Preface

The first edition of *Prealgebra,* which is part of the Lial/Hornsby/Miller/Salzman/Hestwood series, is written for developmental math students who are

- new to algebra,
- relearning the algebra they studied in the past, or
- anxious about their ability to learn algebra.

All too often, math books are written in a rigorous, mathematically correct style, which is satisfying to math instructors but is confusing and frustrating to the students in developmental math courses. The author's main objectives are based on what has worked well for her students over many years of teaching. These main objectives include

- ensuring that students have an understanding of the foundation of algebra, why we use it, and how it works;
- explaining important terminology in everyday English, which is mathematically correct, but written in a friendly style;
- showing *why* something works, without belaboring the point;
- tying the content to the students' experiences and previous knowledge as much as possible; and
- providing well-designed, carefully sequenced exercises that build skills and understanding but are not overwhelming.

The text interweaves arithmetic review and geometry topics, as appropriate, into the algebraic themes of integers, variables and expressions, equations, solving application problems, positive and negative fractions and decimals, proportions, percents, measurements, and graphing.

The recommendations of the **American Mathematical Association of Two Year Colleges** and the **National Council of Teachers of Mathematics** are addressed through

- exercises that build conceptual understanding and place an emphasis on estimation,
- opportunities for students to write about mathematics within exercise sets (designated with ✎),
- collaborative activities within the exercise sets designed for pairs or small groups of students (designated with Ⓖ),
- data presented in the form of graphs, charts, and tables,
- applications that emphasize the use of mathematics in real-world situations, and
- exercises that allow for optional, yet appropriate, use of technology (designated with a 🖩).

Innovative Features

▶ ***A friendly, understandable, mathematically correct writing style*** The author speaks to students as capable adults but understands that they are also novices with regard to algebra. The explanations and definitions are clear, complete, and written in "plain English." Mathematical vocabulary and concepts are introduced carefully, in the context of students' experiences and ideas. Vocabulary is used consistently with frequent, brief reminders of its meaning. Occasional mention of historical connections adds depth. Overall, this style results in a book that students can truly read, relate to, and understand.

▶ ***Early, careful introduction of variables and equations*** Many students see algebra as mysterious and unintelligible. Why use letters instead of numbers? Why solve equations? When would they ever use algebra? Even students who have completed a high school algebra course may not "get the point" of the course. Chapters 2 and 3, in particular, are based on the author's many years of helping students successfully bridge the gap between arithmetic and algebra. Students will see the need for variables, the usefulness of algebraic expressions and formulas, and the power of writing and solving equations. These ideas are then revisited and reinforced throughout Chapters 4–9 as students work with fractions, decimals, proportions, percents, measurement, and graphs.

▶ ***Interweaving of arithmetic review and geometry topics throughout the text*** Developmental math students at the college level are tired of rehashing arithmetic. The backbone of this text is algebra, with necessary arithmetic review in fractions, decimals, ratios, and percents included at points where it is needed. (Skills in whole numbers computation can be assessed with the Diagnostic Test at the beginning of the book. The student then works only the needed sections in Chapter R: Whole Numbers Review.) Geometry topics introduced at appropriate points include perimeter, area, volume, circles, Pythagorean theorem, and similar triangles. A review of lines and angles is included for reference in Appendix C.

▶ ***Using a raised negative sign in Chapters 1–3*** To facilitate the understanding of positive and negative numbers, a red dash in the raised position is used to indicate negative numbers in Chapters 1–3. In Chapter 4, the negative sign is shown in the standard position but is still printed in red. Traditional notation is used in Chapters 5–9.

$^-7-^-3$	$-7-(-3)$	$-7-(-3)$
Chapters 1–3	Chapter 4	Chapters 5–9

The raised negative sign helps students form a clear understanding of the difference between a negative number and the subtraction sign. It also helps students avoid errors related to "forgetting" the negative sign. Using colored chalk during lectures to highlight the raised negative sign is helpful but not necessary. Students generally make the transition to traditional notation without difficulty.

▶ ***Study tips in Chapters 1–3*** Developmental math students need to learn *how* to be good students. The Study Tips pages give specific

Prealgebra

Margaret L. Lial
American River College

Diana L. Hestwood
Minneapolis Community and Technical College

ADDISON-WESLEY

An imprint of Addison Wesley Longman, Inc.

Reading, Massachusetts • Menlo Park, California • New York • Harlow, England
Don Mills, Ontario • Sydney • Mexico City • Madrid • Amsterdam

Manager of Addison Wesley Longman Custom Publishing: Andrea Sheehan
Production Administrator: Kathleen Kollar

Prealgebra

Includes selected sections from Lial/Hestwood's *Prealgebra* (ISBN 0-321-01607-6), and Lial/Hornsby/Miller's *Introductory Algebra,* Sixth Edition (ISBN 0-321-01267-4)

ISBN: 0-201-59751-9

99 00 01 5 4 3 2 1

Contents

suggestions on how to use the textbook effectively, how to retain the material, how to prepare for a test, and how to take a test.

Series Features Common to All Titles in This Series

▶ ***Learning objectives*** Each section begins with clearly stated, numbered objectives, and material in the section is keyed to these objectives. In this way, students and instructors know exactly what is being covered in each section. (example, see p. 207)

▶ ***Margin problems*** Margin problems are found in every section, with answers immediately available at the bottom of the page. These problems allow the student to practice the material covered in the section in preparation for the exercise set that follows. (example, see p. 208)

▶ ***Numbers in the Real World: Collaborative Investigations*** These pages show students how mathematics is used in everyday life. Students read and interpret articles and graphs from newspapers, tables of data from food packaging, labels on consumer goods, and more. These investigations may be completed by individual students but are well-suited as collaborative assignments for pairs or small groups of students, or for open-ended discussion by an entire class. (example, see p. 200 and p. 252)

▶ ***Calculator Tips*** These aids are included at appropriate points to guide students in using their scientific calculators appropriately and effectively. The calculator is used to emphasize such concepts as order of operations, commutativity of addition and multiplication, rounding of decimal numbers, division by zero as undefined, and more. Appropriate calculator exercises are designated in selected sections. (If desired, the *Calculator Tips* and exercises may be omitted, with the exception of finding nonperfect square roots in Section 5.8, Pythagorean Theorem.) (example, see p. 457)

▶ ***Writing exercises*** These exercises ask students to look for patterns, devise their own rules, use estimation to identify reasonable and unreasonable solutions, and extend their thinking to cover new possibilities. (example, see p. 156 and p. 202)

▶ ***Ample and varied exercise sets*** Students of basic mathematics need a large number and variety of exercises. This text meets that need with sequenced exercise sets that help students build upon basic skills to develop more complex skills. (example, see p. 227–230)

▶ ***Opportunity for review*** A group of exercises called *Review and Prepare* appear at the end of each section. These exercises review skills from earlier chapters, particularly ones that will be needed in the upcoming section. Each chapter concludes with a concise *Chapter Summary,* which includes key terms and a quick review, *Chapter Review Exercises* with exercises keyed to individual sections, *Mixed Review Exercises,* and a *Chapter Test*. In addition, following every chapter after Chapter 1, *Cumulative Review Exercises* cover material going back to the first chapter.

▶ ***Web Site*** Instructors and students will have access to a World Wide Web site **(www.mathnotes.com),** where additional support is available.

▶ ***Phonetic spellings*** This feature helps students pronounce key terms correctly. It is especially useful for students who speak English as their second language. A separate *Spanish Glossary* is also available. (example, see p. 194)

▶ ***Answers*** Answers to all margin problems are provided at the bottom of the page on which the problems appear. The *Answer* section at the back of the text provides answers to odd-numbered exercises in numbered sections and answers to *all* chapter test exercises, review exercises, and cumulative review exercises.

Extensive Supplements Package

Our extensive supplements package includes printed and electronic testing materials, solutions, software, and videotapes.

For the Instructor

Annotated Instructor's Edition

This edition provides instructors with immediate access to the answers to every exercise in the text. Each answer is printed in color next to the corresponding text exercise.

Printed Test Bank

The *Printed Test Bank* includes short-answer and multiple-choice versions of a placement test; eight forms of chapter tests for each chapter, including six open-response and two multiple-choice forms; and short-answer and multiple-choice forms of a final examination.

Instructor's Solutions Manual

The *Instructor's Solutions Manual* includes solutions to all of the even-numbered section exercises.

Answer Book

The *Answer Book* includes answers to all exercises.

TestGen EQ with QuizMaster EQ

This test generation software is available in Windows and Macintosh versions and is fully networkable. TestGen EQ's friendly, graphical interface enables instructors to easily view, edit, and add questions; transfer questions to tests; and print tests in a variety of fonts and forms. Search and sort features allow instructors to quickly locate questions and arrange them in a preferred order. Six question formats are available, including short-answer, true-false, multiple-choice, essay, matching, and bimodal formats. A built-in question editor gives instructors power to create graphs, import graphics, insert mathematical symbols and templates, and insert variable numbers or text. Computerized testbanks include algorithmically defined problems organized according to each textbook.

QuizMaster EQ enables instructors to create and save tests using TestGen EQ so that students can take them for practice or a grade on a computer network. Instructors can set preferences for how and when tests are to be administered. QuizMaster EQ automatically grades the exams, stores results on disk, and allows instructors to view or print a variety of reports for individual students, classes, or courses.

InterAct Mathematics Plus—Management System

InterAct Math Plus combines course management and online testing with the features of the basic InterAct Math tutorial software to create an invaluable teaching resource. InterAct Math is available in either Windows or Macintosh versions; consult your Addison Wesley Longman representative for details.

FOR THE STUDENT

Student's Solutions Manual

The *Student's Solutions Manual* contains solutions to every other odd-numbered section exercise and solutions to all margin problems, chapter review exercises, chapter tests, and cumulative review exercises. (ISBN 0-321-02282-3)

InterAct Mathematics Tutorial Software

InterAct Math tutorial software has been developed and designed by professional software engineers working closely with a team of experienced developmental math educators.

InterAct Math tutorial software includes exercises that are linked with every objective in the textbook and require the same computational and problem-solving skills as their companion exercises in the text. Each exercise has an example and an interactive guided solution that are designed to involve students in the solution process and to help them identify precisely where they are having trouble. In addition, the software recognizes common student errors and provides students with appropriate customized feedback.

With its sophisticated answer-recognition capabilities, InterAct Math tutorial software recognizes appropriate forms of the same answer for any kind of input. It also tracks student activity and scores for each section, which can then be printed out. Available for Windows and Macintosh computers, the software is available to qualifying adopters or can be bundled with books for sale to students. (Macintosh: ISBN 0-321-03146-6; Windows: ISBN 0-321-03145-8)

Videotapes

A videotape series, "Real to Reel," has been developed to accompany *Prealgebra*. In a separate lesson for each section in the book, the series covers all objectives, topics, and problem-solving techniques discussed in the text. The video series is available to qualifying adopters. (ISBN 0-321-03144-X)

Spanish Glossary

A separate *Spanish Glossary* is now being offered as part of the supplements package for this textbook series. This book contains the key terms from each of the five texts in the series and their Spanish translations. (ISBN 0-321-01647-5)

Acknowledgments

Writing a textbook is not to be taken lightly. There are so many people whose contributions "behind the scenes" are critical to a quality book. The following professors reviewed part or all of the text. Their comments and suggestions were invaluable.

Jon Becker, *Indiana University—Northwest*

Kenneth Benson, *University of Illinois at Urbana–Champaign*

Rebecca Benson-Beaver, *Valencia Community College*

Sarah Carpenter, *Vincennes University*

Celeste Carter, *Richland College*
Stanley Carter, *Central Missouri State University*
Ky Davis, *Muskingum Area Technical College*
Bob C. Denton, *Orange Coast College*
Lisa Grenier, *Pima Community College—Downtown*
Betty Givan, *Eastern Kentucky University*
Kay Haralson, *Austin Peay State University*
John Heublin, *Kansas State University—Salina*
Nancy Johnson, *Broward Community College—North*
Maryann E. Justinger, *SUNY—Erie*
Steve Kahn, *Anne Arundel Community College*
Robert Kaiden, *Lorain County Community College*
Sharon H. Killian, *Asheville–Buncomb Technical Community College*
Theodore Lai, *Hudson County Community College*
Nenette Loftsgaarden, *University of Montana*
Madeleine N. Mahar, *Pitt Community College*
Michael Montano, *Riverside Community College—City Campus*
Gail C. Phillips, *North Harris College*
Sally Sestini, *Cerritos College*
Patricia A. Stoltenberg, *Sam Houston State University*
Sam J. Tinsley, *Richland College*
Gene Tognazzini, *Fresno City College*
Peter A. Wursthorn, *Capitol Community Technical College*
Kevin Yokoyama, *College of the Redwoods*
Yoshiyuki Yamamoto, *Brevard Community College*

Special thanks go to Barbara Brown, *Anoka Ramsey Community College,* who reviewed each chapter as it was written. She not only critiqued the content, but also provided much needed support and encouragement. Barbara Kurt and Lisa Grenier checked the examples and exercises for accuracy, Elizabeth Fleck did a marvelous job writing the solutions manuals, as Abby Tanenbaum did on the Printed Test Bank, and Linda Russell provided the phonetic spellings.

The folks at Addison Wesley Longman have been patient and supportive throughout the long process of writing, rewriting, and production. My thanks to Rita Ferrandino, who started the project, Jenny Crum, who helped the project over the final hurdles, Kari Heen, who provided support and direction at a critical time, Kathy Manley, who coordinated production, Ron Hampton, who saw that the art was done properly, and Jane Estrella for her work on the supplements. I also appreciate the help of Sandi Goldstein in coordinating reviews and preparing the manuscript.

I would be remiss if I didn't acknowledge Stanley Salzman for his contributions to this book. Also, I want Margaret Lial to know how much her support and encouragement meant to me.

This book is dedicated to my husband, Earl Orf, who is the wind beneath my wings, and to my students at Minneapolis Community and Technical College, from whom I have learned so much.

Diana L. Hestwood

Whole Numbers Computation: Diagnostic Test

This test will check your skills in doing whole numbers computation, using paper and pencil. Each part of the test is keyed to a section in the **Review Chapter,** which follows Chapter 9 in this textbook. Based on your test results, work the appropriate section(s) in the Review Chapter *before* you start Chapter 1.

Addition of whole numbers (Do not use a calculator.)

1. $\begin{array}{r} 368 \\ +\ \ 22 \\ \hline \end{array}$ **2.** $\begin{array}{r} 7093 \\ +\ 6073 \\ \hline \end{array}$ **3.** $\begin{array}{r} 85 \\ +\ 2968 \\ \hline \end{array}$ **4.** $\begin{array}{r} 57{,}208 \\ 915 \\ +\ 59{,}387 \\ \hline \end{array}$

5. 714 + 3728 + 9 + 683,775

1. ______
2. ______
3. ______
4. ______
5. ______

Subtraction of whole numbers (Do not use a calculator.)

1. $\begin{array}{r} 426 \\ -\ \ 76 \\ \hline \end{array}$ **2.** $\begin{array}{r} 3358 \\ -\ 2729 \\ \hline \end{array}$ **3.** $\begin{array}{r} 30{,}602 \\ -\ \ 5708 \\ \hline \end{array}$

4. 4006 − 97 **5.** 679,420 − 88,033

1. ______
2. ______
3. ______
4. ______
5. ______

Multiplication of whole numbers (Do not use a calculator.)

1. $3 \times 3 \times 0 \times 6$ **2.** $\begin{array}{r} 3841 \\ \times\ \ \ 7 \\ \hline \end{array}$ **3.** (520)(3000)

1. ______
2. ______
3. ______

4. ____________________

5. ____________________

6. ____________________

Show your work on Exercises 4–6.

4. $\begin{array}{r} 71 \\ \times\ 26 \\ \hline \end{array}$ **5.** Multiply 359 and 48. **6.** 853×609

Division of whole numbers (Do not use a calculator and show your work.)

1. ____________________

2. ____________________

3. ____________________

4. ____________________

5. ____________________

6. ____________________

7. ____________________

8. ____________________

1. $3\overline{)69}$ **2.** $12 \div 0$ **3.** $\dfrac{25{,}036}{4}$ **4.** $7\overline{)5655}$

5. $52\overline{)1768}$ **6.** $45{,}000 \div 900$ **7.** $38\overline{)2300}$ **8.** $83\overline{)44{,}799}$

Now check your answers on page A-43 in the ***Answers*** *section at the back of the book. Record the number you got correct in each part of the test.*

Addition of whole numbers: ______ correct out of 5.
If you got 0, 1, or 2 correct, work **Section R.1** in the Review Chapter.

Subtraction of whole numbers: ______ correct out of 5.
If you got 0, 1, or 2 correct, work **Section R.2** in the Review Chapter.

Multiplication of whole numbers: ______ correct out of 6.
If you got 0, 1, 2, or 3 correct, work **Section R.3** in the Review Chapter.

Division of whole numbers: ______ correct out of 8.
If you got 0, 1, or 2 correct, work **Sections R.4 and R.5** in the Review Chapter.
If you got 3 or 4 correct, work **Section R.5** in the Review Chapter.

Introduction to Algebra: Integers

1

1.1 Place Value

It would be nice to earn millions of dollars like some of our favorite entertainers or sports stars. But how much is a million? If you received \$1 every second, 24 hours a day, day after day, how many days would it take for you to receive a million dollars? How long to receive a billion dollars? Or a trillion dollars? Make some guesses and write them here.

It would take ______________________ to receive a million dollars.

It would take ______________________ to receive a billion dollars.

It would take ______________________ to receive a trillion dollars.

The answers are at the bottom of the page. Later, in **Section 1.7,** you'll find out how to calculate the answers.

INTERNET

www.mathnotes.com

OBJECTIVES

1. Identify whole numbers.
2. Identify the place value of a digit through hundred trillions.
3. Write a whole number in words or digits.

FOR EXTRA HELP

Tutorial | Tape 1 | SSM, Sec. 1.1

OBJECTIVE 1 First we have to be able to write the number that represents *one million.* We can write *one* as 1. How do we make it 1 *million?* Our number system is a **place value system.** That means that the location, or place, in which a number is written gives it a different value. Using money as an example, you can see that

\$1 is one dollar.
\$10 is ten dollars.
\$100 is one hundred dollars.
\$1000 is one thousand dollars.

Answers: About $11\frac{1}{2}$ days to receive a million dollars; nearly 32 *years* to receive a billion dollars; about 31,710 *years* to receive a trillion dollars.

1. Circle the whole numbers.

0.8	−14	502
$\frac{7}{9}$	3	$\frac{3}{2}$
14	0	$6\frac{4}{5}$
9.082	$-\frac{8}{3}$	60,005

Each time the 1 moved to the left one place, it was worth ten times as much. Can you keep moving it to the left? Yes, as many times as you like. The following chart shows the *value* of each *place.* In other words, you write the 1 in the correct place to represent the number you want to express.

Whole Number Place Value Chart

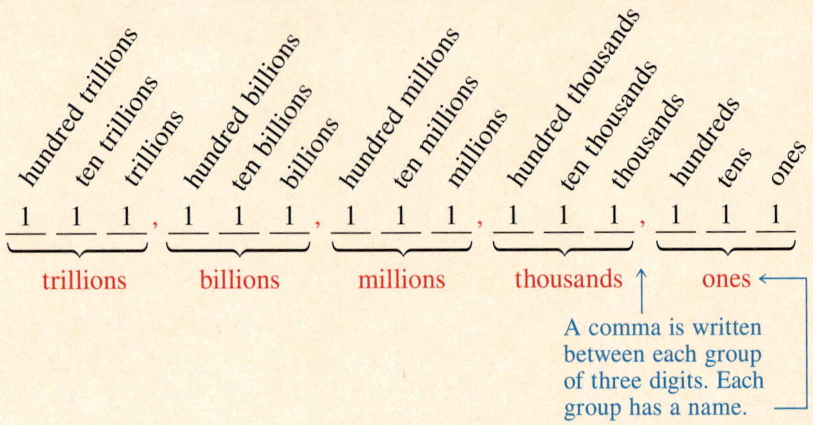

If there had been more room on this page, we could have continued to the left with quadrillions, quintillions, sextillions, septillions, octillions, and more.

Of course, we can use other **digits** besides 1. In our decimal system of writing numbers we can use these ten digits: 0, 1, 2, 3, 4, 5, 6, 7, 8, and 9. In this section we will use the digits to write **whole numbers.**

These are whole numbers.	These are *not* whole numbers.
0 8 37 100 24,014	-6 $\frac{3}{4}$ 7.528 0.3 $5\frac{2}{3}$

EXAMPLE 1 Identifying Whole Numbers

Circle the whole numbers in this list.

75 −4 0 1.5 $\frac{5}{8}$ 300 0.666 $7\frac{1}{2}$ 2

Whole numbers *do* include zero. If we started a list of *all* the whole numbers it would look like this: 0, 1, 2, 3, 4, 5, . . . with the three dots indicating that the list goes on and on. So the whole numbers in this example are: 75, 0, 300, and 2.

◀◀ **WORK PROBLEM 1 AT THE SIDE.**

OBJECTIVE 2 The number of Americans who voted for Bill Clinton for president in 1996 was 45,628,665. There are two 5's in this number, but the value of each 5 is very different. The 5 on the right is in the ones place, so its value is simply 5. But the 5 on the left is worth a great deal more because of the *place* where it is written. Looking back at the place value chart, we see that this 5 is in the millions place, so its value is 5 *million.* It is important to memorize the place value names shown on the chart.

ANSWER

1. 502; 3; 14; 0; 60,005

EXAMPLE 2 Identifying Place Value

Give the place value of each of the 6's in the number of Clinton voters.

45,628,665

Hundred thousands place → the first 6; Hundreds place → the second 6; Tens place → the third 6

WORK PROBLEM 2 AT THE SIDE.

OBJECTIVE 3 To write a whole number in words, or to say it aloud, begin at the left. Write or say the number in each group of three, followed by the name for that group. When you get to the ones group, do *not* include the group name. Hyphens (dashes) are used whenever you write a number from 21 to 99, like twenty-one thousand, or thirty-seven million, or ninety-four billion.

EXAMPLE 3 Writing Numbers in Words

(a) Write 6,058,120 in words.

Start on the left.

6 , 058 , 120

six **million**, fifty-eight **thousand**, one hundred twenty

(million: Group name; thousand: Group name; Do not use group name "ones.")

(b) Write 50,588,000,040,000 in words.

Start on the left.

50 , 588 , 000 , 040 , 000

fifty **trillion**, five hundred eighty-eight **billion**, forty **thousand**

(trillion: Group name; billion: Group name; thousand: Group name; No millions and no ones in this number.)

Note

You often hear people say "AND" when reading a group of three digits. For example, you may hear 120 as "one hundred AND twenty," but this is **not** correct. The word AND is used when reading a DECIMAL POINT, which we do not have here. The correct wording for 120 is "one hundred twenty."

WORK PROBLEM 3 AT THE SIDE.

2. Give the place value of the digit 8 in each number.

(a) 45,628,665

(b) 800,503,622

(c) 428,000,000,000

(d) 2,385,071

3. Write these numbers in words.

(a) 23,605

(b) 400,033,007

(c) 193,080,102,000,000

ANSWERS

2. **(a)** thousands **(b)** hundred millions **(c)** billions **(d)** ten thousands

3. **(a)** twenty-three **thousand,** six hundred five **(b)** four hundred **million,** thirty-three **thousand,** seven **(c)** one hundred ninety-three **trillion,** eighty **billion,** one hundred two **million**

4. Write each number, using digits.

(a) eighteen million, two thousand, three hundred five

(b) two hundred billion, fifty million, six hundred sixteen

(c) five trillion, forty-two billion, nine million

When you're reading a number and want to write it in digits, look for the group names: **trillions, billions, millions,** and **thousands.** Write the number in each group, followed by a comma. Do *not* put a comma at the end of the ones group.

EXAMPLE 4 Writing Numbers in Digits

Write each number, using digits.

(a) five hundred sixteen **thousand**, nine

The first group name is *thousand,* so you need to fill *two groups* of three digits: thousands and ones.

$$\underbrace{\underline{5}\ \underline{1}\ \underline{6}}_{\text{thousands}},\ \underbrace{\underline{0}\ \underline{0}\ \underline{9}}_{\text{ones}}$$

The number is 516,009.

(b) seventy-seven **billion**, thirty **thousand**, five hundred

The first group name is *billion,* so you need to fill *four groups* of three digits: billions, millions, thousands, and ones.

$$\underbrace{\underline{\circ}\ \underline{7}\ \underline{7}}_{\text{billions}},\ \underbrace{\underline{0}\ \underline{0}\ \underline{0}}_{\text{millions}},\ \underbrace{\underline{0}\ \underline{3}\ \underline{0}}_{\text{thousands}},\ \underbrace{\underline{5}\ \underline{0}\ \underline{0}}_{\text{ones}}$$

There are no millions, so fill the millions group with zeros. When writing the number, you can omit the leading zero in the billions group: 77,000,030,500.

◀◀ WORK PROBLEM 4 AT THE SIDE.

ANSWERS

4. **(a)** 18,002,305 **(b)** 200,050,000,616 **(c)** 5,042,009,000,000

35. Six hundred trillion, seventy-one million, four hundred

36. Four hundred forty trillion, thirty-six thousand, one hundred two

G **37.** Here is a group of digits: 6, 0, 9, 1, 5, 0, 7, 1. Using all the digits, arrange them to make the largest possible whole number and the smallest possible whole number. Then write each number in words.

G **38.** Look again at the Whole Number Place Value Chart at the beginning of this section. As you move to the left, each place is worth ten times as much. Computers work on a system where each place is worth two times as much. Complete this place value chart based on 2's.

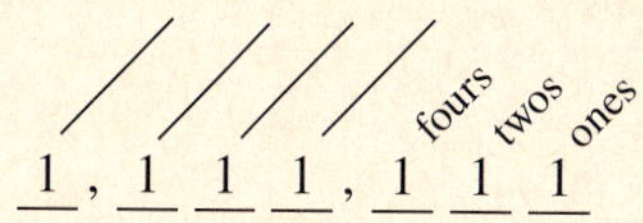

In Exercises 39–46, if the number is given in digits, write it in words. If the number is given in words, write it in digits.

39. In the United States, 6567 couples get married every day.

40. There are 3582 pairs of bowling shoes purchased each day.

41. One hundred one million, two hundred eighty thousand adults are on diets at any one time in the United States.

42. Two million, five thousand Americans suffer from heartburn on any given day.

43. The amount spent each day on exercise equipment in the United States is $2,021,018.

44. Americans spend $434,206,600 daily on toys.

45. People eat twenty-four million, five hundred hot dogs each day.

46. Five hundred twenty-four million servings of cola drinks are consumed every day.

47. Write these numbers in digits and in words.

(a) Your house number or apartment building number.

(b) Your phone number, including the area code.

(c) The approximate cost of your tuition and books for this quarter or semester.

(d) Your zip code.

48. Explain in your own words why our number system is called a place value system. Include an example as part of your explanation.

1.2 *Introduction to Signed Numbers*

OBJECTIVE 1 The whole numbers in **Section 1.1** were either 0 or greater than 0. Numbers *greater* than zero are called *positive numbers.* But many everyday situations involve numbers that are *less* than zero, called *negative numbers.* Here are a few examples.

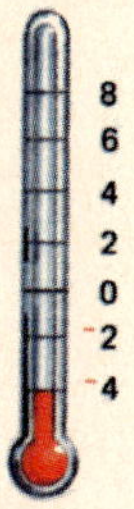

At midnight the temperature dropped to 4 degrees below zero.

⁻4 degrees

Jean had $30 in her checking account. She wrote a check for $40.75. We say she is $10.75 "in the hole," or $10.75 "in the red," or overdrawn by $10.75.

⁻$10.75

The Packers gained 6 yards on the first play. On the second play they lost 9 yards. We can write the results as:

⁺6 yards and **⁻9 yards.**

A plane took off from the airport and climbed to 20,000 feet above sea level. We can write this as

⁺20,000 ft.

A scuba diver dove down to $25\frac{1}{2}$ feet below the surface. We can write this as

⁻$25\frac{1}{2}$ feet.

To write a negative number, put a negative sign (a dash) in front of it: ⁻10. Notice that the negative sign looks exactly like the subtraction sign, as in 5 − 3 = 2. The two signs do *not* mean the same thing (more on that in the next section). To avoid confusion for now, we will write negative signs in red and put them up higher than subtraction signs.

⁻10 means **negative** 10   14 − 10 means 14 **minus** 10

↑ Raised dash

Later, in Chapters 4 and 5, we will start writing negative signs in the traditional way.

Positive numbers can be written two ways:

1. Write a positive sign in front of the number: ⁺2 is positive 2. We will write the sign in the raised position to avoid confusion with the sign for addition, as in 6 + 3 = 9.
2. Do not write any sign. For example, 16 is assumed to be *positive* 16.

OBJECTIVES

1. Write positive and negative numbers used in everyday situations.
2. Graph signed numbers on a number line.
3. Use the < and > symbols to compare integers.
4. Find the absolute value of integers.

FOR EXTRA HELP

Tutorial   Tape 1   SSM, Sec. 1.2

1. Write each negative number with a raised negative sign. Write each positive number in two ways.

   (a) The temperature is $6\frac{1}{2}$ degrees below zero.

   (b) Cameron lost 12 pounds on a diet.

   (c) I deposited \$210.35 in my checking account.

   (d) I wrote too many checks, so my account is overdrawn by \$65.

   (e) The submarine dove to 100 feet below the surface of the sea.

   (f) In this round of the card game, I won 50 points.

2. Graph each set of numbers.

   (a) $^{-}2$ **(b)** 2 **(c)** 0

   (d) $^{-}4$ **(e)** 4

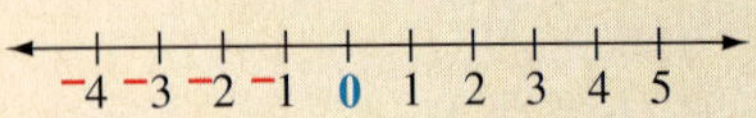

   (f) $^{-}3\frac{1}{2}$ **(g)** $\frac{1}{2}$ **(h)** $^{-}1$ **(i)** 3

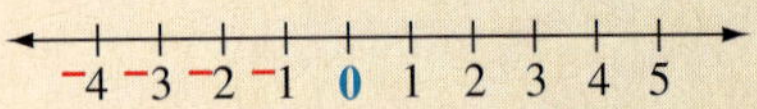

EXAMPLE 1 Writing Positive and Negative Numbers

Write each negative number with a raised negative sign. Write each positive number in two ways.

(a) The river rose to 8 feet above flood stage.

$^{+}8$ or 8

↑ Raised positive sign ↑ No sign

(b) Michael lost \$500 in the stock market.

$^{-}$\$500

↑ Raised negative sign

◀◀ WORK PROBLEM 1 AT THE SIDE.

OBJECTIVE 2 Mathematicians often use a **number line** to show how numbers relate to each other. A number line is like a thermometer turned sideways. Zero is the dividing point between the positive and negative numbers.

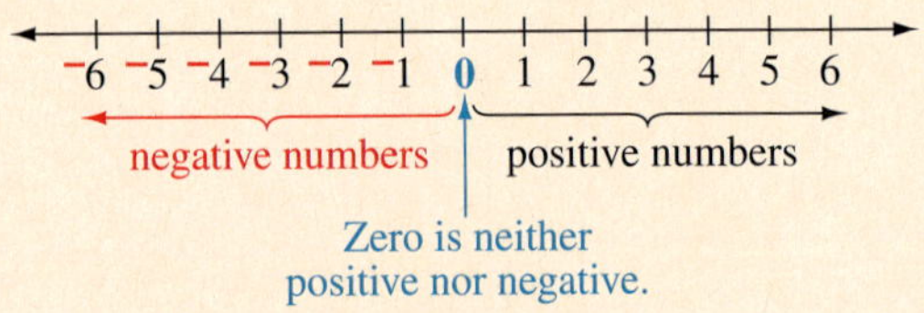

The number line could be shown with positive numbers on the left side of zero instead of the right side. But it helps if everyone draws it the same way, as shown above. This method will also match what you do when graphing points and lines in Chapter 9.

EXAMPLE 2 Graphing Numbers on a Number Line

Graph these numbers on the number line.

(a) $^{-}5$ **(b)** 3 **(c)** $1\frac{1}{2}$ **(d)** 0 **(e)** $^{-}1$

Draw a dot at the correct location for each number.

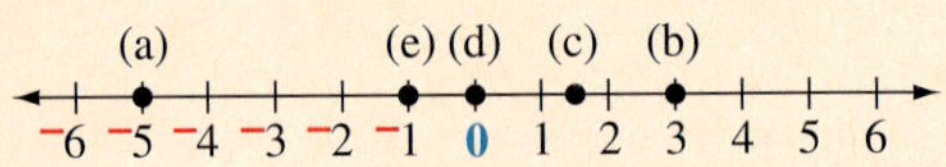

◀◀ WORK PROBLEM 2 AT THE SIDE.

OBJECTIVE 3 In Chapters 4 and 5 you will work with fractions and decimals. For the rest of this chapter, you will work only with **integers** (IN-te-jurs). A list of integers can be written like this:

$$\ldots\ ^{-}6,\ ^{-}5,\ ^{-}4,\ ^{-}3,\ ^{-}2,\ ^{-}1,\ 0,\ 1,\ 2,\ 3,\ 4,\ 5,\ 6,\ \ldots$$

The dots show that the list goes on forever in both directions.

ANSWERS

1. (a) $^{-}6\frac{1}{2}$ degrees
   (b) $^{-}12$ pounds
   (c) \$210.35 or $^{+}$\$210.35
   (d) $^{-}$\$65
   (e) $^{-}100$ feet
   (f) 50 points or $^{+}50$ points
2.

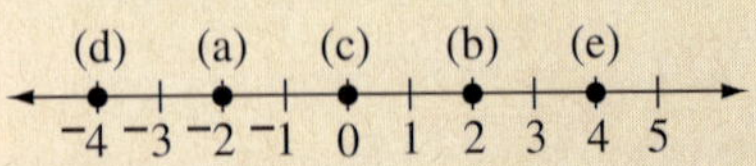

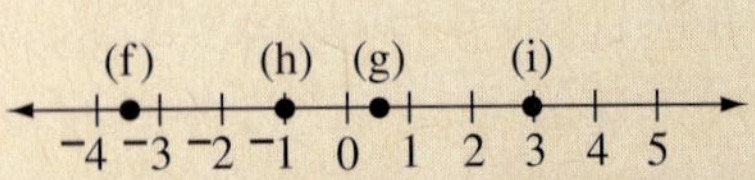

We can use the number line to compare two integers.

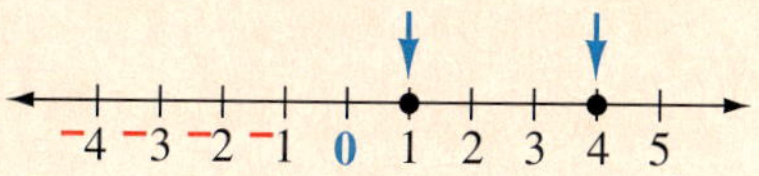

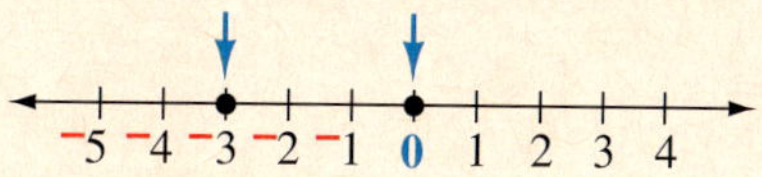

1 is to the left of 4.
1 is less than 4.
Use < to mean "is less than."

1 < 4
1 **is less than** 4

0 is to the right of $^-3$.
0 is greater than $^-3$.
Use > to mean "is greater than."

0 > $^-3$
0 **is greater than** $^-3$

One way to remember which symbol to use is that the "small end of the symbol" points to the "smaller number" (the number that is less).

1 < 4
Smaller number ↑ ↑ Small end of symbol

0 > $^-3$
Small end of symbol ↑ ↑ Smaller number

EXAMPLE 3 Comparing Integers, Using the < and > Symbols

Write < or > between each pair of numbers to make true statements.

(a) 0 __ 2

0 is to the *left* of 2 on the number line, so 0 is *less than* 2. Write 0 < 2.

(b) 1 __ $^-4$

1 is to the *right* of $^-4$, so 1 is *greater than* $^-4$. Write 1 > $^-4$.

(c) $^-4$ __ $^-2$

$^-4$ is to the *left* of $^-2$, so $^-4$ is *less than* $^-2$. Write $^-4$ < $^-2$.

WORK PROBLEM 3 AT THE SIDE.

OBJECTIVE 4 In order to graph a number on the number line, you need to know two things:

1. Which *direction* it is from 0. It can be in a positive direction or a negative direction. You can tell the direction by looking for a positive sign or a negative sign (or no sign, which is positive).
2. How *far* it is from 0. The distance from 0 is the **absolute** (ab-soh-LOOT) **value** of a number.

Absolute value is indicated by two vertical bars. For example,

$|6|$ is read "the **absolute value** of 6."

The absolute value of a number will *always* be positive, because it is the *distance* from zero. A distance is never negative. (You wouldn't say that your living room is $^-16$ feet long.) So absolute value concerns only *how far away* the number is from zero; we don't care which direction it is from zero.

3. Write < or > between each pair of numbers to make true statements.

(a) 5 __ 4

(b) 0 __ 2

(c) $^-3$ __ $^-2$

(d) $^-1$ __ $^-4$

(e) 2 __ $^-2$

(f) $^-5$ __ 1

ANSWERS

3. **(a)** > **(b)** < **(c)** < **(d)** > **(e)** > **(f)** <

EXAMPLE 4 Finding Absolute Value

Find each absolute value.

(a) $|4|$ The distance from 0 to 4 on the number line is 4 spaces.

So $|4| = 4$

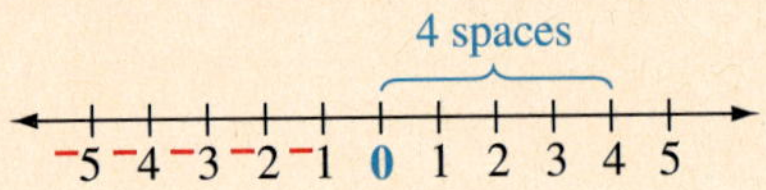

(b) $|{}^{-}4|$ The distance from 0 to $^{-}4$ on the number line is also 4 spaces.

So $|{}^{-}4| = 4$

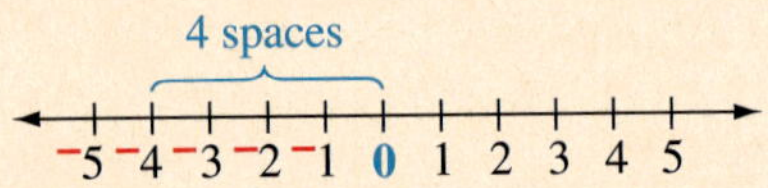

(c) $|0|$ $|0| = 0$ because the distance from 0 to 0 on the number line is 0 spaces.

◀◀ WORK PROBLEM 4 AT THE SIDE.

4. Find each absolute value.

(a) $|13|$

(b) $|{}^{-}7|$

(c) $|0|$

(d) $|{}^{-}350|$

(e) $|6000|$

ANSWERS

4. **(a)** 13 **(b)** 7 **(c)** 0 **(d)** 350 **(e)** 6000

1.2 Exercises

Write each negative number with a raised negative sign. Write each positive number in two ways.

1. Mount Everest, the tallest mountain in the world, rises 29,028 feet above sea level.

2. The bottom of Lake Baykal in central Asia is 6365 feet below the surface of the water.

3. The coldest temperature ever recorded on earth is 128.6 degrees below zero in Antarctica.

4. Normal body temperature is 98.6 degrees Fahrenheit.

5. During the first three plays of the football game, the Trojans lost a total of 18 yards.

6. The Jets gained 25 yards on a pass play.

7. Angelique won \$100 in a prize drawing at the shopping mall.

8. Derice overdrew his checking account by \$37.

9. Keith lost $6\frac{1}{2}$ pounds while he was sick with the flu.

10. The price of Mathtronic stock went up $2\frac{1}{2}$ dollars yesterday.

Graph each set of numbers.

11. $^-3, 3, 0, ^-5$

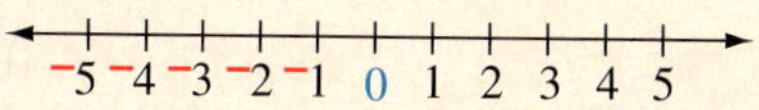

12. $^-2, 2, 0, 5$

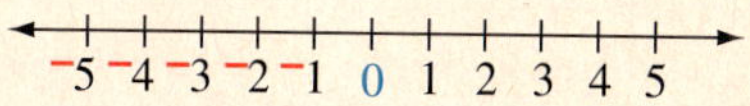

13. $^-1, 4, ^-2, 5$

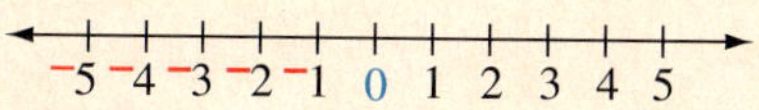

14. $3, ^-4, 1, ^-5$

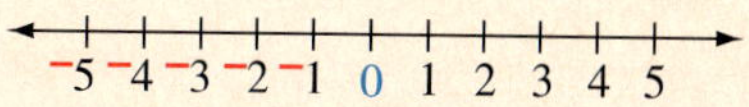

15. $^-4\frac{1}{2}, \frac{1}{2}, 0, ^-8$

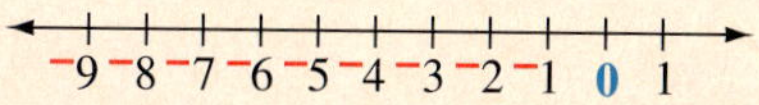

16. $^-7, 1\frac{1}{2}, \frac{^-1}{2}, -9$

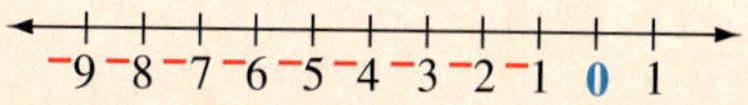

Write < or > between each pair of numbers to make true statements.

17. 10 2 **18.** 6 0 **19.** $^-1$ 0 **20.** $^-3$ $^-1$

21. $^-10$ 2 **22.** $^-9$ 7 **23.** $^-3$ $^-6$ **24.** 0 $^-1$

25. $^-10$ $^-2$ **26.** $^-1$ $^-5$ **27.** 0 $^-8$ **28.** 6 $^-4$

29. 10 $^-2$ **30.** $^-2$ 1 **31.** $^-4$ 4 **32.** 9 $^-9$

Find each absolute value.

33. $|15|$ **34.** $|10|$ **35.** $|^-3|$ **36.** $|^-8|$

37. $|0|$ **38.** $|100|$ **39.** $|200|$ **40.** $|^-99|$

41. $|^-75|$ **42.** $|^-6320|$ **43.** $|^-8042|$ **44.** $|0|$

45. Describe three situations you have encountered where you used negative numbers.

46. Why is it important to have 0 on the number line? What would happen if there were no 0 on the number line?

47. Write in your own words how to find the absolute value of a signed number.

48. Explain why absolute value is never negative.

49. If the distance between two numbers on the number line is 5 spaces and one of the numbers is $^-2$, what is the other number? There are *two* answers; find both. Draw a number line to illustrate your answers.

50. If the distance between two numbers on the number line is 3 spaces and one of the numbers is $^-3$, what is the other number? There are *two* answers; find both. Draw a number line to illustrate your answers.

1.3 *Adding Integers*

Objective 1 Numbers that you are adding are called **addends.** The result is called the **sum.** You can add integers while watching a football game. On each play, you can use a positive integer to stand for the yards gained by your team, and a negative integer for yards lost. Zero indicates no gain or loss. For example,

Our team gained 6 yards on the first play	and	**gained 4 yards on the next play.**		**Total gain of 10 yards**
6	+	**4**	=	**10**
Our team lost 3 yards on the first play	and	**lost 5 yards on the next play.**		**Total loss of 8 yards**
$^-3$	+	**$^-5$**	=	**$^-8$**

A drawing of a football field can help you add integers. Notice how similar the drawing is to a number line. Zero marks your team's starting point.

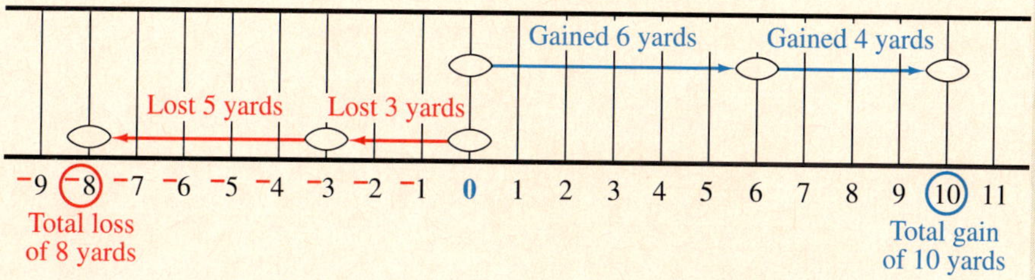

EXAMPLE 1 Using a Number Line to Add Integers

Use a number line to find $^-5 + {}^-4$.

Think of the number line as a football field. Your team starts at zero. On the first play it lost 5 yards. On the next play it lost 4 yards. The total loss is 9 yards.

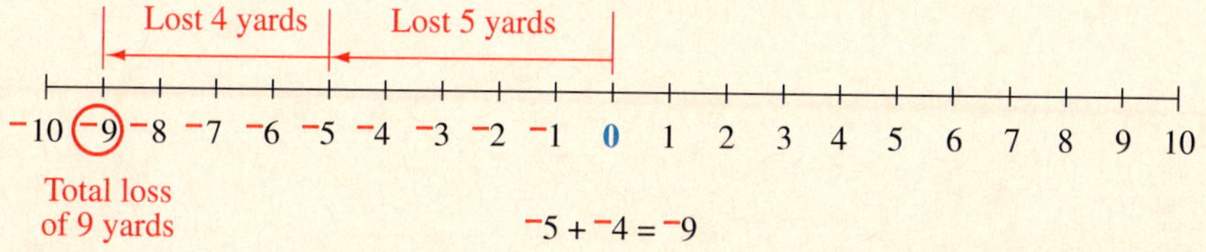

WORK PROBLEM 1 AT THE SIDE.

Do you see a pattern in the margin exercises you just did? The answers to the first two exercises are the same, except for the sign. The same is true for the next two exercises and for the last two exercises. This pattern leads to a rule for adding two integers when the signs are the same.

Adding Two Integers with the Same Sign

Step 1 Add the absolute values of the numbers.

Step 2 Use the common sign as the sign of the sum. If both numbers are positive, the sum is positive. If both numbers are negative, the sum is negative.

Objectives

1. Add integers.
2. Identify properties of addition.

For Extra Help

Tutorial   Tape 1   SSM, Sec. 1.3

1. Find each sum. Use the number line to help you.

(a) $^-2 + {}^-2$

(b) $2 + 2$

(c) $^-10 + {}^-1$

(d) $10 + 1$

(e) $^-3 + {}^-7$

(f) $3 + 7$

Answers

1. (a) $^-4$ **(b)** 4
(c) $^-11$ **(d)** 11
(e) $^-10$ **(f)** 10

2. Find each sum.

(a) $^-6 + {}^-6$

(b) $9 + 7$

(c) $^-5 + {}^-10$

(d) $^-12 + {}^-4$

(e) $13 + 2$

EXAMPLE 2 **Adding Two Integers with the Same Sign**

Add.

(a) $^-8 + {}^-7$

Step 1 Add the absolute values.

$$|^-8| = 8 \quad \text{and} \quad |^-7| = 7$$

Add $8 + 7$ to get 15.

Step 2 Use the common sign as the sign of the sum. Both numbers are negative, so the sum is negative.

$$^-8 + {}^-7 = {}^-15$$

Both negative — Sum is negative.

(b) $3 + 6 = 9$

Both positive — Sum is positive.

WORK PROBLEM 2 AT THE SIDE.

You can also use a number line (or drawing of a football field) to add integers with *different* signs. For example, suppose that your team gained 2 yards on the first play and then lost 7 yards on the next play.

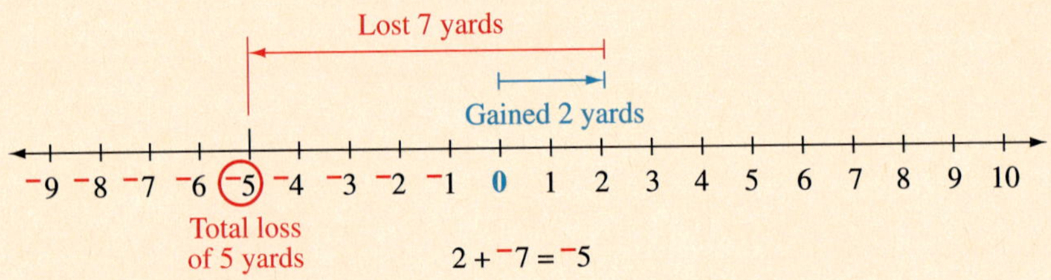

Or, try this one. On the first play your team gained 10 yards, but then it lost 4 yards on the next play.

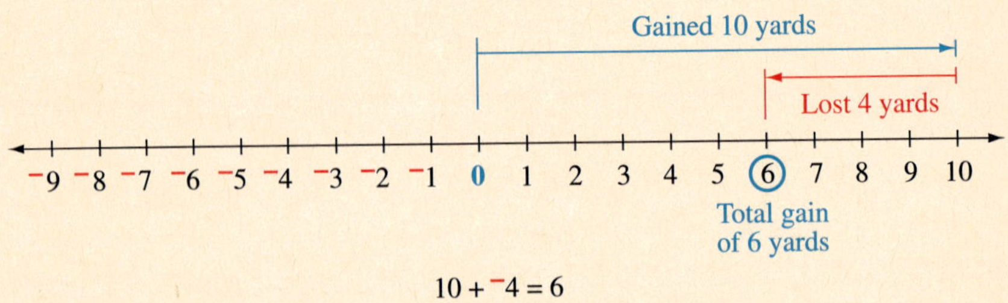

These examples illustrate the rule for adding two integers with unlike, or different, signs.

Adding Two Integers with Unlike Signs

Step 1 Subtract the smaller absolute value from the larger absolute value.

Step 2 Use the sign of the number with the larger absolute value as the sign of the sum.

ANSWERS

2. (a) $^-12$ **(b)** 16 **(c)** $^-15$ **(d)** $^-16$ **(e)** 15

EXAMPLE 3 Adding Two Integers with Unlike Signs

Add.

(a) $^-8 + 3$

Step 1 $|^-8| = 8$ and $|3| = 3$
Subtract $8 - 3$ to get 5.

Step 2 $^-8$ has the larger absolute value and is negative, so the sum is also negative.

$$^-8 + 3 = {}^-5$$

(b) $^-5 + 11$

Step 1 $|^-5| = 5$ and $|11| = 11$
Subtract $11 - 5$ to get 6.

Step 2 11 has the larger absolute value and is positive, so the sum is also positive.

$$^-5 + 11 = {}^+6 \quad \text{or} \quad 6$$

WORK PROBLEM 3 AT THE SIDE.

EXAMPLE 4 Adding Several Integers

A football team has to gain at least 10 yards during four plays in order to keep the ball. Suppose that your college team lost 6 yards on the first play, gained 8 yards on the second play, lost 2 yards on the third play, and gained 7 yards on the fourth play. Did the team gain enough to keep the ball?

When you're adding several integers, work from left to right.

1st play, 2nd play, 3rd play, 4th play

$^-6$ yards + 8 yards + $^-2$ yards + 7 yards — First add $^-6 + 8$.

2 yards + $^-2$ yards + 7 yards — Add $2 + {}^-2$.

0 yards + 7 yards — Add $0 + 7$.

7 yards

No, the team didn't gain enough yards to keep the ball.

WORK PROBLEM 4 AT THE SIDE.

OBJECTIVE 2 In our football model for adding integers, we said that zero indicated no gain or loss, that is, no change in the position of the ball. This example illustrates one of the *properties of addition.* A property of addition is something that applies to all addition problems, regardless of the numbers you use.

Addition Property of 0

Adding zero to any number leaves the number unchanged. Some examples are shown below.

$$0 + 6 = 6 \qquad ^-25 + 0 = {}^-25 \qquad 72{,}399 + 0 = 72{,}399$$

$$0 + {}^-100 = {}^-100$$

3. Find each sum.

(a) $^-3 + 7$

(b) $6 + {}^-12$

(c) $12 + {}^-7$

(d) $^-10 + 2$

(e) $5 + {}^-9$

(f) $^-8 + 9$

4. (a) The temperature was $^-15$ degrees this morning. It rose 21 degrees during the day, then dropped 10 degrees. What is the new temperature?

(b) Andrew had \$60 in his checking account. He wrote a \$20 check for gas and a \$75 check for groceries. Later in the day he deposited an \$85 tax refund in his account. What is the balance in his account?

ANSWERS

3. (a) 4 (b) $^-6$ (c) 5 (d) $^-8$ (e) $^-4$ (f) 1
4. (a) $^-15 + 21 + {}^-10 = {}^-4$ degrees
(b) $60 + {}^-20 + {}^-75 + 85 = \$50$

5. Rewrite each sum, using the commutative property of addition. Check that the sum is unchanged.

(a) $175 + 25 = \_\_ + \_\_$

Both sums are _____.

(b) $7 + {}^{-}37 = \_\_ + \_\_$

Both sums are _____.

(c) ${}^{-}16 + 16 = \_\_ + \_\_$

Both sums are _____.

(d) ${}^{-}9 + {}^{-}41 = \_\_ + \_\_$

Both sums are _____.

ANSWERS

5. (a) $175 + 25 = 25 + 175$; 200
(b) $7 + {}^{-}37 = {}^{-}37 + 7$; ${}^{-}30$
(c) ${}^{-}16 + 16 = 16 + {}^{-}16$; 0
(d) ${}^{-}9 + {}^{-}41 = {}^{-}41 + {}^{-}9$; ${}^{-}50$

Another property of addition is that you can change the *order* of the addends and still get the same sum. For example,

gaining 2 yards, then losing 7 yards, gives a result of ${}^{-}5$ yards.
losing 7 yards, then gaining 2 yards, also gives a result of ${}^{-}5$ yards.

Commutative Property of Addition

Changing the **order** of two addends does not change the sum. Here are some examples.

$84 + 2 = 2 + 84$ Both sums are 86.

${}^{-}10 + 6 = 6 + {}^{-}10$ Both sums are ${}^{-}4$.

EXAMPLE 5 Using the Commutative Property of Addition

Rewrite each sum, using the commutative property of addition. Check that the sum is unchanged.

(a) $65 + 35$

$$65 + 35 = 35 + 65$$

Both sums are 100, so the sum is unchanged.

(b) ${}^{-}20 + {}^{-}30$

$${}^{-}20 + {}^{-}30 = {}^{-}30 + {}^{-}20$$

Both sums are ${}^{-}50$, so the sum is unchanged.

◀◀ WORK PROBLEM 5 AT THE SIDE.

For three addends, parentheses are used to tell you which pair of numbers to add first, as shown here.

$\underbrace{(3 + 4)}_{7} + 2$ First add $3 + 4$.
$\underbrace{7 + 2}_{9}$ Then add $7 + 2$.

$3 + \underbrace{(4 + 2)}_{6}$ First add $4 + 2$.
$\underbrace{3 + 6}_{9}$ Then add $3 + 6$.

Both sums are 9. This example illustrates another property of addition.

Associative Property of Addition

Changing the **grouping** of addends does not change the sum. Some examples are:

$$\underbrace{\underbrace{({}^{-}5 + 5)}_{0} + 8}_{8} = \underbrace{{}^{-}5 + \underbrace{(5 + 8)}_{13}}_{8}$$

Both sums are 8.

$$\underbrace{3 + \underbrace{({}^{-}4 + {}^{-}6)}_{{}^{-}10}}_{{}^{-}7} = \underbrace{\underbrace{(3 + {}^{-}4)}_{{}^{-}1} + {}^{-}6}_{{}^{-}7}$$

Both sums are ${}^{-}7$.

We can use the associative property to make addition problems easier. Notice in the first example in the box above that it is easier to group $^-5 + 5$ (which is 0) and then add 8. In the second example in the box above, it is helpful to group $^-4 + {}^-6$ because the sum is $^-10$, and it is easy to work with multiples of 10.

EXAMPLE 6 Using the Associative Property of Addition

In each addition problem, pick out the two addends that would be easiest to add. Write parentheses around those addends. Then find the sum.

(a) $6 + 9 + {}^-9$

Group $9 + {}^-9$ because the sum is 0.

$$6 + \underbrace{(9 + {}^-9)}$$
$$\underbrace{6 + \quad 0}$$
$$6$$

(b) $17 + 3 + {}^-25$

Group $17 + 3$ because the sum is 20.

$$\underbrace{(17 + 3)} + {}^-25$$
$$\underbrace{20 \quad + {}^-25}$$
$$^-5$$

WORK PROBLEM 6 AT THE SIDE.

6. In each problem, write parentheses around the two addends that would be easiest to add. Then find the sum.

(a) $^-12 + 12 + {}^-19$

(b) $31 + 75 + {}^-75$

(c) $16 + {}^-1 + {}^-9$

(d) $^-8 + 5 + {}^-25$

ANSWERS

6. (a) $\underbrace{({}^-12 + 12)} + {}^-19$
$\underbrace{0 \quad + {}^-19}$
$^-19$

(b) $31 + \underbrace{(75 + {}^-75)}$
$\underbrace{31 + \quad 0}$
31

(c) $16 + \underbrace{({}^-1 + {}^-9)}$
$\underbrace{16 + \quad {}^-10}$
6

(d) $^-8 + \underbrace{(5 + {}^-25)}$
$\underbrace{{}^-8 + \quad {}^-20}$
$^-28$

Study Tips

Using Your Textbook

Do you use your textbook only for homework problems? If so, you're missing out on a lot of help. Answer these questions on a separate sheet of paper. By the time you're done, you'll know how to get the most out of your textbook.

Find the **Whole Number Diagnostic Test** in the front of the textbook.

1. What kinds of problems are on the test?

2. Where are the answers located?

3. What should you do if you get only 3 or 4 of the division problems correct?

Find the **Contents** in the front of the textbook.

4. How many chapters are in the book?

5. What numbering system is used to label the sections of each chapter?

6. What *three* things are listed at the end of *every* chapter?

7. Chapters 2–9 also list "Cumulative Review Exercises" at the end of each chapter. What does the word *cumulative* mean?

8. Use the **Contents**. What is the title of Section 1.6? What page does it start on?

9. Turn to the first page of Section 1.6. What are the objectives for this section? How can you tell where the material for Objective 2 starts?

10. On the first page of Section 1.6, explain the three items shown under *For Extra Help*.

11. Use the **Contents** to find **Chapter 1 Summary**. What page does it start on?

12. What is included in the Summary?

13. How could you use the Summary to prepare for a test?

14. Find **Chapter 1 Review Exercises**. What page does it start on?______ Look at exercises 9–17. How can you tell which section these exercises are from?

15. Find **Chapter 1 Test**. What page does it start on?______ How could you use the Chapter Test to prepare for a test in your math class?

16. Look at the **Contents** to see what is listed *after* Chapter 9. Explain which of these items will be useful to you and why.

17. Find **Answers to Selected Exercises** at the end of the book. What page does it start on?

18. Find the answers for Section 1.3 exercises. Which answers are given?

19. Find the answers for Chapter 1 Review Exercises and Chapter 1 Test. Which answers are given?

20. Find the **Index** in the back of the book. What is the *Index* showing?

21. List the pages where you could find information about *absolute value*. ________________

1.3 Exercises

Add by using the number line.

1. $^{-}2 + 5$

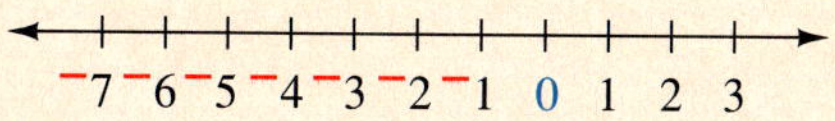

2. $^{-}3 + 4$

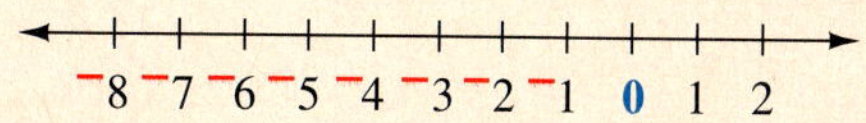

3. $^{-}5 + {}^{-}2$

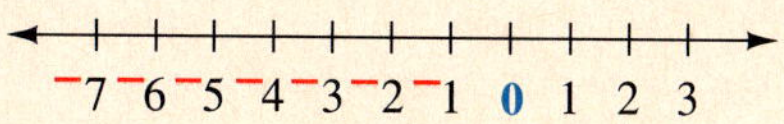

4. $^{-}2 + {}^{-}2$

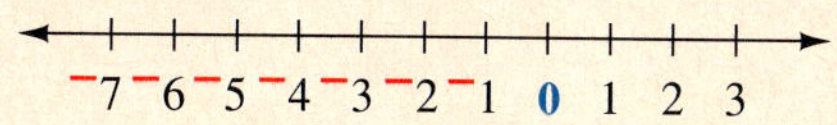

5. $3 + {}^{-}4$

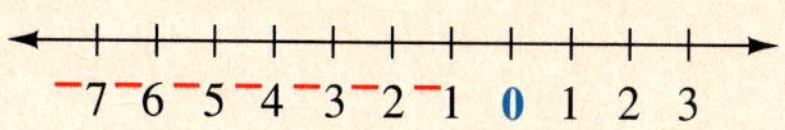

6. $5 + {}^{-}1$

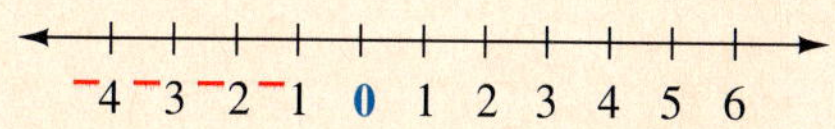

Add.

7. (a) $^{-}5 + {}^{-}5$
(b) $5 + 5$

8. (a) $^{-}9 + {}^{-}9$
(b) $9 + 9$

9. (a) $7 + 5$
(b) $^{-}7 + {}^{-}5$

10. (a) $3 + 6$
(b) $^{-}3 + {}^{-}6$

11. (a) $^{-}25 + {}^{-}25$
(b) $25 + 25$

12. (a) $^{-}30 + {}^{-}30$
(b) $30 + 30$

13. (a) $48 + 110$
(b) $^{-}48 + {}^{-}110$

14. (a) $235 + 21$
(b) $^{-}235 + {}^{-}21$

15. What pattern do you see in your answers to Exercises 7–14? Explain why this pattern occurs.

16. In your own words, explain how to add two integers that have the same sign.

Add.

17. **(a)** $^{-}6 + 8$
(b) $6 + {}^{-}8$

18. **(a)** $^{-}3 + 7$
(b) $3 + {}^{-}7$

19. **(a)** $^{-}9 + 2$
(b) $9 + {}^{-}2$

20. **(a)** $^{-}8 + 7$
(b) $8 + {}^{-}7$

21. **(a)** $20 + {}^{-}25$
(b) $^{-}20 + 25$

22. **(a)** $30 + {}^{-}40$
(b) $^{-}30 + 40$

23. **(a)** $200 + {}^{-}50$
(b) $^{-}200 + 50$

24. **(a)** $150 + {}^{-}100$
(b) $^{-}150 + 100$

25. What pattern do you see in your answers to Exercises 17–24? Explain why this pattern occurs.

26. In your own words, explain how to add two integers that have different signs.

Add.

27. $^{-}8 + 5$

28. $^{-}3 + 2$

29. $^{-}1 + 8$

30. $^{-}4 + 10$

31. $^{-}2 + {}^{-}5$

32. $^{-}7 + {}^{-}3$

33. $6 + {}^{-}5$

34. $11 + {}^{-}3$

35. $4 + {}^{-}12$

36. $9 + {}^{-}10$

37. $^{-}10 + {}^{-}10$

38. $^{-}5 + {}^{-}20$

39. $^{-}17 + 0$

40. $0 + {}^{-}11$

41. $1 + {}^{-}23$

42. $13 + {}^{-}1$

43. $^{-}2 + {}^{-}12 + {}^{-}5$

44. $^{-}16 + {}^{-}1 + {}^{-}3$

45. $8 + 6 + {}^{-}8$

46. $^-5 + 2 + 5$

47. $^-7 + 6 + {}^-4$

48. $^-9 + 8 + {}^-2$

49. $^-3 + {}^-11 + 14$

50. $15 + {}^-7 + {}^-8$

51. $10 + {}^-6 + {}^-3 + 4$

52. $2 + {}^-1 + {}^-9 + 12$

53. $^-7 + 28 + {}^-56 + 3$

54. $4 + {}^-37 + 29 + {}^-5$

Write an addition problem for each situation and find the sum.

55. The football team gained 13 yards on the first play and lost 17 yards on the second play. How many yards did the team gain or lose in all?

56. At penguin breeding grounds on Antarctic islands, temperatures routinely drop to $^-15$ °C. Temperatures in the interior of the continent may drop another 60 °C below that. What is the temperature in the interior?

57. Nick's checking account was overdrawn by \$62. He deposited \$50 in his account. What is the balance in his account?

58. \$88 was stolen from Jay's car. He got \$35 of it back. What was his net loss?

59. Red River flood waters rose 8 feet on Monday, dropped 3 feet on Tuesday, and dropped 1 more foot on Wednesday. What was the new flood level?

60. Marion lost 4 pounds in April, gained 2 pounds in May, and gained 3 pounds in June. How many pounds did she gain or lose in all?

61. While playing a card game, Jeff first lost 20 points, won 75 points, and then lost 55 points. What was his point total?

62. Cynthia had \$100 in her checking account. She wrote a check for \$83 and was charged \$17 for overdrawing her account last month. What is her account balance?

Rewrite each sum, using the commutative property of addition. Show that the sum is unchanged.

63. $^{-}18 + {}^{-}5 = \_\_\_\_ + \_\_\_\_$

Both sums are ____.

64. $^{-}12 + 20 = \_\_\_\_ + \_\_\_\_$

Both sums are ____.

65. $^{-}4 + 15 = \_\_\_\_ + \_\_\_\_$

Both sums are ____.

66. $17 + 1 = \_\_\_\_ + \_\_\_\_$

Both sums are ____.

In each addition problem, write parentheses around the two addends that would be easiest to add. Then find the sum.

67. $6 + {}^{-}14 + 14$

68. $9 + {}^{-}9 + {}^{-}8$

69. $^{-}14 + {}^{-}6 + {}^{-}7$

70. $^{-}18 + 3 + 7$

Ⓖ **71.** Make up three of your own examples that illustrate the addition property of zero.

Ⓖ **72.** Make up three of your own examples that illustrate the associative property of addition. Show that the sum is unchanged.

Ⓖ *Find each sum.*

73. $^{-}7081 + 2965$

74. $^{-}1398 + 3802$

75. $^{-}179 + {}^{-}61 + 8926$

76. $36 + {}^{-}6215 + 428$

77. $86 + {}^{-}99{,}000 + 0 + 2837$

78. $^{-}16{,}719 + 0 + 8878 + {}^{-}14$

1.4 Subtracting Integers

OBJECTIVES

1. Find the opposite of a signed number.
2. Subtract two integers.
3. Combine adding and subtracting of integers.

FOR EXTRA HELP

Tutorial | Tape 1 | SSM, Sec. 1.4

OBJECTIVE 1 Look at how the integers match up on this number line.

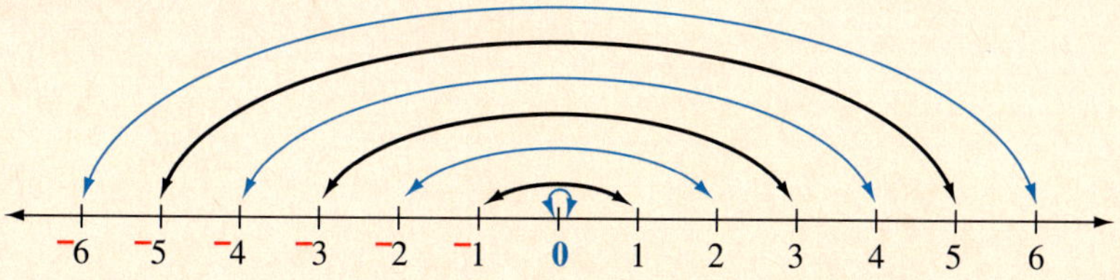

Each integer is matched with its **opposite.** Opposites are the same *distance* from zero on the number line but are on *opposite sides* of zero.

$^{+}2$ is the opposite of $^{-}2$ and $^{-}2$ is the opposite of $^{+}2$.

When you add opposites, the sum is always zero. The opposite of a number is also called its *additive inverse.*

$2 + {}^{-}2 = \mathbf{0}$ and ${}^{-}2 + 2 = \mathbf{0}$

EXAMPLE 1 Finding the Opposite of a Signed Number

Find the opposite (additive inverse) of each number. Show that the sum of the number and its opposite is zero.

(a) 6 — The opposite of 6 is $^{-}6$ and $6 + {}^{-}6 = \mathbf{0}$.

(b) $^{-}10$ — The opposite of $^{-}10$ is 10 and ${}^{-}10 + 10 = \mathbf{0}$.

(c) 0 — The opposite of 0 is 0 and $0 + 0 = \mathbf{0}$.

WORK PROBLEM 1 AT THE SIDE.

1. Find the additive inverse (opposite) of each number. Show that the sum of the number and its additive inverse is zero.

(a) 5

(b) 48

(c) 0

(d) $^{-}1$

(e) $^{-}24$

OBJECTIVE 2 Now that you know how to add integers and how to find opposites, you can subtract integers. Every subtraction problem has the same answer as a related addition problem. The problems below illustrate how to change subtraction problems into addition problems.

$$6 - 2 = 4 \qquad 8 - 3 = 5$$
$$\downarrow \ \downarrow \qquad\qquad \downarrow \ \downarrow$$
$$6 + {}^{-}2 = 4 \qquad 8 + {}^{-}3 = 5$$

Same answer — Same answer

Subtracting Two Integers

To subtract two numbers, *add* the first number to the *opposite* of the second number. Remember to change *two* things:

Step 1 Make one pencil stroke to change the subtraction symbol to an addition symbol.

Step 2 Make a second pencil stroke to change the second number to its opposite. If the second number is positive, change it to negative. If the second number is negative, change it to positive.

Note

When changing a subtraction problem to an addition problem, do *not* make any change in the *first* number. The pattern is

1st number − 2nd number = 1st number + opposite of 2nd number.

ANSWERS

1. (a) $^{-}5$; $5 + {}^{-}5 = 0$
(b) $^{-}48$; $48 + {}^{-}48 = 0$
(c) 0; $0 + 0 = 0$
(d) 1; ${}^{-}1 + 1 = 0$
(e) 24; ${}^{-}24 + 24 = 0$

2. Subtract by changing subtraction to adding the opposite. (Make two pencil strokes.)

(a) $^-6 - 5$

(b) $3 - {}^-10$

(c) $^-8 - {}^-2$

(d) $0 - 10$

(e) $^-4 - {}^-12$

(f) $9 - 7$

EXAMPLE 2 Subtracting Two Integers

Make two pencil strokes to change each subtraction problem into an addition problem. Then find the sum.

(a) Change 10 to $^-10$.

$4 - \mathbf{10} = 4 + {}^-\mathbf{10} = {}^-6$

Change subtraction to addition.

(b) Change $^-6$ to $^+6$.

$^-9 - {}^-\mathbf{6} = {}^-9 + {}^+\mathbf{6} = {}^-3$

Change subtraction to addition.

(c) Change $^-5$ to $^+5$.

$3 - {}^-\mathbf{5} = 3 + {}^+\mathbf{5} = 8$ Make *two* pencil strokes.

Change subtraction to addition.

(d) Change 9 to $^-9$.

$^-2 - \mathbf{9} = {}^-2 + {}^-\mathbf{9} = {}^-11$ Make *two* pencil strokes.

Change subtraction to addition.

◀◀ WORK PROBLEM 2 AT THE SIDE.

OBJECTIVE 3 When a problem involves adding and subtracting more than two signed numbers, first change all the subtractions to additions. Then add from left to right.

3. Simplify.

(a) $6 - 7 + {}^-3$

(b) $^-2 + {}^-3 - {}^-5$

(c) $7 - 7 - 7$

(d) $^-3 - 9 + 4 - {}^-20$

EXAMPLE 3 Combining Addition and Subtraction of Integers

Simplify.

$^-5 - \mathbf{10} - \mathbf{12} + 1$ Change all subtractions to addition. Change 10 to $^-10$. Change 12 to $^-12$

$^-5 + {}^-\mathbf{10} + {}^-\mathbf{12} + 1$ Add from left to right. First add $^-5 + {}^-10$.

$^-15 + {}^-12 + 1$ Then add $^-15 + {}^-12$.

$^-27 + 1$ Finally, add $^-27 + 1$.

$^-26$

◀◀ WORK PROBLEM 3 AT THE SIDE.

Calculator Tip: You can use the *change of sign* key [+/−] or [+○−] on your scientific calculator to enter negative numbers.

To enter $^-5$, press [5] [+/−]. To enter $^+5$, just press [5].

To enter Example 3 above, press the following keys.

5 [+/−] [−] 10 [−] 12 [+] 1 [=] The answer is $^-26$.

(5 [+/−] is $^-5$; [−] means Subtract)

When using a calculator, you do *not* need to change subtraction to addition.

ANSWERS

2. (a) $^-11$ (b) 13 (c) $^-6$ (d) $^-10$ (e) 8 (f) 2

3. (a) $^-4$ (b) 0 (c) $^-7$ (d) 12

Name Date Class

1.4 Exercises

Find the opposite (additive inverse) of each number. Show that the sum of the number and its opposite is zero.

1. 6 **2.** 10 **3.** $^{-}13$

4. $^{-}3$ **5.** 0 **6.** 1

Subtract by changing subtraction to addition.

7. $19 - 5$ **8.** $24 - 11$ **9.** $10 - 12$ **10.** $1 - 8$

11. $7 - 19$ **12.** $2 - 17$ **13.** $^{-}15 - 10$ **14.** $^{-}10 - 4$

15. $^{-}9 - 14$ **16.** $^{-}3 - 11$ **17.** $^{-}3 - {}^{-}8$ **18.** $^{-}1 - {}^{-}4$

19. $6 - {}^{-}14$ **20.** $8 - {}^{-}1$ **21.** $1 - {}^{-}10$ **22.** $6 - {}^{-}1$

23. $^{-}30 - 30$ **24.** $^{-}25 - 25$ **25.** $^{-}16 - {}^{-}16$ **26.** $^{-}20 - {}^{-}20$

27. $13 - 13$ **28.** $19 - 19$ **29.** $0 - 6$ **30.** $0 - 12$

31. **(a)** $3 - {}^{-}5$
(b) $3 - 5$
(c) $^{-}3 - {}^{-}5$
(d) $^{-}3 - 5$

32. **(a)** $9 - 6$
(b) $^{-}9 - 6$
(c) $9 - {}^{-}6$
(d) $^{-}9 - {}^{-}6$

33. **(a)** $4 - 7$
(b) $4 - {}^{-}7$
(c) $^{-}4 - 7$
(d) $^{-}4 - {}^{-}7$

34. **(a)** $8 - {}^{-}2$
(b) $^{-}8 - {}^{-}2$
(c) $8 - 2$
(d) $^{-}8 - 2$

35. Look for a pattern in these pairs of subtractions.

$^-3 - 5 = \_\_$ $\quad$ $^-4 - {^-3} = \_\_$

$5 - {^-3} = \_\_$ $\quad$ $^-3 - {^-4} = \_\_$

Explain what happens when you try to apply the commutative property to subtraction.

36. Recall the addition property of zero. Can zero be used in a subtraction problem without changing the other number? Explain what happens and give several examples.

Simplify.

37. $^-2 - 2 - 2$

38. $^-8 - 4 - 8$

39. $9 - 6 - 3 - 5$

40. $12 - 7 - 5 - 4$

41. $3 - {^-3} - 10 - {^-7}$

42. $1 - 9 - {^-2} - {^-6}$

43. $^-2 + {^-11} - {^-3}$

44. $^-5 - {^-2} + {^-6}$

45. $4 - {^-13} + {^-5}$

46. $6 - {^-1} + {^-10}$

47. $6 + 0 - 12 + 1$

48. $^-10 - 4 + 0 + 18$

49. Find, correct, and explain the mistake made in this subtraction.

$$\begin{array}{c} ^-6 - 6 \\ \downarrow \; \downarrow \\ ^-6 + 6 = 0 \end{array}$$

50. Find, correct, and explain the mistake made in this subtraction.

$$\begin{array}{c} ^-7 - \;\; 5 \\ \downarrow \;\; \downarrow \\ ^+7 + {^-5} = 2 \end{array}$$

Ⓖ *Simplify.*

51. $^-2 + {^-11} + |{^-2}|$

52. $5 - |{^-3}| + 3$

53. $0 - |{^-7} + 2|$

54. $|1 - 8| - |0|$

55. $^-3 - ({^-2} + 4) + {^-5}$

56. $5 - 8 - (6 - 7) + 1$

1.5 *Rounding and Estimating*

One way to get a rough check on an answer is to *round* the numbers in the problem. **Rounding** a number means finding a number that is close to the original number, but easier to work with.

For example, a superintendent of schools in a large city might be discussing the need to build new schools. In making her point, it probably would not be necessary to say that the school district has 152,807 students—it probably would be sufficient to say that there are 153,000 students, or even 150,000 students.

OBJECTIVES

1. Locate the place to which a number is to be rounded.
2. Round integers.
3. Use front end rounding to estimate answers in addition and subtraction.

FOR EXTRA HELP

Tutorial — Tape 2 — SSM, Sec. 1.5

OBJECTIVE 1 The first step in rounding a number is to locate the *place to which the number is to be rounded.*

EXAMPLE 1 Finding the Place to Which a Number Is to Be Rounded

Locate and draw a line under the place to which each number is to be rounded. Then answer the question.

(a) Round −23 to the nearest ten. Is −23 closer to −20 or −30?

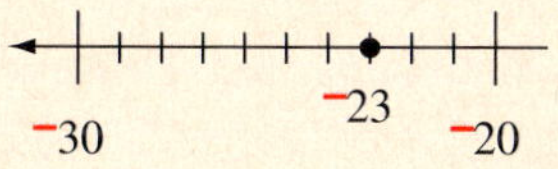

−23 is closer to −20.
↑
Tens place

(b) Round $381 to the nearest hundred. Is it closer to $300 or $400?

$381 is closer to $400.
↑
Hundreds place

(c) Round −54,702 to the nearest thousand. Is it closer to −54,000 or −55,000?

−54,702 is closer to −55,000.
Thousands place ↑

WORK PROBLEM 1 AT THE SIDE.

1. Locate and draw a line under the place to which the number is to be rounded. Then answer the question.

(a) −746 (nearest ten)
Is it closer to −740 or −750? ______

(b) 2412 (nearest thousand)
Is it closer to 2000 or 3000? ______

(c) −89,512 (nearest hundred)
Is it closer to −89,500 or −89,600? ______

(d) 546,325 (nearest ten thousand)
Is it closer to 540,000 or 550,000? ______

OBJECTIVE 2 Use the following rules for rounding integers.

Step 1 Locate the *place* to which the number is to be rounded. Draw a line under that place.

Step 2 Look only at the next digit to the right of the one you underlined. If it is 5 or more, increase the underlined digit by 1. If the next digit to the right is 4 or less, do not change the digit in the underlined place.

Step 3 *Change* all digits to the right of the underlined place to zeros.

Note
If you are rounding a negative number, be careful to write the negative sign in front of the rounded number. For example, −79 rounds to −80.

ANSWERS

1. (a) −746 is closer to −750
(b) 2412 is closer to 2000
(c) −89,512 is closer to −89,500
(d) 546,325 is closer to 550,000

2. Round to the nearest ten.

(a) 34

(b) $^{-}61$

(c) $^{-}683$

(d) 1792

EXAMPLE 2 Using the Rounding Rule for 4 or Less

Round 349 to the nearest hundred.

Step 1 Locate the place to which the number is being rounded. Draw a line under that place.

349
Hundreds place

Step 2 Because the next digit to the right of the underlined place is 4, which is 4 or less, do *not* change the digit in the underlined place.

Next digit is 4 or less.
349
3 remains 3.

Step 3 Change all digits to the right of the underlined place to zeros.

Change to 0.
349 rounded to the nearest hundred is 300.
Leave 3 as 3.

In other words, 349 is closer to 300 than to 400.

WORK PROBLEM 2 AT THE SIDE.

3. Round to the nearest thousand.

(a) 1725

(b) $^{-}6511$

(c) 58,829

(d) $^{-}83,904$

EXAMPLE 3 Using the Rounding Rule for 5 or More

Round 36,833 to the nearest thousand.

Step 1 Find the place to which the number is to be rounded and draw a line under that place.

36,833
Thousands place

Step 2 Because the next digit to the right of the underlined place is 8, which is 5 or more, add 1 to the underlined place.

Next digit is 5 or more.
36,833
Change 6 to 7.

Step 3 All digits to the right of the underlined place are changed to zeros.

Change to 0.
36,833 rounded to the nearest thousand is 37,000.
Change 6 to 7.

In other words, 36,833 is closer to 37,000 than to 36,000.

WORK PROBLEM 3 AT THE SIDE.

ANSWERS

2. **(a)** 30 **(b)** $^{-}60$ **(c)** $^{-}680$ **(d)** 1790

3. **(a)** 2000 **(b)** $^{-}7000$ **(c)** 59,000 **(d)** $^{-}84,000$

EXAMPLE 4 Using Rules for Rounding

(a) Round $^-2382$ to the nearest ten.

Step 1 $^-2382$

Tens place

Step 2 The next digit to the right is 2, which is 4 or less.

Next digit is 4 or less.

$^-2382$

Leave 8 as 8.

Change to 0.

Step 3 $^-2382$ rounds to $^-2380$.

$^-2382$ rounded to the nearest ten is $^-2380$.

(b) Round 13,961 to the nearest hundred.

Step 1 13,961

Hundreds place

Step 2 The next digit to the right is a 6.

Next digit is 5 or more.

13,961

Change 9 to 10; write 0 and carry 1 into thousands place.

3 + carried 1 = 4.

Step 3

Change to 0.

13,961 rounds to 14,000.

Change 39 to 40.

13,961 rounded to the nearest hundred is 14,000.

Note

In Step 2, when you add 1 to the hundreds place, notice that the first three digits increased from 139 to 140.

13,961 rounded to **14,0**00

WORK PROBLEM 4 AT THE SIDE.

EXAMPLE 5 Rounding Large Numbers

(a) Round $^-37,892$ to the nearest ten thousand.

Step 1 $^-37,892$

Ten thousands place

Step 2 The next digit to the right is 7.

Next digit is 5 or more.

$^-37,892$

Change 3 to 4.

Change to 0.

Step 3 $^-37,892$ rounds to $^-40,000$.

Change 3 to 4.

$^-37,892$ rounded to the nearest ten thousand is $^-40,000$.

CONTINUED ON NEXT PAGE

4. Round as shown.

(a) 6036 to the nearest ten

(b) 31,968 to the nearest hundred

(c) 73,077 to the nearest thousand

(d) 4952 to the nearest thousand

(e) 85,949 to the nearest hundred

(f) 40,387 to the nearest thousand

ANSWERS

4. (a) 6040 **(b)** 32,000 **(c)** 73,000 **(d)** 5000 **(e)** 85,900 **(f)** 40,000

(b) Round 528,498,675 to the nearest million.

Next digit is 4 or less.

Step 1 528,498,675
Millions place

Step 2 528,498,675
Leave 8 as 8.

Change to 0.

Step 3 528,498,675 rounds to 528,000,000

528,498,675 rounded to the nearest million is 528,000,000.

◀◀ **WORK PROBLEM 5 AT THE SIDE.**

Objective 3 In many everyday situations, we can round numbers and **estimate** the answer to a problem. For example, suppose that you're thinking about buying a sofa for \$988 and a chair for \$209. You can round the prices and estimate the total cost as \$1000 + \$200 for a total of ≈\$1200. The ≈ symbol means "approximately equal to." The estimated total of ≈\$1200 is close enough to help you decide whether you can afford both items. Of course, when it comes time to pay the bill, you'll want the *exact* total of \$988 + \$209 = \$1197.

Front end rounding is often used to estimate answers. Each number is rounded to the highest possible place, so all the digits become zero except the first digit. Once the numbers have lots of zeros, working with them is easy.

EXAMPLE 6 Using Front End Rounding

Use front end rounding to round each number.

(a) −216

Round to the highest possible place, that is, the left-most digit. In this case, the left-most digit, 2, is in the hundreds place, so round to the nearest hundred.

Next digit is 4 or less.
−216
Leave 2 as 2.

Change to 0.
−216 rounds to −200.
Leave 2 as 2.

Notice that all the digits in the rounded number are zero, except the first digit.

(b) 97,203

The left-most digit, 9, is in the ten thousands place, so round to the nearest ten thousand.

Next digit is 5 or more.
97,203
Change 9 to 10.

Change to 0.
97,203 rounds to 100,000.
Change 9 to 10.
Carry 1 into the hundred thousands place.

Notice that all the digits in the rounded number are zero, except the first digit.

◀◀ **WORK PROBLEM 6 AT THE SIDE.**

5. Round as shown.

(a) −14,679 to the nearest ten thousand

(b) 724,518,715 to the nearest million

(c) −49,900,700 to the nearest million

(d) 306,779,000 to the nearest hundred million

6. Use front end rounding to round each number.

(a) −94

(b) 508

(c) −2522

(d) 9700

(e) 61,888

(f) −963,369

Answers

5. (a) −10,000 **(b)** 725,000,000 **(c)** −50,000,000 **(d)** 300,000,000

6. (a) −90 **(b)** 500 **(c)** −3000 **(d)** 10,000 **(e)** 60,000 **(f)** −1,000,000

EXAMPLE 7 Using Front End Rounding to Estimate an Answer

Use front end rounding to estimate an answer. Also find the exact answer.

Miesha's paycheck showed gross pay of $823. It also listed deductions of $291. What is her net pay after deductions?

estimate: Use front end rounding.

Next digit is 4 or less. — $823 rounds to $800. — Leave 8 as 8.

Next digit is 5 or more. — $291 rounds to $300. — Change 2 to 3.

Use the rounded numbers and subtract to estimate Miesha's net pay.

$800 − $300 = $500 ← Estimate

exact: $823 − $291 = $532 ← Exact

Meisha's paycheck will show the exact amount of $532. Because $532 is fairly close to the estimate of $500, Meisha can quickly see that the amount shown on her paycheck probably is correct. She might also use the estimate when talking to a friend, saying, "My net pay is about $500."

Note
Always estimate the answer first. Then, when you find the exact answer, check that it is close to the estimate. If your exact answer is very far off, rework the problem because you probably made an error.

WORK PROBLEM 7 AT THE SIDE.

Calculator Tip: It's easy to press the wrong key when using a calculator. If you use front end rounding and estimate the answer *before* entering the numbers, you can catch many such mistakes. For example, a student thought that he entered this problem correctly.

7836 [+] 5060 [=] **2776**

Front end rounding gives an estimated answer of 8000 + 5000 = 13,000, which is very different from 2776. Can you figure out which key the student pressed incorrectly? (He pressed [−] instead of [+].)

7. Use front end rounding to estimate an answer. Also find the exact answer. Pao Xiong is a bookkeeper for a small business. The company checking account is overdrawn by $3881. He deposits a check for $2090. What is the balance in the account?

estimate:

exact:

ANSWER

7. *estimate:* −$4000 + $2000 = −$2000
exact: −$3881 + $2090 = −$1791
The account is overdrawn by $1791, which is close to the estimate of −$2000.

Study Tips

Keys to Remembering What You Study

There are three important keys to improving your memory in math. The most successful students use all three.

Immediate Review

Unfortunately, the human memory is *not* very effective. Here is the bad news and good news.

Students who do *not* review	**Students who review immediately and review again later**
• After 20 minutes, they remember only 60% of what they learned. • After 1 day, they remember only 30%.	• They remember 80% of what they learned for long periods of time.

To get information from your *short*-term memory into your *long*-term memory, it needs to be repeated right away after you learn it, and then repeated again later.

Try to schedule some free time right after your math class and immediately do a few homework problems. (If not right after class, then as soon as possible, and definitely before you leave for the day.) Also read your lecture notes and fill in things you didn't have time to write down during class. If you have questions, take time *before you leave* to ask for help from your instructor, or a classmate, or a tutor in your learning center.

Later in the day, review your notes again and finish the homework assignment. This second review reinforces the information in your long-term memory. That way, before a test, you won't have to learn it all over again.

Overlearning

When you are learning many new things, it is easy for the newest concepts to interfere with remembering the ones you learned a week or two earlier. To prevent his problem, build review into *every* study session. After you complete the current homework assignment, rework one or two odd-numbered problems from each of the previous sections in the chapter. Check your answers in the *Answers* section in the back of the book.

Or, if you are farther along in the book, rework several problems from the *Chapter Test* at the end of each chapter. Mark the problems you rework, so that you do different problems during each review. Also try the *Cumulative Review Exercises* at the end of each chapter; they cover all the topics in the book up to that point.

Distributive Learning

Studying math for 3 hours on one night is *not* as effective as studying for 1 hour on each of three nights. Either way the total is 3 hours of studying. Why is the second method better for your memory?

The answer is *time*. Your brain needs time between learning sessions to continue to process the information subconsciously. This is why cramming the night before a math test doesn't work. The best learning, and remembering, happens when you study *every day*, rather than once or twice a week.

Make a Plan

How will *you* improve your math memory? Write down a plan that includes

1. How you will do immediate review after each class.
2. How you will build review into each study session.
3. How you will distribute your studying during each week.

1.5 Exercises

Round each number. Write your answers using the ≈ *symbol to mean "approximately equal to."*

1. 623 to the nearest ten

2. 206 to the nearest ten

3. ⁻1085 to the nearest ten

4. ⁻2439 to the nearest ten

5. 7862 to the nearest hundred

6. 6746 to the nearest hundred

7. ⁻86,813 to the nearest hundred

8. ⁻17,211 to the nearest hundred

9. 42,495 to the nearest hundred

10. 18,273 to the nearest hundred

11. ⁻5996 to the nearest hundred

12. ⁻8451 to the nearest hundred

13. 15,758 to the nearest hundred

14. 28,065 to the nearest hundred

15. ⁻78,499 to the nearest thousand

16. ⁻14,314 to the nearest thousand

17. 5847 to the nearest thousand

18. 49,706 to the nearest thousand

19. 53,182 to the nearest thousand

20. 13,124 to the nearest thousand

21. 595,008 to the nearest ten thousand

22. 725,182 to the nearest ten thousand

23. ⁻8,906,422 to the nearest million

24. ⁻13,713,409 to the nearest million

25. 139,610,000 to the nearest million

26. 609,845,500 to the nearest million

Use front end rounding to round each number.

27. Tyrone drove his truck 31,500 miles last year.

28. Ezra bought a used car with 67,300 miles on it.

29. From winter to summer the average temperature drops 56 degrees.

30. The flood waters fell 42 inches yesterday.

31. Jan earned $9942 working part time.

32. Carol deposited $285 in her checking account.

33. 60,950,000 Americans go to a video store each week.

34. 97,475,000 U.S. households have at least one TV.

35. The submarine will dive to 255 feet below the surface of the ocean.

36. DeAnne lost $1352 in the stock market.

37. The population of Alaska is 597,497 people.

38. The population of California is 31,905,000 people.

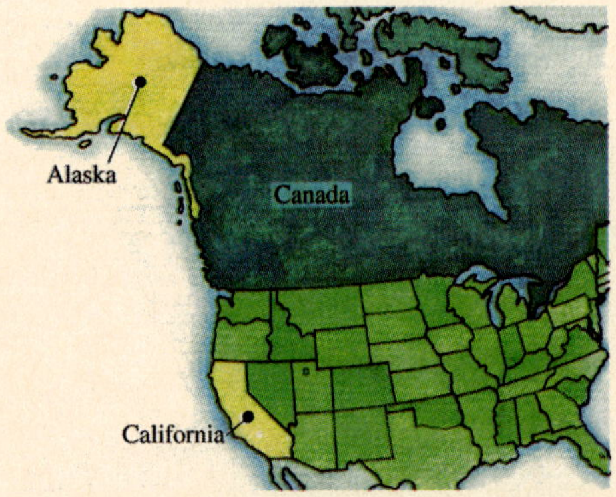

39. Explain in your own words how to do front end rounding. Also show two examples of numbers and how you round them.

40. Describe two situations in your own life when you might use rounded numbers. Describe two situations in which exact numbers are important.

Ⓖ *Use your estimation skills to pick the most reasonable answer for each addition. Do **not** solve the problems. Circle your choices.*

41. $^-42 + 89$

estimate: ______ + ______ = ______

exact: 131 $^-47$ 47

42. $^-66 + 25$

estimate: ______ + ______ = ______

exact: $^-91$ $^-41$ $^-21$

43. $16 + {}^-97$

estimate: ______ + ______ = ______

exact: $^-81$ $^-113$ $^-41$

44. $58 + {}^-19$

estimate: ______ + ______ = ______

exact: 39 $^-39$ $^-77$

45. $^-273 + {}^-399$

estimate:

exact: $^-126$ $^-672$ $^-992$

46. $^-311 + {}^-582$

estimate:

exact: 893 $^-271$ $^-893$

47. $3081 + 6826$

estimate:

exact: 3745 9907 15,907

48. $4904 + 1181$

estimate:

exact: 3723 9025 6085

Ⓖ *Change subtractions to addition. Then use your estimation skills to pick the most reasonable answer for each problem. Circle your choices.*

49. $23 - 81$

estimate:

exact: 58 104 $^-58$

50. $72 - 84$

estimate:

exact: 12 $^-12$ $^-156$

51. $^-39 - 39$

estimate:

exact: $^-78$ 0 78

52. $^-91 - 91$

estimate:

exact: 0 182 $^-182$

53. $^-106 + 34 - {}^-72$

estimate:

exact: $^-143$ 68 0

54. $52 - {}^-87 - 139$

estimate:

exact: 0 $^-104$ $^-174$

First use front end rounding to estimate the answer to each application problem. Then find the exact answer.

55. The community has raised \$52,882 for the homeless shelter. If the amount needed for the shelter is \$78,650, how much more needs to be collected?

estimate:
exact:

56. A truck weighs 9250 pounds when empty. After being loaded with firewood, it weighs 21,375 pounds. What is the weight of the firewood?

estimate:
exact:

57. Dorene Cox decided to establish a budget. She will spend \$485 for rent, \$325 for food, \$320 for child care, \$182 for transportation, and \$150 for other expenses, and she will put the remainder in savings. If her monthly take-home pay is \$1920, find her monthly savings.

estimate:
exact:

58. Jared Ueda had \$2874 in his checking account. He wrote checks for \$308 for auto repairs, \$580 for child support, and \$778 for an insurance payment. Find the amount remaining in his account.

estimate:
exact:

59. In a laboratory experiment, a mixture started at a temperature of $^-102$ degrees. First the temperature was raised 37 degrees and then raised 52 degrees. What was the final temperature?

estimate:
exact:

60. A scuba diver was photographing fish at 65 feet below the surface of the lagoon. She swam up 24 feet and then swam down 49 feet. What was her final depth?

estimate:
exact:

61. A "riding type" lawn mower costs \$525 more than a self-propelled lawn mower. If a self-propelled lawn mower costs \$380, find the cost of a "riding type" mower.

estimate:
exact:

62. The price of the least expensive rear bagging lawn mower used in a recent test was \$175. If this was \$475 less than the most expensive model used in the test, find the price of the most expensive mower.

estimate:
exact:

1.6 *Multiplying Integers*

OBJECTIVES

1. Use a raised dot or parentheses to express multiplication.
2. Multiply integers.
3. Identify properties of multiplication.
4. Estimate answers to application problems involving multiplication.

FOR EXTRA HELP

Tutorial | Tape 2 | SSM, Sec. 1.6

OBJECTIVE 1 In arithmetic we usually use "×" when writing multiplication problems. But in algebra, we use a raised dot or parentheses to show multiplication. The numbers being multiplied are called **factors** and the answer is called the **product.**

ARITHMETIC

$3 \times 5 = 15$ (3 and 5: Factors; 15: Product)

ALGEBRA

$3 \cdot 5 = 15$ or $3(5) = 15$ or $(3)(5) = 15$ (3 and 5: Factors; 15: Product)

EXAMPLE 1 Expressing Multiplication in Algebra

Rewrite each multiplication in three different ways, using a dot or parentheses. Also identify the factors and the product.

(a) 10×7

Rewrite it as $10 \cdot 7$ (Raised dot) or $10(7)$ or $(10)(7)$.

The factors are 10 and 7. The product is 70.

(b) 4×80

Rewrite it as $4 \cdot 80$ or $4(80)$ or $(4)(80)$.

The factors are 4 and 80. The product is 320.

WORK PROBLEM 1 AT THE SIDE.

1. Rewrite each multiplication in three different ways using a dot or parentheses. Also identify the factors and the product.

(a) 100×6

(b) 7×12

Note

Parentheses are used to show several different things in algebra. When we discussed the associative property earlier in this chapter, we used parentheses in this way.

$$6 + \underbrace{(9 + {}^{-}9)}_{} \leftarrow \text{Parentheses show which numbers to add first.}$$
$$\underbrace{6 + \quad 0}_{6}$$

Now we are using parentheses to tell you to multiply, as in 3(5) or (3)(5).

OBJECTIVE 2 Suppose that our football team gained 5 yards on the first play, gained 5 yards again on the second play, and gained 5 yards again on the third play. We can add to find the result.

5 yards + 5 yards + 5 yards = 15 yards

A quick way to add the same number several times is to multiply.

Our team made 3 plays	and	gained 5 yards each time.		Our team gained a total of 15 yards.
3	•	5	=	15

ANSWERS

1. (a) 100 • 6 or 100(6) or (100)(6)
The factors are 100 and 6; the product is 600.
(b) 7 • 12 or 7(12) or (7)(12)
The factors are 7 and 12; the product is 84.

Here are the rules for multiplying two integers.

Multiplying Two Integers

If the factors have *different signs,* the product is *negative.* For example,

$$^{-}2 \cdot 6 = {}^{-}12 \qquad \text{and} \qquad 4 \cdot {}^{-}5 = {}^{-}20.$$

If the factors have the *same sign,* the product is *positive.* For example,

$$7 \cdot 3 = 21 \qquad \text{and} \qquad {}^{-}3 \cdot {}^{-}10 = 30.$$

There are several ways to illustrate these rules. First we'll continue with football. Remember, you are interested in the results for *our* team. We will designate **our team** with a **positive sign** and **their team** with a **negative sign**.

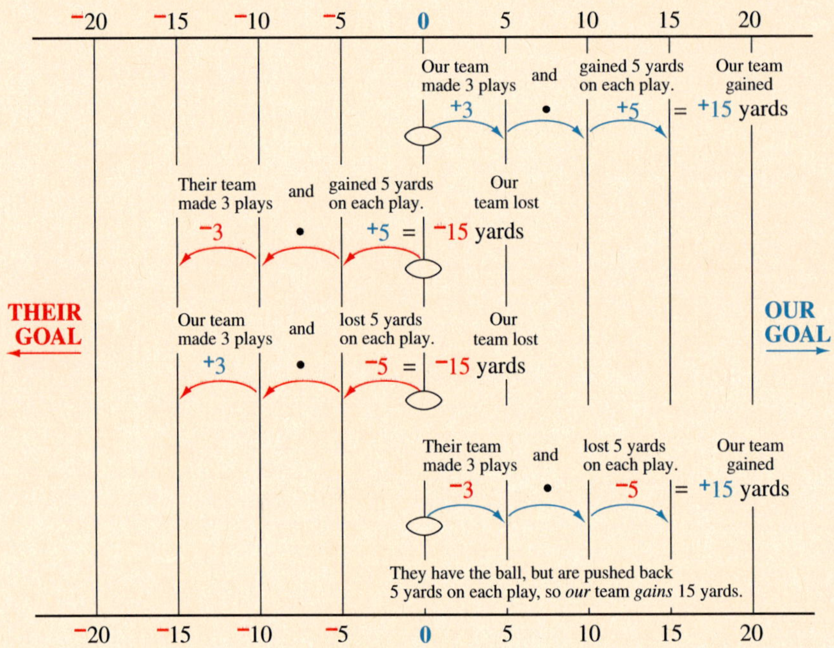

Here is a summary of the football examples.

When both factors have the *same* sign, the product is *positive.*

Both positive

$$3 \cdot 5 = 15$$

$$^{-}3 \cdot {}^{-}5 = 15$$

Both negative

Product is positive.

When the factors have *different* signs, the product is *negative.*

$$^{-}3 \cdot 5 = {}^{-}15$$

$$3 \cdot {}^{-}5 = {}^{-}15$$

Product is negative.

There is another way to look at these multiplication rules. In mathematics, the rules or patterns must always be consistent.

Look for a pattern in this list of products

Blue numbers decrease by 1. Red numbers decrease by 2.

$$4 \cdot 2 = 8$$
$$3 \cdot 2 = 6$$
$$2 \cdot 2 = 4$$
$$1 \cdot 2 = 2$$
$$0 \cdot 2 = 0$$
$$^{-}1 \cdot 2 = ?$$

To keep the red pattern going, replace the ? with a number that is 2 *less than* 0, which is $^{-}2$. This pattern illustrates that the product of two numbers with *different* signs is *negative.*

Look for a pattern in this list of products

Blue numbers decrease by 1. Red numbers increase by 2.

$$4 \cdot {}^{-}2 = {}^{-}8$$
$$3 \cdot {}^{-}2 = {}^{-}6$$
$$2 \cdot {}^{-}2 = {}^{-}4$$
$$1 \cdot {}^{-}2 = {}^{-}2$$
$$0 \cdot {}^{-}2 = 0$$
$$^{-}1 \cdot {}^{-}2 = ?$$

To keep the red pattern going, replace the ? with a number that is 2 *more than* 0, which is $^{+}2$. This pattern illustrates that the product of two numbers with the *same* sign is *positive.*

EXAMPLE 2 Multiplying Two Integers

(a) $^{-}2 \cdot 8 = {}^{-}16$ The factors have *different signs,* so the product is *negative.*

($^{-}2$: Negative; 8: Positive)

(b) $^{-}10\,({}^{-}6) = 60$ The factors have the *same sign,* so the product is *positive.*

(Both negative)

(c) $(9)({}^{-}11) = {}^{-}99$ The factors have *different signs,* so the product is *negative.*

(9: Positive; $^{-}11$: Negative)

WORK PROBLEM 2 AT THE SIDE.

Sometimes there are more than two factors in a multiplication problem. If there are parentheses around two of the factors, multiply them first. If there aren't any parentheses, start at the left and work with two factors at a time.

EXAMPLE 3 Multiplying Several Factors

Multiply.

(a) $^{-}3 \cdot (4 \cdot {}^{-}5)$ Parentheses tell you to multiply $4 \cdot {}^{-}5$ first. The factors have different signs, so the product is negative.

$^{-}3 \cdot {}^{-}20$ Then multiply $^{-}3 \cdot {}^{-}20$. Both factors have the same sign so the product is positive.

60

(b) $^{-}2 \cdot {}^{-}2 \cdot {}^{-}2$ There are no parentheses, so multiply $^{-}2 \cdot {}^{-}2$ first. The factors have the same sign, so the product is positive.

$4 \cdot {}^{-}2$ Then multiply $4 \cdot {}^{-}2$. The factors have different signs, so the product is negative.

$^{-}8$

---

2. Multiply.

(a) $7({}^{-}2)$

(b) $^{-}5 \cdot {}^{-}5$

(c) $^{-}1(14)$

(d) $10 \cdot 6$

(e) $({}^{-}4)({}^{-}9)$

ANSWERS

2. (a) $^{-}14$ (b) 25 (c) $^{-}14$ (d) 60 (e) 36

Note

In Example 3(b) you may be tempted to think that the final product will be *positive* because all the factors have the *same sign*. Be careful to work with just two factors at a time and keep track of the sign at each step.

Calculator Tip: You can use the *change of sign* key for multiplication and division, just as you did for adding and subtracting. To enter Example 3(b) on your calculator, press the following keys.

2 [+/−] [×] 2 [+/−] [×] 2 [+/−] [=] The answer is $^-8$.

$^-2$ $\quad$ $^-2$ $\quad$ $^-2$

◀◀ WORK PROBLEM 3 AT THE SIDE.

OBJECTIVE 3 Addition involving zero is unusual because adding zero does *not* change the number. For example, $7 + 0$ is still 7. (See **Section 1.3**.) But what happens in multiplication? Let's use our football team as an example.

Our team made 3 plays	and	didn't gain or lose yards on any play.		Altogether our team didn't gain or lose any yards.
3	•	0	=	0

This example illustrates one of the properties of multiplication.

Multiplication Property of 0

Multiplying any number by 0 gives a product of 0. Some examples are shown below.

$$^-16 \cdot 0 = 0 \qquad 0 \cdot 5 = 0 \qquad 32{,}977 \cdot 0 = 0$$

So, can you multiply a number by something that will *not* change the number?

$$6 \cdot ? = 6 \qquad ^-12 \cdot ? = {}^-12 \qquad ? \cdot 5876 = 5876$$

The number 1 can replace the ? in each example. These examples illustrate another property of multiplication.

Multiplication Property of 1

Multiplying a number by 1 leaves the number unchanged. Some examples are shown here.

$$6 \cdot 1 = 6 \qquad ^-12 \cdot 1 = {}^-12 \qquad 1 \cdot 5876 = 5876$$

EXAMPLE 4 Using Properties of Multiplication

Multiply. Then name the property illustrated by each example.

(a) $0 \cdot {}^-48 = 0$ Illustrates the multiplication property of 0.

(b) $615 \cdot 1 = 615$ Illustrates the multiplication property of 1.

◀◀ WORK PROBLEM 4 AT THE SIDE.

3. Multiply.

(a) $5 \cdot (^-10 \cdot 2)$

(b) $^-1 \cdot 8 \cdot {}^-5$

(c) $^-3 \cdot {}^-2 \cdot {}^-4$

(d) $^-2 \cdot (7 \cdot {}^-3)$

(e) $^-1 \cdot {}^-1 \cdot {}^-1$

4. Multiply. Then name the property illustrated by each example.

(a) $819 \cdot 0$

(b) $1 \cdot {}^-90$

(c) $25 \cdot 1$

(d) $0 \cdot {}^-75$

ANSWERS

3. **(a)** $^-100$ **(b)** 40 **(c)** $^-24$ **(d)** 42 **(e)** $^-1$

4. **(a)** 0; multiplication property of 0
(b) $^-90$; multiplication property of 1
(c) 25; multiplication property of 1
(d) 0; multiplication property of 0

When adding, we said that changing the *order* of the addends did not change the sum (commutative property of addition). We also found that changing the *grouping* of addends did not change the sum (associative property of addition). These same ideas apply to multiplication.

Commutative Property of Multiplication

Changing the *order* of two factors does not change the product.

Associative Property of Multiplication

Changing the *grouping* of factors does not change the product.

EXAMPLE 5 Using the Commutative and Associative Properties

Show that the product is unchanged and name the property that is illustrated by each of the following.

(a) $\underbrace{{}^-7 \cdot {}^-4}_{28} = \underbrace{{}^-4 \cdot {}^-7}_{28}$ Both products are 28.

This illustrates the commutative property of multiplication.

(b) $5 \cdot \underbrace{(10 \cdot 2)}_{20} = \underbrace{(5 \cdot 10)}_{50} \cdot 2$

$\underbrace{5 \cdot 20}_{100} = \underbrace{50 \cdot 2}_{100}$ Both products are 100.

This illustrates the associative property of multiplication.

WORK PROBLEM 5 AT THE SIDE.

Now that you are familiar with multiplication and addition, we can look at a property that involves both operations.

Distributive Property

Multiplication distributes over addition. For example,

$$3(6 + 2) = 3 \cdot 6 + 3 \cdot 2.$$

What is the **distributive** (dis-TRIB-yoo-tiv) **property** really saying? Notice that there is an understood multiplication symbol between the 3 and the parentheses. To "distribute" the 3 means to multiply 3 times each number inside the parentheses.

Understood to be *multiplying* by 3

$3(6 + 2)$

$3 \cdot (6 + 2)$

Using the distributive property,

$3 \cdot (6 + 2)$ can be rewritten as $3 \cdot 6 + 3 \cdot 2$.

Check that the product is unchanged. Either way the product is 24.

5. Show that the product is unchanged and name the property that is illustrated in each case.

(a) $({}^-3 \cdot {}^-3) \cdot {}^-2 = {}^-3 \cdot ({}^-3 \cdot {}^-2)$

(b) $11 \cdot 8 = 8 \cdot 11$

(c) $0 \cdot {}^-15 = {}^-15 \cdot 0$

(d) $4 \cdot ({}^-1 \cdot {}^-5) = (4 \cdot {}^-1) \cdot {}^-5$

ANSWERS

5. **(a)** ${}^-18 = {}^-18$; associative property of multiplication
(b) $88 = 88$; commutative property of multiplication
(c) $0 = 0$; commutative property of multiplication
(d) $20 = 20$; associative property of multiplication

6. Rewrite each of the following using the distributive property. Show that the product is unchanged.

(a) $3(8 + 7)$

(b) $10(6 + {}^{-}9)$

(c) ${}^{-}6(4 + 4)$

7. Use front end rounding to estimate an answer. Also find the exact answer.

An average of 27,095 baseball fans attended each of the 81 home games during the season. What was the total home game attendance for the season?

estimate:

exact:

EXAMPLE 6 Using the Distributive Property

Rewrite each of the following using the distributive property. Show that the product is unchanged.

(a) $4(3 + 7)$

$$4(3+7) = 4 \cdot 3 + 4 \cdot 7$$
$$4 \cdot (10) = 12 + 28$$
$$40 = 40$$

Both products are 40.

(b) ${}^{-}2({}^{-}5 + 1)$

$${}^{-}2({}^{-}5+1) = {}^{-}2 \cdot {}^{-}5 + {}^{-}2 \cdot 1$$
$${}^{-}2 \cdot ({}^{-}4) = 10 + {}^{-}2$$
$$8 = 8$$

Both products are 8.

WORK PROBLEM 6 AT THE SIDE.

OBJECTIVE 4 Front end rounding can be used to estimate answers in multiplication, just as we did when adding and subtracting (see **Section 1.5**). Once the numbers have been rounded so that there are lots of zeros, we can use the multiplication shortcut described in the Review chapter (see **Section R.3**). As a brief review, look at the pattern in these examples.

${}^{-}3 \cdot 2$ is ${}^{-}6$.

$${}^{-}30 \cdot 200 = {}^{-}6000$$

Total of 3 zeros — Write 3 zeros after the ${}^{-}6$.

${}^{-}2 \cdot {}^{-}5$ is 10.

$${}^{-}2000 \cdot {}^{-}5000 = 10{,}000{,}000$$

Total of 6 zeros — Write 6 zeros after the 10.

EXAMPLE 7 Using Front End Rounding to Estimate an Answer

Use front end rounding to estimate an answer. Also find the exact answer.

Last year the Video Land store had to replace 392 defective videos at a cost of \$19 each. How much money did the store lose on defective videos? (*Hint:* Because it's a loss, use a negative number.)

estimate: Use front end rounding: 392 rounds to 400 and ${}^{-}\$19$ rounds to ${}^{-}\$20$. Use the rounded numbers and multiply to estimate the total amount of money lost.

$4 \cdot {}^{-}2$ is ${}^{-}8$.

$$400 \cdot {}^{-}\$20 = {}^{-}\$8000 \quad \text{Estimate}$$

Total of 3 zeros — Write 3 zeros after the ${}^{-}8$.

exact: $392 \cdot {}^{-}\$19 = {}^{-}\$7448$

Because ${}^{-}\$7448$ is fairly close to the estimate of ${}^{-}\$8000$, you can see that ${}^{-}\$7448$ probably is correct. The store manager could also use the estimate to say, "We lost about \$8000 on defective videos last year."

WORK PROBLEM 7 AT THE SIDE.

ANSWERS

6. (a) $3 \cdot 8 + 3 \cdot 7$; both products are 45.
(b) $10 \cdot 6 + 10 \cdot {}^{-}9$; both products are ${}^{-}30$.
(c) ${}^{-}6 \cdot 4 + {}^{-}6 \cdot 4$; both products are ${}^{-}48$.

7. *estimate:* $30{,}000 \cdot 80 = 2{,}400{,}000$ fans
exact: $27{,}095 \cdot 81 = 2{,}194{,}695$ fans

Name Date Class

1.6 Exercises

Multiply.

1. **(a)** $9 \cdot 7$

(b) ${}^{-}9 \cdot {}^{-}7$

(c) ${}^{-}9 \cdot 7$

(d) $9 \cdot {}^{-}7$

2. **(a)** ${}^{-}6 \cdot 9$

(b) $6 \cdot {}^{-}9$

(c) ${}^{-}6 \cdot {}^{-}9$

(d) $6 \cdot 9$

3. **(a)** $7({}^{-}8)$

(b) ${}^{-}7(8)$

(c) $7(8)$

(d) ${}^{-}7({}^{-}8)$

4. **(a)** $8(6)$

(b) ${}^{-}8({}^{-}6)$

(c) ${}^{-}8(6)$

(d) $8({}^{-}6)$

5. ${}^{-}5 \cdot 7$

6. ${}^{-}10 \cdot 2$

7. $({}^{-}5)(9)$

8. $({}^{-}9)(4)$

9. $3({}^{-}6)$

10. $8({}^{-}9)$

11. $10({}^{-}5)$

12. $5({}^{-}11)$

13. $({}^{-}1)(40)$

14. $(75)({}^{-}1)$

15. ${}^{-}56 \cdot 1$

16. $1 \cdot {}^{-}87$

17. ${}^{-}8({}^{-}4)$

18. ${}^{-}3({}^{-}9)$

19. $11 \cdot 7$

20. $4 \cdot 25$

21. $25 \cdot 0$

22. $0 \cdot 30$

23. ${}^{-}19({}^{-}7)$

24. ${}^{-}21({}^{-}3)$

25. ${}^{-}13({}^{-}1)$

26. ${}^{-}1({}^{-}31)$

27. $(0)({}^{-}25)$

28. $({}^{-}50)(0)$

29. ${}^{-}4 \cdot {}^{-}6 \cdot 2$

30. ${}^{-}9 \cdot 3 \cdot {}^{-}3$

31. ${}^{-}4 \cdot {}^{-}2 \cdot {}^{-}7$

32. ${}^{-}6 \cdot {}^{-}2 \cdot {}^{-}3$

33. $5 \cdot {}^{-}8 \cdot 4$

34. $5 \cdot 4 \cdot {}^{-}6$

Writing Calculator Small Group

35. In your own words, explain the difference between the commutative and associative properties of multiplication. Show an example of each.

36. A student did this multiplication:

$$^{-}3 \cdot {}^{-}3 \cdot {}^{-}3 = 27.$$

He knew that $3 \cdot 3 \cdot 3$ is 27. Since all the factors have the same sign, he made the product positive. Do you agree with his reasoning? Explain.

37. Write three numerical examples for each of these situations:
(a) a positive number multiplied by $^{-}1$
(b) a negative number multiplied by $^{-}1$

Now write a rule that explains what happens when you multiply a signed number by $^{-}1$.

38. Do these multiplications.

$$^{-}2 \cdot {}^{-}2 = \underline{\qquad}$$

$$^{-}2 \cdot {}^{-}2 \cdot {}^{-}2 = \underline{\qquad}$$

$$^{-}2 \cdot {}^{-}2 \cdot {}^{-}2 \cdot {}^{-}2 = \underline{\qquad}$$

$$^{-}2 \cdot {}^{-}2 \cdot {}^{-}2 \cdot {}^{-}2 \cdot {}^{-}2 = \underline{\qquad}$$

Describe the pattern in the products. Then find the next three products without doing any multiplication.

Rewrite each multiplication, using the stated property. Show that the product is unchanged.

39. Distributive property
$9(^{-}3 + 5)$

40. Distributive property
$^{-}6(4 + {}^{-}5)$

41. Commutative property
$25 \cdot 8$

42. Commutative property
$^{-}7 \cdot {}^{-}11$

43. Associative property
$^{-}3 \cdot (^{-}2 \cdot {}^{-}5)$

44. Associative property
$(5 \cdot 5) \cdot 10$

First use front end rounding to estimate the answer to each application problem. Then find the exact answer.

45. Alliette receives $324 per week for doing child care in her home. How much income will she have for an entire year? There are 52 weeks in a year.

estimate:
exact:

46. Enrollment at our community college has increased by 875 students each of the last four semesters. What is the total increase?

estimate:
exact:

47. A new computer software store had losses of $9950 during each month of its first year. What was the total loss for the year?

estimate:
exact:

48. A long distance phone company estimates that it is losing 95 customers each week. How many customers will it lose in a year?

estimate:
exact:

49. Tuition at the state university is $182 per credit for undergraduates. How much tuition will Wei Chen pay for 13 credits?

estimate:
exact:

50. Pat ate a dozen crackers as a snack. Each cracker had 17 calories. How many calories did Pat eat?

estimate:
exact:

51. There are 24 hours in one day. How many hours are in one year (365 days)?

estimate:
exact:

52. There are 5280 feet in one mile. How many feet are in 17 miles?

estimate:
exact:

Ⓖ *Simplify.*

53. $^-8 \cdot |^-8 \cdot 8|$

54. $^-7 \cdot |7| \cdot |^-7|$

55. $^-37 \cdot {}^-1 \cdot 85 \cdot 0$

56. $^-1 \cdot 9732 \cdot {}^-1 \cdot {}^-1$

57. $|6 - 7| \cdot {}^-355{,}299$

58. $987 \cdot {}^-65{,}432 \cdot |9 - 9|$

Ⓖ *Each of these application problems requires several steps and may involve addition and subtraction as well as multiplication.*

59. Each of Maurice's four cats needs a $24 rabies shot and a $29 shot to prevent respiratory infections. There will also be one $35 "office charge." What will be the total amount of his bill?

60. In Ms. Zubero's math class there are six tests of 100 points each, eight quizzes of six points each, and 20 homework assignments of five points each. There are also four "bonus points" on each test. What is the total number of possible points?

61. There is a 3-degree drop in temperature for every thousand feet that an airplane climbs into the sky. If the temperature on the ground is 50 degrees, what will be the temperature when the plane reaches an altitude of 24,000 feet?

62. An unmanned submarine descends to 150 feet below the surface of the ocean. Then it continues to go deeper, taking a water sample every 25 feet. What is its depth when it takes the 15th sample?

1.7 Dividing Integers

OBJECTIVE 1 In arithmetic we usually use $\overline{)}$ to write division problems so that we can do the problem by hand. Calculators use the ÷ symbol for division. In algebra we usually show division by using a fraction bar or the ÷ symbol. The answer to a division problem is called the **quotient.**

ARITHMETIC

8 ← Quotient
Divisor → $2\overline{)16}$ ← Dividend

CALCULATOR AND ALGEBRA

Dividend ↓, Divisor ↓: $16 \div 2 = 8$ (↑ Quotient)

Dividend → $\dfrac{16}{2} = 8$ ← Quotient; Divisor → 2

For every division problem, you can write a related multiplication problem. (See **Section R.4**.) Because of this relationship, the rules for dividing integers are the same as the rules for multiplying integers. For example,

$$\frac{16}{8} = 2 \quad \text{because} \quad 2 \cdot 8 = 16.$$

$$\frac{^-16}{^-8} = 2 \quad \text{because} \quad 2 \cdot {^-8} = {^-16}.$$

$$\frac{^-16}{8} = {^-2} \quad \text{because} \quad {^-2} \cdot 8 = {^-16}.$$

$$\frac{16}{^-8} = {^-2} \quad \text{because} \quad {^-2} \cdot {^-8} = 16.$$

Dividing Two Integers

If the numbers have *different signs,* the quotient is *negative.* For example,

$$\frac{^-18}{3} = {^-6} \quad \text{and} \quad \frac{40}{^-5} = {^-8}.$$

If the numbers have the *same sign,* the quotient is *positive.* For example,

$$\frac{^-30}{^-6} = 5 \quad \text{and} \quad \frac{48}{8} = 6.$$

EXAMPLE 1 Dividing Two Integers

Divide.

(a) $\dfrac{^-20}{5}$ Numbers have different signs, so the quotient is negative. $\dfrac{^-20}{5} = {^-4}$

(b) $\dfrac{^-24}{^-4}$ Numbers have the same sign, so the quotient is positive. $\dfrac{^-24}{^-4} = 6$

(c) $60 \div {^-2}$ Numbers have different signs, so the quotient is negative. $60 \div {^-2} = {^-30}$

WORK PROBLEM 1 AT THE SIDE.

OBJECTIVES

1. Divide integers.
2. Identify properties of division.
3. Combine multiplying and dividing of integers.
4. Estimate answers to application problems involving division.
5. Interpret remainders in division application problems.

FOR EXTRA HELP

Tutorial | Tape 2 | SSM, Sec. 1.7

1. Divide.

(a) $\dfrac{40}{^-8}$

(b) $\dfrac{49}{7}$

(c) $\dfrac{^-32}{4}$

(d) $\dfrac{^-10}{^-10}$

(e) $^-81 \div 9$

(f) $^-100 \div {^-50}$

ANSWERS

1. (a) $^-5$ (b) 7 (c) $^-8$ (d) 1 (e) $^-9$ (f) 2

2. Divide. Then state the property illustrated by each division.

(a) $\frac{-12}{0}$

(b) $\frac{0}{39}$

(c) $\frac{-9}{1}$

(d) $\frac{21}{21}$

Answers

2. **(a)** undefined; division by 0 is undefined.
(b) 0; 0 divided by any nonzero number is 0.
(c) -9; any number divided by 1 is the number.
(d) 1; a nonzero number divided by itself is 1.

Objective 2 You have seen that 0 and 1 are used in special ways in addition and multiplication. This is also true in division.

Examples			Pattern (Division Property)
$\frac{5}{5} = 1$	$\frac{-18}{-18} = 1$	$\frac{-793}{-793} = 1$	When a nonzero number is divided by itself, the quotient is 1.
$\frac{5}{1} = 5$	$\frac{-18}{1} = -18$	$\frac{-793}{1} = -793$	When a number is divided by 1, the quotient is the number.
$\frac{0}{5} = 0$	$\frac{0}{-18} = 0$	$\frac{0}{-793} = 0$	When 0 is divided by any other number (except 0) the quotient is 0.
$\frac{5}{0}$ is undefined.	$\frac{-18}{0}$ is undefined.		Division by 0 is undefined. There is no answer.

The most surprising property is that division by zero cannot be done. Let's review the reason for that by rewriting this division problem as a related multiplication problem.

$\frac{-18}{0} = ?$ can be written as the multiplication $? \cdot 0 = -18$.

It you thought the answer to $\frac{-18}{0}$ should be 0, try replacing ? with 0. It doesn't work in the related multiplication problem! Try replacing ? with any number you like. The result in the related multiplication problem is always 0 instead of -18. That is how we know that dividing by zero cannot be done. Mathematicians say that it is *undefined.*

EXAMPLE 2 Using the Properties of Division

Divide. Then state the property illustrated by each example.

(a) $\frac{-312}{-312} = 1$ Any nonzero number divided by itself is 1.

(b) $\frac{75}{1} = 75$ Any number divided by 1 is the number.

(c) $\frac{0}{-19} = 0$ 0 divided by any nonzero number is 0.

(d) $\frac{48}{0}$ is undefined. Division by 0 is undefined.

Calculator Tip: Try Examples 2(c) and 2(d) on your calculator. You can use the change of sign key to enter -19.

0 [÷] 19 [+/_] [=] Answer is 0.
(19 [+/_] is -19)

48 [÷] 0 [=] Calculator shows "Error" or "E" for error because it cannot divide by 0.

◀◀ WORK PROBLEM 2 AT THE SIDE.

OBJECTIVE 3 When a problem involves both multiplying and dividing, first check to see if there are any parentheses. Do what is inside parentheses first. Then start at the left and work toward the right, using two numbers at a time.

EXAMPLE 3 Combining Multiplication and Division of Integers

Simplify.

(a) $6 \cdot {}^-10 \div ({}^-3 \cdot 2)$ — Work inside parentheses first. Do ${}^-3 \cdot 2$.

$6 \cdot {}^-10 \div {}^-6$ — Start at the left. Do $6 \cdot {}^-10$.

${}^-60 \div {}^-6$ — Finally, do ${}^-60 \div {}^-6$.

10

(b) ${}^-24 \div {}^-2 \cdot 4 \div {}^-6$ — There are no parentheses, so start at the left. First do ${}^-24 \div {}^-2$.

$12 \cdot 4 \div {}^-6$ — Next do $12 \cdot 4$.

$48 \div {}^-6$ — Finally, do $48 \div {}^-6$.

${}^-8$

(c) ${}^-50 \div {}^-5 \div {}^-2$ — There are no parentheses, so start at the left. First do ${}^-50 \div {}^-5$. The signs are the same, so the quotient is positive.

$10 \div {}^-2$ — Now do $10 \div {}^-2$. The signs are different, so the quotient is negative.

${}^-5$

WORK PROBLEM 3 AT THE SIDE.

OBJECTIVE 4 Front end rounding can be used to estimate answers in division just as you did when multiplying (see **Section 1.6**). Once the numbers have been rounded so that there are lots of zeros, you can use the division shortcut described in the Review chapter (see **Section R.5**). As a brief review, look at the pattern in these examples.

$400\not{0} \div {}^-5\not{0} = 400 \div {}^-5 = {}^-80$

Drop one zero from both dividend and divisor.

${}^-6\not{0}\not{0}\not{0} \div {}^-3\not{0}\not{0}\not{0} = {}^-6 \div {}^-3 = 2$

Drop 3 zeros from both dividend and divisor.

3. Simplify.

(a) $60 \div {}^-3 \div 4 \cdot {}^-5$

(b) ${}^-6 \cdot ({}^-16 \div {}^-8) \cdot 2$

(c) ${}^-8 \cdot 10 \div 4 \cdot {}^-3 \div {}^-6$

(d) $56 \div {}^-8 \div {}^-1$

ANSWERS

3. **(a)** 25 **(b)** ${}^-24$ **(c)** ${}^-10$ **(d)** 7

4. First use front end rounding to estimate an answer. Then find the exact answer.

Laurie and Chuck Struthers lost \$2724 on their stock investments last year. What was their average loss each month?

estimate:
exact:

EXAMPLE 4 Using Front End Rounding to Estimate an Answer in Division

First use front end rounding to estimate an answer. Then find the exact answer.

During a 24-hour laboratory experiment, the temperature of a solution dropped 96 degrees. What was the average drop in temperature each hour?

estimate: Use front end rounding: $^{-}96$ degrees rounds to $^{-}100$ degrees and 24 hours rounds to 20 hours. To estimate the average, divide the rounded number of degrees by the rounded number of hours.

$^{-}100$ degrees $\div$ 20 hours $=$ $^{-}5$ degrees each hour ← Estimate

exact: $^{-}96$ degrees $\div$ 24 hours $=$ $^{-}4$ degrees each hour

Because $^{-}4$ degrees is close to the estimate of $^{-}5$ degrees, you can see that $^{-}4$ degrees probably is correct.

Calculator Tip: The answer in Example 4 "came out even." In other words, the quotient was an integer. Suppose that the drop in temperature had been 97 degrees. Do the division on your calculator.

97 [+/−] [÷] 24 [=]   Calculator shows −4.041666667

(97 [+/−] is $^{-}97$)

The quotient is *not* an integer. We will work with numbers like these in Chapter 5, Decimals.

WORK PROBLEM 4 AT THE SIDE.

OBJECTIVE 5 In arithmetic, division problems often have a remainder, as shown below.

$$\begin{array}{r} 14 \text{ R10} \\ 25\overline{)360} \\ \underline{25} \\ 110 \\ \underline{100} \\ 10 \end{array} \leftarrow \text{Remainder}$$

But what does **R10** really mean? Let's look at this same problem by using money amounts.

EXAMPLE 5 Interpreting Remainders in Division Applications

Divide; then interpret the remainder in each of these applications.

(a) The math department at Lake Community College has \$360 in its budget to buy calculators for the math lab. If the calculators cost \$25 each, how many can be purchased? How much money will be left over?

We can solve this problem by using the same division as shown above. But this time we can decide what the remainder really means.

$$\begin{array}{r} 14 \\ \$25\overline{)\$360} \\ \underline{25} \\ 110 \\ \underline{100} \\ \$10 \end{array}$$

14 ← Number of calculators purchased
Cost of one calculator → \$25
\$360 ← Budget
\$10 ← Money left over

The department can buy 14 calculators. There will be \$10 left over.

CONTINUED ON NEXT PAGE

ANSWER

4. *estimate:* $^{-}$\$3000 $\div$ 10 months = $^{-}$\$300 each month
exact: $^{-}$\$2724 $\div$ 12 months = $^{-}$\$227 each month

(b) Luke's son is going on a Scout camping trip. There are 135 Scouts. Luke is renting tents that sleep 4 people each. How many tents should he rent?

We again use division to solve the problem. There is a remainder, but this time it must be interpreted differently than in the calculator example.

```
                    33  ← Number of tents with 4 Scouts each
Each tent holds → 4)135 ← Total number of Scouts
                    12
                    15
                    12
                     3  ← Scouts left over
```

If Luke rents 33 tents, 3 Scouts will have to sleep out in the rain. He must rent 34 tents to accommodate all the Scouts. (One tent will have only 3 Scouts in it.)

Calculator Tip: You can use your calculator to solve Example 5(a). Recall that digits on the *right* side of the decimal point show *part* of one whole. You cannot order *part* of one calculator, so ignore those digits and use only the *whole number part* of the quotient.

```
                     Write whole number →  14
360 [÷] 25 [=] 14.4                     25)360
               └──────────────────────→    350   Subtract 360 − 350
[AC] 14 [×] 25 [=] 350                      10     to get remainder.
 ↑
Clear 14.4 from calculator.
```

WORK PROBLEM 5 AT THE SIDE.

5. Divide; then interpret the remainder in each of these applications.

(a) Chad and Martha are baking cookies for a fund raiser. They baked 116 cookies and are putting them into packages of a dozen each. How many packages will they have for the fundraiser? How many cookies will be left over for them to eat?

(b) Coreen is a dispatcher for a bus company. A group of 249 senior citizens is going to a baseball game. If the buses will each hold 44 people, how many buses should she send to pick up the seniors?

Answers

5. (a) 9 packages, with 8 cookies left over to eat

(b) 6 buses, because 5 buses would leave 29 seniors standing on the curb

Numbers in the *Real World*

collaborative investigations

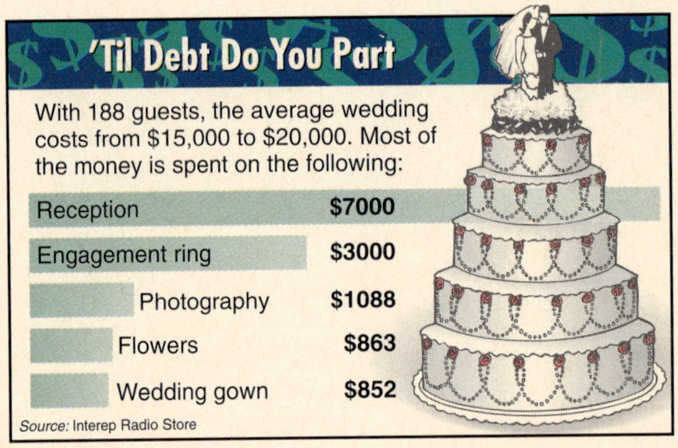

1. What is the total of the expenses shown in the graph?

2. This total is how much less than the "average wedding cost"?

3. What are some of the other expenses that could make up the difference? List three or four possibilities.

4. If you budget $7000 for your wedding reception, and the cost per person is $37, how many guests can you invite without going over your budget?
How much of your budgeted amount will be left over?

5. If you budget $4500, and the cost per person is $37, how many guests can you invite, and how much of your budgeted amount will be left over?

6. If you budget $1000?

7. Show how the last three answers are similar to doing a division problem where the quotient includes a remainder.

8. Different people will pick different amounts to budget for a reception. But if the price per person is $37 and as many guests are invited as possible, what is the maximum amount of money, in whole dollars, that could be left over? Explain your answer.

Name Date Class

1.7 Exercises

Divide.

1. **(a)** $14 \div 2$ **(b)** $^{-}14 \div {}^{-}2$ **(c)** $14 \div {}^{-}2$ **(d)** $^{-}14 \div 2$

2. **(a)** $^{-}18 \div {}^{-}3$ **(b)** $18 \div 3$ **(c)** $^{-}18 \div 3$ **(d)** $18 \div {}^{-}3$

3. **(a)** $^{-}42 \div 6$ **(b)** $^{-}42 \div {}^{-}6$ **(c)** $42 \div {}^{-}6$ **(d)** $42 \div 6$

4. **(a)** $45 \div 5$ **(b)** $45 \div {}^{-}5$ **(c)** $^{-}45 \div {}^{-}5$ **(d)** $^{-}45 \div 5$

5. **(a)** $\frac{35}{35}$ **(b)** $\frac{35}{1}$ **(c)** $\frac{^{-}13}{1}$ **(d)** $\frac{^{-}13}{^{-}13}$

6. **(a)** $\frac{^{-}23}{1}$ **(b)** $\frac{^{-}23}{^{-}23}$ **(c)** $\frac{17}{1}$ **(d)** $\frac{17}{17}$

7. **(a)** $\frac{0}{50}$ **(b)** $\frac{50}{0}$ **(c)** $\frac{^{-}11}{0}$ **(d)** $\frac{0}{^{-}11}$

8. **(a)** $\frac{^{-}85}{0}$ **(b)** $\frac{0}{^{-}85}$ **(c)** $\frac{6}{0}$ **(d)** $\frac{0}{6}$

9. $\frac{^{-}8}{2}$ **10.** $\frac{^{-}14}{7}$ **11.** $\frac{21}{^{-}7}$ **12.** $\frac{30}{^{-}6}$

13. $\frac{^{-}54}{^{-}9}$ **14.** $\frac{^{-}48}{^{-}6}$ **15.** $\frac{55}{^{-}5}$ **16.** $\frac{70}{^{-}7}$

17. $\frac{^{-}28}{0}$ **18.** $\frac{^{-}40}{0}$ **19.** $\frac{14}{^{-}1}$ **20.** $\frac{25}{^{-}1}$

21. $\frac{^{-}20}{^{-}2}$ **22.** $\frac{^{-}80}{^{-}4}$ **23.** $\frac{^{-}48}{^{-}12}$ **24.** $\frac{^{-}30}{^{-}15}$

25. $\frac{^{-}18}{18}$ **26.** $\frac{50}{^{-}50}$ **27.** $\frac{0}{^{-}9}$ **28.** $\frac{0}{^{-}4}$

29. $\frac{^{-}573}{^{-}3}$ **30.** $\frac{^{-}580}{^{-}5}$ **31.** $\frac{163{,}672}{^{-}328}$ **32.** $\frac{^{-}69{,}496}{1022}$

Writing Calculator Small Group

Simplify.

33. $^-60 \div 10 \div {}^-3$

34. $36 \div {}^-4 \div 3$

35. $^-64 \div {}^-8 \div {}^-2$

36. $^-72 \div {}^-9 \div {}^-4$

37. $100 \div {}^-5 \cdot {}^-2$

38. $^-80 \div 4 \cdot {}^-5$

39. $48 \div 3 \cdot (12 \div {}^-4)$

40. $^-2 \cdot ({}^-3 \cdot {}^-7) \div 7$

41. $^-5 \div {}^-5 \cdot {}^-10 \div {}^-2$

42. $^-9 \cdot 4 \div {}^-36 \cdot 50$

43. $64 \cdot 0 \div {}^-8 \cdot 10$

44. $^-88 \div {}^-8 \div {}^-11 \cdot 0$

45. Explain whether or not division is commutative like multiplication. Start by doing these two divisions on your calculator: $2 \div 1$ and $1 \div 2$.

46. Explain whether or not division is associative like multiplication. Start by doing these two divisions: $(12 \div 6) \div 2$ and $12 \div (6 \div 2)$.

47. Explain what is different and what is similar about multiplying and dividing signed numbers.

48. In your own words, describe at least three division properties. Include examples to illustrate each property.

49. Write three numerical examples for each of these situations:

(a) a negative number divided by $^-1$

(b) a positive number divided by $^-1$

Now write a rule that explains what happens when you divide a signed number by $^-1$.

50. Explain why $\frac{0}{^-3}$ and $\frac{^-3}{0}$ do not give the same result.

Solve these application problems by using addition, subtraction, multiplication, or division. First use front end rounding to estimate the answer. Then find the exact answer.

51. The greatest ocean depth is 36,198 feet below sea level. If an unmanned research sub dives to that depth in 18 equal steps, how far does it dive in each step?

estimate:
exact:

52. Our college enrollment dropped by 3245 students over the last 11 years. What was the average drop in enrollment each year?

estimate:
exact:

53. When Ashwini discovered that her checking account was overdrawn by $238, she quickly transferred $450 from her savings to her checking account. What is the balance in her checking account?

estimate:
exact:

54. The Tigers offensive team lost a total of 48 yards during the first half of the football game. During the second half they gained 191 yards. How many yards did they gain or lose during the entire game?

estimate:
exact:

55. The foggiest place in the United States is Cape Disappointment, Washington. It is foggy there an average of 106 days each year. How many days is it not foggy each year?

estimate:
exact:

56. The number of cellular phone users worldwide in 1993 was 34 million. The number is expected to reach 298 million users by 2001. What increase in the number of users is expected during this 8-year period?

estimate:
exact:

57. A plane descended an average of 730 feet each minute during a 37-minute landing. How far did the plane descend during the landing?

estimate:
exact:

58. A discount store found that 174 items were lost to shoplifting last month. The average value of each item was $24. What was the total loss due to shoplifting?

estimate:
exact:

59. Mr. and Mrs. Martinez drove on the Interstate for five hours and traveled 315 miles. What was the average number of miles they drove each hour?

estimate:
exact:

60. Rochelle has a 48-month car loan for $9072. How much is her monthly payment?

estimate:
exact:

Ⓖ *Find the exact answer in Exercises 61–66. Solving these problems requires more than one step.*

61. Clarence bowled four games and had scores of 143, 190, 162, and 177. What was his average score? (*Hint:* To find the average, add all the scores and divide by the number of scores.)

62. Sheila kept track of her grocery expenses for six weeks. The amounts she spent were $84, $111, $82, $110, $98, and $79. What was the average weekly cost of her groceries?

63. On the back of an oatmeal box it says that one serving weighs 40 grams and that there are 13 servings in the box. On the front of the box it says that the weight of the contents is 510 grams. What is the difference in the total weight on the front and the back of the box?

64. A 2000-calorie-per-day diet recommends that you eat no more than 65 grams of fat. If each gram of fat is 9 calories, how many calories can you consume in other types of food?

65. Stephanie had $302 in her checking account. She wrote a $116 check for day care and a $548 check for rent. She also deposited her $347 paycheck. What is the balance in her account?

66. Gary started a new checking account with a $500 deposit. The bank charged him $18 to print his checks. He also wrote a $193 check for car repairs and a $289 check to his credit card company. What is the balance in his account?

Divide; then interpret the remainder in each of these applications.

67. A cellular phone company is offering 1000 free minutes of air time to new subscribers. How many hours of free time will a new subscriber receive?

68. Ralph's chickens laid 263 eggs. How many cartons holding a dozen eggs each can he fill?

69. Nikki is catering a large party. If one pie will serve 8 guests, how many pies should she make for 100 guests?

70. Flood victims are being given temporary shelter in a hotel. Each room can hold 5 people. How many rooms are needed for 163 people?

71. The Mathtronic company has budgeted $135,000 for new computers. How many can be purchased if the price is $2788 each?

72. A college has received a $250,000 donation to be used for scholarships. How many $3500 scholarships can be given to students?

G *Simplify.*

73. $|{}^{-}8| \div {}^{-}4 \cdot |{}^{-}5| \cdot |1|$

74. ${}^{-}6 \cdot |{}^{-}3| \div |9| \cdot {}^{-}2$

75. ${}^{-}6 \cdot {}^{-}8 \div ({}^{-}5 - {}^{-}5)$

76. ${}^{-}9 \div {}^{-}9 \cdot ({}^{-}9 \div 9) \div (12 - 13)$

Problem Solving with Your Calculator: Look back at the first page of this chapter. You guessed how many days it would take to receive a million dollars if you got $1 each second. Here's how to use your calculator to get the answer. If you get $1 per second, it would take 1,000,000 seconds to receive $1,000,000. Enter

1000000	[÷] 60	[=] 16666.66667	[÷] 60	[=] 277.7777778
	There are 60 seconds in one minute.	About 16,667 minutes	There are 60 minutes in one hour.	About 278 hours

[÷] 24	[=] 11.57407407
There are 24 hours in one day.	About $11\frac{1}{2}$ days (11.5 is equivalent to $11\frac{1}{2}$.)

Now use your calculator to find how long it takes to receive a billion dollars. Start by entering 1000000000. Then follow the pattern shown above. You will need to do one more division step to get the number of years. (Assume that there are 365 days in one year.) Your answer should be 31.70979198 (which rounds to 32 years).

1.8 Exponents and Order of Operations

OBJECTIVES

1. Write repeated factors, using exponents.
2. Simplify expressions containing exponents.
3. Use the order of operations.
4. Simplify expressions with fraction bars.

FOR EXTRA HELP

Tutorial | Tape 2 | SSM, Sec. 1.8

OBJECTIVE 1 An **exponent** (EX-poh-nent) is a quick way to write repeated multiplication. Here is an example.

$$2 \cdot 2 \cdot 2 \cdot 2 \cdot 2 \quad \text{can be written} \quad 2^5$$

(2 is the Base; 5 is the Exponent)

The *base* is the number being multiplied over and over, and the exponent tells how many times to use the number as a factor. This is called *exponential notation* or *exponential form.*

To simplify 2^5, actually do the multiplication.

$$2^5 = 2 \cdot 2 \cdot 2 \cdot 2 \cdot 2 = 32$$

Here are some more examples, using 2 as the base.

$2 = 2^1$ is read "2 to the **first power**."

$2 \cdot 2 = 2^2$ is read "2 to the **second power**" or, more commonly, "2 **squared**."

$2 \cdot 2 \cdot 2 = 2^3$ is read "2 to the **third power**" or, more commonly, "2 **cubed**."

$2 \cdot 2 \cdot 2 \cdot 2 = 2^4$ is read "2 to the **fourth power**."

$2 \cdot 2 \cdot 2 \cdot 2 \cdot 2 = 2^5$ is read "2 to the **fifth power**."

and so on.

We usually, don't write an exponent of 1, so if no exponent is shown, you can assume that it is 1. For example, 6 is actually 6^1, and 4 is actually 4^1.

Note

Exponents can also be negative numbers, zero, and fractions—for example, 2^{-3}, 2^0, and $2^{1/2}$. You will learn more about these exponents in other math courses.

EXAMPLE 1 Using Exponents

Write these multiplications using exponents. Also indicate how to read the exponential form.

(a) $5 \cdot 5 \cdot 5$ Can be written as 5^3, which is read "5 cubed" or "5 to the third power."

(b) $4 \cdot 4$ Can be written as 4^2, which is read "4 squared" or "4 to the second power."

(c) 7 Can be written as 7^1, which is read "7 to the first power."

WORK PROBLEM 1 AT THE SIDE.

1. Write each multiplication, using exponents. Indicate how to read the exponential form.

(a) $3 \cdot 3 \cdot 3 \cdot 3$

(b) $6 \cdot 6$

(c) 9

(d) $2 \cdot 2 \cdot 2 \cdot 2 \cdot 2 \cdot 2$

OBJECTIVE 2 Exponents are also used with signed numbers, as shown here.

$({}^{-}3)^2 = {}^{-}3 \cdot {}^{-}3 = 9$ The factors have the same sign, so the product is positive.

$({}^{-}4)^3 = {}^{-}4 \cdot {}^{-}4 \cdot {}^{-}4$ Multiply two numbers at a time.

$16 \cdot {}^{-}4$ First ${}^{-}4 \cdot {}^{-}4$ is positive 16.

${}^{-}64$ Then $16 \cdot {}^{-}4$ is ${}^{-}64$.

ANSWERS

1. (a) 3^4 is read "3 to the fourth power."
 (b) 6^2 is read "6 squared" or "6 to the second power."
 (c) 9^1 is read "9 to the first power."
 (d) 2^6 is read "2 to the sixth power."

2. Simplify.

(a) $(^-2)^3$

(b) $(^-6)^2$

(c) $2^4 \cdot (^-3)^2$

(d) $(^-4)^2 \cdot 3^3$

Work with exponents before you do other multiplications, as shown below. Notice that the exponent applies only to the *first* thing to its *left*.

Exponent applies only to the 2. $2^3 \cdot 5 \cdot 4^2$ Exponent applies only to the 4.

2 • 2 • 2 is 8. → $\underbrace{8 \cdot 5} \cdot 16$ ← 4 • 4 is 16.

$\underbrace{40 \cdot 16}$

640

EXAMPLE 2 Using Exponents with Negative Numbers

Simplify.

(a) $(^-5)^2 = {^-5} \cdot {^-5} = 25$

(b) $(^-5)^3 = \underbrace{^-5 \cdot {^-5}} \cdot {^-5}$

$\underbrace{25 \cdot {^-5}}$

$^-125$

(c) $(^-2)^4 = {^-2} \cdot {^-2} \cdot {^-2} \cdot {^-2} = 16$

(d) $(^-3)^2 \cdot 2^3 = \underbrace{^-3 \cdot {^-3}} \cdot \underbrace{2 \cdot 2 \cdot 2}$

$\underbrace{9 \cdot 8}$

72

Calculator Tip: Use the exponent key $\boxed{y^x}$ to enter exponents. To enter $(^-5)^2$, press the following keys.

5 $\boxed{+/-}$ $\boxed{y^x}$ 2 $\boxed{=}$ Answer is 25.

($^-5$) ↑ Exponent

To enter 5^8 press

5 $\boxed{y^x}$ 8 $\boxed{=}$ Answer is 390,625.

↑ Base ↑ Exponent

◀◀ WORK PROBLEM 2 AT THE SIDE.

OBJECTIVE 3 In **Sections 1.4** and **1.6** you worked examples that mixed either addition and subtraction or multiplication and division. In those situations you worked from left to right. Example 3 is a review.

EXAMPLE 3 Working from Left to Right

Simplify.

(a) $\underbrace{^-8 - {^-6}} + {^-11}$ Do additions and subtractions from left to right.

$\underbrace{^-2 + {^-11}}$

$^-13$

CONTINUED ON NEXT PAGE

ANSWERS

2. **(a)** $^-2 \cdot {^-2} \cdot {^-2} = {^-8}$
(b) $^-6 \cdot {^-6} = 36$
(c) $16 \cdot 9 = 144$
(d) $16 \cdot 27 = 432$

(b) $\underbrace{^{-}15 \div {}^{-}3} \cdot 6$ Do multiplications and divisions from left to right.

$\underbrace{5 \quad \cdot 6}$

30

WORK PROBLEM 3 AT THE SIDE.

Now we're ready to do problems that use a mix of the four operations, parentheses, and exponents. Let's start with a simple example: $4 + 2 \cdot 3$.

If we work from left to right	If we multiply first
$\underbrace{4 + 2} \cdot 3$	$4 + \underbrace{2 \cdot 3}$
$\underbrace{6 \quad \cdot 3}$	$\underbrace{4 + \quad 6}$
18	10

To be sure that everyone gets the same answer to a problem like this, mathematicians have agreed to do things in a certain order. The following order of operations shows that multiplying is done ahead of adding, so the correct answer is 10.

Order of Operations

Step 1 Work inside **parentheses** or other grouping symbols.

Step 2 Simplify expressions with **exponents.**

Step 3 Do the remaining **multiplications and divisions** as they occur from left to right.

Step 4 Do the remaining **additions and subtractions** as they occur from left to right.

Calculator Tip: Enter the example above in your calculator.

4 [+] 2 [×] 3 [=]

Which answer do you get? If you have a scientific calculator, it automatically uses the order of operations and multiplies first to get the correct answer of 10. Some standard, four-function calculators may *not* have the order of operations built into them and will give the *incorrect* answer of 18.

3. Simplify.

(a) $^{-}9 + {}^{-}15 - 3$

(b) $^{-}4 - 2 + {}^{-}6$

(c) $3 \cdot {}^{-}4 \div {}^{-}6$

(d) $^{-}18 \div 9 \cdot {}^{-}4$

ANSWERS

3. **(a)** $^{-}27$ **(b)** $^{-}12$
(c) 2 **(d)** 8

4. Simplify.

(a) $8 + (14 \div 2) \cdot 6$

(b) $4(1) + 8(9 - 2)$

(c) $3(5 + 1) + 20 \div 4$

EXAMPLE 4 Using the Order of Operations with Whole Numbers

Simplify.

$9 + 3(20 - 4) \div 8$ — Work inside parentheses first: 20 − 4 is 16. Bring down the other numbers and signs that you haven't used.

$9 + 3(16) \div 8$ — Look for exponents: none. Move from left to right, looking for multiplying and dividing.

$9 + 3(16) \div 8$ — Yes, here is multiplying: 3(16) is 48.

$9 + 48 \div 8$ — Here is dividing. 48 ÷ 8 is 6. There is no other multiplying or dividing, so look for adding and subtracting.

$9 + 6$ — Add last: 9 + 6 is 15.

15

◀◀ WORK PROBLEM 4 AT THE SIDE.

EXAMPLE 5 Using Order of Operations with Integers

Simplify.

(a) $^-8 \cdot (7 - 5) - 9$ — Work inside parentheses first: 7 − 5 is 2. Bring down the other numbers and signs you haven't used.

$^-8 \cdot (2) - 9$ — Look for exponents: none. Move from left to right, looking for multiplying and dividing.

$^-8 \cdot (2) - 9$ — Here is multiplying. $^-8 \cdot 2$ is $^-16$. No other multiplying or dividing, so look for adding and subtracting.

$^-16 - 9$ — Change subtracting to adding. Change 9 to its opposite.

$^-16 + {}^-9$ — Add $^-16 + {}^-9$.

$^-25$

CONTINUED ON NEXT PAGE

Answers

4. (a) 50 **(b)** 60 **(c)** 23

(b) $3 + 2(6 - 8) \cdot (15 \div 3)$ — Work inside first set of parentheses. Change $6 - 8$ to $6 + {}^-8$ to get ${}^-2$.

$3 + 2\ ({}^-2)\ \cdot (15 \div 3)$ — Work inside second set of parentheses: $15 \div 3$ is 5.

$3 + 2\ ({}^-2) \cdot 5$ — Multiply and divide from left to right. First multiply $2({}^-2)$ to get ${}^-4$.

$3 + {}^-4 \cdot 5$ — Then multiply ${}^-4 \cdot 5$ to get ${}^-20$.

$3 + {}^-20$ — Add last: $3 + {}^-20$ is ${}^-17$.

${}^-17$

WORK PROBLEM 5 AT THE SIDE.

EXAMPLE 6 Using Exponents and Order of Operations

Simplify.

(a) $4^2 - ({}^-3)^2$ — The only parentheses are around ${}^-3$, but no work can be done inside these parentheses.

$4^2 - ({}^-3)^2$ — Work with exponents: $4^2 = 4 \cdot 4 = 16$, and $({}^-3)^2 = {}^-3 \cdot {}^-3 = 9$.

$16 - 9$ — There is no multiplying or dividing, so add and subtract: $16 - 9$ is 7.

7

(b) $({}^-4)^3 - (4 - 6)^2 \cdot {}^-3$ — Work inside parentheses: $4 - 6$ becomes $4 + {}^-6$, which is ${}^-2$.

$({}^-4)^3 - ({}^-2)^2 \cdot {}^-3$ — Work with exponents next: $({}^-4)^3$ is ${}^-4 \cdot {}^-4 \cdot {}^-4 = {}^-64$, and $({}^-2)^2$ is ${}^-2 \cdot {}^-2 = 4$.

${}^-64 - 4 \cdot {}^-3$ — Look for multiplying and dividing. Multiply $4 \cdot {}^-3$.

${}^-64 - {}^-12$ — Change subtraction to addition. Change ${}^-12$ to its opposite.

${}^-64 + {}^+12$ — Add: ${}^-64 + 12$ is ${}^-52$.

${}^-52$

WORK PROBLEM 6 AT THE SIDE.

5. Simplify.

(a) $2 + 40 \div ({}^-5 + 3)$

(b) ${}^-5 \cdot 5 - (15 + 5)$

(c) $({}^-24 \div 2) + (15 - 3)$

(d) ${}^-3 \cdot (2 - 8) - 5 \cdot (4 - 3)$

(e) $3 \cdot 3 - (10 \cdot 3) \div 5$

(f) $6 - (2 + 7) \div ({}^-4 + 1)$

6. Simplify.

(a) $2^3 - 3^2$

(b) $({}^-4)^2 - 3^2 \cdot (5 - 2)$

(c) $({}^-3)^3 + (3 - 9)^2$

ANSWERS

5. **(a)** ${}^-18$ **(b)** ${}^-45$ **(c)** 0 **(d)** 13 **(e)** 3 **(f)** 9

6. **(a)** ${}^-1$ **(b)** ${}^-11$ **(c)** 9

7. Simplify.

(a) $\dfrac{{}^{-}3 \cdot 2^3}{{}^{-}10 - 6 + 8}$

(b) $\dfrac{({}^{-}10)({}^{-}5)}{{}^{-}6 \div 3 \cdot 5}$

(c) $\dfrac{6 + 18 \div ({}^{-}2)}{(1 - 10) \div 3}$

(d) $\dfrac{6^2 - 3^2 \cdot 4}{5 + (3 - 7)^2}$

ANSWERS

7. (a) $\dfrac{{}^{-}24}{{}^{-}8} = 3$ **(b)** $\dfrac{50}{{}^{-}10} = {}^{-}5$

(c) $\dfrac{{}^{-}3}{{}^{-}3} = 1$ **(d)** $\dfrac{0}{21} = 0$

OBJECTIVE 4 A fraction bar indicates division, as in $\frac{{}^{-}6}{2}$ which means ${}^{-}6 \div 2$. In an expression like

$$\frac{{}^{-}5 + 3^2}{16 - 7(2)}$$

the fraction bar also acts as a grouping symbol, like parentheses. It tells us to do the work in the numerator (above the bar) and then the work in the denominator (below the bar). The last step is to divide the results.

$$\frac{{}^{-}5 + 3^2}{16 - 7(2)} \rightarrow \frac{{}^{-}5 + 9}{16 - 14} \rightarrow \frac{4}{2} \rightarrow 4 \div 2 = 2$$

The final result is 2.

EXAMPLE 7 Fraction Bars and Order of Operations

Simplify $\dfrac{{}^{-}8 + (4 - 6) \cdot 5}{4 - 4^2 \div 8}$.

First do the work in the numerator.

$${}^{-}8 + \underbrace{(4 - 6)} \cdot 5 \qquad \text{Work inside the parentheses.}$$

$${}^{-}8 + \underbrace{{}^{-}2 \cdot 5} \qquad \text{Multiply.}$$

$$\underbrace{{}^{-}8 + ({}^{-}10)} \qquad \text{Add.}$$

Numerator → $^{-}18$

Now do the work in the denominator.

$$4 - \underbrace{4^2} \div 8 \qquad \text{There are no parentheses; use the exponent.}$$

$$4 - \underbrace{16 \div 8} \qquad \text{Divide.}$$

$$\underbrace{4 - 2} \qquad \text{Subtract.}$$

Denominator → 2

The last step is the division.

Numerator → / Denominator → $\dfrac{{}^{-}18}{2} = {}^{-}9$

WORK PROBLEM 7 AT THE SIDE.

Name ______ Date ______ Class ______

1.8 Exercises

Complete this table.

	Exponential Form	Factored Form	Simplified	Read as
1.	4^3		64	
2.	10^2		100	
3.		$2 \cdot 2 \cdot 2 \cdot 2 \cdot 2 \cdot 2 \cdot 2$		
4.		$3 \cdot 3 \cdot 3 \cdot 3 \cdot 3$		
5.		$5 \cdot 5 \cdot 5 \cdot 5$		
6.		$2 \cdot 2 \cdot 2 \cdot 2 \cdot 2 \cdot 2$		
7.				7 squared
8.				6 cubed
9.				10 to the first power
10.				4 to the fourth power

Simplify.

11. **(a)** 10^1 **(b)** 10^2 **(c)** 10^3 **(d)** 10^4

12. **(a)** 5^1 **(b)** 5^2 **(c)** 5^3 **(d)** 5^4

13. **(a)** 4^1 **(b)** 4^2 **(c)** 4^3 **(d)** 4^4

14. **(a)** 3^1 **(b)** 3^2 **(c)** 3^3 **(d)** 3^4

15. 5^{10}

16. 4^9

17. 2^{12}

18. 3^{10}

19. $(^-2)^2$

20. $(^-4)^2$

21. $(^-5)^2$

22. $(^-10)^2$

23. $(^-4)^3$

24. $(^-2)^3$

25. $(^-3)^4$

26. $(^-2)^4$

27. $(^-10)^3$

28. $(^-5)^3$

29. 1^4

30. 1^5

Writing Calculator Small Group

31. $3^3 \cdot 2^2$

32. $4^2 \cdot 5^2$

33. $(^{-}5)^2 \cdot 3^2$

34. $3^2 \cdot (^{-}2)^2$

35. $(^{-}5)^3 \cdot 6^1$

36. $7^1 \cdot (^{-}4)^3$

37. $(^{-}2)^4 \cdot {}^{-}2$

38. ${}^{-}6 \cdot (^{-}6)^2$

39. Evaluate this series.

$(^{-}2)^2 =$ $(^{-}2)^6 =$

$(^{-}2)^3 =$ $(^{-}2)^7 =$

$(^{-}2)^4 =$ $(^{-}2)^8 =$

$(^{-}2)^5 =$ $(^{-}2)^9 =$

What pattern do you see in the sign of the answers?

40. Explain why it is important to have rules for the order of operations. Why do you think our "natural instinct" is to just work from left to right?

Simplify.

41. $6 + 3 \cdot {}^{-}4$

42. $10 - 30 \div 2$

43. ${}^{-}1 + 15 + {}^{-}7 \cdot 2$

44. $9 + {}^{-}5 + 2 \cdot {}^{-}2$

45. $10 - 7^2$

46. $5 - 5^2$

47. $2 - {}^{-}5 + 3^2$

48. $6 - {}^{-}9 + 2^3$

49. $3 + 5(6 - 2)$

50. $4 + 3(8 - 3)$

51. ${}^{-}7 + 6(8 - 14)$

52. ${}^{-}3 + 5(9 - 12)$

53. $2(^{-}3 + 5) - (9 - 12)$

54. $3(2 - 7) - (^{-}5 + 1)$

55. ${}^{-}5(7 - 13) \div {}^{-}10$

56. ${}^{-}4(9 - 17) \div {}^{-}8$

57. $9 \div ({}^{-}3)^2 + {}^{-}1$

58. ${}^{-}48 \div ({}^{-}4)^2 + 3$

59. $2 - {}^{-}5 \cdot ({}^{-}2)^3$

60. $1 - {}^{-}10 \cdot ({}^{-}3)^3$

61. ${}^{-}2({}^{-}7) + 3(9)$

62. $4({}^{-}2) + {}^{-}3({}^{-}5)$

63. $30 \div {}^{-}5 - 36 \div {}^{-}9$

64. $8 \div {}^{-}4 - 42 \div {}^{-}7$

65. $2 \cdot 5 - 3 \cdot 4 + 5 \cdot 3$

66. $9 \cdot 3 - 6 \cdot 4 + 3 \cdot 7$

67. $4 \cdot 3^2 + 7(3 + 9) - {}^{-}6$

68. $5 \cdot 4^2 - 6(1 + 4) - {}^{-}3$

69. $({}^{-}4)^2 \cdot (7 - 9)^2 \div 2^3$

70. $({}^{-}5)^2 \cdot (9 - 17)^2 \div ({}^{-}10)^2$

71. $\dfrac{{}^{-}1 + 5^2 - {}^{-}3}{{}^{-}6 - 9 + 12}$

72. $\dfrac{{}^{-}6 + 3^2 - {}^{-}7}{7 - 9 - 3}$

73. $\dfrac{^-2 \cdot 4^2 - 4(6 - 2)}{^-4(8 - 13) \div {^-5}}$

74. $\dfrac{3 \cdot 3^2 - 5(9 - 2)}{8(6 - 9) \div {^-3}}$

75. $\dfrac{2^3 \cdot (^-2 - 5) + 4(^-1)}{4 + 5(^-6 \cdot 2) + (5 \cdot 11)}$

76. $\dfrac{3^3 + (^-1 - 2) \cdot 4 - 25}{^-4 + 4(3 \cdot 5) + (^-6 \cdot 9)}$

Ⓖ *Simplify.*

77. $5^2(9 - 11)(^-3) \cdot (^-3)^3$

78. $4^2(13 - 17)(^-2) \cdot (^-2)^3$

79. $|^-12| \div 4 + 2 \cdot |(^-2)^3| \div 4$

80. $6 - |2 - 3 \cdot 4| + (^-5)^2 \div 5^2$

81. $\dfrac{^-9 + 18 \div {^-3}(^-6)}{32 - 4(12) \div 3 \cdot 2}$

82. $\dfrac{^-20 - 15(^-4) - {^-40}}{14 + 27 \div 3 \cdot {^-2} - {^-4}}$

Chapter 1 Summary

Key Terms

1.1	**place value system**	A number system in which the location, or place, where a digit is written gives it a different value.
	digits	The 10 digits in our number system are 0, 1, 2, 3, 4, 5, 6, 7, 8, and 9.
	whole numbers	The whole numbers are 0, 1, 2, 3, and so on.
1.2	**number line**	A number line is like a thermometer turned sideways. It is used to show how numbers relate to each other.
	integers	Integers are the whole numbers and all their opposites (negative whole numbers).
	absolute value	The absolute value of a number is its distance from zero. Absolute value is indicated by two vertical bars and is always positive (or zero) but never negative.
1.3	**addends**	In addition, the numbers being added are called addends.
	sum	The answer to an addition problem is called the sum.
1.4	**opposite**	The opposite of a number is the same distance from zero on the number line but on the opposite side of zero. It is also called the additive inverse because a number plus its opposite equals zero.
1.5	**rounding**	Rounding a number means finding a number that is close to the original number but easier to work with.
	estimate	The use of rounded numbers to get an approximate answer, or estimate.
	front end rounding	Front end rounding is rounding numbers to the highest possible place, so all the digits become zero except the first digit.
1.6	**factors**	In multiplication, the numbers being multiplied are called factors.
	product	The answer to a multiplication problem is called the product.
	distributive property	Multiplication distributes over addition. For example, $3(6 + 2) = 3 \cdot 6 + 3 \cdot 2$.
1.7	**quotient**	The answer to a division problem is called the quotient.
1.8	**exponent**	An exponent tells how many times a number is used as a factor in repeated multiplication.

Quick Review

Concepts	*Examples*
1.1 Reading and Writing Whole Numbers Do not use the word "and" with whole numbers. Commas separate groups of three digits. The first few group names are ones, thousands, millions, billions, trillions.	Write **3**, **008**, **160** in words. **three million, eight thousand**, **one hundred sixty** Write this number, using digits: twenty **billion,** sixty-five **thousand,** eighteen. __ 2 0 , 0 0 0 , 0 6 5 , 0 1 8 billions millions thousands ones

Concepts	*Examples*
1.2 Graphing Signed Numbers Place a dot at the correct location on the number line.	Graph (a) $^-4$ (b) 0 (c) $^-1$ (d) $\frac{1}{2}$ (e) 2 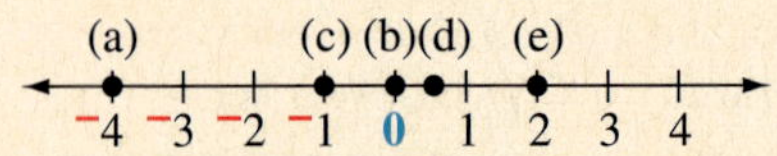
1.2 Comparing Integers When you are comparing two integers, the one that is farther to the left on the number line is less than the other. Use the $<$ symbol for "is less than" and the $>$ symbol for "is greater than."	Write $<$ or $>$ between each pair of numbers to make a true statement. $^-3 < {^-2}$ $4 > {^-4}$ $^-3$ **is less than** $^-2$ because $^-3$ is to the *left* of $^-2$ on the number line. 4 **is greater than** $^-4$ because 4 is to the *right* of $^-4$ on the number line.
1.2 Finding the Absolute Value of a Number Find the distance on the number line from 0 to the number. The absolute value is always positive (or zero) but never negative.	Find each absolute value. $\lvert {^-5} \rvert = 5$ because $^-5$ is 5 steps away from 0 on the number line. $\lvert 3 \rvert = 3$ because 3 is 3 steps away from 0 on the number line.
1.3 Adding Two Integers When both integers have the *same sign,* add the absolute values and use the common sign as the sign of the sum. When the integers have *different signs,* subtract the smaller absolute value from the larger absolute value. Use the sign of the number with the larger absolute value as the sign of the sum.	Add. **(a)** $^-6 + {^-7}$ Add the absolute values. $\lvert {^-6} \rvert = 6$ and $\lvert {^-7} \rvert = 7$ Add $6 + 7 = 13$ and use the common sign as the sign of the sum: $^-6 + {^-7} = {^-13}$. **(b)** $^-10 + 4$ Subtract the smaller absolute value from the larger. $\lvert {^-10} \rvert = 10$ and $\lvert 4 \rvert = 4$ Subtract $10 - 4 = 6$; the number with the larger absolute value is negative, so the sum is negative: $^-10 + 4 = {^-6}$.

Concepts	*Examples*
1.3 Using Properties of Addition Addition Property of 0: Adding zero to any number leaves the number unchanged. Commutative Property of Addition: Changing the *order* of two addends does not change the sum. Associative Property of Addition: Changing the *grouping* of addends does not change the sum.	Name the property illustrated by each case. **(a)** $^-16 + 0 = {}^-16$ **(b)** $4 + 10 = 10 + 4$ Both sums are 14. **(c)** $2 + (^-6 + 1) = (2 + {}^-6) + 1$ Both sums are $^-3$. **(a)** Addition property of 0. **(b)** Commutative property of addition. **(c)** Associative property of addition.
1.4 Subtracting Two Integers To subtract two numbers, add the first number to the opposite of the second number. *Step 1* Make one pencil stroke to change the subtraction symbol to an addition symbol. *Step 2* Make a second pencil stroke to change the second number to its opposite.	Subtract. **(a)** $7 - {}^-2$ → $7 + {}^+2 = 9$ Change subtraction to addition. Change $^-2$ to its opposite, $^+2$. **(b)** $^-9 - 12$ → $^-9 + {}^-12 = {}^-21$ Change subtraction to addition. Change 12 to its opposite, $^-12$.
1.5 Rounding Integers *Step 1* Draw a line under the place to which the number is to be rounded. *Step 2* Look only at the next digit to the right of the underlined place. If it is 5 or more, increase the underlined digit by 1. If it is 4 or less, do not change the digit in the underlined place. *Step 3* Change all digits to the right of the underlined place to zeros.	Round 36,833 to the nearest thousand. 36,833 rounds to 37,000 Next digit is 5 or more. Thousands. Changed to 0's. Change 6 to 7. Round $^-3582$ to the nearest ten. $^-3582$ rounds to $^-3580$ Next digit is 4 or less. Tens. Changed to 0. Leave 8 as 8.
1.5 Front End Rounding Round to the highest possible place so that all the digits become zero except the first digit.	Use front end rounding. 97,203 rounds to 100,000 Next digit is 5 or more. Changed to 0. Change 9 to 10. Carry 1 into hundred thousands place

Concepts	Examples
1.5 Estimating Answers in Addition and Subtraction	
Use front end rounding to round the numbers in a problem. Then add or subtract the rounded numbers to estimate the answer.	First use front end rounding to estimate the answer. Then find the exact answer. The temperature was 48 degrees below zero. During the morning it rose 21 degrees. What was the new temperature? *estimate:* $^{-}50 + 20 = {}^{-}30$ degrees *exact:* $^{-}48 + 21 = {}^{-}27$ degrees
1.6 Multiplying Two Integers	
If the factors have *different signs,* the product is *negative.* If the factors have the *same sign,* the product is *positive.*	Multiply. **(a)** $^{-}5(6) = {}^{-}30$ The factors have different signs, so the product is negative. **(b)** $(^{-}10)(^{-}2) = 20$ The factors have the same sign, so the product is positive.
1.6 Using Properties of Multiplication	
Multiplication property of 0: Multiplying any number by 0 gives a product of 0. Multiplication property of 1: Multiplying a number by 1 leaves the number unchanged. Commutative property of multiplication: Changing the *order* of two factors does not change the product. Associative property of multiplication: Changing the *grouping* of factors does not change the product. Distributive property: Multiplication distributes over addition.	Name the property illustrated by each case. **(a)** $^{-}49 \cdot 0 = 0$ **(b)** $1(675) = 675$ **(c)** $^{-}8 \cdot 2 = 2 \cdot {}^{-}8$ Both products are $^{-}16$. **(d)** $(^{-}3 \cdot {}^{-}2) \cdot 4 = {}^{-}3 \cdot (^{-}2 \cdot 4)$ Both products are 24. **(e)** $5(2 + 4) = 5 \cdot 2 + 5 \cdot 4$ Both products are 30. **(a)** Multiplication property of 0. **(b)** Multiplication property of 1. **(c)** Commutative property of multiplication. **(d)** Associative property of multiplication. **(e)** Distributive property.
1.6 Estimating Answers in Multiplication	
First use front end rounding. Then multiply the rounded numbers using a shortcut: Multiply the nonzero digits in each factor; count the total number of zeros in the two factors and write that number of zeros in the product.	First use front end rounding to estimate the answer. Then find the exact answer. At a PTA fund raiser, Lionel sold 96 photo albums at \$22 each. How much money did he take in? *estimate:* $100 \cdot \$20 = \$2000$ *exact:* $96 \cdot \$22 = \$2112$

Concepts	*Examples*
1.7 Dividing Two Integers Use the same rules as for multiplying two integers. If the numbers have *different signs,* the quotient is *negative.* If the numbers have the *same sign,* the quotient is *positive.*	Divide. $\frac{^{-}24}{6} = {^{-}4}$ Numbers have different signs, so the quotient is negative. $^{-}72 \div {^{-}8} = 9$ Numbers have the same sign, so the quotient is positive. $\frac{50}{^{-}5} = {^{-}10}$ Numbers have different signs, so the quotient is negative.
1.7 Using Properties of Division When a nonzero number is divided by itself, the quotient is 1. When a number is divided by 1, the quotient is the number. When 0 is divided by any other number (except 0) the quotient is 0. Division by 0 is undefined. There is no answer.	State the property illustrated by each case. **(a)** $\frac{^{-}4}{^{-}4} = 1$ **(b)** $\frac{65}{1} = 65$ **(c)** $\frac{0}{9} = 0$ **(d)** $\frac{^{-}10}{0}$ undefined The examples are in the same order as the properties listed at the left. Note that division is *not* commutative or associative.
1.7 Estimating Answers in Division First use front end rounding. Then divide the rounded numbers, using a shortcut: Drop the same number of zeros in both the divisor and the dividend.	First use front end rounding to estimate the answer. Then find the exact answer. Joan has one year to pay off a \$1020 loan. What is her monthly payment? *estimate:* \$1000 ÷ 10 = \$100 *exact:* \$1020 ÷ 12 = \$85
1.7 Interpret Remainders in Division In some situations the remainder tells you how much is left over. In other situations, you must increase the quotient by 1 in order to accommodate the "left over."	Divide; then interpret the remainder. Each chemistry student needs 35 milliliters of acid for an experiment. How many students can be served from a bottle holding 500 milliliters of acid? **14** → 14 students served 35)500 490 **10** → 10 milliliters of acid left over
1.8 Using Exponents An exponent tells how many times a number is used as a factor in repeated multiplication. An exponent applies only to its base (the first thing to the left of the exponent).	Simplify. Exponent **(a)** $2^5 = 2 \cdot 2 \cdot 2 \cdot 2 \cdot 2 = 32$ **(b)** $({^{-}3})^2 = {^{-}3} \cdot {^{-}3} = 9$

Concepts	*Examples*
1.8 Order of Operations Mathematicians have agreed to follow this order. *Step 1* Work inside parentheses or other grouping symbols. *Step 2* Simplify expressions with exponents. *Step 3* Do the remaining multiplications and divisions as they occur from left to right. *Step 4* Do the remaining additions and subtractions as they occur from left to right.	Simplify. $(^-2)^4 + 3(^-4 - {}^-2)$ Work inside parentheses. $(^-2)^4 + 3(^-2)$ Work with exponents. $16 + 3(^-2)$ Multiply. $16 + {}^-6$ Add. 10
1.8 Using Fraction Bars with Order of Operations When there is a fraction bar, do all the work in the numerator. Then do all the work in the denominator. Finally, divide numerator by denominator.	Simplify. $\frac{^-10 + 4^2 - 6}{2 + 3(1 - 4)} = \frac{^-10 + 16 - 6}{2 + 3(^-3)} = \frac{0}{^-7} = 0$

Name ____________________ Date ____________ Class ____________

CHAPTER 1 REVIEW EXERCISES

If you need help with any of these review exercises, look in the section indicated in the brackets.

[1.1] **1.** Circle the whole numbers: 86 2.831 -4 0 $\frac{2}{3}$ 35,600

Write these numbers in words.

2. 806

3. 319,012

4. 60,003,200

5. 15,749,000,000,006

Write these numbers, using digits.

6. Five hundred four thousand, one hundred

7. Six hundred twenty million, eighty thousand

8. Ninety-nine billion, seven million, three hundred fifty-six

[1.2] **9.** Graph these numbers: $^-3\frac{1}{2}$, 2, $^-5$, 0

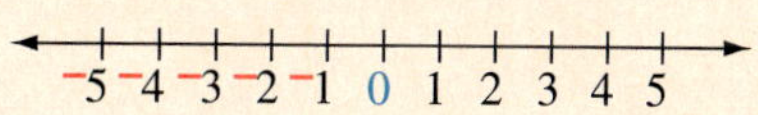

Write < or > between each pair of numbers to make true statements.

10. 0 ___ $^-4$

11. $^-3$ ___ $^-1$

12. 2 ___ $^-2$

13. $^-2$ ___ 1

Find each absolute value.

14. $|^-5|$

15. $|9|$

16. $|0|$

17. $|^-125|$

[1.3] *Add.*

18. $^-9 + 8$

19. $^-8 + {}^-5$

20. $16 + {}^-19$

21. $^-4 + 4$

22. $6 + {}^-5$

23. $^-12 + {}^-12$

24. $0 + {}^-7$

25. $^-16 + 19$

26. $9 + {}^-4 + {}^-8 + 3$

27. $^-11 + {}^-7 + 5 + {}^-4$

Writing Calculator Small Group

[1.4] *Find the opposite (additive inverse) of each number. Show that the sum of the number and its opposite is zero.*

28. $^-5$

29. 18

Subtract by changing subtraction to addition.

30. $5 - 12$

31. $24 - 7$

32. $^-12 - 4$

33. $4 - {}^-9$

34. $^-12 - {}^-30$

35. $^-8 - 14$

36. $^-6 - {}^-6$

37. $^-10 - 10$

38. $^-8 - {}^-7$

39. $0 - 3$

40. $1 - {}^-13$

41. $15 - 0$

Simplify.

42. $3 - 12 - 7$

43. $^-7 - {}^-3 + 7$

44. $4 + {}^-2 - 0 - 10$

45. $^-12 - 12 + 20 - {}^-4$

[1.5] *Round each number. Write your answers using the ≈ symbol.*

46. $^-205$ to the nearest ten

47. 59,499 to the nearest thousand

48. 85,066,000 to the nearest million

49. $^-2963$ to the nearest hundred

50. $^-7,063,885$ to the nearest ten thousand

51. 399,712 to the nearest thousand

Use front end rounding to round each number.

52. The combined weight loss of 10 dieters was 197 pounds.

53. The land on the shore of the Dead Sea in the Middle East is 1312 feet below sea level.

54. There are 362,000,000 Yellow Pages directories published in the United States each year.

55. There were 9,150,700,000 purchases made in 1996 by people who used the Yellow Pages.

[1.6] *Multiply.*

56. $^{-}6(9)$

57. $(^{-}7)(^{-}8)$

58. $10(^{-}10)$

59. $^{-}45 \cdot 0$

60. $^{-}1(^{-}24)$

61. $17 \cdot 1$

62. $4(^{-}12)$

63. $(^{-}5)(^{-}25)$

64. $^{-}3 \cdot {}^{-}4 \cdot {}^{-}3$

65. $^{-}5(2) \cdot {}^{-}5$

66. $^{-}8 \cdot {}^{-}1(^{-}9)$

[1.7] *Simplify.*

67. $\frac{^{-}63}{^{-}7}$

68. $\frac{70}{^{-}10}$

69. $\frac{^{-}15}{0}$

70. $^{-}100 \div {}^{-}20$

71. $18 \div {}^{-}1$

72. $\frac{0}{12}$

73. $\frac{^{-}30}{^{-}2}$

74. $\frac{^{-}35}{35}$

75. $^{-}40 \div {}^{-}4 \div {}^{-}2$

76. $^{-}18 \div 3 \cdot {}^{-}3$

77. $0 \div {}^{-}10 \cdot 5 \div 5$

78. Divide; then interpret the remainder. It took 1250 hours to build the set for a new play. How many work days of 8 hours each did it take to build the set?

[1.8] *Simplify.*

79. 10^4

80. 2^5

81. 3^3

82. $(^{-}4)^2$

83. $(^{-}5)^3$

84. 8^1

85. $6^2 \cdot 3^2$

86. $(^{-}2)^3 \cdot 5^2$

87. $^{-}30 \div 6 - 4 \cdot 5$

88. $6 + 8(2 - 3)$

89. $16 \div 4^2 + (^{-}6 + 9)^2$

90. $^{-}3(4) - 2(5) + 3(^{-}2)$

91. $\frac{^{-}10 + 3^2 - {}^{-}9}{3 - 10 - 1}$

92. $\frac{^{-}1(1 - 3)^3 + 12 \div 4}{^{-}5 + 24 \div 8 \cdot 2(6 - 6) + 5}$

MIXED REVIEW EXERCISES

Name the property illustrated by each case.

93. $^-3 + (5 + 1) = (^-3 + 5) + 1$

94. $^-7(2) = 2(^-7)$

95. $0 + 19 = 19$

96. $^-42 \cdot 0 = 0$

97. $2(^-6 + 4) = 2 \cdot {^-6} + 2 \cdot 4$

98. $(^-6 \cdot 3) \cdot {^-1} = {^-6} \cdot (3 \cdot {^-1})$

First use front end rounding to estimate each answer. Then find the exact answer.

99. Last year 192 Elvis jukeboxes were sold at a price of \$11,900 each. What was the total value of the jukeboxes?

estimate:
exact:

100. Chad had \$185 in his checking account. He deposited his \$428 paycheck and then wrote a \$706 check for car repairs. What is the balance in his account?

estimate:
exact:

101. Georgia's car used 24 gallons of gas on her 840-mile vacation trip. What was the average number of miles she drove on each gallon of gas?

estimate:
exact:

102. When inventory was taken at Mathtronic Company, 19 calculators and 12 computer modems were missing. Each calculator is worth \$39 and each modem is worth \$85. What is the total value of the missing items?

estimate:
exact:

Elena Sanchez opened a shop that does alterations and designs custom clothing. Use the table of her income and expenses to answer Exercises 103–106.

Month	Income	Expenses	Profit or Loss
Jan.	\$2400	\$3100	
Feb.	\$1900	\$2000	
Mar.	\$2500	\$1800	
Apr.	\$2300	\$1400	
May	\$1600	\$1600	
June	\$1900	\$1200	

103. Complete the table by finding Elena's profit or loss for each month.

104. Which month had the greatest loss? Which month had the greatest profit?

105. What was Elena's average monthly income?

106. What was the average monthly amount of expenses?

Name ______________________ Date ______________ Class ______________

Chapter 1 Test

1. Write this number in words: 20,008,307

2. Write this number using digits:
   thirty billion, seven hundred thousand, five

3. Graph the numbers 3, $^{-}2$, 0, $\frac{^{-}1}{2}$

4. Write < or > between each pair of numbers to make a true statement.

   0 _____ $^{-}3$ $\qquad$ $^{-}2$ _____ $^{-}1$

5. Find $|10|$ and $|^{-}14|$

Add, subtract, multiply, or divide.

6. $3 - 9$
7. $^{-}12 + 7$
8. $\frac{^{-}28}{^{-}4}$
9. $^{-}1(40)$
10. $^{-}5 - {^{-}15}$
11. $(^{-}8)(^{-}8)$
12. $^{-}25 + {^{-}25}$
13. $\frac{17}{0}$
14. $^{-}30 - 30$
15. $\frac{50}{^{-}10}$
16. $5 \cdot {^{-}9}$
17. $0 - {^{-}6}$

Simplify.

18. $^{-}35 \div 7 \cdot {^{-}5}$
19. $^{-}15 - {^{-}8} + 7$
20. $3 - 7(^{-}2) - 8$
21. $(^{-}4)^2 \cdot 2^3$
22. $\frac{5^2 - 3^2}{(4)(^{-}2)}$
23. $^{-}2(^{-}4 + 10) + 5 \cdot 4$
24. $^{-}3 + (^{-}7 - {^{-}10}) + 4(6 - 10)$
25. Explain how an exponent is used. Include two examples.

1. ____________________

2. ____________________

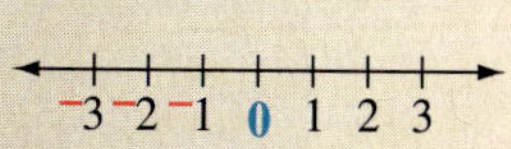

3. ____________________

4. ____________________

5. ____________________

6. ____________________
7. ____________________
8. ____________________
9. ____________________
10. ____________________
11. ____________________
12. ____________________
13. ____________________
14. ____________________
15. ____________________
16. ____________________
17. ____________________

18. ____________________
19. ____________________
20. ____________________
21. ____________________
22. ____________________
23. ____________________
24. ____________________

25. ____________________

26. ____________

26. Explain the commutative and associative properties of addition. Also give an example to illustrate each property.

Round each number.

27. ____________

27. 851 to the nearest hundred.

28. ____________

28. 36,420,498,725 to the nearest million.

29. ____________

29. 349,812 to the nearest thousand.

First use front end rounding to estimate the answer to each of these application problems. Then find the exact answer.

30. *estimate:* ____________
exact: ____________

30. In 1975, the number of Japanese cars imported to the United States was 807,931. In 1996, the number was 726,940 cars. What was the decrease in the number of imported Japanese cars?

31. *estimate:* ____________
exact: ____________

31. Lorene had $184 in her checking account. She deposited her $293 paycheck and then wrote a $506 check for tuition. What is the balance in her account?

32. *estimate:* ____________
exact: ____________

32. The Cardinals football team had a bad year. It lost a total of 1140 yards in 12 games. What was the average loss in each game?

33. *estimate:* ____________
exact: ____________

33. One kind of cereal has 220 calories in each serving. Another kind has 110 calories in each serving. During a month with 31 days, how many calories would you save by eating the second kind of cereal each morning for breakfast?

Divide; then interpret the remainder.

34. ____________

34. Anthony has a part-time job as a shipping clerk. He is sending 1276 pounds of books to a bookstore. Each shipping carton can safely hold 48 pounds. What is the minimum number of cartons he will need?

Understanding Variables and Solving Equations

2

2.1 Introduction to Variables

OBJECTIVE 1 You probably know that algebra uses letters, especially the letter x. But why use letters when numbers are easier to understand? Here is an example.

Suppose that you run your college bookstore. When deciding how many books to order for a certain class, you first find out the class limit, that is, the maximum number of students allowed in the class. You will need at least that many books. But you decide to order 5 extra copies for emergencies.

Rule for ordering books: Order the class limit + 5 extra

How many books would you order for a prealgebra class with a limit of 25 students?

Class limit ↓ Extra ↓

$25 + 5$ You would order 30 prealgebra books.

How many books would you order for a geometry class that allows 40 students to register?

Class limit ↓ Extra ↓

$40 + 5$ You would order 45 geometry books.

You could set up a table to keep track of the number of books to order for various classes.

Class	Rule for Ordering Books: Class Limit + 5 Extra	Number of Books to Order
Prealgebra	$25 + 5$	30
Geometry	$40 + 5$	45
College algebra	$35 + 5$	40
Calculus 1	$50 + 5$	55

INTERNET

www.mathnotes.com

OBJECTIVES

1. Identify variables, constants, and expressions.
2. Evaluate variable expressions for given replacement values.
3. Write properties of operations using variables.
4. Understand the use of exponents with variables.

FOR EXTRA HELP

Tutorial Tape 3 SSM, Sec. 2.1

1. Write an expression for this rule. Identify the variable and the constant.

Order the class limit plus 15 extra books because it is a very large class.

A shorthand way to write your rule would be

$$c + 5$$

↑ c stands for class limit.

You can't write your rule by using just numbers because the class limit varies, or changes, depending on which class you're talking about. So you use a letter, called a **variable** (VAIR-ee-uh-bul) to represent the part of the rule that varies. Notice the similarity in the words "**vari**es" and "**vari**able." When part of a rule does *not* change, it is called a **constant.**

The constant, or the part of the rule that does *not* change. ↓

$$c + 5$$

↑ The variable, or the part of the rule that varies or changes.

$c + 5$ is called an **expression** (eks-PRESH-un). It expresses (tells) the rule for ordering books. You could use any letter you like for the variable part of the expression, such as $x + 5$, or $n + 5$, and so on. But one suggestion is to use a letter that reminds you of what it stands for. In this situation, the letter "**c**" reminds us of "**c**lass limit."

EXAMPLE 1 Writing Expressions and Identifying Variables and Constants

Write an expression for this rule. Identify the variable and the constant.

Order the class limit minus 10 books because some students will buy used books.

Constant (10)

$$c - 10$$

Variable (c)

◀◀ **WORK PROBLEM 1 AT THE SIDE.**

OBJECTIVE 2 When you need to figure out how many books to order for a particular class, you use a specific value for the class limit, like 25 students in prealgebra. Then you **evaluate the expression,** that is, you follow the rule.

ORDERING BOOKS FOR A PREALGEBRA CLASS

$c + 5$	Expression (rule for ordering books) is $c + 5$. Replace c with 25, the class limit for prealgebra.
$25 + 5$	Follow the rule. Add $25 + 5$.
30	Order 30 prealgebra books.

EXAMPLE 2 Evaluating an Expression

Use this rule for ordering books: Order the class limit minus 10. The expression is $c - 10$.

(a) Evaluate the expression when the class limit is 32.

$c - 10$	Replace c with 32.
$32 - 10$	Follow the rule. Subtract to find $32 - 10$.
22	Order 22 books.

CONTINUED ON NEXT PAGE

ANSWER

1. The expression is $c + 15$. The variable is c, and the constant is 15.

(b) Evaluate the expression when the class limit is 48.

$c - 10$ Replace c with 48.

$\underbrace{48 - 10}$ Follow the rule. Subtract to find $48 - 10$.

38 Order 38 books.

WORK PROBLEM 2 AT THE SIDE.

In any career you choose, there will be many useful "rules" that need to be written using letters (variables) because part of the rule changes depending on the situation. This is one reason why algebra is such a powerful tool. Here is another example.

Suppose that you work in a landscaping business. You are putting a fence around a square-shaped garden. Each side of the garden is 6 feet long. How much fencing material should you bring to finish the job? You could add the lengths of the four sides.

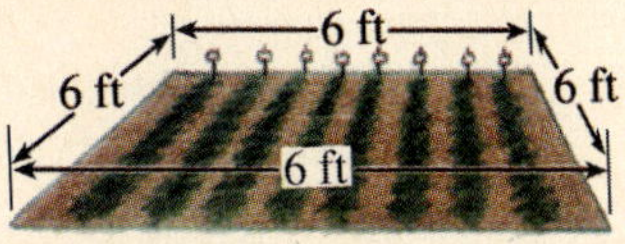

6 feet + 6 feet + 6 feet + 6 feet = 24 feet of fencing

Or, recall that multiplication is a quick way to do repeated addition. There are 4 sides, so multiply by 4.

4 • 6 feet = 24 feet of fencing

So the rule for figuring the amount of fencing for a square garden is

4 • length of one side.

Other jobs may require fencing for larger or smaller square shapes. The following table shows how much fencing you will need.

Length of one side of square shape	Expression (rule) to find total amount of fencing: 4 • length of one side	Total amount of fencing needed
6 feet	4 • 6 feet	24 feet
9 feet	4 • 9 feet	36 feet
10 feet	4 • 10 feet	40 feet
3 feet	4 • 3 feet	12 feet

The expression (rule) can be written in shorthand form as

Length of one side
↓
$4 \cdot s$
Coefficient ↑ ↑ Variable

The number part in a multiplication expression is called the numerical coefficient, or just the **coefficient** (koh-uh-FISH-ent). We usually don't write multiplication dots in expressions, so we do the following.

$4 \cdot s$ is written as $4s$

You can use the expression $4s$ any time you need to know the perimeter of a square shape, that is, the total distance around all four sides of the square.

2. Use this expression for ordering books: $c + 3$.

(a) Evaluate the expression when the class limit is 25.

(b) Evaluate the expression when the class limit is 60.

ANSWERS

2. (a) $25 + 3$ is 28; order 28 books
(b) $60 + 3$ is 63; order 63 books

3. **(a)** Evaluate the expression $4s$ when the length of one side of a square table is 3 feet.

(b) Evaluate the expression $4s$ when the length of one side of a square park is 7 miles.

4. Evaluate the expression $100 + \frac{a}{2}$ when the age of the person is 40.

Note

If an expression involves adding, subtracting, or dividing, then you **do** have to write $+$, $-$, or $\div$. It is only multiplication that is understood without writing a symbol.

$4 + s$	$4 - s$	$4 \div s$	$4s$
Add s.	Subtract s.	Divide by s.	Multiply by s.

EXAMPLE 3 Evaluating an Expression with Multiplication

The expression (rule) for finding the perimeter of a square shape is $4s$. Evaluate the expression when the length of one side of a square parking lot is 30 yards.

$4s$ — Replace s with 30 yards.

$4 \cdot \mathbf{30\ yards}$ — There is no operation symbol between the 4 and the s, so it is understood to be multiplication.

120 yards — Total distance (perimeter) around the lot.

◀◀ WORK PROBLEM 3 AT THE SIDE.

Some expressions (rules) involve several different steps. An expression for finding the approximate systolic blood pressure of a person of a certain age is shown below.

$$100 + \frac{a}{2}$$ ← a: Age of person (the variable)

Remember that a fraction bar means division, so $\frac{a}{2}$ is the person's age divided by 2. You also need to remember the order of operations, which means doing division before addition.

EXAMPLE 4 Evaluating Expressions with Several Steps

Evaluate the expression $100 + \frac{a}{2}$ when the age of the person is 24.

$100 + \frac{a}{2}$ — Replace a with 24, the age of the person.

$100 + \frac{\mathbf{24}}{2}$ — Follow the rule using order of operations. First divide: $24 \div 2$ is 12.

$100 + 12$ — Now add: $100 + 12$ is 112.

112 — The approximate systolic blood pressure is 112.

◀◀ WORK PROBLEM 4 AT THE SIDE.

ANSWERS

3. **(a)** $4 \cdot 3$ feet; 12 feet
(b) $4 \cdot 7$ miles; 28 miles

4. $100 + \frac{40}{2}$ Replace a with 40.

$100 + 20$

120 Approximate systolic blood pressure.

Calculator Tip: If you like to fish, you can use an expression (rule) like the one below to find the approximate weight (in pounds) of a fish you catch. Measure the length (in inches) of the fish and then use the correct expression for that type of fish. For a northern pike, the weight expression is shown.

Variable (length of fish) → $\dfrac{l^3}{3600}$

where l is the length of the fish in inches.

To evaluate this expression for a fish that is 43 inches long, follow the rule by calculating

$\dfrac{43^3}{3600}$ Replace l with 43, the length of the fish in inches.

In the numerator, you can multiply 43 • 43 • 43 or use the y^x key. Then divide by 3600.

Enter 43 y^x 3 ÷ 3600 = **Calculator shows 22.08527778**

(43: Base; 3: Exponent)

The fish weighs about 22 pounds.

Now use the expression to find the approximate weight of a northern pike that is 37 inches long. (Answer: about 14 pounds.)

Notice that variables are used on your calculator keys. On the y^x key, y represents the base and x represents the exponent. You evaluated y^x by entering 43 as the base and 3 as the exponent for the first fish. Then you evaluated y^x by entering 37 as the base and 3 as the exponent for the second fish.

Some expressions (rules) involve several variables. For example, if you bowl three games and want to know your average score, you can use this expression.

$\dfrac{t}{g}$ — t ← Total score for all games (variable); g ← Number of games (variable)

EXAMPLE 5 Evaluating Expressions with Two Variables

(a) Find your average score if you bowl three games and your total score for all three games is 378.

Use the expression (rule) for finding your average score.

$\dfrac{t}{g}$ Replace t with your total score of 378. Replace g with 3, the number of games.

$\dfrac{378}{3}$ Follow the rule. Divide 378 by 3.

126 Your average score is 126.

CONTINUED ON NEXT PAGE

5. (a) Use the expression for finding your average bowling score. Evaluate the expression if your total score for 4 games is 532.

(b) Complete this table.

Value of x	Value of y	Expression $x - y$
16	10	
3	7	
8	0	

(b) Complete these tables to show how to evaluate each expression.

Value of x	Value of y	Expression (Rule) $x + y$
2	5	$2 + 5$ is 7
$^-6$	4	$\_ + \_$ is
0	16	is

Value of x	Value of y	Expression (Rule) xy
2	5	$2 \cdot 5$ is 10
$^-6$	4	$\cdot$ is
0	16	is

The expression (rule) is to *add* the two variables. So the completed table is:

Value of x	Value of y	Expression (Rule) $x + y$
2	5	$2 + 5$ is 7
$^-6$	4	$^-6 + 4$ is $^-2$
0	16	$0 + 16$ is 16

The expression (rule) is to *multiply* the two variables. We know that it's multiplication because there is no operation symbol between the x and y. So the completed table is:

Value of x	Value of y	Expression (Rule) xy
2	5	$2 \cdot 5$ is 10
$^-6$	4	$^-6 \cdot 4$ is $^-24$
0	16	$0 \cdot 16$ is 0

◀◀ WORK PROBLEM 5 AT THE SIDE.

OBJECTIVE 3 Now you can use variables as a shorthand way to express the properties you learned about in **Sections 1.3, 1.6,** and **1.7**. We'll use the letters a and b to represent numbers.

COMMUTATIVE PROPERTY OF ADDITION

$$a + b = b + a$$

COMMUTATIVE PROPERTY OF MULTIPLICATION

$$a \cdot b = b \cdot a$$

To get specific examples, you can pick values for a and b. For example, if a is $^-3$, replace every a with $^-3$. If b is 5, replace every b with 5.

$$\underbrace{^-3 + 5}_{2} = \underbrace{5 + {}^-3}_{2}$$

Both sums are 2.

$$\underbrace{^-3 \cdot 5}_{^-15} = \underbrace{5 \cdot {}^-3}_{^-15}$$

Both products are $^-15$.

Of course, you could pick many different values for a and b, because the commutative "rule" will always work for adding any two numbers or multiplying any two numbers.

ANSWERS

5. (a) $\frac{532}{4}$; average score is 133.

(b) $16 - 10$ is 6.
$3 - 7$ is $^-4$.
$8 - 0$ is 8.

EXAMPLE 6 Writing Properties of Operations Using Variables

Use the variable b to state this property: When a number is divided by 1, the quotient is the number.

Use the letter b to represent any number.

$$\frac{b}{1} = b$$

WORK PROBLEM 6 AT THE SIDE.

OBJECTIVE 4 In **Section 1.8** we used an exponent as a quick way to write repeated multiplication. For example,

$$\underbrace{3 \cdot 3 \cdot 3 \cdot 3 \cdot 3}_{\text{3 is used as a factor 5 times.}} \quad \text{can be written} \quad 3^5$$

(3 is the Base; 5 is the Exponent.)

The meaning of an exponent remains the same when a variable (a letter) is the base.

$$\underbrace{c \cdot c \cdot c \cdot c \cdot c}_{c\text{ is used as a factor 5 times.}} \quad \text{can be written} \quad c^5$$

(c is the Base; 5 is the Exponent.)

m^2 means $m \cdot m$ — m is used as a factor 2 times.

x^4y^3 means $x^4 \cdot y^3$ or $\underbrace{x \cdot x \cdot x \cdot x}_{x^4} \cdot \underbrace{y \cdot y \cdot y}_{y^3}$

$7b^2$ means $7 \cdot b \cdot b$ — The exponent applies *only* to b.

$^-4xy^2z$ means $^-4 \cdot x \cdot y \cdot y \cdot z$ — The exponent applies *only* to y.

EXAMPLE 7 Understanding Exponents Used with Variables

Rewrite each expression without exponents.

(a) y^6 can be written as $\underbrace{y \cdot y \cdot y \cdot y \cdot y \cdot y}_{y\text{ is used as a factor 6 times}}$

(b) $12bc^3$ can be written as $12 \cdot b \cdot \underbrace{c \cdot c \cdot c}_{c^3}$ — Coefficient is 12. The exponent applies only to c.

(c) $^-2m^2n^4$ can be written as $^-2 \cdot \underbrace{m \cdot m}_{m^2} \cdot \underbrace{n \cdot n \cdot n \cdot n}_{n^4}$ — Coefficient is $^-2$.

WORK PROBLEM 7 AT THE SIDE.

---

6. (a) Use the variable a to state this property: Multiplying any number by 0 gives a product of 0.

(b) Use the variables a, b, and c to state the associative property of addition: Changing the grouping of addends does not change the sum.

7. Rewrite each expression without exponents.

(a) x^5

(b) $4a^2b^2$

(c) $^-10xy^3$

(d) s^4tu^2

ANSWERS

6. (a) $0 \cdot a = 0$ or $a \cdot 0 = 0$
(b) $(a + b) + c = a + (b + c)$

7. (a) $x \cdot x \cdot x \cdot x \cdot x$
(b) $4 \cdot a \cdot a \cdot b \cdot b$
(c) $^-10 \cdot x \cdot y \cdot y \cdot y$
(d) $s \cdot s \cdot s \cdot s \cdot t \cdot u \cdot u$

To evaluate an expression with exponents, multiply all the factors.

EXAMPLE 8 Evaluating Expressions with Exponents

Evaluate each expression.

(a) x^2 when x is $^-3$

$$x^2 \text{ means } x \cdot x \quad \text{Replace each } x \text{ with } {}^-3.$$

$$\underbrace{{}^-3 \cdot {}^-3}_{9} \quad \text{Multiply } {}^-3 \text{ times } {}^-3.$$

So x^2 becomes $(^-3)^2$, which is $(^-3)(^-3)$, or 9.

(b) x^3y when x is $^-4$ and y is $^-10$

$$x^3y \text{ means } x \cdot x \cdot x \cdot y \quad \text{Replace } x \text{ with } {}^-4, \text{ and replace } y \text{ with } {}^-10.$$

$$\underbrace{{}^-4 \cdot {}^-4}_{16} \cdot {}^-4 \cdot {}^-10 \quad \text{Multiply two factors at a time.}$$

$$\underbrace{16 \cdot {}^-4}_{^-64} \cdot {}^-10$$

$$\underbrace{{}^-64 \cdot {}^-10}_{640}$$

So x^3y becomes $(^-4)^3\,(^-10)$, which is $(^-4)(^-4)(^-4)(^-10)$, or 640.

(c) $^-5ab^2$ when a is 5 and b is 3

$$^-5ab^2 \text{ means } {}^-5 \cdot a \cdot b \cdot b \quad \text{Replace } a \text{ with 5, and replace } b \text{ with 3.}$$

$$\underbrace{{}^-5 \cdot 5}_{^-25} \cdot 3 \cdot 3 \quad \text{Multiply two factors at a time.}$$

$$\underbrace{{}^-25 \cdot 3}_{^-75} \cdot 3$$

$$\underbrace{{}^-75 \cdot 3}_{^-225}$$

So $^-5ab^2$ becomes $^-5(5)(3)^2$, which is $^-5(5)(3)(3)$, or $^-225$.
↑Coefficient is $^-5$.↑

◀◀ WORK PROBLEM 8 AT THE SIDE.

8. Evaluate each expression.

(a) y^3 when y is $^-5$

(b) r^2s^2 when r is 6 and s is 3

(c) $10xy^2$ when x is 4 and y is $^-3$

(d) $^-3c^4$ when c is 2

Answers

8. **(a)** $(^-5)^3$ is $^-125$.
(b) $(6)^2\,(3)^2$ is 324.
(c) $10(4)(^-3)^2$ is 360.
(d) $^-3(2)^4$ is $^-48$.

2.1 Exercises

Identify the parts of each expression. Choose from these labels: variable, constant, and coefficient.

1. $c + 4$

2. $d + 6$

3. $5h$

4. $3s$

5. $^{-}3 + m$

6. $^{-}4 + n$

7. $2c + 10$

8. $6b + 1$

9. $x - y$

10. xy

11. $^{-}6g$

12. $^{-}10k$

Evaluate each expression.

13. The expression (rule) for ordering robes for the graduation ceremony is $g + 10$, where g is the number of graduates. Evaluate the expression when
(a) there are 654 graduates.
(b) there are 208 graduates.

14. The expression (rule) for ordering individual milk cartons at the daycare center is $n + 4$, where n is the number of children. Evaluate the expression when
(a) there are 41 children.
(b) there are 35 children.

15. The expression for finding the perimeter of a triangle with sides of equal length is $3s$, where s is the length of one side. Evaluate the expression when
(a) the length of one side is 11 inches.
(b) the length of one side is 3 feet.

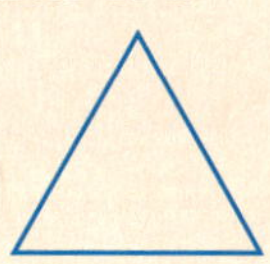

16. The expression for finding the perimeter of a pentagon with sides of equal length is $5s$, where s is the length of one side. Evaluate the expression when
(a) the length of one side is 25 meters.
(b) the length of one side is 8 inches.

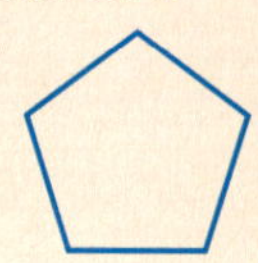

Writing Calculator Small Group

17. The expression for ordering brushes for an art class is $3c - 5$, where c is the class limit. Evaluate the expression when
(a) the class limit is 12.
(b) the class limit is 16.

18. The expression for ordering doughnuts for the office staff is $2n - 4$, where n is the number of people at work. Evaluate the expression when
(a) there are 13 people at work.
(b) there are 18 people at work.

19. The expression for figuring a student's average test score is $\frac{p}{t}$, where p is the total points earned on all the tests and t is the number of tests. Evaluate the expression when
(a) 332 points were earned on 4 tests.
(b) there were 7 tests and 637 points earned.

20. The expression for deciding how many buses are needed for a group trip is $\frac{p}{b}$, where p is the total number of people and b is the number of people that one bus will hold. Evaluate the expression when
(a) 176 people are going on a trip and one bus holds 44 people.
(b) A bus holds 36 people and 72 people are going on a trip.

Complete each table by evaluating the expressions.

21.

Value of x	Expression $x + x + x + x$	Expression $4x$
$^-2$	$^-2 + {^-2} + {^-2} + {^-2}$ is $^-8$	$4 \cdot {^-2}$ is $^-8$
12		
0		
$^-5$		

22.

Value of y	Expression $3y$	Expression $y + 2y$
$^-6$	$3 \cdot {^-6}$ is $^-18$	$^-6 + 2 \cdot {^-6}$ is $^-6 + {^-12}$, or $^-18$
10		
$^-3$		
0		

23.

Value of x	Value of y	Expression $^-2x + y$
3	7	$^-2 \cdot 3 + 7$ is $^-6 + 7$, or 1
$^-4$	5	
$^-6$	$^-2$	
0	$^-8$	

24.

Value of x	Value of y	Expression ^-2xy
3	7	$^-2 \cdot 3 \cdot 7$ is $^-42$
$^-4$	5	
$^-6$	$^-2$	
0	$^-8$	

25. Explain the words *variable* and *expression.*

26. Explain the words *coefficient* and *constant.*

Use the variable b to express each of these properties.

27. Multiplying a number by 1 leaves the number unchanged.

28. Adding zero to any number leaves the number unchanged.

29. Any number divided by zero is undefined.

30. Multiplication distributes over addition. (Use *a*, *b*, and *c* as the variables.)

Rewrite each expression without exponents.

31. c^6

32. d^7

33. x^4y^3

34. c^2d^5

35. $^-3a^3b$

36. $^-8m^2n$

37. $9xy^2$

38. $5ab^4$

39. $^-2c^5d$

40. $^-4x^3y$

41. a^3bc^2

42. x^2yz^6

Evaluate each expression when r is $^{-}3$, *s is* 2, *and t is* $^{-}4$.

43. t^2

44. r^2

45. rs^3

46. s^4t

47. $3rs$

48. $6st$

49. $^{-}2s^2t^2$

50. $^{-}4rs^4$

51. $r^2s^5t^3$

52. $r^3s^4t^2$

53. $^{-}10r^5s^7$

54. $^{-}5s^6t^5$

G *Evaluate each expression when x is* 4, *y is* $^{-}2$, *and z is* $^{-}6$.

55. $|xy| + |xyz|$

56. $x + |y^2| + |xz|$

57. $\dfrac{z^2}{^{-}3y + z}$

58. $\dfrac{y^2}{x + 2y}$

Review and Prepare

Add, or subtract by changing subtraction to addition. (For help, see ***Sections 1.3*** *and* ***1.4****.)*

59. $^{-}9 + 12 + {^{-}4}$

60. $10 + {^{-}9} + 3$

61. $2 - 7$

62. $5 - 6$

63. $^{-}3 - 8 + 11$

64. $2 + 8 - 12$

65. $^{-}7 + 1 - {^{-}10}$

66. $^{-}1 - 15 + 16$

2.2 Simplifying Expressions

Objectives

1. Combine like terms, using the distributive property.
2. Simplify expressions.
3. Use the distributive property to multiply.

For Extra Help

Tutorial — Tape 3 — SSM, Sec. 2.2

Objective 1 In **Section 2.1,** the expression for ordering math textbooks was $c + 5$. This expression was simple and easy to use. Sometimes expressions are *not* written in the simplest possible way. For example:

Evaluate this expression when c is 20.

$c + 5$ — Replace c with 20.

$20 + 5$ — Add $20 + 5$.

25

Evaluate this expression when c is 20.

$2c - 10 - c + 15$ — Replace c with 20.

$2 \cdot 20 - 10 - 20 + 15$ — Multiply $2 \cdot 20$.

$40 - 10 - 20 + 15$ — Change subtraction to adding the opposite.

$40 + {}^-10 + {}^-20 + 15$ — Add from left to right.

$30 + {}^-20 + 15$

$10 + 15$

25

These two expressions are actually equivalent. When you evaluate them, the final result is the same, but it takes a lot more work when you use the right-hand expression. To save a lot of work, you need to learn how to **simplify expressions.** Then you can rewrite $2c - 10 - c + 15$ in the simplest way possible, which is $c + 5$.

The basic idea in simplifying expressions is to *combine,* or *add,* like terms. Each addend in an expression is a **term.** Here are two examples.

$c + 5$ — the + separates the terms; Two terms

$6x^2 + {}^-2xy + 8$ — the + signs separate the terms; Three terms

In $6x^2 + {}^-2xy + 8$, the 8 is the *constant term.* There are also two variable terms in the expression: $6x^2$ is a variable term, and ${}^-2xy$ is a variable term. A **variable term** has a number part (coefficient) and a letter part (variable).

$6x^2$ — Coefficient is 6. Variable part is x^2.

${}^-2xy$ — Coefficient is $^-2$. Variable part is xy.

If no coefficient is shown, it is assumed to be 1. Remember from **Section 1.6** that multiplying any number by 1 does *not* change the number.

c can be written $1 \cdot c$ or just $1c$

Coefficient is understood to be 1. Variable part is c.

Also, ${}^-c$ can be written ${}^-1 \cdot c$. The coefficient of ${}^-c$ is understood to be ${}^-1$.

Like Terms

Like terms are terms with exactly the same variable parts (the same letters and exponents). The coefficients can be different.

LIKE TERMS		UNLIKE TERMS	
$5x$ and $3x$	Variable parts match; both are x.	$3x$ and $3x^2$	Variable parts do *not* match; exponents are different.
$^-6y^3$ and y^3	Variable parts match; both are y^3.	^-2x and ^-2y	Variable parts do *not* match; letters are different.
$4a^2b$ and $5a^2b$	Variable parts match; both are a^2b.	a^3b and a^2b	Variable parts do *not* match; exponents are different.
$^-8$ and 4	There are no variable parts; numbers are like terms.	^-8c and 4	Variable parts do *not* match; one term has a variable part, but the other term does not.

EXAMPLE 1 Identifying Like Terms and Their Coefficients

List the like terms in each expression. Then identify the coefficients of the like terms.

(a) $^-5x + {}^-5x^2 + 3xy + x + {}^-5$

The like terms are ^-5x and x.

The coefficient of ^-5x is $^-5$, and the coefficient of x is understood to be 1.

(b) $2yz^2 + 2y^2z + {}^-3y^2z + 2 + {}^-6yz$

The like terms are $2y^2z$ and $^-3y^2z$.

The coefficients are 2 and $^-3$.

(c) $10ab + 12 + {}^-10a + 12b + {}^-6$

The like terms are 12 and $^-6$.

The like terms are constants.

WORK PROBLEM 1 AT THE SIDE.

The distributive property (see **Section 1.6**) can be used "in reverse" to combine like terms. For example,

$$\underbrace{3 \cdot x}_{3x} + \underbrace{4 \cdot x}_{4x} \quad \text{can be written as} \quad \underbrace{\underbrace{(3+4)}_{7} \cdot x}_{7x}$$

Thus $3x + 4x$ can be written in simplified form as $7x$. To check, evaluate each expression when x is 2.

$$3x + 4x \rightarrow \underbrace{\underbrace{3 \cdot 2}_{6} + \underbrace{4 \cdot 2}_{8}}_{14} \qquad 7x \rightarrow \underbrace{7 \cdot 2}_{14}$$

Both results are 14, so the expressions are equivalent. But you can see how much easier it is to work with $7x$, the simplified expression.

1. List the like terms in each expression. Then identify the coefficients of the like terms.

(a) $3b^2 + {}^-3b + 3 + b^3 + b$

(b) $^-4xy + 4x^2y + {}^-4xy^2 + {}^-4 + 4$

(c) $5r^2 + 2r + {}^-2r^2 + 5 + 5r^3$

(d) $^-10 + {}^-x + {}^-10x + {}^-x^2 + {}^-10y$

ANSWERS

1. (a) ^-3b and b; the coefficients are $^-3$ and 1.
(b) $^-4$ and 4; they are constants.
(c) $5r^2$ and $^-2r^2$; the coefficients are 5 and $^-2$.
(d) ^-x and ^-10x; the coefficients are $^-1$ and $^-10$.

Note

Notice that $3x + 4x$ is simplified to $7x$, **not** to $7x^2$

Variable part is unchanged. Do not change x to x^2.

Combining Like Terms

Step 1 If there are variable terms with no coefficient, write in the understood 1.

Step 2 If there are any subtractions, change each one to adding the opposite.

Step 3 Find like terms (the variable parts match).

Step 4 Add the coefficients (number parts) of like terms. **The variable part stays the same.**

EXAMPLE 2 Combining Like Terms

Combine like terms.

(a) $2x + 4x + x$

$2x + 4x + x$ No coefficient; write understood 1. There are no subtractions to change.

$2x + 4x + 1x$ Find like terms: $2x$, $4x$, and $1x$ are like, so add the coefficients, $2 + 4 + 1$.

$(2 + 4 + 1)x$ The variable part, x, stays the same.

$7x$

Therefore $2x + 4x + x$ can be written $7x$.

(b) $^-3y^2 - 8y^2$

$^-3y^2 - 8y^2$ Both coefficients are shown. Change subtraction to adding the opposite.

$^-3y^2 + {}^-8y^2$ Find like terms: $^-3y^2$ and $^-8y^2$ are like, so add the coefficients, $^-3 + {}^-8$.

$(^-3 + {}^-8)y^2$ The variable part, y^2, stays the same.

$^-11y^2$

Therefore $^-3y^2 - 8y^2$ can be written $^-11y^2$.

WORK PROBLEM 2 AT THE SIDE.

OBJECTIVE 2 When simplifying expressions, be careful to combine only *like* terms—those having variable parts that match. You *cannot* combine terms if the variable parts are different.

2. Combine like terms.

(a) $10b + 4b + 10b$

(b) $y^3 + 8y^3$

(c) $^-7n - n$

(d) $3c - 5c - 4c$

(e) $^-9xy + xy$

(f) $^-4p^2 - 3p^2 + 8p^2$

(g) $ab - ab$

ANSWERS

2. **(a)** $24b$
(b) $9y^3$
(c) ^-8n
(d) ^-6c
(e) ^-8xy
(f) $1p^2$, or just p^2
(g) 0 because $1ab + {}^-1ab$ is $(1 + {}^-1)ab$, or $0ab$, and zero times anything is zero.

3. Simplify.

(a) $3b^2 + 4d^2 + 7b^2$

(b) $4a + b - 6a + b$

(c) $^-6x + 5 + 6x + 2$

(d) $2y - 7 - y + 7$

(e) $^-3x - 5 + 12 + 10x$

ANSWERS

3. (a) $10b^2 + 4d^2$
(b) $^-2a + 2b$
(c) 7
(d) y
(e) $7x + 7$

EXAMPLE 3 Simplifying Expressions

Simplify.

(a) $6xy + 2y + 3xy$

The like terms are $6xy$ and $3xy$. We can use the commutative property to rewrite the expression so that the like terms are next to each other. This helps to organize our work.

$6xy + 3xy + 2y$	Combine like terms only.
$(6 + 3)xy + 2y$	Add the coefficients, 6 + 3. The variable part, xy, stays the same.
$9xy + 2y$	Keep writing the term that was *not* combined. It is still part of the expression.

The simplified expression is $9xy + 2y$.

(b) Here is the expression from the first page in this section.

$2c - 10 - c + 15$	Write the understood 1 as the coefficient of c.
$2c - 10 - 1c + 15$	Change subtractions to adding the opposite.
$2c + {}^-10 + {}^-1c + 15$	Rewrite the expression so that like terms are next to each other.
$2c + {}^-1c + {}^-10 + 15$	Combine $2c + {}^-1c$. Also combine $^-10 + 15$.
$(2 + {}^-1)c + 5$	
$1c + 5$	

The simplified expression is

$$1c + 5 \quad \text{or just} \quad c + 5.$$

$1c$ is the same as c.

Note

In this book, when combining like terms, we will usually write the variable terms in alphabetical order. A constant term (number only) will be written last. So, in Examples 3(a) and 3(b) above, the preferred and alternative ways of writing the expressions are as follows.

Simplified expression is $9xy + 2y$ (alphabetical order). However, by the commutative property, $2y + 9xy$ is also correct.

Simplified expression is $c + 5$ (constant written last). However, by the commutative property, $5 + c$ is also correct.

◀◀ WORK PROBLEM 3 AT THE SIDE.

We can use the associative property of multiplication to simplify an expression such as $4(3x)$.

$$4(3x) \quad \text{can be written as} \quad 4 \cdot (3 \cdot x)$$

↑↑ Understood multiplications

Using the associative property, we can regroup the factors.

$4 \cdot (3 \cdot x)$ can be written as $\underbrace{(4 \cdot 3)}\cdot x$ To simplify, multiply $4 \cdot 3$.

$\underbrace{12 \cdot x}$ Write $12 \cdot x$ without the multiplication dot.

$12x$

The simplified expression is $12x$.

EXAMPLE 4 Simplifying Multiplication Expressions

Simplify.

(a) $5(10y)$

Use the associative property.

$5 \cdot (10 \cdot y)$ can be written as $\underbrace{(5 \cdot 10)} \cdot y$ Multiply $5 \cdot 10$.

$\underbrace{50 \cdot y}$ Write $50 \cdot y$ without the multiplication dot.

$50y$

So, $5(10y)$ simplifies to $50y$.

(b) $^-6(3b)$

Use the associative property.

$^-6(3b)$ can be written as $(^-6 \cdot 3)b$

^-18b

So, $^-6(3b)$ simplifies to ^-18b.

(c) $^-4(^-2x^2)$

Use the associative property.

$^-4(^-2x^2)$ can be written as $(^-4 \cdot {}^-2)x^2$

$8x^2$

So, $^-4(^-2x^2)$ simplifies to $8x^2$.

WORK PROBLEM 4 AT THE SIDE. ▶▶

OBJECTIVE 3 The distributive property can also be used to simplify expressions such as $3(x + 5)$. You *cannot* add the terms inside the parentheses because x and 5 are *not* like terms. But notice the understood multiplication dot between the 3 and the parentheses.

$$3(x + 5)$$
$$\downarrow$$
$$3 \cdot (x + 5)$$

Thus you can distribute multiplication over addition, as you did in **Section 1.6.** That is, multiply 3 times each term inside the parentheses.

$3 \cdot (x + 5)$ can be written as $\underbrace{3 \cdot x} + \underbrace{3 \cdot 5}$

$3x + 15$

So, $3(x + 5)$ simplifies to $3x + 15$.

(stays as addition)

4. Simplify.

(a) $7(4c)$

(b) $^-3(5y^3)$

(c) $20(^-2a)$

(d) $^-10(^-x)$

ANSWERS

4. (a) $28c$ **(b)** $^-15y^3$ **(c)** ^-40a **(d)** $10x$

Multiplication also distributes over subtraction.

$4 \cdot (y - 2)$ can be written as $\underbrace{4 \cdot y}_{4y} - \underbrace{4 \cdot 2}_{8}$

So, $4(y - 2)$ simplifies to $4y - 8$.
↑ stays as subtraction ↑

Notice that we did *not* need to change subtraction to adding the opposite.

EXAMPLE 5 Using the Distributive Property

Simplify.

(a) $6(y - 4)$ can be written as $\underbrace{6 \cdot y}_{6y} - \underbrace{6 \cdot 4}_{24}$

stays as subtraction

So, $6(y - 4)$ simplifies to $6y - 24$.

(b) $5(3x + 2)$ can be written as

$$\underbrace{5 \cdot 3x}_{} + \underbrace{5 \cdot 2}_{}$$
$$\underbrace{5 \cdot 3} \cdot x + 10$$
$$\underbrace{15 \cdot x} + 10$$
$$15x + 10$$

So, $5(3x + 2)$ simplifies to $15x + 10$.

(c) $^-2(4a + 3)$ can be written as

$$\underbrace{^-2 \cdot 4a} + \underbrace{^-2 \cdot 3}$$
$$\underbrace{^-2 \cdot 4} \cdot a + {}^-6$$
$$\underbrace{^-8 \cdot a} + {}^-6$$
$$^-8a + {}^-6$$

Now we will use the definition of subtraction "in reverse" to rewrite $^-8a + {}^-6$.

Change $^-6$ to its opposite, $^+6$.

Write $^-8a + {}^-6$ as $^-8a - 6$

Change addition to subtraction.

Think back to the way you changed subtraction to adding the opposite. Here you are "working backwards." From now on, whenever addition is followed by a negative number, we will change it to subtracting a positive number.

$^-8a + {}^-6$
↕ ↕ Equivalent expressions
$^-8a - 6$

So, $^-2(4a + 3)$ simplifies to $^-8a - 6$.

◀◀ WORK PROBLEM 5 AT THE SIDE.

5. Simplify.

(a) $7(a + 10)$

(b) $3(x - 3)$

(c) $4(2y + 6)$

(d) $^-5(3b + 2)$

(e) $^-8(c + 4)$

ANSWERS

5. (a) $7a + 70$ **(b)** $3x - 9$ **(c)** $8y + 24$ **(d)** $^-15b - 10$ **(e)** $^-8c - 32$

Sometimes you need to do several steps to simplify an expression.

EXAMPLE 6 Simplifying More Complex Expressions

Simplify: $8 + 3(x - 2)$.

$8 + 3(x - 2)$ Do *not* add 8 + 3. Use the distributive property because multiplying is done *before* adding.

$8 + \underbrace{3 \cdot x} - \underbrace{3 \cdot 2}$ Do the multiplications.

$8 + 3x - 6$ Rewrite so that like terms are next to each other.

$3x + \underbrace{8 - 6}$ Subtract to find 8 − 6 or change to adding $8 + {}^-6$.

$3x + 2$

The simplified expression is $3x + 2$.

Note

Do *not* add 8 + 3 as the first step in Example 6 above. Remember that the order of operations tells you to add *last*.

WORK PROBLEM 6 AT THE SIDE.

6. Simplify.

(a) ${}^-4 + 5(y + 1)$

(b) $2(3w + 4) - 5$

(c) $5(6x - 2) + 3x$

(d) $21 + 7(a^2 - 3)$

(e) ${}^-y + 3(2y + 5) - 18$

ANSWERS

6. (a) $5y + 1$ **(b)** $6w + 3$ **(c)** $33x - 10$ **(d)** $7a^2$ **(e)** $5y - 3$

NUMBERS IN THE *Real World*

collaborative investigations

Expressions

An expression for the number of words in a child's vocabulary is given below.

$60A - 900$ where A is the age in months

1. Evaluate the expression for a child who is 20 months old. How many words are in the child's vocabulary?

2. Evaluate the expression for a child who is 2 years old. (How many months are in two years?)

3. How many words does a child learn between the ages of 20 months and 2 years?

4. Evaluate the expression for a child who is 3 years old.

5. How many words does a child learn between the ages of 2 years and 3 years?

6. Evaluate the expression for a child who is 15 months old. Explain why the answer does, or does not, make sense.

7. Evaluate the expression for a child who is 12 months old. Explain why the answer does, or does not, make sense.

You can estimate how many miles away a thunderstorm is by counting the number of seconds from the time you see a lightning flash until you hear the thunder. Then use this expression.

$\frac{s}{5}$ where s is the number of seconds

8. Evaluate this expression for 15 seconds, for 10 seconds, and for 5 seconds. How many miles away is the storm?

9. Using the answers you just calculated, how could you estimate the distance when the time is $2\frac{1}{2}$ seconds?

Traffic engineers have to decide how long to have the red, yellow, and green lights showing on a traffic signal. They use this expression to find the number of seconds that the yellow light should be on.

$\frac{5v}{100} + 1$ where v is the speed limit in miles per hour

10. Evaluate this expression for speed limits of 20, 40, and 60 miles per hour. How many seconds should the yellow light be on?

11. Using the answers you just calculated, how could you estimate the yellow light time for speed limits of 30 and 50 miles per hour?

Sometimes multiple choice tests are scored in such a way as to discourage guessing. The number of correct answers is reduced by some portion of the number of wrong answers. Here is an expression for finding the score on a multiple choice test.

$$R - \frac{W}{n-1}$$

where R is the number of right answers, W is the number of wrong answers, and n is the number of choices for each question.

12. Evaluate this expression for a 50-item test with 4 choices for each item. The student got 38 items right and 12 items wrong. What is the student's score?

13. Explain what happens when you evaluate the expression for a student who got all 50 items correct.

14. Explain what happens when you evaluate the expression for a student who got 11 items correct and 39 items wrong.

2.2 Exercises

Circle the like terms in each expression. Then identify the coefficients of the like terms.

1. $2b^2 + 2b + 2b^3 + b^2 + 6$

2. $3x + x^3 + 3x^2 + 3 + 2x^3$

3. $^-x^2y + {}^-xy + 2xy + {}^-2xy^2$

4. $ab^2 + {}^-a^2b + 2ab + {}^-3a^2b$

5. $7 + 7c + 3 + 7c^3 + {}^-4$

6. $4d + {}^-5 + 1 + {}^-5d^2 + 4$

Simplify these expressions.

7. $6r + 6r$

8. $4t + 10t$

9. $x^2 + 5x^2$

10. $9y^3 + y^3$

11. $p - 5p$

12. $n - 3n$

13. $^-2a^3 - a^3$

14. $^-10x^2 - x^2$

15. $c - c$

16. $b^2 - b^2$

17. $9xy + xy - 9xy$

18. $r^2s - 7r^2s + 7r^2s$

19. $5t^4 + 7t^4 - 6t^4$

20. $10mn - 9mn + 3mn$

21. $y^2 + y^2 + y^2 + y^2$

22. $a + a + a$

23. $^-x - 6x - x$

24. $^-y - y - 3y$

Simplify by combining like terms. Write your answers with the variables in alphabetical order and any constant term last.

25. $8a + 4b + 4a$

26. $6x + 5y + 4y$

27. $6 + 8 + 7rs$

28. $10 + 2c^2 + 15$

29. $a + ab^2 + ab^2$

30. $n + mn + n$

31. $6x + y - 8x + y$

32. $d + 3c - 7c + 3d$

33. $8b^2 - a^2 - b^2 + a^2$

34. $5ab - ab + 3a^2b - 4ab$

35. $^-x^3 + 3x - 3x^2 + 2$

36. $a^2b - 2ab - ab^3 + 3a^3b$

37. $^-9r + 6t - s - 5r + s + t - 6t + 5s - r$

38. $^-x - 3y + 4z + x - z + 5y - 8x - y$

Simplify, using multiplication.

39. $3(10a)$

40. $8(4b)$

41. $^-4(2x^2)$

42. $^-7(3b^3)$

43. $5(^-4y^3)$

44. $2(^-6x)$

45. $^-9(^-2cd)$

46. $^-6(^-4rs)$

47. $7(3a^2bc)$

48. $4(2xy^2z^2)$

49. $^-12(^-w)$

50. $^-10(^-k)$

Use the distributive property to simplify these expressions.

51. $6(b + 6)$ **52.** $5(a + 3)$ **53.** $7(x - 1)$

54. $4(y - 4)$ **55.** $3(7t + 1)$ **56.** $8(2c + 5)$

57. $^{-}2(5r + 3)$ **58.** $^{-}5(6z + 2)$ **59.** $^{-}9(k + 4)$

60. $^{-}3(p + 7)$ **61.** $50(m - 6)$ **62.** $25(n - 1)$

Simplify these expressions.

63. $10 + 2(4y + 3)$ **64.** $4 + 7(x^2 + 3)$ **65.** $6(a^2 - 2) + 15$

66. $5(b - 4) + 25$ **67.** $2 + 9(m - 4)$ **68.** $6 + 3(n - 8)$

69. $^{-}5(k + 5) + 5k$ **70.** $^{-}7(p + 2) + 7p$ **71.** $4(6x - 3) + 12$

72. $6(3y - 3) + 18$ **73.** $5 + 2(3n + 4) - n$ **74.** $8 + 8(4z + 5) - z$

75. $^{-}p + 6(2p - 1) + 5$ **76.** $^{-}k + 3(4k - 1) + 2$

77. Explain the difference between *simplifying* an expression and *evaluating* an expression.

78. Simplify each expression. Are the answers equivalent? Explain why or why not.

$$5(3x + 2) \qquad\qquad 5(2 + 3x)$$

79. Explain what makes two terms *like* terms. Include several examples in your explanation.

80. Explain how to combine like terms. Include an example in your explanation.

81. Explain and correct the error made by a student who simplified this expression.

$$\underbrace{^{-}2x + 7x} + 8$$
$$5x^2 \qquad + 8$$

82. Explain and correct the error made by a student who simplified this expression;

$$\underbrace{^{-}10a + 6a} - \quad 7 + 2$$
$$\downarrow \quad \downarrow$$
$$^{-}4a \qquad + \underbrace{^{-}7 + 2}$$
$$^{-}4a \qquad + \quad ^{-}5$$
$$4a - 5$$

Simplify.

83. $^{-}4(3y) - 5 + 2(5y + 7)$

84. $6(^{-}3x) - 9 + 3(^{-}2x + 6)$

85. $^{-}10 + 4(^{-}3b + 3) + 2(6b - 1)$

86. $12 + 2(4a - 4) + 4(^{-}2a - 1)$

87. $^{-}5(^{-}x + 2) + 8(^{-}x) + 3(^{-}2x - 2) + 16$

88. $^{-}7(^{-}y) + 6(y - 1) + 3(^{-}2y) + 6 - y$

Review and Prepare

Find the opposite of each number. Show that the sum of the number and its opposite is zero. *(For help, see* ***Section 1.4****.)*

89. 7 **90.** 5 **91.** $^{-}12$ **92.** $^{-}10$

2.3 *Solving Equations Using Addition*

OBJECTIVES

1. Determine whether a given number is the solution of an equation.
2. Solve equations, using the addition property of equality.
3. Simplify equations and then use the addition property of equality.

FOR EXTRA HELP

Tutorial — Tape 3 — SSM, Sec. 2.3

Now you are ready for a look at the "heart" of algebra, writing and solving **equations** (ee-KWAY-zhuhns). *Writing* an equation is a way to show the relationship between what you *know* about a problem and what you *don't* know. Then, *solving* the equation is a way to figure out the part that you didn't know and answer your question. The questions you can answer by writing and solving equations are as varied as the careers people choose. A zoo keeper can solve an equation that answers the question of how long to incubate the egg of a particular tropical bird. An aerobics instructor can solve an equation that answers the question of how hard a certain person should exercise for maximum benefit.

OBJECTIVE 1 Let's start with the example from the beginning of this chapter: ordering textbooks for math classes. The expression we used to order books was $c + 5$, where c was the class limit (the maximum number of students allowed in the class). Suppose that 30 prealgebra books were ordered. What is the class limit for prealgebra? To answer this question, write an equation showing the relationship between what you know and what you don't know.

$$c + 5 = 30$$

You don't know the class limit. (c)
You do know the total number of books ordered. (30)
You do know that 5 extra books were ordered. (5)

Note
An equation has an equal sign. Notice the similarity in the words **equa**tion and **equal**. An expression does *not* have an equal sign.

The equal sign in an equation is like the balance point on a playground teeter-totter, or seesaw. To have a true equation, the two sides must balance.

These equations balance, so we can use the = sign.

$$6 + 8 = 14$$
$$10 = 5 \cdot 2$$
$$3 \cdot 2 = 5 + 1$$

These equations do *not* balance, so we write $\neq$ to mean "not equal to."

$$6 + 8 \neq 15$$
$$10 \neq 4 \cdot 2$$
$$4 + 5 \neq 5 \cdot 4$$

When an equation has a variable, we **solve the equation** by finding a number that can replace the variable and make the equation balance. For the textbook example:

$$c + 5 = 30$$

What number can replace c so that the equation balances?

Try replacing c with **15**. $\quad 15 + 5 \neq 30 \quad$ Does *not* balance: $15 + 5$ is only 20.

Try replacing c with **40**. $\quad 40 + 5 \neq 30 \quad$ Does *not* balance: $40 + 5$ is more than 30.

Try replacing c with **25**. $\quad 25 + 5 = 30 \quad$ Balances: $25 + 5$ is 30.

1. Which of these numbers, 95, 65, or 70, is the solution for the equation $c + 15 = 80$?

The **solution** is **25** because **25** is the *only* number that makes the equation balance. By solving the equation, you have answered the question about the class limit. The class limit is 25.

Note

Most of the equations that you will solve in this book have only one solution, that is, one number that makes the equation balance. In Chapter 9, and in other algebra courses, you will solve equations that have two or more solutions.

EXAMPLE 1 Identifying Solutions to Equations

Which of these numbers, 70, 40, or 60, is the solution for the equation $c - 10 = 50$?

Replace c with each of the numbers. The one that makes the equation balance is the solution.

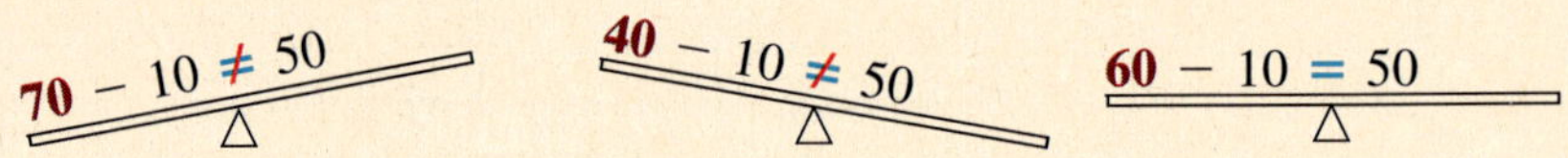

Does *not* balance: $70 - 10$ is more than 50.

Does *not* balance: $40 - 10$ is only 30.

Balances: $60 - 10$ is 50.

The solution is 60 because, when c is 60, the equation balances.

WORK PROBLEM 1 AT THE SIDE.

OBJECTIVE 2 When you solved the book ordering equation, $c + 5 = 30$, you could just look at the equation and think, "What number, plus 5, would balance with 30?" You could easily see that c had to be 25. Not all equations can be solved this easily, so you'll need some tools for the harder ones. The first tool is called the **addition property of equality.**

Addition Property of Equality

If $a = b$, then $a + c = b + c$. In other words, you may add the same number to both sides of an equation and still keep it balanced.

Think of the teeter-totter. If there are 3 children of the same size on each side, it will balance. If 2 more children climb onto the left side, the only way to keep the balance is to have 2 more children climb onto the right side as well.

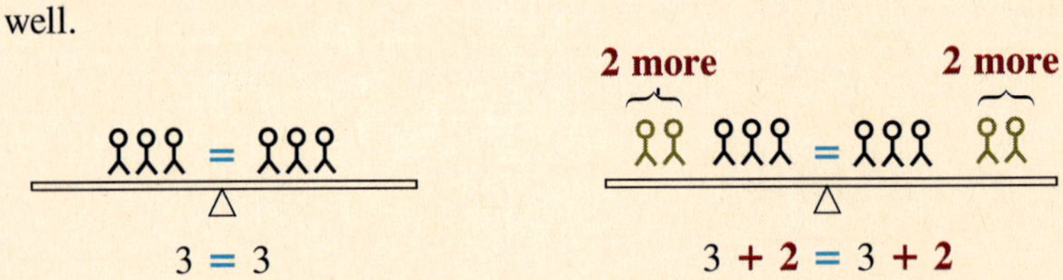

All the tools you will learn to use with equations have one goal.

ANSWER

1. 65 is the solution.

Goal in Solving an Equation

The goal is to end up with the variable (letter) on one side of the equal sign balancing a number on the other side. We work on the original equation until we get:

variable = number or number = variable

Once we have arrived at that point, the number balancing the variable is the solution to the original equation.

EXAMPLE 2 Using the Addition Property of Equality

Solve each equation and check the solution.

(a) $c + 5 = 30$

You want to get the variable, c, by itself on the left side of the equal sign. To do that, add the *opposite* of 5, which is $^{-}5$. Then $5 + {}^{-}5$ will be 0.

$$\begin{array}{rcl} c + 5 &=& 30 \\ {}^{-}5 & & {}^{-}5 \\ \hline c + 0 &=& 25 \end{array}$$

Add $^{-}5$ to the left side.

To keep the balance, add $^{-}5$ to the right side also.

$30 + {}^{-}5$ is 25.

$5 + {}^{-}5$ is 0.

Recall that adding 0 to any number leaves the number unchanged, so $c + 0$ is c. Now we have

$$\underbrace{c + 0}_{} = 25$$
$$c = 25$$

and c balances with 25, so, as we already knew, the solution is 25.

Check the solution by replacing c with 25 *in the original equation.*

$$c + 5 = 30$$
$$\underbrace{25 + 5}_{} = 30$$
$$30 = 30 \quad \text{Balances}$$

Because the equation balances when we use 25 to replace the variable, we know that **25 is the correct solution**. If it had *not* balanced, we would need to rework the problem, find our error, and correct it.

(b) $^{-}5 = x - 3$

We want the variable, x, by itself on the right side of the equal sign. (Remember, it doesn't matter which side of the equal sign the variable is on, just so it ends up by itself.) To see what number to add, we change the subtraction to adding the opposite.

$^{-}5 = x - 3$ — Change subtraction to adding the opposite.

$^{-}5 = x + {}^{-}3$ — To get x by itself on the right side, add the opposite of $^{-}3$, which is $^{+}3$. Then $^{-}3 + {}^{+}3$ is 0.

$^{+}3 \qquad {}^{+}3$ — To keep the balance, add $^{+}3$ to the left side also; $^{-}5 + {}^{+}3$ is $^{-}2$.

$^{-}2 = \underbrace{x + 0}_{}$ — Adding 0 to x leaves x unchanged.

$^{-}2 = x$ — x balances with $^{-}2$, so $^{-}2$ is the solution.

CONTINUED ON NEXT PAGE

2. Solve each equation and check each solution.

(a) $12 = y + 5$

Check

(b) $b - 2 = {}^-6$

Check

ANSWERS

2. (a) $y = 7$

Check $12 = y + 5$; $12 = \underbrace{7 + 5}$; Balances $12 = 12$

(b) $b = {}^-4$

Check $b - 2 = {}^-6$; $\underbrace{{}^-4 + {}^-2} = {}^-6$; Balances ${}^-6 = {}^-6$

We check the solution by replacing x with $^-2$ in the original equation. If the equation balances when we use $^-2$, we know that it is the correct solution. If the equation does *not* balance when we use $^-2$, we made an error and need to try solving the equation again.

Check

${}^-5 = x - 3$	Replace x with $^-2$.
${}^-5 = {}^-2 - 3$	Change subtraction to adding the opposite.
${}^-5 = \underbrace{{}^-2 + {}^-3}$	
${}^-5 = {}^-5$	Balances.

The equation balances, so **$^-2$ is the correct solution.**

Note

When checking the solution to Example 2(b), we ended up with ${}^-5 = {}^-5$. Notice that $^-5$ is **not** the solution. The solution is $^-2$, the number used to replace x in the original equation.

WORK PROBLEM 2 AT THE SIDE.

OBJECTIVE 3 Sometimes you can simplify the expression on one or both sides of the equal sign. Doing so will make it easier to solve the equation.

EXAMPLE 3 Simplifying Before Solving an Equation

Solve each equation and check each solution.

(a) $y + 8 = 3 - 7$

You cannot simplify $y + 8$ because y and 8 are not like terms.	$y + 8 = 3 - 7$	Simplify the right side of the equation by changing subtraction to adding the opposite.
	$y + 8 = \underbrace{3 + {}^-7}$	Add $3 + {}^-7$.
To get y by itself, add the opposite of 8, which is $^-8$.	$y + 8 = {}^-4$	
	$\quad {}^-8 \qquad {}^-8$	To keep the balance add $^-8$ to the right side also.
$8 + {}^-8$ is 0.	$\underbrace{y + 0} = {}^-12$	${}^-4 + {}^-8$ is $^-12$.
	$y = {}^-12$	

The solution is $^-12$. Now check the solution.

Check

	$y + 8 = 3 - 7$	Go back to the original equation and replace y with $^-12$.
Add ${}^-12 + 8$.	$\underbrace{{}^-12 + 8} = 3 - 7$	Change $3 - 7$ to $3 + {}^-7$.
	${}^-4 = \underbrace{3 + {}^-7}$	Add $3 + {}^-7$.
	${}^-4 = {}^-4$	Balances.

The equation balances, so **$^-12$ is the correct solution.**

CONTINUED ON NEXT PAGE

(b) $^{-}2 + 2 = {}^{-}4b - 6 + 5b$

Simplify the left side by adding $^{-}2 + 2$.	$\underbrace{^{-}2 + 2} = {}^{-}4b - 6 + 5b$	Simplify the right side by changing subtraction to adding the opposite.
	$0 = {}^{-}4b + {}^{-}6 + 5b$	Find like terms.
	$0 = \underbrace{^{-}4b + 5b} + {}^{-}6$	Combine $^{-}4b + 5b$.
To keep the balance, add 6 to this side also.	$0 = 1b + {}^{-}6$	To get $1b$ by itself, add the opposite of $^{-}6$, which is 6.
	$\underline{6} = \underline{\phantom{1b + {}}6}$	
	$6 = \underbrace{1b + 0}$	
	$6 = 1b$	$1b$ is equivalent to b.
	$6 = b$	

The solution is 6.

Check	$^{-}2 + 2 = {}^{-}4b - 6 + 5b$	Go back to the original equation and replace each b with 6.
Add $^{-}2 + 2$.	$\underbrace{^{-}2 + 2} = \underbrace{^{-}4 \cdot 6} - 6 + \underbrace{5 \cdot 6}$	Do multiplications first.
	$0 = \underbrace{^{-}24 + {}^{-}6} + 30$	Change subtraction to adding the opposite.
	$0 = \underbrace{^{-}30 + 30}$	Add from left to right.
	$0 = 0$	Balances.

The equation balances, so **6 is the correct solution**.

Note

When checking a solution, always go back to the *original* equation. That way you will catch any errors you made when simplifying each side of the equation.

WORK PROBLEM 3 AT THE SIDE.

3. Simplify each side of the equation when possible. Then solve the equation and check the solution.

(a) $2 - 8 = k - 2$

Check

(b) $4r + 1 - 3r = {}^{-}8 + 11$

Check

Answers

3. (a) $k = {}^{-}4$

Check $2 - 8 = k - 2$

$\underbrace{2 + {}^{-}8} = \underbrace{^{-}4 + {}^{-}2}$

Balances $^{-}6 = {}^{-}6$

(b) $r = 2$

Check $4r + 1 - 3r = {}^{-}8 + 11$

$\underbrace{4 \cdot 2} + 1 - \underbrace{3 \cdot 2} = \underbrace{^{-}8 + 11}$

$\underbrace{8 + 1} - 6 = 3$

$\underbrace{9 + {}^{-}6} = 3$

Balances $3 = 3$

Study Tips

Preparing for a Test

You've probably taken one test already in your math course. Did you do as well as you wanted to? If not, there are some things you can do to improve your next test score. Think about how you prepared for the last test you took and answer each of these questions honestly.

Studying for the Test

Did you

Yes	No	Read your notes *every* day *right* after class and *again* later?
Yes	No	Do your homework the same day as it was assigned?
Yes	No	Start reviewing *at least* three days before the test?
Yes	No	Take a 10-minute break after each hour of studying?
Yes	No	Make study cards for new vocabulary and drill yourself on them each day?
Yes	No	Rework some problems from every set of exercises?
Yes	No	Rework errors on quizzes and homework assignments corrected by your teacher?
Yes	No	Practice doing a sample test in the same amount of time as allowed on the real test?

Before the Test

Did you

Yes	No	Get a good night's sleep?
Yes	No	Eat a high-energy meal with protein about two hours before the test?
Yes	No	Avoid caffeine and high-fat or sugary snacks before the test?
Yes	No	Do some walking or moderate exercise shortly before the test?
Yes	No	Arrive on time?
Yes	No	Come in with a positive attitude?
Yes	No	Ignore what other students were saying right before the test?
Yes	No	Bring a calculator (if allowed) and a pencil (not pen)?
Yes	No	Do a minute or two of deep breathing or other relaxation?

Did You Know...

- Fatigue causes poor memory recall.
- Exercise reduces stress and prevents "blanking out" on a test. Exercise increases clear thinking.
- Just 3 or 4 ounces of protein increases the amount of the chemical tyrosine in your brain, which improves alertness, accuracy, and motivation.

Make a Plan

How will *you* improve the way you prepare for tests? Look at each question that you answered "No." Pick three or four of them that you can change to "Yes." Write down your plans. For more help, see the Study Tips pages in Chapters 1 and 3.

2.3 Exercises

In each list of numbers, find the one that is a solution to the given equation.

1. $n - 50 = 8$

58, 42, 60

2. $r - 20 = 5$

15, 30, 25

3. $^-6 = y + 10$

$^-4$, $^-16$, 16

4. $^-4 = x + 13$

17, $^-17$, $^-9$

5. $t + 12 = 0$

0, $^-12$, $^-24$

6. $b - 8 = 0$

8, 0, $^-8$

Solve each equation and check each solution.

7. $p + 5 = 9$ **Check** $p + 5 = 9$ ↓

8. $a + 3 = 12$ **Check** $a + 3 = 12$ ↓

9. $8 = r - 2$ **Check** $8 = r - 2$ ↓

10. $3 = b - 5$ **Check** $3 = b - 5$ ↓

11. $^-5 = n + 3$ **Check**

12. $^-1 = a + 8$ **Check**

13. $^-4 + k = 14$ **Check**

14. $^-9 + y = 7$ **Check**

15. $y - 6 = 0$ **Check**

16. $k - 15 = 0$ **Check**

17. $7 = r + 13$ **Check**

18. $12 = z + 19$ **Check**

19. $x - 12 = {}^{-}1$ **Check**

20. $m - 3 = {}^{-}9$ **Check**

21. ${}^{-}5 = {}^{-}2 + t$ **Check**

22. ${}^{-}1 = {}^{-}10 + w$ **Check**

A solution is given for each equation. Show how to check the solution. If the solution is correct, leave it. If the solution is not correct, solve the equation and check your solution.

23. $z - 5 = 3$
Solution is ${}^{-}2$.
Check $z - 5 = 3$
↓

24. $x - 9 = 4$
Solution is 13.
Check $x - 9 = 4$
↓

25. $7 + x = {}^{-}11$
Solution is ${}^{-}18$.
Check

26. $2 + k = {}^{-}7$
Solution is ${}^{-}5$.
Check

27. ${}^{-}10 = {}^{-}10 + b$
Solution is 10.
Check

28. $0 = {}^{-}14 + a$
Solution is 0.
Check

Simplify each side of the equation when possible. Then solve each equation and check each solution.

29. $c - 4 = {}^-8 + 10$ **Check**

30. $b - 8 = 10 - 6$ **Check**

31. ${}^-1 + 4 = y - 2$ **Check**

32. $2 + 3 = k - 4$ **Check**

33. $10 + b = {}^-14 - 6$ **Check**

34. $1 + w = {}^-8 - 8$ **Check**

35. $t - 2 = 3 - 5$ **Check**

36. $p - 8 = {}^-10 + 2$ **Check**

37. $10z - 9z = {}^-15 + 8$ **Check**

38. $2r - r = 5 - 10$ **Check**

39. ${}^-5w + 2 + 6w = {}^-4 + 9$ **Check**

40. ${}^-2t + 4 + 3t = 6 - 7$ **Check**

Solve each equation. Show the steps you used.

41. $^{-}3 - 3 = 4 - 3x + 4x$

42. $^{-}5 - 5 = {}^{-}2 - 6b + 7b$

43. $^{-}3 + 7 - 4 = {}^{-}2a + 3a$

44. $6 - 11 + 5 = {}^{-}8c + 9c$

45. $y - 75 = {}^{-}100$

46. $a - 200 = {}^{-}100$

47. $^{-}x + 3 + 2x = 18$

48. $^{-}s + 2s - 4 = 13$

49. $82 = {}^{-}31 + k$

50. $^{-}5 = 72 + w$

51. $^{-}2 + 11 = 2b - 9 - b$

52. $^{-}6 + 7 = 2h - 1 - h$

53. $r - 6 = 7 - 10 - 8$

54. $m - 5 = 2 - 9 + 1$

55. ${}^{-}14 = n + 91$

56. $66 = x - 28$

57. ${}^{-}9 + 9 = 5 + h$

58. $18 - 18 = 6 + p$

59. Explain the difference between an expression and an equation. Give two examples of each.

60. Explain the addition property of equality.

61. A student did this work when solving an equation. Do you agree that the solution is ${}^{-}7$? Explain why or why not.

$$\begin{aligned}
\underbrace{{}^{-}8 + 1} &= x + 7 \\
{}^{-}7 &= x + 7 \\
\underline{{}^{-}7} &= \underline{\quad {}^{-}7} \\
{}^{-}14 &= \underbrace{x + 0} \\
{}^{-}14 &= x
\end{aligned}$$

Check

$$\begin{aligned}
{}^{-}8 + 1 &= x + 7 \\
&\quad \downarrow \\
\underbrace{{}^{-}8 + 1} &= \underbrace{{}^{-}14 + 7} \\
{}^{-}7 &= {}^{-}7
\end{aligned}$$

Balances, so
${}^{-}7$ is the solution

62. A student did this work when solving an equation. Show how to check the solution. If the solution does not check, find and correct the errors.

$$\begin{aligned}
{}^{-}3 - 6 &= n - 5 \\
\underbrace{{}^{-}3 + 6} &= n - 5 \\
3 &= n - 5 \\
\underline{{}^{-}5} & \quad \underline{\quad {}^{-}5} \\
{}^{-}2 &= \underbrace{n + 0} \\
{}^{-}2 &= n
\end{aligned}$$

Ⓖ **63.** Write two *different* equations that have $^{-}2$ as the solution. Be sure that you have to use the addition property of equality to solve the equations. Show how to solve each equation.

Ⓖ **64.** Follow the directions in Exercise 63, but this time write two equations that have 0 as the solution.

Ⓖ *Solve each equation. Show your work.*

65. $^{-}17 - 1 + 26 - 38 = {}^{-}3 - m - 8 + 2m$

66. $19 - 38 - 9 + 11 = {}^{-}t - 6 + 2t - 6$

67. $^{-}6x + 2x + 6 + 5x = |0 - 9| - |{}^{-}6 + 5|$

68. $^{-}h - |{}^{-}9 - 9| + 8h - 6h = {}^{-}12 - |{}^{-}5 + 0|$

Review and Prepare

Multiply or divide. (For help, see ***Sections 1.6*** *and* ***1.7****.)*

69. (a) $^{-}6(7)$ **(b)** $^{-}6(^{-}7)$ **(c)** $6(^{-}7)$

70. (a) $9(8)$ **(b)** $^{-}9(8)$ **(c)** $^{-}9(^{-}8)$

71. (a) $\frac{24}{^{-}8}$ **(b)** $\frac{24}{8}$ **(c)** $\frac{^{-}24}{^{-}8}$

72. (a) $\frac{^{-}35}{^{-}7}$ **(b)** $\frac{35}{^{-}7}$ **(c)** $\frac{^{-}35}{7}$

73. (a) $\frac{^{-}5}{0}$ **(b)** $\frac{0}{^{-}5}$

74. (a) $\frac{0}{10}$ **(b)** $\frac{10}{0}$

75. (a) $\frac{15}{^{-}1}$ **(b)** $\frac{^{-}4}{^{-}4}$

76. (a) $\frac{11}{11}$ **(b)** $\frac{3}{^{-}1}$

2.4 *Solving Equations Using Division*

Objectives

1 Solve equations, using the division property of equality.

2 Simplify equations and then use the division property of equality.

3 Solve equations such as $^{-}x = 5$.

For Extra Help

Tutorial — Tape 3 — SSM, Sec. 2.4

Objective 1 In **Section 2.1** you worked with the expression for finding the perimeter of a square-shaped garden, that is, finding the total distance around all four sides of the garden:

$4s$, where s is the length of one side of the square.

Suppose you know that 24 feet of fencing were used around a square-shaped garden. What was the length of one side of the garden? To answer this question, write an equation showing the relationship between what you know and what you don't know.

You don't know the length of one side.

You do know that there are 4 sides. → $4\,s = 24$ ← You do know the perimeter.

To solve the equation, what number can replace s so that the equation balances? You can see that s is 6 feet.

$4 \cdot \mathbf{6} = 24$ — Balances: $4 \cdot 6$ is exactly 24.

The **solution** is **6 feet** because **6** is the *only* number that makes the equation balance. You have answered the question about the length of one side: The length is 6 feet.

There is a tool that you can use to solve equations such as $4s = 24$. It is the **division property of equality.**

Division Property of Equality

If $a = b$ then $\dfrac{a}{c} = \dfrac{b}{c}$ as long as c is not 0.

In other words, you may divide both sides of an equation by the same nonzero number and still keep it balanced.

In **Section 2.3,** you saw that *adding* the same number to both sides of an equation kept it balanced. (We could also have *subtracted* the same number from both sides because subtraction is defined as adding the opposite.) Now we're saying that you can *divide* both sides by the same number. (In Chapter 4 we'll *multiply* both sides by the same number.)

Equality Principle for Solving Equations

As long as you do the *same* thing to *both* sides of an equation, the balance is maintained and you still have a true equation. (The only exception is that you cannot divide by 0.)

EXAMPLE 1 Using the Division Principle of Equality

Solve each equation and check each solution.

(a) $4s = 24$

As with any equation, the goal is to get the variable by itself on one side of the equal sign. On the left side we have $4s$, which means $4 \cdot s$. The variable is multiplied by 4. Division is the opposite of multiplication, so dividing by 4 can be used to "undo" multiplying by 4.

The result of $4 \cdot s \div 4$ is just s.

To see how this method works, replace s with several specific values.

Evaluate $4 \cdot s \div 4$ when s is $^{-}3$.

$$4 \cdot {}^{-}3 \div 4$$
$$^{-}12 \div 4$$
$$^{-}3$$

The result is the original value of s.

Evaluate $4 \cdot s \div 4$ when s is 25.

$$4 \cdot 25 \div 4$$
$$100 \div 4$$
$$25$$

The result is the original value of s.

In algebra we usually write division using a fraction bar.

Divide $4s$ by 4. The fraction bar indicates division: $4s \div 4$ is s. → $\dfrac{4s}{4} = \dfrac{24}{4}$ To keep the balance, divide the right side by 4 also: $24 \div 4$ is 6.

$$s = 6$$

So, as we already knew, 6 is the solution. Check the solution by replacing s with 6 in the original equation.

Check

$$4s = 24$$
$$4 \cdot 6 = 24$$
$$24 = 24 \quad \text{Balances}$$

The equation balances, so **6 is the correct solution**.

(b) $42 = {}^{-}6w$

On the right side of the equation, the variable is multiplied by $^{-}6$. To undo the multiplication, divide by $^{-}6$.

To keep the balance, divide by $^{-}6$ on the left side also. $\dfrac{42}{^{-}6} = \dfrac{^{-}6w}{^{-}6}$ Use division to undo multiplication: $^{-}6 \cdot w \div {}^{-}6$ is w.

$$^{-}7 = w$$

The solution is $^{-}7$.

Check

$$42 = {}^{-}6w \quad \text{Replace } w \text{ with } {}^{-}7.$$
$$42 = {}^{-}6 \cdot {}^{-}7$$
$$42 = 42 \quad \text{Balances}$$

The equation balances, so **$^{-}7$ is the correct solution**.

◀◀ WORK PROBLEM 1 AT THE SIDE.

---

1. Solve each equation and check each solution.

(a) $4s = 44$

Check

(b) $27 = {}^{-}9p$

Check

(c) $^{-}40 = {}^{-}5x$

Check

(d) $7t = {}^{-}70$

Check

ANSWERS

1. (a) $s = 11$

Check $4s = 44$; $4 \cdot 11 = 44$

Balances $44 = 44$

(b) $p = {}^{-}3$

Check $27 = {}^{-}9p$; $27 = {}^{-}9 \cdot {}^{-}3$

Balances $27 = 27$

(c) $x = 8$

Check $^{-}40 = {}^{-}5x$; $^{-}40 = {}^{-}5 \cdot 8$

Balances $^{-}40 = {}^{-}40$

(d) $t = {}^{-}10$

Check $7t = {}^{-}70$; $7 \cdot {}^{-}10 = {}^{-}70$

Balances $^{-}70 = {}^{-}70$

Note

Be careful to divide both sides by the *same* number as the coefficient of the variable term. In Example 1(b), the coefficient of ^-6w is $^-6$, so divide both sides by $^-6$. (Do **not** use the *opposite* of $^-6$, which is 6. Use the opposite only when you're *adding* the same number to both sides.)

OBJECTIVE 2 You can sometimes simplify the expression on one or both sides of the equal sign, as you did in **Section 2.3.**

EXAMPLE 2 Simplifying Before Solving an Equation

Solve each equation and check each solution.

(a) $4y - 7y = {}^-12$

Simplify the left side by combining like terms. Change subtraction to adding the opposite.

$$4y - 7y = {}^-12$$

The right side cannot be simplified.

$$\underbrace{4y + {}^-7y} = {}^-12$$

Divide by the coefficient, which is $^-3$.

$$\frac{^-3y}{^-3} = \frac{^-12}{^-3}$$

To keep the balance, divide by $^-3$ on the right side also.

$$y = 4$$

$^-12 \div {}^-3$ is 4.

The solution is 4.

Check

$$4y - 7y = {}^-12$$

Go back to the original equation and replace each y with 4.

Do multiplications first.

$$\underbrace{4 \cdot 4} - \underbrace{7 \cdot 4} = {}^-12$$

Change subtraction to adding the opposite.

$$16 - 28 = {}^-12$$

$$\underbrace{16 + {}^-28} = {}^-12$$

$$^-12 = {}^-12$$

Balances

The equation balances, so **4 is the correct solution.**

(b) $3 - 10 + 7 = h + 7h$

Change subtraction to adding the opposite. Write in the understood 1 as the coefficient of h.

$$3 - 10 + 7 = h + 7h$$

Add from left to right. Combine like terms.

$$\underbrace{3 + {}^-10} + 7 = \underbrace{1h + 7h}$$

$$\underbrace{^-7 + 7} = 8h$$

To keep the balance, divide by 8 on the left side also: $0 \div 8$ is 0. Divide by the coefficient, which is 8.

$$\frac{0}{8} = \frac{8h}{8}$$

$$0 = h$$

The solution is 0.

Check

$$3 - 10 + 7 = h + 7h$$

Go back to the original equation and replace each h with 0.

$$\underbrace{3 + {}^-10} + 7 = 0 + \underbrace{7 \cdot 0}$$

Do multiplication first.

$$\underbrace{^-7 + 7} = \underbrace{0 + 0}$$

$$0 = 0$$

Balances.

The equation balances, so **0 is the correct solution.**

WORK PROBLEM 2 AT THE SIDE.

---

2. Simplify each side of the equation when possible. Then solve the equation and check the solution.

(a) $^-28 = {}^-6n + 10n$

Check

(b) $p - 14p = {}^-2 + 18 - 3$

Check

ANSWERS

2. (a) $n = {}^-7$

Check

$$^-28 = {}^-6n + 10n$$
$$^-28 = \underbrace{^-6 \cdot {}^-7} + \underbrace{10 \cdot {}^-7}$$
$$^-28 = \underbrace{42 + {}^-70}$$
$$^-28 = {}^-28$$

Balances

(b) $p = {}^-1$

Check

$$p - 14p = \underbrace{^-2 + 18} - 3$$
$$^-1 - \underbrace{14(^-1)} = \underbrace{16 - 3}$$
$$^-1 - {}^-14 = 13$$
$$\underbrace{^-1 + {}^+14} = 13$$
$$13 = 13$$

Balances

3. Solve each equation. Check each solution.

(a) $^{-}k = {}^{-}12$

Check

(b) $7 = {}^{-}t$

Check

(c) $^{-}m = {}^{-}20$

Check

Answers

3. (a) $k = 12$

Check $^{-}1k = {}^{-}12$

$^{-}1 \cdot 12 = {}^{-}12$

Balances $^{-}12 = {}^{-}12$

(b) $t = {}^{-}7$

Check $7 = {}^{-}1t$

$7 = {}^{-}1 \cdot {}^{-}7$

Balances $7 = 7$

(c) $m = 20$

Check $^{-}1m = {}^{-}20$

$^{-}1 \cdot 20 = {}^{-}20$

Balances $^{-}20 = {}^{-}20$

Objective 3 It may look like there's nothing more you can do to the equation

$$^{-}x = 5$$

but ^{-}x is *not* the same as x. So the variable is not by itself on one side of the equal sign. To see this, write in the understood $^{-}1$ as the coefficient of ^{-}x.

$$^{-}x = 5 \quad \text{can be written} \quad {}^{-}1x = 5$$

We want the coefficient of x to be $^{+}1$, not $^{-}1$. To accomplish that, we can divide both sides by the coefficient of x, which is $^{-}1$.

$$\frac{^{-}1x}{^{-}1} = \frac{5}{^{-}1}$$ Divide both sides by $^{-}1$. On the left side, $^{-}1 \div {}^{-}1$ is 1. On the right side, $5 \div {}^{-}1$ is $^{-}5$.

$$1x = {}^{-}5$$

Now x is by itself on one side of the equal sign and has a coefficient of $^{+}1$. The solution is $^{-}5$.

Check $^{-}x = 5$ Go back to the original equation. Write in the understood $^{-}1$ as the coefficient of ^{-}x.

$^{-}1x = 5$ Replace x with $^{-}5$.

$^{-}1 \cdot {}^{-}5 = 5$

$5 = 5$ Balances

Note

As the last step in solving an equation, do **not** leave a negative sign in front of a variable. For example, do *not* leave $^{-}y = {}^{-}8$. Write in the understood $^{-}1$ as the coefficient, so that

$$^{-}y = {}^{-}8 \quad \text{is written as} \quad {}^{-}1y = {}^{-}8.$$

Then divide both sides by $^{-}1$ to get $y = 8$. The solution is 8.

WORK PROBLEM 3 AT THE SIDE.

Name Date Class

2.4 Exercises

Solve each equation and check each solution.

1. $6z = 12$ **Check** $6z = 12$ ↓

2. $8k = 24$ **Check** $8k = 24$ ↓

3. $48 = 12r$ **Check**

4. $99 = 11m$ **Check**

5. $3y = 0$ **Check**

6. $5a = 0$ **Check**

7. $^{-}7k = 70$ **Check**

8. $^{-}6y = 36$ **Check**

9. $^{-}54 = {}^{-}9r$ **Check**

10. $^{-}36 = {}^{-}4p$ **Check**

11. $^{-}25 = 5b$ **Check**

12. $^{-}70 = 10x$ **Check**

Simplify where possible. Then solve each equation and check each solution.

13. $2r = {}^{-}7 + 13$ **Check** $2r = \underbrace{{}^{-}7 + 13}$ ↓

14. $6y = 28 - 4$ **Check** $6y = \underbrace{28 - 4}$ ↓

15. $^{-}12 = 5p - p$ **Check**

16. $20 = z - 11z$ **Check**

Solve each equation. Show your work.

17. $3 - 28 = 5a$

18. $^{-}55 + 7 = 8n$

19. $x - 9x = 80$

20. $4c - c = {}^{-}27$

21. $13 - 13 = 2w - w$

22. $^{-}11 + 11 = 8t - 7t$

23. $3t + 9t = 20 - 10 + 26$

24. $6m + 6m = 40 + 20 - 12$

25. $0 = {}^{-}9t$

26. $^{-}10 = 10b$

27. $^{-}14m + 8m = 6 - 60$

28. $7w - 14w = 1 - 50$

29. $100 - 96 = 31y - 35y$

30. $150 - 139 = 20x - 9x$

Use multiplication to simplify the side of the equation with the variable. Then solve each equation.

31. $3(2z) = {}^-30$

32. $2(4k) = 16$

33. $50 = {}^-5(5p)$

34. $60 = 4({}^-3a)$

35. ${}^-2({}^-4k) = 56$

36. ${}^-5(4r) = {}^-80$

37. ${}^-90 = {}^-10({}^-3b)$

38. ${}^-90 = {}^-5({}^-2y)$

Solve each equation.

39. ${}^-x = 32$

40. ${}^-c = 23$

41. ${}^-2 = {}^-w$

42. ${}^-75 = {}^-t$

43. ${}^-n = {}^-50$

44. ${}^-x = {}^-1$

45. $10 = {}^-p$

46. $100 = {}^-k$

47. Look again at the solutions to Exercises 39–46. Describe any pattern you see. Then write a rule for solving equations with a negative sign in front of the variable, such as $^-x = 5$.

48. Explain the division property of equality.

49. Explain and correct the error made by a student who solved this equation.

$$3x = \underbrace{16 - 1}$$
$$\frac{3x}{^-3} = \frac{15}{^-3}$$
$$x = {^-5}$$

50. Write two *different* equations that have $^-4$ as the solution. Be sure that you have to use the division property of equality to solve the equations. Show how to solve each equation.

Solve each equation. Show your work.

51. $89 - 116 = {^-4}({^-4}y) - 9(2y) + y$

52. $58 - 208 = {^-b} + 8({^-3}b) + 5({^-5}b)$

53. $^-37(14x) + 28(21x) = |72 - 72| + |{^-166} + 96|$

54. $6a - 10a - 3(2a) = |{^-25} - 25| - 5(8)$

Review and Prepare

Simplify. (For help, see ***Section 1.8.****)*

55. $5 - 3 \cdot {^-4}$

56. $^-4 + 6 \div 2$

57. $^-8 + 2(0 - 5)$

58. $1 - 6({^-2} + 4)$

59. $^-36 \div 9 \cdot 2({^-4} + 1)$

60. $^-6 - 4 + 7({^-1} + 4)$

2.5 *Solving Equations with Several Steps*

OBJECTIVE 1 To solve some equations, you need to use both the addition property of equality (see **Section 2.3**) and the division property of equality (see **Section 2.4**). Here are the steps.

Solving Equations by Using the Addition and Division Properties

Step 1 Add the same amount to both sides of the equation so that the variable term (the variable and its coefficient) ends up by itself on one side of the equal sign.

Step 2 Divide both sides by the coefficient of the variable term to find the solution.

Step 3 Check the solution by going back to the *original* equation.

EXAMPLE 1 Solving Equations with Several Steps

Solve the equation and check the solution: $5m + 1 = 16$.

Step 1 Get the variable term by itself on one side of the equal sign. The variable term is $5m$. Adding $^-1$ to the left side of the equation will leave $5m$ by itself. To keep the balance, add $^-1$ to the right side also.

$$\begin{aligned} 5m + 1 &= 16 \\ {}^-1 \quad & \quad {}^-1 \\ \underbrace{5m + 0} &= 15 \\ 5m \quad &= 15 \end{aligned}$$

Step 2 Divide both sides by the coefficient of the variable term. In $5m$, the coefficient is 5. So divide both sides by 5.

$$\frac{5m}{5} = \frac{15}{5}$$

$$m = 3$$

Step 3 Check the solution by going back to the original equation.

$$\begin{aligned} 5m + 1 &= 16 \\ \underbrace{5(3)} + 1 &= 16 \\ \underbrace{15 + 1} &= 16 \\ 16 &= 16 \end{aligned}$$

Use the original equation and replace m with 3.

Balances

The equation balances, so **3 is the correct solution**.

WORK PROBLEM 1 AT THE SIDE.

So far, variable terms have appeared on just one side of the equal sign. But some equations start with variable terms on both sides. In that case, you can use the addition property of equality to add the same *variable term* to both sides of the equation, just as you have added the same *number* to both sides.

OBJECTIVES

1. Solve equations that require use of both the addition and division properties of equality.
2. Solve equations by using the distributive, addition, and division properties.

FOR EXTRA HELP

Tutorial | Tape 3 | SSM, Sec. 2.5

1. Solve each equation. Check each solution.

(a) $2r + 7 = 13$ **Check**

(b) $20 = 6y - 4$ **Check**

(c) $^-10z - 9 = 11$ **Check**

ANSWERS

1. (a) $r = 3$

Check $\underbrace{2(3)} + 7 = 13$

$\underbrace{6 + 7} = 13$

Balances $13 = 13$

(b) $y = 4$

Check $20 = \underbrace{6(4)} - 4$

$20 = \underbrace{24 - 4}$

Balances $20 = 20$

(c) $z = {}^-2$

Check $\underbrace{^-10(^-2)} - 9 = 11$

$\underbrace{20 - 9} = 11$

Balances $11 = 11$

2. Solve each equation *two* ways. First keep the variable term on the left side when you solve. Then solve again, keeping the variable term on the right side.

(a) $3y - 1 = 2y + 7$

$3y - 1 = 2y + 7$

(b) $3p - 2 = p - 6$

$3p - 2 = p - 6$

First decide whether to keep the variable term on the left side, or to keep the variable term on the right side. It doesn't matter which one you keep; just pick one side or the other. Then use the addition property to "get rid of" the variable term on the *other* side by adding its opposite.

EXAMPLE 2 Solving Equations with Variable Terms on Both Sides

Solve the equation and check the solution: $2k - 2 = 5k - 11$.

First let's keep $2k$, the variable term on the left side. That means we need to "get rid of" $5k$ on the right side. We can do that by adding the opposite of $5k$, which is ^-5k.

To keep the balance, add ^-5k to the left side also. Write ^-5k underneath $2k$, *not* underneath 2.

Write ^-5k underneath $5k$, *not* underneath 11. $5k + {}^-5k$ is $0k$, or just 0.

$$\begin{array}{rcl} 2k - 2 &=& 5k - 11 \\ \underline{^-5k \qquad} & & \underline{^-5k \qquad} \\ ^-3k - 2 &=& 0 - 11 \end{array}$$

Change subtractions to adding the opposite.

$$^-3k + {}^-2 = 0 + {}^-11$$

To get ^-3k by itself, add 2 to both sides.

$$\begin{array}{rcl} ^-3k + {}^-2 &=& 0 + {}^-11 \\ \underline{\qquad 2} & & \underline{\qquad 2} \\ ^-3k + 0 &=& 0 + {}^-9 \end{array}$$

Divide both sides by $^-3$, the coefficient of the variable term.

$$\frac{^-3k}{^-3} = \frac{^-9}{^-3}$$

$$k = 3$$

Suppose that, in the first step, you decided to keep $5k$ on the right side and "get rid of" $2k$ on the left side. Let's see what happens.

Add ^-2k to both sides.

$$\begin{array}{rcl} 2k - 2 &=& 5k - 11 \\ \underline{^-2k \qquad} & & \underline{^-2k \qquad} \\ 0 - 2 &=& 3k - 11 \end{array}$$

Change subtractions to adding the opposite.

$$0 + {}^-2 = 3k + {}^-11$$

To get $3k$ by itself, add 11 to both sides.

$$\begin{array}{rcl} 0 + {}^-2 &=& 3k + {}^-11 \\ \underline{\qquad 11} & & \underline{\qquad 11} \\ 0 + 9 &=& 3k + 0 \end{array}$$

Divide both sides by 3.

$$\frac{9}{3} = \frac{3k}{3}$$

$$3 = k$$

The two solutions are the same. In both cases, k balances with 3. Notice that you used the addition principle *twice:* once to "get rid of" a variable term and once to "get rid of" a number. You could have done those steps in the reverse order without changing the result.

Note

More than one sequence of steps will work to solve complicated equations. The basic approach is the following:

- Simplify each side of the equation, if possible.
- Get the variable term by itself on one side of the equal sign and a number by itself on the other side.
- Divide both sides by the coefficient of the variable term.

◀◀ WORK PROBLEM 2 AT THE SIDE.

ANSWERS

2. (a) $y = 8$ and $y = 8$
(b) $p = {}^-2$ and $p = {}^-2$

OBJECTIVE 2 If an equation contains parentheses, check to see whether you can use the distributive property to remove them.

EXAMPLE 3 Solving Equations by Using the Distributive Property

Solve the equation and check the solution: $^{-}6 = 3(y - 2)$.

We can use the distributive property to simplify the right side of the equation. Recall from **Section 2.2** that

$$3(y - 2) \text{ can be written as } \underbrace{3 \cdot y}_{3y} - \underbrace{3 \cdot 2}_{6}$$

So now the original equation $^{-}6 = 3(y - 2)$ looks like this.

$^{-}6 = 3y - 6$ Change subtraction to adding the opposite.

$^{-}6 = 3y + {}^{-}6$ To get $3y$ by itself, add 6 to both sides.

$\underline{\;6\;} \qquad \underline{\qquad\quad 6}$

$\dfrac{0}{3} = \dfrac{3y}{3}$ Divide both sides by 3, the coefficient of $3y$.

$0 = y$

The solution is 0.

Check: $^{-}6 = 3(y - 2)$ Go back to the original equation and replace y with 0.

$^{-}6 = 3(0 - 2)$ Change subtraction to addition.

$^{-}6 = 3(\underbrace{0 + {}^{-}2}_{})$ Follow the order of operations; work inside parentheses first.

$^{-}6 = \underbrace{3({}^{-}2)}_{}$

$^{-}6 = {}^{-}6$ Balances.

The equation balances, so **0 is the solution**.

WORK PROBLEM 3 AT THE SIDE.

3. Solve each equation. Check each solution.

(a) $^{-}12 = 4(y - 1)$

(b) $5(m + 4) = 20$

(c) $6(t - 2) = 18$

Here is a summary of all the steps you can use to solve an equation.

Solving Equations

Step 1 If possible, use the **distributive property** to remove parentheses.

Step 2 **Combine** any like terms on the left side of the equation. Combine any like terms on the right side of the equation.

Step 3 **Add** the same amount to both sides of the equation so that the variable term ends up by itself on one side of the equal sign and a number is by itself on the other side. You may have to do this step more than once.

Step 4 **Divide** both sides by the coefficient of the variable term to find the solution.

Step 5 **Check** your solution by going back to the original equation. Replace the variable with your solution. Follow the order of operations to complete the calculations. If the two sides of the equation balance, your solution is correct.

ANSWERS

3. (a) $y = {}^{-}2$ **(b)** $m = 0$ **(c)** $t = 5$

4. Solve each equation. Check each solution.

(a) $3(b + 7) = 2b - 1$

(b) $6 - 2n = 14 + 4(n - 5)$

EXAMPLE 4 Solving Equations

Solve this equation and check the solution: $8 + 5(m + 2) = 6 + 2m$.

Step 1 Use the distributive property on the left side.

$$8 + 5(m + 2) = 6 + 2m$$

Step 2 Combine like terms on the left side.

$$8 + 5m + 10 = 6 + 2m$$

No like terms on the right side.

$$5m + 18 = 6 + 2m$$

Step 3 Add ^-2m to both sides.

$$\begin{array}{r} 5m + 18 = 6 + 2m \\ \underline{^-2m \qquad\quad ^-2m} \\ 3m + 18 = 6 + 0 \\ 3m + 18 = 6 \end{array}$$

Step 3 To get $3m$ by itself, add $^-18$ to both sides.

$$\begin{array}{r} 3m + 18 = 6 \\ \underline{^-18 \quad ^-18} \\ 3m + 0 = {^-12} \end{array}$$

Step 4 Divide both sides by 3, the coefficient of the variable term $3m$.

$$\frac{3m}{3} = \frac{^-12}{3}$$

$$m = {^-4}$$

The solution is $^-4$.

Step 5 **Check:**

$$8 + 5(m + 2) = 6 + 2m$$

Replace each m with $^-4$.

$$8 + 5(^-4 + 2) = 6 + 2(^-4)$$

$$8 + 5(^-2) = 6 + {^-8}$$

$$8 + {^-10} = {^-2}$$

$$^-2 = {^-2} \quad \text{Balances}$$

The equation balances, so **$^-4$ is the correct solution.**

◀◀ WORK PROBLEM 4 AT THE SIDE.

Answers

4. **(a)** $b = {^-22}$ **(b)** $n = 2$

2.5 Exercises

Solve each equation and check each solution.

1. $7p + 5 = 12$ **Check** $7p + 5 = 12$
↓

2. $6k + 3 = 15$ **Check** $6k + 3 = 15$
↓

3. $2 = 8y - 6$ **Check**

4. $10 = 11p - 12$ **Check**

5. $^{-}3m + 1 = 1$ **Check**

6. $^{-}4k + 5 = 5$ **Check**

7. $28 = {}^{-}9a + 10$ **Check**

8. $75 = {}^{-}10w + 25$ **Check**

9. $^{-}5x - 4 = 16$ **Check**

10. $^{-}12b - 3 = 21$ **Check**

*Solve each equation **two** ways. First keep the variable term on the left side when you solve it. Then solve it again, keeping the variable term on the right side. Finally, check your solution.*

11. $6p - 2 = 4p + 6$ $\qquad$ $6p - 2 = 4p + 6$ $\qquad$ **Check** $6p - 2 = 4p + 6$

$\downarrow \qquad \downarrow$

12. $5y - 5 = 2y + 10$ $\qquad$ $5y - 5 = 2y + 10$ $\qquad$ **Check** $5y - 5 = 2y + 10$

$\downarrow \qquad \downarrow$

13. $^-2k - 6 = 6k + 10$ $\qquad$ $^-2k - 6 = 6k + 10$ $\qquad$ **Check**

14. $5x + 4 = {}^-3x - 4$ $\qquad$ $5x + 4 = {}^-3x - 4$ $\qquad$ **Check**

15. $^-18 + 7a = 2a + 7$ $\qquad$ $^-18 + 7a = 2a + 7$ $\qquad$ **Check**

16. $^-9 + 2z = 9z + 12$ $^-9 + 2z = 9z + 12$ **Check**

Use the distributive property to help you solve each equation.

17. $8(w - 2) = 32$

18. $9(b - 4) = 27$

19. $^-10 = 2(y + 4)$

20. $^-3 = 3(x + 6)$

21. $^-4(t + 2) = 12$

22. $^-5(k + 3) = 25$

23. $6(x - 5) = {}^-30$

24. $7(r - 7) = {}^-49$

25. $^-12 = 12(h - 2)$

26. $^-11 = 11(c - 3)$

27. $0 = {}^-2(y + 2)$

28. $0 = {}^-9(b + 1)$

Solve each equation. Show the steps you used.

29. $6m + 18 = 0$

30. $8p - 40 = 0$

31. $6 = 9w - 12$

32. $8 = 8h + 24$

33. $5x = 3x + 10$

34. $7n = {}^{-}2n - 36$

35. $2a + 11 = 8a - 7$

36. $r - 10 = 10r + 8$

37. $7 - 5b = 28 + 2b$

38. $1 - 8t = {}^{-}9 - 3t$

39. ${}^{-}20 + 2k = k - 4k$

40. $6y - y = {}^{-}16 + y$

41. $10(c - 6) + 4 = 2 + c - 58$

42. $8(z + 7) - 6 = z + 60 - 10$

43. $^{-}18 + 13y + 3 = 3(5y - 1) - 2$

44. $3 + 5h - 9 = 4(3h + 4) - 1$

45. $6 - 4n + 3n = 20 - 35$

46. $^{-}19 + 8 = 6p - 7p - 5$

47. $6(c - 2) = 7(c - 6)$

48. $^{-}3(5 + x) = 4(x - 2)$

49. $^{-}5(2p + 2) - 7 = 3(2p + 5)$

50. $4(3m - 6) = 72 + 3(m - 8)$

51. $^{-}6b - 4b + 7b = 10 - b + 3b$

52. $w + 8 - 5w = {}^{-}w - 15w + 11w$

53. Solve $^{-}2t - 10 = 3t + 5$. Show each step you take while solving it. Next to each step, write a sentence that explains what you did in that step. Be sure to tell when you used the addition property of equality and when you used the division property of equality.

54. Explain the distributive property. Show two examples of using the distributive property to remove parentheses in an expression.

55. Here is one student's solution to an equation. Show how to check the solution. If the solution doesn't check, explain the error and correct it.

$$\begin{array}{rcl} ^{-}8 + 4a & = & 2a + 2 \\ ^{-}2a & & ^{-}2a \\ \hline ^{-}10 + 4a & = & \underbrace{0 + 2} \\ ^{-}10 + 4a & = & 2 \\ 10 & & 10 \\ \hline \underbrace{0 + 4a} & = & 12 \\ \dfrac{4a}{4} & = & \dfrac{12}{4} \\ a & = & 3 \end{array}$$

56. Here is one student's solution to an equation. Show how to check the solution. If the solution doesn't check, explain the error and correct it.

$$\begin{array}{rcl} 2(x + 4) & = & ^{-}16 \\ 2x + 4 & = & ^{-}16 \\ ^{-}4 & & ^{-}4 \\ \hline \underbrace{2x + 0} & = & ^{-}20 \\ \dfrac{2x}{2} & = & \dfrac{^{-}20}{2} \\ x & = & ^{-}10 \end{array}$$

Review and Prepare

Simplify. (For help, see ***Section 1.8****.)*

57. $6 \div 3 + 4 \cdot 0 - 3(6)$

58. $5 \cdot 1 - 6 \cdot 2 + 7(^{-}1)$

59. $3(0 - 3) - 10(2 - 3)$

60. $(^{-}2 + 7) + 5(^{-}5 + 1)$

61. $\dfrac{^{-}18 \div 3 \cdot 2 \div 4}{3(2 - {}^{-}2) - 12}$

62. $\dfrac{9 - 9 + 6(^{-}1 + 1)}{^{-}3 + 2(^{-}6 - 4)}$

KEY TERMS

2.1	**variable**	A variable is a letter that represents a number that varies or changes, depending on the situation.
	constant	A constant is a number that is added or subtracted in an expression. It does not vary. For example, 5 is the constant in the expression $c + 5$.
	expression	An expression expresses, or tells, the rule for doing something. It is a combination of operations on variables and numbers.
	evaluate the expression	To evaluate an expression, replace each variable with specific values (numbers) and follow the order of operations.
	coefficient	The number part in a multiplication expression is the coefficient. For example, 4 is the coefficient in the expression $4s$.
2.2	**simplify expressions**	To simplify an expression, write it in a simpler way by combining all the like terms.
	term	Each addend in an expression is a term.
	variable term	A variable term has a number part (called the coefficient) multiplied by a variable part (a letter). An example is $4s$.
	like terms	Like terms are terms with exactly the same variable parts (the same letters and exponents). The coefficients may be different.
2.3	**equations**	An equation has an equal sign. It shows the relationship between what is known about a problem and what isn't known.
	solve the equation	To solve an equation, find a number that can replace the variable and make the equation balance.
	solution	A solution of an equation is a number that can replace the variable and make the equation balance.
	addition property of equality	The addition property of equality states that adding the same quantity to both sides of an equation will keep it balanced.
	check the solution	To check the solution of an equation, go back to the original equation and replace the variable with the solution. If the equation balances, the solution is correct.
2.4	**division property of equality**	The division property of equality states that dividing both sides of an equation by the same nonzero number will keep it balanced.

QUICK REVIEW

Concepts	*Examples*
2.1 Evaluating Expressions Replace each variable with the specified value. Then follow the order of operations to simplify the expression.	The expression for ordering textbooks for two pre-algebra classes is $2c + 10$, where c is the class limit. Evaluate the expression when the class limit is 24. $2c + 10$ Replace c with 24. $2 \cdot 24 + 10$ Multiply first. $48 + 10$ Add last. 58 Order 58 books.
2.1 Using Exponents with Variables An exponent next to a variable tells how many times to use the variable as a factor in multiplication.	Rewrite $^{-}6x^4$ without exponents. $^{-}6x^4$ can be written as $^{-}6 \cdot x \cdot x \cdot x \cdot x$ Coefficient is $^{-}6$. x is used as a factor 4 times.

Concepts	*Examples*
2.1 Evaluating Expressions with Exponents Rewrite the expression without exponents, replace each variable with the specified value, and multiply all the factors.	Evaluate x^3y when x is $^{-}4$ and y is 5. x^3y means $x \cdot x \cdot x \cdot y$ Replace x with $^{-}4$ and y with 5. $^{-}4 \cdot {^{-}4} \cdot {^{-}4} \cdot 5$ $16 \cdot {^{-}4} \cdot 5$ $^{-}64 \cdot 5$ $^{-}320$
2.2 Identifying Like Terms Like terms have *exactly* the same letters and exponents. The coefficients can be different.	List the like terms in this expression. Then identify the coefficients of the like terms. $^{-}3b + {^{-}3b^2} + 3ab + b + 3$ The like terms are $^{-}3b$ and b. The coefficient of $^{-}3b$ is $^{-}3$, and the coefficient of b is understood to be 1.
2.2 Combining Like Terms If there are any variable terms without coefficients, write in the understood 1. Also change any subtractions to adding the opposite. Then find like terms and add their coefficients, keeping the variable part the same. Be careful to combine (add) *like terms* only.	Simplify $4x^2 - 10 + x^2 + 15$. $4x^2 - 10 + 1x^2 + 15$ Write understood 1. $4x^2 + {^{-}10} + 1x^2 + 15$ Change subtraction to adding the opposite. $4x^2 + 1x^2 + {^{-}10} + 15$ Combine $4x^2 + 1x^2$. Also combine $^{-}10 + 15$. The variable part stays the same. $(4 + 1)x^2 + 5$ $5x^2 + 5$ The simplified expression is $5x^2 + 5$.
2.2 Simplifying Multiplication Expressions Use the associative property to rewrite the expression so that the two number parts can be multiplied. The variable part stays the same.	Simplify $^{-}7(5k)$. Use the associative property of multiplication. $^{-}7 \cdot (5 \cdot k)$ can be written as $(^{-}7 \cdot 5) \cdot k$ $^{-}35 \cdot k$ $^{-}35\,k$ The simplified expression is $^{-}35k$.
2.2 Using the Distributive Property Multiplication distributes over addition and over subtraction. Be careful to multiply *every* term inside the parentheses by the number outside the parentheses.	Simplify. **(a)** $6(w - 4)$ can be written as $6 \cdot w - 6 \cdot 4$ $6w - 24$ The simplified expression is $6w - 24$. **(b)** $^{-}3(2b + 5)$ can be written as $^{-}3 \cdot 2b + {^{-}3} \cdot 5$ $^{-}6b + {^{-}15}$ Use the definition of subtraction "in reverse" to write $^{-}6b + {^{-}15}$ as $^{-}6b - 15$. The simplified expresssion is $^{-}6b - 15$.

Concepts	*Examples*
2.3 Solving and Checking Equations Using the Addition Property of Equality	
If possible, simplify the expression on one or both sides of the equal sign. Next, to get the variable by itself on one side of the equal sign, add the same number to both sides. Finally, check the solution by going back to the original equation and replacing the variable with the solution. If the equation balances, the solution is correct.	Solve the equation and check the solution. $\underbrace{^{-}5 + 8}_{} = 9 + r$ Simplify the left side by adding $^{-}5 + 8$. $3 = 9 + r$ To get r by itself, add the opposite of 9, which is $^{-}9$, to both sides. $\underline{^{-}9} \quad \underline{^{-}9}$ $^{-}6 = \underbrace{0 + r}_{}$ $^{-}6 = r$ The solution is $^{-}6$. **Check** $^{-}5 + 8 = 9 + r$ Use the original equation and replace r with $^{-}6$. $\underbrace{^{-}5 + 8}_{3} = \underbrace{9 + {}^{-}6}_{3}$ $3 = 3$ Balances The equation balances, so $^{-}6$ is the correct solution (**not** 3).
2.4 Solving and Checking Equations Using the Division Property of Equality	
If possible, simplify the expression on one or both sides of the equal sign. Next, to get the variable by itself on one side of the equal sign, divide both sides by the coefficient of the variable term. Finally, check the solution by going back to the original equation and replacing the variable with the solution. If the equation balances, the solution is correct.	Solve the equation and check the solution. Simplify the left side. Change subtraction to adding the opposite. $2h - 6h = \underbrace{18 + 22}_{}$ Simplify the right side. Add $18 + 22$. $\underbrace{2h + {}^{-}6h}_{} = 40$ Divide by $^{-}4$, the coefficient of $^{-}4h$. $\dfrac{^{-}4h}{^{-}4} = \dfrac{40}{^{-}4}$ Also divide 40 by $^{-}4$ to keep the balance. $h = {}^{-}10$ The solution is $^{-}10$. **Check** $2h - 6h = \underbrace{18 + 22}_{}$ Replace h with $^{-}10$. $\underbrace{2(^{-}10)}_{} - \underbrace{6(^{-}10)}_{} = 40$ $^{-}20 - {}^{-}60 = 40$ $\underbrace{^{-}20 + {}^{+}60}_{}$ $40 = 40$ Balances The equation balances, so $^{-}10$ is the correct solution.

Concepts	*Examples*

2.4 Solving Equations Such as $^-x = 5$

As the last step in solving an equation, do **not** leave a negative sign in front of the variable, such as $^-x = 5$, because ^-x is **not** the same as x. Divide both sides by $^-1$, the understood coefficient of ^-x.

Solve the equation and check the solution.

$$9 = {}^-n$$

Write the understood $^-1$ as the coefficient of n.

$9 = {}^-n$ can be written as $9 = {}^-\mathbf{1}n$

Now divide both sides by $^-1$.

$$\frac{9}{^-\mathbf{1}} = \frac{^-1n}{^-\mathbf{1}}$$

$$^-9 = n$$

The solution is $^-9$.

Check $9 = {}^-n$ Write understood $^-1$.

$9 = {}^-\mathbf{1}n$ Replace n with $^-9$.

$9 = \underbrace{^-1(^-9)}$

$9 = 9$ Balances

The equation balances, so $^-\mathbf{9}$ is the correct solution.

2.5 Solving Equations with Several Steps

Step 1 If possible, use the distributive property to remove parentheses.

Step 2 Combine any like terms on the left side of the equal sign. Combine any like terms on the right side of the equal sign.

Step 3 Add the same amount to both sides of the equation so that the variable term ends up by itself on one side of the equal sign, and a number is by itself on the other side. You may have to do this step more than once.

Step 4 Divide both sides by the coefficient of the variable term to find the solution.

Step 5 Check the solution by going back to the original equation. Replace the variable with the solution. If the equation balances, the solution is correct.

Solve this equation and check the solution.

Step 1 $3 + 2(y + 8) = 5y + 4$

Step 2 $\mathbf{3} + 2y + \mathbf{16} = 5y + 4$

$2y + \mathbf{19} = 5y + 4$

Step 3
$$\begin{array}{rcl} 2y + 19 &=& 5y + 4 \\ \underline{^-\mathbf{2y}} && \underline{^-\mathbf{2y}} \\ 0 + 19 &=& 3y + 4 \end{array}$$

Step 3
$$\begin{array}{rcl} 0 + 19 &=& 3y + 4 \\ \underline{{}^-\mathbf{4}} && \underline{{}^-\mathbf{4}} \\ \underbrace{0 + 15} &=& \underbrace{3y + 0} \end{array}$$

Step 4
$$\frac{15}{\mathbf{3}} = \frac{3y}{\mathbf{3}}$$

$$5 = y$$

The solution is 5.

Check $3 + 2(y + 8) = 5y + 4$

$3 + 2(\underbrace{\mathbf{5} + 8}) = \underbrace{5(\mathbf{5})} + 4$

$3 + \underbrace{2(13)} = \underbrace{25 + 4}$

$\underbrace{3 + 26} = 29$

$29 = 29$

The equation balances, so **5 is the correct solution**.

Name Date Class

CHAPTER 2 REVIEW EXERCISES

If you need help with any of these review exercises, look in the section indicated in the brackets.

[2.1] **1.** Identify the variable and the constant in this expression.

$k + 3$

2. The expression for ordering test tubes for a chemistry lab is $4c + 10$, where c is the class limit. Evaluate the expression when

(a) the class limit is 15.

(b) the class limit is 24.

3. Rewrite each expression without exponents.

(a) x^2y^4

(b) $5ab^3$

4. Evaluate each expression when m is 2, n is $^-3$, and p is 4.

(a) n^2

(b) n^3

(c) $^-4mp^2$

(d) $5m^4n^2$

[2.2] *Simplify.*

5. $ab + ab^2 + 2ab$

6. $^-3x + 2y - x - 7$

7. $^-8(^-2g^3)$

8. $4(3r^2t)$

9. $5(k + 2)$

10. $^-2(3b + 4)$

11. $3(2y - 4) + 12$

12. $^-4 + 6(4x + 1) - 4x$

13. Write an expression with four terms that *cannot* be simplified.

[2.3] *Solve each equation and check each solution.*

14. $16 + n = 5$ **Check**

15. $^-4 + 2 = 2a - 6 - a$ **Check**

[2.4] *Solve each equation. Show the steps you used.*

16. $48 = {}^-6m$

17. $k - 5k = {}^-40$

18. $^-17 + 11 + 6 = 7t$

19. $^-2p + 5p = 3 - 21$

20. $^-30 = 3(^-5r)$

21. $12 = {}^-h$

[2.5] *Solve each equation. Show the steps you used.*

22. $12w - 4 = 8w + 12$

23. $0 = {}^-4(c + 2)$

24. $6 + 10n - 18 = 8(3n + 4) - 2$

MIXED REVIEW EXERCISES

Solve each equation. Show the steps you used.

25. $12 + 7a = 4a - 3$

26. $^-2(p - 3) = {}^-14$

27. $10y = 6y + 20$

28. $2m - 7m = 5 - 20$

29. $20 = 3x - 7$

30. $b + 6 = 3b - 8$

31. $z + 3 = 0$

32. $3(2n - 1) = 3(n + 3)$

33. $^-4 + 46 = 7(^-3t + 6)$

34. $6 + 10d - 19 = 2(3d + 4) - 1$

35. $^-4(3b + 9) = 24 + 3(2b - 8)$

Chapter 2 Test

1. Identify the parts of this expression: $^-7w + 6$.
Choose from these labels: variable, constant term, coefficient.

1. ______

2. The expression for buying hot dogs for the company picnic is $3a + 2c$, where a is the number of adults and c is the number of children. Evaluate the expression when there are 45 adults and 21 children.

2. ______

Rewrite each expression without exponents.

3. x^5y^3

4. $4ab^4$

3. ______

4. ______

5. Evaluate $^-2s^2t$ when s is $^-5$ and t is 4.

5. ______

Simplify each expression.

6. $3w^3 - 8w^3 + w^3$

7. $xy - xy$

6. ______

7. ______

8. $^-6c - 5 + 7c + 5$

9. $3m^2 - 3m + 3mn$

8. ______

9. ______

10. $^-10(4b^2)$

11. $^-5(^-3k)$

10. ______

11. ______

12. $7(3t + 4)$

13. $^-4(a + 6)$

12. ______

13. ______

14. $^-8 + 6(x - 2) + 5$

15. $^-9b - c - 3 + 9 + 2c$

14. ______

15. ______

16. ______

17. ______

18. ______

19. ______

20. ______

21. ______

22. ______

23. ______

24. ______

25. ______

26. ______

Solve each equation and check each solution.

16. $^{-}4 = x - 9$ **Check**

17. $^{-}7w = 77$ **Check**

18. $^{-}p = 14$ **Check**

19. $^{-}15 = {}^{-}3(a + 2)$ **Check**

Solve each equation. Show the steps you used.

20. $6n + 8 - 5n = {}^{-}4 + 4$

21. $5 - 20 = 2m - 3m$

22. $^{-}2x + 2 = 5x + 9$

23. $3m - 5 = 7m - 13$

24. $2 + 7b - 44 = {}^{-}3b + 12 + 9b$

25. $3c - 24 = 6(c - 4)$

26. Write an equation that requires the addition property of equality to solve it and the solution is $^{-}4$. Then write a different equation that requires the division property of equality to solve it and the solution is $^{-}4$. Show how to solve each equation.

CUMULATIVE REVIEW EXERCISES CHAPTERS 1–2

1. Write this number in words.

306,000,004,210

2. Write this number, using digits.

eight hundred million, sixty-six thousand

3. Write < or > between each pair of numbers to make true statements.

$^{-}3$ __ $^{-}10$ $^{-}1$ __ 0

4. Name the property illustrated by each example.

(a) $^{-}6 + 2 = 2 + {}^{-}6$ **(b)** $0 \cdot 25 = 0$

(c) $5({}^{-}6 + 4) = 5 \cdot {}^{-}6 + 5 \cdot 4$

5. (a) Round 9047 to the nearest hundred.

(b) Round 289,610 to the nearest thousand.

Simplify.

6. $0 - 8$

7. $|{}^{-}6| + |4|$

8. $^{-}3({}^{-}10)$

9. $({}^{-}5)^2$

10. $\frac{{}^{-}42}{{}^{-}6}$

11. $^{-}19 + 19$

12. $({}^{-}4)^3$

13. $\frac{{}^{-}14}{0}$

14. $^{-}5 \cdot 12$

15. $^{-}20 - 20$

16. $\frac{45}{{}^{-}5}$

17. $^{-}50 + 25$

18. $^{-}10 + 6(4 - 7)$

19. $\frac{{}^{-}20 - 3({}^{-}5) + 16}{({}^{-}4)^2 - 3^3}$

First use front end rounding to estimate the answer to each application problem. Then find the exact answer.

20. One Siberian tiger was tracked for 22 days while it was searching for food. It traveled 616 miles. What was the average distance it traveled each day?

estimate:
exact:

21. The temperature in Siberia got down to $^{-}48$ degrees one night. The next day the temperature rose 23 degrees. What was the daytime temperature?

estimate:
exact:

22. Doug owned 52 shares of Mathtronic stock that had a total value of $2132. Yesterday the value of each share dropped $8. What are his shares worth now?

estimate:
exact:

23. Ikuko's monthly rent is $552 plus $35 for parking. How much will she spend for rent and parking in one year?

estimate:
exact:

24. Rewrite $^{-}4ab^3c^2$ without exponents.

25. Evaluate $3xy^3$ when x is $^{-}5$ and y is $^{-}2$.

Simplify.

26. $3h - 7h + 5h$

27. $c^2d - c^2d$

28. $4n^2 - 4n + 6 - 8 + n^2$

29. $^{-}10(3b^2)$

30. $7(4p - 4)$

31. $3 + 5(^{-}2w^2 - 3) + w^2$

Solve each equation and check each solution.

32. $3x = x - 8$ **Check**

33. $^{-}44 = {}^{-}2 + 7y$ **Check**

34. $2k - 5k = {}^{-}21$ **Check**

35. $m - 6 = {}^{-}2m + 6$ **Check**

Solve each equation. Show the steps you used.

36. $4 - 4x = 18 + 10x$

37. $18 = {}^{-}r$

38. $^{-}8b - 11 + 7b = b - 1$

39. $^{-}2(t + 1) = 4(1 - 2t)$

40. $5 + 6y - 23 = 5(2y + 8) - 10$

Solving Application Problems

3

3.1 Perimeter

OBJECTIVE 1 If you have ever studied geometry, you probably used several different formulas such as $P = 2l + 2w$ and $A = lw$. A **formula** (FOR-mew-lah) is just a shorthand way of writing a rule for solving a particular type of problem. A formula uses variables (letters) and it has an equal sign, so it is an equation. That means you can use the equation-solving techniques you learned in Chapter 2 to work with formulas.

But let's start at the beginning. Geometry was developed centuries ago when people needed a way to measure land. The name *geometry* comes from the Greek words *ge,* meaning earth, and *metron,* meaning measure. Today we still use geometry to measure land. It is also important in architecture, construction, navigation, art and design, physics, chemistry, and astronomy. You can use it at home when you buy carpet or wallpaper, hang a picture, or do home repairs. In this chapter you'll learn about two basic ideas, perimeter and area. Other geometry concepts will appear in later chapters.

In **Section 2.1,** you found the **perimeter** (per-IM-it-er) of a square garden.

Perimeter

The distance around the outside edges of any flat shape is called the perimeter of the shape.

To review, a **square** has four sides that are all the same length. Also, the sides meet to form 90° (90 degree) angles. This means that the sides form "square corners." (For more review of angles, see **Appendix C.**) Two examples of squares are shown below.

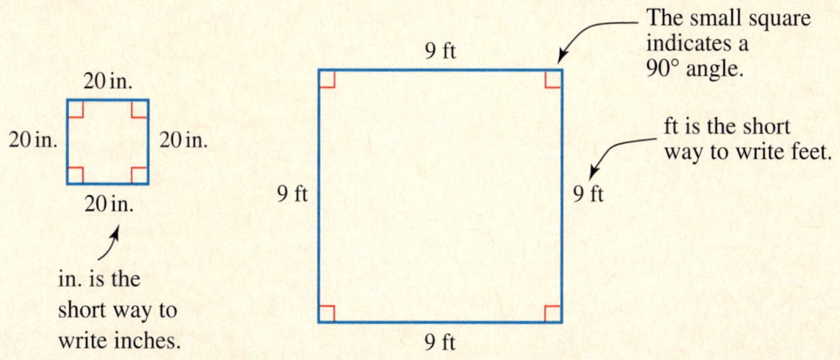

INTERNET
www.mathnotes.com

OBJECTIVES

1. Use the formula for perimeter of a square to find the perimeter or the length of one side.
2. Use the formula for perimeter of a rectangle to find the perimeter, the length, or the width.
3. Find the perimeter of parallelograms, triangles, and irregular shapes.

FOR EXTRA HELP

Tutorial | Tape 4 | SSM, Sec. 3.1

1. Find the perimeter of each square, using the appropriate formula.

(a) the 20 in. square shown on the previous page

(b)

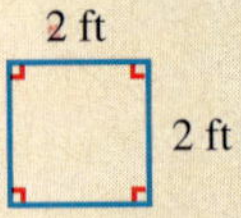

(c) a square measuring 14 miles on each side (*Hint:* First draw a sketch of the square and label each side with its length.)

2. Use the perimeter of each square and the appropriate formula to find the length of one side. Then check your solution by drawing a square, labeling each side, and finding the perimeter.

(a) Perimeter is 28 in.

(b) Perimeter is 100 ft.

(c) Perimeter is 64 cm.

ANSWERS

1. (a) $P = 80$ in. **(b)** $P = 8$ ft

(c) $P = 56$ miles

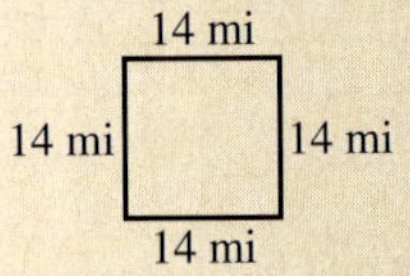

2. (a) $s = 7$ in.

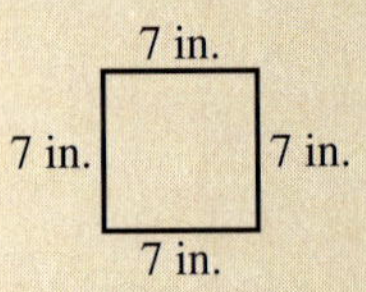

$P = 7 \text{ in.} + 7 \text{ in.} + 7 \text{ in.} + 7 \text{ in.}$
$= 28$ in.

(b) $s = 25$ ft

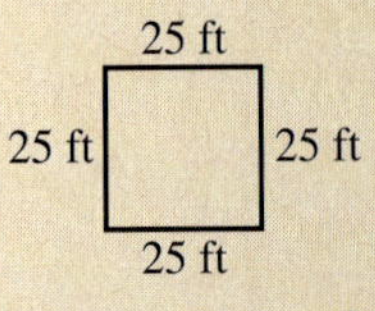

$P = 25 \text{ ft} + 25 \text{ ft} + 25 \text{ ft} + 25 \text{ ft}$
$= 100$ ft

(c) $s = 16$ cm

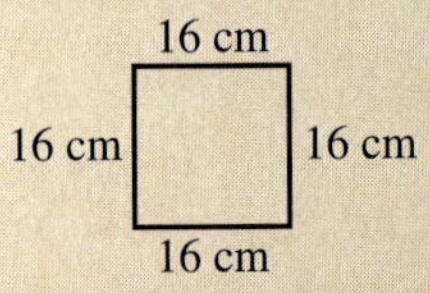

$P = 16 \text{ cm} + 16 \text{ cm} + 16 \text{ cm}$
$+ 16 \text{ cm} = 64$ cm

To find the *perimeter* of the right-hand square on the previous page (the distance around the square), you could add 9 ft + 9 ft + 9 ft + 9 ft to get 36 ft. A shorter way is to multiply the length of one side times 4, because all 4 sides are the same length.

Finding the Perimeter of a Square

Perimeter of a square = side + side + side + side

or, $P = 4 \cdot \text{side}$

$P = 4s$

EXAMPLE 1 Finding the Perimeter of a Square

Find the perimeter of a square that measures 9 ft on each side.

Use the formula for perimeter of a square, $P = 4s$. You do know that for this particular square, the value of s is 9 ft.

$P = 4s$ — Formula for perimeter of a square.

$P = 4 \cdot \mathbf{9\ ft}$ — Replace s with 9 ft. Multiply 4 times 9 ft.

$P = 36$ ft — Write 36 ft; ft is the unit of measure.

The perimeter of the square is 36 ft. Notice that this answer matches the result we got when adding the four sides.

WORK PROBLEM 1 AT THE SIDE.

EXAMPLE 2 Finding the Length of One Side of a Square

If the perimeter of a square is 40 cm, find the length of one side. (Note: cm is the short way to write *centimeters.*)

Use the formula for perimeter of a square, $P = 4s$. This time you know that the value of P (the perimeter) is 40 cm.

$P = 4s$ — Formula for perimeter of a square.

$\mathbf{40\ cm} = 4s$ — Replace P with 40 cm.

$\dfrac{40 \text{ cm}}{4} = \dfrac{4s}{4}$ — To get the variable by itself on the right side, divide both sides by 4.

$10 \text{ cm} = s$

The length of one side of the square is 10 cm.

Check Check the solution by drawing a square with each side 10 cm. The perimeter is 10 cm + 10 cm + 10 cm + 10 cm = 40 cm. This result matches the perimeter given in the problem.

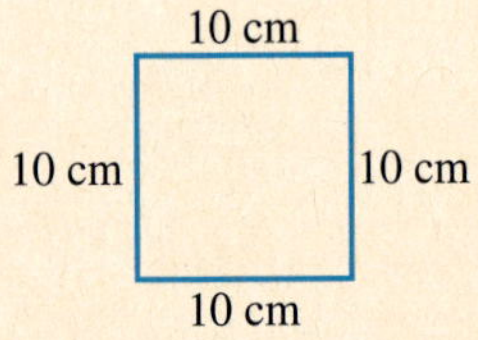

WORK PROBLEM 2 AT THE SIDE.

OBJECTIVE 2 A **rectangle** is a figure with four sides that intersect to form 90° angles. Each set of opposite sides is parallel and congruent (has the same length).

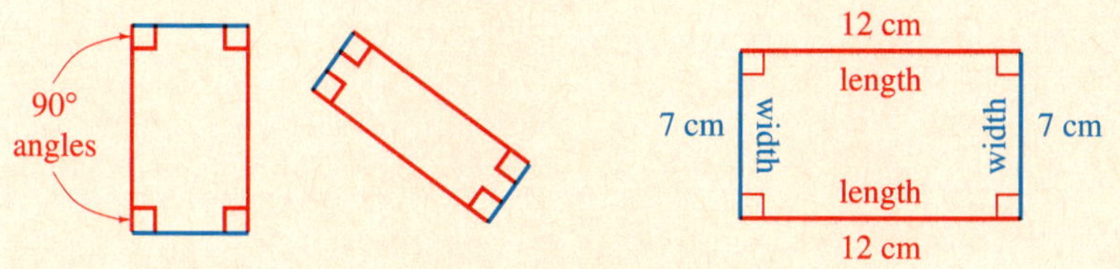

Each longer side of a rectangle is called the length (l) and each shorter side is called the width (w).

Look at the rectangle above with the lengths of the sides labeled. To find the perimeter (distance around), you could add **12 cm** + **12 cm** + **7 cm** + **7 cm** to get 38 cm. Because the two long sides are both 12 cm, and the two short sides are both 7 cm, you can also use this formula.

Finding the Perimeter of a Rectangle

Perimeter of a rectangle = length + length + width + width

$$P = (2 \bullet \text{length}) + (2 \bullet \text{width})$$

$$P = 2l + 2w$$

EXAMPLE 3 Finding the Perimeter of a Rectangle

Find the perimeter of this rectangle.

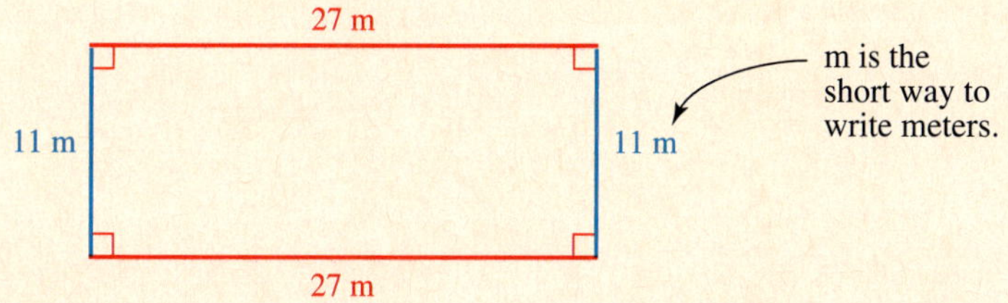

The length is **27 m**, and the width is **11 m**.

$P = 2l + 2w$ — Replace l with 27 m and w with 11 m.

$P = \underbrace{2 \bullet 27\text{ m}} + \underbrace{2 \bullet 11\text{ m}}$ — Do the multiplications first.

$P = 54\text{ m} + 22\text{ m}$ — Add last.

$P = 76\text{ m}$

The perimeter of the rectangle (the distance you would walk around the outside edges of the rectangle) is 76 m.

Check To check the solution, add the lengths of the four sides.

$$P = 27\text{ m} + 27\text{ m} + 11\text{ m} + 11\text{ m}$$

$P = 76$ m ← Matches the solution above

WORK PROBLEM 3 AT THE SIDE.

3. Find the perimeter of each rectangle by using the appropriate formula. Check your solutions by adding the lengths of the four sides.

(a)

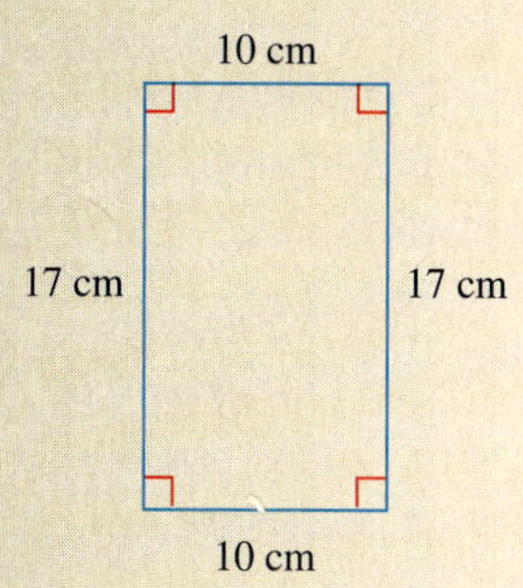

(b) 6 m wide and 11 m long (*Hint:* First draw a sketch of the rectangle.)

(c)

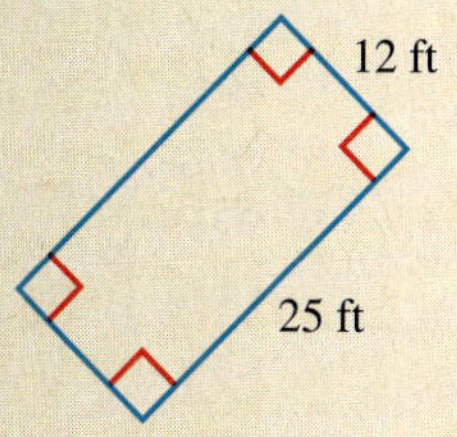

ANSWERS

3. (a) $P = 54$ cm
Check 17 cm + 17 cm + 10 cm + 10 cm = 54 cm

(b)

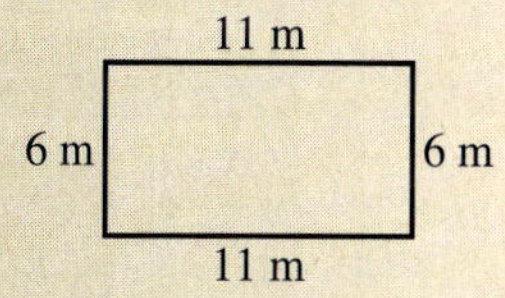

$P = 34$ m
Check 11 m + 11 m + 6 m + 6 m = 34 m

(c) $P = 74$ ft
Check 25 ft + 25 ft + 12 ft + 12 ft = 74 ft

4. Use the perimeter of each rectangle and the appropriate formula to find the length or width. Draw a sketch of each rectangle and use it to check your solution.

(a) The perimeter of a rectangle is 36 in. and the width is 8 in. Find the length.

(b) A rectangle has a width of 4 cm. The perimeter is 32 cm. Find the length.

(c) A rectangle with a perimeter of 14 ft has a length of 6 ft. Find the width.

EXAMPLE 4 Finding the Length or Width of a Rectangle

If the perimeter of a rectangle is 20 ft and the width is 3 ft, find the length.

First draw a sketch of the rectangle and label the widths as 3 ft.

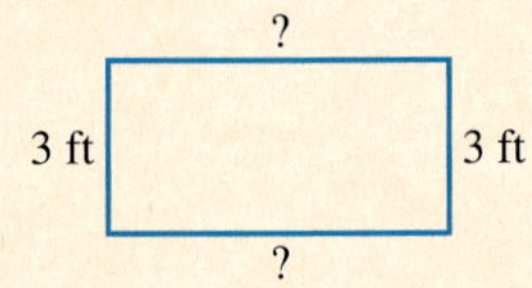

Then use the formula for perimeter of a rectangle, $P = 2l + 2w$. The value of P is 20 ft and the value of w is 3 ft.

$P = 2l + 2w$ — Formula for perimeter of a rectangle.

$20 \text{ ft} = 2l + \underbrace{2 \cdot 3 \text{ ft}}$ — Replace P with 20 ft and w with 3 ft. Simplify the right side by multiplying $2 \cdot 3$ ft.

$$\begin{array}{rcl} 20 \text{ ft} & = & 2l + 6 \text{ ft} \\ {}^{-}6 \text{ ft} & & \quad\;\; {}^{-}6 \text{ ft} \\ \hline 14 \text{ ft} & = & \underbrace{2l + 0} \end{array}$$

To get $2l$ by itself, add ${}^{-}6$ ft to both sides: $6 + {}^{-}6$ is 0.

$$\frac{14 \text{ ft}}{2} = \frac{2l}{2}$$

To get l by itself, divide both sides by 2.

$$7 \text{ ft} = l$$

The length is 7 ft.

Check To check the solution, put the length measurements on your sketch. Then add the four measurements.

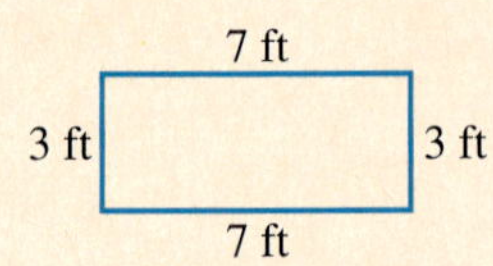

$$P = 7 \text{ ft} + 7 \text{ ft} + 3 \text{ ft} + 3 \text{ ft}$$

$$P = 20 \text{ ft}$$

A perimeter of 20 ft matches the information in the original problem, so 7 ft is the correct length of the rectangle.

WORK PROBLEM 4 AT THE SIDE.

OBJECTIVE 3 A **parallelogram** is a four-sided figure with opposite sides parallel. Some examples are shown below. Notice that opposite sides have the same length.

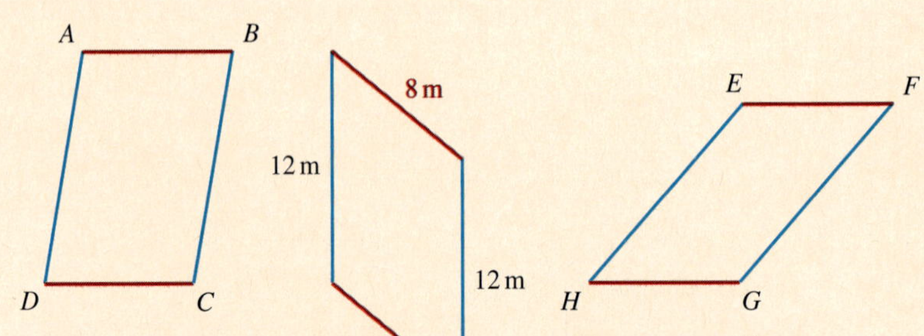

$\overline{AB}$ parallel to $\overline{DC}$ and $\overline{AD}$ parallel to $\overline{BC}$

$\overline{EF}$ parallel to $\overline{HG}$ and $\overline{EH}$ parallel to $\overline{FG}$

Perimeter is the distance around a figure, so the easiest way to find the perimeter of a parallelogram is to add the lengths of the four sides.

ANSWERS

4. (a) $l = 10$ in.

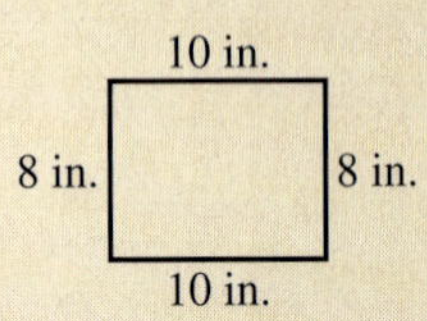

Check 10 in. + 10 in. + 8 in. + 8 in. = 36 in.

(b) $l = 12$ cm

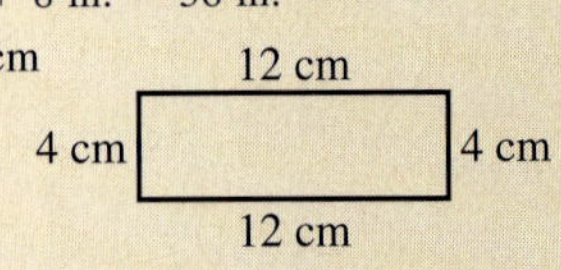

Check 12 cm + 12 cm + 4 cm + 4 cm = 32 cm

(c) $w = 1$ ft

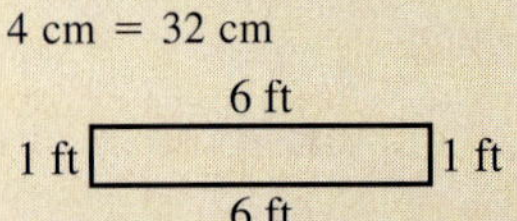

Check 6 ft + 6 ft + 1 ft + 1 ft = 14 ft

EXAMPLE 5 Finding the Perimeter of a Parallelogram

Find the perimeter of the middle parallelogram on the previous page.

$$P = \mathbf{12\ m} + \mathbf{12\ m} + \mathbf{8\ m} + \mathbf{8\ m}$$

$$P = 40 \text{ m}$$

WORK PROBLEM 5 AT THE SIDE.

A **triangle** is a figure with exactly three sides. Some examples are shown below.

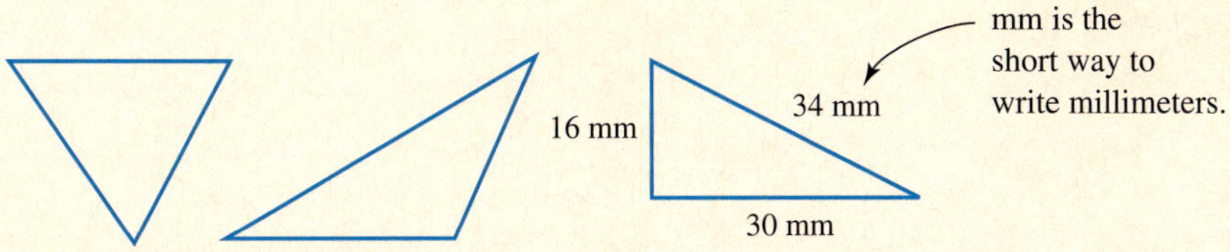

To find the perimeter of a triangle (the distance around the edges), add the lengths of the three sides.

EXAMPLE 6 Finding the Perimeter of a Triangle

The perimeter of the triangle above on the right is

$$P = 16 \text{ mm} + 30 \text{ mm} + 34 \text{ mm}$$

$$P = 80 \text{ mm}$$

WORK PROBLEM 6 AT THE SIDE.

As with any other shape, you can find the perimeter (distance around) of an irregular shape by adding the lengths of the sides.

EXAMPLE 7 Finding Perimeter of an Irregular Shape

A room has the shape shown here.

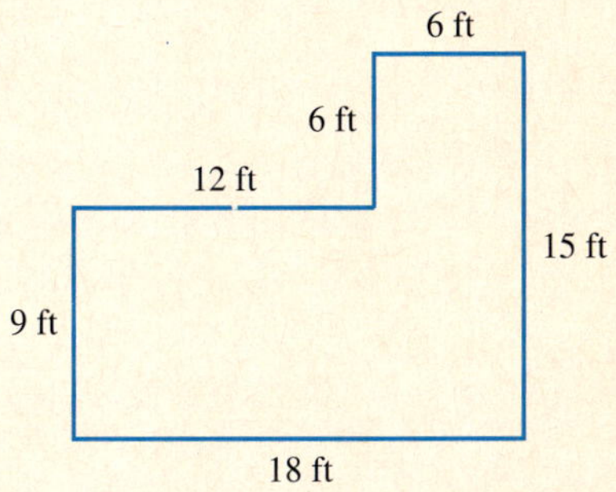

Suppose you want to put a new baseboard (wooden strip) along the base of all the walls. How much material do you need?

CONTINUED ON NEXT PAGE

5. Find the perimeter of each parallelogram.

(a)

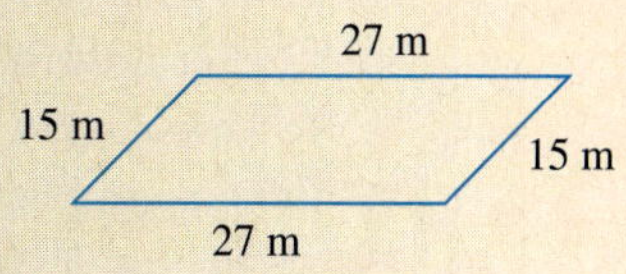

(b)

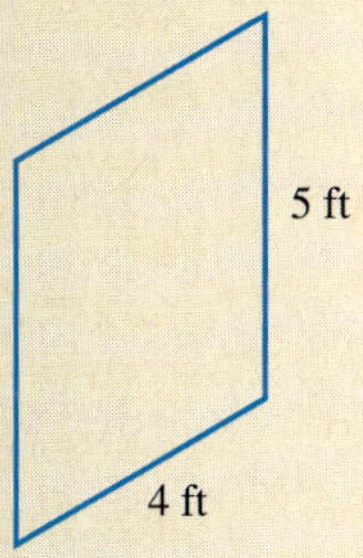

6. Find the perimeter of each triangle.

(a)

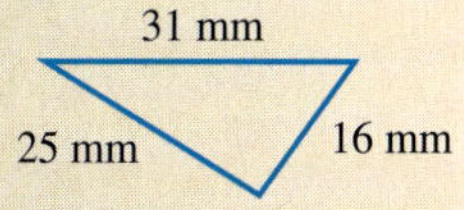

(b) A triangle with sides that each measure 5 in. Draw a sketch of the triangle and label the length of each side.

ANSWERS

5. **(a)** $P = 84$ m **(b)** $P = 18$ ft

6. **(a)** $P = 72$ mm
(b) $P = 15$ in.

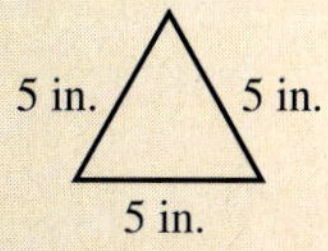

7. How much fencing will be needed to go around a flower bed with the measurements shown below?

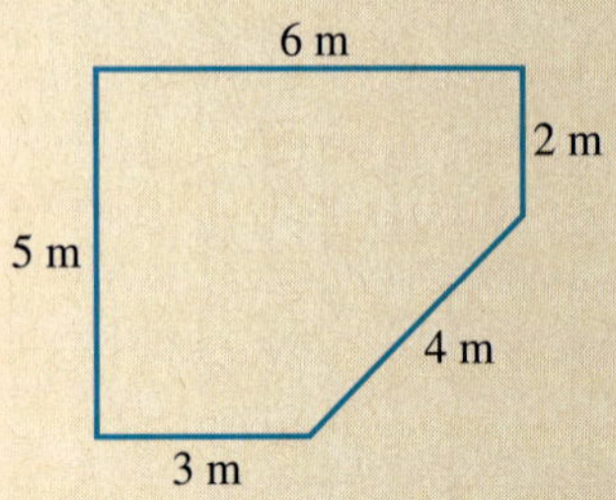

Find the perimeter of the room by adding the lengths of the sides.

$$P = 9 \text{ ft} + 12 \text{ ft} + 6 \text{ ft} + 6 \text{ ft} + 15 \text{ ft} + 18 \text{ ft}$$

$$P = 66 \text{ ft}$$

You need 66 feet of baseboard material.

WORK PROBLEM 7 AT THE SIDE.

Answer

7. 20 m of fencing are needed.

3.1 Exercises

Find the perimeter of each square, using the appropriate formula.

1.

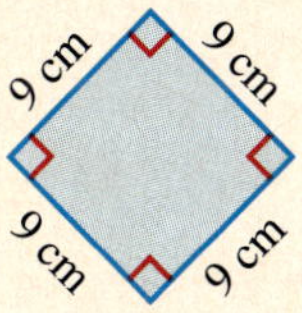

2.

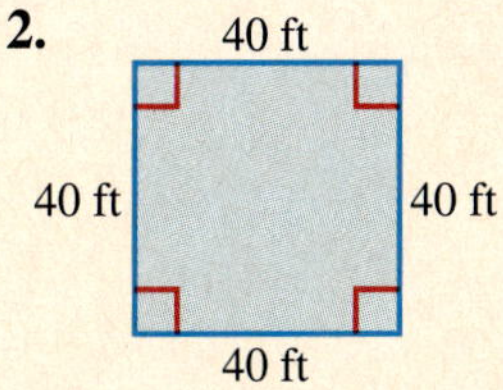

3.

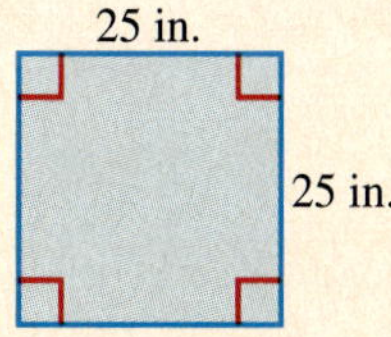

4.

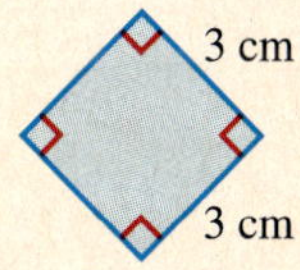

In Exercises 5–8, draw a sketch of each figure and label the lengths of the sides. Then find the perimeter. (Sketches may vary; show your sketches to your instructor.)

5. A square park measuring 1 mile on each side.

6. A square garden measuring 4 meters on each side.

7. A 22 mm square postage stamp.

8. A 10 in. square piece of cardboard.

For the given perimeter of each square, find the length of one side using the appropriate formula.

9. Perimeter is 120 ft.

10. Perimeter is 52 cm.

11. Perimeter is 4 mm.

12. Perimeter is 20 miles.

13. A square parking lot with a perimeter of 92 yards.

14. A square building with a perimeter of 144 meters.

15. A square closet with a perimeter of 8 ft.

16. A square bedroom with a perimeter of 44 ft.

Find the perimeter of each rectangle, using the appropriate formula. Check your solutions by adding the lengths of the four sides.

17.

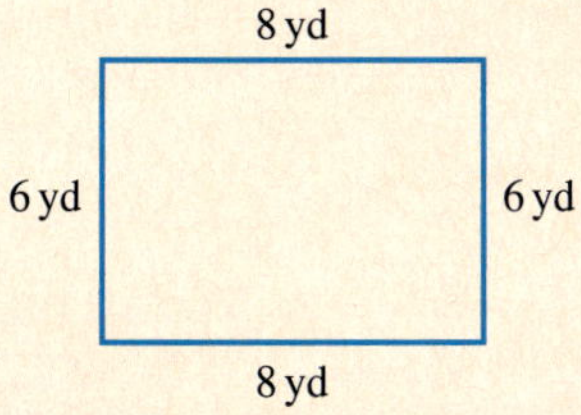

18.

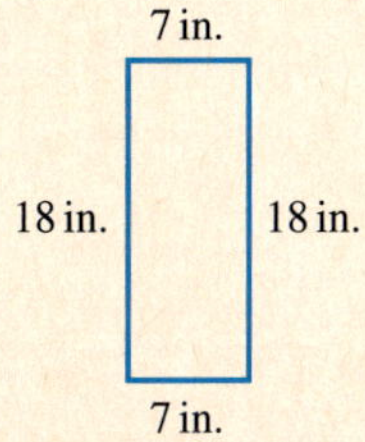

19.

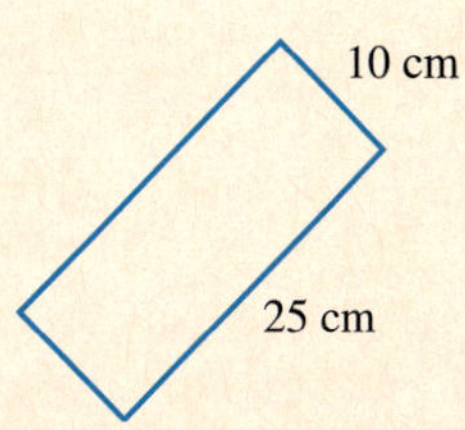

20.

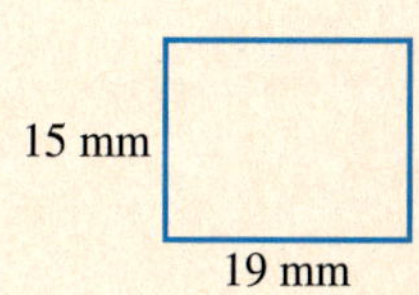

Draw a sketch of each rectangle and label the lengths of the sides. Then find the perimeter by using the appropriate formula. (Sketches may vary; show your sketches to your instructor.)

21. A rectangular living room 20 ft long by 16 ft wide.

22. A rectangular placemat 45 cm long by 30 cm wide.

23. An 8 in. by 5 in. rectangular piece of paper.

24. A 2 ft by 3 ft rectangular window.

For each rectangle, you are given the perimeter and either the length or width. Find the unknown measurement by using the appropriate formula. Draw a sketch of each rectangle and use it to check your solution. (Show your sketches to your instructor.)

25. The perimeter is 30 cm and the width is 6 cm.

26. The perimeter is 48 yards and the length is 14 yards.

27. The length is 4 miles and the perimeter is 10 miles.

28. The width is 8 meters and the perimeter is 34 meters.

29. A 6 ft long rectangular table has a perimeter of 16 ft.

30. A 13 in. wide rectangular picture frame has a perimeter of 56 in.

31. A rectangular door 1 meter wide has a perimeter of 6 meters.

32. A rectangular house 33 ft long has a perimeter of 118 ft.

Find the perimeter of each shape.

33.

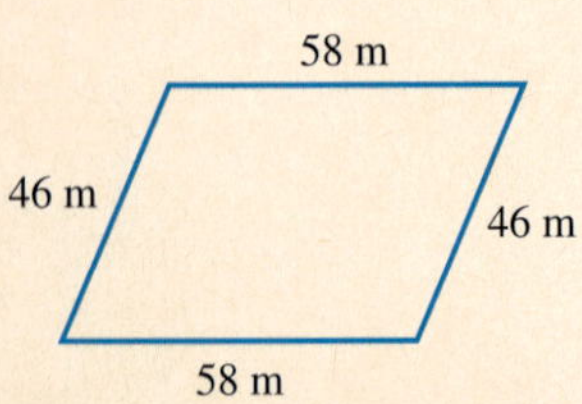

34.

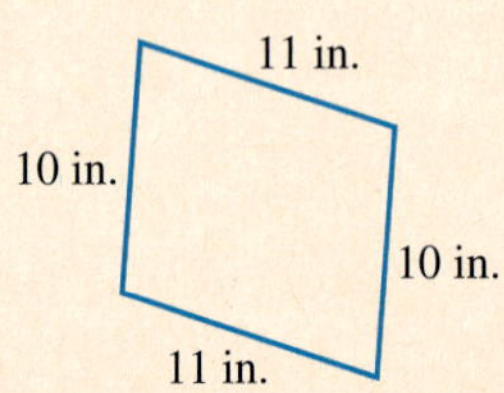

35. Parallelogram

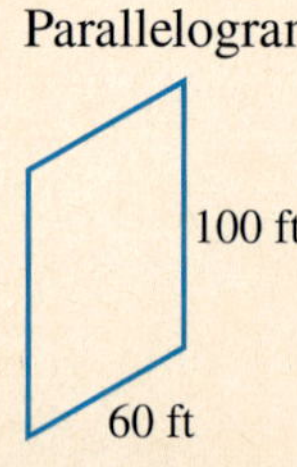

36. Parallelogram

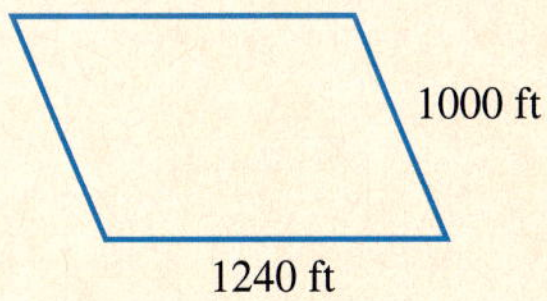

37.

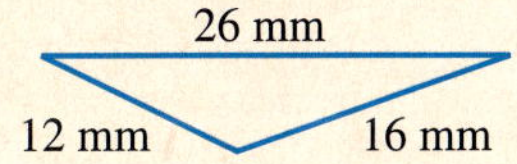

38.

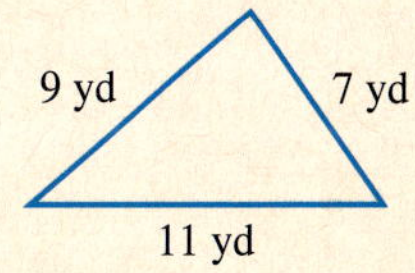

39.

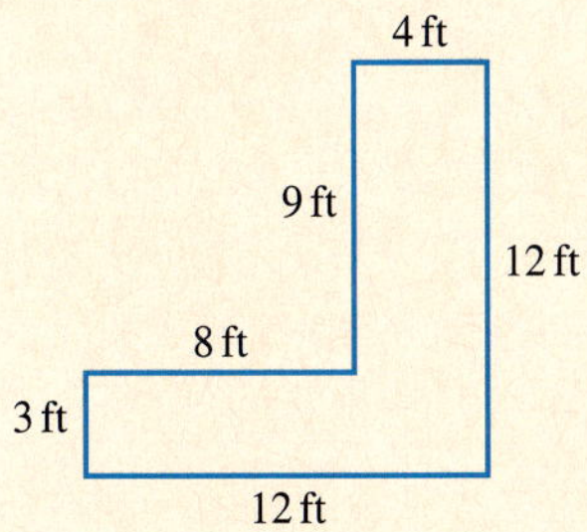

40.

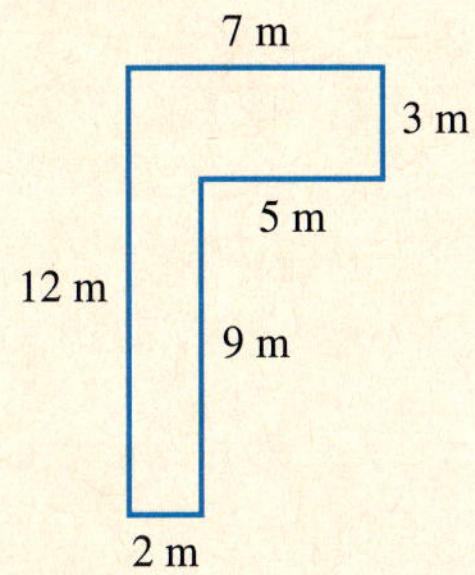

41.

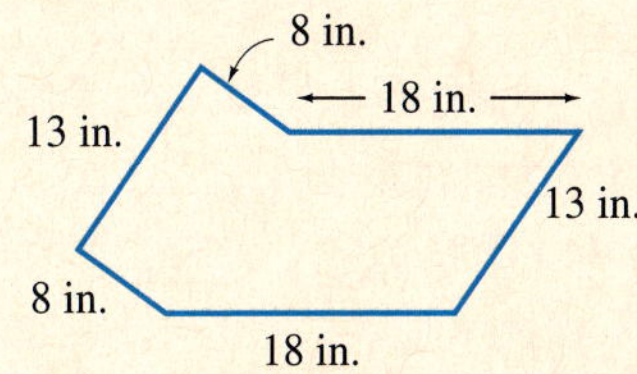

42.

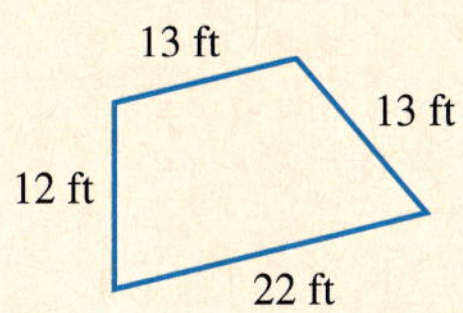

43.

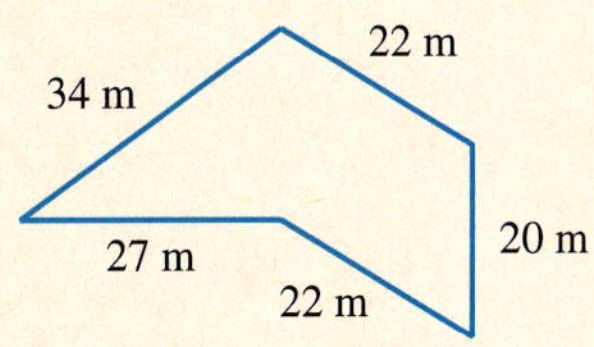

44.

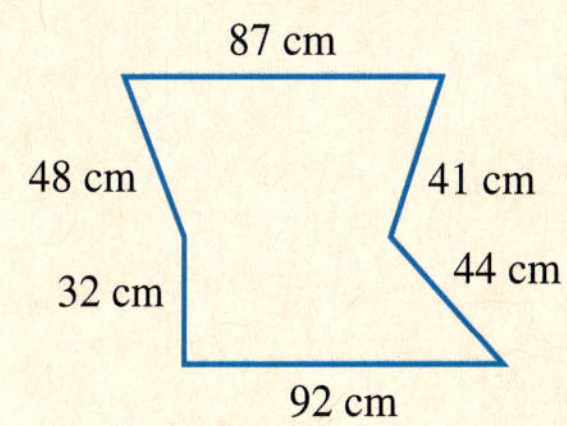

For each shape, you are given the perimeter and the lengths of all sides except one. Find the length of the unlabeled side.

45. Perimeter is 115 cm.

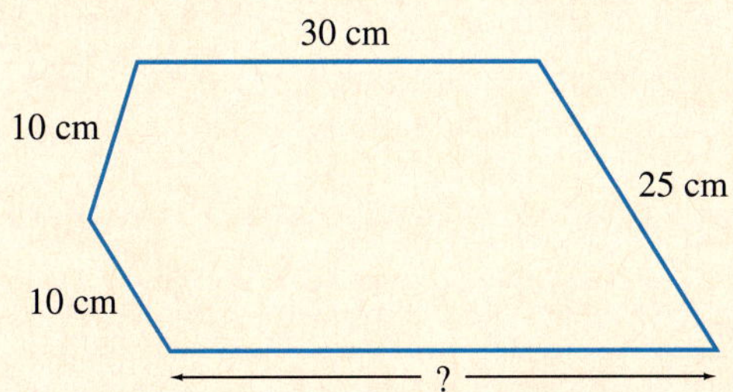

46. Perimeter is 63 in.

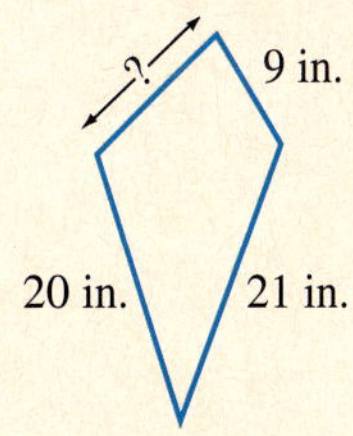

G *For each irregular figure, first find the length of the unlabeled side. Then find the perimeter.*

47.

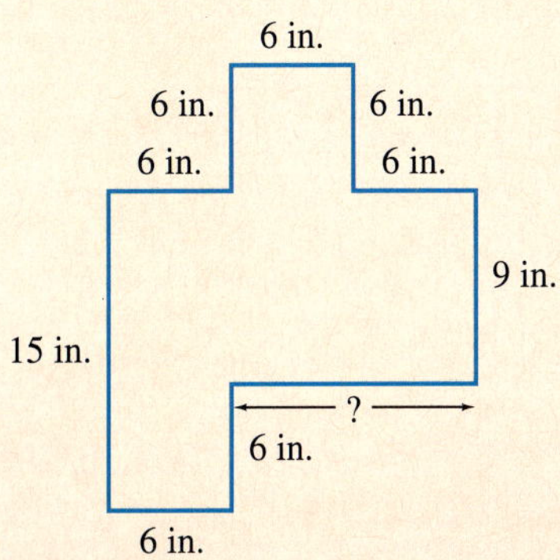

48.

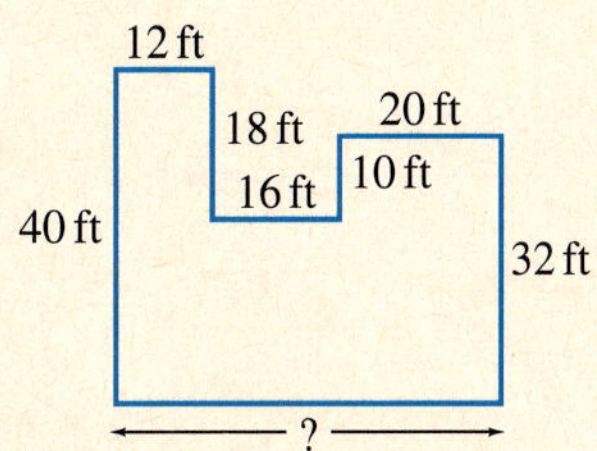

49. **(a)** Draw sketches of four *different* rectangles, each with a perimeter of 20 ft. Label the lengths of the sides on your sketches.
(b) Show how to find the perimeter of each of your rectangles, using the appropriate formula.

50. **(a)** Describe what happens to the perimeter of a square when the length of one side is doubled. Draw sketches of squares, with sides labeled, to illustrate your description.
(b) Describe what happens to the perimeter of a square when the length of one side is tripled.

51. In an *equilateral* triangle, all sides have the same length.
(a) Draw sketches of four different equilateral triangles, label the lengths of the sides, and find the perimeter.
(b) Write a "shortcut" rule (a formula) for finding the perimeter of an equilateral triangle.
(c) Will your formula work for other kinds of triangles that are not equilateral? Explain why or why not.

52. Be sure that you have done Exercise 51 before you do this one.
(a) Draw a sketch of a figure with five sides of equal length. Write a "shortcut" rule (a formula) for finding the perimeter of this shape.
(b) Draw a sketch of a figure with six sides of equal length. Write a formula for finding the perimeter of the shape.
(c) Write a formula for finding the perimeter of a shape with 10 sides of equal length.

Review and Prepare

Simplify. (For help, see ***Sections 1.8*** *and* ***2.1****.)*

53. 6^2

54. 9^2

55. $(^-2)^3$

56. $(^-5)^3$

57. Write x^2 without exponents.

58. Write b^4 without exponents.

59. Evaluate t^2 when t is 4.

60. Evaluate w^3 when w is $^-3$.

3.2 Area

OBJECTIVES

1. Use the formula for area of a rectangle to find the area, the length, or the width.
2. Use the formula for area of a square to find the area or the length of one side.
3. Use the formula for area of a parallelogram to find the area, the base, or the height.
4. Solve application problems involving perimeter and area of rectangles, squares, or parallelograms.

FOR EXTRA HELP

Tutorial | Tape 4 | SSM, Sec. 3.2

OBJECTIVE 1

The Difference Between Perimeter and Area

Perimeter is the **distance around the outside edges** of a flat shape. **Area** is the amount of **surface inside** a flat shape.

The *perimeter* of a rectangle is the distance around the *outside edges*. The *area* of a rectangle is the amount of surface *inside* the rectangle. We measure area by finding the number of squares of a certain size needed to cover the surface inside the rectangle. Think of covering the floor of a rectangular living room with carpet. Carpet is measured in square yards, that is, square pieces that measure 1 yard along each side. Here is a drawing of a living room floor.

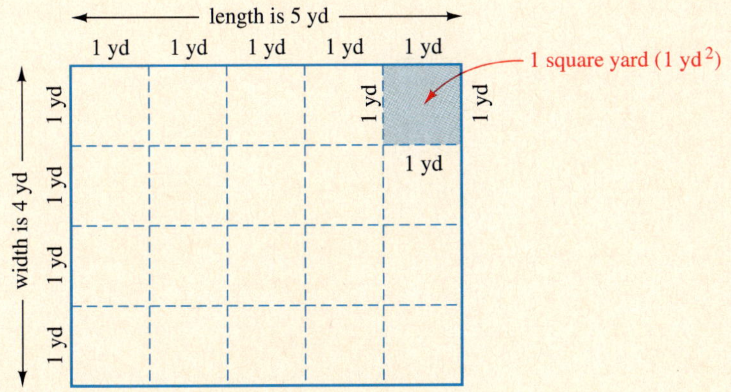

You can see from the drawing that it takes 20 squares to cover the floor. We say that the area of the floor is 20 *square yards.* A short way to write square yards is yd^2.

20 **square yards** can be written as 20 $\mathbf{yd^2}$

To find the number of squares, you can count them, or you can multiply the number of squares in the length (5) times the number of squares in the width (4) to get 20. The formula is:

Finding the Area of a Rectangle

$$\text{Area of a rectangle} = \text{length} \cdot \text{width}$$

$$A = lw$$

Remember to use square units when measuring area.

Squares of other sizes can be used to measure area. For smaller areas, you might use these:

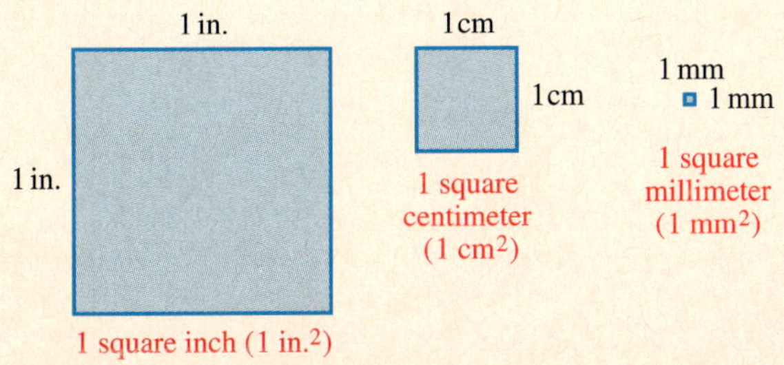

Actual-size drawings

1. Find the area of each rectangle, using the appropriate formula.

(a)

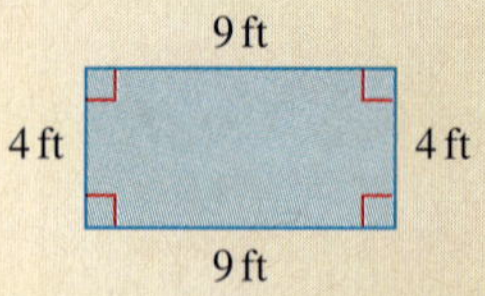

(b) a rectangle that is 35 mi long and 20 mi wide (First make a sketch of the rectangle.)

(c) a rectangular patio that measures 3 m by 2 m (First make a sketch of the patio.)

Other sizes of squares that are often used to measure area are listed here, but they are too large to draw on this page.

1 square meter (1 m^2)	1 square foot (1 ft^2)
1 square kilometer (1 km^2)	1 square yard (1 yd^2)
	1 square mile (1 mi^2)

Note

The raised 2 in 4^2 means that you multiply 4 • 4 to get 16. The raised 2 in cm^2 or yd^2 is a short way to write the word "square." It means that you multiplied cm times cm or yd times yd. Recall that a short way to write $x \cdot x$ is x^2. Similarly, cm • cm is cm^2. When you see 5 cm^2, say "five square centimeters." Do *not* multiply 5 • 5.

EXAMPLE 1 Finding the Area of a Rectangle

Find the area of each rectangle.

(a)

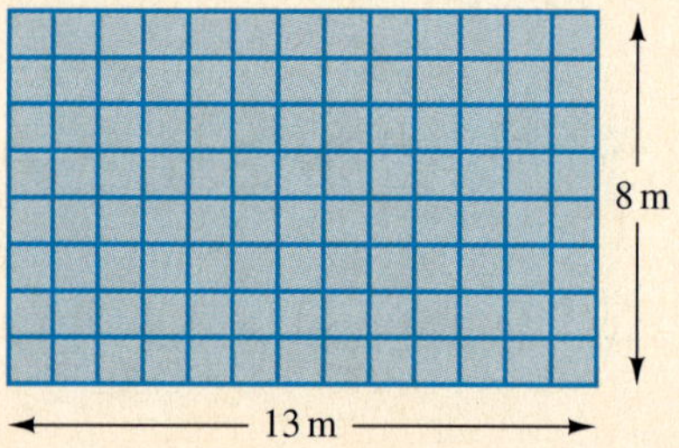

The length of this rectangle is 13 m and the width is 8 m. Use the formula $A = lw$.

$$A = l \cdot w$$ Replace l with 13 m and w with 8 m.

$$A = 13 \text{ m} \cdot 8 \text{ m}$$ Multiply 13 times 8 to get 104. Multiply m times m to get m^2.

$$A = 104 \text{ m}^2$$

The area of the rectangle is 104 m^2. If you count the number of squares in the sketch, you will also get 104 m^2. (Each square in the sketch is 1 m by 1m or 1 m^2.)

(b) A rectangle measuring 7 cm by 21 cm.

First make a sketch of the rectangle. The length is 21 cm (the longer measurement) and the width is 7 cm. Then use the formula for area of a rectangle, $A = lw$.

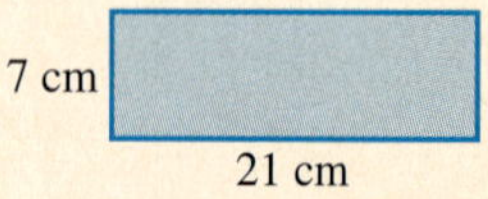

$$A = l \cdot w$$

$$A = 21 \text{ cm} \cdot 7 \text{ cm}$$ Multiply 21 • 7 to get 147. Multiply cm • cm to get cm^2.

$$A = 147 \text{ cm}^2$$

The area of the rectangle is 147 cm^2.

Note

The units for *area* will always be *square* units (cm^2, m^2, yd^2, mi^2, and so on). The units for *perimeter* will always be *linear* units (cm, m, yd, mi, and so on; not square units).

WORK PROBLEM 1 AT THE SIDE.

ANSWERS

1. (a) $A = 36 \text{ ft}^2$

(b)

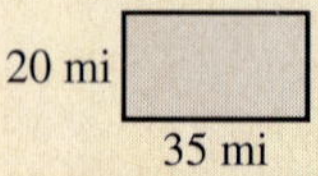

$A = 700 \text{ mi}^2$

(c)

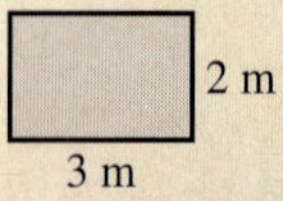

$A = 6 \text{ m}^2$

EXAMPLE 2 Finding the Length or Width of a Rectangle

If the area of a rectangular rug is 12 yd^2 and the length is 4 yd, find the width.

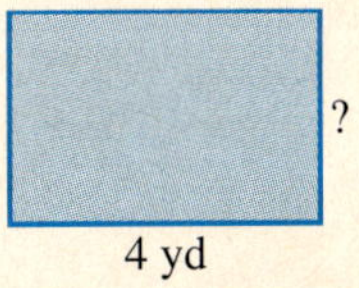

First draw a sketch of the rug and label the length as 4 yd.

Use the formula for area of a rectangle, $A = lw$.
The value of A is 12 yd^2 and the value of l is 4 yd.

$$A = l \cdot w$$
Replace A with 12 yd^2 and replace l with 4 yd.

$$12\ yd^2 = 4\ yd \cdot w$$
To get w by itself, divide both sides by 4 yd.

$$\frac{12\ yd \cdot \cancel{yd}}{4\ \cancel{yd}} = \frac{4\ yd \cdot w}{4\ yd}$$
On the left side, rewrite yd^2 as yd • yd. Then $\frac{yd}{yd}$ is 1, so they "cancel out."

$$3\ yd = w$$
On the left side, 12 yd ÷ 4 is 3 yd.

The width of the rug is 3 yd.

Check To check the solution, put the width measurement on your sketch. Then use the area formula.

$$A = l \cdot w$$
$$A = 4\ yd \cdot 3\ yd$$
$$A = 12\ yd^2$$

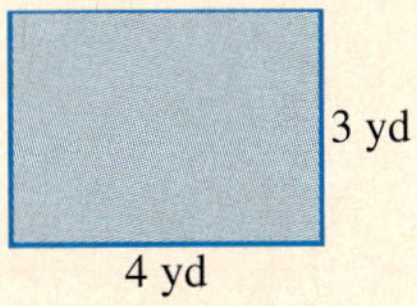

An area of 12 yd^2 matches the information in the original problem. So 3 yd is the correct width of the rug.

WORK PROBLEM 2 AT THE SIDE.

OBJECTIVE 2 As with a rectangle, you can multiply length times width to find the area (surface inside) of a square. Because the length and the width are the same in a square, the formula is written:

Finding the Area of a Square

Area of a square = side • side

$$A = s \cdot s$$
$$A = s^2$$

Remember to use square units when measuring area.

EXAMPLE 3 Finding the Area of a Square

Find the area of a square sign that is 4 ft on each side.

Use the formula for area of a square, $A = s^2$.

$$A = s^2$$
Remember that s^2 means $s \cdot s$.

$$A = s \cdot s$$
Replace s with 4 ft.

$$A = 4\ ft \cdot 4\ ft$$
Multiply 4 • 4 to get 16.

$$A = 16\ ft^2$$
Multiply ft • ft to get ft^2.

The area of the sign is 16 ft^2.

CONTINUED ON NEXT PAGE

2. Use the area of each rectangle and the appropriate formula to find the length or width. Draw a sketch of each rectangle and use it to check your solution.

(a) The area of a microscope slide is 12 cm^2, and the length is 6 cm. Find the width.

(b) A child's play lot is 10 ft wide and has an area of 160 ft^2. Find the length.

(c) A hallway is 31 m long and has an area of 93 m^2. Find the width of the hall.

ANSWERS

2. (a) $w = 2$ cm

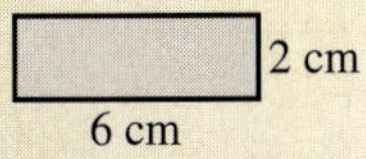

Check $A = 6$ cm • 2 cm
$A = 12\ cm^2$
Matches original problem.

(b) $l = 16$ ft

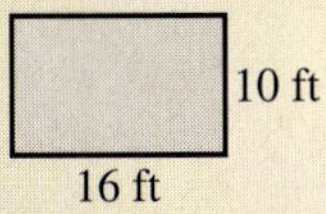

Check $A = 16$ ft • 10 ft
$A = 160\ ft^2$
Matches original problem.

(c) $w = 3$ m

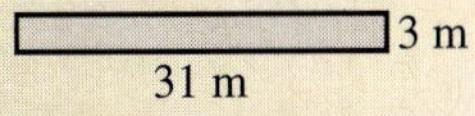

Check $A = 31$ m • 3 m
$A = 93\ m^2$
Matches original problem.

3. Find the area of each square, using the appropriate formula. Make a sketch of each square.

(a) a 12 in. square piece of fabric

(b) a square township 7 mi on a side

(c) a square earring measuring 20 mm on each side

4. Given the area of each square, find the length of one side by inspection.

(a) The area of a square-shaped nature center is 16 mi^2.

(b) A square floor has an area of 100 m^2.

(c) A square clock face has an area of 81 $in.^2$

Answers

3. (a) 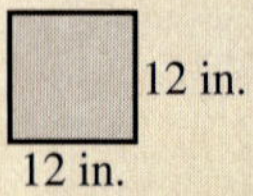

$A = 144 \text{ in}^2$

(b) 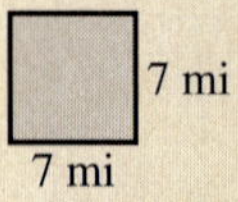

$A = 49 \text{ mi}^2$

(c) 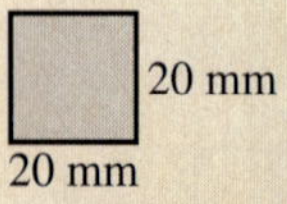

$A = 400 \text{ mm}^2$

4. (a) $s = 4$ mi **(b)** $s = 10$ m **(c)** $s = 9$ in.

Note

Be careful! s^2 means $s \cdot s$. It does **not** mean $2 \cdot s$. In this example s is 4 ft, so $(4 \text{ ft})^2$ is $4 \text{ ft} \cdot 4 \text{ ft} = 16 \text{ ft}^2$. It is **not** $2 \cdot 4 \text{ ft} = 8 \text{ ft}$.

Check Check the solution by drawing a square with each side 4 ft. You can multiply length (4 ft) times width (4 ft), as for a rectangle. So the area is $4 \text{ ft} \cdot 4 \text{ ft}$, or 16 ft^2. This result matches the solution we got by using the formula $A = s^2$.

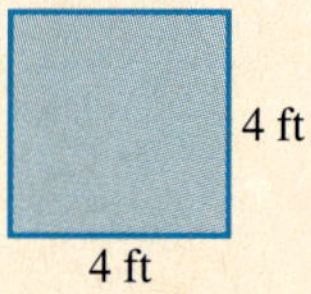

◀◀ WORK PROBLEM 3 AT THE SIDE.

EXAMPLE 4 Finding the Length of One Side of a Square

If the area of a square township is 49 mi^2, what is the length of one side of the township?

Use the formula for area of a square, $A = s^2$. The value of A is 49 mi^2.

$A = s^2$ — Replace A with 49 mi^2.

$49 \text{ mi}^2 = s^2$ — To get s by itself, we have to "undo" the squaring of s. This is called *finding the square root* (more on square roots in Chapter 5).

$49 \text{ mi}^2 = s \cdot s$ — For now, solve by inspection. Ask, what number times itself gives 49?

$49 \text{ mi}^2 = 7 \text{ mi} \cdot 7 \text{ mi}$ — $7 \cdot 7$ is 49, so $7 \text{ mi} \cdot 7 \text{ mi}$ is 49 mi^2.

The length of one side of the township is 7 miles. Notice how this result matches the information about the township in Margin Exercise 3(b).

◀◀ WORK PROBLEM 4 AT THE SIDE.

Objective 3 To find the area of a parallelogram, first draw a dashed line inside the figure, as shown here.

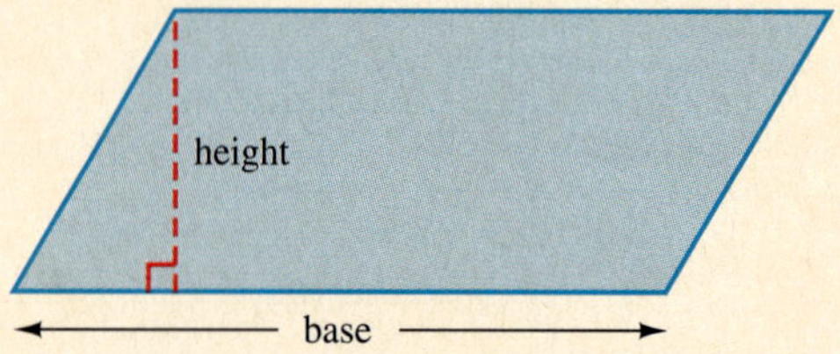

Try this yourself by tracing this parallelogram onto a piece of paper.

The length of the dashed line is the *height* of the parallelogram. It forms a 90° angle (a square corner) with the base. A 90° angle is also called a *right angle.* The height is the shortest distance between the base and the opposite side.

Now cut off the triangle created on the left side of the parallelogram and move it to the right side, as shown below.

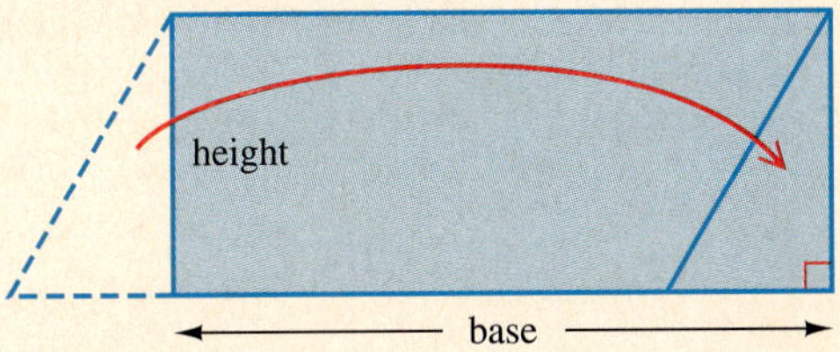

You have made the parallelogram into a rectangle. You can see that the area of the parallelogram and the rectangle are the same. The area of the rectangle is *length* times *width*. For the parallelogram, the area translates into *base* times *height*.

Finding the Area of a Parallelogram

Area of a parallelogram = base • height

$$A = bh$$

Remember to use square units when measuring area.

EXAMPLE 5 Finding the Area of a Parallelogram

Find the area of each parallelogram.

(a)

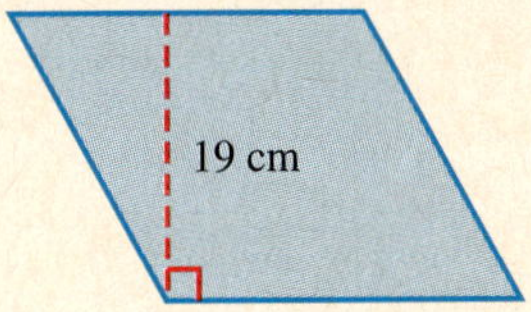

(b)

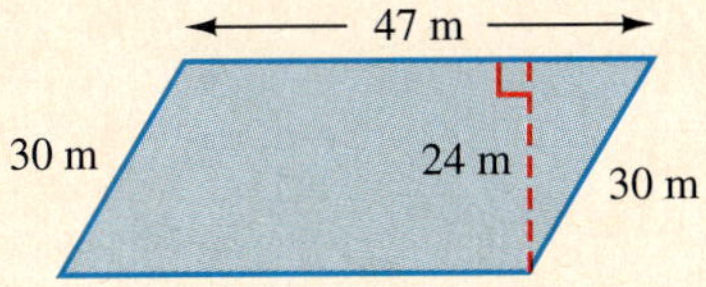

(a) The base is 24 cm and the height is 19 cm. The formula for the area is $A = bh$.

$A = b \cdot h$	Replace b with 24 cm and h with 19 cm.
$A = 24 \text{ cm} \cdot 19 \text{ cm}$	Multiply 24 • 19 to get 456.
$A = 456 \text{ cm}^2$	Multiply cm • cm to get cm^2.

The area of the parallelogram is 456 cm^2.

(b) Use the formula for area of a parallelogram, $A = bh$.

$A = b \cdot h$	Replace b with 47 m and h with 24 m.
$A = 47 \text{ m} \cdot 24 \text{ m}$	Multiply 47 • 24 to get 1128.
$A = 1128 \text{ m}^2$	Multiply m • m to get m^2.

Notice that the 30 m sides are *not* used in finding the area. But you would use them when finding the *perimeter* of the parallelogram.

WORK PROBLEM 5 AT THE SIDE.

EXAMPLE 6 Finding the Base or Height of a Parallelogram

If the area of a parallelogram is 24 ft^2 and the base is 6 ft, find the height.

First draw a sketch of the parallelogram and label the base as 6 ft.

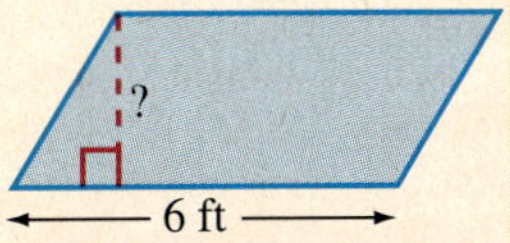

Use the formula for the area of a parallelogram, $A = bh$.

CONTINUED ON NEXT PAGE

5. Find the area of each parallelogram.

(a)

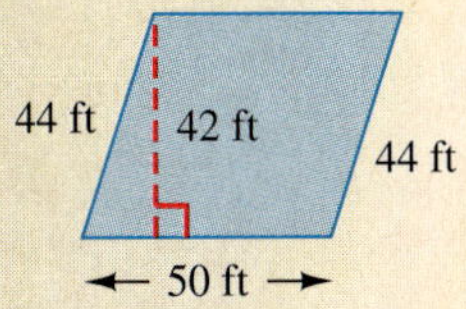

(b)

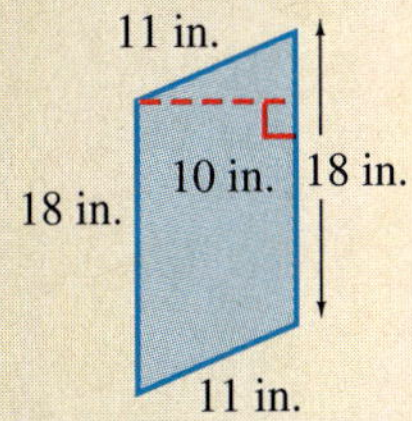

(c) a parallelogram with base 8 cm and height 1 cm.

ANSWERS

5. **(a)** $A = 2100 \text{ ft}^2$ **(b)** $A = 180 \text{ in.}^2$ **(c)** $A = 8 \text{ cm}^2$

6. Use the area of each parallelogram and the appropriate formula to find the base or height. Draw a sketch of each parallelogram and use it to check your solution.

(a) The area of a parallelogram is 140 in.^{2} and the base is 14 in. Find the height.

(b) A parallelogram has an area of 4 yd^2. The height is 1 yd. Find the base.

The value of A is 24 ft^2, and the value of b is 6 ft.

$A = b \cdot h$ — Replace A with 24 ft^2 and b with 6 ft.

$24\text{ ft}^2 = 6\text{ ft} \cdot h$ — To get h by itself, divide both sides by 6 ft.

$\dfrac{24\text{ ft} \cdot \cancel{\text{ft}}}{6\,\cancel{\text{ft}}} = \dfrac{6\text{ ft} \cdot h}{6\text{ ft}}$ — On the left side, rewrite ft^2 as ft • ft. Then $\frac{\text{ft}}{\text{ft}}$ is 1, so they "cancel out."

$4\text{ ft} = h$ — On the left side, 24 ft ÷ 6 is 4 ft.

The height of the parallelogram is 4 ft.

Check To check the solution, put the height measurement on your sketch. Then use the area formula.

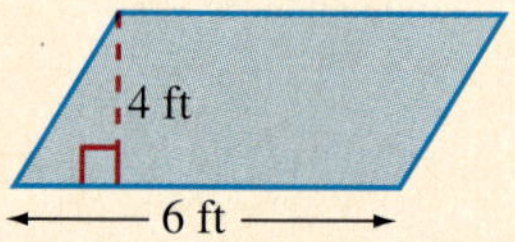

$$A = b \cdot h$$
$$A = 6\text{ ft} \cdot 4\text{ ft}$$
$$A = 24\text{ ft}^2$$

An area of 24 ft^2 matches the information in the original problem. So 4 ft is the correct height of the parallelogram.

WORK PROBLEM 6 AT THE SIDE.

Objective 4 When you are solving problems, first decide whether you need to find the perimeter or the area.

EXAMPLE 7 Solving Application Problems Involving Perimeter or Area

A group of neighbors is fixing up a playground area for their children. The rectangular lot is 22 yd by 16 yd. If chain-link fencing costs \$6 per yard, how much will they spend to put a fence around the lot?

First draw a sketch of the rectangular lot and label the lengths of the sides. The fence will go around the edges of the lot, so you need to find the *perimeter* of the lot.

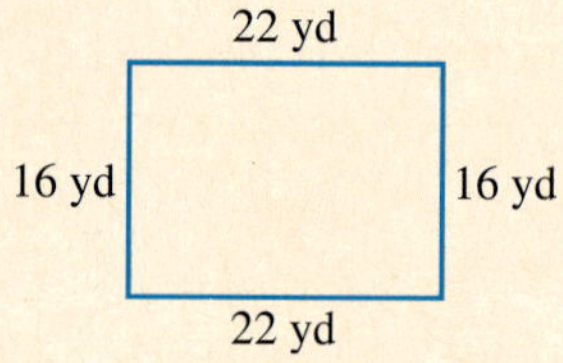

$P = 2l + 2w$ — Formula for perimeter of a rectangle.

$$P = 2 \cdot 22\text{ yd} + 2 \cdot 16\text{ yd}$$
$$P = 44\text{ yd} + 32\text{ yd}$$
$$P = 76\text{ yd}$$

The perimeter of the lot is 76 yd, so the neighbors need to buy 76 yd of fencing. The cost of the fencing is \$6 *per yard,* which means \$6 *for 1 yard.* To find the cost for 76 yd, multiply \$6 • 76. The neighbors will spend \$456 on the fence.

WORK PROBLEM 7 AT THE SIDE.

7. If sod costs \$3 per square yard, how much will the neighbors in Example 7 spend to cover the playground with grass?

Answers

6. (a) $h = 10$ in.

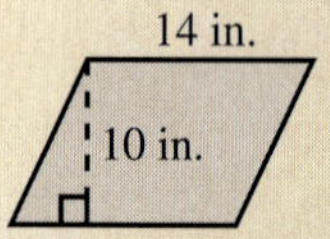

Check $A = 14\text{ in.} \cdot 10\text{ in.}$
$A = 140\text{ in.}^2$
Matches original problem.

(b) $b = 4$ yd

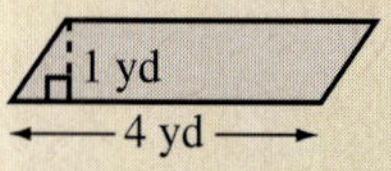

Check $A = 4\text{ yd} \cdot 1\text{ yd}$
$A = 4\text{ yd}^2$
Matches original problem.

7. \$1056 (Find the area by multiplying 22 yd • 16 yd to get 352 yd^2. Then multiply \$3 • 352 to get \$1056.)

3.2 Exercises

Find the area of each rectangle, square, or parallelogram using the appropriate formula.

1.
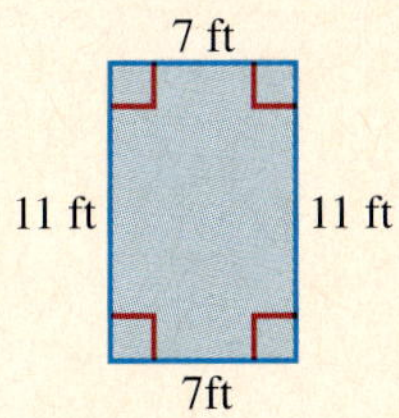

2.
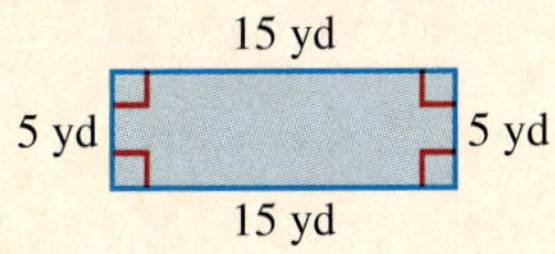

3.
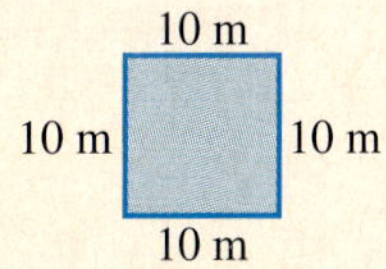

4.
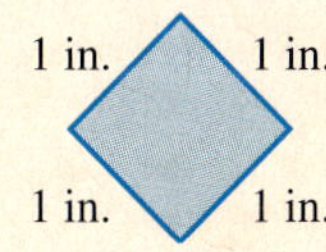

5.
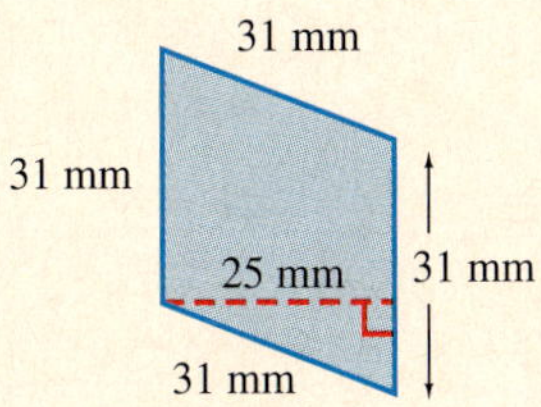

6.
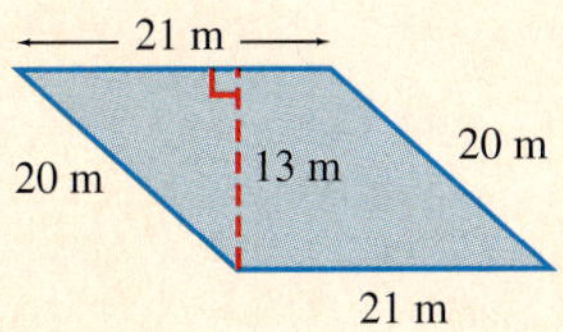

7.

8.
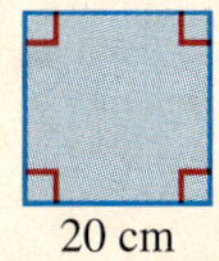

In Exercises 9–16, first draw a sketch of the shape and label the lengths of the sides or base and height. Then find the area. (Sketches may vary; show your sketches to your instructor.)

9. A rectangular calculator that measures 15 cm by 7 cm.

10. A rectangular board is 8 ft long and 2 ft wide.

11. A parallelogram with height of 9 ft and base of 8 ft.

12. A parallelogram measuring 18 mm on the base and 3 mm on the height.

13. A fire burned a 25-mile-square forest.

14. An 11 in. square pillow.

15. A piece of window glass 1 m on each side.

16. A table 12 ft long by 3 ft wide.

Use the area of each rectangle and either its length or width, and the appropriate formula, to find the other measurement. Draw a sketch of each rectangle and use it to check your solution. (Sketches may vary; show your sketches to your instructor.)

17. The area of a desk is 18 ft^2, and the width is 3 ft. Find its length.

18. The area of a classroom is 630 ft^2, and the length is 30 ft. Find its width.

19. A football field is 100 yd long and has an area of 5300 yd^2. Find its width.

20. A hockey field is 60 yd wide and has an area of 6000 yd^2. Find its length.

21. A 154 $in.^2$ photo has a width of 11 in. Find its length.

22. A 15 $in.^2$ recipe card has a width of 3 in. Find its length.

Given the area of each square, find the length of one side by inspection.

23. A square floor has an area of 36 m^2.

24. A square stamp has an area of 9 cm^2.

25. The area of a square sign is 4 ft^2.

26. The area of a square piece of metal is 64 $in.^2$.

Use the area of each parallelogram and either its base or height, and the appropriate formula, to find the other measurement. Draw a sketch of each parallelogram and use it to check your solution. (Sketches may vary; show your sketches to your instructor.)

27. The area is 500 cm^2, and the base is 25 cm. Find the height.

28. The area is 1500 m^2, and the height is 30 m. Find the base.

29. The height is 13 in. and the area is 221 $in.^2$. Find the base.

30. The base is 19 cm, and the area is 114 cm^2. Find the height.

31. The base is 9 m, and the area is 9 m^2. Find the height.

32. The area is 25 mm^2, and the height is 5 mm. Find the base.

33. In your own words, describe the difference between perimeter and area. Then make a drawing of a square and a rectangle, label the sides with measurements, and show the steps for finding the perimeter and area of each figure.

(G) **34.** Suppose that you had 16 feet of fencing. Draw three different square or rectangular garden plots that would use exactly 16 feet of fencing; label the lengths of the sides. What shape plot would have the greatest area?

Find both the perimeter and the area of each rectangle, square, or parallelogram.

35.

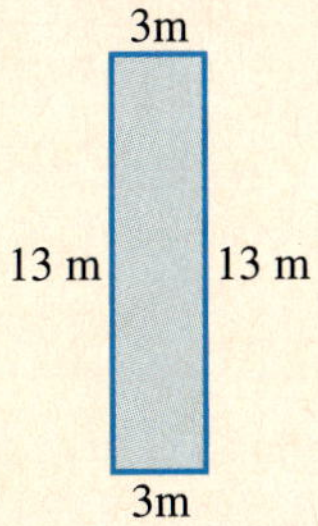

36.

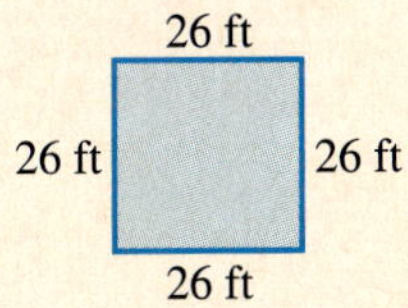

37.

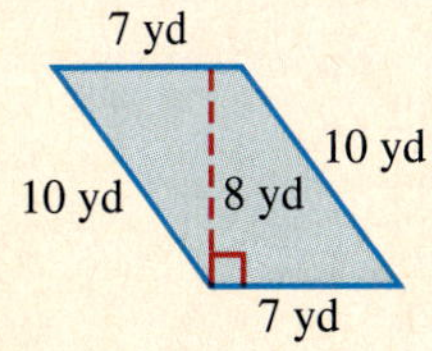

38.

39.

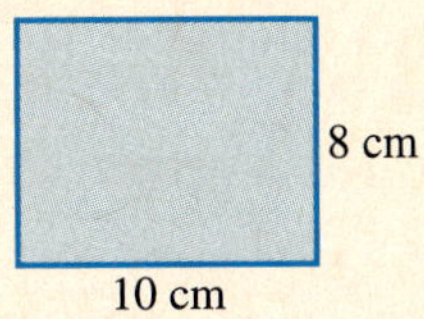

40.

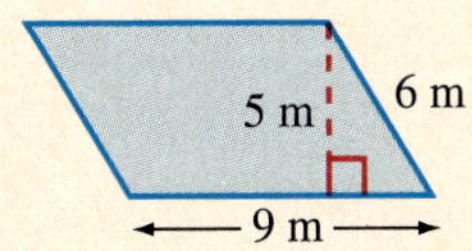

*Explain and correct the **two** errors made by students in Exercises 41 and 42.*

41.

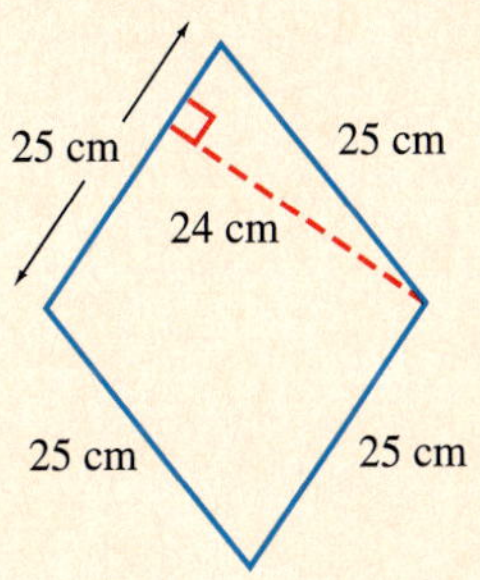

$P = 25 \text{ cm} + 24 \text{ cm} + 25 \text{ cm} + 25 \text{ cm}$
$\quad + 25 \text{ cm}$
$P = 124 \text{ cm}^2$

42.

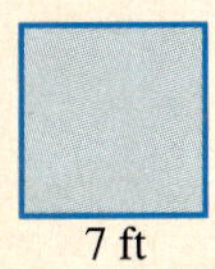

$A = s^2$
$A = 2 \cdot 7 \text{ ft}$
$A = 14 \text{ ft}$

Solve each application problem. Sometimes you will need to find the perimeter and sometimes the area. Draw a sketch for each problem and label the sketch with the appropriate measurements. (Sketches may vary; show your sketches to your instructor.)

43. Tyra's kitchen is 4 m wide and 5 m long. She is pasting a decorative strip that costs \$6 per meter around the top edge of all the walls. How much will she spend?

44. The Wang's family room measures 20 ft by 25 ft. They are covering the floor with square tiles that measure 1 ft on a side and cost \$1 each. How much will they spend on tile?

45. Mr. and Mrs. Gomez are buying carpet for their square-shaped bedroom that is 5 yd wide. The carpet is \$23 per square yard and padding and installation is another \$6 per square yard. How much will they spend in all?

46. A page in this book measures about 27 cm from top to bottom and 21 cm from side to side. Find the perimeter and the area of the page.

47. Find the area of Colin's L-shaped office. (*Hint:* Cut the office into two rectangles. Find the area of each rectangle and then add the areas.)

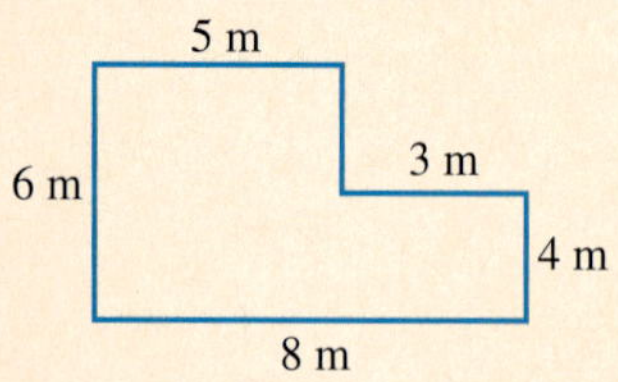

48. Find the area of this piece of land. (*Hint:* Cut the land into two pieces. Find the area of each piece and then add the areas.)

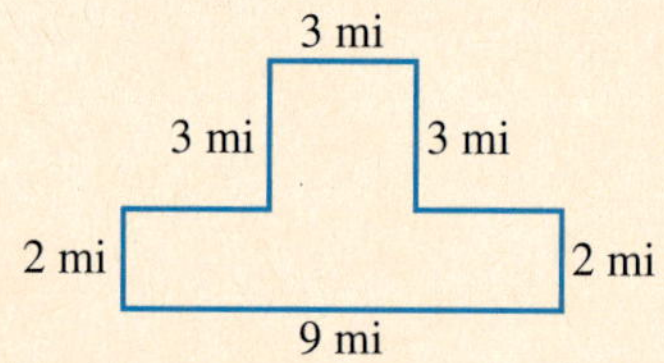

49. A lot is 124 ft by 172 ft. County rules require that nothing be built on land within 12 ft of any edge of the lot. Draw a sketch of the lot, showing the land that cannot be built on. What is the area of the land that cannot be built on?

50. Find the cost of fencing needed for this rectangular field. Fencing along the country roads costs \$4.25 per foot. Fencing for the other two sides costs \$2.75 per foot.

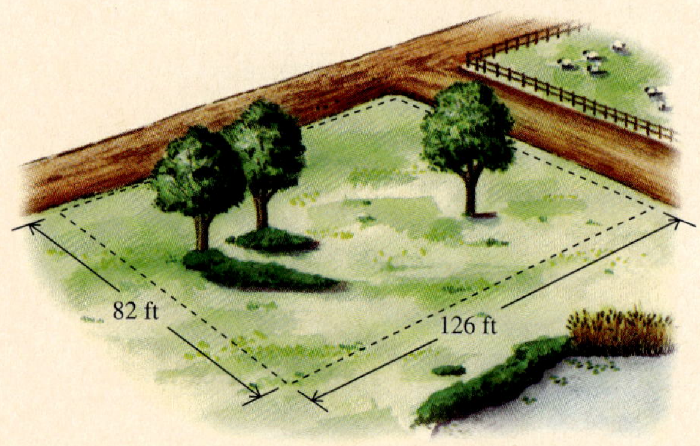

Review and Prepare

Solve each equation. Show your work. (For help, see ***Sections 2.3*** *and* ***2.4****.)*

51. $y + 15 - 11 = {}^{-}2 + 8$

52. $5 + d - 12 = 3 - 7$

53. $4n - 8 - 10 = {}^{-}30$

54. $3x - 6 + 9 = {}^{-}42$

55. ${}^{-}60 + 5 = 2x - 7x$

56. $10 - 20 = {}^{-}3m + 8m$

3.3 Application Problems with One Unknown Quantity

OBJECTIVE 1 In **Sections 3.1** and **3.2** you worked with applications involving perimeter and area. You were able to use well-known rules (formulas) to set up equations that could be solved. However, you will encounter many problems for which no formula is available. Then you need to analyze the problem and translate the words into an equation that fits the particular situation. We'll start by translating word phrases into algebraic expressions.

EXAMPLE 1 Translating Word Phrases into Algebraic Expressions

Write in symbols by using x as the variable.

WORDS	ALGEBRA
a number **plus** 2	$x + 2$ or $2 + x$
the **sum** of 8 and a number	$8 + x$ or $x + 8$
5 **more than** a number	$x + 5$ or $5 + x$
$^{-}35$ **added to** a number	$^{-}35 + x$ or $x + {}^{-}35$
a number **increased by** 6	$x + 6$ or $6 + x$
9 **less than** a number	$x - 9$
a number **subtracted from** 3	$3 - x$
3 **subtracted from** a number	$x - 3$
a number **decreased by** 4	$x - 4$
10 **minus** a number	$10 - x$

Note

Recall that addition can be done in any order, so $x + 2$ gives the same result as $2 + x$. This is *not* true in subtraction, so be careful. $10 - x$ does *not* give the same result as $x - 10$.

WORK PROBLEM 1 AT THE SIDE.

EXAMPLE 2 Translating Word Phrases into Algebraic Expressions

Write in symbols by using x as the variable.

WORDS	ALGEBRA
8 **times** a number	$8x$
the **product** of 12 and a number	$12x$
double a number (meaning "2 times")	$2x$
the **quotient** of $^{-}6$ and a number	$\frac{^{-}6}{x}$
a number **divided by** 10	$\frac{x}{10}$
15 **subtracted from** 4 **times** a number	$4x - 15$
the result **is**	$=$

WORK PROBLEM 2 AT THE SIDE.

OBJECTIVES

1. Translate word phrases into algebraic expressions.
2. Translate sentences into equations.
3. Solve application problems with one unknown quantity.

FOR EXTRA HELP

Tutorial | Tape 4 | SSM, Sec. 3.3

1. Write in symbols by using x as the variable.

(a) 15 less than a number

(b) 12 more than a number

(c) a number increased by 13

(d) a number minus 8

(e) 10 plus a number

(f) a number subtracted from 6

(g) 6 subtracted from a number

2. Write in symbols by using x as the variable.

(a) double a number

(b) the product of $^{-}8$ and a number

(c) the quotient of 15 and a number

(d) 5 times a number subtracted from 30

ANSWERS

1. (a) $x - 15$ **(b)** $x + 12$ or $12 + x$ **(c)** $x + 13$ or $13 + x$ **(d)** $x - 8$ **(e)** $10 + x$ or $x + 10$ **(f)** $6 - x$ **(g)** $x - 6$

2. (a) $2x$ **(b)** $^{-}8x$ **(c)** $\frac{15}{x}$ **(d)** $30 - 5x$ (**not** $5x - 30$)

3. Translate each sentence into an equation and solve it. Check your solution by going back to the words in the original problem.

(a) If 3 times a number is added to 4, the result is 19. Find the number.

(b) If $^{-}6$ times a number is added to 5, the result is $^{-}13$. Find the number.

ANSWERS

3. (a) $3x + 4 = 19$

$x = 5$

Check $\underbrace{3 \cdot 5}_{15} + 4$ does equal 19; $\underbrace{15 + 4}_{19}$

(b) $^{-}6x + 5 = {}^{-}13$

$x = 3$

Check $\underbrace{^{-}6 \cdot 3}_{^{-}18} + 5$ does equal $^{-}13$; $\underbrace{^{-}18 + 5}_{^{-}13}$

OBJECTIVE 2 The next example shows you how to translate a sentence into an equation that you can solve.

EXAMPLE 3 Translating Sentences into Equations

If 5 times a number is added to 11, the result is 26. Find the number.

Let x represent the unknown number.

Use the information in the problem to write an equation.

$$\underbrace{\text{5 times a number}}_{5x}\ \underbrace{\text{added to}}_{+}\ \underset{11}{\text{11}}\ \underset{=}{\text{is}}\ \underset{26}{\text{26.}}$$

Next, solve the equation.

$$\begin{aligned} 5x + 11 &= 26 \\ {}^{-}11 &\quad {}^{-}11 \\ \underbrace{5x + 0} &= 15 \\ \frac{5x}{5} &= \frac{15}{5} \\ x &= 3 \end{aligned}$$

To get $5x$ by itself, add $^{-}11$ to both sides.

To get x by itself, divide both sides by 5.

The number is 3.

Check Go back to the words of the original problem.

$$\underset{5}{\text{If 5}}\ \underset{\cdot}{\text{times}}\ \underbrace{\text{a number}}_{3}\ \underbrace{\text{is added to}}_{+}\ \underset{11}{\text{11,}}\ \underbrace{\text{the result is}}_{=}\ \underset{26}{\text{26.}}$$

Does $5 \cdot 3 + 11$ really equal 26? Yes it does. So 3 is the correct solution because it "works" when you put it back into the original problem.

WORK PROBLEM 3 AT THE SIDE.

OBJECTIVE 3 Now you are ready to tackle application problems. The steps you will use are summarized below.

Solving Application Problems

Step 1 Read the problem once to see what it is about. Read it carefully a second time. As you read, make a sketch or write word phrases that identify the known and the unknown parts of the problem.

Step 2(a) If there is one unknown quantity, choose a variable to represent it. Write down what your variable represents.

Step 2(b) If there is more than one unknown quantity, choose a variable to represent "the thing you know the least about." Then write variable expression(s) to show the relationship of the other unknown quantities to the first one.

Step 3 Write an equation, using your sketch or word phrases as the guide.

Step 4 Solve the equation.

Step 5 Answer the question in the problem and label your answer(s).

Step 6 Go back to the original statement of the problem. Check whether your answer fits all the facts given in the problem. If it does, you are done. If it doesn't, start again at Step 1.

EXAMPLE 4 Solving Application Problems with One Unknown Quantity

Heather had put some money aside in an envelope for household expenses. Yesterday she took out \$20 for groceries. Today a friend paid back a loan and Heather put the \$34 in the envelope. Now she has \$43 in the envelope. How much was in the envelope at the start?

Step 1 Read the problem once. It is about money in an envelope. Read it a second time and write word phrases.

Unknown: amount of money in the envelope at the start.
Known: took out \$20; put in \$34; ended up with \$43.

Step 2(a) There is only one unknown quantity. Let m represent the money at the start.

Step 3 Write an equation, using the phrases you wrote as a guide.

money at the start	took out \$20	put in \$34	ended up with \$43
m	$-$ \$20	+ \$34	= \$43

Step 4 Solve the equation.

$$\begin{aligned} m - 20 + 34 &= 43 && \text{Change subtraction to adding the opposite.} \\ m + {}^{-}20 + 34 &= 43 && \text{Simplify the left side.} \\ m + 14 &= 43 && \text{To get } m \text{ by itself, add } {}^{-}14 \text{ to both sides.} \\ {}^{-}14 & \quad {}^{-}14 \\ m + 0 &= 29 \\ m &= 29 \end{aligned}$$

Step 5 Answer the question, "How much was in the envelope at the start?" **Solution: \$29**

Step 6 **Check** the solution by going back to the original problem and inserting the solution.

Started with \$29 in the envelope.
Took out \$20, so \$29 − \$20 = \$9 in the envelope.
Put in \$34, so \$9 + \$34 = \$43
Now has \$43. ← matches ↑

The correct solution is \$29 because it "works" when put back in the original problem.

WORK PROBLEM 4 AT THE SIDE.

EXAMPLE 5 Solving Application Problems with One Unknown Quantity

Three friends each put in the same amount of money to buy a gift. After they spent \$2 for a card and \$31 for the gift, they had \$6 left. How much money had each friend put in originally?

Step 1 The problem is about 3 friends buying a gift.

Unknown: amount of money each friend contributed.
Known: 3 friends put in money; spent \$2 and \$31; had \$6 left.

Step 2(a) There is only one unknown quantity. Let m represent the amount of money each friend contributed.

CONTINUED ON NEXT PAGE

4. Some people got on an empty bus at its first stop. At the second stop, 3 people got on. At the third stop, 5 more people got on. At the fourth stop, 10 people got off, but 4 people were still on the bus. How many people got on at the first stop? Show your work for each of the six problem-solving steps.

ANSWER

4. *Step 1*
Unknown: number of people who got on at first stop.
Known: 3 got on; 5 got on; 10 got off; 4 people still on bus.

Step 2(a)
Let p be people who got on at first stop. (You may use any letter you like as the variable.)

Step 3 $p + 3 + 5 - 10 = 4$

Step 4
$$\begin{aligned} p + 3 + 5 + {}^{-}10 &= 4 \\ p + {}^{-}2 &= 4 \\ {}^{+}2 & \quad {}^{+}2 \\ p + 0 &= 6 \\ p &= 6 \end{aligned}$$

Step 5
6 people got on at the first stop.

Step 6
6 got on at first stop.
3 got on at 2nd stop: 6 + 3 = 9
5 got on at 3rd stop: 9 + 5 = 14
10 got off at 4th stop: 14 − 10 = 4
4 people are left. ← matches ↑

5. Five donors each gave the same amount of money to a college to use for scholarships. From the money, scholarships of $1250, $900, and $850 were given to students; $250 was left. How much money did each donor give to the college? Show your work for each of the six problem-solving steps.

Step 3

number of friends		amount each friend put in		spent on card		spent on gift		left over
3	•	m	−	\$2	−	\$31	=	\$6

To see why this is multiplication, think of an example. If each friend put in \$10, how much money would there be? 3 • \$10, or \$30

Step 4

$3m - 2 - 31 = 6$ Change subtractions to adding the opposite.

$3m + {}^{-}2 + {}^{-}31 = 6$ Simplify the left side.

$3m + {}^{-}33 = 6$ To get $3m$ by itself, add 33 to both sides.

$ {}^{+}33 \quad {}^{+}33$

$3m + 0 = 39$

$\dfrac{3m}{3} = \dfrac{39}{3}$ To get m by itself, divide both sides by 3.

$m = 13$

Step 5 Each friend put in \$13.

Step 6 **Check** the solution by putting it back in the original problem.

3 friends each put in \$13, so 3 • \$13 = \$39
Spent \$2, spent \$31, so \$39 − \$2 − \$31 = \$6
Had \$6 left ← matches ↑

\$13 is the correct solution because it "works."

◀◀ **WORK PROBLEM 5 AT THE SIDE.**

EXAMPLE 6 Solving More Complex Application Problems with One Unknown Quantity

Michael has completed 5 less than three times as many lab experiments as David. If Michael has completed 13 experiments, how many experiments has David completed?

Step 1 The problem is about the number of experiments done by two students.

Unknown: number of experiments David did.
Known: Michael did 5 less than 3 times the number David did; Michael did 13.

Step 2(a) Let n represent the number of experiments David did.

Step 3

The number Michael did	is	5 less than 3 times David's number.
13	=	$3n - 5$

Step 4

$13 = 3n - 5$ Change subtraction to adding the opposite.

$13 = 3n + {}^{-}5$ To get $3n$ by itself, add 5 to both sides.

${}^{+}5 \qquad {}^{+}5$

$18 = 3n + 0$

$\dfrac{18}{3} = \dfrac{3n}{3}$ To get n by itself, divide both sides by 3.

$6 = n$

CONTINUED ON NEXT PAGE

ANSWER

5. *Step 1*
Unknown: money given by each donor.
Known: 5 donors; gave out \$1250, \$900, \$850; \$250 left.

Step 2(a)
Let m be each donor's money.

Step 3
$5 \cdot m - \$1250 - \$900 - \$850 = \$250$

Step 4
$5m + {}^{-}1250 + {}^{-}900 + {}^{-}850 = 250$
$5m + {}^{-}3000 = 250$
$ {}^{+}3000 \quad {}^{+}3000$
$5m + 0 = 3250$
$\dfrac{5m}{5} = \dfrac{3250}{5}$
$m = 650$

Step 5 Each donor gave \$650.

Step 6 5 donors each gave \$650, so
5 • \$650 = \$3250
Gave out \$1250, \$900, \$850, so
\$3250 − 1250 − 900 − 850 = \$250
Had \$250 left ← matches ↑

Step 5 David did 6 experiments.

Step 6 **Check** the solution by putting it back in the original problem.

3 times David's number $3 \cdot 6 = 18$
less 5 $18 - 5 = 13$
Michael did 13 ← matches → (13)

The correct solution is: David did 6 experiments.

WORK PROBLEM 6 AT THE SIDE.

6. Susan donated \$10 more than twice what LuAnn donated. If Susan donated \$22, how much did LuAnn donate?

Show your work for each of the six problem-solving steps.

ANSWER

6. *Step 1*
Unknown: LuAnn's donation.
Known: Susan donated \$10 more than twice what LuAnn donated; Susan donated \$22.

Step 2(a)
Let d be LuAnn's donation.

Step 3
$2d + \$10 = \$22$
(or $\$10 + 2d = \$22$)

Step 4

$$\begin{aligned} 2d + 10 &= 22 \\ -10 &\quad -10 \\ 2d + 0 &= 12 \\ \frac{2d}{2} &= \frac{12}{2} \\ d &= 6 \end{aligned}$$

Step 5
LuAnn donated \$6.

Step 6
\$10 more than twice \$6 is
$\$10 + 2 \cdot \$6 = \$22.$ ←
Susan donated \$22. ← matches

NUMBERS IN THE *Real World*

collaborative investigations

Formulas

In **Section 3.1**, you found the perimeter of a triangle by adding the lengths of the three sides. A formula for this is shown here.

$$P = a + b + c$$

where P is the perimeter and a, b, and c are the lengths of the sides.

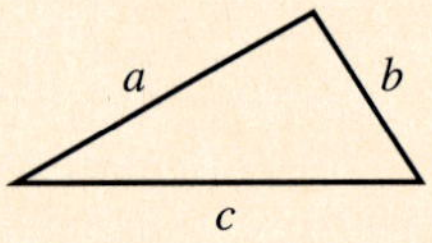

1. Use the formula to find the length of the third side in this triangle. Replace P with 33 ft, replace a with 12 ft, and replace b with 7 ft. Then solve the equation.

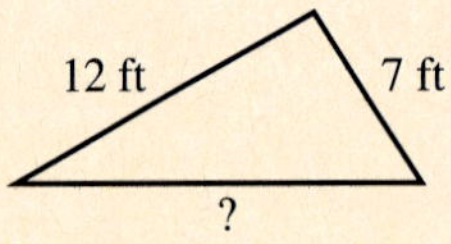

Perimeter is 33 ft.

Use the formula to find the length of the third side in each of these triangles.

2.

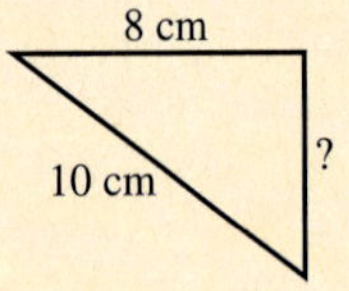

Perimeter is 24 cm.

3.

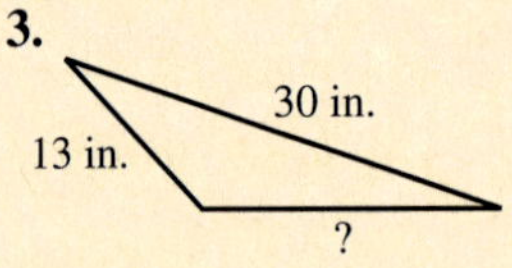

Perimeter is 64 in.

4. Describe *in words* what you were doing as you found the length of the third side. What things were being added? What were you doing with the perimeter?

5. Bonus Question: How can you simplify the equation when *two* sides of the triangle are the *same* length?

Use the simplified equation to find the unknown lengths in these triangles.

6.

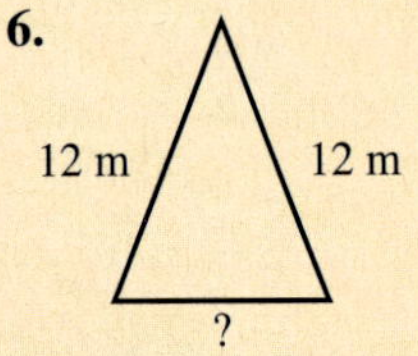

Perimeter is 33 m.

7.

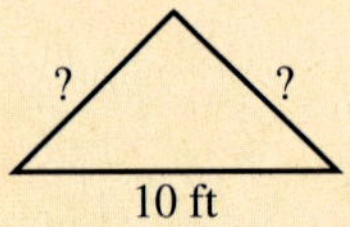

Perimeter is 24 ft.

A formula that has many uses is $d = rt$, called the distance formula. If you are driving a car, then

d is the *distance* you travel (how many miles)
r is the *rate* (how fast you are driving)
t is the *time* (how many hours you drive)

For example, suppose you are driving at a rate of 65 miles per hour. If you drive 3 hours, how far will you have traveled? Use the formula.

$$d = rt$$
$$d = (65)(3)$$
$$d = 195$$

You will have traveled 195 miles.

Now use the formula to solve these problems.

8. You traveled 232 miles. If you were driving at a rate (speed) of 58 miles per hour, how long did you drive?

9. You drove for five hours and traveled 315 miles. At what rate were you driving?

In Chapter 2 you worked with an expression for the vocabulary of a child. Now we can use this formula.

$$V = 60A - 900$$

where V is the number of vocabulary words and A is the age of the child in months.

Use the formula to find the age of each child.

10. The child's vocabulary is 180 words.

11. The child's vocabulary is 1440 words.

12. The child's vocabulary is 60 words.

13. The child's vocabulary is 0 words.

14. Does the answer to Question 13 make sense? Are there other ages that would also have a vocabulary of 0 words?

15. Describe *in words* what you were doing as you found each child's age.

3.3 Exercises

Write an algebraic expression, using x as the variable.

1. 14 plus a number

2. the sum of a number and $^{-}8$

3. $^{-}5$ added to a number

4. 16 more than a number

5. 20 minus a number

6. a number decreased by 25

7. 9 less than a number

8. a number subtracted from $^{-}7$

9. Subtract 4 from a number.

10. 3 fewer than a number

11. $^{-}6$ times a number

12. the product of $^{-}3$ and a number

13. double a number

14. a number times 10

15. a number divided by 2

16. 4 divided by a number

17. twice a number added to 8

18. five times a number plus 5

19. 10 fewer than seven times a number

20. 12 less than six times a number

21. the sum of twice a number and the number

22. triple a number subtracted from the number

Translate each sentence into an equation and solve it. Check your solution by going back to the words in the original problem.

23. If four times a number is decreased by 2, the result is 26. Find the number.

24. The sum of 8 and five times a number is 53. Find the number.

25. If a number is added to twice a number, the result is $^{-}15$. What is the number?

26. If a number is subtracted from three times the number, the result is $^{-}8$. What is the number?

27. If the product of some number and 5 is increased by 12, the result is seven times the number. Find the number.

28. If eight times a number is subtracted from eleven times the number, the result is $^{-}9$. Find the number.

29. When three times a number is subtracted from 30, the result is 2 plus the number. What is the number?

30. When twice a number is decreased by 8, the result is the number increased by 7. Find the number.

Solve each application problem. Use the six problem-solving steps you learned in this section.

31. Ricardo gained 15 pounds over the winter. He went on a diet and lost 28 pounds. Then he regained 5 pounds and weighed 177 pounds. How much did he weigh originally?

32. Mr. Chee deposited $80 into his checking account. He then wrote a $23 check for gas and a $90 check for his child's day care. The balance in his account was $67. How much was in his account before he made the deposit?

33. There were 18 cookies in Magan's cookie jar. While she was busy in another room, her children ate some of the cookies. Magan bought three dozen cookies and added them to the jar. At that point she had 49 cookies in the jar. How many cookies did her children eat?

34. The Greens had a 20-pound bag of bird seed in their garage. Mice got into the bag and ate some of it. They bought an 8-pound bag of seed and put all the seed in a metal container. They now have 24 pounds of seed. How much did the mice eat?

35. A college bookstore ordered six boxes of red pens. The store sold 32 red pens last week and 35 red pens this week. Five pens were left on the shelf. How many pens were in each box?

36. The manager of an apartment complex had 11 packages of lightbulbs on hand. He replaced 29 burned out bulbs in hallway lights and 7 bulbs in the party room. Eight bulbs were left. How many bulbs were in each package?

37. The 14 music club members each paid the same amount for dues. The club also earned $340 selling magazine subscriptions. They spent $575 to organize a jazz festival. Now their bank account is overdrawn by $25. How much did each member pay in dues?

38. A local charity received a donation of eight cartons filled with cans of soup. The charity gave out 100 cans of soup yesterday and 92 cans today before running out. How many cans were in each carton?

39. When 75 is subtracted from four times Tamu's age, the result is Tamu's age. How old is Tamu?

40. If three times Linda's age is decreased by 36, the result is twice Linda's age. How old is Linda?

41. While shopping for clothes, Consuelo spent \$3 less than twice what Brenda spent. Consuelo spent \$81. How much did Brenda spend?

42. Dennis weighs 184 pounds. His weight is 2 pounds less than six times his child's weight. How much does his child weigh?

43. Paige bought five bags of candy for Halloween. Forty-eight children visited her home and she gave each child three pieces of candy. At the end of the night she still had one bag of candy. How many pieces of candy were in each bag?

44. A restaurant ordered four packages of paper napkins. Yesterday they used up one package, and today they used up 140 napkins. Two packages plus 60 napkins remain. How many napkins are in each package?

Review and Prepare

Solve each equation. Show the steps you use. (For help, see ***Section 2.5.****)*

45. $k + 45 + k = {}^{-}55 + 800$

46. $1000 = a + 100 + a - 300 + a$

47. $82 = 2(n + 3) + 2n$

48. $3(n - 4) + 6 = {}^{-}18$

3.4 *Application Problems with Two Unknown Quantities*

OBJECTIVE

1 Solve application problems with two unknown quantities.

FOR EXTRA HELP

Tutorial | Tape 4 | SSM, Sec. 3.4

OBJECTIVE 1 In the preceding section, the problems had only one unknown quantity. As a result, we used Step 2(a) rather than Step 2(b) in the problem solving steps. For easy reference, we repeat them here.

Solving Application Problems

Step 1 Read the problem once to see what it is about. Read it carefully a second time. As you read, make a sketch or write word phrases that identify the known and the unknown parts of the problem.

Step 2(a) If there is one unknown quantity, choose a variable to represent it. Write down what your variable represents.

Step 2(b) If there is more than one unknown quantity, choose a variable to represent "the thing you know the least about." Then write variable expression(s) to show the relationship of the other unknown quantities to the first one.

Step 3 Write an equation, using your sketch or word phrases as the guide.

Step 4 Solve the equation.

Step 5 Answer the question in the problem and label your answer(s).

Step 6 Go back to the original statement of the problem. Check whether your answer fits all the facts given in the problem. If it does, you are done. If it doesn't, start again at Step 1.

Now you are ready to solve problems with two unknown quantities.

EXAMPLE 1 Solving Application Problems with Two Unknown Quantities

Last month, Sheila worked 72 hours more than Russell. Together they worked a total of 232 hours. Find the number of hours each person worked last month.

Step 1 The problem is about the number of hours worked by Sheila and by Russell.

Unknowns: hours worked by Sheila;
hours worked by Russell.

Known: Sheila worked 72 hours more than Russell;
232 hours total for Sheila and Russell.

Step 2(b) There are *two* unknowns. You know the *least* about the hours worked by Russell, so let h represent Russell's hours.

Sheila worked 72 hours more than Russell, so her hours are $h + 72$, that is, Russell's hours (h) plus 72 more.

Step 3

hours worked by Russell		hours worked by Sheila		total hours worked
h	$+$	$h + 72$	$=$	232

CONTINUED ON NEXT PAGE

1. In a day of work, Keonda made \$12 more than her daughter. Together they made \$182. Find the amount that each person made. (*Hint:* Which amount do you know the *least* about, Keonda's or her daughter's? Let m be that amount.) Use the six problem-solving steps.

Step 4

$$\underbrace{h + h} + 72 = 232$$
Simplify the left side by combining like terms.

$$2h + 72 = 232$$
$$\underline{-72 \quad -72}$$
To get $2h$ by itself, add $^-72$ to both sides.

$$\underbrace{2h + 0} = 160$$

$$\frac{2h}{2} = \frac{160}{2}$$
To get h by itself, divide both sides by 2.

$$h = 80$$

Step 5 Because h represents Russell's hours, and the solution of the equation is $h = 80$, Russell worked 80 hours.

h + 72 represents Sheila's hours. Replace h with 80.
↓
80 + 72 = 152, so Sheila worked 152 hours.

The final answer is: Russell worked 80 hours and Sheila worked 152 hours.

Step 6 **Check** the solution by putting both numbers back in the original problem.

"Sheila worked 72 hours more than Russell."
Sheila's 152 hours are 72 more than Russell's 80 hours, so the solution checks.

$$\begin{array}{r} 152 \\ -\ 72 \\ \hline 80 \end{array}$$

"Together they worked a total of 232 hours."
Sheila's 152 hours + Russell's 80 hours = 232 hours so the solution checks.

$$\begin{array}{r} 152 \\ +\ 80 \\ \hline 232 \end{array}$$

We've answered the question correctly because 80 hours and 152 hours fit all the facts given in the problem.

Note
Check the solution to an application problem by putting the numbers back in the *original* problem. If they do *not* work, solve the problem a different way.

◀◀ WORK PROBLEM 1 AT THE SIDE.

EXAMPLE 2 Solving a Geometry Application with Two Unknown Quantities

The length of a rectangle is 2 cm more than the width. The perimeter is 68 cm. Find the length and width.

Step 1 The problem is about a rectangle. Make a sketch of a rectangle.

Unknown(s): length of the rectangle; width of the rectangle
Known: the length is 2 cm more than the width; the perimeter is 68 cm

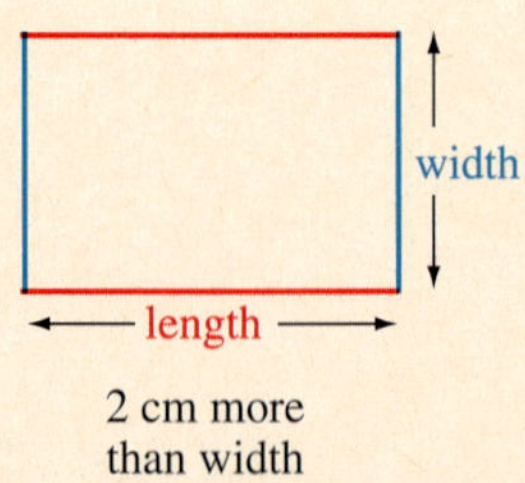

CONTINUED ON NEXT PAGE

ANSWER

1. Daughter made m.
Keonda made $m + 12$.
$m + m + 12 = 182$
Daughter made \$85.
Keonda made \$97.
Check \$97 − \$85 = \$12 and \$97 + \$85 = \$182.

Step 2(b) There are *two* unknowns. You know the least about the width, so let x represent the **width**.

The length is 2 cm more than the width, so the **length** is $x + 2$.

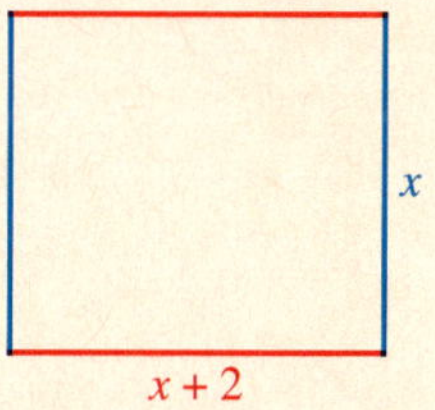

Step 3 Use the formula for perimeter of a rectangle, $P = 2l + 2w$ to help you write the equation.

$$P = 2\,l + 2\,w$$
$$68 = 2(x + 2) + 2 \cdot x$$

Replace l with $(x + 2)$. Replace w with x.

Step 4

$$68 = 2(x + 2) + 2x$$ Use the distributive property.

$$68 = 2x + 4 + 2x$$ Combine like terms.

$$68 = 4x + 4$$ To get $4x$ by itself, add $^-4$ to both sides.

$$\underline{^-4 \qquad\quad ^-4}$$
$$64 = 4x + 0$$

$$\frac{64}{4} = \frac{4x}{4}$$ To get x by itself, divide both sides by 4.

$$16 = x$$

Step 5 x represents the width, and $x = 16$, so the width is 16 cm.

$x + 2$ represents the length. Replace x with 16.

$16 + 2 = 18$, so the length is 18 cm. ← The label for both answers is cm.

Step 6 **Check** the solution by putting the measurements on your sketch and going back to the original problem.

"The length of a rectangle is 2 cm more than the width."

18 cm is 2 cm more than 16 cm, so the solution checks.

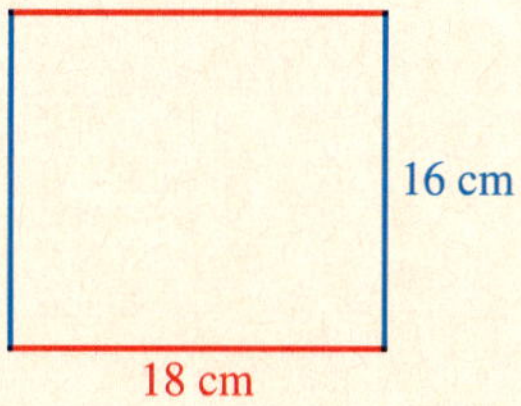

"The perimeter is 68 cm."

$$P = \underbrace{2 \cdot 18 \text{ cm}} + \underbrace{2 \cdot 16 \text{ cm}}$$
$$P = 36 \text{ cm} + 32 \text{ cm}$$
$$P = 68 \text{ cm}$$

Matches perimeter given in the original problem, so the solution checks.

WORK PROBLEM 2 AT THE SIDE.

2. Make a drawing to help solve this problem. The length of Ann's rectangular garden plot is 3 yd more than the width. She used 22 yd of fencing around the edge. Find the length and the width of the garden, using the six problem-solving steps.

ANSWER

2.

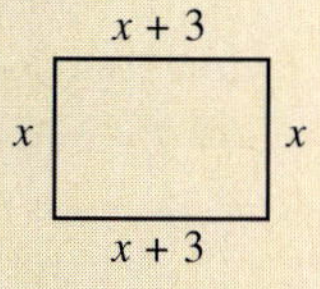

width is x
length is $x + 3$

$22 = 2(x + 3) + 2 \cdot x$
width is 4 yd
length is 7 yd
Check 7 yd is 3 yd more than 4 yd.
$P = 2 \cdot 7$ yd $+ 2 \cdot 4$ yd $= 22$ yd
Matches perimeter given in the original problem.

Study Tips

Taking a Test

The study tips in Chapter 2 talked about how to prepare for a test. But even with good preparation, you may not do as well as you could on the actual test. Here are some ideas about ways to maximize your test score. Think about the last test you took and answer each question honestly.

During the Test

Did you

Yes	No	Write memorized formulas and key ideas on a corner of the test paper before starting to work the problems?
Yes	No	Scan the entire test and plan your time accordingly?
Yes	No	Work easy problems first, to build confidence?
Yes	No	Replace negative self-talk with positive self-talk?
Yes	No	Read all directions carefully and circle significant words?
Yes	No	Leave problems that you couldn't work and come back to them?
Yes	No	Show your work, neatly, so you could earn partial credit?
Yes	No	Check that answers to application problems were reasonable and made sense?
Yes	No	*Rework* problems (not just look at them) to catch careless errors?
Yes	No	Change your answers *only* when you could find and explain the error?
Yes	No	Close your eyes and take a few deep breaths as soon as the tension began to build?
Yes	No	Ignore students who finished early?

Did You Know...

- If your writing becomes larger, smaller, or you are pressing down harder, you may be getting anxious. Anxiety causes shallow breathing which leads to confusion and inability to concentrate. Deep breathing will calm you down.
- If you are stuck on a problem, write down *anything* you can think of that might relate to the problem. The answer may come to you as you work on it, or at least you may get partial credit. But don't spend too long on one problem. Your subconscious mind will work on it while you go on to the rest of the test.
- If you just "look over" your work, your mind can easily make the same mistake again without noticing it. Reworking the problem from the beginning forces you to rethink it. If possible, use a different method to solve the problem the second time.

Make a Plan

How will *you* improve the way you take tests? Look at each question that you answered "No." Pick two or three of them that you can change to a "Yes" on the next test you take. Write down your plans.

3.4 Exercises

Solve each application problem by using the six problem-solving steps you learned in this section.

1. My sister is 9 years older than I am. The sum of our ages is 51. Find our ages.

2. Ed and Marge were candidates for city council. Marge won, with 93 more votes than Ed. The total number of votes cast in the election was 587. Find the number of votes received by each candidate.

3. Last year, Lien earned \$1500 more than her husband. Together they earned \$37,500. How much did each of them earn?

4. A \$149,000 estate is to be divided between two charities so that one charity receives \$18,000 less than the other. How much will each charity receive?

5. Jason paid five times as much for his computer as he did for his printer. He paid a total of \$1320 for both items. What did each item cost?

6. The attendance at the Saturday night baseball game was three times the attendance at Sunday's game. In all, 56,000 fans attended the games. How many fans were at each game?

7. A board is 78 cm long. Rosa cut the board into two pieces, with one piece 10 cm longer than the other. Find the length of both pieces. (*Hint:* Make a sketch of the board.)

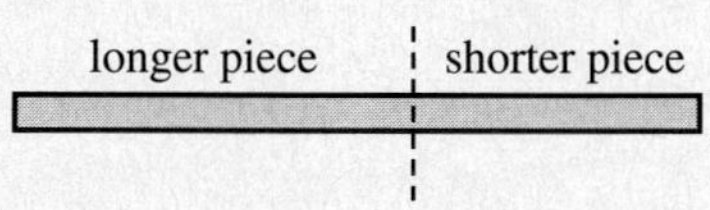

Which piece do you know the least about? Let x be the length of that piece.

8. A rope is 21 m long. Marcos cut it into two pieces so that one piece is 3 m longer than the other. Find the length of each piece.

9. A wire is cut into two pieces, with one piece 7 ft shorter than the other. The wire was 31 ft long before it was cut. How long was each piece?

10. A 90 cm pipe is cut into two pieces so that one piece is 6 cm shorter than the other. Find the length of each piece.

G **11.** A fence is 706 m long. It is to be cut into three parts. Two parts are the same length, and the third part is 25 m longer than each of the other two. Find the length of each part.

G **12.** A wooden railing is 82 m long. It is to be divided into four pieces. Three pieces will be the same length, and the fourth piece will be 2 m longer than each of the other three. Find the length of each piece.

In Exercises 13–18 use the formula for the perimeter of a rectangle, $P = 2l + 2w$. Make a sketch to help you solve each problem.

13. The perimeter of a rectangle is 48 m. The width is 5 m. Find the length.

14. The length of a rectangle is 27 cm, and the perimeter is 74 cm. Find the width of the rectangle.

15. A rectangular dog pen is twice as long as it is wide. The perimeter of the pen is 36 ft. Find the length and the width of the pen.

16. A new city park is a rectangular shape. The length is triple the width. It will take 240 meters of fencing to go around the park. Find the length and width of the park.

17. The length of a rectangular jewelry box is 3 inches more than twice the width. The perimeter is 36 inches. Find the length and the width.

18. The perimeter of a rectangular house is 122 ft. The width is 5 ft less than the length. Find the length and the width.

G **19.** A photograph measures 8 in. by 10 in. Earl put it in a frame that is 2 in. wide. Find the area of the frame and its outside perimeter. Make a sketch.

G **20.** Barb had a 16 in. by 20 in. photograph. She cropped 3 in. off each side of the photo. What is the perimeter and the area of the cropped photo? Make a sketch.

Review and Prepare

For help, see ***Section 1.2***.

21. Graph these numbers on the number line: $^-1$, 0, $^-3$, 1.

22. Find each absolute value.
(a) $|^-8|$ **(b)** $|0|$ **(c)** $|16|$ **(d)** $|2 - 9|$

KEY TERMS

3.1	**formula**	Formulas are well-known rules for solving common types of problems. They are written in a shorthand form that uses variables.
	perimeter	Perimeter is the distance around the outside edges of a flat shape. It is measured in linear units of in., ft, yd, mm, cm, m, and so on.
	square	A square is a figure with four sides that are all the same length and meet to form 90° angles (square corners).
	rectangle	A rectangle is a four-sided figure with all sides meeting at 90° angles. The opposite sides are the same length.
	parallelogram	A parallelogram is a four-sided figure with both pairs of opposite sides parallel.
	triangle	A triangle is a figure with exactly three sides.
3.2	**area**	Area is the surface inside a two-dimensional (flat) shape. It is measured by determining the number of squares of a certain size needed to cover the surface inside the shape. Some of the commonly used units for measuring area are square inches (in.^{2}), square feet (ft^2), square yards (yd^2), square centimeters (cm^2), and square meters (m^2).

KEY FORMULAS

Perimeter of a square: $P = 4s$
Perimeter of a rectangle: $P = 2l + 2w$
Area of a rectangle: $A = lw$
Area of a square: $A = s^2$
Area of a parallelogram: $A = bh$

QUICK REVIEW

Concepts	*Examples*
3.1 Finding Perimeter To find the perimeter of *any* shape, add the lengths of the sides. Perimeter is measured in linear units (cm, m, ft, yd, and so on). Or, for some shapes, there are formulas that you can use. Perimeter of a square: $P = 4s$ Perimeter of a rectangle: $P = 2l + 2w$	Find the perimeter of each figure. 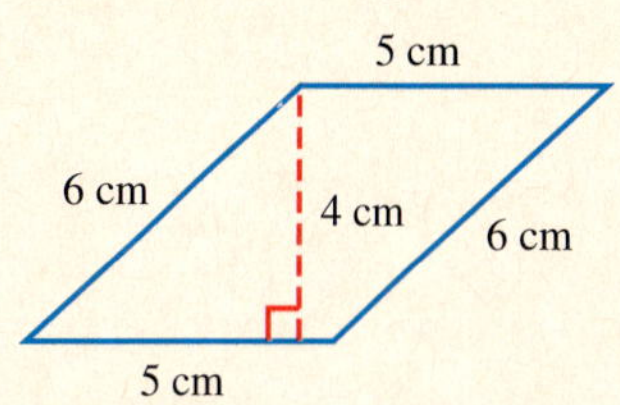 $P = 5\text{ cm} + 6\text{ cm} + 5\text{ cm} + 6\text{ cm}$ $P = 22\text{ cm}$ 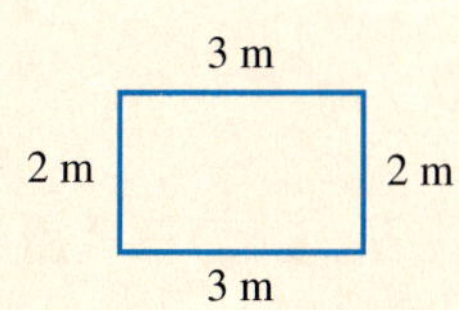 $P = 2 \cdot l + 2 \cdot w$ $P = 2 \cdot 3\text{ m} + 2 \cdot 2\text{ m}$ $P = 6\text{ m} + 4\text{ m}$ $P = 10\text{ m}$
3.1 Finding the Length of One Side of a Square If you know the perimeter of a square, use the formula $P = 4s$. Replace P with the value for the perimeter and solve the equation for s.	If the perimeter of a square room is 44ft, find the length of one side. $P = 4s$ Replace P with 44 ft. $44\text{ ft} = 4s$ $\frac{44\text{ ft}}{4} = \frac{4s}{4}$ Divide both sides by 4. $11\text{ ft} = s$ The length of one side is 11 ft.

Concepts	*Examples*

3.1 Finding the Length or Width of a Rectangle

Concepts: If you know the perimeter of a rectangle and either its width or length, use the formula $P = 2l + 2w$. Replace P and either l or w with the values that you know. Then solve the equation.

Examples: The width of a rectangular rug is 8 ft. The perimeter is 36 ft. Find the length.

$$P = 2l + 2w$$ Replace P with 36 ft and w with 8 ft.

$$36 \text{ ft} = 2l + 2 \cdot 8 \text{ ft}$$

$$36 \text{ ft} = 2l + 16 \text{ ft}$$ Add $^-16$ ft to both sides.

$$\underline{^-16 \text{ ft}} \qquad \underline{^-16 \text{ ft}}$$

$$20 \text{ ft} = 2l + 0$$

$$\frac{20 \text{ ft}}{2} = \frac{2l}{2}$$ Divide both sides by 2.

$$10 \text{ ft} = l$$ The length is 10 ft.

3.2 Finding Area

Concepts: Use the appropriate formula. Remember to measure area in square units (cm^2, m^2, ft^2, yd^2, and so on).

Area of a rectangle: $A = lw$,
where l is the length and w is the width.

Area of a square: $A = s^2$ which means $s \cdot s$,
where s is the length of one side.

Area of a parallelogram: $A = bh$,
where b is the base and h is the height.

Examples: Find the area of each figure.

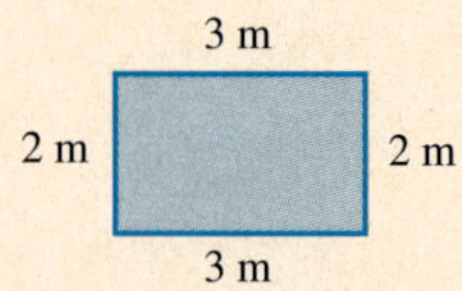

$A = lw$
$A = 3 \text{ m} \cdot 2 \text{ m}$
$A = 6 \text{ m}^2$

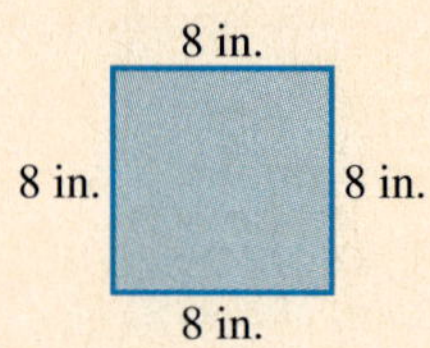

$A = s^2$
$A = s \cdot s$
$A = 8 \text{ in.} \cdot 8 \text{ in.}$
$A = 64 \text{ in.}^2$

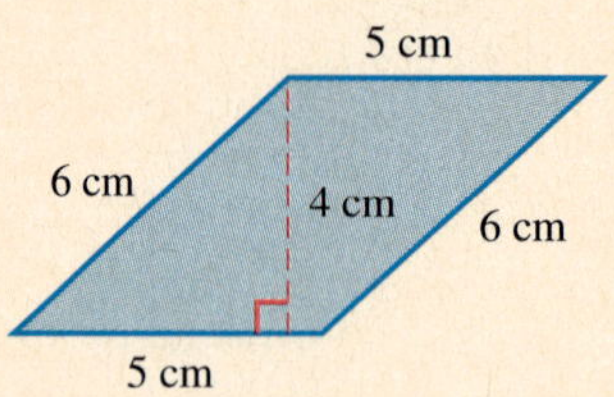

$A = bh$
$A = 5 \text{ cm} \cdot 4 \text{ cm}$
$A = 20 \text{ cm}^2$

3.2 Finding the Unknown Length in a Rectangle or Parallelogram

Concepts: If you know the area of a rectangle or parallelogram and one of the measurements, use the appropriate area formula (see above). Replace A and one of the other variables with the values that you know. Then solve the equation.

Examples: The area of a parallelogram is 72 yd^2, and its height is 9 yd. Find the base.

$$A = bh$$ Replace A with 72 yd^2 and h with 9 yd.

$$72 \text{ yd}^2 = b \cdot 9 \text{ yd}$$

$$\frac{72 \text{ yd} \cdot \cancel{\text{yd}}}{9\ \cancel{\text{yd}}} = \frac{b \cdot 9 \text{ yd}}{9 \text{ yd}}$$ Divide both sides by 9 yd.

$$8 \text{ yd} = b$$

The base of the parallelogram is 8 yd.

Concepts	*Examples*
3.2 Finding the Length of One Side of a Square If you know the area of a square, use the formula $A = s^2$. Replace A with the value that you know. Then solve the equation by asking, "What number, times itself, gives the value of A?"	A square ceiling has an area of 100 ft^2. What is the length of each side of the ceiling? $A = s^2$ Replace A with 100 ft^2. $\mathbf{100\ ft^2} = s^2$ Rewrite s^2 as $s \cdot s$. $100\ ft^2 = s \cdot s$ Ask, "What number times itself gives 100?" $100\ ft^2 = 10\ ft \cdot 10\ ft$ Each side of the ceiling is 10 ft long.
3.3 Translating Sentences into Equations Use x (or any other letter) as the variable to symbolize the operations described by the words in the sentence. Then solve the equation. Check the solution by putting it back in the original problem.	If 10 is subtracted from three times a number, the result is 14. Let x represent the unknown number. $3x - 10 = 14$ $3x + {}^-10 = 14$ Add 10 to both sides. $\quad\ {}^+10 \quad {}^+10$ $3x + 0 = 24$ $\frac{3x}{3} = \frac{24}{3}$ Divide both sides by 3. $x = 8$ The number is 8. **Check:** If 10 is subtracted from three times 8, do you get 14? Yes, $3 \cdot 8 - 10 = 14$.
3.3 Solving Application Problems with One Unknown Quantity Use the six problem-solving steps outlined in Section 3.3. In abbreviated form they are:	Denise had some money in her purse this morning. She gave \$15 to her daughter and paid \$4 to park in the lot at work. At that point she still had \$27. How much was in her purse this morning?
Step 1 Identify what is known and what is unknown.	*Step 1* Unknown: money in purse this morning. Known: took out \$15; took out \$4; still had \$27.
Step 2 Choose a variable to represent the unknown quantity.	*Step 2(a)* Let m represent the money in her purse this morning.
Step 3 Write an equation.	*Step 3* $m - \$15 - \$4 = \$27$
Step 4 Solve the equation.	*Step 4* $m - 15 - 4 = 27$ $m + {}^-15 + {}^-4 = 27$ $m + {}^-19 = 27$ $\quad\ {}^+19 \quad {}^+19$ $m + 0 = 46$ $m = 46$
Step 5 Answer the question.	*Step 5* She had \$46 in her purse this morning.
Step 6 Check the solution.	*Step 6* Started with \$46. Took out \$15, so \$46 − \$15 = \$31. Took out \$4, so \$31 − \$4 = \$27. Had \$27 left. ← matches ↑

Concepts	*Examples*
3.4 Solving Application Problems with Two Unknown Quantities	
Use the six problem-solving steps outlined in Section 3.3. Because there are *two* unknown quantities, choose a variable to represent "the thing you know the least about." Then write variable expressions to show the relationship of the other unknown quantities to the first one.	Last week, Brian earned \$50 more than twice what Dan earned. How much did each person earn if the total for both of them was \$254? *Step 1* Unknowns: Brian's earnings; Dan's earnings. Known: Brian earned \$50 more than twice what Dan earned; the sum of their earnings was \$254. *Step 2(b)* You know the least about Dan's earnings, so let m represent Dan's earnings. Brian earned \$50 more than twice what Dan earned, so Brian's earnings are $2m$ + \$50. *Step 3* $\underbrace{m + 2m} + 50 = 254$ *Step 4* $3m + 50 = 254$ $\quad\quad\;\; {}^{-}50 \quad {}^{-}50$ $\underbrace{3m + 0} = 204$ $\frac{3m}{3} = \frac{204}{3}$ $m = 68$ *Step 5* m represents Brian's earnings, so Brian earned \$68. $2m$ + \$50 represents Dan's earnings, and 2 • \$68 + \$50 is \$186. *Step 6* Is \$186 actually \$50 more than twice \$68? Yes, the solution checks. Does \$68 + \$186 = \$254? Yes, the solution checks.

Name Date Class

CHAPTER 3 REVIEW EXERCISES

If you need help with any of these review exercises, look in the section indicated in the brackets.

[3.1] *In Exercises 1–4, find the perimeter of each shape.*

1. Square

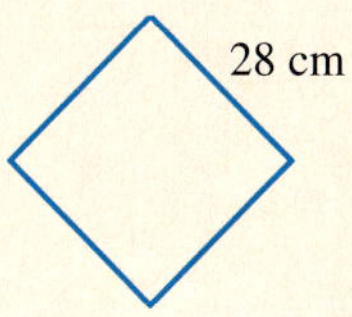

2. Rectangle

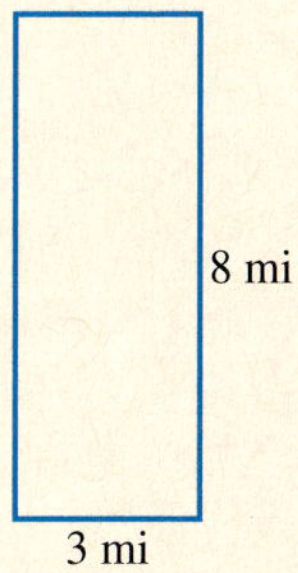

3. Parallelogram

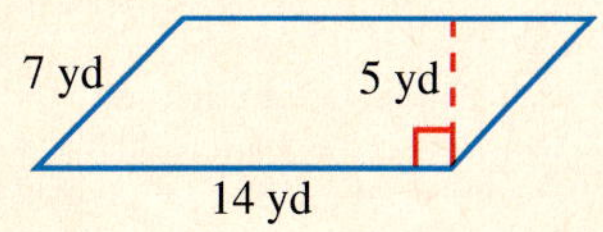

4.

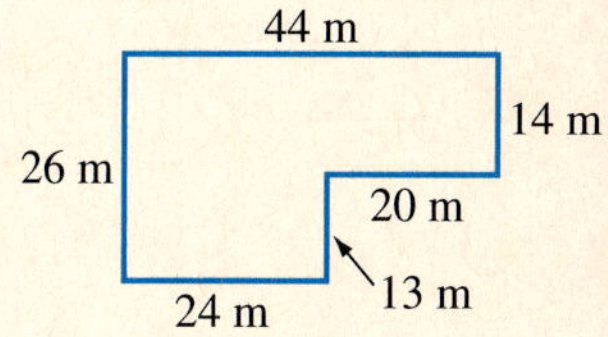

5. A square card table has a perimeter of 12 ft. Find the length of one side of the table.

6. A rectangular playground has a perimeter of 128 m. Find its length if it is 31 m wide.

7. A rectangular painting that is 21 in. long has a perimeter of 72 in. What is the width of the painting?

[3.2] *In Exercises 8–10, draw a sketch of each shape and label the lengths of the sides or the base and height. Then find the area, using the appropriate formula.*

8. A tablecloth that measures 5 ft by 8 ft.

9. A 25 m square dance floor.

10. A parallelogram-shaped lot with a base of 16 yd and a height of 13 yd.

11. A concrete patio that is 14 ft long has an area of 126 ft^2. Find its width.

12. A parallelogram has an area of 88 cm^2. If the base is 11 cm, what is the height?

13. The area of a square piece of land is 100 mi^2. What is the length of one side?

[3.3] *Write in symbols, using x as the variable.*

14. A number subtracted from 57.

15. The sum of 15 and twice a number.

16. The product of $^-9$ and a number.

Writing Calculator Small Group

Translate each sentence into an equation and solve it.

17. The sum of four times a number and 6 is $^{-}30$. What is the number?

18. When twice a number is subtracted from 10, the result is 4 plus the number. Find the number.

[3.3–3.4] *Use the six problem-solving steps to solve each problem.*

19. Grace wrote a \$600 check for her rent. Then she deposited her \$750 paycheck and a \$75 tax refund into her account. The new balance was \$309. How much was in her account before she wrote the rent check?

20. Yoku ordered four boxes of candles for his restaurant. One candle was put on each of the 25 tables. There were 23 candles left. How many candles were originally in each box?

21. \$1000 in prize money in an essay contest is being split between Reggie and Donald. Donald should get \$300 more than Reggie. How much will each man receive?

22. A rectangular photograph is twice as long as it is wide. The perimeter of the photograph is 84 cm. Find the length and the width of the photograph.

MIXED REVIEW EXERCISES

Use the information in the advertisement to do Exercises 23–26.

Now on SALE

Build Your Own Dog Pen

Kit #1 includes 20 feet of fencing.

Kit #2 includes 36 feet of fencing.

23. Anthony made a square dog pen using Kit #2.

(a) What was the length of each side of the pen?

(b) What was the area of the pen?

24. First draw sketches of two *different* rectangular dog pens that you could build using all the fencing in Kit #1. Label the lengths of the sides. Then find the area of each pen.

25. Timotha bought Kit #2. But she used some of the fencing around her garden, so she went back and bought Kit #1. Now she has 41 ft of fencing for a dog pen. How much fencing did she use around her garden? Use the six problem-solving steps.

26. Diana bought Kit #2. The pen she built had a length that was 2 ft more than the width. Find the length and width of the pen. Use the six problem-solving steps.

Chapter 3 Test

Find the perimeter of each figure.

1.

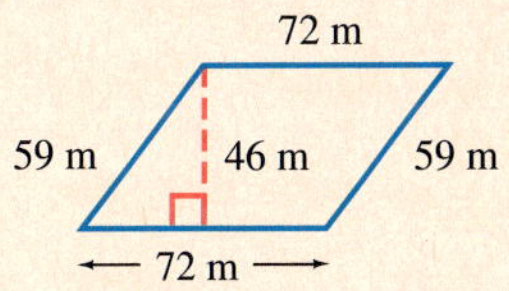

2.

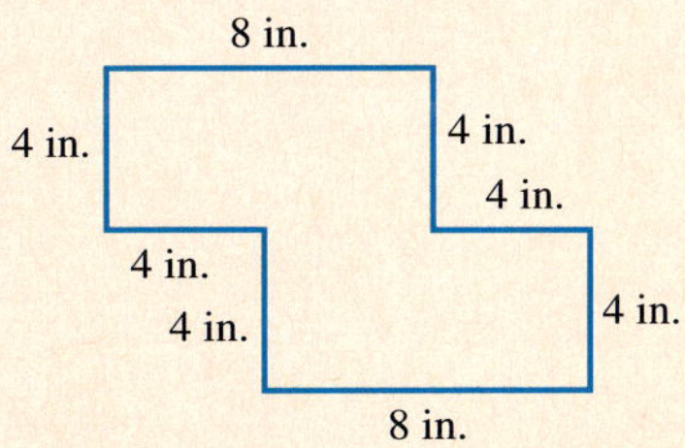

3. A square wetland 3 mi on a side.

4. A mirror that measures 2 ft by 4 ft.

5.

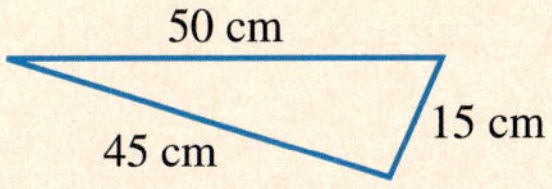

Find the area of each figure.

6.

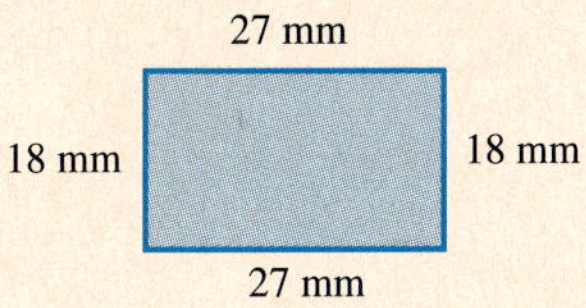

7.

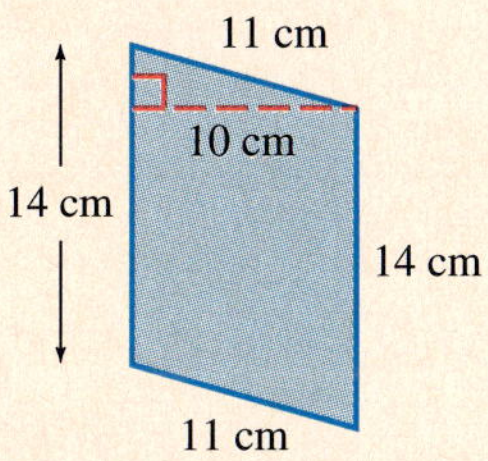

8. A rectangular animal preserve is 55 mi wide and 68 mi long.

9. A square room measures 6 m on each side.

Solve each problem by using the appropriate formula.

10. A square table has a perimeter of 12 ft. Find the length of one side.

11. The Mercado family has 34 ft of fencing to put around a garden plot. The plot is rectangular in shape. If it is 6 ft wide, find the length of the plot.

12. The area of a parallelogram is 65 $in.^2$, and the base is 13 in. What is the height of the parallelogram?

1. ____________
2. ____________
3. ____________
4. ____________
5. ____________
6. ____________
7. ____________
8. ____________
9. ____________
10. ____________
11. ____________
12. ____________

13. ____________________

13. A rectangular postage stamp has a length of 4 cm and an area of 12 cm^2. Find its width.

14. ____________________

14. A square bulletin board has an area of 16 ft^2. How long is each side of the bulletin board?

15. ____________________

15. Explain the difference between ft and ft^2. For which types of problems might you use each of these units?

Translate each sentence into an equation and solve it.

16. ____________________

17. ____________________

16. If 40 is added to four times a number, the result is zero. Find the number.

17. When 7 times a number is decreased by 23, the result is the number plus 7. What is the number?

Solve each application problem, using the six problem-solving steps.

18. ____________________

18. Josephine had \$43 in her wallet. Her son took some of it when he went to the grocery store. Josephine found \$16 in her desk drawer and put it in her wallet. She counted \$44 in the wallet. How much money did her son take to the store?

19. ____________________

19. Ray is 39 years old. His age is 4 years more than five times his daughter's age. How old is his daughter?

20. ____________________

20. A board is 118 cm long. Karin cut it into two pieces, with one piece 4 cm longer than the other. Find the length of both pieces.

21. ____________________

21. The perimeter of a rectangular building is 420 feet. The length is four times as long as the width. Find the length and the width. Make a drawing to help you solve this problem.

22. ____________________

22. Marcella and her husband Tim spent a total of 19 hours redecorating their living room. Tim spent 3 hours less time than Marcella. How long did each person work on the room?

Name ____________________ Date __________ Class __________

CUMULATIVE REVIEW EXERCISES CHAPTERS 1–3

1. Write this number in words.
4,000,206,300

2. Write this number, using digits.
seventy million, five thousand, four hundred eighty-nine.

3. Write $<$ or $>$ between each pair of numbers to make true statements.

$^-7$ __ $^-1$ $\qquad$ 0 __ $^-5$

4. Name the property illustrated by each equation.

(a) $1 \cdot 97 = 97$ $\quad$ **(b)** $^-10 + 0 = {}^-10$

(c) $(3 \cdot {}^-7) \cdot 6 = 3 \cdot ({}^-7 \cdot 6)$

5. (a) Round 3795 to the nearest ten.

(b) Round 493,662 to the nearest ten thousand.

Simplify.

6. $^-12 - 12$

7. $^-3(^-9)$

8. $|7| - |^-10|$

9. $^-40 \div 2 \cdot 5$

10. $3 - 8 + 10$

11. $\dfrac{0}{^-6}$

12. $^-8 + 5(2 - 3)$

13. $(^-3)^2 + 4^2$

14. $4 - 3(^-6 \div 3) + 7(0 - 6)$

15. $\dfrac{4 - 2^3 + 5^2 - 3(^-2)}{^-1(3) - 6(^-2) - 9}$

16. Rewrite $10w^2xy^4$ without exponents.

17. Evaluate $^-6cd^3$ when c is 5 and d is $^-2$.

Simplify.

18. $^-4k + k + 5k$

19. $m^2 + 2m + 2m^2$

20. $xy^3 - xy^3$

21. $5(^-4a)$

22. $^-8 + x + 5 - 2x^2 - x$

23. $^-3(4n + 3) + 10$

Solve each equation and check each solution.

24. $6 - 20 = 2x - 9x$ $\qquad$ **Check**

25. $^-5y = y + 6$ $\qquad$ **Check**

Writing $\quad$ Calculator $\quad$ Small Group

Solve each equation. Show the steps you used.

26. $3b - 9 = 19 - 4b$

27. $^-16 - h + 2 = h - 10$

28. $^-5(2x + 4) = 3x - 20$

29. $6 + 4(a + 8) = {}^-8a + 5 + a$

Find the perimeter and area of each shape.

30.

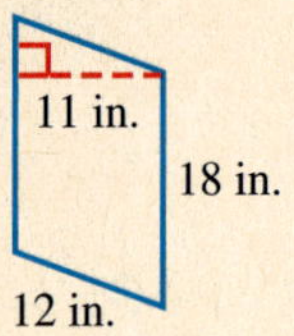

31. A square plaza is 15 m on a side.

32. A rectangular piece of plywood is 4 ft wide and 8 ft long.

Translate each sentence into an equation and solve it.

33. When $^-50$ is added to five times a number, the result is zero. What is the number?

34. When three times a number is subtracted from 10, the result is two times the number. Find the number.

Solve each application problem, using the six problem-solving steps.

35. Some people were waiting in line at the bank. Three people made deposits and left. Six more people got in line. Then two more people were helped, but five people were still in line. How many were in the line originally?

36. Twelve soccer players each paid the same amount toward a team trip. Expenses for the trip were \$2200, and now the team bank account is overdrawn by \$40. How much did each player originally pay?

37. A group of 192 students was divided into two smaller groups. One smaller group was three time the size of the other. How many students were in each smaller group?

38. A rectangular swimming pool is 14 ft longer than it is wide. If the perimeter of the pool is 92 ft, find the length and the width.

Rational Numbers: Positive and Negative Fractions

4

4.1 Introduction to Fractions

OBJECTIVE 1 In Chapters 1–3 you worked with integers. Recall that a list of integers can be written as follows.

$$\ldots\ ^{-}6,\ ^{-}5,\ ^{-}4,\ ^{-}3,\ ^{-}2,\ ^{-}1,\ 0,\ 1,\ 2,\ 3,\ 4,\ 5,\ 6\ \ldots$$

The dots show that the list goes on forever.

Now we will work with *fractions* (FRAK-shuns).

Fraction

A **fraction** is a number of the form $\frac{a}{b}$ where a and b are integers and b is not 0.

One use for fractions is situations in which we need a number that is between two integers. For example,

the recipe uses $\frac{2}{3}$ cup of milk,
$\frac{2}{3}$ is between 0 and 1, and
$\frac{2}{3}$ is a fraction because it is of the form $\frac{a}{b}$ and 2 and 3 are integers.

The number $\frac{2}{3}$ is a fraction that represents 2 of 3 equal parts. Read $\frac{2}{3}$ as "two thirds."

EXAMPLE 1 Identifying Fractions

Use fractions to represent the shaded portions and the unshaded portions of the figures shown.

(a) The figure on the left has 3 equal parts. The 2 shaded parts are represented by the fraction $\frac{2}{3}$. The *un*shaded part is $\frac{1}{3}$ of the figure.

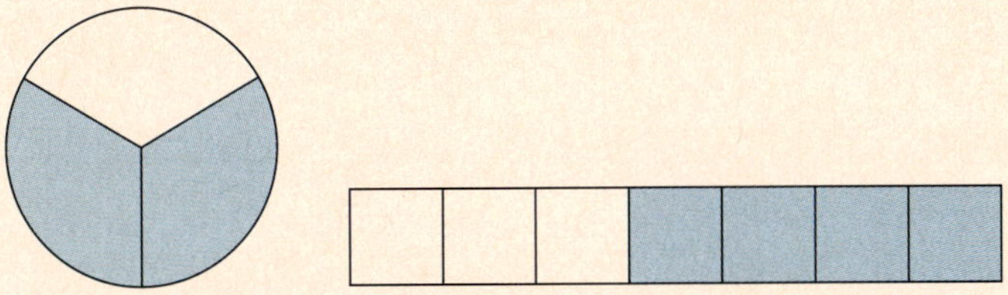

(b) The 4 shaded parts of the 7-part figure on the right are represented by the fraction $\frac{4}{7}$. The *un*shaded part is $\frac{3}{7}$ of the figure.

WORK PROBLEM 1 AT THE SIDE.

INTERNET
www.mathnotes.com

OBJECTIVES

1. Use a fraction to show which part of a whole is shaded.
2. Define *numerator, denominator, proper fraction,* and *improper fraction.*
3. Graph positive and negative fractions on a number line.
4. Find the absolute value of a fraction.
5. Write equivalent fractions.

FOR EXTRA HELP

Tutorial — Tape 5 — SSM, Sec. 4.1

1. Write fractions for the shaded portions and the unshaded portions of each figure.

(a)

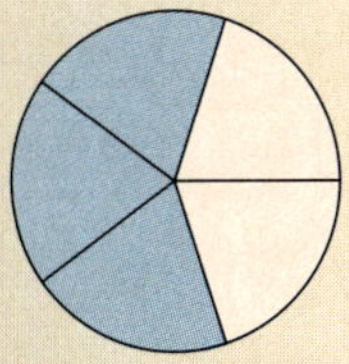

(b)

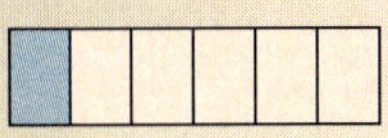

(c)

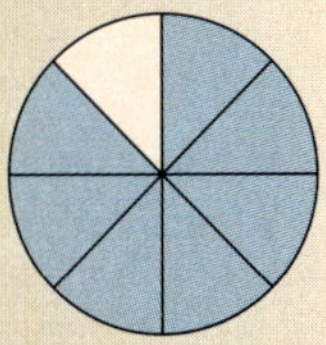

ANSWERS

1. (a) $\frac{3}{5}$; $\frac{2}{5}$ **(b)** $\frac{1}{6}$; $\frac{5}{6}$ **(c)** $\frac{7}{8}$; $\frac{1}{8}$

2. Write fractions for the shaded portions of the figures.

(a)

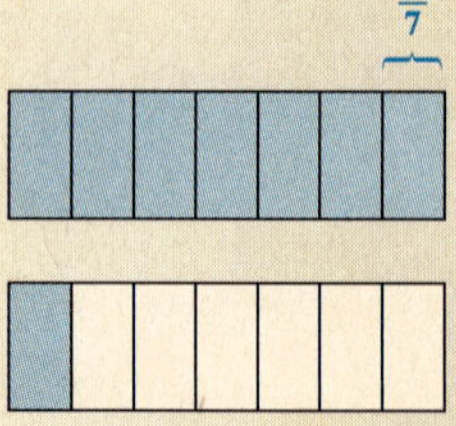

(b)

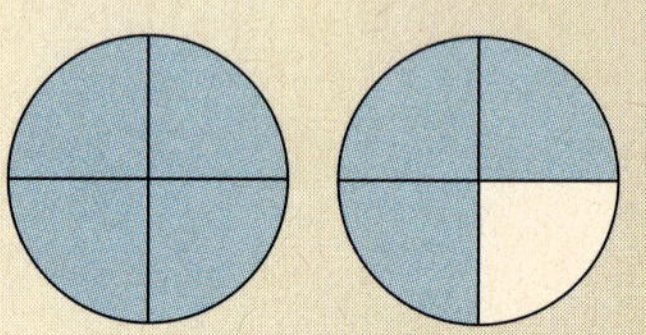

3. Identify the numerator and the denominator. Draw a picture with shaded parts to show each fraction. Your drawings may vary, but they should have the correct number of shaded parts.

(a) $\frac{2}{3}$

(b) $\frac{1}{4}$

(c) $\frac{8}{5}$

(d) $\frac{5}{2}$

ANSWERS

2. **(a)** $\frac{8}{7}$ **(b)** $\frac{7}{4}$

3. **(a)** N: 2; D: 3

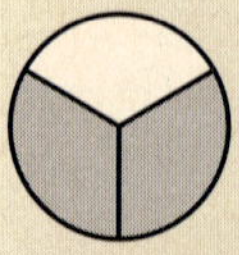

(b) N: 1; D: 4

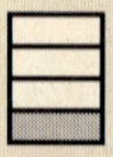

(c) N: 8; D: 5

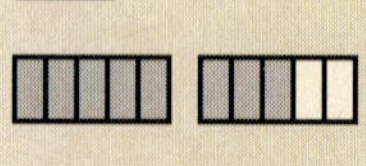

(d) N: 5; D: 2

Fractions can also be used to represent more than one whole object.

EXAMPLE 2 Representing Fractions Greater Than 1

Use a fraction to represent the shaded parts of the figures.

(a)

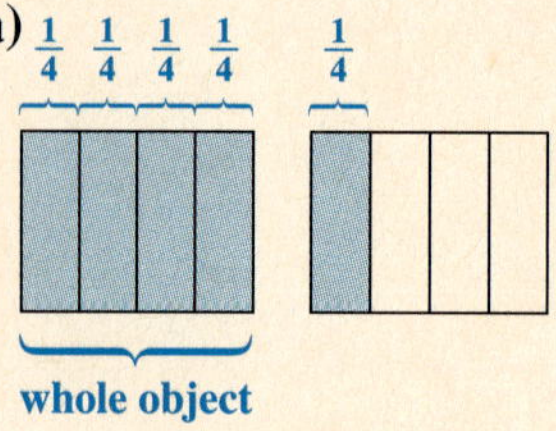

An area equal to 5 of the $\frac{1}{4}$ parts is shaded, so $\frac{5}{4}$ is shaded.

(b)

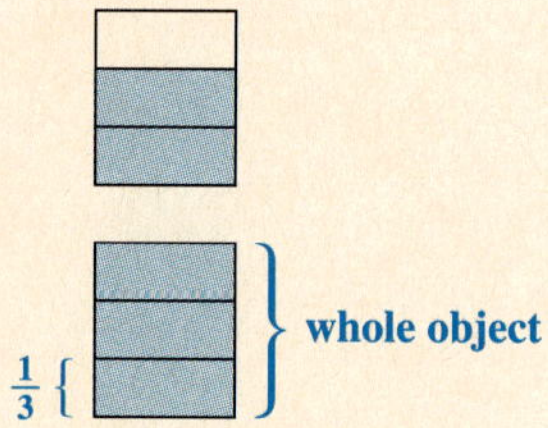

An area equal to 5 of the $\frac{1}{3}$ parts is shaded, so $\frac{5}{3}$ is shaded.

◀◀ WORK PROBLEM 2 AT THE SIDE.

OBJECTIVE 2 In the fraction $\frac{2}{3}$, the number 2 is the **numerator** (NOOM-er-ay-ter), and 3 is the **denominator** (di-NAHM-in-ay-ter). The bar between the numerator and the denominator is the *fraction bar.*

Fraction bar → $\frac{2}{3}$ ← Numerator / ← Denominator

The Numerator and Denominator

The denominator of a fraction shows the number of equivalent parts in the whole, and the numerator shows how many parts are being considered.

Note

Recall that a fraction bar, —, is one of the division symbols and that division by 0 is undefined. Therefore a fraction with a denominator of 0 is also undefined.

EXAMPLE 3 Identifying Numerator and Denominator

Identify the numerator and denominator in each fraction.

(a) $\frac{5}{9}$

$\frac{5}{9}$ 5 ← Numerator; 9 ← Denominator

(b) $\frac{11}{7}$

$\frac{11}{7}$ 11 ← Numerator; 7 ← Denominator

◀◀ WORK PROBLEM 3 AT THE SIDE.

Fractions are sometimes called *proper* or *improper* fractions.

Proper and Improper Fractions

If the numerator of a fraction is *smaller* than the denominator, the fraction is a **proper fraction.**

If the numerator is *greater than or equal to* the denominator, the fraction is an **improper fraction.**

Proper Fractions	*Improper Fractions*
$\frac{1}{2}, \frac{5}{11}, \frac{35}{36}$	$\frac{9}{7}, \frac{126}{125}, \frac{7}{7}$

EXAMPLE 4 Classifying Types of Fractions

(a) Identify all proper fractions in this list.

$$\frac{3}{4}, \frac{5}{9}, \frac{17}{5}, \frac{9}{7}, \frac{3}{3}, \frac{12}{25}, \frac{1}{9}, \frac{5}{3}$$

Proper fractions have a numerator that is *smaller* than the denominator. The proper fractions in the list are

$\frac{3}{4}$, ← 3 is smaller than 4 $\quad \frac{5}{9}, \quad \frac{12}{25}, \quad$ and $\quad \frac{1}{9}$.

(b) Identify all improper fractions in the list in part (a).

Improper fractions have a numerator that is *equal to or greater* than the denominator. The improper fractions in the list are

$\frac{17}{5}$, ← 17 is greater than 5 $\quad \frac{9}{7}, \quad \frac{3}{3}, \quad$ and $\quad \frac{5}{3}$.

WORK PROBLEM 4 AT THE SIDE.

OBJECTIVE 3 Sometimes we need *negative* numbers that are between two integers. For example,

the change in the price of the stock is $-\frac{3}{4}$ dollar, and $-\frac{3}{4}$ is between 0 and $^{-}1$.

Graphing numbers on a number line helps us see the difference between $\frac{3}{4}$ and $-\frac{3}{4}$. Both represent 3 out of 4 equal parts, but they are in opposite directions from zero on the number line. For $\frac{3}{4}$, divide the distance from 0 to 1 into 4 equal parts. Then start at 0, count over 3 parts, and make a dot. For $-\frac{3}{4}$, repeat the same process between 0 and $^{-}1$.

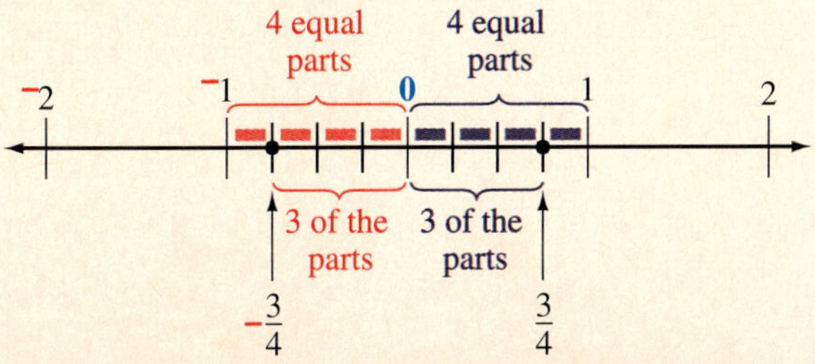

4. From the following group of fractions:

$$\frac{3}{4}, \frac{8}{7}, \frac{5}{7}, \frac{6}{6}, \frac{1}{2}, \frac{2}{1}$$

(a) list all proper fractions.

(b) list all improper fractions.

ANSWERS

4. (a) $\frac{3}{4}, \frac{5}{7}, \frac{1}{2}$

(b) $\frac{8}{7}, \frac{6}{6}, \frac{2}{1}$

5. Graph each fraction on the number line.

(a) $\frac{2}{4}$

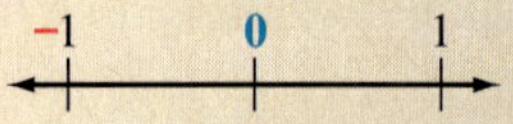

(b) $\frac{1}{2}$

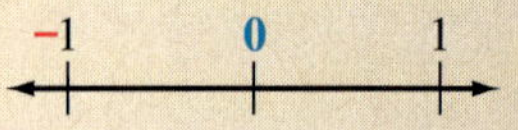

(c) $-\frac{2}{3}$

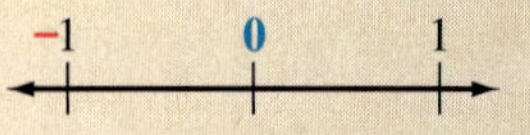

Note

In Chapters 1–3 we used a raised negative sign to help you avoid confusion between negative numbers and subtraction. Now you are ready to start writing the negative sign in the more traditional way. In this chapter, the negative sign will still be red, but it will be centered on the number instead of raised: for example, -2 instead of $^{-}2$. When the negative sign might be confused with the sign for subtraction, we will write parentheses around the negative number. Here is an example.

$3 - (-2)$ means 3 **minus** (**negative** 2)

For fractions, the negative sign will be written in front of the fraction bar: for example, $-\frac{3}{4}$. As with integers, the negative sign tells you that a fraction is *less than 0*; it is to the *left* of zero on the number line. When there is *no* sign in front of a fraction, the fraction is assumed to be positive. For example, $\frac{3}{4}$ is assumed to be $+\frac{3}{4}$. It is to the *right* of zero on the number line.

EXAMPLE 5 Graphing Positive and Negative Fractions

Graph each fraction on the number line.

(a) $\frac{2}{5}$

There is *no* sign in front of $\frac{2}{5}$, so it is *positive.* Because $\frac{2}{5}$ is between 0 and 1, we divide that space into 5 equal parts. Then we start at 0 and count to the right 2 parts.

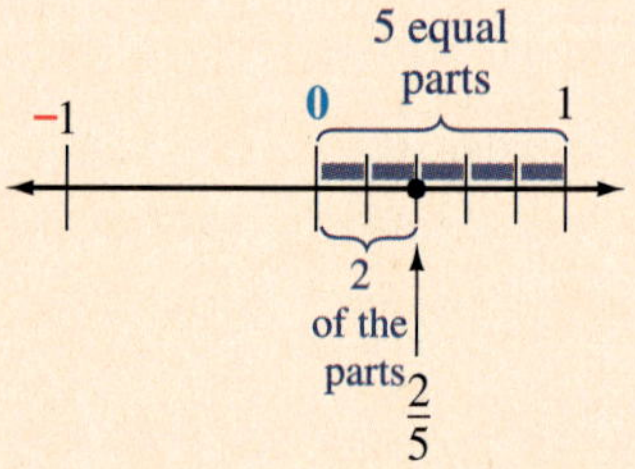

(b) $-\frac{4}{5}$

The fraction is *negative,* so it is between 0 and -1. We divide that space into 5 equal parts. Then we start at 0 and count to the left 4 parts.

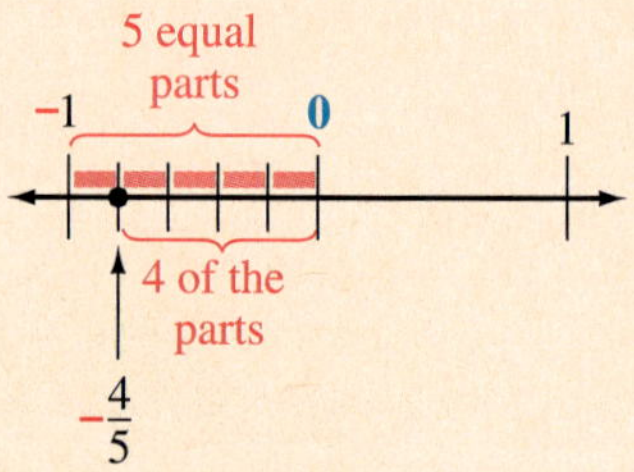

◀◀ WORK PROBLEM 5 AT THE SIDE.

Objective 4 In **Section 1.2** we said that the absolute value of a number was its distance from zero on the number line. Two vertical bars indicate absolute value, as shown.

$\left|-\frac{3}{4}\right|$ is read "the absolute value of negative three-fourths."

As with integers, the absolute value of fractions will *always* be positive (or zero) because it is the *distance* from zero on the number line.

Answers

5. (a)

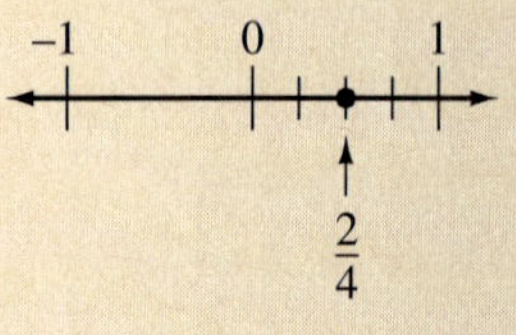

(b)

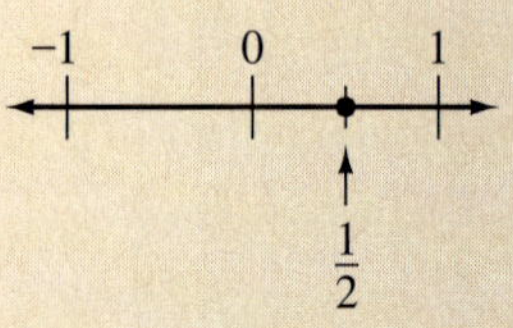

(c)

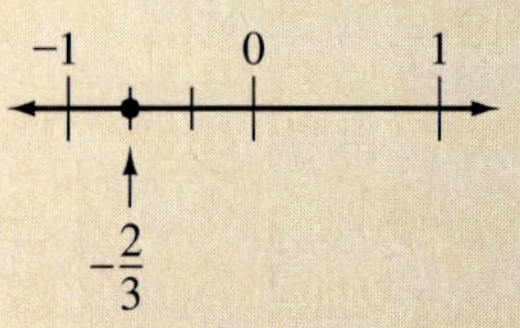

EXAMPLE 6 Finding Absolute Value of a Fraction

Find each absolute value: $\left|\frac{1}{2}\right|$ and $\left|-\frac{1}{2}\right|$

The distance from 0 to $\frac{1}{2}$ on the number line is $\frac{1}{2}$ space, so $\left|\frac{1}{2}\right| = \frac{1}{2}$.
The distance from 0 to $-\frac{1}{2}$ is also $\frac{1}{2}$ space, so $\left|-\frac{1}{2}\right| = \frac{1}{2}$.

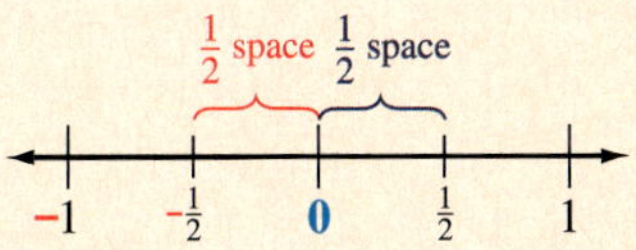

WORK PROBLEM 6 AT THE SIDE.

OBJECTIVE 5 You may have noticed in margin exercises 5(a) and (b) that $\frac{2}{4}$ and $\frac{1}{2}$ were at the same point on the number line. Both of them were halfway between 0 and 1. There are actually *many* different names for this point. We illustrate some of them here.

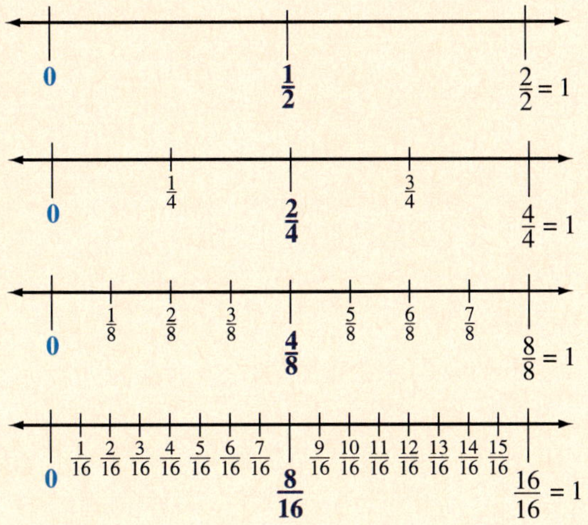

That is, $\frac{8}{16} = \frac{4}{8} = \frac{2}{4} = \frac{1}{2}$. If you have ever used a standard ruler with inches divided into sixteenths, you have probably already noticed that these distances are the same. Although the fractions look different, they all name the same point between 0 and 1. In other words, they have the same value. We say that they are *equivalent* (ee-KWIV-uh-lent) *fractions.*

Equivalent Fractions

Fractions that represent the same number (the same point on a number line) are **equivalent fractions.**

Drawing number lines is tedious, so we usually find equivalent fractions by multiplying or dividing both the numerator and denominator by the same number. We can use some of the fractions that we just graphed to illustrate this method.

$$\frac{1}{2} = \frac{1 \cdot 2}{2 \cdot 2} = \frac{2}{4}$$

Multiply both numerator and denominator by 2.

$$\frac{8}{16} = \frac{8 \div 4}{16 \div 4} = \frac{2}{4}$$

Divide both numerator and denominator by 4.

---

6. Find each absolute value.

(a) $\left|-\frac{3}{4}\right|$

(b) $\left|\frac{5}{8}\right|$

(c) $|0|$

ANSWERS

6. **(a)** $\frac{3}{4}$ **(b)** $\frac{5}{8}$ **(c)** 0

7. (a) Write $\frac{2}{5}$ as an equivalent fraction with a denominator of 20.

(b) Write $-\frac{21}{28}$ as an equivalent fraction with a denominator of 4.

Answers

7. (a) $\frac{2}{5} = \frac{2 \cdot 4}{5 \cdot 4} = \frac{8}{20}$

(b) $-\frac{21}{28} = -\frac{21 \div 7}{28 \div 7} = -\frac{3}{4}$

Writing Equivalent Fractions

If a, b, and c are numbers (and b and c are not zero), then,

$$\frac{a}{b} = \frac{a \cdot c}{b \cdot c} \quad \text{or} \quad \frac{a}{b} = \frac{a \div c}{b \div c}.$$

In other words, if the numerator and denominator of a fraction are multiplied or divided by the *same* nonzero number, the result is an equivalent fraction.

EXAMPLE 7 Writing Equivalent Fractions

(a) Write $-\frac{1}{2}$ as an equivalent fraction with a denominator of 16. In other words, $-\frac{1}{2} = -\frac{?}{16}$. The original denominator is 2. *Multiplying* 2 times 8 gives 16, the new denominator. To write an equivalent fraction, multiply *both* the numerator and denominator by 8.

$$-\frac{1}{2} = -\frac{1 \cdot 8}{2 \cdot 8} = -\frac{8}{16}$$

Keep the negative sign.

So $-\frac{1}{2}$ is equivalent to $-\frac{8}{16}$.

(b) Write $\frac{12}{15}$ as an equivalent fraction with a denominator of 5. In other words, $\frac{12}{15} = \frac{?}{5}$. The original denominator is 15. *Dividing* 15 by 3 gives 5, the new denominator. To write an equivalent fraction, divide *both* the numerator and denominator by 3.

$$\frac{12}{15} = \frac{12 \div 3}{15 \div 3} = \frac{4}{5}$$

So $\frac{12}{15}$ is equivalent to $\frac{4}{5}$.

WORK PROBLEM 7 AT THE SIDE.

Look back at the set of four number lines on the previous page. Notice that there are many different names for 1.

$$\frac{2}{2} = \mathbf{1} \qquad \frac{4}{4} = \mathbf{1} \qquad \frac{8}{8} = \mathbf{1} \qquad \frac{16}{16} = \mathbf{1}$$

Because a fraction bar is a symbol for division, you can think of $\frac{2}{2}$ as $2 \div 2$, which equals 1. Similarly, $\frac{4}{4}$ is $4 \div 4$, which also is 1, and so on. These examples illustrate one of the division properties from **Section 1.7**.

Division Properties

If a is any number (except 0), then $\frac{a}{a} = 1$. For example, $\frac{6}{6} = 1$ and $\frac{-4}{-4} = 1$. In other words, when a nonzero number is divided by itself, the result is 1.

Also recall that when any number is divided by 1, the result is the number. That is, $\frac{a}{1} = a$. For example, $\frac{6}{1} = 6$ and $-\frac{12}{1} = -12$.

EXAMPLE 8 Using Division to Simplify Fractions

Simplify each fraction by dividing the numerator by the denominator.

(a) $\frac{5}{5}$ Think of $\frac{5}{5}$ as $5 \div 5$. The result is 1, so $\frac{5}{5} = 1$.

(b) $\frac{12}{4}$ Think of $\frac{12}{4}$ as $12 \div 4$. The result is 3, so $\frac{12}{4} = 3$.

(c) $-\frac{8}{1}$ Think of $-\frac{8}{1}$ as $-8 \div 1$. The result is -8, so $-\frac{8}{1} = -8$.

Keep the negative sign.

WORK PROBLEM 8 AT THE SIDE.

8. Simplify each fraction by dividing the numerator by the denominator.

(a) $\frac{10}{10}$

(b) $-\frac{3}{1}$

(c) $\frac{8}{2}$

(d) $-\frac{25}{5}$

ANSWERS

8. **(a)** 1 **(b)** -3 **(c)** 4 **(d)** -5

NUMBERS IN THE *Real World*

collaborative investigations

People who make quilts often base their designs on a block cut into 4, 9, 16, or 25 squares. Then, they select the colors for the various pieces in the block. Finally, they calculate how much fabric of each color to buy, based on the fractional part of the block in that color. For each quilt design shown, figure out what fraction of the block is red, what fraction is blue, and so on for each color.

Use the following blocks to create and color your own quilt patterns. Then indicate what fractional part of the block is in each color.

Find the next two numbers in this pattern: 4, 9, 16, 25,
Explain how the pattern works.

4.1 Exercises

Write the fractions that represent the shaded and unshaded portions of each figure.

1. 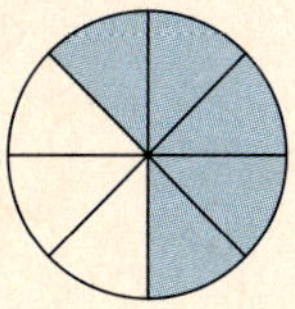

2. 

3. 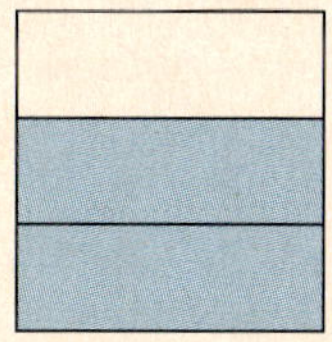

4. 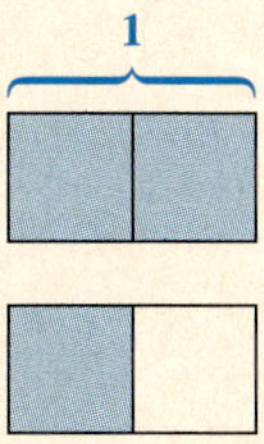

5. 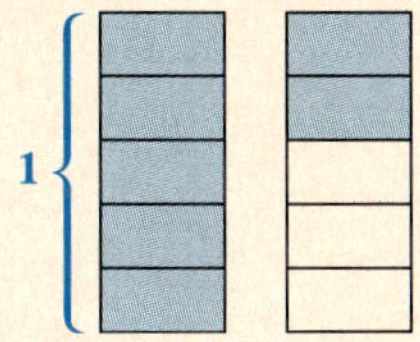

6. 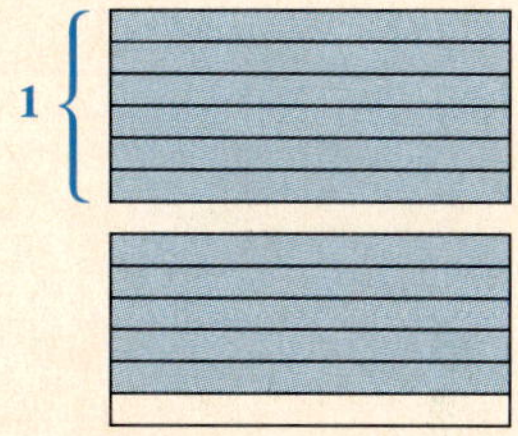

7. What fraction of these 11 coins are dimes? What fraction are pennies? What fraction are nickels?

8. What fraction of these six recording stars are men? What fraction are women? What fraction are wearing headbands?

9. In an American Sign Language (A.S.L.) class of 25 students, 8 are hearing impaired. What fraction of the students are hearing impaired?

10. Of 35 motorcycles in the parking lot, 17 are Harley Davidsons. What fraction of the motorcycles are *not* Harley Davidsons?

11. Of 71 cars making up a freight train, 58 are boxcars. What fraction of the cars are *not* boxcars?

12. A college cheerleading squad has 12 members. If 5 of the cheerleaders are sophomores and the rest are freshmen, find the fraction of the members that are freshmen.

Identify the numerator and denominator in each fraction.

	Numerator	***Denominator***		***Numerator***	***Denominator***
13. $\frac{3}{4}$	______	______	**14.** $\frac{5}{8}$	______	______
15. $\frac{12}{7}$	______	______	**16.** $\frac{8}{3}$	______	______

List the proper and improper fractions in each group of numbers.

	Proper	***Improper***
17. $\frac{8}{5}, \frac{1}{3}, \frac{5}{8}, \frac{6}{6}, \frac{12}{2}, \frac{7}{16}$	______	______
18. $\frac{1}{6}, \frac{5}{8}, \frac{15}{14}, \frac{11}{9}, \frac{7}{7}, \frac{3}{4}$	______	______
19. $\frac{3}{4}, \frac{3}{2}, \frac{5}{5}, \frac{9}{11}, \frac{7}{15}, \frac{19}{18}$	______	______
20. $\frac{12}{12}, \frac{15}{11}, \frac{13}{12}, \frac{11}{8}, \frac{17}{17}, \frac{19}{12}$	______	______

21. Write a fraction of your own choice. Label the parts of the fraction and write a sentence describing what each part represents. Draw a figure with shaded parts showing your fraction.

22. Give one example of a proper fraction and one example of an improper fraction. What determines whether a fraction is proper or improper? Draw figures with shaded parts showing these fractions.

Graph each pair of fractions on the number line.

23. $\frac{1}{4}, -\frac{1}{4}$

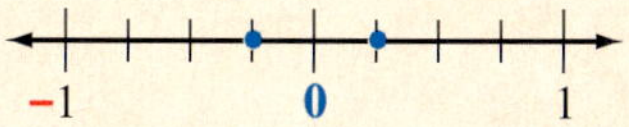

24. $-\frac{1}{3}, \frac{1}{3}$

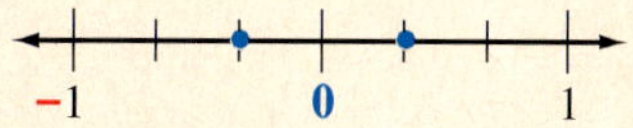

25. $-\frac{3}{5}, \frac{3}{5}$

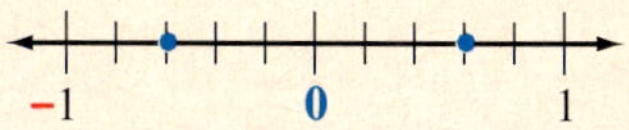

26. $\frac{5}{6}, -\frac{5}{6}$

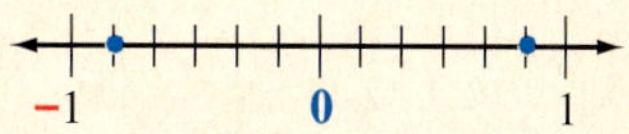

25. $\frac{7}{8}, -\frac{7}{8}$

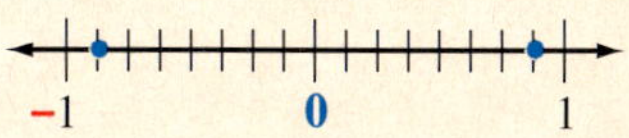

28. $-\frac{3}{4}, \frac{3}{4}$

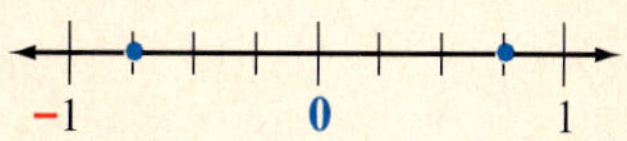

Write a positive or negative fraction to describe each situation.

29. The baby lost $\frac{3}{4}$ pound in weight while she was sick.

30. Greta needed $\frac{1}{3}$ cup of brown sugar for the cookie recipe.

31. The price of Mathtronic stock went up $\frac{1}{2}$ dollar per share.

32. The mice who were on an experimental diet lost an average of $\frac{1}{4}$ ounce.

33. Marcel cut $\frac{5}{8}$ in. from the bottom of the door so that it wouldn't scrape the floor.

34. The Brown's driveway is $\frac{3}{10}$ mile long.

Find each absolute value.

35. $\left|-\frac{2}{5}\right|$

36. $\left|-\frac{2}{3}\right|$

37. $\left|\frac{9}{10}\right|$

38. $|0|$

39. Rewrite each fraction as an equivalent fraction with a denominator of 24.

(a) $\frac{1}{2} = \underline{\quad}$ (b) $\frac{1}{3} = \underline{\quad}$ (c) $\frac{2}{3} = \underline{\quad}$ (d) $\frac{1}{4} = \underline{\quad}$ (e) $\frac{3}{4} = \underline{\quad}$

(f) $\frac{1}{6} = \underline{\quad}$ (g) $\frac{5}{6} = \underline{\quad}$ (h) $\frac{1}{8} = \underline{\quad}$ (i) $\frac{3}{8} = \underline{\quad}$ (j) $\frac{5}{8} = \underline{\quad}$

40. Rewrite each fraction as an equivalent fraction with a denominator of 36.

(a) $\frac{1}{2} = \underline{\quad}$ (b) $\frac{1}{3} = \underline{\quad}$ (c) $\frac{2}{3} = \underline{\quad}$ (d) $\frac{1}{4} = \underline{\quad}$ (e) $\frac{3}{4} = \underline{\quad}$

(f) $\frac{1}{6} = \underline{\quad}$ (g) $\frac{5}{6} = \underline{\quad}$ (h) $\frac{1}{9} = \underline{\quad}$ (i) $\frac{4}{9} = \underline{\quad}$ (j) $\frac{8}{9} = \underline{\quad}$

41. Rewrite each fraction as an equivalent fraction with a denominator of 3.

(a) $-\frac{2}{6} = \underline{\quad}$ (b) $-\frac{4}{6} = \underline{\quad}$ (c) $-\frac{12}{18} = \underline{\quad}$ (d) $-\frac{6}{18} = \underline{\quad}$ (e) $-\frac{200}{300} = \underline{\quad}$

(f) Write two more fractions that are equivalent to $-\frac{1}{3}$ and two more fractions equivalent to $-\frac{2}{3}$.

42. Rewrite each fraction as an equivalent fraction with a denominator of 4.

(a) $-\frac{2}{8} = \underline{\quad}$ (b) $-\frac{6}{8} = \underline{\quad}$ (c) $-\frac{15}{20} = \underline{\quad}$ (d) $-\frac{50}{200} = \underline{\quad}$ (e) $-\frac{150}{200} = \underline{\quad}$

(f) Write two more fractions that are equivalent to $-\frac{1}{4}$ and two more fractions equivalent to $-\frac{3}{4}$.

Use your calculator to help you solve Exercises 43–50.

43. Write $\frac{3}{8}$ as an equivalent fraction with a denominator of 3912.

44. Write $\frac{7}{9}$ as an equivalent fraction with a denominator of 5472.

45. Is $-\frac{697}{3485}$ equivalent to $-\frac{1}{2}$, $-\frac{1}{3}$, or $-\frac{1}{5}$?

46. Is $-\frac{817}{4902}$ equivalent to $-\frac{1}{4}$, $-\frac{1}{6}$, or $-\frac{1}{8}$?

Find a number to replace the ? that will make the two fractions equivalent.

47. $\frac{1183}{2028} = \frac{?}{12}$

48. $\frac{2775}{6105} = \frac{?}{11}$

49. $\frac{13}{?} = \frac{1157}{1335}$

50. $\frac{9}{?} = \frac{891}{1584}$

51. Explain how to write equivalent fractions. Show one example in which the new denominator is larger than the original denominator, and one example in which the new denominator is smaller.

52. Explain how to find the absolute value of a fraction. Draw a number line to illustrate your explanation and include a positive fraction and a negative fraction as examples.

53. Can you write $\frac{3}{5}$ as an equivalent fraction with a denominator of 18? Explain why or why not. If not, what denominators could you use instead of 18?

54. Can you write $\frac{3}{4}$ as an equivalent fraction with a denominator of 0? Explain why or why not.

Simplify each fraction by dividing the numerator by the denominator.

55. $\frac{10}{1}$

56. $\frac{9}{9}$

57. $-\frac{16}{16}$

58. $-\frac{7}{1}$

59. $-\frac{18}{3}$

60. $-\frac{40}{4}$

61. $\frac{24}{8}$

62. $\frac{42}{6}$

63. $\frac{12}{12}$

64. $-\frac{5}{5}$

65. $\frac{14}{7}$

66. $\frac{8}{2}$

There are many correct ways to draw the answers for Exercises 67–82, so ask your instructor to check your work.

67. Shade $\frac{3}{5}$ of this figure. What fraction is unshaded?

68. Shade $\frac{5}{6}$ of this figure. What fraction is unshaded?

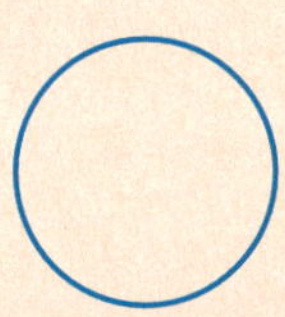

69. Shade $\frac{3}{8}$ of this figure. What fraction is unshaded?

70. Shade $\frac{1}{3}$ of this figure. What fraction is unshaded?

71. Shade $\frac{7}{4}$ of this figure. What fraction is unshaded?

72. Shade $\frac{6}{5}$ of this figure. What fraction is unshaded?

73. Shade $\frac{4}{3}$ of this figure. What fraction is unshaded?

74. Shade $\frac{11}{8}$ of this figure. What fraction is unshaded?

75. Shade $\frac{6}{6}$ of this figure. What fraction is unshaded?

76. Shade $\frac{10}{10}$ of this figure. What fraction is unshaded?

77. Shade $\frac{10}{5}$ of this figure.

78. Shade $\frac{8}{4}$ of this figure.

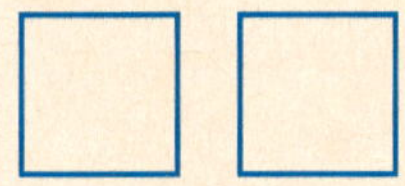

79. Draw a group of seven faces. Draw some of the faces smiling and some sad. What fraction of your faces are smiling? What fraction are sad?

80. Draw a group of eight apples. Draw some of the apples with a stem and some of them with a bite taken out. What fraction of your apples have stems? What fraction have a bite taken out?

81. Draw a group of figures. Make $\frac{1}{10}$ of the figures circles, $\frac{6}{10}$ of the figures squares, and $\frac{3}{10}$ of the figures triangles. Then shade $\frac{1}{6}$ of the squares and $\frac{2}{3}$ of the triangles.

82. Write a group of capital letters. Make $\frac{4}{9}$ of the letters A's, $\frac{2}{9}$ of the letters B's, and $\frac{3}{9}$ of the letters C's. Then draw a line under $\frac{3}{4}$ of the A's and $\frac{1}{2}$ of the B's.

Review and Prepare

Write each expression without exponents. (For help, see ***Section 1.8****.)*

83. x^4 **84.** y^5 **85.** a^5b^2 **86.** m^2n^3

87. $6b^3$ **88.** $10r^4$ **89.** $12xy^2$ **90.** $2ab^6$

4.2 *Writing Fractions in Lowest Terms*

OBJECTIVES

1. Define *lowest terms.*
2. Write a fraction in lowest terms using common factors.
3. Write a number as a product of prime factors.
4. Write a fraction in lowest terms, using prime factorization.
5. Write fractions with variables in lowest terms.

FOR EXTRA HELP

Tutorial | Tape 5 | SSM, Sec. 4.2

OBJECTIVE 1 You can see from these drawings that $\frac{1}{2}$ and $\frac{4}{8}$ are different names for the same amount of pizza.

$\frac{1}{2}$ of the pizza has pepperoni on it.

$\frac{4}{8}$ of the pizza has pepperoni on it.

You saw in the last section that $\frac{1}{2}$ and $\frac{4}{8}$ are equivalent fractions. But we say that the fraction $\frac{1}{2}$ is in *lowest terms* because the numerator and denominator have no *common factor* other than 1. That means that 1 is the only number that divides evenly into both 1 and 2. However, the fraction $\frac{4}{8}$ is *not* in lowest terms because its numerator and denominator have a common factor of 4. That means 4 will divide evenly into both 4 and 8.

Note

Recall that factors are numbers being multiplied to give a product. Here is an example.

$1 \cdot 4 = 4$, so 1 and 4 are factors of 4
$2 \cdot 4 = 8$, so 2 and 4 are factors of 8

4 is a factor of both 4 and 8, so 4 is a *common factor* of those numbers.

Writing a Fraction in Lowest Terms

A fraction is written in **lowest terms** when the numerator and denominator have no common factor other than 1. Examples are $\frac{1}{3}$, $\frac{3}{4}$, $\frac{2}{5}$, and $\frac{7}{10}$. When you work with fractions, always write the final answer in lowest terms.

EXAMPLE 1 Understanding Lowest Terms

Are the following fractions in lowest terms?

(a) $\frac{3}{8}$

The numerator and denominator have no common factor other than 1, so the fraction is in lowest terms.

(b) $\frac{21}{36}$

The numerator and denominator have a common factor of 3, so the fraction is *not* in lowest terms.

WORK PROBLEM 1 AT THE SIDE.

1. Are the following fractions in lowest terms? If not, find a common factor of the numerator and denominator (other than 1).

(a) $\frac{2}{3}$

(b) $-\frac{8}{10}$

(c) $-\frac{9}{11}$

(d) $\frac{15}{20}$

ANSWERS

1. (a) Yes **(b)** No; 2 is a common factor. **(c)** Yes **(d)** No; 5 is a common factor.

2. Write in lowest terms.

(a) $\frac{5}{10}$

(b) $\frac{9}{12}$

(c) $-\frac{24}{30}$

(d) $\frac{15}{40}$

(e) $-\frac{50}{90}$

OBJECTIVE 2 We will show you two common methods for writing a fraction in lowest terms. The first method, dividing by a common factor, works best when the numerator and denominator are small numbers.

EXAMPLE 2 Using Common Factors to Write Fractions in Lowest Terms

Write each fraction in lowest terms.

(a) $\frac{20}{24}$

The largest common factor of 20 and 24 is 4. Divide both numerator and denominator by **4.**

$$\frac{20}{24} = \frac{20 \div 4}{24 \div 4} = \frac{5}{6}$$

(b) $\frac{30}{50} = \frac{30 \div 10}{50 \div 10} = \frac{3}{5}$ Divide both numerator and denominator by 10.

(c) $-\frac{24}{42} = -\frac{24 \div 6}{42 \div 6} = -\frac{4}{7}$ Divide both numerator and denominator by 6. Keep the negative sign.

(d) $\frac{60}{72}$

Suppose we made an error and thought that 4 was the largest common factor of 60 and 72. Dividing by 4 would give

$$\frac{60}{72} = \frac{60 \div 4}{72 \div 4} = \frac{15}{18}.$$

But $\frac{15}{18}$ is not in lowest terms because 15 and 18 have a common factor of 3. Divide by 3.

$$\frac{15}{18} = \frac{15 \div 3}{18 \div 3} = \frac{5}{6} \leftarrow \text{Lowest terms}$$

The fraction $\frac{60}{72}$ could have been written in lowest terms in one step by dividing by 12, the largest common factor of 60 and 72.

$$\frac{60}{72} = \frac{60 \div 12}{72 \div 12} = \frac{5}{6} \leftarrow \text{Same answer as above}$$

Either way works. Just keep dividing until the fraction is in lowest terms.

This method of writing a fraction in lowest terms by dividing by a common factor is summarized in the following steps.

Dividing by a Common Factor to Write Fractions in Lowest Terms

Step 1 Find the largest number that will divide evenly into both the numerator and denominator. This number is a **common factor.**

Step 2 **Divide** both numerator and denominator by the common factor.

Step 3 **Check** to see if the new fraction has any common factors (besides 1). If it does, repeat Steps 2 and 3. If the only common factor is 1, the fraction is in lowest terms.

ANSWERS

2. (a) $\frac{1}{2}$ (b) $\frac{3}{4}$ (c) $-\frac{4}{5}$ (d) $\frac{3}{8}$ (e) $-\frac{5}{9}$

WORK PROBLEM 2 AT THE SIDE.

OBJECTIVE 3 In Example 2(d) the largest common factor of 60 and 72 was difficult to see quickly. You can handle a problem like that by writing the numerator and denominator as a product of *prime numbers.*

Prime Numbers

A **prime number** is a whole number that has exactly *two different* factors, itself and 1.

The number 3 is a prime number because it can be divided evenly only by itself and 1. The number 8 is *not* a prime number. The number 8 is a *composite number* because it can be divided evenly by 2 and 4, as well as by itself and 1.

Composite Numbers

A number with a factor other than itself or 1 is called a **composite** (kahm-PAHZ-it) **number.**

Note

A prime number has *only two* different factors, itself and 1. The number 1 is *not* a prime number because it does not have *two different* factors; the only factor of 1 is 1. Likewise, 0 is *not* a prime number, and neither 0 nor 1 are composite numbers.

EXAMPLE 3 Finding Prime Numbers

Which of the following numbers are prime?

2 5 8 11 15

The number 8 can be divided by 4 and 2, so it is *not* prime. Also, 15 is *not* prime because 15 can be divided by 5 and 3. All the other numbers in the list are prime. Each of these numbers is divisible only by itself and 1.

WORK PROBLEM 3 AT THE SIDE.

For reference, here are the primes smaller than 100.

2, 3, 5, 7, 11,
13, 17, 19, 23, 29,
31, 37, 41, 43, 47,
53, 59, 61, 67, 71,
73, 79, 83, 89, 97

Note

All prime numbers are odd numbers except the number 2. Be careful though, because *all odd numbers are not prime numbers.* For example, 9, 15, and 21 are odd numbers that are *not* prime numbers.

The *prime factorization* of a number can be especially useful when working with fractions.

Prime Factorization

A **prime factorization** of a number is a factorization in which every factor is a prime number.

3. Which of the following are prime numbers?

1, 2, 3, 4, 7, 9, 13, 19, 25, 29

ANSWER

3. 2, 3, 7, 13, 19, 29

4. Find the prime factorization of each number.

(a) 8

(b) 42

(c) 90

(d) 100

(e) 81

EXAMPLE 4 Factoring by Using the Division Method

(a) Find the prime factorization of 48.

All prime factors	$2\overline{)48}$	← Divide 48 by 2 (first prime).
	$2\overline{)24}$	← Divide 24 by 2.
	$2\overline{)12}$	← Divide 12 by 2.
	$2\overline{)6}$	← Divide 6 by 2.
	$3\overline{)3}$	← Divide 3 by 3.
	1	← Continue to divide until the quotient is 1.

Because all factors (divisors) are prime, the prime factorization of 48 is

$$2 \cdot 2 \cdot 2 \cdot 2 \cdot 3.$$

Check by multiplying the factors to see if the product is 48. Yes, $2 \cdot 2 \cdot 2 \cdot 2 \cdot 3$ does equal 48.

Note

You may write the factors in any order because multiplication is commutative and associative. So you could write the factorization of 48 as $3 \cdot 2 \cdot 2 \cdot 2 \cdot 2$. We will show the factors from smallest to largest in our examples.

(b) Find the prime factorization of 225.

All prime factors	$3\overline{)225}$	← 225 is not divisible by 2; use 3.
	$3\overline{)75}$	← Divide 75 by 3.
	$5\overline{)25}$	← 25 is not divisible by 3; use 5.
	$5\overline{)5}$	← Divide 5 by 5.
	1	← Quotient is 1.

So $225 = 3 \cdot 3 \cdot 5 \cdot 5$.

Note

When you're using the division method of factoring, the last quotient is 1. The 1 is never used as a prime factor because 1 is neither prime nor composite. Besides, 1 times any number is the number itself.

◀◀ WORK PROBLEM 4 AT THE SIDE.

Another method of factoring uses what is called a *factor tree.*

EXAMPLE 5 Factoring by Using a Factor Tree

Find the prime factorization of each number.

(a) 60

Try to divide by the first prime, 2. Write the factors under the 60. Circle the 2, because it is a prime.

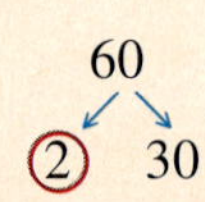

CONTINUED ON NEXT PAGE

ANSWERS

4. (a) $2 \cdot 2 \cdot 2$ (b) $2 \cdot 3 \cdot 7$ (c) $2 \cdot 3 \cdot 3 \cdot 5$ (d) $2 \cdot 2 \cdot 5 \cdot 5$ (e) $3 \cdot 3 \cdot 3 \cdot 3$

Try dividing 30 by 2. Write the factors under the 30.

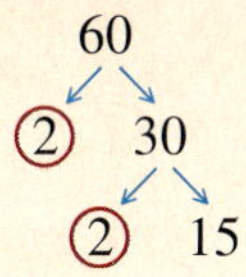

Because 15 cannot be evenly divided by 2, you should try dividing 15 by the next prime number, 3.

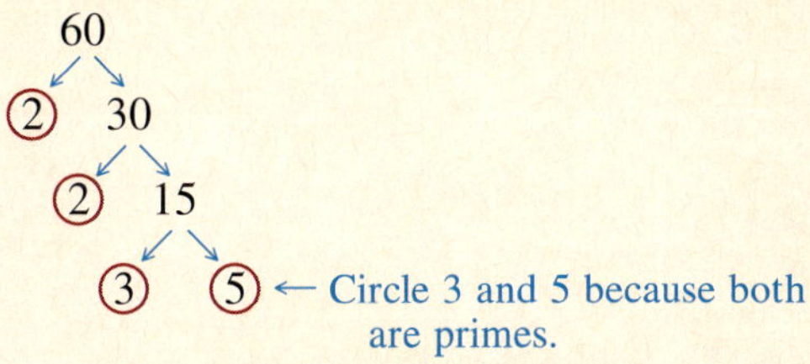

No uncircled factors remain, so you have found the prime factorization (the circled factors).

$$60 = \mathbf{2 \cdot 2 \cdot 3 \cdot 5}$$

(b) 72

Divide by 2.

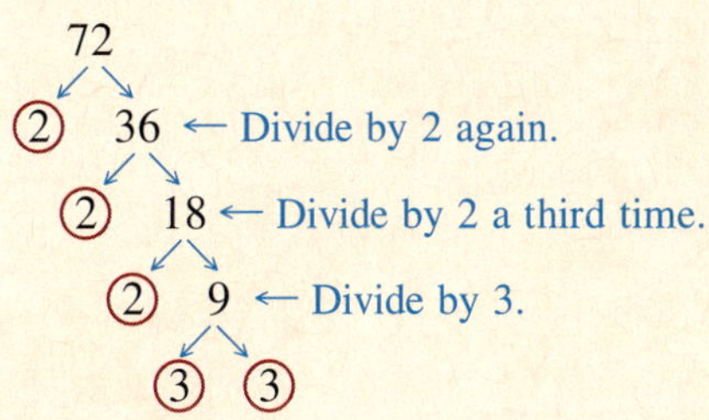

So 72 = **2 • 2 • 2 • 3 • 3**.

(c) 45

Because 45 cannot be divided evenly by 2, try dividing by the next prime, 3.

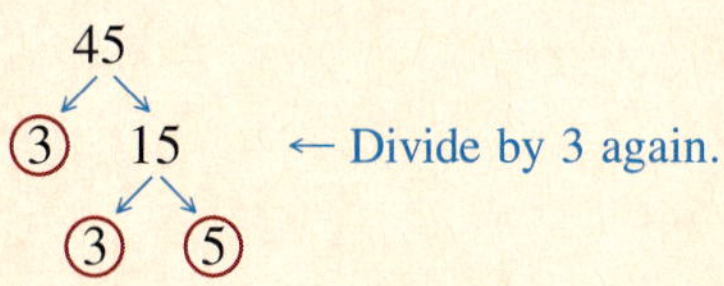

So 45 = **3 • 3 • 5**.

Note

Here is a reminder about the quick way to see whether a number can be divided evenly by 2, 3, or 5.

A number is divisible by 2 if the ones digit is 0, 2, 4, 6, or 8. For example, 30, 512, 76, and 3018 are all divisible by 2.

A number is divisible by 3 if the *sum* of the digits is divisible by 3. For example, 129 is divisible by 3 because 1 + 2 + 9 = 12 and 12 is divisible by 3.

A number is divisible by 5 if it has 0 or 5 in the ones place. For example, 85, 610, and 1725 are all divisible by 5.

See **Section R.4** for more information.

WORK PROBLEM 5 AT THE SIDE.

5. Complete each factor tree and give the prime factorization.

(a)

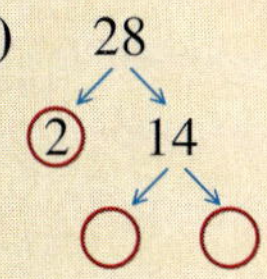

(b)

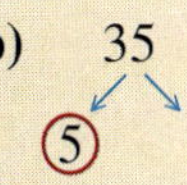

(c)

ANSWERS

5. (a)

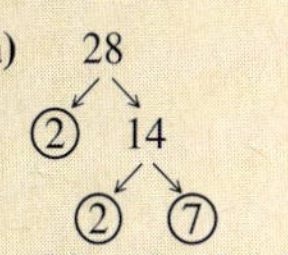

28 = 2 • 2 • 7

(b)

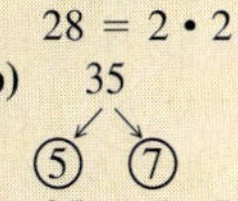

35 = 5 • 7

(c)

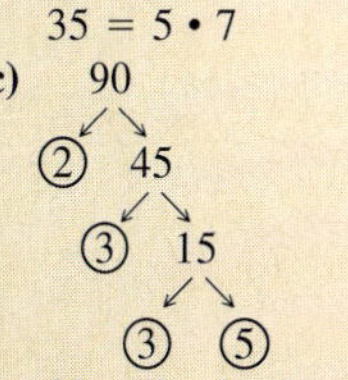

90 = 2 • 3 • 3 • 5

Calculator Tip: You can use your calculator to find the prime factorization of a number. Here is an example that uses 539.

Try dividing 539 by the first prime number, 2.

539 [÷] **2** [=] 269.5 Does not divide evenly.

Try dividing by the next prime number, 3.

539 [÷] **3** [=] 179.6666667 Does not divide evenly.

Keep trying the next prime numbers until you find one that divides evenly.

539 [÷] **5** [=] 107.8 Does not divide evenly.
539 [÷] **7** [=] 77 Divides evenly.

Once you have found that 7 works, try using it again.

77 [÷] **7** [=] 11 Divides evenly.

Because 11 is prime, you're finished. The prime factorization of 539 is $7 \cdot 7 \cdot 11$.

Now try factoring 2431. (The answer is at the bottom of this page.)

OBJECTIVE 4 Now you can use prime factorization to write fractions in lowest terms. This is a good method to use when the numerator and denominator are larger numbers.

EXAMPLE 6 Using Prime Factorization to Write Fractions in Lowest Terms

(a) Write $\frac{20}{35}$ in lowest terms.

20 can be written as $2 \cdot 2 \cdot 5$ (prime factors)
35 can be written as $5 \cdot 7$ (prime factors)

$$\frac{20}{35} = \frac{2 \cdot 2 \cdot 5}{5 \cdot 7}$$

The numerator and denominator have 5 as a common factor. Dividing both numerator and denominator by 5 will give an equivalent fraction.

$$\frac{20}{35} = \frac{2 \cdot 2 \cdot 5}{5 \cdot 7} = \frac{2 \cdot 2 \cdot 5}{7 \cdot 5} = \frac{2 \cdot 2 \cdot \boxed{5 \div 5}}{7 \cdot \boxed{5 \div 5}} = \frac{2 \cdot 2 \cdot 1}{7 \cdot 1} = \frac{4}{7}$$

Multiplication is commutative.

Any number divided by itself is 1.

$\frac{20}{35}$ is written in lowest terms as $\frac{4}{7}$.

To shorten the work, you may use slashes to indicate the divisions. For example, the work on $\frac{20}{35}$ can be shown as follows.

$$\frac{20}{35} = \frac{2 \cdot 2 \cdot \overset{1}{\cancel{5}}}{\underset{1}{\cancel{5}} \cdot 7}$$

Slashes indicate $5 \div 5$, and 1 is the result.

CONTINUED ON NEXT PAGE

Calculator Tip Answer: **2431 = 11 • 13 • 17**

(b) Write $\frac{60}{72}$ in lowest terms.

Use the prime factorizations of 60 and 72 from Example 5(a) and 5(b).

$$\frac{60}{72} = \frac{2 \cdot 2 \cdot 3 \cdot 5}{2 \cdot 2 \cdot 2 \cdot 3 \cdot 3}$$

This time there are three common factors. Use slashes to show the three divisions.

2 ÷ 2 is 1. 2 ÷ 2 is 1. 3 ÷ 3 is 1.

$$\frac{60}{72} = \frac{\overset{1}{\cancel{2}} \cdot \overset{1}{\cancel{2}} \cdot \overset{1}{\cancel{3}} \cdot 5}{\underset{1}{\cancel{2}} \cdot \underset{1}{\cancel{2}} \cdot 2 \cdot \underset{1}{\cancel{3}} \cdot 3} = \frac{5}{6}$$

5 ← Multiply $1 \cdot 1 \cdot 1 \cdot 5$ to get 5.
6 ← Multiply $1 \cdot 1 \cdot 2 \cdot 1 \cdot 3$ to get 6.

(c) $\frac{18}{90}$

$$\frac{18}{90} = \frac{\overset{1}{\cancel{2}} \cdot \overset{1}{\cancel{3}} \cdot \overset{1}{\cancel{3}}}{\underset{1}{\cancel{2}} \cdot \underset{1}{\cancel{3}} \cdot \underset{1}{\cancel{3}} \cdot 5} = \frac{1}{5}$$

1 ← Multiply $1 \cdot 1 \cdot 1$ to get 1.
5 ← Multiply $1 \cdot 1 \cdot 1 \cdot 5$ to get 5.

Note

In Example 6(c), all factors of the numerator divided out. But $1 \cdot 1 \cdot 1$ is still 1, so the final answer is $\frac{1}{5}$ (not 5).

This method of writing a fraction in lowest terms is summarized as follows.

Using Prime Factorization to Write Fractions in Lowest Terms

Step 1 Write the **prime factorization** of both numerator and denominator.

Step 2 Use slashes to show where you are **dividing** the numerator and denominator by any common factors.

Step 3 **Multiply** the remaining factors in the numerator and in the denominator.

WORK PROBLEM 6 AT THE SIDE.

OBJECTIVE 5 Fractions may have variables in the numerator or denominator. Examples are shown below.

$$\frac{6}{2x} \qquad \frac{3xy}{9xy} \qquad \frac{4b^3}{8ab} \qquad \frac{7ab^2}{n^2}$$

You can use prime factorization to write these fractions in lowest terms.

6. Use the method of prime factorization to write each fraction in lowest terms.

(a) $\frac{16}{48}$

(b) $\frac{28}{60}$

(c) $\frac{74}{111}$

(d) $\frac{124}{340}$

ANSWERS

6. (a) $\frac{1}{3}$ (b) $\frac{7}{15}$ (c) $\frac{2}{3}$ (d) $\frac{31}{85}$

7. Write each fraction in lowest terms.

(a) $\frac{5c}{15}$

(b) $\frac{10x^2}{8x^2}$

(c) $\frac{9a^3}{11b^3}$

(d) $\frac{6m^2n}{9n^2}$

ANSWERS

7. (a) $\frac{c}{3}$ (b) $\frac{5}{4}$ (c) already in lowest terms (d) $\frac{2m^2}{3n}$

EXAMPLE 7 Writing Fractions with Variables in Lowest Terms

(a) $\frac{6}{2x}$ ← Prime factors of 6 are $2 \cdot 3$. ← $2x$ means $2 \cdot x$.

$$\frac{6}{2x} = \frac{\overset{1}{\cancel{2}} \cdot 3}{\underset{1}{\cancel{2}} \cdot x} = \frac{3}{x}$$ ← $1 \cdot 3$ is 3. ← $1 \cdot x$ is x.

$3xy$ means $3 \cdot x \cdot y$.

(b) $$\frac{3xy}{9xy} = \frac{3 \cdot x \cdot y}{3 \cdot 3 \cdot x \cdot y} = \frac{\overset{1}{\cancel{3}} \cdot \overset{1}{\cancel{x}} \cdot \overset{1}{\cancel{y}}}{\underset{1}{\cancel{3}} \cdot 3 \cdot \underset{1}{\cancel{x}} \cdot \underset{1}{\cancel{y}}} = \frac{1}{3}$$

The prime factors of 9 are $3 \cdot 3$.

b^3 means $b \cdot b \cdot b$.

(c) $$\frac{4b^3}{8ab} = \frac{2 \cdot 2 \cdot b \cdot b \cdot b}{2 \cdot 2 \cdot 2 \cdot a \cdot b} = \frac{\overset{1}{\cancel{2}} \cdot \overset{1}{\cancel{2}} \cdot \overset{1}{\cancel{b}} \cdot b \cdot b}{\underset{1}{\cancel{2}} \cdot \underset{1}{\cancel{2}} \cdot 2 \cdot a \cdot \underset{1}{\cancel{b}}} = \frac{b^2}{2a}$$ ← $b \cdot b$ is b^2. ← $2 \cdot a$ is $2a$.

The prime factors of 8 are $2 \cdot 2 \cdot 2$.

(d) $$\frac{7ab^2}{n^2} = \frac{7 \cdot a \cdot b \cdot b}{n \cdot n}$$ There are no common factors.

$\frac{7ab^2}{n^2}$ is already in lowest terms.

◀◀ **WORK PROBLEM 7 AT THE SIDE.**

4.2 Exercises

Label each number as prime or composite.

1. 9 2 8 5 11 10 21

2. 12 3 7 6 15 13 25

Find the prime factorization of each number.

3. 6

4. 12

5. 20

6. 30

7. 25

8. 18

9. 36

10. 56

11. 44

12. 68

13. 88

14. 64

15. 75

16. 80

17. Give a definition of both a composite number and a prime number. Give three examples of each. Which whole numbers are neither prime nor composite?

18. With the exception of the number 2, all prime numbers are odd numbers. Nevertheless, all odd numbers are not prime numbers. Explain why these statements are true.

Write each numerator and denominator as a product of prime factors. Then use the prime factorization to write the fraction in lowest terms.

19. $\frac{8}{16}$

20. $\frac{6}{8}$

21. $\frac{32}{48}$

22. $\frac{9}{27}$

23. $\frac{14}{21}$

24. $\frac{20}{32}$

25. $\frac{36}{42}$

26. $\frac{22}{33}$

27. $\frac{63}{70}$

28. $\frac{72}{80}$

29. $\frac{27}{45}$

30. $\frac{36}{63}$

31. $\frac{12}{18}$

32. $\frac{63}{90}$

33. $\frac{35}{40}$

34. $\frac{36}{48}$

35. $\frac{90}{180}$

36. $\frac{16}{64}$

Ⓖ **37.** $\frac{210}{315}$

Ⓖ **38.** $\frac{96}{192}$

Ⓖ **39.** $\frac{429}{495}$

Ⓖ **40.** $\frac{135}{182}$

Write your answers in lowest terms.

41. There are 60 minutes in an hour.
- **(a)** What fraction of an hour is 15 minutes?
- **(b)** What fraction of an hour is 30 minutes?
- **(c)** What fraction of an hour is 6 minutes?
- **(d)** What fraction of an hour is 60 minutes?

42. There are 24 hours in a day.
- **(a)** What fraction of a day is 8 hours?
- **(b)** What fraction of a day is 18 hours?
- **(c)** What fraction of a day is 12 hours?
- **(d)** What fraction of a day is 3 hours?

43. SueLynn's monthly income is $1500.
- **(a)** She spends $500 on rent. What fraction of her income is spent on rent?
- **(b)** She spends $300 on food. What fraction of her income is spent on food?
- **(c)** What fraction of her income is left for other expenses?

44. There are 10,000 students at Minneapolis Community and Technical College.
- **(a)** 8000 of the students receive some form of financial aid. What fraction of the students receive financial aid?
- **(b)** 6000 of the students are women. What fraction are women?
- **(c)** What fraction of the students are men?

45. Explain the error in this problem and correct it.

$$\frac{9}{36} = \frac{\cancel{3} \cdot \cancel{3}}{2 \cdot 2 \cdot \cancel{3} \cdot \cancel{3}} = 4$$

46. Explain the error in this problem and correct it.

$$\frac{9}{16} = \frac{9 \div 3}{16 \div 4} = \frac{3}{4}$$

47. Explain how you could use your calculator to find the prime factorization of 437. Then find the prime factorization.

48. The text lists all the prime numbers less than 100. Use the divisibility rules and your calculator to find at least five prime numbers between 100 and 150.

Write each fraction in lowest terms.

49. $\frac{16c}{40}$

50. $\frac{36}{54a}$

51. $\frac{20x}{35x}$

52. $\frac{21n}{28n}$

53. $\frac{18r^2}{15rs}$

54. $\frac{18ab}{48b^2}$

55. $\frac{6m}{42mn^2}$

56. $\frac{10g^2}{90g^2h}$

57. $\frac{9x^2}{16y^2}$

58. $\frac{5rst}{8st}$

59. $\frac{7xz}{9xyz}$

60. $\frac{6a^3}{23b^3}$

61. $\frac{21k^3}{6k^2}$

62. $\frac{16x^3}{12x^4}$

63. $\frac{13a^2bc^3}{39a^2bc^3}$

64. $\frac{22m^3n^4}{55m^3n^4}$

65. $\frac{14c^2d}{14cd^2}$

66. $\frac{19rs}{19s^3}$

Ⓖ **67.** $\frac{210ab^3cd^2}{35b^2c^2de^2}$

Ⓖ **68.** $\frac{81w^4xy^2z}{300wy^4z^3}$

Review and Prepare

Multiply or divide, as indicated. (For help, see ***Sections 1.6*** *and* ***1.7****.)*

69. **(a)** $7(-4)$ **(b)** $-3 \cdot -9$ **(c)** $-8(10)$

70. **(a)** $-5(-4)$ **(b)** $-6 \cdot 9$ **(c)** $20(-5)$

71. **(a)** $-12 \div 0$ **(b)** $0 \div 8$ **(c)** $\frac{-9}{-1}$

72. **(a)** $0 \div (-3)$ **(b)** $\frac{7}{1}$ **(c)** $15 \div 0$

4.3 *Multiplying and Dividing Signed Fractions*

Objectives

1. Multiply signed fractions.
2. Multiply fractions that involve variables.
3. Divide signed fractions.
4. Divide fractions that involve variables.
5. Solve application problems involving multiplying and dividing fractions.

For Extra Help

Tutorial | Tape 5 | SSM, Sec. 4.3

Objective 1 Suppose that you give $\frac{1}{3}$ of your candy bar to your friend Ann. Then Ann gives $\frac{1}{2}$ of her share to Tim. How much of the bar does Tim get to eat?

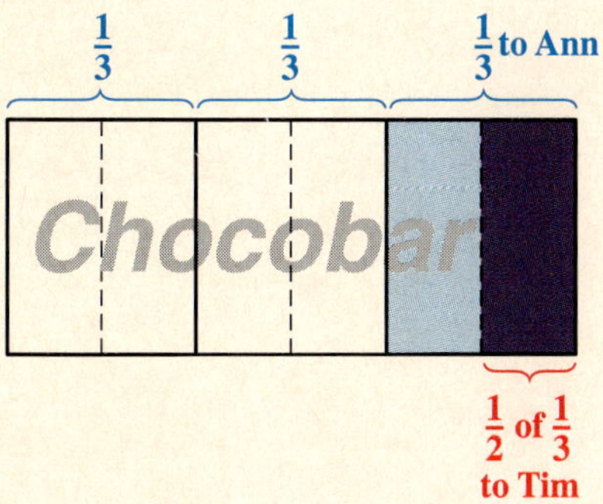

A sketch of the candy bar shows that Tim will get $\frac{1}{6}$ of the bar.

Tim's share is $\frac{1}{2}$ **of** $\frac{1}{3}$ candy bar. When used with fractions, the word **of** indicates multiplication.

$$\frac{1}{2} \textbf{ of } \frac{1}{3} \quad \text{means} \quad \frac{1}{2} \bullet \frac{1}{3}.$$

Tim's share is $\frac{1}{6}$ bar, so $\frac{1}{2} \bullet \frac{1}{3} = \frac{1}{6}$.
This example illustrates the rule for multiplying fractions.

Multiplying Fractions

If a, b, c, and d are numbers (but b and d are not zero), then

$$\frac{a}{b} \bullet \frac{c}{d} = \frac{a \bullet c}{b \bullet d}.$$

In other words, multiply the numerators and multiply the denominators.

When we apply this rule to find Tim's part of the bar, we get

$$\frac{1}{2} \bullet \frac{1}{3} = \frac{1 \bullet 1}{2 \bullet 3} = \frac{1}{6}. \quad \begin{matrix} \leftarrow \text{Multiply numerators.} \\ \leftarrow \text{Multiply denominators.} \end{matrix}$$

EXAMPLE 1 Multiplying Signed Fractions

Multiply.

(a) $-\frac{5}{8} \bullet -\frac{3}{4}$ Recall that the product of two negative numbers is a positive number.

Multiply the numerators and multiply the denominators.

$$-\frac{5}{8} \bullet -\frac{3}{4} = \frac{5 \bullet 3}{8 \bullet 4} = \frac{15}{32} \quad \text{Lowest terms.}$$

The product of two negative numbers is positive.

The answer is in lowest terms because 15 and 32 have no common factor other than 1.

CONTINUED ON NEXT PAGE

1. Multiply.

(a) $-\frac{3}{4} \cdot \frac{1}{2}$

(b) $\left(-\frac{2}{5}\right)\left(-\frac{2}{3}\right)$

(c) $\frac{3}{4}\left(\frac{3}{8}\right)$

(b) $\left(\frac{4}{7}\right)\left(-\frac{2}{5}\right) = -\frac{4 \cdot 2}{7 \cdot 5} = -\frac{8}{35}$ Recall that the product of a negative number and a positive number is negative.

◀◀ **WORK PROBLEM 1 AT THE SIDE.**

Sometimes the result won't be in lowest terms. For example, find $\frac{3}{10}$ of $\frac{5}{6}$.

$$\frac{3}{10} \text{ of } \frac{5}{6} \quad \text{means} \quad \frac{3}{10} \cdot \frac{5}{6} = \frac{3 \cdot 5}{10 \cdot 6} = \frac{15}{60} \quad \text{Not in lowest terms.}$$

Now write $\frac{15}{60}$ in lowest terms.

$$\frac{15}{60} = \frac{\overset{1}{\cancel{3}} \cdot \overset{1}{\cancel{5}}}{2 \cdot 2 \cdot \underset{1}{\cancel{3}} \cdot \underset{1}{\cancel{5}}} = \frac{1}{4} \quad \text{Lowest terms}$$

You used prime factorization in **Section 4.2** to write fractions in lowest terms. You can also use it when multiplying fractions. Writing the prime factors of the original fractions and dividing out common factors *before* multiplying usually saves time. If you divide out *all* the common factors, the result will automatically be in lowest terms. Let's see how that works when finding $\frac{3}{10}$ of $\frac{5}{6}$.

$$\frac{3}{\mathbf{10}} \cdot \frac{5}{\mathbf{6}} = \frac{3 \cdot 5}{\mathbf{2 \cdot 5 \cdot 2 \cdot 3}} = \frac{\overset{1}{\cancel{3}} \cdot \overset{1}{\cancel{5}}}{2 \cdot \underset{1}{\cancel{5}} \cdot 2 \cdot \underset{1}{\cancel{3}}} = \frac{1}{4}$$

3 and 5 are already prime. Write 10 as 2 • 5. Write 6 as 2 • 3. Divide common factors. Same result as above.

Note

When you are working with fractions, always write the final result in lowest terms. Visualizing $\frac{15}{60}$ is hard. But when $\frac{15}{60}$ is written as $\frac{1}{4}$, working with it is much easier. Lowest terms is the simplest way to write a fraction.

EXAMPLE 2 Using Prime Factorization to Multiply Fractions

(a) Multiply $-\frac{8}{5}\left(\frac{5}{12}\right)$.

Multiplying a negative number times a positive number gives a negative product.

$$-\frac{8}{5}\left(\frac{5}{12}\right) = -\frac{2 \cdot 2 \cdot 2 \cdot 5}{5 \cdot 2 \cdot 2 \cdot 3} = -\frac{\overset{1}{\cancel{2}} \cdot \overset{1}{\cancel{2}} \cdot 2 \cdot \overset{1}{\cancel{5}}}{\underset{1}{\cancel{5}} \cdot \underset{1}{\cancel{2}} \cdot \underset{1}{\cancel{2}} \cdot 3} = -\frac{2}{3} \quad \text{Lowest terms}$$

Write 8 as 2 • 2 • 2. 5 is already prime. Write 12 as 2 • 2 • 3. Negative product

CONTINUED ON NEXT PAGE

Answers

1. (a) $-\frac{3}{8}$ (b) $\frac{4}{15}$ (c) $\frac{9}{32}$

(b) Find $\frac{2}{9}$ of $\frac{15}{16}$.

Recall that, when used with fractions, *of* indicates multiplication.

2 is already prime. Write 15 as 3 • 5. Write 9 as 3 • 3. Write 16 as 2 • 2 • 2 • 2.

$$\frac{2}{9} \cdot \frac{15}{16} = \frac{2 \cdot 3 \cdot 5}{3 \cdot 3 \cdot 2 \cdot 2 \cdot 2 \cdot 2} = \frac{\overset{1}{\cancel{2}} \cdot \overset{1}{\cancel{3}} \cdot 5}{\underset{1}{\cancel{3}} \cdot 3 \cdot \underset{1}{\cancel{2}} \cdot 2 \cdot 2 \cdot 2} = \frac{5}{24} \quad \text{Lowest terms}$$

WORK PROBLEM 2 AT THE SIDE.

Calculator Tip: If your scientific calculator has a fraction key [a b/c] you can do calculations with fractions. You'll also need the change of sign key [+/−] to enter negative fractions, as you did to enter negative integers in **Section 1.4**.

Start by entering several different fractions. Clear your calculator after each one.

To enter $\frac{3}{4}$, press 3 [a b/c] 4. The display will show [3⌟4]. (← Fraction bar)

To enter $-\frac{9}{10}$, press 9 [a b/c] 10 [+/−]. The display will show [−9⌟10].

Try entering a fraction that is *not* in lowest terms. As soon as you press an operation key, such as [×] or [÷], the calculator will automatically show the fraction in lowest terms. Suppose that you start to enter the multiplication problem $\frac{4}{16} \cdot \frac{2}{3}$. Press 4 [a b/c] 16 [×]. The display shows [1⌟4], or $\frac{1}{4}$, which is $\frac{4}{16}$ in lowest terms. The calculator will always show fractions in lowest terms.

Let's check the result of Example 2(a): Multiply $-\frac{8}{5}\left(\frac{5}{12}\right)$ by pressing

8 [a b/c] 5 [+/−] [×] 5 [a b/c] 12 [=]. The display shows [−2⌟3]. ← $-\frac{2}{3}$

(8 [a b/c] 5 [+/−] is $-\frac{8}{5}$; 5 [a b/c] 12 is $\frac{5}{12}$)

Now try Example 2(b): Find $\frac{2}{9}$ of $\frac{15}{16}$. (Did you get $\frac{5}{24}$?)

There are some limitations to the calculations that you can do using the fraction key.

Try entering the fraction $\frac{9}{1000}$. What happens? (You can't enter denominators >999.)

Try doing this multiplication: $\frac{7}{10} \cdot \frac{3}{100}$. The result should be $\frac{21}{1000}$. What happens?
(The answer is given in decimal form because the denominator is >999.)

Note

The fraction key is useful for *checking* your work. But knowing the rules for fraction computation is important because you'll need them when fractions involve variables. You *cannot* enter fractions such as $\frac{3x}{5}$ or $\frac{9}{m^2}$ on your calculator (see Example 4, coming up shortly).

---

2. Use prime factorization to multiply these fractions.

(a) $\frac{15}{28} \cdot -\frac{6}{5}$

(b) $\frac{12}{7}\left(\frac{7}{24}\right)$

(c) $\left(-\frac{11}{18}\right)\left(-\frac{9}{20}\right)$

Answers

2. (a) $-\frac{3 \cdot \overset{1}{\cancel{5}} \cdot \overset{1}{\cancel{2}} \cdot 3}{2 \cdot \underset{1}{\cancel{2}} \cdot 7 \cdot \underset{1}{\cancel{5}}} = -\frac{9}{14}$

(b) $\frac{\overset{1}{\cancel{2}} \cdot \overset{1}{\cancel{2}} \cdot \overset{1}{\cancel{3}} \cdot \overset{1}{\cancel{7}}}{\underset{1}{\cancel{7}} \cdot \underset{1}{\cancel{2}} \cdot \underset{1}{\cancel{2}} \cdot 2 \cdot \underset{1}{\cancel{3}}} = \frac{1}{2}$

(c) $\frac{11 \cdot \overset{1}{\cancel{3}} \cdot \overset{1}{\cancel{3}}}{2 \cdot \underset{1}{\cancel{3}} \cdot \underset{1}{\cancel{3}} \cdot 2 \cdot 2 \cdot 5} = \frac{11}{40}$

3. Use prime factorization to find these products.

(a) $\frac{3}{4}$ of 36

(b) $-10 \cdot \frac{2}{5}$

(c) $-\frac{7}{8} \cdot -24$

4. Use prime factorization to find these products.

(a) $\frac{2c}{5} \cdot \frac{c}{4}$

(b) $\left(\frac{m}{6}\right)\left(\frac{9}{m^2}\right)$

(c) $\frac{w^2}{y} \cdot \frac{x^2y}{w}$

ANSWERS

3. (a) $\frac{3 \cdot \cancel{2} \cdot \cancel{2} \cdot 3 \cdot 3}{\cancel{2} \cdot \cancel{2} \cdot 1} = \frac{27}{1} = 27$

(b) $-\frac{2 \cdot \cancel{5} \cdot 2}{1 \cdot \cancel{5}} = -\frac{4}{1} = -4$

(c) $\frac{7 \cdot \cancel{2} \cdot \cancel{2} \cdot \cancel{2} \cdot 3}{\cancel{2} \cdot \cancel{2} \cdot \cancel{2}} = \frac{21}{1} = 21$

4. (a) $\frac{\cancel{2} \cdot c \cdot c}{5 \cdot \cancel{2} \cdot 2} = \frac{c^2}{10}$

(b) $\frac{\cancel{m} \cdot \cancel{3} \cdot 3}{2 \cdot \cancel{3} \cdot \cancel{m} \cdot m} = \frac{3}{2m}$

(c) $\frac{\cancel{w} \cdot w \cdot x \cdot x \cdot \cancel{y}}{\cancel{y} \cdot \cancel{w}} = \frac{wx^2}{1} = wx^2$

EXAMPLE 3 Multiplying a Fraction and an Integer

Find $\frac{2}{3}$ of 6.

We can write 6 in fraction form as $\frac{6}{1}$. Recall that $\frac{6}{1}$ means $6 \div 1$, which is 6. So we can write any integer a as $\frac{a}{1}$.

$$\frac{2}{3} \text{ of } 6 \quad \text{means} \quad \frac{2}{3} \cdot \frac{6}{1} = \frac{2 \cdot 2 \cdot \cancel{3}}{\cancel{3} \cdot 1} = \frac{4}{1} = 4.$$

($4 \div 1$ is 4)

◀◀ WORK PROBLEM 3 AT THE SIDE.

OBJECTIVE 2 The multiplication method involving the use of prime factors also works when there are variables in the numerators and/or denominators of the fractions.

EXAMPLE 4 Multiplying Fractions with Variables

Multiply.

(a) $\frac{3x}{5} \cdot \frac{2}{9x}$

$3x$ means $3 \cdot x$.

$$\frac{3x}{5} \cdot \frac{2}{9x} = \frac{3 \cdot x \cdot 2}{5 \cdot 3 \cdot 3 \cdot x} = \frac{\cancel{3} \cdot \cancel{x} \cdot 2}{5 \cdot \cancel{3} \cdot 3 \cdot \cancel{x}} = \frac{2}{15}$$

$\frac{x}{x}$ is 1.

The prime factors of 9 are $3 \cdot 3$, so $9x$ is $3 \cdot 3 \cdot x$.

(b) $\left(\frac{3y}{4x}\right)\left(\frac{2x^2}{y}\right)$

$2x^2$ means $2 \cdot x \cdot x$.

$$\left(\frac{3y}{4x}\right)\left(\frac{2x^2}{y}\right) = \frac{3 \cdot y \cdot 2 \cdot x \cdot x}{2 \cdot 2 \cdot x \cdot y} = \frac{3 \cdot \cancel{y} \cdot \cancel{2} \cdot \cancel{x} \cdot x}{2 \cdot \cancel{2} \cdot \cancel{x} \cdot \cancel{y}} = \frac{3x}{2}$$

The prime factors of 4 are $2 \cdot 2$, so $4x$ is $2 \cdot 2 \cdot x$.

◀◀ WORK PROBLEM 4 AT THE SIDE.

OBJECTIVE 3 To divide fractions, we will rewrite division problems as multiplication problems. Recall that you rewrote subtraction problems as addition problems in **Section 1.4**. For subtraction, you left the first number as it was, and changed the second number to it's opposite. This time, for division, you will also leave the first number (the dividend) as it is but change the second number (the divisor) to its *reciprocal* (ree-SIP-ruh-kul).

Reciprocal of a Fraction

Two numbers are **reciprocals** of each other if their product is 1. The reciprocal of the fraction $\frac{a}{b}$ is $\frac{b}{a}$ because

$$\frac{a}{b} \cdot \frac{b}{a} = \frac{\overset{1}{\cancel{a}} \cdot \overset{1}{\cancel{b}}}{\underset{1}{\cancel{b}} \cdot \underset{1}{\cancel{a}}} = \frac{1}{1} = 1.$$

Notice that you "flip" or "invert" a fraction to find its reciprocal. Here are some examples.

Number	Reciprocal	Reason
$\frac{1}{6}$	$\frac{6}{1}$	Because $\frac{1}{6} \cdot \frac{6}{1} = \frac{6}{6} = 1$
$-\frac{2}{5}$	$-\frac{5}{2}$	Because $\left(-\frac{2}{5}\right)\left(-\frac{5}{2}\right) = \frac{10}{10} = 1$
4 Think of 4 as $\frac{4}{1}$.	$\frac{1}{4}$	Because $4 \cdot \frac{1}{4} = \frac{4}{1} \cdot \frac{1}{4} = \frac{4}{4} = 1$

Note

Every number has a reciprocal except 0. Why not 0? Recall that a number times its reciprocal equals 1. But that doesn't work for 0.

$$0 \cdot (\text{reciprocal}) = 1$$

Put any number here. When you multiply it by 0, you get 0, never 1.

Dividing Fractions

If a, b, c, and d are numbers (but b, c, and d are not zero), then we have the following.

$$\frac{a}{b} \div \frac{c}{d} = \frac{a}{b} \cdot \frac{d}{c}$$

reciprocals

In other words, change division to multiplying by the reciprocal of the divisor.

Now you can find the quotient for $\frac{2}{3} \div \frac{1}{6}$. Rewrite it as a multiplication problem and then use the steps for multiplying fractions.

Change division to multiplication.

$$\frac{2}{3} \div \frac{1}{6} = \frac{2}{3} \cdot \frac{6}{1} = \frac{2 \cdot 2 \cdot \overset{1}{\cancel{3}}}{\underset{1}{\cancel{3}} \cdot 1} = \frac{4}{1} = 4$$

The reciprocal of $\frac{1}{6}$ is $\frac{6}{1}$.

Let's see if a quotient of 4 makes sense by comparing division of fractions to division of whole numbers.

$12 \div 3$ is asking, "How many 3's are in 12?"

$\frac{2}{3} \div \frac{1}{6}$ is asking, "How many $\frac{1}{6}$'s are in $\frac{2}{3}$?"

This figure illustrates $\frac{2}{3} \div \frac{1}{6}$.

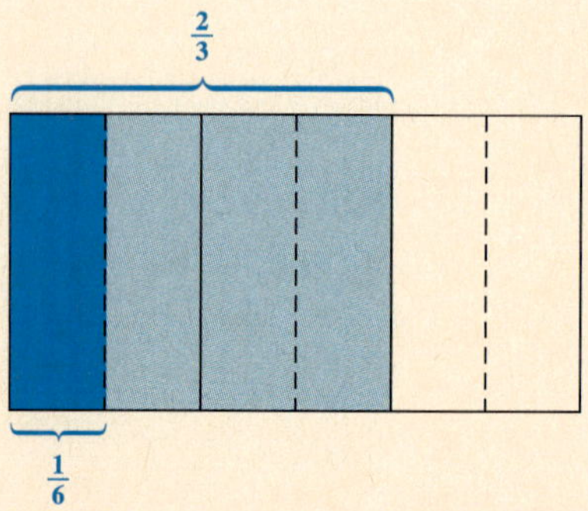

How many $\frac{1}{6}$'s are in $\frac{2}{3}$?
There are 4 of the $\frac{1}{6}$ pieces in $\frac{2}{3}$.
So $\frac{2}{3} \div \frac{1}{6} = 4$.

As a final check on this method of dividing, try changing $12 \div 3$ into a multiplication problem. You know that the quotient should be 4.

$$12 \div 3 = 12 \cdot \frac{1}{3} = \frac{12}{1} \cdot \frac{1}{3} = \frac{2 \cdot 2 \cdot \overset{1}{\cancel{3}} \cdot 1}{1 \cdot \underset{1}{\cancel{3}}} = \frac{4}{1} = 4 \leftarrow \text{The quotient we expected}$$

reciprocals

EXAMPLE 5 Dividing Signed Fractions

Rewrite each division problem as a multiplication problem. Then multiply.

(a) $$\frac{3}{10} \div \frac{4}{5} = \frac{3}{10} \cdot \frac{5}{4} = \frac{3 \cdot \overset{1}{\cancel{5}}}{2 \cdot \underset{1}{\cancel{5}} \cdot 2 \cdot 2} = \frac{3}{8}$$

reciprocals

Note

When multiplying fractions, you don't always have to factor the numerator and denominator completely into prime numbers. In part (a) above, if you notice that 5 is a common factor of the numerator and denominator, you can write

$$\frac{3}{10} \cdot \frac{5}{4} = \frac{3 \cdot \overset{1}{\cancel{5}}}{2 \cdot \cancel{5} \cdot 4} = \frac{3}{8}.$$

Factor 10 into 2 • 5.

Leave 4 as it is.

If no common factors are obvious to you, then write out the complete prime factorization to help you find the common factors.

CONTINUED ON NEXT PAGE

(b) $2 \div \left(-\frac{1}{3}\right)$

First notice that the numbers have different signs. In a division problem, different signs mean that the quotient is negative. Then write 2 in fraction form as $\frac{2}{1}$.

$$2 \div \left(-\frac{1}{3}\right) = \frac{2}{1} \cdot \left(-\frac{3}{1}\right) = -\frac{2 \cdot 3}{1 \cdot 1} = -\frac{6}{1} = -6$$

reciprocals

Negative product

(c) $-\frac{3}{4} \div (-8) = -\frac{3}{4} \cdot \left(-\frac{1}{8}\right) = \frac{3 \cdot 1}{4 \cdot 8} = \frac{3}{32}$

reciprocals

No common factor to divide out

The quotient is positive because both numbers in the problem were negative. When signs match, the quotient is positive.

(d) $\frac{9}{16} \div 0$

This is *undefined,* as dividing by 0 was undefined for integers (see **Section 1.7**). Recall that 0 does *not* have a reciprocal, so you can't change the division to multiplying by the reciprocal of the divisor.

(e) $0 \div \frac{9}{16} = 0 \cdot \frac{16}{9} = 0$ Recall that zero divided by any number gives a result of zero.

reciprocals

WORK PROBLEM 5 AT THE SIDE.

OBJECTIVE 4 The method for dividing fractions also works when there are variables in the numerators and/or denominators of the fractions.

EXAMPLE 6 Dividing Fractions with Variables

(a) $\frac{x^2}{y} \div \frac{x}{3y} = \frac{x^2}{y} \cdot \frac{3y}{x} = \frac{\overset{1}{\cancel{x}} \cdot x \cdot 3 \cdot \overset{1}{\cancel{y}}}{\underset{1}{\cancel{y}} \cdot \underset{1}{\cancel{x}}} = \frac{3x}{1} = 3x$

reciprocals

(b) $\frac{8b}{5} \div b^2 = \frac{8b}{5} \cdot \frac{1}{b^2} = \frac{8 \cdot \overset{1}{\cancel{b}} \cdot 1}{5 \cdot \underset{1}{\cancel{b}} \cdot b} = \frac{8}{5b}$

Write b^2 as $\frac{b^2}{1}$.
The reciprocal is $\frac{1}{b^2}$.

WORK PROBLEM 6 AT THE SIDE.

5. Rewrite each division problem as a multiplication problem. Then multiply.

(a) $-\frac{3}{4} \div \frac{5}{8}$

(b) $0 \div \left(-\frac{7}{12}\right)$

(c) $\frac{5}{6} \div 10$

(d) $-9 \div \left(-\frac{9}{16}\right)$

(e) $\frac{2}{5} \div 0$

6. Divide.

(a) $\frac{c^2d^2}{4} \div \frac{c^2d}{4}$

(b) $\frac{20}{7h} \div \frac{5h}{7}$

(c) $\frac{n}{8} \div mn$

ANSWERS

5. (a) $-\frac{3}{4} \cdot \frac{8}{5} = -\frac{3 \cdot 2 \cdot \overset{1}{\cancel{4}}}{\underset{1}{\cancel{4}} \cdot 5} = -\frac{6}{5}$

(b) $0 \cdot \left(-\frac{12}{7}\right) = 0$

(c) $\frac{5}{6} \cdot \frac{1}{10} = \frac{\overset{1}{\cancel{5}} \cdot 1}{6 \cdot 2 \cdot \underset{1}{\cancel{5}}} = \frac{1}{12}$

(d) $-\frac{9}{1} \cdot \left(-\frac{16}{9}\right) = \frac{\overset{1}{\cancel{9}} \cdot 16}{1 \cdot \underset{1}{\cancel{9}}} = \frac{16}{1} = 16$

(e) undefined; can't be written as multiplication because 0 doesn't have a reciprocal

6. (a) d **(b)** $\frac{4}{h^2}$ **(c)** $\frac{1}{8m}$

7. Look for indicator words or draw sketches to help you with these problems.

(a) How many times can a $\frac{2}{3}$ quart spray bottle be filled before 18 quarts of window cleaner are used up?

(b) A retiring police officer will receive $\frac{5}{8}$ of her highest annual salary as retirement income. If her highest annual salary is \$48,000, how much will she receive as retirement income?

Answers

7. (a) $18 \div \frac{2}{3} = \frac{18}{1} \cdot \frac{3}{2} = \frac{\overset{1}{\cancel{2}} \cdot 9 \cdot 3}{1 \cdot \underset{1}{\cancel{2}}} = \frac{27}{1} = 27$

The bottle can be filled 27 times.

(b) $\frac{5}{8} \cdot \frac{48{,}000}{1} = \frac{5 \cdot \overset{1}{\cancel{8}} \cdot 6000}{\underset{1}{\cancel{8}} \cdot 1} = \frac{30{,}000}{1} = 30{,}000$

The officer will receive \$30,000.

Objective 5 When you're solving application problems, some indicator words are used to suggest multiplication and some are used to suggest division.

Indicator Words for Multiplication	Indicator Words for Division
product	per
double	each
triple	goes into
times	divided by
twice	divided into
of (when it follows a fraction)	divided equally

Look for these indicator words in the following examples. However, you won't always find an indicator word. Then, you need to think through the problem to decide what to do. Sometimes, drawing a sketch of the situation described in the problem will help you decide which operation to use.

EXAMPLE 7 Using Indicator Words and Sketches to Solve Application Problems

(a) Lois gives $\frac{1}{10}$ **of** her income to her church. Last month she earned \$1980. How much of that did she give to her church?

Notice the word **of**. Because the word **of** *follows the fraction* $\frac{1}{10}$, it indicates multiplication.

$$\frac{1}{10} \text{ of } 1980 = \frac{1}{10} \cdot \frac{1980}{1} = \frac{1 \cdot \overset{1}{\cancel{10}} \cdot 198}{\underset{1}{\cancel{10}} \cdot 1} = \frac{198}{1} = 198$$

Lois gave \$198 to her church.

(b) The apparel design class is making infant snowsuits to give to a local shelter. A fabric store donated 12 yd of fabric for the project. If one snowsuit needs $\frac{2}{3}$ yd of fabric, how many suits can the class make?

The word **of** appears in the second sentence: "A fabric store donated 12 yd **of** fabric." But there is *no fraction* next to the word **of**, so it is *not* an indicator to multiply. Let's try a sketch. There is a 12 yd piece of fabric. One snowsuit will use $\frac{2}{3}$ yd. The question is, how many $\frac{2}{3}$ yd pieces can be cut from the 12 yards?

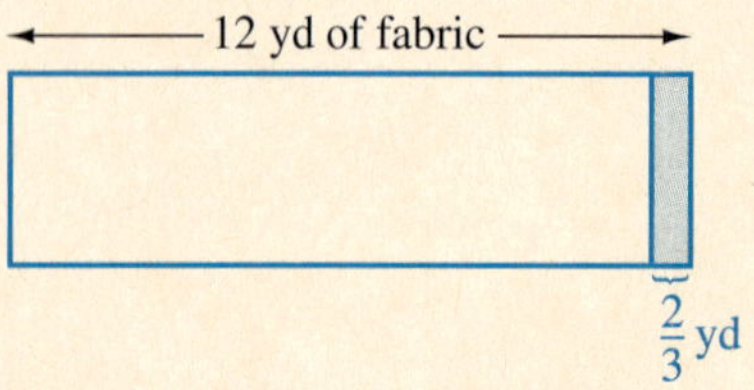

Cutting 12 yards into equal size pieces indicates division. How many $\frac{2}{3}$'s are in 12?

$$12 \div \frac{2}{3} = \frac{12}{1} \cdot \frac{3}{2} = \frac{\overset{1}{\cancel{2}} \cdot 6 \cdot 3}{1 \cdot \underset{1}{\cancel{2}}} = \frac{18}{1} = 18$$

reciprocals

The class can make 18 snowsuits.

WORK PROBLEM 7 AT THE SIDE.

4.3 Exercises

Multiply. Write the products in lowest terms.

1. $-\frac{3}{8} \cdot \frac{1}{2}$

2. $\left(\frac{2}{3}\right)\left(-\frac{5}{7}\right)$

3. $\left(-\frac{3}{8}\right)\left(-\frac{12}{5}\right)$

4. $\frac{4}{9} \cdot \frac{12}{7}$

5. $\frac{21}{30}\left(\frac{5}{7}\right)$

6. $\left(-\frac{6}{11}\right)\left(-\frac{22}{15}\right)$

7. $10\left(-\frac{3}{5}\right)$

8. $-20\left(\frac{3}{4}\right)$

9. $\frac{4}{9}$ of 81

10. $\frac{2}{3}$ of 48

11. $\frac{3x}{4} \cdot \frac{5}{xy}$

12. $\frac{2}{5a^2} \cdot \frac{a}{8}$

Divide. Write the quotients in lowest terms.

13. $\frac{1}{6} \div \frac{1}{3}$

14. $-\frac{1}{2} \div \frac{2}{3}$

15. $-\frac{3}{4} \div \left(-\frac{5}{8}\right)$

16. $\frac{7}{10} \div \frac{2}{5}$

17. $6 \div \left(-\frac{2}{3}\right)$

18. $-7 \div \left(-\frac{1}{4}\right)$

19. $-\frac{2}{3} \div 4$

20. $\frac{5}{6} \div (-15)$

21. $\frac{11c}{5d} \div 3c$

22. $8x^2 \div \frac{4x}{7}$

23. $\frac{ab^2}{c} \div \frac{ab}{c}$

24. $\frac{mn}{6} \div \frac{n}{3m}$

25. **(a)** Find each product. Then state what property these problems illustrate.

$\frac{3}{8} \cdot \frac{2}{5} =$ _____ $\frac{2}{5} \cdot \frac{3}{8} =$ _____

(b) Find each quotient. Then state what these problems illustrate.

$6 \div \frac{1}{2} =$ _____ $\frac{1}{2} \div 6 =$ _____

26. **(a)** Find each product. Then state what property these problems illustrate.

$\frac{7}{16} \cdot 1 =$ _____ $1 \cdot \left(-\frac{11}{12}\right) =$ _____

(b) Find each quotient. Then state what property these problems illustrate.

$-\frac{5}{4} \div \left(-\frac{5}{4}\right) =$ _____ $\frac{7}{10} \div \frac{7}{10} =$ _____

27. Explain the error in each of these calculations and correct it.

(a) $\frac{3}{4} \cdot \frac{8}{9} = \frac{3 \cdot 8}{4 \cdot 9} = \frac{24}{36}$

(b) $\frac{2}{3} \div 4 = \frac{2}{3} \cdot \frac{4}{1} = \frac{2 \cdot 4}{3 \cdot 1} = \frac{8}{3}$

(c) $\frac{\overset{1}{\cancel{3}}}{10} \div \frac{1}{\underset{1}{\cancel{3}}} = \frac{1}{10}$

(d) $\frac{1}{2} \div 0 = 0$

28. Explain the error in each of these calculations and correct it.

(a) $\frac{3}{14} \cdot \frac{7}{9} = \frac{\cancel{3} \cdot \cancel{7}}{2 \cdot \cancel{7} \cdot \cancel{3} \cdot 3} = 6$

(b) $\frac{5}{6} \div \frac{10}{9} = \frac{6}{5} \cdot \frac{10}{9} = \frac{2 \cdot \overset{1}{\cancel{3}} \cdot 2 \cdot \overset{1}{\cancel{5}}}{\underset{1}{\cancel{5}} \cdot \underset{1}{\cancel{3}} \cdot 3} = \frac{4}{3}$

(c) $8 \cdot \frac{2}{3} = \frac{8}{1} \cdot \frac{3}{2} = \frac{\overset{1}{\cancel{2}} \cdot 4 \cdot 3}{1 \cdot \underset{1}{\cancel{2}}} = \frac{12}{1} = 12$

(d) $0 \div \frac{1}{2} = \frac{1}{2}$

29. Your friend missed class and is confused about how to divide fractions. Write a short explanation for your friend.

30. Mary spilled coffee on her math homework, and part of one problem is covered up.

$$\frac{3}{\blacksquare} \div \frac{4}{5}$$

She knows the answer given in the back of the book is $\frac{3}{4}$. Describe how to find the missing number.

Find each product or quotient. Write all answers in lowest terms.

31. $\frac{4}{5} \div 3$

32. $\left(-\frac{20}{21}\right)\left(-\frac{14}{15}\right)$

33. $-\frac{3}{8}\left(\frac{3}{4}\right)$

34. $-\frac{8}{17} \div \frac{4}{5}$

35. $\frac{3}{5}$ of 35

36. $\frac{2}{3} \div (-6)$

37. $-9 \div \left(-\frac{3}{5}\right)$

38. $\frac{7}{8} \cdot \frac{25}{21}$

39. $\frac{12}{7} \div 0$

40. $\frac{5}{8}$ of (-48)

41. $\left(\frac{11}{2}\right)\left(-\frac{5}{6}\right)$

42. $\frac{3}{4} \div \frac{3}{16}$

43. $\frac{4}{7}$ of $14b$

44. $\frac{ab}{6} \div \frac{b}{9}$

45. $\frac{12}{5} \div 4d$

46. $\frac{18}{7} \div 2t$

47. $\frac{x^2}{y} \div \frac{w}{2y}$

48. $\frac{5}{6}$ of $18w$

Solve each application problem.

49. Joyce Chen wants to make vests to sell at a craft fair. Each vest requires $\frac{3}{4}$ yd of material. She has 36 yd of material. Find the number of vests she can make.

50. Al is helping Tim make a mahogany lamp table for Jill's birthday. Find the area of the top of the table if it is $\frac{4}{5}$ yd long by $\frac{3}{8}$ yd wide.

51. Parking lot A is $\frac{1}{4}$ mile long and $\frac{3}{16}$ mile wide, and parking lot B is $\frac{3}{8}$ mile long and $\frac{1}{8}$ mile wide. Which parking lot has the larger area?

52. The Sweepstakes Lottery pays out $\frac{7}{8}$ of the total revenue to 14 top winners. What fraction of the total revenue does each winner receive?

53. How many $\frac{1}{8}$-ounce eye drop dispensers can be filled with 10 ounces of eye drops?

54. Todd estimates that it will cost him \$12,400, including living expenses, to attend college for one year. If he must earn $\frac{3}{4}$ of the cost and borrow the balance, find the amount that he must earn and the amount he must borrow.

55. A motorcycle race course is $\frac{4}{3}$ mile wide by 6 miles long. Find the area of the race course.

56. Ms. Shaffer has a piece of property with an area that is $\frac{9}{10}$ acre. She wishes to divide it into three equal parts for her children. How many acres of land will each child get?

57. At the Garlic Festival Fun Run, $\frac{5}{12}$ of the runners are women. If there are 780 runners, how many are women? How many are men?

58. There are 25 New England residents on *Forbes* magazine's list of the 400 wealthiest Americans. Two-fifths of the New Englanders had fortunes of more than \$700 million. How many New Englanders had fortunes of more than \$700 million?

Ⓖ *The following table shows the earnings for the Gomez family last year and the circle graph shows how they spent their earnings. Use this information to solve Exercises 59–62.*

Month	*Earnings*	*Month*	*Earnings*
January	\$3050	July	\$3160
February	\$2875	August	\$2355
March	\$3325	September	\$2780
April	\$3020	October	\$3675
May	\$2880	November	\$3310
June	\$3265	December	\$4305

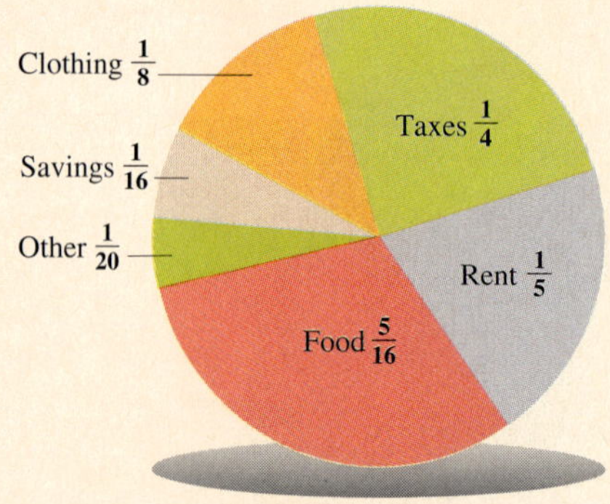

59. **(a)** What was the family's total income for the year?
(b) Find the amount of the family's rent for the year.

60. **(a)** How much did the family pay in taxes during the year?
(b) How much more did the family spend on taxes than on rent?

61. How much did the family spend for food and clothing last year?

62. Find the amount the family saved during the year.

63. Pam Trizlia has a small pickup truck that can carry $\frac{2}{3}$ cord of firewood. Find the number of trips needed to deliver 6 cords of wood.

64. A dog bed is $\frac{7}{8}$ yd by $\frac{10}{9}$ yd. Find its area.

Review and Prepare

*Add, or subtract by changing subtraction to addition. (For help, see **Sections 1.3** and **1.4**.)*

65. $9 - 10$

66. $3 - 7$

67. $-11 + 5$

68. $-3 + 4$

69. $-2 - 3$

70. $-10 - 4$

71. $9 + (-5)$

72. $12 + (-10)$

4.4 Adding and Subtracting Signed Fractions

OBJECTIVE 1 You probably remember learning something about "common denominators" in other math classes. When fractions have the *same* denominator, we say that they have a *common* denominator, which makes them **like fractions.** When fractions have different denominators, they are called **unlike fractions.** Here are some examples.

LIKE FRACTIONS	UNLIKE FRACTIONS
$\frac{3}{4}$ and $-\frac{7}{4}$ (Common denominator)	$\frac{2}{9}$ and $\frac{2}{8}$ (Different denominators)
$\frac{6}{x}$ and $\frac{y}{x}$ (Common denominator)	$\frac{a^2}{3}$ and $\frac{3}{a}$ (Different denominators)

You can add or subtract fractions *only* when they have a common denominator. To see why, let's look at more pizzas.

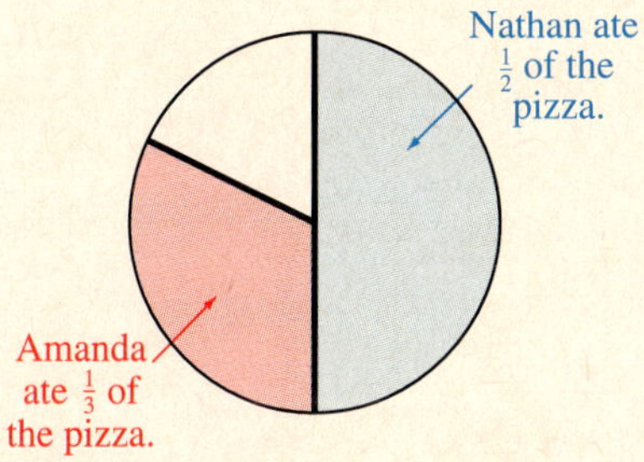

What fraction of the pizza has been eaten? We can't write a fraction until the pizza is cut into pieces of *equal* size. That's what the denominator of a fraction tells us: the number of *equal* size pieces in the pizza.

Now the pizza is cut into 6 *equal* pieces, and we can find out how much was eaten.

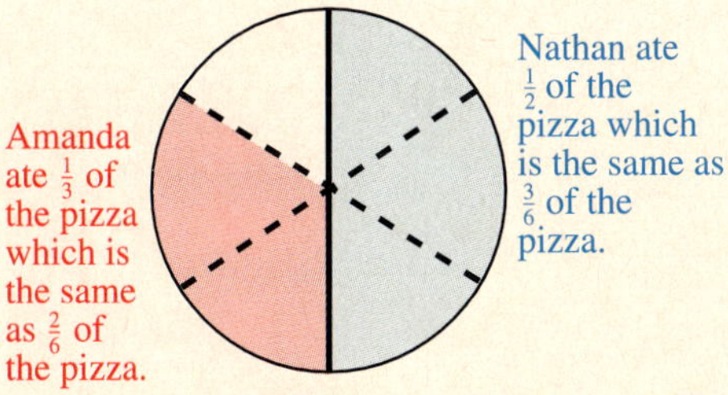

$$\frac{1}{2} + \frac{1}{3}$$
$$\downarrow \qquad \downarrow$$
$$\frac{3}{6} + \frac{2}{6}$$
$$= \frac{5}{6}$$

5 ← Number of pieces eaten
6 ← Number of equal pieces in the pizza

Adding and Subtracting Like Fractions

You can add or subtract fractions **only** when they have a common denominator. If a, b, and c are numbers (and b is not 0), then

$$\frac{a}{b} + \frac{c}{b} = \frac{a + c}{b} \quad \text{and} \quad \frac{a}{b} - \frac{c}{b} = \frac{a - c}{b}.$$

In other words, add or subtract the numerators and write the result over the common denominator. Then check to be sure that the answer is in lowest terms.

OBJECTIVES

1. Add and subtract like fractions.
2. Find the lowest common denominator for unlike fractions.
3. Add and subtract unlike fractions.
4. Add and subtract unlike fractions that contain variables.

FOR EXTRA HELP

1. Find each sum or difference.

(a) $\frac{1}{6} + \frac{5}{6}$

(b) $-\frac{11}{12} + \frac{5}{12}$

(c) $-\frac{2}{9} - \frac{3}{9}$

(d) $\frac{8}{ab} + \frac{3}{ab}$

ANSWERS

1. (a) $\frac{6}{6} = 1$ (b) $-\frac{1}{2}$ (c) $-\frac{5}{9}$ (d) $\frac{11}{ab}$

EXAMPLE 1 Adding and Subtracting Like Fractions

Find each sum or difference.

(a) $\frac{1}{8} + \frac{3}{8}$

These are *like* fractions because they have a *common denominator of 8*. So they are ready to be added. Add the numerators and write the sum over the common denominator.

$$\frac{1}{8} + \frac{3}{8} = \frac{1+3}{8} = \frac{4}{8}$$

Common denominator

Now write $\frac{4}{8}$ in lowest terms.

$$\frac{4}{8} = \frac{\overset{1}{\cancel{2}} \cdot \overset{1}{\cancel{2}}}{\underset{1}{\cancel{2}} \cdot \underset{1}{\cancel{2}} \cdot 2} = \frac{1}{2}$$

The sum, in lowest terms, is $\frac{1}{2}$.

Note

Add **only** the numerators. **Do not add the denominators.** In part (a) above we **kept the common denominator.**

$$\frac{1}{8} + \frac{3}{8} = \frac{1+3}{8} \quad \textbf{not} \quad \frac{1}{8} + \frac{3}{8} = \cancel{\frac{1+3}{8+8} = \frac{4}{16}}$$

Incorrect

To help you understand why we add *only* the numerators, think of $\frac{1}{8}$ as $1(\frac{1}{8})$ and $\frac{3}{8}$ as $3(\frac{1}{8})$. Then we use the distributive property.

$$\frac{1}{8} + \frac{3}{8} = 1\left(\frac{1}{8}\right) + 3\left(\frac{1}{8}\right) = (1+3)\left(\frac{1}{8}\right) = 4\left(\frac{1}{8}\right) = \frac{4}{8} = \frac{1}{2}$$

Use the distributive property.

(b) $-\frac{3}{5} + \frac{4}{5} = \frac{-3+4}{5} = \frac{1}{5}$ Lowest terms

Common denominator

Rewrite subtraction as adding the opposite.

(c) $\frac{3}{10} - \frac{7}{10} = \frac{3-7}{10} = \frac{3+(-7)}{10} = \frac{-4}{10}$ or $-\frac{4}{10}$

Common denominator

Now write $-\frac{4}{10}$ in lowest terms.

$$-\frac{4}{10} = -\frac{\overset{1}{\cancel{2}} \cdot 2}{\underset{1}{\cancel{2}} \cdot 5} = -\frac{2}{5}$$ Lowest terms

(d) $\frac{5}{x^2} - \frac{2}{x^2} = \frac{5-2}{x^2} = \frac{3}{x^2}$

Common denominator

◀◀ WORK PROBLEM 1 AT THE SIDE.

OBJECTIVE 2 When we first tried to add the pizza eaten by Nathan and Amanda, we could *not* do so because the pizza was *not cut into pieces of the same size.* So we rewrote $\frac{1}{2}$ and $\frac{1}{3}$ as equivalent fractions that both had 6 as the common denominator.

$$\frac{1}{2} = \frac{1 \cdot 3}{2 \cdot 3} = \frac{3}{6}$$
$$\frac{1}{3} = \frac{1 \cdot 2}{3 \cdot 2} = \frac{2}{6}$$

Common denominator

Then we could add the fractions, because the pizza was cut into pieces of *equal* size.

$$\frac{1}{2} + \frac{1}{3} = \frac{3}{6} + \frac{2}{6} = \frac{3+2}{6} = \frac{5}{6}$$ Lowest terms

In other words, when you want to add or subtract unlike fractions, the first thing you must do is rewrite them so that they have a common denominator.

Finding a Common Denominator for Unlike Fractions

To find a common denominator for two unlike fractions, find a number that is evenly divisible by *both* of the original denominators. For example, the common denominator for $\frac{1}{2}$ and $\frac{1}{3}$ is 6 because 2 goes into 6 evenly and 3 goes into 6 evenly.

Notice that 12 is also a common denominator for $\frac{1}{2}$ and $\frac{1}{3}$ because 2 and 3 both go into 12 evenly.

$$\frac{1}{2} = \frac{1 \cdot 6}{2 \cdot 6} = \frac{6}{12}$$
$$\frac{1}{3} = \frac{1 \cdot 4}{3 \cdot 4} = \frac{4}{12}$$

Common denominator

Now that the fractions have a common denominator, we can add them.

$$\frac{1}{2} + \frac{1}{3} = \frac{6}{12} + \frac{4}{12} = \frac{6+4}{12} = \frac{10}{12}$$ Not in lowest terms but $$\frac{10}{12} = \frac{\overset{1}{\cancel{2}} \cdot 5}{\underset{1}{\cancel{2}} \cdot 6} = \frac{5}{6}$$ Same result as above

Both 6 and 12 worked as common denominators for adding $\frac{1}{2}$ and $\frac{1}{3}$, but using the smaller number saved some work. You should always try to find the smallest common denominator. If you don't for some reason, you can still work the problem—but it may take you longer. You'll have to divide out some common factors at the end in order to write the answer in lowest terms.

Least Common Denominator

The **least common denominator** (LCD) for two fractions is the *smallest* positive number divisible by both denominators of the original fractions. In other words, both 6 and 12 are common denominators for $\frac{1}{2}$ and $\frac{1}{3}$, but 6 is smaller, so it is the LCD.

There are several ways to find the LCD. When the original denominators are small numbers, you can often find the LCD by inspection. HINT: always check to see if the larger denominator will work as the LCD.

EXAMPLE 2 Finding the LCD by Inspection

(a) Find the LCD for $\frac{2}{3}$ and $\frac{1}{9}$.

Check to see if 9 (the larger denominator) will work as the LCD. Is 9 divisible by 3 (the other denominator)? Yes, so 9 is the LCD for $\frac{2}{3}$ and $\frac{1}{9}$.

(b) Find the LCD for $\frac{5}{8}$ and $\frac{5}{6}$.

Check to see if 8 (the larger denominator) will work. No, 8 is not evenly divisible by 6. So start checking numbers that are multiples of 8, that is, 16, 24, and 32. Notice that 24 will work because it is evenly divisible by 8 and by 6.

The LCD for $\frac{5}{8}$ and $\frac{5}{6}$ is 24.

WORK PROBLEM 2 AT THE SIDE.

For larger denominators, you can use prime factorization to find the LCD. Factor each denominator completely into prime numbers. Then use the factors to build the LCD.

EXAMPLE 3 Using Prime Factors to Find the LCD

(a) What is the LCD for $\frac{7}{12}$ and $\frac{13}{18}$?

Write 12 and 18 as the product of prime factors. Then use enough prime factors in the LCD to "cover" both 12 and 18.

$$12 = 2 \cdot 2 \cdot 3 \qquad 18 = 2 \cdot 3 \cdot 3$$

Factors of 12

$$\text{LCD} = 2 \cdot 2 \cdot 3 \cdot 3 = 36$$

Factors of 18

Check whether 36 is divisible by 12 (yes) and by 18 (yes). Notice that we did *not* have to repeat the factors that 12 and 18 have in common. If we had used *all* the 2's and 3's, we would get a common denominator, but not the *smallest* one.

(b) What is the LCD for $\frac{11}{15}$ and $\frac{9}{70}$?

$$15 = 3 \cdot 5 \qquad 70 = 2 \cdot 5 \cdot 7$$

Factors of 15

$$\text{LCD} = 3 \cdot 5 \cdot 2 \cdot 7 = 210$$

Factors of 70

Check whether 210 is divisible by 15 (yes) and divisible by 70 (yes). So 210 is the LCD for $\frac{11}{15}$ and $\frac{9}{70}$.

WORK PROBLEM 3 AT THE SIDE.

2. Find the LCD for each pair of fractions.

(a) $\frac{3}{5}$ and $\frac{3}{10}$

(b) $\frac{1}{2}$ and $\frac{2}{5}$

(c) $\frac{3}{4}$ and $\frac{1}{6}$

(d) $\frac{5}{6}$ and $\frac{7}{18}$

3. Use prime factorization to find the LCD for each pair of fractions.

(a) $\frac{1}{10}$ and $\frac{13}{14}$

(b) $\frac{5}{12}$ and $\frac{17}{20}$

(c) $\frac{7}{15}$ and $\frac{7}{9}$

ANSWERS

2. **(a)** 10 **(b)** 10 **(c)** 12 **(d)** 18

3. **(a)** $10 = 2 \cdot 5$; $14 = 2 \cdot 7$; LCD $= 2 \cdot 5 \cdot 7 = 70$ (10 above, 14 below)

(b) $12 = 2 \cdot 2 \cdot 3$; $20 = 2 \cdot 2 \cdot 5$; LCD $= 2 \cdot 2 \cdot 3 \cdot 5 = 60$ (12 above, 20 below)

(c) $15 = 3 \cdot 5$; $9 = 3 \cdot 3$; LCD $= 3 \cdot 3 \cdot 5 = 45$ (15 above, 9 below)

OBJECTIVE 3 Here are the steps for adding or subtracting unlike fractions. The key idea is that you must rewrite the fractions so that they have a common denominator before you can add or subtract them.

Adding and Subtracting Unlike Fractions

Step 1 Find the LCD, the smallest number divisible by both denominators in the problem.

Step 2 Rewrite each original fraction as an equivalent fraction whose denominator is the LCD.

Step 3 Add or subtract the numerators of the like fractions. Keep the common denominator.

Step 4 Write the sum or difference in lowest terms.

EXAMPLE 4 Adding or Subtracting Unlike Fractions

Find each sum or difference.

(a) $\frac{1}{5} + \frac{3}{10}$

Step 1 The larger denominator (10) is the LCD.

Step 2 $\frac{1}{5} = \frac{1 \cdot 2}{5 \cdot 2} = \frac{2}{10} \leftarrow \text{LCD}$ and $\frac{3}{10}$ already has the LCD.

Step 3 Add the numerators. Write the sum over the common denominator.

$$\frac{1}{5} + \frac{3}{10} = \frac{2}{10} + \frac{3}{10} = \frac{2+3}{10} = \frac{5}{10}$$

Step 4 Write $\frac{5}{10}$ in lowest terms.

$$\frac{5}{10} = \frac{\overset{1}{\cancel{5}}}{2 \cdot \underset{1}{\cancel{5}}} = \frac{1}{2} \quad \text{Lowest terms}$$

(b) $\frac{3}{4} - \frac{5}{6}$

Step 1 The LCD is 12.

Step 2 $\frac{3}{4} = \frac{3 \cdot 3}{4 \cdot 3} = \frac{9}{12} \leftarrow \text{LCD}$ and $\frac{5}{6} = \frac{5 \cdot 2}{6 \cdot 2} = \frac{10}{12} \leftarrow \text{LCD}$

Step 3 Subtract the numerators. Write the difference over the common denominator.

$9 + (-10)$ is -1

$$\frac{3}{4} - \frac{5}{6} = \frac{9}{12} - \frac{10}{12} = \frac{9 - 10}{12} = \frac{-1}{12} \text{ or } -\frac{1}{12}$$

Step 4 $-\frac{1}{12}$ is in lowest terms.

CONTINUED ON NEXT PAGE

4. Find each sum or difference. Write all answers in lowest terms.

(a) $-\frac{2}{3} + \frac{1}{6}$

(b) $\frac{1}{12} - \frac{5}{6}$

(c) $3 - \frac{4}{5}$

(d) $\frac{9}{16} + \left(-\frac{5}{12}\right)$

ANSWERS

4. (a) $-\frac{1}{2}$ **(b)** $-\frac{3}{4}$ **(c)** $\frac{11}{5}$ **(d)** $\frac{7}{48}$

(c) $-\frac{5}{12} + \frac{5}{9}$

Step 1 Use prime factorization to find the LCD.

$$12 = 2 \cdot 2 \cdot 3 \qquad 9 = 3 \cdot 3 \qquad \text{LCD} = \underbrace{2 \cdot 2 \cdot 3}_{12} \cdot 3 = 36$$

(The factors $3 \cdot 3$ in the LCD come from 9.)

Step 2 $-\frac{5}{12} = -\frac{5 \cdot 3}{12 \cdot 3} = -\frac{15}{36}$ and $\frac{5}{9} = \frac{5 \cdot 4}{9 \cdot 4} = \frac{20}{36}$

Step 3 Add the numerators. Keep the common denominator.

$$-\frac{5}{12} + \frac{5}{9} = -\frac{15}{36} + \frac{20}{36} = \frac{-15 + 20}{36} = \frac{5}{36}$$

Step 4 $\frac{5}{36}$ is in lowest terms.

(d) $4 - \frac{2}{3}$

Step 1 Think of 4 as $\frac{4}{1}$. The LCD for $\frac{4}{1}$ and $\frac{2}{3}$ is 3, the larger denominator.

Step 2 $\frac{4}{1} = \frac{4 \cdot 3}{1 \cdot 3} = \frac{12}{3}$ and $\frac{2}{3}$ already has the LCD.

Step 3 Subtract the numerators. Keep the common denominator.

$$\frac{4}{1} - \frac{2}{3} = \frac{12}{3} - \frac{2}{3} = \frac{12 - 2}{3} = \frac{10}{3}$$

Step 4 $\frac{10}{3}$ is in lowest terms.

WORK PROBLEM 4 AT THE SIDE.

OBJECTIVE 4 We use the same steps to add or subtract unlike fractions with variables in the numerators or denominators.

EXAMPLE 5 Adding or Subtracting Unlike Fractions with Variables

Find each sum or difference.

(a) $\frac{1}{4} + \frac{b}{5}$

Step 1 The LCD is 20.

Step 2 $\frac{1}{4} = \frac{1 \cdot 5}{4 \cdot 5} = \frac{5}{20}$ and $\frac{b}{5} = \frac{b \cdot 4}{5 \cdot 4} = \frac{4b}{20}$

Step 3 $\frac{1}{4} + \frac{b}{5} = \frac{5}{20} + \frac{4b}{20} = \frac{5 + 4b}{20}$ ← Add the numerators. ← Keep the common denominator.

Step 4 $\frac{5 + 4b}{20}$ is in lowest terms.

CONTINUED ON NEXT PAGE

Note

In part (a), we could *not* add 5 + 4b in the numerator of the answer because 5 and 4b are *not* like terms. We *could* add 5b + 4b but *not* 5 + 4b.

↖ ↗ Variable parts match.

(b) $\frac{2}{3} - \frac{6}{x}$

Step 1 The LCD is $3 \cdot x$, or $3x$.

Step 2 $\frac{2}{3} = \frac{2 \cdot x}{3 \cdot x} = \frac{2x}{3x}$ and $\frac{6}{x} = \frac{6 \cdot 3}{x \cdot 3} = \frac{18}{3x}$

Step 3 $\frac{2}{3} - \frac{6}{x} = \frac{2x}{3x} - \frac{18}{3x} = \frac{2x - 18}{3x}$ ← Keep the common denominator.

Step 4 $\frac{2x - 18}{3x}$ is in lowest terms.

WORK PROBLEM 5 AT THE SIDE. ▶▶

5. Find each sum or difference.

(a) $\frac{5}{6} - \frac{h}{2}$

(b) $\frac{7}{t} + \frac{3}{5}$

(c) $\frac{4}{x} - \frac{8}{3}$

Answers

5. (a) $\frac{5 - 3h}{6}$ **(b)** $\frac{35 + 3t}{5t}$ **(c)** $\frac{12 - 8x}{3x}$

NUMBERS IN THE *Real World*

collaborative investigations

The time signature at the beginning of a piece of music looks like a fraction. Commonly used time signatures are $\frac{2}{4}$, $\frac{3}{4}$, $\frac{4}{4}$, and $\frac{6}{8}$. Musicians use the time signature to tell how long to hold each note. The values of different notes can be written as fractions:

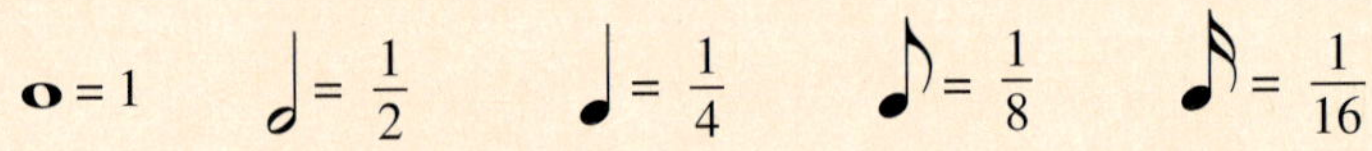

Music is divided into measures. In $\frac{4}{4}$ time, each measure contains notes that add up to $\frac{4}{4}$ (or 1). In $\frac{2}{4}$ time the notes in each measure add up to $\frac{2}{4}$ (or $\frac{1}{2}$), and so on for $\frac{3}{4}$ time and $\frac{6}{8}$ time.

Write one or more notes in each measure to make it add up to its time signature. Use as many different kinds of notes as possible.

Below are excerpts from "Jingle Bells" and "The Star-Spangled Banner." Divide each line of music into measures.

4.4 Exercises

Find each sum or difference. Write all answers in lowest terms.

1. $\frac{3}{4} + \frac{1}{8}$

2. $\frac{1}{3} + \frac{1}{2}$

3. $-\frac{1}{14} + \left(-\frac{3}{7}\right)$

4. $-\frac{2}{9} + \frac{2}{3}$

5. $\frac{2}{3} - \frac{1}{6}$

6. $\frac{5}{12} - \frac{1}{4}$

7. $\frac{3}{8} - \frac{3}{5}$

8. $\frac{1}{3} - \frac{3}{5}$

9. $-\frac{5}{8} + \frac{1}{12}$

10. $-\frac{13}{16} + \frac{13}{16}$

11. $-\frac{7}{20} - \frac{5}{20}$

12. $-\frac{7}{9} - \frac{5}{6}$

13. $0 - \frac{7}{18}$

14. $-\frac{7}{8} + 3$

15. $2 - \frac{6}{7}$

16. $5 - \frac{2}{5}$

17. $-\frac{1}{2} + \frac{3}{24}$

18. $\frac{7}{10} + \frac{7}{15}$

19. $\frac{1}{5} + \frac{c}{3}$

20. $\frac{x}{4} + \frac{2}{3}$

21. $\frac{5}{m} - \frac{1}{2}$

22. $\frac{2}{9} - \frac{4}{y}$

23. $\frac{3}{b^2} + \frac{5}{b^2}$

24. $\frac{10}{xy} - \frac{7}{xy}$

Ⓖ *Simplify.*

25. $\frac{c}{7} + \frac{3}{b}$

26. $\frac{2}{x} - \frac{y}{5}$

27. $-\frac{4}{c^2} - \frac{d}{c}$

28. $-\frac{1}{n} + \frac{m}{n^2}$

29. $-\frac{11}{42} - \frac{11}{70}$

30. $\frac{7}{45} - \frac{7}{20}$

31. A key step in adding or subtracting unlike fractions is to rewrite the fractions so that they have the least common denominator. Explain why this step is necessary.

32. Explain how to write a fraction with an indicated denominator. As part of your explanation, show how to change $\frac{3}{4}$ to an equivalent fraction having 12 as a denominator.

33. **(a)** Find each sum. Then state the property illustrated by these problems.

$-\frac{2}{3} + \frac{3}{4} =$ ____ $\quad \frac{3}{4} + \left(-\frac{2}{3}\right) =$ ____

(b) Find each difference. Then explain what these problems illustrate.

$\frac{5}{6} - \frac{1}{2} =$ ____ $\quad \frac{1}{2} - \frac{5}{6} =$ ____

34. **(a)** Find each sum. Then think about the properties you learned when adding integers. Which of those properties do these problems illustrate?

$-\frac{7}{12} + \frac{7}{12} =$ ____ $\quad \frac{3}{5} + \left(-\frac{3}{5}\right) =$ ____

(b) Find each product. Then state the property that these problems illustrate.

$-\frac{2}{3} \cdot \frac{9}{10} =$ ____ $\quad \frac{9}{10} \cdot \left(-\frac{2}{3}\right) =$ ____

35. Ⓖ Explain the error in each calculation and correct it.

(a) $\frac{3}{4} + \frac{2}{5} = \frac{3+2}{4+5} = \frac{5}{9}$

(b) $\frac{5}{6} - \frac{4}{9} = \frac{5}{18} - \frac{4}{18} = \frac{5-4}{18} = \frac{1}{18}$

36. Ⓖ Explain the error in each calculation and correct it.

(a) $-\frac{1}{4} + \frac{7}{12} = -\frac{3}{12} + \frac{7}{12} = \frac{-3+7}{12} = \frac{4}{12}$

(b) $\frac{3}{10} - \frac{1}{4} = \frac{3-1}{10-4} = \frac{2}{6} = \frac{1}{3}$

Solve each application problem.

37. A forester planted $\frac{5}{12}$ acre in seedlings in the morning and $\frac{11}{12}$ acre in the afternoon. If $\frac{7}{12}$ acre of seedlings were destroyed by an unexpected frost, how many acres of seedlings remained?

38. How much fencing will be needed to enclose this rectangular wildflower preserve?

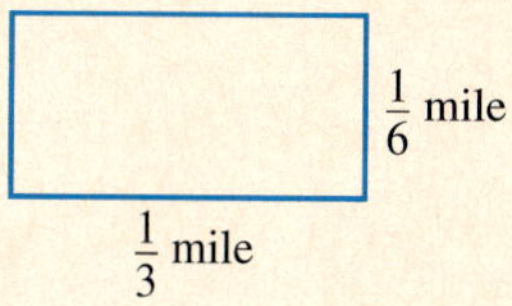

39. The owner of Racy's Feed Store ordered $\frac{1}{3}$ cubic yard of corn, $\frac{3}{8}$ cubic yard of oats, and $\frac{1}{4}$ cubic yard of washed medium mesh gravel. Find the total cubic yards of products ordered.

40. A flower grower purchased $\frac{9}{10}$ acre of land one year and $\frac{3}{10}$ acre the next year. She then sold $\frac{7}{10}$ acre of land. How much land does she now have?

41. The warrior princess, Xena, urges local peasants to fortify their village against a warlord's attack. Find the perimeter of (distance around) the village.

42. When installing cabinets, Pam Phelps must be certain that the proper type and size of mounting hardware is used. Find the total length of the bolt below.

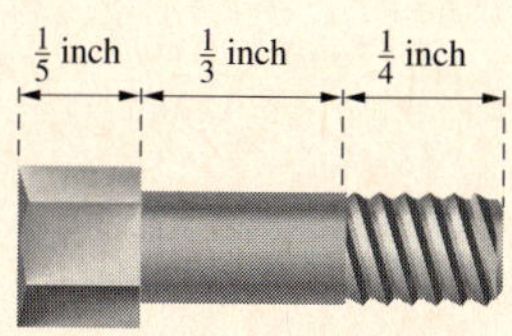

43. A hydraulic jack contains $\frac{7}{8}$ gallon of hydraulic fluid. A cracked seal resulted in a loss of $\frac{1}{6}$ gallon of fluid in the morning and another $\frac{1}{3}$ gallon in the afternoon. Find the amount of fluid remaining.

44. Adrian Ortega drives a tanker for the British Petroleum Company. He leaves the refinery with his tanker filled to $\frac{7}{8}$ of capacity. If he delivers $\frac{1}{4}$ of the tank capacity at the first stop and $\frac{1}{3}$ of the tank capacity at the second stop, find the fraction of the tanker's capacity remaining.

Refer to the circle graph to answer Exercises 45–48.

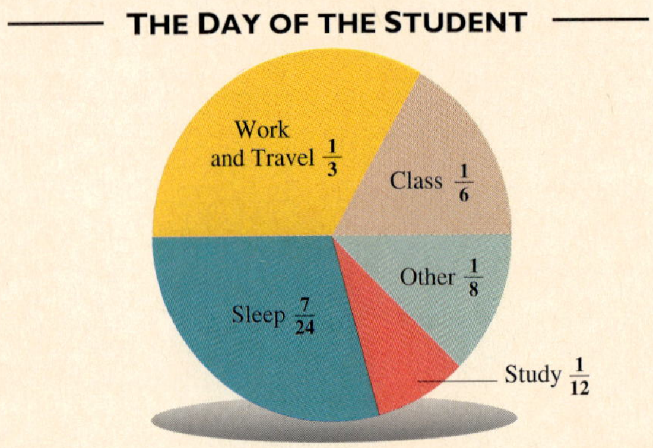

45. What fraction of the day was spent in class and study?

46. What fraction of the day was spent in work and travel and other?

47. In which activity was the greatest amount of time spent? How many hours did this activity take?

48. In which activity was the least amount of time spent? How many hours did this activity take?

49. A hazardous waste dump site will require $\frac{7}{8}$ mile of security fencing. The site has four sides with three of the sides measuring $\frac{1}{4}$ mile, $\frac{1}{6}$ mile, and $\frac{3}{8}$ mile. Find the length of the fourth side.

50. Chakotay is fitting a turquoise stone into a bear claw pendant. Find the diameter of the hole in the pendant. (The diameter is the distance across the center of the hole.)

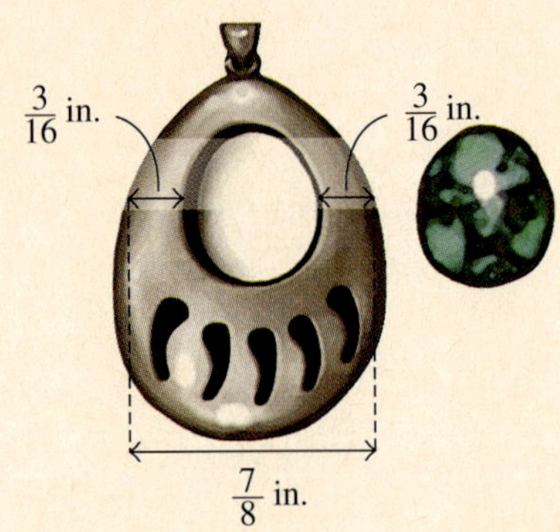

Review and Prepare

First use front end rounding to estimate each answer. Then find the exact answer. (For help, see ***Section 1.5****.)*

51. Elena's $518 paycheck had three deductions: $48, $23, and $96. What was her net pay?

estimate:

exact:

52. Mathtronic stock started the day at $93 per share. By noon it had dropped $17, but then it rose $25. What was the final price per share?

estimate:

exact:

53. There are 12 inches in 1 foot. How many inches are in 47 feet?

estimate:

exact:

54. There are 128 fluid ounces in 1 gallon. How many fluid ounces are in 35 gallons?

estimate:

exact:

4.5 Problem Solving: Mixed Numbers and Estimating

Objectives

1. Identify mixed numbers and graph them on a number line.
2. Rewrite mixed numbers as improper fractions, or the reverse.
3. Estimate the answer and multiply or divide mixed numbers.
4. Estimate the answer and add or subtract mixed numbers.
5. Solve application problems containing mixed numbers.

For Extra Help

Tutorial | Tape 6 | SSM, Sec. 4.5

Objective 1 When a fraction and a whole number are written together the result is a **mixed number.** For example, the mixed number

$$3\frac{1}{2} \quad \text{represents} \quad 3 + \frac{1}{2},$$

or 3 wholes and $\frac{1}{2}$ of a whole. Read $3\frac{1}{2}$ as "three and one half."

One common use of mixed numbers is to measure things. Examples include:

Juan worked $5\frac{1}{2}$ hours. The box weighs $2\frac{3}{4}$ pounds.

The park is $1\frac{7}{10}$ miles long. Add $1\frac{2}{3}$ cups of flour.

EXAMPLE 1 Illustrating Mixed Numbers with Diagrams and Number Lines

As this diagram shows, the mixed number $3\frac{1}{2}$ is equivalent to the improper fraction $\frac{7}{2}$.

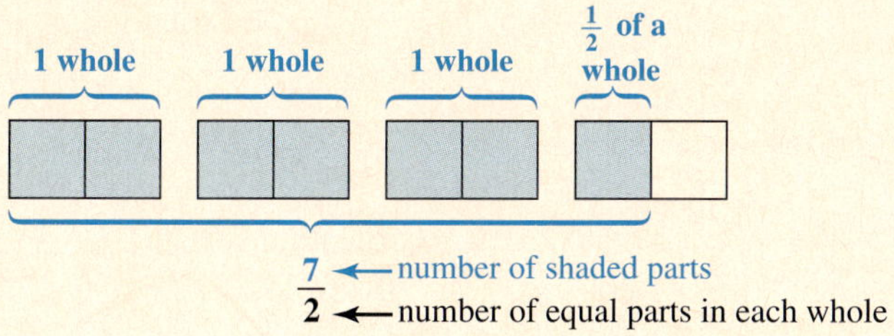

We can also use a number line to show mixed numbers, as in this graph of $3\frac{1}{2}$ and $-3\frac{1}{2}$.

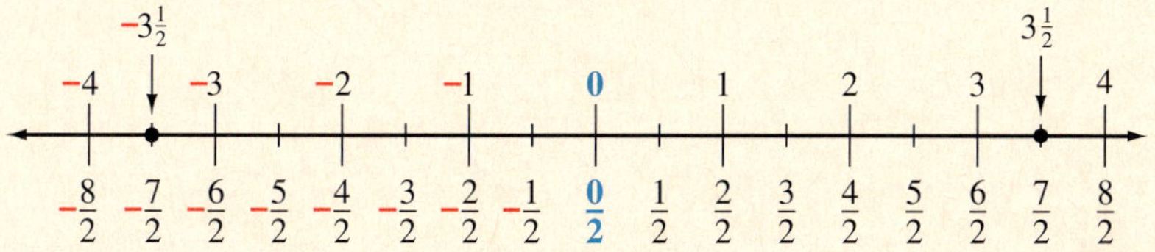

The number line shows that

$$3\frac{1}{2} \text{ is equivalent to } \frac{7}{2}$$

$$-3\frac{1}{2} \text{ is equivalent to } -\frac{7}{2}.$$

Note

$3\frac{1}{2}$ represents $3 + \frac{1}{2}$.

$-3\frac{1}{2}$ represents $-3 + (-\frac{1}{2})$, which can also be written as $-3 - \frac{1}{2}$.

In algebra we usually work with the improper fraction form of mixed numbers, especially for negative mixed numbers. However, positive mixed numbers are frequently used in daily life, so it's important to know how to work with them. For example, we usually say $3\frac{1}{2}$ inches rather than $\frac{7}{2}$ inches.

1. (a) Use these diagrams to write $1\frac{2}{3}$ as an improper fraction.

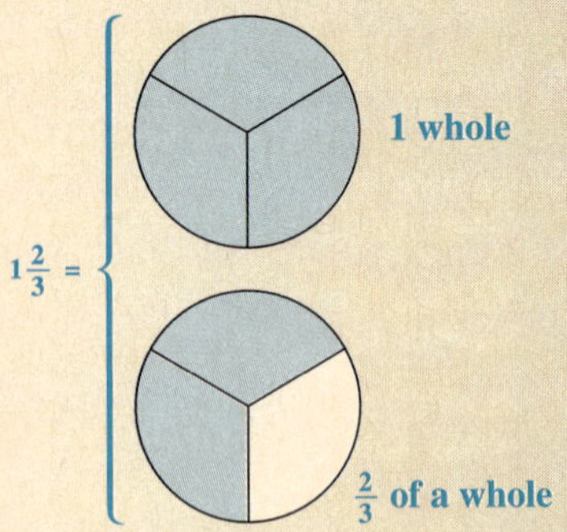

(b) Now graph $1\frac{2}{3}$ and $-1\frac{2}{3}$ on this number line.

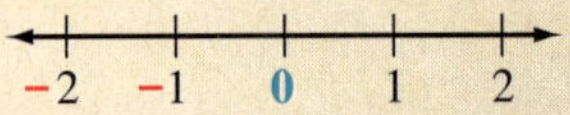

(c) Use these diagrams to write $2\frac{1}{4}$ as an improper fraction.

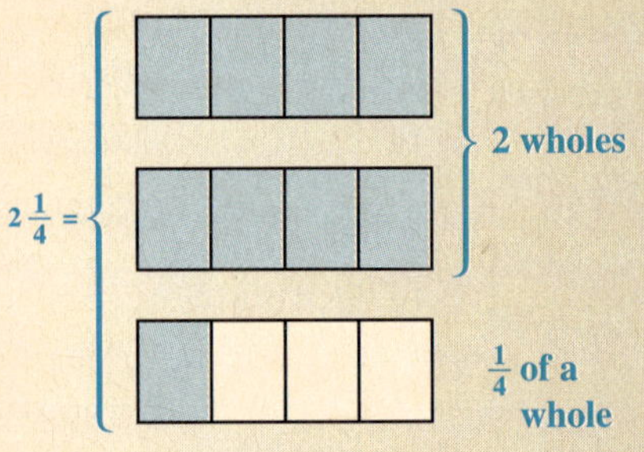

(d) Now graph $2\frac{1}{4}$ and $-2\frac{1}{4}$ on this number line.

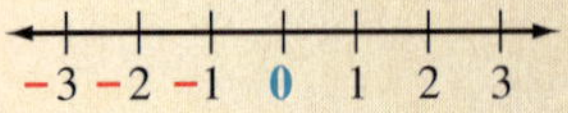

ANSWERS

1. (a) $\frac{5}{3}$ **(b)**

(c) $\frac{9}{4}$ **(d)**

◀◀ **WORK PROBLEM 1 AT THE SIDE.**

OBJECTIVE 2 You can use the following steps to write $3\frac{1}{2}$ as an improper fraction without drawing a diagram or a number line.

Step 1 Multiply 3 and 2.

$$3\frac{1}{2} \quad 3 \cdot 2 = 6$$

Step 2 Add 1 to the product.

$$3\frac{1}{2} \quad 3 \cdot 2 = 6 + 1 = 7$$

Step 3 Use 7 (from Step 2) as the numerator and 2 as the denominator.

$$3\frac{1}{2} = \frac{7}{2}$$

Same denominator

To see why this method works, recall that $3\frac{1}{2}$ represents $3 + \frac{1}{2}$. Let's add $3 + \frac{1}{2}$.

$$3 + \frac{1}{2} = \frac{3}{1} + \frac{1}{2} = \frac{6}{2} + \frac{1}{2} = \frac{6+1}{2} = \frac{7}{2}$$

Common denominator — Same result as above

In summary, use the following steps to *write a mixed number as an improper fraction.*

Writing a Mixed Number as an Improper Fraction

Step 1 **Multiply** the denominator of the fraction times the whole number.

Step 2 **Add** to this product the numerator of the fraction.

Step 3 Write the result of Step 2 as the **numerator** and keep the original **denominator.**

EXAMPLE 2 Writing Mixed Numbers as Improper Fractions

Write $7\frac{2}{3}$ as an improper fraction (numerator greater than denominator).

Step 1 $7\frac{2}{3} \quad 7 \cdot 3 = 21$ Multiply 7 and 3.

CONTINUED ON NEXT PAGE

Step 2 $7\frac{2}{3}$ $7 \cdot 3 = 21 + 2 = 23$ Add 2.

Step 3 $7\frac{2}{3} = \frac{23}{3}$

Same denominator

WORK PROBLEM 2 AT THE SIDE.

We used *multiplication* for the first step in writing a mixed number as an improper fraction. To work in *reverse,* writing an improper fraction as a mixed number, we will use *division.* Recall that the fraction bar is a symbol for division.

Writing an Improper Fraction as a Mixed Number

Write an *improper fraction* as a mixed number by dividing the numerator by the denominator. The quotient is the whole number (of the mixed number), the remainder is the numerator of the fraction part, and the denominator remains the same.

Always check to be sure that the fraction part of the mixed number is in lowest terms. Then the mixed number is in *simplest form.*

EXAMPLE 3 Writing Improper Fractions as Mixed Numbers

Write each improper fraction as an equivalent mixed number.

(a) $\frac{17}{5}$

Divide 17 by 5.

$$\begin{array}{r} 3 \\ 5\overline{)17} \\ \underline{15} \\ 2 \end{array}$$

3 ← Whole number part

2 ← Remainder

The quotient **3** is the whole number part of the mixed number. The remainder **2** is the numerator of the fraction and the denominator remains as **5.**

$$\frac{17}{5} = 3\frac{2}{5}$$

2 ← Remainder

Same denominator

Let's look at a drawing of $\frac{17}{5}$ to check our work.

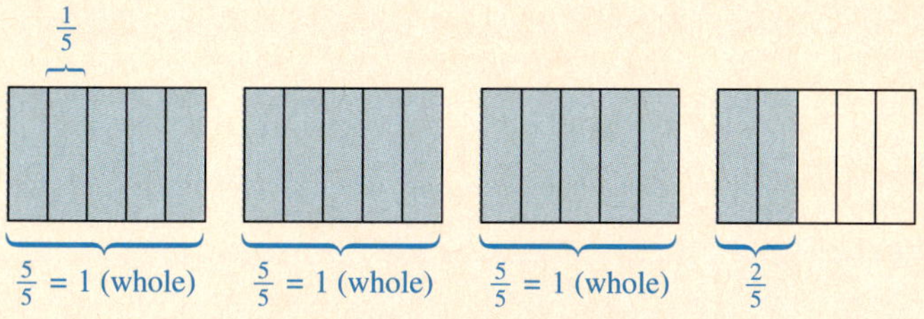

CONTINUED ON NEXT PAGE

2. Write each mixed number as an equivalent improper fraction.

(a) $3\frac{2}{3}$

(b) $4\frac{7}{10}$

(c) $5\frac{3}{4}$

(d) $8\frac{5}{6}$

ANSWERS

2. (a) $\frac{11}{3}$ **(b)** $\frac{47}{10}$ **(c)** $\frac{23}{4}$ **(d)** $\frac{53}{6}$

3. Write each improper fraction as an equivalent mixed number in simplest form.

(a) $\frac{5}{2}$

(b) $\frac{14}{4}$

(c) $\frac{33}{5}$

(d) $\frac{58}{10}$

4. Round each mixed number to the nearest whole number.

(a) $2\frac{3}{4}$

(b) $6\frac{3}{8}$

(c) $4\frac{2}{3}$

(d) $1\frac{7}{10}$

(e) $3\frac{1}{2}$

(f) $5\frac{4}{9}$

Answers

3. **(a)** $2\frac{1}{2}$ **(b)** $3\frac{1}{2}$ **(c)** $6\frac{3}{5}$ **(d)** $5\frac{4}{5}$

4. **(a)** 3 **(b)** 6 **(c)** 5 **(d)** 2 **(e)** 4 **(f)** 5

In other words,

$$\frac{17}{5} = \frac{5}{5} + \frac{5}{5} + \frac{5}{5} + \frac{2}{5}$$
$$= \underbrace{1 + 1 + 1}_{} + \frac{2}{5}$$
$$= 3 + \frac{2}{5}$$
$$= 3\frac{2}{5}$$

(b) $\frac{26}{4}$

Divide 26 by 4.

$$4\overline{)26} = 6 \text{ R } 2 \quad (24) \qquad \text{so} \qquad \frac{26}{4} = 6\frac{2}{4} = 6\frac{1}{2} \quad \text{Simplest form}$$

Write $\frac{2}{4}$ in lowest terms.

You could write $\frac{26}{4}$ in lowest terms first.

$$\frac{26}{4} = \frac{\overset{1}{\cancel{2}} \cdot 13}{\underset{1}{\cancel{2}} \cdot 2} = \frac{13}{2} \qquad \text{or} \qquad 2\overline{)13} = 6 \text{ R } 1 \quad (12) \qquad \text{so} \qquad \frac{13}{2} = 6\frac{1}{2} \quad \text{Same result as above}$$

◀◀ WORK PROBLEM 3 AT THE SIDE.

Objective 3 Once you have rewritten mixed numbers as improper fractions, you can use the steps you learned in **Section 4.3** to multiply and divide. However, it's a good idea to estimate the answer before you start any other work.

EXAMPLE 4 Rounding Mixed Numbers

To estimate answers, first round each mixed number to the nearest whole number. If the numerator is *half* of the denominator or *more,* round up the whole number part. If the numerator is *less* than half the denominator, leave the whole number as it is.

(a) Round $1\frac{5}{8}$ ← 5 is more than 4. ← Half of 8 is 4. $1\frac{5}{8}$ rounds up to 2

(b) Round $3\frac{2}{5}$ ← 2 is less than $2\frac{1}{2}$. ← Half of 5 is $2\frac{1}{2}$. $3\frac{2}{5}$ rounds to 3

◀◀ WORK PROBLEM 4 AT THE SIDE.

Multiplying and Dividing Mixed Numbers

Step 1 **Rewrite** each mixed number as an improper fraction.

Step 2 **Multiply** or **divide** the improper fractions.

Step 3 Write the answer in lowest terms and change it to a mixed number or whole number where possible. This step gives you an answer that is in **simplest form.**

EXAMPLE 5 Estimating the Answer and Multiplying Mixed Numbers

First, round the numbers and estimate the answer. Then find the exact answer. Write all answers in simplest form.

(a) $2\frac{1}{2} \cdot 3\frac{1}{5}$

Estimate the answer by rounding the mixed numbers.

$2\frac{1}{2}$ rounds to 3 and $3\frac{1}{5}$ rounds to 3

$3 \cdot 3 = 9$ Estimated answer

To find the exact answer, first rewrite each mixed number as an improper fraction.

Step 1 $2\frac{1}{2} = \frac{5}{2}$ and $3\frac{1}{5} = \frac{16}{5}$

Next, multiply.

$$2\frac{1}{2} \cdot 3\frac{1}{5} = \underbrace{\frac{5}{2} \cdot \frac{16}{5} = \frac{\overset{1}{\cancel{5}} \cdot \overset{1}{\cancel{2}} \cdot 8}{\underset{1}{\cancel{2}} \cdot \underset{1}{\cancel{5}}}}_{\text{Step 2}} = \underbrace{\frac{8}{1} = 8}_{\text{Step 3}} \quad \text{Simplest form}$$

The estimated answer is 9 and the exact answer is 8, so the exact answer is reasonable.

(b) $3\frac{5}{8} \cdot 4\frac{4}{5}$

First, round each mixed number and estimate the answer.

$3\frac{5}{8}$ rounds to 4 and $4\frac{4}{5}$ rounds to 5.

$4 \cdot 5 = 20$ Estimated answer

Now find the exact answer.

$$3\frac{5}{8} \cdot 4\frac{4}{5} = \underbrace{\frac{29}{8} \cdot \frac{24}{5}}_{\text{Step 1}} = \underbrace{\frac{29 \cdot 3 \cdot \overset{1}{\cancel{8}}}{\underset{1}{\cancel{8}} \cdot 5}}_{\text{Step 2}} = \underbrace{\frac{87}{5} = 17\frac{2}{5}}_{\text{Step 3}} \quad \text{Simplest form}$$

$17\frac{2}{5}$ is close to the estimated answer of 20.

WORK PROBLEM 5 AT THE SIDE.

5. First, round the numbers and estimate the answer. Then find the exact answer. Write answers in simplest form.

(a) $2\frac{1}{4} \cdot 7\frac{1}{3}$

Rounded

_______ • _______

= _______ *estimate*

= _______ *exact*

(b) $4\frac{1}{2} \cdot 1\frac{2}{3}$

Rounded

_______ • _______

= _______ *estimate*

= _______ *exact*

(c) $3\frac{3}{5} \cdot 4\frac{4}{9}$

Rounded

_______ • _______

= _______ *estimate*

= _______ *exact*

(d) $3\frac{1}{5} \cdot 5\frac{3}{8}$

Rounded

_______ • _______

= _______ *estimate*

= _______ *exact*

ANSWERS

5. **(a)** estimate: $2 \cdot 7 = 14$; exact: $16\frac{1}{2}$
(b) estimate: $5 \cdot 2 = 10$; exact: $7\frac{1}{2}$
(c) estimate: $4 \cdot 4 = 16$; exact: 16
(d) estimate: $3 \cdot 5 = 15$; exact: $17\frac{1}{5}$

6. First, round the numbers and estimate the answer. Then find the exact answer. Write answers in simplest form.

(a) $6\frac{1}{4} \div 3\frac{1}{3}$

Rounded

_______ ÷ _______

= _______ *estimate*

= _______ *exact*

(b) $3\frac{3}{8} \div 2\frac{4}{7}$

Rounded

_______ ÷ _______

= _______ *estimate*

= _______ *exact*

(c) $8 \div 5\frac{1}{3}$

Rounded

_______ ÷ _______

= _______ *estimate*

= _______ *exact*

(d) $4\frac{1}{2} \div 6$

Rounded

_______ ÷ _______

= _______ *estimate*

= _______ *exact*

ANSWERS

6. **(a)** estimate: $6 \div 3 = 2$; exact: $1\frac{7}{8}$
(b) estimate: $3 \div 3 = 1$; exact: $1\frac{5}{16}$
(c) estimate: $8 \div 5 = 1\frac{3}{5}$; exact: $1\frac{1}{2}$
(d) estimate: $5 \div 6 = \frac{5}{6}$; exact: $\frac{3}{4}$

EXAMPLE 6 Estimating the Answer and Dividing Mixed Numbers

First round the numbers and estimate the answer. Then find the exact answer. Write answers in simplest form.

(a) $3\frac{3}{5} \div 1\frac{1}{2}$

To estimate the answer, round each mixed number to the nearest whole number.

$$3\frac{3}{5} \div 1\frac{1}{2}$$

Rounded

$$4 \div 2 = 2 \quad \text{Estimate}$$

To find the exact answer, first rewrite each mixed number as an improper fraction.

$$3\frac{3}{5} \div 1\frac{1}{2} = \frac{18}{5} \div \frac{3}{2}$$

Now rewrite the problem as multiplying by the reciprocal of $\frac{3}{2}$.

$$\frac{18}{5} \div \frac{3}{2} = \frac{18}{5} \cdot \frac{2}{3} = \frac{\overset{1}{\cancel{3}} \cdot 6 \cdot 2}{5 \cdot \underset{1}{\cancel{3}}} = \frac{12}{5} = 2\frac{2}{5} \quad \text{Simplest form}$$

reciprocals

The estimate was 2, so an exact answer of $2\frac{2}{5}$ is reasonable.

(b) $4\frac{3}{8} \div 5$

First round the numbers and estimate the answer.

$$4\frac{3}{8} \div 5$$

Rounded

$$4 \div 5 \quad \text{Write } 4 \div 5 \text{ using a fraction bar.} \quad \frac{4}{5} \text{ Estimate}$$

Find the exact answer.

Write 5 as $\frac{5}{1}$.

$$4\frac{3}{8} \div 5 = \frac{35}{8} \div \frac{5}{1} = \frac{35}{8} \cdot \frac{1}{5} = \frac{\overset{1}{\cancel{5}} \cdot 7 \cdot 1}{8 \cdot \underset{1}{\cancel{5}}} = \frac{7}{8} \quad \text{Simplest form}$$

reciprocals

The estimate was $\frac{4}{5}$, so the exact answer of $\frac{7}{8}$ is reasonable. They are both less than 1.

WORK PROBLEM 6 AT THE SIDE.

OBJECTIVE 4 The steps you learned for adding and subtracting fractions in **Section 4.4** will also work for mixed numbers: Just rewrite the mixed numbers as equivalent improper fractions. Again, it is a good idea to estimate the answer before you start any other work.

EXAMPLE 7 Estimating the Answer and Adding or Subtracting Mixed Numbers

First estimate the answer. Then add or subtract to find the exact answer. Write answers in simplest form.

(a) $2\frac{3}{8} + 3\frac{3}{4}$

To estimate an answer, round each mixed number to the nearest whole number.

$$2\frac{3}{8} + 3\frac{3}{4}$$
$$\downarrow \qquad \downarrow$$
$$2 + 4 = 6 \quad \text{Estimate}$$

To find the exact answer, first rewrite each mixed number as an equivalent improper fraction.

$$2\frac{3}{8} + 3\frac{3}{4} = \frac{19}{8} + \frac{15}{4}$$

You can't add fractions until they have a common denominator. The LCD for $\frac{19}{8}$ and $\frac{15}{4}$ is 8. Rewrite $\frac{15}{4}$ as an equivalent fraction with a denominator of 8.

$$\frac{19}{8} + \frac{\mathbf{15}}{\mathbf{4}} = \frac{19}{8} + \frac{\mathbf{30}}{\mathbf{8}} = \frac{19 + 30}{8} = \frac{49}{8} = 6\frac{1}{8} \quad \text{Simplest form}$$

Common denominator

The estimate was 6, so an exact answer of $6\frac{1}{8}$ is reasonable.

(b) $4\frac{2}{3} - 2\frac{4}{5}$

Round each number and estimate the answer.

$$4\frac{2}{3} - 2\frac{4}{5}$$
$$\downarrow \qquad \downarrow$$
$$5 - 3 = 2 \quad \text{Estimate}$$

Rewrite the mixed numbers as improper fractions and subtract.

$$4\frac{2}{3} - 2\frac{4}{5} = \frac{14}{3} - \frac{14}{5} = \frac{70}{15} - \frac{42}{15} = \frac{70 - 42}{15} = \frac{28}{15} = 1\frac{13}{15} \quad \text{Simplest form}$$

LCD is 15.

The estimate was 2, so an exact answer of $1\frac{13}{15}$ is reasonable.

CONTINUED ON NEXT PAGE

7. First, round the numbers and estimate the answer. Then add or subtract to find the exact answer.

(a) $5\frac{1}{3} - 2\frac{5}{6}$

$\downarrow \quad \downarrow$

__ − __ = __ *estimate*

(b) $\frac{3}{4} + 3\frac{1}{8}$

$\downarrow \quad \downarrow$

__ + __ = __ *estimate*

(c) $6 - 3\frac{4}{5}$

$\downarrow \quad \downarrow$

__ − __ = __ *estimate*

(c) $5 - 1\frac{3}{8}$

$$5 - 1\frac{3}{8}$$

$$\downarrow \quad \downarrow$$

$$5 - 1 = 4 \quad \text{Estimate}$$

Write 5 as $\frac{5}{1}$.

$$5 - 1\frac{3}{8} = \frac{5}{1} - \frac{11}{8} = \frac{40}{8} - \frac{11}{8} = \frac{29}{8} = 3\frac{5}{8} \quad \text{Simplest form}$$

LCD is 8.

The estimate was 4, so an exact answer of $3\frac{5}{8}$ is reasonable.

WORK PROBLEM 7 AT THE SIDE.

Problem Solving with Your Calculator: In some situations the method of rewriting mixed numbers as improper fractions may result in very large numerators. Consider this example.

Last year Hue's child was $48\frac{3}{8}$ inches tall. This year the child is $51\frac{1}{4}$ inches tall. How much has the child grown?

First, estimate the answer by rounding each mixed number to the nearest whole number.

$$51\frac{1}{4} - 48\frac{3}{8}$$

$$\downarrow \quad \downarrow$$

$$51 - 48 = 3 \text{ inches} \quad \text{Estimate}$$

Rewrite the mixed numbers as improper fractions.

Rewrite $\frac{205}{4}$ as $\frac{410}{8}$.

$$51\frac{1}{4} - 48\frac{3}{8} = \frac{205}{4} - \frac{387}{8} = \frac{410}{8} - \frac{387}{8} = \frac{410 - 387}{8} = \frac{23}{8} = 2\frac{7}{8}$$

LCD is 8.

You can also use the fraction key [a b/c] to solve this problem.

To enter $51\frac{1}{4}$, press

51 [a b/c] 1 [a b/c] 4 [a b/c]. The display shows [51 ⌞ 1⌟4].

(Whole number) (Numerator) (Denominator)

Then press [−] 48 [a b/c] 3 [a b/c] 8 [=]. The display shows [2 ⌞ 7⌟8].

(Subtract) ($48\frac{3}{8}$) ($2\frac{7}{8}$)

Either way the exact answer is $2\frac{7}{8}$ inches, which is close to the estimate of 3 inches.

Another efficient method for handling large mixed numbers is to rewrite them in decimal form. You will learn how to do that in Chapter 5.

ANSWERS

7. (a) $5 - 3 = 2$; $2\frac{1}{2}$
(b) $1 + 3 = 4$; $3\frac{7}{8}$
(c) $6 - 4 = 2$; $2\frac{1}{5}$

OBJECTIVE 5 Rounding mixed numbers to the nearest whole number can also help you decide whether to solve an application problem by adding, subtracting, multiplying, or dividing.

EXAMPLE 8 Solving Application Problems with Mixed Numbers

First, estimate the answer to each application problem. Then find the exact answer.

(a) Gary needs to haul $15\frac{3}{4}$ **tons** of sand to a construction site. His truck can carry $2\frac{1}{4}$ **tons**. How many trips will he need to make?

First round each mixed number to the nearest whole number:

$$15\frac{3}{4} \text{ rounds to } 16 \quad \text{and} \quad 2\frac{1}{4} \text{ rounds to } 2.$$

Now read the problem again, *using the rounded numbers:* Gary needs to haul **16 tons** of sand to a construction site. His truck can carry **2 tons**. How many trips will he need to make? Using the rounded numbers in the problem makes it easier to see that you need to *divide.*

$$16 \div 2 = 8 \text{ trips} \qquad \text{Estimate}$$

To find the exact answer, use the original mixed numbers and divide.

$$15\frac{3}{4} \div 2\frac{1}{4} = \frac{63}{4} \div \frac{9}{4} = \frac{63}{4} \cdot \frac{4}{9} = \frac{7 \cdot \overset{1}{\cancel{9}} \cdot \overset{1}{\cancel{4}}}{\underset{1}{\cancel{4}} \cdot \underset{1}{\cancel{9}}} = \frac{7}{1} = 7 \qquad \text{Simplest form}$$

reciprocals

Gary needs to make 7 trips to haul all the sand. This result is close to the estimate of 8 trips.

(b) Zenitia worked $3\frac{5}{6}$ **hours** on Monday and $6\frac{1}{2}$ **hours** on Tuesday. How much longer did she work on Tuesday than on Monday?

First, round each mixed number to the nearest whole number:

$$3\frac{5}{6} \text{ rounds to } 4 \quad \text{and} \quad 6\frac{1}{2} \text{ rounds to } 7.$$

Now read the problem again, *using the rounded numbers*: Zenitia worked **4 hours** on Monday and **7 hours** on Tuesday. How much longer did she work on Tuesday than on Monday? Using the rounded numbers in the problem makes it easier to see that you need to *subtract.*

$$7 - 4 = 3 \text{ hours} \qquad \text{Estimate}$$

To find the exact answer, use the original mixed numbers and subtract.

$$6\frac{1}{2} - 3\frac{5}{6} = \frac{13}{2} - \frac{23}{6} = \frac{39}{6} - \frac{23}{6} = \frac{39-23}{6} = \frac{16}{6} = 2\frac{4}{6} = 2\frac{2}{3}$$

LCD is 6. Write answer in simplest form.

Zenitia worked $2\frac{2}{3}$ hours longer on Tuesday. This result is close to the estimate of 3 hours.

WORK PROBLEM 8 AT THE SIDE.

8. First, round the numbers and estimate the answer to each problem. Then find the exact answer.

(a) Richard's son grew $3\frac{5}{8}$ inches last year and $2\frac{1}{4}$ inches this year. How much has his height increased over the two years?

estimate:

exact:

(b) Ernestine used $2\frac{1}{2}$ packages of chocolate chips in her cookie recipe. Each package has $5\frac{1}{2}$ ounces of chips. How many ounces of chips did she use in the recipe?

estimate:

exact:

ANSWERS

8. (a) $4 + 2 = 6$ inches; $5\frac{7}{8}$ inches
(b) $3 \cdot 6 = 18$ ounces; $13\frac{3}{4}$ ounces

NUMBERS IN THE *Real World*

collaborative investigations

Math teachers attending conferences in Dallas, Texas, and Tampa Bay, Florida, received the following information about hotel rates:

	Distance from Convention Center	Single	Double/ Twin	Triple	Quad
Dallas, Texas, February 12–14					
Wyndham Anatole	3 blocks	$135	$145	$155	$165
Wilson World Market Center	$3\frac{1}{2}$ blocks	$105	$105	$105	$105
Tampa Bay, Florida, March 5–7					
Wyndham Harbor Island	3 blocks	$153	$168	$183	$198
Hyatt Regency	2 blocks	$132	$142	$152	$162
Club Hotel	$2\frac{1}{2}$ blocks	$99	$99	$109	$119

1. The double/twin rate is for 2 people sharing a room. So each person pays $\frac{1}{2}$ of the cost. Multiply the double/twin rate by $\frac{1}{2}$ to find each person's cost at each hotel.

2. Which hotel has the cheapest double/twin rate?

3. Do you see a shortcut for finding $\frac{1}{2}$ of a number? Describe the shortcut.

4. Why does multiplying by $\frac{1}{2}$ give the same result as dividing by 2?

5. The triple rate is for three people. What fraction of the room cost does each person pay?

6. The quad rate is for four people. What fraction of the room cost does each person pay?

7. Multiply each triple rate by $\frac{1}{3}$ and each quad rate by $\frac{1}{4}$ to find each person's cost.

8. Which hotel has the cheapest triple rate? The cheapest quad rate?

9. What shortcut could you use for finding $\frac{1}{3}$ of a number? For finding $\frac{1}{4}$ of a number?

10. Name a city you would like to visit. How could you get information on room rates at various hotels in that city?

4.5 Exercises

First, round the mixed numbers to the nearest whole number and estimate each answer. Then add, subtract, multiply, or divide to find the exact answer. Write all answers in simplest form.

1. *exact*

$2\frac{1}{4} \cdot 3\frac{1}{2}$

estimate

____ • ____ = ____

2. *exact*

$1\frac{1}{2} \cdot 3\frac{3}{4}$

estimate

____ • ____ = ____

3. *exact*

$3\frac{1}{4} \div 2\frac{5}{8}$

estimate

____ ÷ ____ = ____

4. *exact*

$2\frac{1}{4} \div 1\frac{1}{8}$

estimate

____ ÷ ____ = ____

5. *exact*

$3\frac{2}{3} + 1\frac{5}{6}$

estimate

____ + ____ = ____

6. *exact*

$4\frac{4}{5} + 2\frac{1}{3}$

estimate

____ + ____ = ____

7. *exact*

$4\frac{1}{4} - \frac{7}{12}$

estimate

____ − ____ = ____

8. *exact*

$10\frac{1}{3} - 6\frac{5}{6}$

estimate

____ − ____ = ____

9. *exact*

$5\frac{2}{3} \div 6$

estimate

____ ÷ ____ = ____

10. *exact*

$1\frac{7}{8} \div 6\frac{1}{4}$

estimate

____ ÷ ____ = ____

11. *exact*

$8 - 1\frac{4}{5}$

estimate

____ − ____ = ____

12. *exact*

$7 - 3\frac{3}{10}$

estimate

____ − ____ = ____

Find the perimeter and the area of each square or rectangle. Write all answers in simplest form.

13.

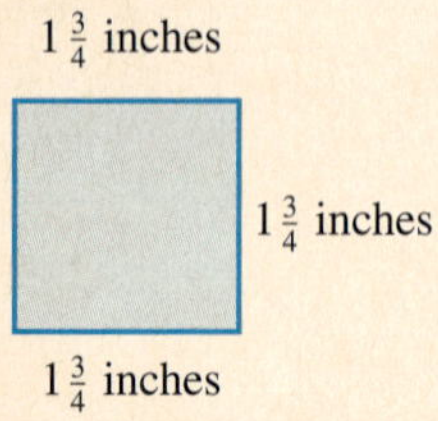

14.

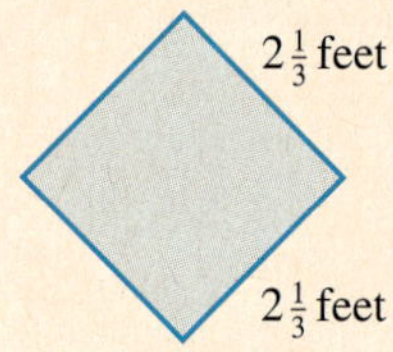

15.

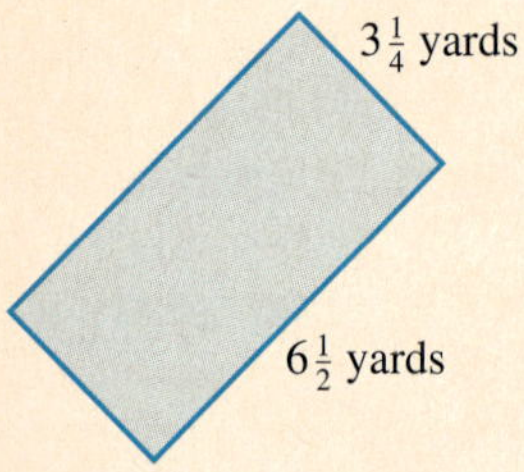

16.

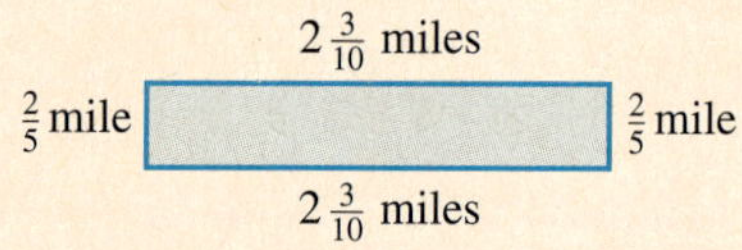

First, estimate the answer to each application problem. Then find the exact answer. Write all answers in simplest form.

17. A carpenter has two pieces of oak trim. One piece of trim is $12\frac{1}{2}$ ft long and the other is $8\frac{2}{3}$ ft in length. How many feet of oak trim does he have in all?

estimate:

exact:

18. On Monday, $5\frac{3}{4}$ tons of cans were recycled, and, on Tuesday, $9\frac{3}{5}$ tons were recycled. How many tons were recycled on these two days?

estimate:

exact:

19. The directions for mixing an insect spray say to use $1\frac{3}{4}$ ounces of chemical in each gallon of water. How many ounces of chemical should be mixed with $12\frac{1}{2}$ gallons of water?

estimate:

exact:

20. Shirley Cicero wants to make 16 holiday wreaths to sell at the craft fair. Each wreath requires $2\frac{1}{4}$ yd of ribbon. How many yards does she need?

estimate:

exact:

21. The Boy Scout troop has volunteered to pick up trash along a 4 mile stretch of highway. So far they have done $1\frac{7}{10}$ miles. How much do they have left to do?

estimate:

exact:

22. The gas tank on a Jeep Cherokee has a capacity of $21\frac{3}{8}$ gallons. Scott started with a full tank and then used $8\frac{1}{2}$ gallons of gasoline. Find the number of gallons that remain.

estimate:

exact:

23. Suppose that a dress requires $2\frac{3}{4}$ yd of material. How much material would be needed for 7 dresses?

estimate:

exact:

24. A cookie recipe uses $\frac{2}{3}$ cup brown sugar. How much brown sugar is needed to make $2\frac{1}{2}$ times the original recipe?

estimate:

exact:

25. A landscaper has $9\frac{5}{8}$ cubic yards of peat moss in a truck. If she unloads $1\frac{1}{2}$ cubic yards at the first stop and 3 cubic yards at the second stop, how much peat moss remains in the truck?

estimate:

exact:

26. Marv bought 10 yd of Italian silk fabric. He used $3\frac{7}{8}$ yd to make a jacket. How much fabric is left for other sewing projects?

estimate:

exact:

27. Melissa worked $18\frac{3}{4}$ hours over the last five days. If she worked the same amount each day, how long was she at work each day?

estimate:

exact:

28. Michael is cutting a $10\frac{1}{2}$ ft board into shelves for a bookcase. Each shelf will be $1\frac{3}{4}$ ft long. How many shelves can he cut?

estimate:

exact:

29. The head keeper at Jurassic Park had to transfer several dinosaurs to new enclosures. A 3-ton stegosaurus was moved on Monday, and a $7\frac{1}{2}$-ton adult tyrannosaurus and a $2\frac{3}{8}$-ton juvenile were moved on Wednesday. Find the total number of tons of dinosaurs moved during the week.

estimate:

exact:

30. Find the length of the indented section on this board.

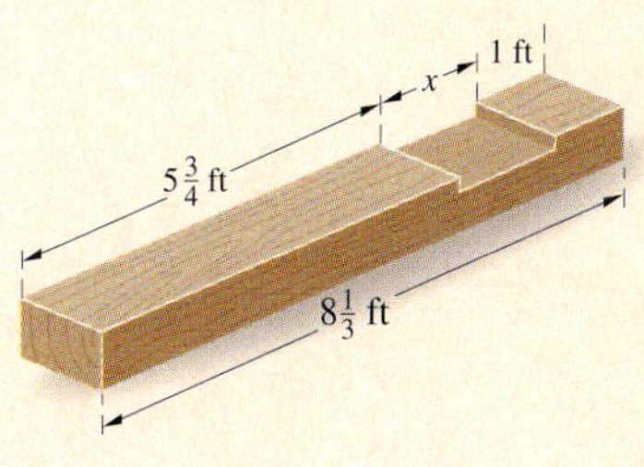

estimate:

exact:

First, estimate the answer to each application problem. Then use your calculator to find the exact answer.

31. A craftsperson must attach a lead strip around all four sides of a stained glass window before it is installed. Find the length of lead stripping needed for the window shown.

estimate:

exact:

32. To complete a custom order, Zak Morten of Home Depot must find the number of inches of brass trim needed to go around the four sides of the lamp base plate shown. Find the length of brass trim needed.

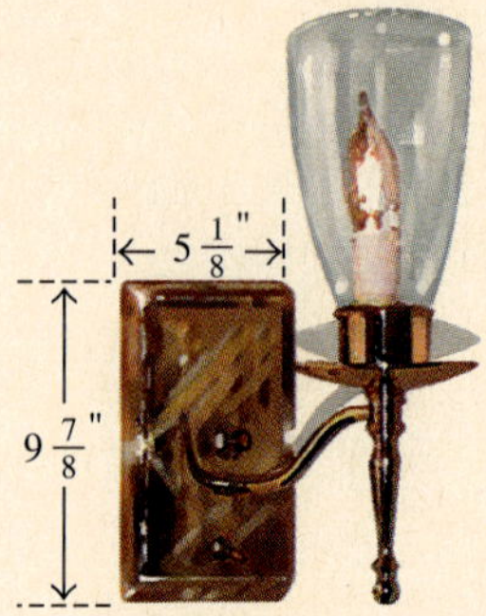

estimate:

exact:

33. A fishing boat anchor requires $10\frac{3}{8}$ pounds of steel. Find the number of anchors that can be manufactured with 25,730 pounds of steel.

estimate:

exact:

34. Each apartment requires $62\frac{1}{2}$ square yards of carpet. Find the number of apartments that can be carpeted with 6750 square yards of carpet.

estimate:

exact:

35. Ms. Nishimoto bought 12 shares of stock at $\$18\frac{3}{4}$ per share, 24 shares at $\$36\frac{3}{8}$ per share, and 16 shares at $\$74\frac{1}{8}$ per share. Her broker charged her a commission of \$12. Find the total amount that she paid.

estimate:

exact:

36. Three sides of a parking lot are $108\frac{1}{4}$ ft, $162\frac{3}{8}$ ft, and $143\frac{1}{2}$ ft. If the distance around the lot is $518\frac{3}{4}$ ft, find the length of the fourth side.

estimate:

exact:

Review and Prepare

Simplify. (For help, see ***Section 1.8****.)*

37. $(-2)^3$

38. $(-5)^2$

39. $(-4)^2 \cdot 3^2$

40. $(-3)^3 \cdot 2^3$

41. $3 + 7(-4)$

42. $5 + 2(6)$

43. $-6 + (2 - 5)^2$

44. $-8 + (7 - 9)^2$

4.6 Exponents, Order of Operations, and Complex Fractions

OBJECTIVES

1. Use exponents with fractions.
2. Use the order of operations with fractions.
3. Simplify complex fractions.

FOR EXTRA HELP

Tutorial | Tape 6 | SSM, Sec. 4.6

OBJECTIVE 1 You have used exponents as a quick way to write repeated multiplication of integers and variables. Here are two examples.

$$\underbrace{(-3)^2}_{\text{Base } -3,\ \text{Exponent } 2} = \underbrace{(-3)\cdot(-3)}_{\text{Two factors of } -3} = 9 \quad \text{and} \quad \underbrace{x^3}_{\text{Base } x,\ \text{Exponent } 3} = \underbrace{x\cdot x\cdot x}_{\text{Three factors of } x}$$

The meaning of an exponent remains the same when a fraction is the base.

EXAMPLE 1 Using Exponents with Fractions

Simplify.

(a) $\left(-\frac{1}{2}\right)^3$

The exponent indicates that there are three factors of $-\frac{1}{2}$.

$$\left(-\frac{1}{2}\right)^3 = \left(-\frac{1}{2}\right)\left(-\frac{1}{2}\right)\left(-\frac{1}{2}\right)$$ Three factors of $-\frac{1}{2}$. Watch the signs carefully. Multiply $\left(-\frac{1}{2}\right)\left(-\frac{1}{2}\right)$ to get $\frac{1}{4}$.

$$= \frac{1}{4}\left(-\frac{1}{2}\right)$$ Now multiply $\frac{1}{4}\left(-\frac{1}{2}\right)$ to get $-\frac{1}{8}$.

$$= -\frac{1}{8}$$ The product is negative.

(b) $\left(\frac{3}{4}\right)^2 \cdot \left(\frac{2}{3}\right)^3$

$$\left(\frac{3}{4}\right)^2 \cdot \left(\frac{2}{3}\right)^3 = \left(\frac{3}{4}\cdot\frac{3}{4}\right)\cdot\left(\frac{2}{3}\cdot\frac{2}{3}\cdot\frac{2}{3}\right)$$ Two factors of $\frac{3}{4}$. Three factors of $\frac{2}{3}$.

$$= \frac{\overset{1}{\cancel{3}}\cdot\overset{1}{\cancel{3}}\cdot\overset{1}{\cancel{2}}\cdot\overset{1}{\cancel{2}}\cdot\overset{1}{\cancel{2}}}{\underset{1}{\cancel{2}}\cdot\underset{1}{\cancel{2}}\cdot\underset{1}{\cancel{2}}\cdot 2\cdot\underset{1}{\cancel{3}}\cdot\underset{1}{\cancel{3}}\cdot 3}$$ Divide out all common factors.

$$= \frac{1}{6}$$

WORK PROBLEM 1 AT THE SIDE.

1. Simplify.

(a) $\left(-\frac{3}{5}\right)^2$

(b) $\left(\frac{1}{3}\right)^4$

(c) $\left(-\frac{2}{3}\right)^3\left(\frac{1}{2}\right)^2$

(d) $\left(-\frac{1}{2}\right)^2\left(\frac{1}{4}\right)^2$

OBJECTIVE 2 The order of operations that you used in **Section 1.8** for integers also applies to fractions.

ANSWERS

1. (a) $\frac{9}{25}$ (b) $\frac{1}{81}$ (c) $-\frac{2}{27}$ (d) $\frac{1}{64}$

2. Simplify.

(a) $\frac{1}{3} - \frac{5}{9}\left(\frac{3}{4}\right)$

(b) $-\frac{3}{4} + \left(-\frac{1}{2}\right)^2 \div \frac{2}{3}$

(c) $\frac{12}{5} - \frac{1}{6}\left(3 - \frac{3}{5}\right)$

Order of Operations

Step 1 Work inside **parentheses** or other grouping symbols.

Step 2 Simply expressions with **exponents.**

Step 3 Do the remaining multiplications and divisions as they occur from left to right.

Step 4 Do the remaining additions and subtractions as they occur from left to right.

EXAMPLE 2 Using the Order of Operations with Fractions

Simplify.

(a) $-\frac{1}{3} + \frac{1}{2}\left(\frac{4}{5}\right)$ There is no work to be done inside the parentheses. There are no exponents, so start with Step 3, multiplying and dividing.

$-\frac{1}{3} + \frac{1 \cdot 4}{2 \cdot 5}$ Multiply.

$-\frac{1}{3} + \frac{4}{10}$ Now add. The LCD is 30. $-\frac{1}{3} = -\frac{10}{30}$ and $\frac{4}{10} = \frac{12}{30}$

$\frac{-10 + 12}{30}$ Add the numerators.

$\frac{2}{30}$ Write $\frac{2}{30}$ in lowest terms $\frac{\overset{1}{\cancel{2}}}{\underset{1}{\cancel{2}} \cdot 15}$

$\frac{1}{15}$ The answer is in lowest terms.

(b) $-2 + \left(\frac{1}{4} - \frac{3}{2}\right)^2$ Work inside parentheses. The LCD for $\frac{1}{4}$ and $\frac{3}{2}$ is 4. Rewrite $\frac{3}{2}$ as $\frac{6}{4}$ and subtract.

$-2 + \left(\frac{1 - 6}{4}\right)^2$

$-2 + \left(\frac{-5}{4}\right)^2$ Simplify the term with the exponent. Multiply $(-\frac{5}{4})(-\frac{5}{4})$. Signs match, so the product is positive.

$-2 + \left(\frac{25}{16}\right)$ Add last. Write -2 as $-\frac{2}{1}$.

$-\frac{2}{1} + \frac{25}{16}$ The LCD is 16. Rewrite $-\frac{2}{1}$ as $-\frac{32}{16}$.

$\frac{-32 + 25}{16}$ Add the numerators.

$-\frac{7}{16}$ The answer is in lowest terms.

WORK PROBLEM 2 AT THE SIDE.

ANSWERS

2. (a) $-\frac{1}{12}$ (b) $-\frac{3}{8}$ (c) 2

OBJECTIVE 3 We have used both the symbol ÷ and a fraction bar to indicate division. For example,

Indicates division → $\frac{6}{2}$ can be written as $6 \div 2$. ← Indicates division

That means we could write $-\frac{4}{5} \div \left(-\frac{3}{10}\right)$ using a fraction bar instead of ÷.

$$-\frac{4}{5} \div \left(-\frac{3}{10}\right) \text{ can be written as } \frac{-\frac{4}{5}}{-\frac{3}{10}}.$$

(÷ — Indicates division; fraction bar ← Indicates division)

The result looks a bit complicated, and its name reflects that fact. We call it a *complex fraction.*

Complex Fractions

A **complex fraction** is a fraction in which the numerator and/or denominator contain one or more fractions.

EXAMPLE 3 Simplifying Complex Fractions

Simplify. $\frac{-\frac{4}{5}}{-\frac{3}{10}}$

Rewrite the complex fraction using the ÷ symbol for division. Then follow the steps for dividing fractions.

$$\frac{-\frac{4}{5}}{-\frac{3}{10}} = -\frac{4}{5} \div -\frac{3}{10} = -\frac{4}{5} \cdot -\frac{10}{3} = \frac{4 \cdot 2 \cdot \overset{1}{\cancel{5}}}{\underset{1}{\cancel{5}} \cdot 3} = \frac{8}{3} \text{ or } 2\frac{2}{3}$$

reciprocal

The quotient is positive because the numbers in the problem had matching signs (both were negative).

WORK PROBLEM 3 AT THE SIDE.

3. Simplify.

(a) $\frac{-\frac{3}{5}}{\frac{9}{10}}$

(b) $\frac{6}{\frac{3}{4}}$

(c) $\frac{-\frac{15}{16}}{-5}$

ANSWERS

3. (a) $-\frac{2}{3}$ (b) 8 (c) $\frac{3}{16}$

NUMBERS IN THE *Real World*

collaborative investigations

Many food packages have useful recipes on them. The side of a corn starch box is shown below.

1. Suppose you work at a daycare center, or run one in your home. You decide to mix up 3 pounds of Fun-Time Dough for the children. The recipe on the box makes how many pounds?

You will need to multiply each ingredient by some number that will give you 3 pounds of dough. Multiplying by 2 would give you 4 pounds of dough. What should you multiply by to get 3 pounds of dough?

—FARMER'S—

CORN STARCH

FAVORITE RECIPES

Fun-Time Dough

$1\frac{1}{2}$ cups Farmer's Corn Starch
$\frac{1}{2}$ cup flour
2 cups water
2 tsp cream of tartar
1 cup salt
1 T. vegetable oil

Mix all ingredients together in saucepan. Cook over medium heat, stirring constantly, until mixture gathers on the stirring spoon and forms dough. This will take about 6 minutes. Dump onto waxed paper until cool enough to handle and knead to form a pliable mass. Store in covered container or plastic bag. Food coloring may be added to make different colors.
Makes about 2 lbs. of Fun-Time Dough.

Great Gravy

3 T. bacon fat or meat drippings
2 T. Farmer's Corn Starch
$1\frac{1}{2}$ cups water
$\frac{3}{4}$ tsp. salt
$\frac{1}{8}$ tsp. pepper

Blend fat and Farmer's Corn Starch over low heat until it is a rich brown color, stirring constantly. Gradually add water, salt and pepper. Heat to boiling over direct heat and then boil gently 2 minutes, stirring constantly. Makes $1\frac{1}{2}$ cups.

— Satisfaction Guaranteed —

Now figure out how much of each ingredient you should use to make 3 pounds of dough. Show your work and record your answers

corn starch **flour**

water **cream of tartar**

salt **vegetable oil**

2. Suppose you decide to use the recipe on the box to make gravy for Thanksgiving dinner. How much gravy does the recipe make?

If you need 4 cups of gravy, what will you do to the recipe amounts?

Figure out the amount of each ingredient needed to make 4 cups of gravy. Show your work and record your answers below.

4.6 Exercises

Simplify.

1. $\left(-\frac{3}{4}\right)^2$

2. $\left(-\frac{4}{5}\right)^2$

3. $\left(\frac{2}{5}\right)^3$

4. $\left(\frac{1}{4}\right)^3$

5. $\left(-\frac{1}{3}\right)^3$

6. $\left(-\frac{3}{5}\right)^3$

7. $\left(\frac{1}{2}\right)^5$

8. $\left(\frac{1}{3}\right)^4$

9. $\left(\frac{7}{10}\right)^2$

10. $\left(\frac{8}{9}\right)^2$

11. $\left(-\frac{6}{5}\right)^2$

12. $\left(-\frac{8}{7}\right)^2$

13. $\frac{15}{16}\left(\frac{4}{5}\right)^3$

14. $-8\left(-\frac{3}{8}\right)^2$

15. $\left(\frac{1}{3}\right)^4\left(\frac{9}{10}\right)^2$

16. $\left(\frac{4}{5}\right)^2\left(\frac{1}{2}\right)^6$

17. $\left(-\frac{3}{2}\right)^3\left(-\frac{2}{3}\right)^2$

18. $\left(\frac{5}{6}\right)^2\left(-\frac{2}{5}\right)^3$

19. Evaluate this series of examples. Explain the pattern you see in the sign of the answers.

$\left(-\frac{1}{2}\right)^2 = \underline{\qquad}$ $\left(-\frac{1}{2}\right)^6 = \underline{\qquad}$

$\left(-\frac{1}{2}\right)^3 = \underline{\qquad}$ $\left(-\frac{1}{2}\right)^7 = \underline{\qquad}$

$\left(-\frac{1}{2}\right)^4 = \underline{\qquad}$ $\left(-\frac{1}{2}\right)^8 = \underline{\qquad}$

$\left(-\frac{1}{2}\right)^5 = \underline{\qquad}$ $\left(-\frac{1}{2}\right)^9 = \underline{\qquad}$

20. Several drops of ketchup fell on Ron's homework. Explain how he can figure out what number is covered by each drop. Be careful. More than one number may work.

(a) ()$^2 = \frac{4}{9}$ **(b)** ()$^3 = -\frac{1}{27}$

Simplify.

21. $\frac{1}{5} - \frac{7}{10}(6)$

22. $\frac{2}{9} - 4\left(\frac{5}{6}\right)$

23. $\left(\frac{4}{3} \div \frac{8}{3}\right) + \left(-\frac{3}{4} \cdot \frac{1}{4}\right)$

24. $\left(-\frac{1}{3} \cdot \frac{3}{5}\right) + \left(\frac{3}{4} \div \frac{1}{4}\right)$

25. $-\frac{3}{10} \div \frac{3}{5}\left(-\frac{2}{3}\right)$

26. $5 \div \left(-\frac{10}{3}\right)\left(-\frac{4}{9}\right)$

27. $\frac{8}{3}\left(\frac{1}{4} - \frac{1}{2}\right)^2$

28. $\frac{1}{3}\left(\frac{4}{5} - \frac{3}{10}\right)^3$

29. $-\frac{3}{8} + \frac{2}{3}\left(-\frac{2}{3} + \frac{1}{6}\right)$

30. $\frac{1}{6} + 4\left(\frac{2}{5} - \frac{7}{10}\right)$

31. $2\left(\frac{1}{3}\right)^3 - \frac{2}{9}$

32. $8\left(-\frac{3}{4}\right)^2 + \frac{3}{2}$

Ⓖ **33.** $\left(-\frac{2}{3}\right)^3\left(\frac{1}{8} - \frac{1}{2}\right) - \frac{2}{3}\left(\frac{1}{8}\right)$

Ⓖ **34.** $\left(\frac{3}{5}\right)^2\left(\frac{5}{9} - \frac{2}{3}\right) \div \left(-\frac{1}{5}\right)^2$

35. A square operation key on a calculator is $\frac{3}{8}$ in. on each side. What is the area of the key? Use the formula $A = s^2$.

36. A square lot for sale in the country is $\frac{3}{10}$ mile on a side. Find the area of the lot by using the formula $A = s^2$.

37. A rectangular parking lot at the megamall is $\frac{7}{10}$ mile long and $\frac{1}{4}$ mile wide. How much fencing is needed to enclose the lot? Use the formula $P = 2l + 2w$.

38. A computer chip in a rectangular shape is $\frac{7}{8}$ in. long and $\frac{5}{16}$ in. wide. An insulating strip must be put around all sides of the chip. Find the length of the strip by using the formula $P = 2l + 2w$.

Simplify.

39. $\dfrac{-\frac{7}{9}}{-\frac{7}{36}}$

40. $\dfrac{\frac{15}{32}}{-\frac{5}{64}}$

41. $\dfrac{-15}{\frac{6}{5}}$

42. $\dfrac{-6}{-\frac{5}{8}}$

43. $\dfrac{\frac{4}{7}}{8}$

44. $\dfrac{-\frac{11}{5}}{3}$

Ⓖ **45.** $\dfrac{\left(\frac{2}{5}\right)^2}{\left(-\frac{4}{3}\right)^2}$

Ⓖ **46.** $\dfrac{\left(-\frac{5}{6}\right)^2}{\left(\frac{1}{2}\right)^3}$

Ⓖ **47.** $\dfrac{\frac{5}{6}+\frac{2}{3}}{2\frac{2}{5}}$

Ⓖ **48.** $\dfrac{\frac{1}{2}+\frac{3}{4}}{3\frac{1}{3}}$

Ⓖ **49.** $\dfrac{4\frac{1}{2}}{\frac{1}{2}-\frac{3}{4}}$

Ⓖ **50.** $\dfrac{1\frac{2}{3}}{\frac{3}{10}-\frac{4}{5}}$

Review and Prepare

Solve each equation and check each solution. (For help, see ***Section 2.5****.)*

51. $-9 = 2y + 7$ **Check**

52. $12 = 3c - 6$ **Check**

53. $300 - 4n = 100$ **Check**

54. $200 + 8a = 232$ **Check**

4.7 Problem Solving: Equations Containing Fractions

OBJECTIVES

1. Use the multiplication property of equality to solve equations containing fractions.
2. Use both the addition and multiplication properties of equality to solve equations containing fractions.
3. Solve application problems using equations containing fractions

FOR EXTRA HELP

Tutorial — Tape 6 — SSM, Sec. 4.7

OBJECTIVE 1 In **Section 2.4** you used the division property of equality to solve an equation such as $4s = 24$. The division property says that you may divide both sides of an equation by the same nonzero number and that the equation will still be balanced. Now that you have some experience with fractions, let's look again at how the division property works.

4s means 4 • s

$$4s = 24$$

$$\frac{4 \cdot s}{4} = \frac{24}{4} \quad \text{Divide both sides by 4.}$$

$$\frac{\overset{1}{\cancel{4}} \cdot s}{\underset{1}{\cancel{4}}} = 6 \quad \text{Divide out the common factor of 4.}$$

$$s = 6$$

Because multiplication and division are related to each other, we can also *multiply* both sides of an equation by the same nonzero number and keep it balanced.

Multiplication and Division Properties of Equality

If $a = b$, then $a \cdot c = b \cdot c$. This is the **multiplication property of equality.** Also $\frac{a}{c} = \frac{b}{c}$ as long as c is not zero (division property of equality). In other words, you may multiply or divide both sides of an equation by the same nonzero number and it will still be balanced.

EXAMPLE 1 Using the Multiplication Property of Equality

Solve each equation and check each solution.

(a) $\frac{1}{2}b = 5$

As in Chapter 2, you want the variable by itself on one side of the equal sign. In this example, you want to end up with $1b$, not $\frac{1}{2}b$, on the left side. (Recall that $1b$ is equivalent to b.)

In **Section 4.3** you learned that the product of a number and its reciprocal is 1. Thus multiplying $\frac{1}{2}$ by $\frac{2}{1}$ will give us the desired coefficient of 1.

$$\frac{1}{2}b = 5 \quad \text{Multiply both sides by } \tfrac{2}{1} \text{ (the reciprocal of } \tfrac{1}{2}\text{).}$$

On the left side, use the associative property to regroup the factors.

$$\frac{\mathbf{2}}{\mathbf{1}}\left(\frac{1}{2}b\right) = \frac{\mathbf{2}}{\mathbf{1}}(5) \quad \text{On the right side, 5 is equivalent to } \tfrac{5}{1}.$$

$$\left(\frac{2}{1} \cdot \frac{1}{2}\right)b = \frac{2}{1}\left(\frac{5}{1}\right)$$

$$\left(\frac{\overset{1}{\cancel{2}}}{1} \cdot \frac{1}{\underset{1}{\cancel{2}}}\right)b = \frac{10}{1}$$

1b is equivalent to b.

$$1b = 10$$

$$b = 10$$

CONTINUED ON NEXT PAGE

Once you understand the process, you don't have to show every step. Here is a shorthand solution of the same problem.

$$\frac{1}{2}b = 5$$

$$\frac{\overset{1}{\cancel{2}}}{1}\left(\frac{1}{\underset{1}{\cancel{2}}}b\right) = \frac{2}{1}(5)$$

$$b = 10$$

The solution is 10. Check the solution by going back to the original equation.

Check $\frac{1}{2}b = 5$ Replace b with 10 in the original equation.

$\frac{1}{2}(10) = 5$ Multiply on the left side: $\frac{1}{2}(10)$ is $\frac{1}{2} \bullet \frac{10}{1}$, or $\frac{1}{2} \bullet \frac{2 \bullet 5}{1}$.

$$\frac{1 \bullet \overset{1}{\cancel{2}} \bullet 5}{\underset{1}{\cancel{2}} \bullet 1} = 5$$

$5 = 5$ Balances, so 10 is the correct solution.

(b) $12 = -\frac{3}{4}x$

$12 = -\frac{3}{4}x$ Multiply both sides by $-\frac{4}{3}$ (the reciprocal of $-\frac{3}{4}$. The reciprocal of a negative number is also negative.

$-\frac{4}{3}(12) = -\frac{\overset{1}{\cancel{4}}}{\underset{1}{\cancel{3}}}\left(-\frac{\overset{1}{\cancel{3}}}{\underset{1}{\cancel{4}}}x\right)$ On the left side $-\frac{4}{3}(12)$ is $-\frac{4}{3} \bullet \frac{12}{1}$, or $-\frac{4 \bullet \overset{1}{\cancel{3}} \bullet 4}{\underset{1}{\cancel{3}} \bullet 1} = -16$

$-16 = x$

The solution is -16. Check the solution by going back to the original equation.

Check $12 = -\frac{3}{4}x$ Replace x with -16.

$12 = -\frac{3}{4}(-16)$ The product of two negative numbers is positive. $-\frac{3}{4}(-16)$ is $\frac{3}{4} \bullet \frac{16}{1}$, or $\frac{3}{4} \bullet \frac{4 \bullet 4}{1}$.

$$12 = \frac{3 \bullet \overset{1}{\cancel{4}} \bullet 4}{\underset{1}{\cancel{4}} \bullet 1}$$

$12 = 12$ Balances, so -16 is the correct solution.

(c) $-\frac{2}{5}n = -\frac{1}{3}$

$-\frac{\overset{1}{\cancel{5}}}{\underset{1}{\cancel{2}}}\left(-\frac{\overset{1}{\cancel{2}}}{\underset{1}{\cancel{5}}}n\right) = \left(-\frac{5}{2}\right)\left(-\frac{1}{3}\right)$ Multiply both sides by $-\frac{5}{2}$ (the reciprocal of $-\frac{2}{5}$).

$n = \frac{5 \bullet 1}{2 \bullet 3}$ The product of two negative numbers is positive.

$$n = \frac{5}{6}$$

CONTINUED ON NEXT PAGE

Check $-\frac{2}{5}n = -\frac{1}{3}$ Replace n with $\frac{5}{6}$.

$\left(-\frac{2}{5}\right)\left(\frac{5}{6}\right) = -\frac{1}{3}$ Multiply on the left side.

$-\frac{\overset{1}{\cancel{2}} \cdot \overset{1}{\cancel{5}}}{\underset{1}{\cancel{5}} \cdot \underset{1}{\cancel{2}} \cdot 3} = -\frac{1}{3}$

$-\frac{1}{3} = -\frac{1}{3}$ Balances, so $\frac{5}{6}$ is the correct solution.

WORK PROBLEM 1 AT THE SIDE.

OBJECTIVE 2 In **Section 2.5** you used both the addition and division properties of equality to solve equations. Now you can use both the addition and multiplication properties.

EXAMPLE 2 Using the Addition and Multiplication Properties of Equality

Solve each equation and check each solution.

(a) $\frac{1}{3}c + 5 = 7$

The first step is to get the variable term $\frac{1}{3}c$ by itself on the left side of the equal sign. Recall that to "get rid of" the 5 on the left side, add the opposite of 5, which is -5.

$$\frac{1}{3}c + 5 = 7 \quad \text{Add } -5 \text{ to both sides.}$$

$$-5 = -5$$

$$\frac{1}{3}c + 0 = 2$$

$$\frac{1}{3}c = 2$$

$$\frac{\overset{1}{\cancel{3}}}{1}\left(\frac{1}{\underset{1}{\cancel{3}}}c\right) = \left(\frac{3}{1}\right)2 \quad \text{Multiply both sides by } \tfrac{3}{1} \text{ (the reciprocal of } \tfrac{1}{3}\text{).}$$

$$c = 6$$

The solution is 6. Check the solution by going back to the original equation.

Check $\frac{1}{3}c + 5 = 7$ Replace c in the original equation with 6.

$\frac{1}{3}(6) + 5 = 7$

$2 + 5 = 7$

$7 = 7$ Balances, so 6 is the correct solution.

(b) $-3 = \frac{2}{3}y + 7$

To get the variable term $\frac{2}{3}y$ by itself on the right side, add -7 to both sides.

CONTINUED ON NEXT PAGE

1. Solve each equation. Check each solution.

(a) $\frac{1}{6}m = 3$ **Check**

(b) $\frac{3}{2}a = -9$ **Check**

(c) $\frac{3}{14} = -\frac{2}{7}x$ **Check**

ANSWERS

1. (a) $m = 18$ **Check** $\frac{1}{6}m = 3$; $\frac{1}{6}(18) = 3$; Balances $3 = 3$

(b) $a = -6$ **Check** $\frac{3}{2}a = -9$; $\frac{3}{2}(-6) = -9$; Balances $-9 = -9$

(c) $x = -\frac{3}{4}$ **Check** $\frac{3}{14} = -\frac{2}{7}x$; $\frac{3}{14} = -\frac{2}{7}\left(-\frac{3}{4}\right)$; Balances $\frac{3}{14} = \frac{3}{14}$

$$-3 = \frac{2}{3}y + 7 \quad \text{Add } -7 \text{ to both sides.}$$
$$\underline{-7} \qquad \underline{-7}$$
$$-10 = \frac{2}{3}y + 0$$
$$\frac{3}{2}(-10) = \frac{\overset{1}{\cancel{3}}}{\underset{1}{\cancel{2}}}\left(\frac{\overset{1}{\cancel{2}}}{\underset{1}{\cancel{3}}}y\right) \quad \text{Multiply both sides by } \tfrac{3}{2} \text{ (the reciprocal of } \tfrac{2}{3}\text{).}$$
$$-15 = y$$

Check $-3 = \frac{2}{3}y + 7$ Replace y with -15.
$$-3 = \frac{2}{3}(-15) + 7$$
$$-3 = -10 + 7$$
$$-3 = -3$$
Balances, so -15 is the correct solution.

WORK PROBLEM 2 AT THE SIDE.

Objective 3 One of the expressions you worked with in **Section 2.1** told how to find a person's approximate systolic blood pressure.

$$100 + \frac{a}{2} \quad \leftarrow a \text{ represents the age of the person.}$$

EXAMPLE 3 Solving Application Problems by Using Equations with Fractions

Suppose you know that your friend's systolic blood pressure is 116 (and that your friend has normal blood pressure). Use this information and the expression for finding systolic blood pressure to write an equation. Then solve the equation to find your friend's age.

$$100 + \frac{a}{2} = 116 \quad \text{Add } -100 \text{ to both sides.}$$
$$\underline{-100} \qquad \underline{-100}$$
$$0 + \frac{a}{2} = 16$$
$$\frac{a}{2} = 16$$
$$\frac{1}{2}a = 16$$

$\frac{1}{2}a$ is equivalent to $\frac{a}{2}$ because $\frac{1}{2}a$ is $\frac{1}{2} \cdot \frac{a}{1}$

$$\frac{\overset{1}{\cancel{2}}}{1}\left(\frac{1}{\underset{1}{\cancel{2}}}a\right) = \frac{2}{1}(16) \quad \text{Multiply both sides by } \tfrac{2}{1} \text{ (the reciprocal of } \tfrac{1}{2}\text{).}$$
$$a = 32$$

Your friend is 32 years old.

WORK PROBLEM 3 AT THE SIDE.

---

2. Solve each equation. Check each solution.

(a) $18 = \frac{4}{5}x + 2$ **Check**

(b) $\frac{1}{4}h - 5 = 1$ **Check**

(c) $\frac{4}{3}r + 4 = -8$ **Check**

3. A woman's systolic blood pressure is 111. Write an equation and solve it to find her age, using the expression for systolic blood pressure from Example 3. (Assume that the woman has normal blood pressure.)

Answers

2. (a) $x = 20$ **Check** $18 = \frac{4}{5}x + 2$
$18 = \frac{4}{5}(20) + 2$
$18 = 16 + 2$
Balances $18 = 18$

(b) $h = 24$ **Check** $\frac{1}{4}h - 5 = 1$
$\frac{1}{4}(24) - 5 = 1$
$6 - 5 = 1$
Balances $1 = 1$

(c) $r = -9$ **Check** $\frac{4}{3}r + 4 = -8$
$\frac{4}{3}(-9) + 4 = -8$
$-12 + 4 = -8$
Balances $-8 = -8$

3. $100 + \frac{a}{2} = 111$ The woman is 22 years old.

Name Date Class

4.7 Exercises

Solve each equation and check each solution.

1. $\frac{1}{3}a = 10$ **Check**

2. $7 = \frac{1}{5}y$ **Check**

3. $-20 = \frac{5}{6}b$ **Check**

4. $-\frac{4}{9}w = 16$ **Check**

5. $-\frac{7}{2}c = -21$ **Check**

6. $-25 = \frac{5}{3}x$ **Check**

7. $\frac{9}{16} = \frac{3}{4}m$ **Check**

8. $\frac{5}{12}k = \frac{15}{16}$ **Check**

9. $\frac{3}{10} = -\frac{1}{4}d$ **Check**

10. $-\frac{7}{8}h = -\frac{1}{6}$ **Check**

11. $\frac{1}{6}n + 7 = 9$ **Check**

12. $3 + \frac{1}{4}p = 5$ **Check**

13. $-10 = \frac{5}{3}r + 5$ **Check**

14. $0 = 6 + \frac{3}{2}t$ **Check**

15. $\frac{3}{8}x - 9 = 0$ **Check**

16. $\frac{1}{3}s - 10 = -5$ **Check**

Ⓖ *Solve each equation. Show the steps you used.*

17. $7 - 2 = \frac{1}{5}y + 2 - 6$

18. $0 - 8 = \frac{1}{10}k - 8 + 5$

19. $-3 + 7 + \frac{2}{3}n = -10 + 2$

20. $-\frac{2}{5}m - 3 = 6 - 10 - 5$

21. $3x + \frac{1}{2} = \frac{3}{4}$

22. $4y + \frac{1}{3} = \frac{7}{9}$

23. $\frac{3}{10} = -4b - \frac{1}{5}$

24. $\frac{5}{6} = -3c - \frac{2}{3}$

25. Check the solution given for each equation. If a solution doesn't check, find and correct the error.

(a) $\frac{1}{6}x + 1 = -2$

$x = 18$

(b) $-\frac{3}{2} = \frac{9}{4}k$

$k = -\frac{2}{3}$

26. Check the solution given for each equation. If a solution doesn't check, find and correct the error.

(a) $-\frac{3}{4}y = -\frac{5}{8}$

$y = \frac{5}{6}$

(b) $16 = -\frac{7}{3}w + 2$

$w = 6$

Ⓖ 27. Write two different equations that have 8 as a solution. Write your equations with a fraction as the coefficient of the variable term.

Ⓖ 28. Write two different equations that have -12 as the solution. Write your equations with a fraction as the coefficient of the variable term.

In Exercises 29–32, write an equation and solve it to find each person's age. Use this expression for the approximate systolic blood pressure: $100 + \frac{a}{2}$, *where a is the person's age. Assume that all the people have normal blood pressure.*

29. A man has systolic blood pressure of 109. How old is he?

30. A man has systolic blood pressure of 118. How old is he?

31. A woman has systolic blood pressure of 122. How old is she?

32. A woman has systolic blood pressure of 113. How old is she?

Ⓖ *In Exercises 33–36, write an equation and solve it to find each person's height. Use this expression for the recommended weight of an adult: $\frac{11}{2}h - 220$, where h is the person's height in inches. Assume that all the people are at their recommended weight.*

33. A man weighs 209 pounds. What is his height in inches?

34. A woman weighs 110 pounds. What is her height in inches?

35. A woman weighs 132 pounds. What is her height in inches?

36. A man weighs 176 pounds. What is his height in inches?

Review and Prepare

Find the perimeter and area of each figure. (For help, see ***Sections 3.1*** *and* ***3.2****.)*

37. Rectangle

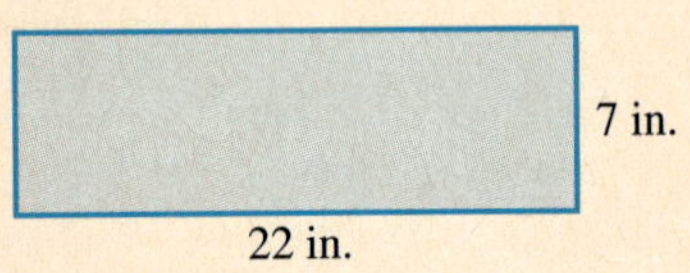

38. Square

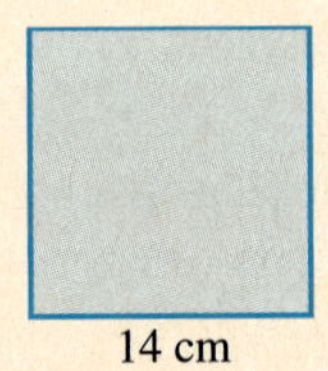

39. Parallelogram

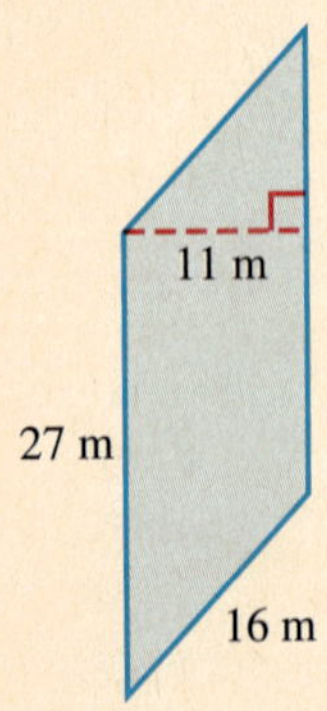

40. Parallelogram

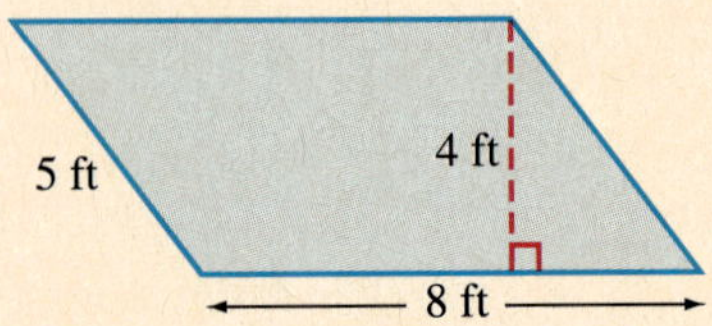

4.8 Geometry Applications: Area and Volume

OBJECTIVES

1. Find the area of a triangle.
2. Find the volume of a rectangular solid.
3. Find the volume of a pyramid.

FOR EXTRA HELP

OBJECTIVE 1 In **Section 3.1** you worked with triangles, which are figures that have exactly three sides. You found the perimeter of a triangle by adding the lengths of the three sides. Now you are ready to find the area of a triangle (the amount of surface inside the triangle).

You can find the *height* of a triangle by measuring the distance from one corner of the triangle to the opposite side (the base). The height line must be *perpendicular* to the base, that is, it must form a right angle with the base. Sometimes you have to extend the base line in order to draw the height perpendicular to it.

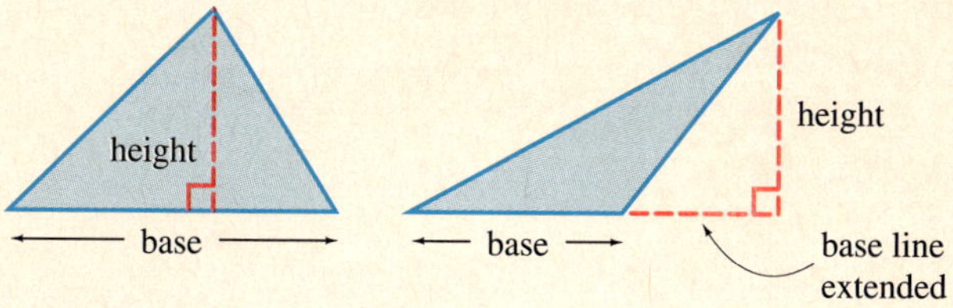

If you cut out two identical triangles and turn one upside down, you can fit them together to form a parallelogram.

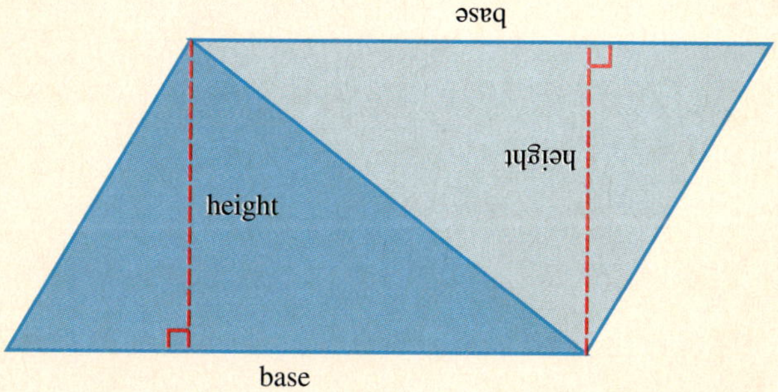

Recall from **Section 3.2** that the area of the parallelogram is base times height. Because each triangle is *half* of the parallelogram, the area of one triangle is

$$\frac{1}{2} \text{ of base times height.}$$

Use the following formula to find the area of a triangle.

Finding the Area of a Triangle

$$\text{Area of triangle} = \frac{1}{2} \cdot \text{base} \cdot \text{height}$$

$$A = \frac{1}{2}bh$$

Remember to use square units when measuring area.

1. Find the area of each triangle.

(a)

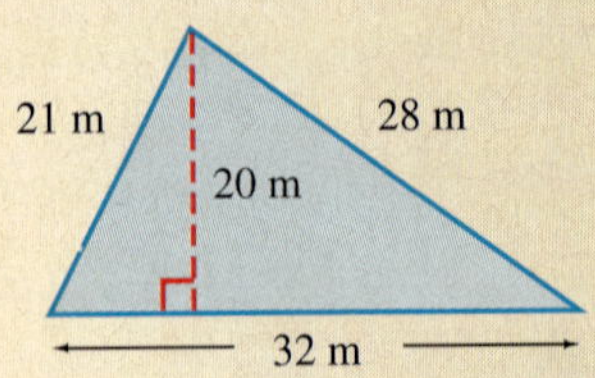

(b)

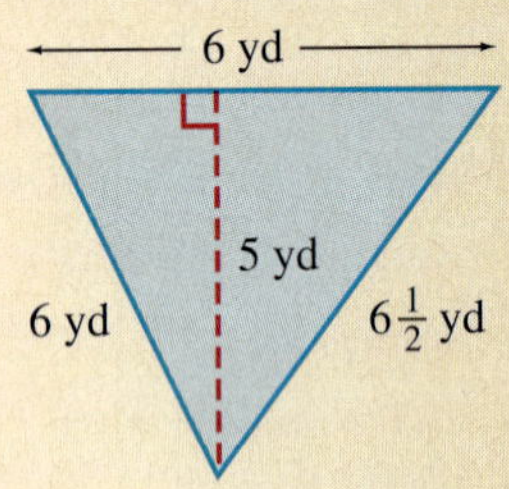

(c)

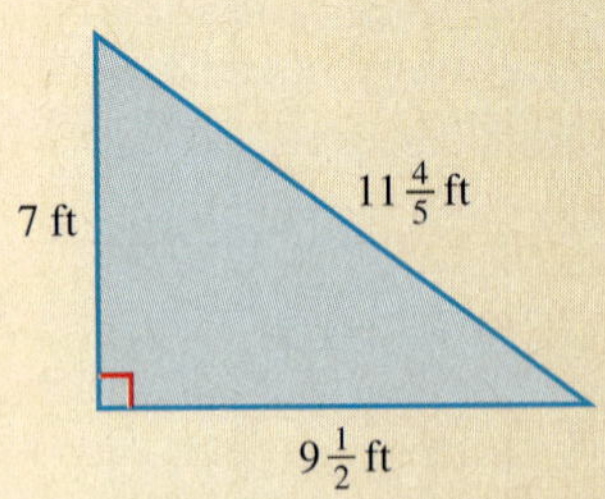

ANSWERS

1. (a) 320 m^2 **(b)** 15 yd^2
(c) $\frac{133}{4}$ ft^2 or $33\frac{1}{4}$ ft^2

EXAMPLE 1 Finding the Area of a Triangle

Find the area of each triangle.

(a)

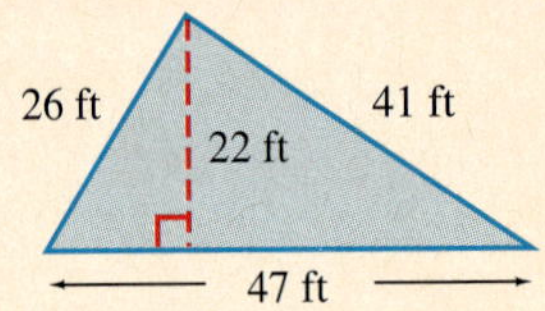

The base is 47 ft and the height is 22 ft. You do *not* need the 26 ft or 41 ft sides to find the area.

$$A = \frac{1}{2} \cdot b \cdot h$$ Replace b with 47 ft and h with 22 ft.

$$A = \frac{1}{2} \cdot 47 \text{ ft} \cdot 22 \text{ ft}$$

$$A = \frac{1}{2} \cdot \frac{47 \text{ ft}}{1} \cdot \frac{22 \text{ ft}}{1}$$ Divide out the common factor of 2.

$$A = \frac{1 \cdot 47 \text{ ft} \cdot \overset{1}{\cancel{2}} \cdot 11 \text{ ft}}{\underset{1}{\cancel{2}} \cdot 1 \cdot 1}$$ Multiply 47 • 11 to get 517. Multiply ft • ft to get ft^2.

$$A = 517 \text{ ft}^2$$

(b)

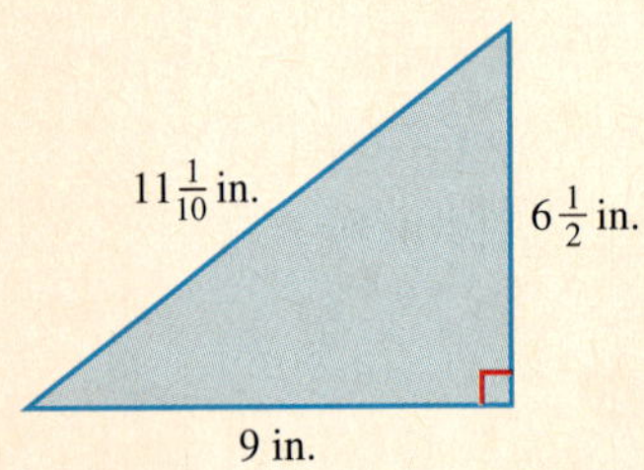

Because two sides of the triangle are perpendicular to each other, use those sides as the base and the height. (Remember that the height line must be perpendicular to the base line.)

$$A = \frac{1}{2}bh$$ Replace b with 9 in. and h with $6\frac{1}{2}$ in.

$$A = \frac{1}{2} \cdot 9 \text{ in.} \cdot 6\frac{1}{2} \text{ in.}$$ Write 9 in. and $6\frac{1}{2}$ in. as improper fractions.

$$A = \frac{1}{2} \cdot \frac{9 \text{ in.}}{1} \cdot \frac{13 \text{ in.}}{2}$$

$$A = \frac{1 \cdot 9 \text{ in.} \cdot 13 \text{ in.}}{2 \cdot 1 \cdot 2}$$ ← Multiply 9 • 13 to get 117 and in. • in. to get in.^{2}.

$$A = \frac{117}{4} \text{ in.}^2 \quad \text{or} \quad 29\frac{1}{4} \text{ in.}^2$$

◀◀ WORK PROBLEM 1 AT THE SIDE.

EXAMPLE 2 Using the Concept of Area

Find the area of the shaded part in this figure.

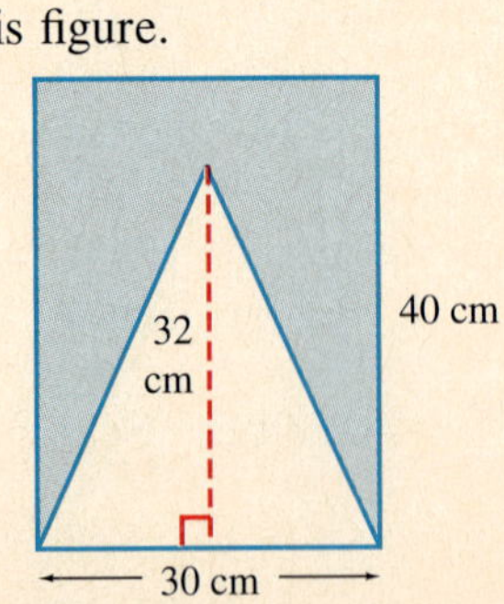

The *entire* figure is a rectangle.

$$A = lw$$

$$A = 30 \text{ cm} \cdot 40 \text{ cm} = 1200 \text{ cm}^2$$

CONTINUED ON NEXT PAGE

The *un*shaded part is a triangle.

$$A = \frac{1}{2}bh$$

$$A = \frac{1}{2} \cdot \frac{30 \text{ cm}}{1} \cdot \frac{32 \text{ cm}}{1}$$

$$A = \frac{1 \cdot \overset{1}{\cancel{2}} \cdot 15 \text{ cm} \cdot 32 \text{ cm}}{\underset{1}{\cancel{2}} \cdot 1 \cdot 1}$$

$$A = 480 \text{ cm}^2$$

Subtract to find the area of the shaded part.

$$A = \overbrace{1200 \text{ cm}^2}^{\text{Entire area}} - \overbrace{480 \text{ cm}^2}^{\text{Unshaded part}} = \overbrace{720 \text{ cm}^2}^{\text{Shaded part}}$$

WORK PROBLEM 2 AT THE SIDE.

OBJECTIVE 2 A shoe box and a cereal box are examples of three-dimensional (or solid) figures. The three dimensions are length, width, and height. (A rectangle or square is a two-dimensional figure. The two dimensions are length and width.) If we want to know how much the shoe box will hold, we find its *volume*. We measure volume by seeing how many cubes of a certain size will fill the space inside the box. Three sizes of *cubic units* are shown here.

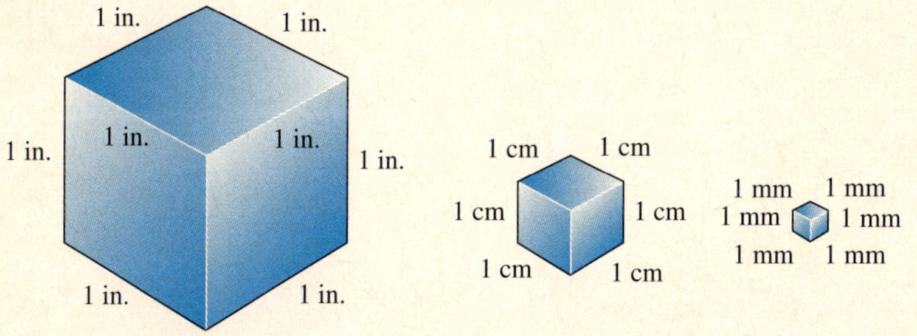

1 cubic inch **(1 in.^{3})** **1 cubic centimeter** **(1 cm^3)** **1 cubic millimeter** **(1 mm^3)**

Notice that all the edges of a cube have the same length. Some other sizes of cubes that are used to measure volume are 1 cubic foot (1 ft^3), 1 cubic yard (1 yd^3), and 1 cubic meter (1 m^3).

Note

The raised 3 in 4^3 means that you multiply $4 \cdot 4 \cdot 4$ to get 64. The raised 3 in cm^3 or ft^3 is a short way to write the word "cubic." It means that you multiplied cm times cm times cm, or ft times ft times ft. Recall that a short way to write $x \cdot x \cdot x$ is x^3. Similarly, cm • cm • cm is cm^3. When you see 5 cm^3, say "five cubic centimeters." Do *not* multiply $5 \cdot 5 \cdot 5$.

Volume

Volume is a measure of the space inside a solid shape. The volume of a solid is how many cubic units it takes to fill the solid.

Use the following formula for finding the volume of rectangular solids (box-like shapes).

2. Find the area of the shaded part in this figure.

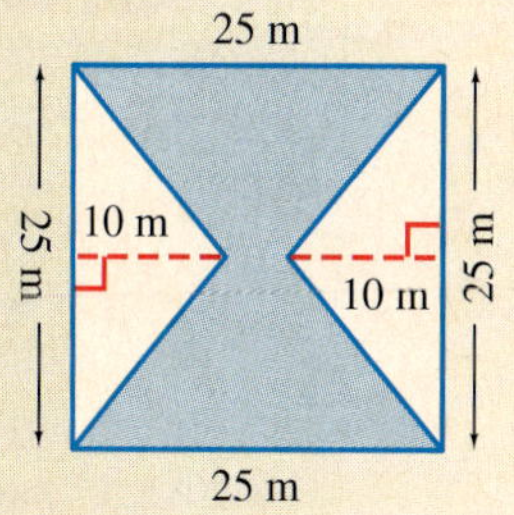

ANSWER

2. $A = 625 \text{ m}^2 - 125 \text{ m}^2 - 125 \text{ m}^2 = 375 \text{ m}^2$

3. Find the volume of each rectangular solid.

(a)

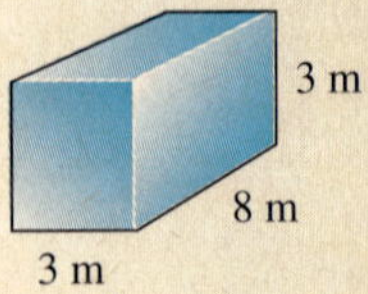

(b) length $6\frac{1}{4}$ ft, width $3\frac{1}{2}$ ft, height 2 ft

Finding the Volume of Rectangular Solids

Volume of rectangular solid = length • width • height

$$V = lwh$$

Remember to use cubic units when measuring volume.

EXAMPLE 3 Finding the Volume of a Rectangular Solid

Find the volume of each box.

(a)

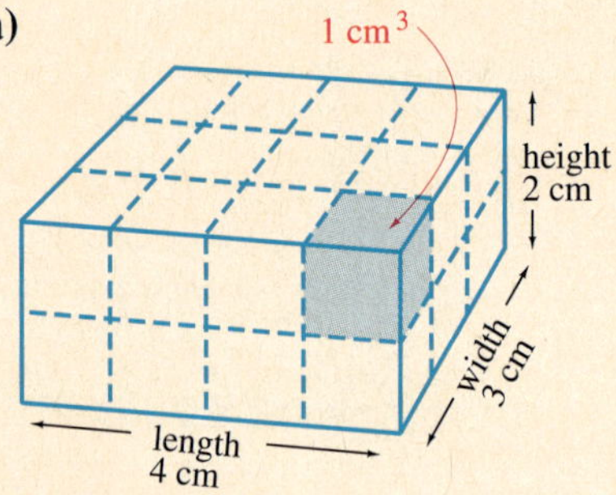

Each cube that fits in the box is 1 cubic centimeter (1 cm^3). To find the volume, you can count the number of cubes.

Bottom layer has 12 cubes. Top layer has 12 cubes. } Total of 24 cubes (24 cm^3)

Or you can use the formula for rectangular solids.

$V = l \cdot w \cdot h$

$V = 4\text{ cm} \cdot 3\text{ cm} \cdot 2\text{ cm}$ Multiply 4 • 3 • 2 to get 24.

$V = 24\text{ cm}^3$ Multiply cm • cm • cm to get cm^3.

(b)

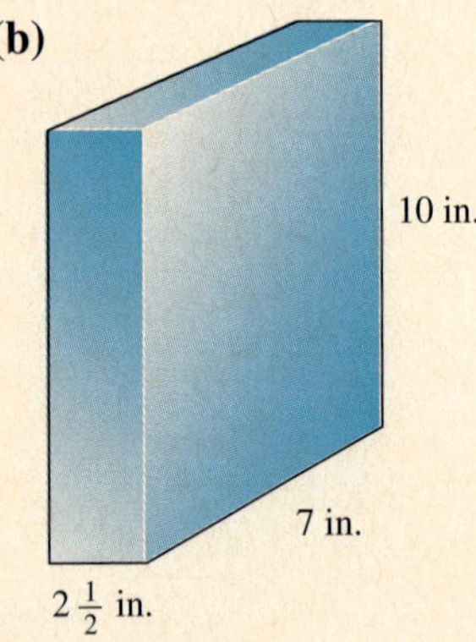

Use the formula $V = lwh$

$V = 7\text{ in.} \cdot 2\frac{1}{2}\text{ in.} \cdot 10\text{ in.}$ Write each measurement as an improper fraction.

$$V = \frac{7\text{ in.}}{1} \cdot \frac{5\text{ in.}}{2} \cdot \frac{10\text{ in.}}{1}$$

$$V = \frac{7\text{ in.} \cdot 5\text{ in.} \cdot \overset{1}{\cancel{2}} \cdot 5\text{ in.}}{1 \cdot \underset{1}{\cancel{2}} \cdot 1}$$ Divide out the common factor of 2.

$V = 175\text{ in.}^3$

WORK PROBLEM 3 AT THE SIDE.

OBJECTIVE 3 A pyramid is a solid shape like the one shown here.

The base of a pyramid may be a square or a rectangle.

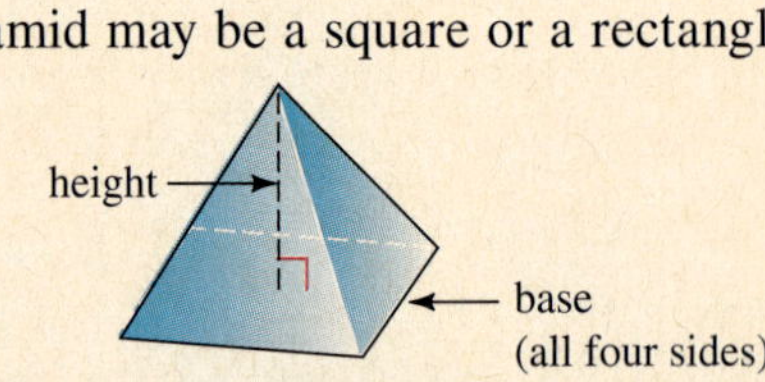

pyramid

The height is the distance from the base to the highest point of the pyramid.

ANSWERS

3. (a) 72 m^3 (b) $\frac{175}{4}$ ft^3 or $43\frac{3}{4}$ ft^3

Use this formula to find the volume of a pyramid.

Finding the Volume of a Pyramid

$$\text{Volume of pyramid} = \frac{1}{3} \cdot B \cdot h$$

$$V = \frac{1}{3} Bh$$

where B is the area of the square or rectangular base of the pyramid and h is the height of the pyramid. Remember to use cubic units when measuring volume.

EXAMPLE 4 Finding the Volume of a Pyramid

Find the volume of the pyramid.

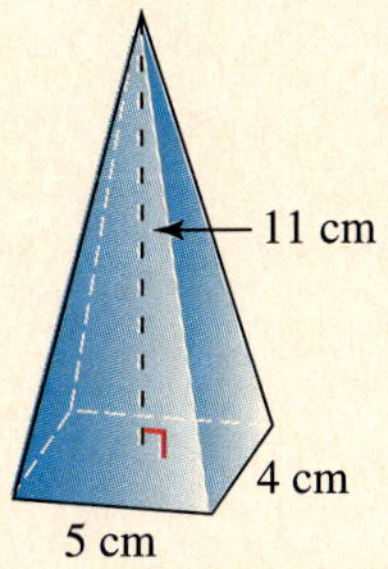

First find the area of the rectangular base by multiplying its length times its width.

$$B = 5 \text{ cm} \cdot 4 \text{ cm}$$

$$\mathbf{B = 20 \text{ cm}^2}$$

Next, find the volume.

$$V = \frac{1}{3} Bh$$ Replace B with 20 cm^2 and h with 11 cm.

$$V = \frac{1}{3} \cdot 20 \text{ cm}^2 \cdot 11 \text{ cm}$$

$$V = \frac{1}{3} \cdot \frac{20 \text{ cm}^2}{1} \cdot \frac{11 \text{ cm}}{1}$$

$$V = \frac{1 \cdot 20 \text{ cm}^2 \cdot 11 \text{ cm}}{3 \cdot 1 \cdot 1}$$ There are no common factors to divide out.

$$V = \frac{220}{3} \text{ cm}^3 \quad \text{or} \quad 73\frac{1}{3} \text{ cm}^3$$

WORK PROBLEM 4 AT THE SIDE.

4. Find the volume of a pyramid with a square base measuring 10 ft by 10 ft and a height of 6 ft.

Answer

4. $V = 200 \text{ ft}^3$

NUMBERS IN THE *Real World*

collaborative investigations

One of the ways people invest their money is to buy stock in a company. The price of each share of stock changes frequently, and many newspapers list the prices. You could buy stock in hundreds of different companies. The listing here shows the price for just five companies.

Market Review

	52 Week			Net
	High	Low	Close	Chg
AT&T	43 3/4	30 3/4	31 1/4	+1/2
AppleC	28 7/8	15 1/8	17 1/2	−3/8
MusicLd	5 1/8	11/16	1 1/4	−1/8
RenoAir	14	6 3/8	8 1/8	+3/4
Spiegel	13 1/4	5 7/8	6	−1/8

1. Stock prices use fractions to show parts of a dollar. For example:

$\frac{1}{4} = \$0.25$ $\frac{1}{2} = \$0.50$ $\frac{3}{4} = \$0.75$

From this information, figure out the value of 1/8, 3/8, 5/8, and 7/8.

The first two numbers after each company name show the highest and lowest price for the stock over the last 52 weeks (one year). Then comes the current price, ("Close") and finally, how the current price has changed from yesterday ("Net Chg"). Use the fractions or mixed numbers in the list to answer these questions.

2. What is the difference between the highest and lowest price for RenoAir during the last year?

3. How does the current price of RenoAir compare to its lowest price during the past year?

4. Use your estimating skills to find the company that had the greatest difference between its highest and lowest price during the past year. Then find the exact difference for that company.

5. How much would it cost to buy 10 shares of MusicLd stock at the current price?

6. How much would it have cost to buy 10 shares of MusicLd at the lowest price during the past year?

7. If AT&T stock is 1/2 dollar higher than it was yesterday (+1/2), what was its price yesterday? Find the price of each of the other stocks yesterday.

8. If you have $100 to invest in one of these companies, how many shares could you buy of each company's stock? You can only buy whole shares, not parts of a share. How much would be left over? (Ignore broker fees.)

4.8 Exercises

Find the area of each triangle.

1.

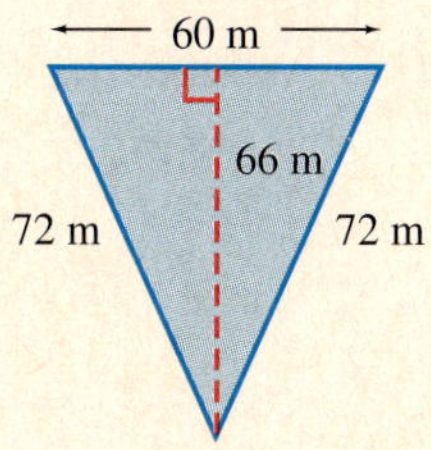

2.

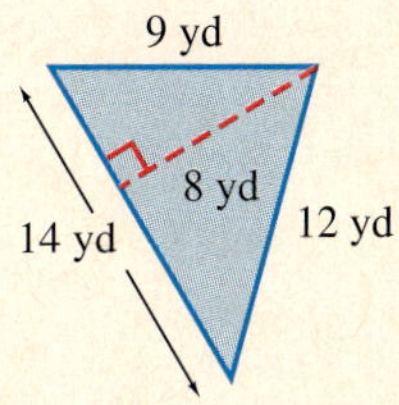

3.

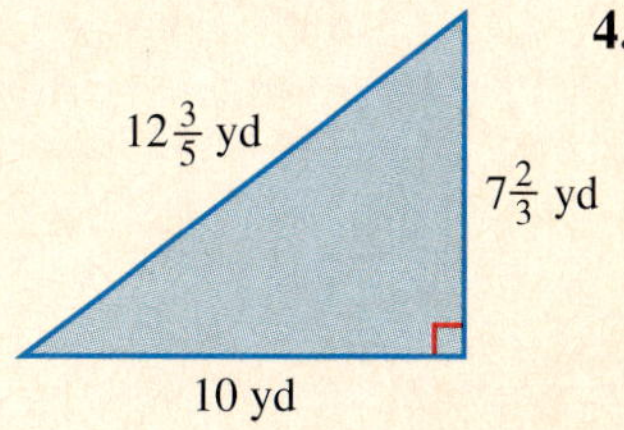

4.

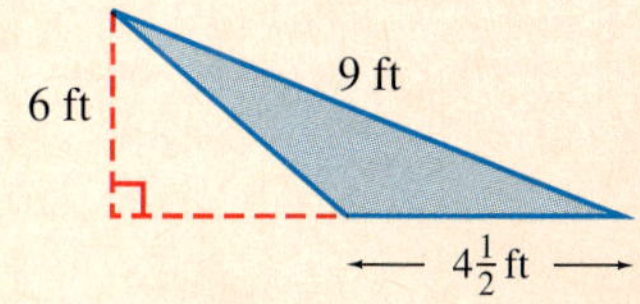

Find the shaded area in each figure.

5.

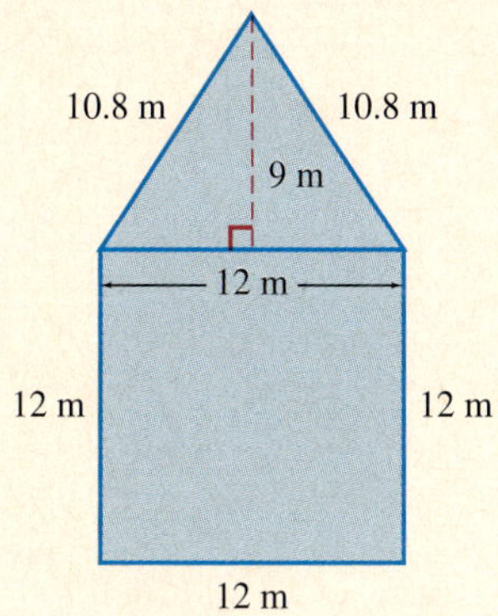

6.

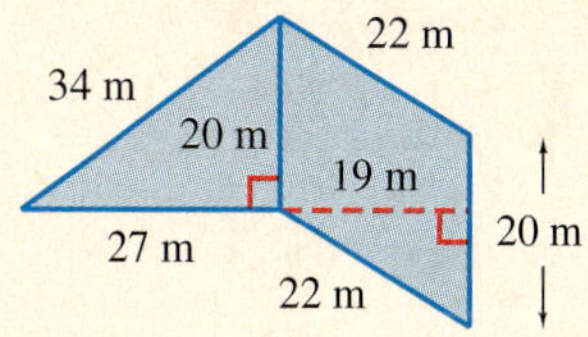

7.

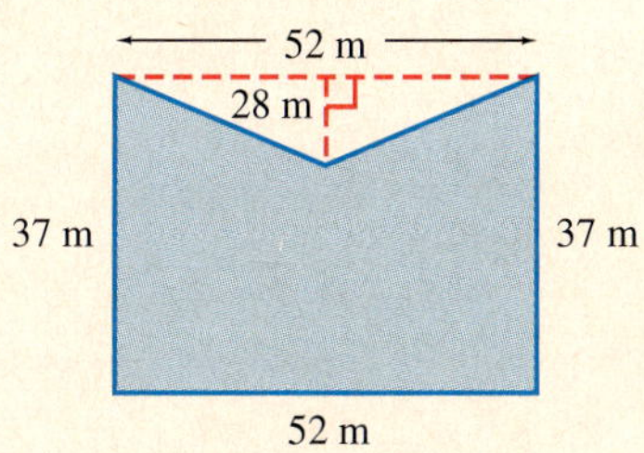

8.

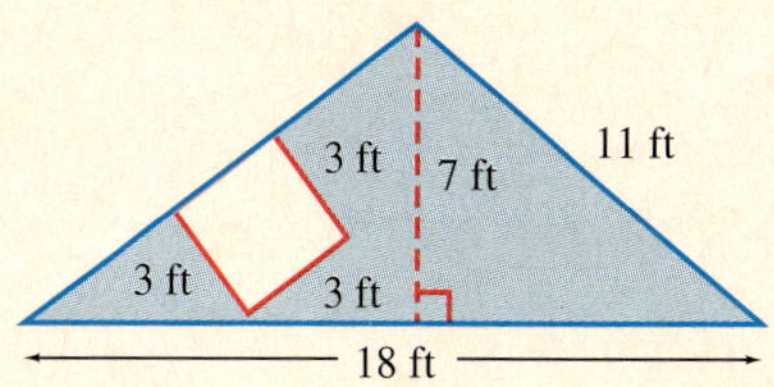

9. Explain the difference between perimeter, area, and volume.

10. Explain where the $\frac{1}{2}$ comes from in the formula for area of a triangle.

Solve each application problem.

11. A triangular tent flap measures $3\frac{1}{2}$ ft along the base and has a height of $4\frac{1}{2}$ ft. How much canvas is needed to make the flap?

12. A wooden sign in the shape of a right triangle has perpendicular sides measuring $1\frac{1}{2}$ m and 1 m. How much surface area does the sign have?

13. A triangular space between three streets has the measurements shown. How much new curbing will be needed to go around the space? How much sod will be needed to cover the space?

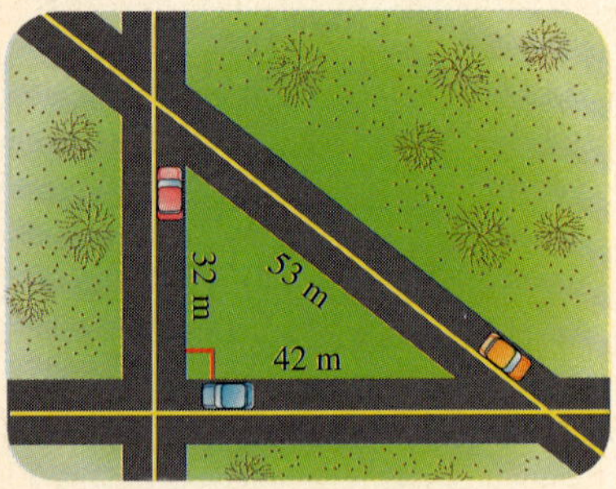

14. Each gable end of a new house has a span of 36 ft and a rise of $9\frac{1}{2}$ ft. What is the total area of both gable ends of the house?

Find the volume of each figure.

15.

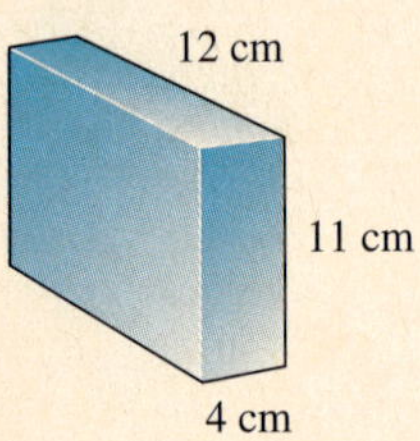

16.

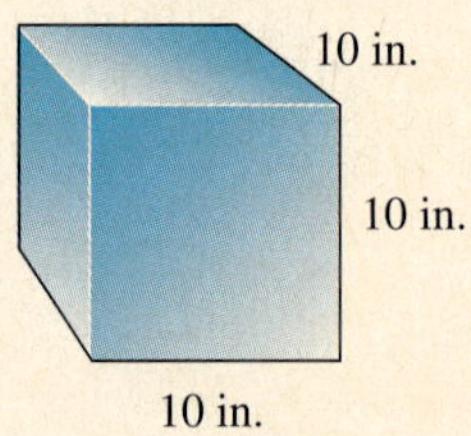

17.

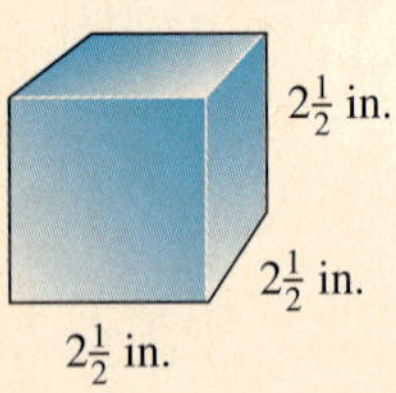

18.

19.

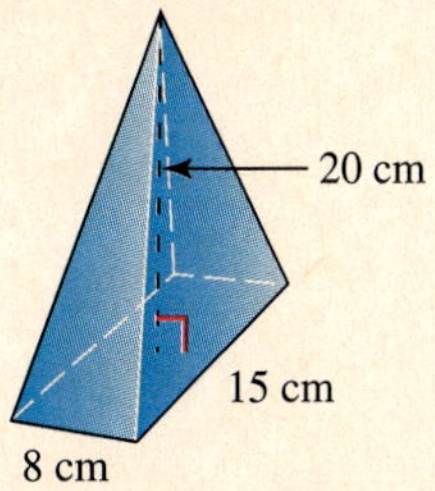

20.

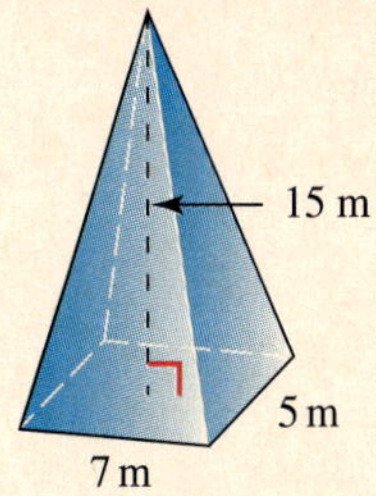

21. A box to hold pencils measures 3 in. by 8 in. by $\frac{3}{4}$ in. high. Find the volume of the box.

22. A train is being loaded with shipping crates. Each one is 12 ft long, 8 ft wide, and $2\frac{1}{4}$ ft high. How much space will each crate take?

23. One of the ancient stone pyramids in Egypt has a square base that measures 145 m on each side. The height is 93 m. What is the volume of the pyramid?

24. A cardboard model of an ancient stone pyramid has a square base that is $10\frac{3}{8}$ in. on each side. The height is $6\frac{1}{2}$ in. Find the volume of the model.

25. Find the volume of the object.

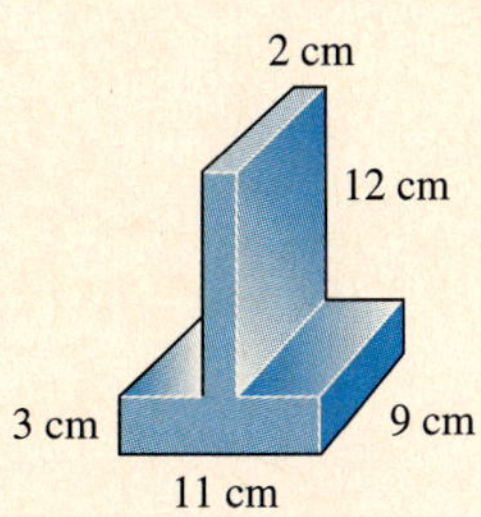

26. Find the volume of the shaded part. (*Hint:* Notice the hole that goes through the center of the shape.)

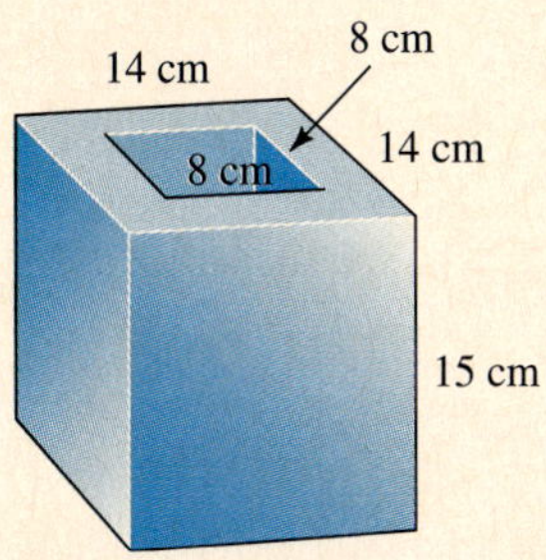

G *The following are some answers from a student's test. The number part of each answer is correct, but some of the units are not. Find the errors, explain what is wrong, and correct the errors.*

27. There are two errors in this list of answers.

$$A = 135^2 \text{ ft}$$
$$V = 76 \text{ yd}^3$$
$$P = 5\tfrac{1}{2} \text{ in.}^2$$

28. There are two errors in this list of answers.

$$P = 215 \text{ cm}$$
$$A = 8\tfrac{1}{2} \text{ m}$$
$$V = 98^3 \text{ in.}$$

Review and Prepare

Give the place value of the digit 5 in each number. (For help, see ***Section 1.1****.)*

29. 3502

30. 60,751

31. 245,617

32. 1,250,897

33. 539,410

34. 83,015

35. Write this number in words: 60,003,508

36. Write this number in words: 207,000,300,007

KEY TERMS

4.1	**fraction**	A fraction is a number of the form $\frac{a}{b}$, where a and b are integers and b is not zero.
	numerator	The top number in a fraction is the numerator. It shows how many of the equal parts are being considered.
	denominator	The bottom number in a fraction is the denominator. It shows how many equal parts are in the whole.
	proper fraction	In a proper fraction, the numerator is smaller than the denominator. The fraction is less than 1.
	improper fraction	In an improper fraction, the numerator is greater than or equal to the denominator. The fraction is greater than or equal to 1.
	equivalent fractions	Equivalent fractions have the same value even though they look different. When graphed on a number line, they are names for the same point.
4.2	**lowest terms**	A fraction is written in lowest terms when its numerator and denominator have no common factor other than 1.
	prime number	A prime number is a whole number that has exactly two different factors, itself and 1. The first few prime numbers are 2, 3, 5, 7, 11, 13, and 17.
	composite number	A composite number has at least one factor other than itself or 1. Examples are 4, 6, 9, and 10. The numbers 0 and 1 are neither prime nor composite.
	prime factorization	In a prime factorization, every factor is a prime number. For example, the prime factorization of 24 is $2 \cdot 2 \cdot 2 \cdot 3$.
4.3	**reciprocal**	Two numbers are reciprocals of each other if their product is 1. The reciprocal of $\frac{a}{b}$ is $\frac{b}{a}$ because $\frac{a}{b} \cdot \frac{b}{a} = 1$.
4.4	**like fractions**	Like fractions have the same denominator.
	unlike fractions	Unlike fractions have different denominators.
	least common denominator	The least common denominator (LCD) for two fractions is the smallest positive number that is evenly divisible by both denominators.
4.5	**mixed number**	A mixed number is a fraction and a whole number written together. It represents the sum of the whole number and the fraction. For example, $3 + \frac{1}{2}$ is written as $3\frac{1}{2}$.
4.6	**complex fraction**	A complex fraction is a fraction in which the numerator and/or denominator contain one or more fractions.
4.7	**multiplication property of equality**	The multiplication property of equality states that you may multiply both sides of an equation by the same nonzero number and still keep it balanced.
4.8	**volume**	Volume is a measure of the space inside a solid shape. Volume is measured in cubic units, such as in.^{3}, ft^3, yd^3, mm^3, cm^3, and so on.

KEY FORMULAS

Area of a triangle: $A = \frac{1}{2}bh$

Volume of a rectangular solid: $V = lwh$

Volume of a pyramid: $V = \frac{1}{3}Bh$

QUICK REVIEW

Concepts	*Examples*
4.1 Understanding Fraction Terminology The *numerator* is the top number. The *denominator* is the bottom number. In a *proper fraction* the numerator is smaller than the denominator. In an *improper fraction* the numerator is greater than or equal to the denominator.	Proper fractions $\frac{2}{3}, \frac{3}{4}, \frac{15}{16}, \frac{1}{8}$ ← Numerator ← Denominator Improper fractions $\frac{17}{8}, \frac{19}{12}, \frac{11}{2}, \frac{5}{3}, \frac{7}{7}$
4.1 Writing Equivalent Fractions Multiply or divide the numerator and denominator by the same nonzero number. The result is an equivalent fraction.	$\frac{1}{2} = \frac{1 \cdot \mathbf{8}}{2 \cdot \mathbf{8}} = \frac{8}{16}$ ← Equivalent to $\frac{1}{2}$ $-\frac{12}{15} = -\frac{12 \div \mathbf{3}}{15 \div \mathbf{3}} = -\frac{4}{5}$ ← Equivalent to $-\frac{12}{15}$
4.2 Finding Prime Factorizations A prime factorization of a number shows the number as the product of prime numbers. The first few prime numbers are 2, 3, 5, 7, 11, 13, and 17. You can use a division method or a factor tree to find the prime factorization.	Division Method: Prime Factorization of 24 $\mathbf{2}\overline{)24}$ ← Divide 24 by 2, the first prime. $\mathbf{2}\overline{)12}$ ← Divide 12 by 2. $\mathbf{2}\overline{)6}$ ← Divide 6 by 2. $\mathbf{3}\overline{)3}$ ← Divide 3 by 3. 1 ← Continue to divide until the quotient is 1. $24 = \mathbf{2 \cdot 2 \cdot 2 \cdot 3}$ Factor Tree Method 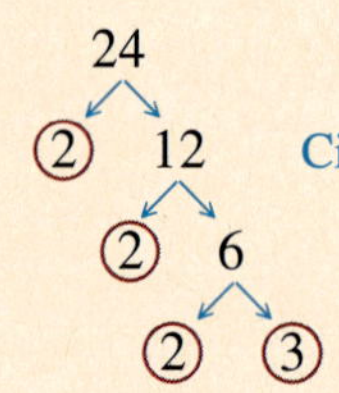Circle each prime number. $24 = \mathbf{2 \cdot 2 \cdot 2 \cdot 3}$
4.2 Writing Fractions in Lowest Terms Write the prime factorization of both numerator and denominator. Divide out all common factors, using slashes to show the division. Multiply any remaining factors in the numerator and in the denominator.	$\frac{18}{90} = \frac{\overset{1}{\cancel{2}} \cdot \overset{1}{\cancel{3}} \cdot \overset{1}{\cancel{3}}}{\underset{1}{\cancel{2}} \cdot \underset{1}{\cancel{3}} \cdot \underset{1}{\cancel{3}} \cdot 5} = \frac{1}{5}$ $\frac{2b^3}{8ab} = \frac{\overset{1}{\cancel{2}} \cdot \overset{1}{\cancel{b}} \cdot b \cdot b}{\underset{1}{\cancel{2}} \cdot 2 \cdot 2 \cdot a \cdot \underset{1}{\cancel{b}}} = \frac{b^2}{4a}$
4.3 Multiplying Fractions Multiply the numerators and multiply the denominators. The product must be written in lowest terms. One way to do so is to write each original numerator and each original denominator as the product of primes and divide out any common factors before multiplying.	$\left(-\frac{7}{10}\right)\left(\frac{5}{6}\right) = -\frac{7 \cdot \overset{1}{\cancel{5}}}{\underbrace{2 \cdot \underset{1}{\cancel{5}}} \cdot \underbrace{2 \cdot 3}} = -\frac{7}{12}$ $\frac{3x^2}{5} \cdot \frac{2}{9x} = \frac{\overset{1}{\cancel{3}} \cdot \overset{1}{\cancel{x}} \cdot x \cdot 2}{5 \cdot \underset{1}{\cancel{3}} \cdot 3 \cdot \underset{1}{\cancel{x}}} = \frac{2x}{15}$

Concepts	*Examples*

4.3 Dividing Fractions

Rewrite the division problem as multiplying by the reciprocal of the divisor. In other words, the first number (dividend) stays the same and the second number (divisor) is changed to its reciprocal. Then use the steps for multiplying fractions. The quotient must be in lowest terms. Division by 0 is undefined.

$$2 \div \left(-\frac{1}{3}\right) = \frac{2}{1} \cdot \left(-\frac{3}{1}\right) = -\frac{2 \cdot 3}{1 \cdot 1} = -6$$

reciprocals

$$\frac{x^2}{y^2} \div \frac{x}{3y} = \frac{x^2}{y^2} \cdot \frac{3y}{x} = \frac{\overset{1}{\cancel{x}} \cdot x \cdot 3 \cdot \overset{1}{\cancel{y}}}{\underset{1}{\cancel{y}} \cdot y \cdot \underset{1}{\cancel{x}}} = \frac{3x}{y}$$

reciprocals

4.4 Adding and Subtracting Like Fractions

You can add or subtract fractions *only* when they have the *same* denominator. Add or subtract the numerators and write the result over the common denominator. The result must be in lowest terms.

$$\frac{3}{10} - \frac{7}{10} = \frac{3-7}{10} = \frac{-4}{10} \text{ or } -\frac{4}{10}$$

Write $-\frac{4}{10}$ in lowest terms.

$$-\frac{4}{10} = -\frac{\overset{1}{\cancel{2}} \cdot 2}{\underset{1}{\cancel{2}} \cdot 5} = -\frac{2}{5}$$

$$\frac{5}{a} + \frac{7}{a} = \frac{5+7}{a} = \frac{12}{a}$$

4.4 Finding the Lowest Common Denominator (LCD)

Write the prime factorization of each denominator. Then use enough prime factors in the LCD to "cover" both denominators.

What is the LCD for $\frac{5}{12}$ and $\frac{5}{18}$?

$$12 = 2 \cdot 2 \cdot 3$$
$$18 = 2 \cdot 3 \cdot 3$$

$$\text{LCD} = 2 \cdot 2 \cdot 3 \cdot 3 = 36$$

(12 covers $2 \cdot 2 \cdot 3$; 18 covers $2 \cdot 3 \cdot 3$)

4.4 Adding and Subtracting Unlike Fractions

Find the LCD. Rewrite each original fraction as an equivalent fraction whose denominator is the LCD. Then add or subtract the numerators and keep the common denominator. Be sure that the final result is in lowest terms.

$$-\frac{5}{12} + \frac{7}{9}$$

The LCD is 36.

Rewrite: $-\frac{5}{12} = -\frac{5 \cdot 3}{12 \cdot 3} = -\frac{15}{36}$

Rewrite: $\frac{7}{9} = \frac{7 \cdot 4}{9 \cdot 4} = \frac{28}{36}$

Add: $-\frac{15}{36} + \frac{28}{36} = \frac{-15+28}{36} = \frac{13}{36}$ Lowest terms

$$\frac{2}{3} - \frac{6}{x}$$

The LCD is $3 \cdot x$ or $3x$.

Rewrite: $\frac{2}{3} = \frac{2 \cdot x}{3 \cdot x} = \frac{2x}{3x}$

Rewrite: $\frac{6}{x} = \frac{6 \cdot 3}{x \cdot 3} = \frac{18}{3x}$

Subtract: $\frac{2x}{3x} - \frac{18}{3x} = \frac{2x - 18}{3x}$ Lowest terms

Concepts	*Examples*
4.5 Mixed Numbers and Improper Fractions **Changing mixed numbers to improper fractions** Multiply denominator by whole number, add numerator, and place over denominator. **Changing improper fractions to mixed numbers** Divide numerator by denominator and place remainder over denominator.	**Mixed to improper** $7\frac{2}{3} = \frac{23}{3}$ ← $3 \times 7 + 2$ Same denominator **Improper to mixed** $\frac{17}{5} = 3\frac{2}{5}$ Same denominator
4.5 Multiplying Mixed Numbers First round the numbers and estimate the answer. Then follow these steps to find the exact answer. **1. Rewrite** each mixed number as an improper fraction. **2. Multiply.** **3.** Write the answer in lowest terms and change the answer to a mixed number if desired. Then the answer is in simplest form.	*Estimate* $1\frac{3}{5} \cdot 3\frac{1}{3}$ Rounded $2 \cdot 3 = 6$ *Exact* $1\frac{3}{5} \cdot 3\frac{1}{3} = \frac{8}{5} \cdot \frac{10}{3}$ $= \frac{8 \cdot 2 \cdot \cancel{5}^{1}}{\cancel{5}_{1} \cdot 3}$ $= \frac{16}{3} = 5\frac{1}{3}$ Close to estimate
4.5 Dividing Mixed Numbers First round the numbers and estimate the answer. Then follow these steps to find the exact answer. **1. Rewrite** each mixed number as an improper fraction. **2. Divide** (Rewrite as multiplication by the reciprocal of the divisor.) **3.** Write the answer in lowest terms and change the answer to a mixed number if desired. Then the answer is in simplest form.	*Estimate* $3\frac{3}{4} \div 2\frac{2}{5}$ Rounded $4 \div 2 = 2$ *Exact* $3\frac{3}{4} \div 2\frac{2}{5} = \frac{15}{4} \div \frac{12}{5}$ Reciprocals $= \frac{15}{4} \cdot \frac{5}{12}$ $= \frac{\cancel{3}^{1} \cdot 5 \cdot 5}{4 \cdot \cancel{3}_{1} \cdot 4}$ $= \frac{25}{16} = 1\frac{9}{16}$ Close to estimate
4.5 Adding and Subtracting Mixed Numbers First estimate the answer. Then rewrite the mixed numbers as improper fractions and follow the steps for adding and subtracting fractions. Write the answer in simplest form.	$2\frac{3}{8} + 3\frac{3}{4}$ $2 + 4 = 6$ Estimate $2\frac{3}{8} + 3\frac{3}{4} = \frac{19}{8} + \frac{15}{4} = \frac{19}{8} + \frac{30}{8} = \frac{19 + 30}{8}$ $= \frac{49}{8} = 6\frac{1}{8}$ Close to estimate

Concepts	Examples
4.6 Exponents and Order of Operations The meaning of an exponent is the same for fractions as it is for integers. An exponent is a way to write repeated multiplication. The order of operations is also the same for fractions as for integers. **1.** Work inside parentheses or other grouping symbols. **2.** Simplify expressions with exponents. **3.** Do the remaining multiplications and divisions as they occur from left to right. **4.** Do the remaining additions and subtractions as they occur from left to right.	$\left(-\frac{2}{3}\right)^2$ means $\left(-\frac{2}{3}\right)\left(-\frac{2}{3}\right) = \frac{2 \cdot 2}{3 \cdot 3} = \frac{4}{9}$. The product of two negative numbers is positive. Simplify. $-\frac{2}{3} + 3\left(\frac{1}{4}\right)^2$ Cannot work inside parentheses. Use exponent: $\frac{1}{4} \cdot \frac{1}{4}$ is $\frac{1}{16}$. $-\frac{2}{3} + 3\left(\frac{1}{16}\right)$ Multiply next: $3(\frac{1}{16})$ is $\frac{3}{1} \cdot \frac{1}{16} = \frac{3}{16}$. $-\frac{2}{3} + \frac{3}{16}$ Add last. The LCD is 48. $-\frac{32}{48} + \frac{9}{48}$ Rewrite $-\frac{2}{3}$ as $-\frac{32}{48}$. Rewrite $\frac{3}{16}$ as $\frac{9}{48}$. $\frac{-32 + 9}{48}$ Add the numerators. Keep the common denominator. $-\frac{23}{48}$ The answer is in lowest terms.
4.6 Simplifying Complex Fractions Recall that the fraction bar indicates division. Rewrite the complex fraction using the ÷ symbol for division. Then follow the steps for dividing fractions.	Simplify. $\frac{-\frac{4}{5}}{10}$ Rewrite as $-\frac{4}{5} \div 10$. $-\frac{4}{5} \div 10 = -\frac{4}{5} \cdot \frac{1}{10} = -\frac{\overset{1}{\cancel{2}} \cdot 2 \cdot 1}{5 \cdot \underset{1}{\cancel{2}} \cdot 5} = -\frac{2}{25}$ (10 and $\frac{1}{10}$: reciprocals)
4.7 Solving Equations Containing Fractions If necessary, add the same number to both sides of the equation so that the variable term is by itself on one side of the equal sign. Then multiply both sides by the reciprocal of the coefficient of the variable term. To check your solution, go back to the original equation and replace the variable with your solution. If the equation balances, your solution is correct. If not, rework the problem.	Solve the equation. Check the solution. $\frac{1}{3}b + 6 = 10$ Add −6 to both sides. $\quad\;\; -6 \quad -6$ $\frac{1}{3}b + 0 = 4$ $\frac{1}{3}b = 4$ $\frac{\overset{1}{\cancel{3}}}{1}\left(\frac{1}{\underset{1}{\cancel{3}}}b\right) = \frac{3}{1}(4)$ Multiply both sides by $\frac{3}{1}$ (the reciprocal of $\frac{1}{3}$). $b = 12$ **Check** $\frac{1}{3}b + 6 = 10$ Replace b with 12. $\frac{1}{3}(12) + 6 = 10$ $4 + 6 = 10$ $10 = 10$ Balances, so 12 is the correct solution.

Concepts	*Examples*

4.8 Finding the Area of a Triangle

Use this formula to find the area of a triangle.

$$A = \frac{1}{2}bh$$

where b is the base and h is the height.

Remember that area is measured in **square units**.

Find the area of this triangle.

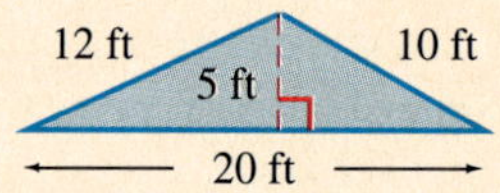

$$A = \frac{1}{2}\mathbf{bh}$$

$$A = \frac{1}{2} \cdot \mathbf{20\ ft} \cdot \mathbf{5\ ft}$$

$$A = \frac{1}{2} \cdot \frac{20\ \text{ft}}{1} \cdot \frac{5\ \text{ft}}{1}$$

$$A = \frac{1 \cdot \overset{1}{\cancel{2}} \cdot 10\ \text{ft} \cdot 5\ \text{ft}}{\underset{1}{\cancel{2}} \cdot 1 \cdot 1}$$

$$A = 50\ \mathbf{ft^2}$$ Measure area in square units.

4.8 Finding the Volume of a Rectangular Solid

Use this formula to find the volume of box-like solids.

$$\text{Volume} = \text{length} \cdot \text{width} \cdot \text{height}$$

$$V = lwh$$

Volume is measured in **cubic units**.

Find the volume of this box.

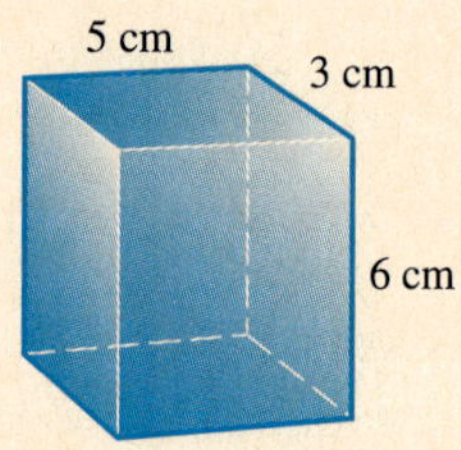

$$V = \mathbf{l \cdot w \cdot h}$$

$$V = \mathbf{5\ cm} \cdot \mathbf{3\ cm} \cdot \mathbf{6\ cm}$$

$$V = 90\ \mathbf{cm^3}$$

4.8 Finding the Volume of a Pyramid

Use this formula to find the volume of a pyramid.

$$V = \frac{1}{3} \cdot B \cdot h$$

$$V = \frac{1}{3}Bh$$

where B is the area of the square or rectangular base and h is the height of the pyramid.

Volume is measured in **cubic units**.

Find the volume of a pyramid with a square base 2 cm by 2 cm and a height of 6 cm.

Area of square base = 2 cm • 2 cm = **4 cm²**

$$V = \frac{1}{3} \cdot \mathbf{B} \cdot h$$

$$V = \frac{1}{3} \cdot \frac{\mathbf{4\ cm^2}}{\mathbf{1}} \cdot \frac{6\ \text{cm}}{1}$$

$$V = \frac{1 \cdot 4\ \text{cm}^2 \cdot \overset{1}{\cancel{3}} \cdot 2\ \text{cm}}{\underset{1}{\cancel{3}} \cdot 1 \cdot 1}$$

$$V = 8\ \mathbf{cm^3}$$

CHAPTER 4 REVIEW EXERCISES

If you need help with any of these review exercises, look in the section indicated in the brackets.

[4.1] **1.** What fraction of these figures are squares? What fraction are triangles?

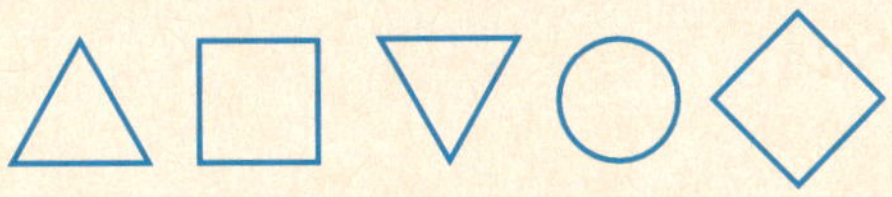

2. Write fractions to represent the shaded and unshaded portions of this figure.

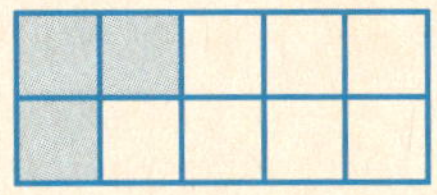

3. Graph $-\frac{1}{2}$ and $1\frac{1}{2}$ on the number line.

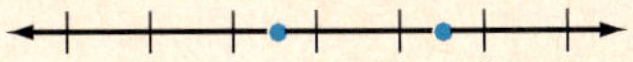

4. Simplify each fraction.

(a) $-\frac{20}{5}$ **(b)** $\frac{8}{1}$ **(c)** $-\frac{3}{3}$

[4.2] *Write each fraction in lowest terms.*

5. $\frac{28}{32}$

6. $\frac{54}{90}$

7. $\frac{16}{25}$

8. $\frac{15x^2}{40x}$

9. $\frac{7a^3}{35a^3b}$

10. $\frac{12mn^2}{21m^3n}$

[4.3] *Multiply or divide. Write all answers in lowest terms.*

11. $-\frac{3}{8} \div (-6)$

12. $\frac{2}{5}$ of (-30)

13. $\frac{4}{9}\left(\frac{2}{3}\right)$

14. $\frac{7}{3x^3} \cdot \frac{x^2}{14}$

15. $\frac{ab}{5} \div \frac{b}{10a}$

16. $\frac{18}{7} \div 3k$

[4.4] *Add or subtract. Write all answers in lowest terms.*

17. $-\frac{5}{12} + \frac{5}{8}$

18. $\frac{2}{3} - \frac{4}{5}$

19. $4 - \frac{5}{6}$

20. $\frac{7}{9} + \frac{13}{18}$

21. $\frac{n}{5} + \frac{3}{4}$

22. $\frac{3}{10} - \frac{7}{y}$

[4.5] *First round the mixed numbers to the nearest whole number and estimate each answer. Then find the exact answer.*

23. *exact*

$2\frac{1}{4} \div 1\frac{5}{8}$

estimate

____ ÷ ____ = ____

24. *exact*

$7\frac{1}{3} - 4\frac{5}{6}$

estimate

____ − ____ = ____

25. *exact*

$1\frac{3}{4} + 2\frac{3}{10}$

estimate

____ + ____ = ____

[4.6] *Simplify.*

26. $\left(-\frac{3}{4}\right)^3$

27. $\left(\frac{2}{3}\right)^2\left(-\frac{1}{2}\right)^4$

28. $\frac{2}{5} + \frac{3}{10}(-4)$

29. $-\frac{5}{8} \div \left(-\frac{1}{2}\right)\left(\frac{14}{15}\right)$

30. $\dfrac{\frac{5}{8}}{\frac{1}{16}}$

31. $\dfrac{\frac{8}{9}}{-6}$

[4.7] *Solve each equation. Show the steps you used.*

32. $-12 = -\frac{3}{5}w$

33. $18 + \frac{6}{5}r = 0$

34. $3x - \frac{2}{3} = \frac{5}{6}$

[4.8] *Find the area of the triangle and the volume of each solid.*

35.

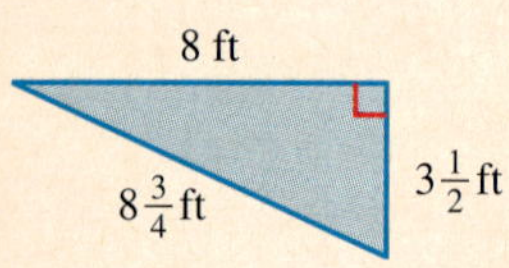

36.

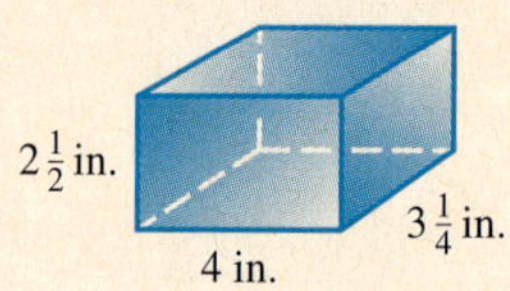

37.

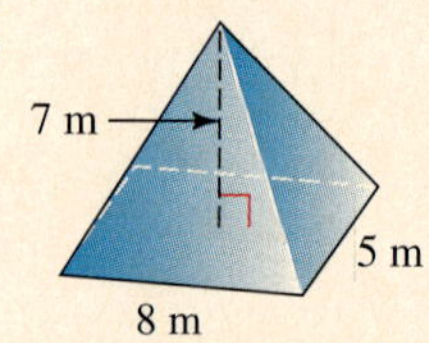

MIXED REVIEW EXERCISES

Solve each application problem.

38. A chili recipe that makes 10 servings uses $2\frac{1}{2}$ pounds of meat. How much meat will be in each serving? How much meat would be needed to make 30 servings?

39. Yanli worked as a math tutor for $4\frac{1}{2}$ hours on Monday, $2\frac{3}{4}$ hours on Tuesday, and $3\frac{2}{3}$ hours on Friday. How much longer did she work on Monday than on Friday? How many hours did she work in all?

40. There are 60 children in the daycare center. If $\frac{1}{5}$ of the children are preschoolers, $\frac{2}{3}$ of the children are toddlers, and the rest are infants, find the number of children in each age group.

41. A rectangular city park is $\frac{3}{4}$ mile long and $\frac{3}{10}$ mile wide. Find the perimeter and area of the park.

Name Date Class

1. Write fractions to represent the shaded and unshaded portions of the figure.

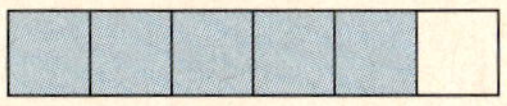

2. Graph $-\frac{2}{3}$ and $2\frac{1}{3}$ on the number line.

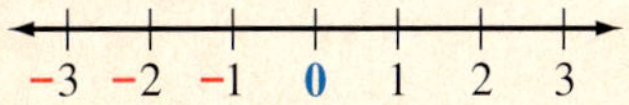

1. ______

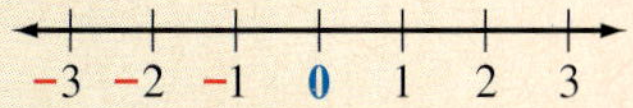

2. ______

Write each fraction in lowest terms.

3. $\frac{21}{84}$

4. $\frac{25}{54}$

5. $\frac{6a^2b}{9b^2}$

3. ______

4. ______

5. ______

Add, subtract, multiply, or divide, as indicated. Write all answers in lowest terms.

6. $\frac{1}{6} + \frac{7}{10}$

7. $-\frac{3}{4} \div \frac{3}{8}$

8. $\frac{5}{8} - \frac{4}{5}$

9. $(-20)\left(-\frac{7}{10}\right)$

10. $\frac{\frac{4}{9}}{-6}$

11. $4 - \frac{7}{8}$

12. $-\frac{2}{9} + \frac{2}{3}$

13. $\frac{21}{24}\left(\frac{9}{14}\right)$

14. $\frac{12x}{7y} \div 3x$

15. $\frac{6}{n} - \frac{1}{4}$

16. $\frac{2}{3} + \frac{a}{5}$

17. $\frac{5}{9b^2} \cdot \frac{b}{10}$

18. Simplify.
$\left(-\frac{1}{2}\right)^3\left(\frac{2}{3}\right)^2$

19. Simplify.
$\frac{1}{6} + 4\left(\frac{2}{5} - \frac{7}{10}\right)$

6. ______

7. ______

8. ______

9. ______

10. ______

11. ______

12. ______

13. ______

14. ______

15. ______

16. ______

17. ______

18. ______

19. ______

20. *estimate:* *exact:* ____________

21. *estimate:* *exact:* ____________

22. ____________

23. ____________

24. ____________

25. ____________

26. ____________

27. ____________

28. ____________

29. ____________

30. ____________

31. ____________

32. ____________

First round the numbers and estimate each answer. Then find the exact answer. Write your answers in simplest form.

20. $4\frac{4}{5} \div 1\frac{1}{8}$

21. $3\frac{2}{5} - 1\frac{9}{10}$

Solve each equation. Show the steps you used.

22. $7 = \frac{1}{5}d$

23. $-\frac{3}{10}t = \frac{9}{14}$

24. $0 = \frac{1}{4}b - 2$

25. $\frac{4}{3}x + 7 = -13$

Find the area of each triangle.

26.

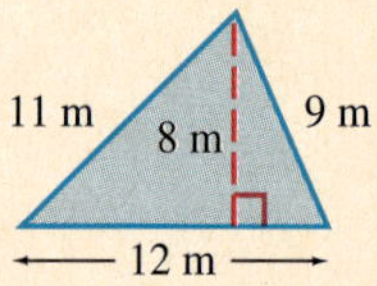

27.

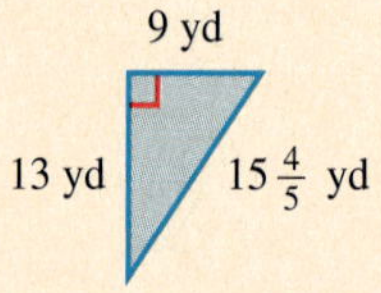

Find the volume of each solid.

28.

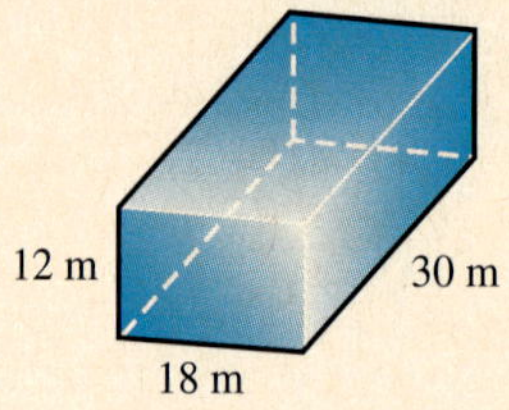

29.

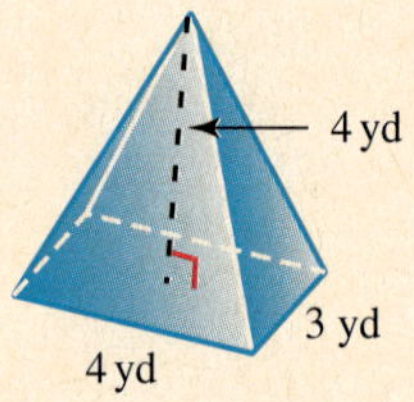

Solve each application problem.

30. Ann-Marie Sargent is training for an upcoming wheelchair race. She rides $4\frac{5}{6}$ hours on Monday, $6\frac{2}{3}$ hours on Tuesday, and $3\frac{1}{4}$ hours on Wednesday. How many hours did she spend in all? How many more hours did she train on Tuesday than on Monday?

31. A new vaccine is synthesized at the rate of $2\frac{1}{2}$ ounces per day. How long will it take to synthesize $8\frac{3}{4}$ ounces?

32. There are 8448 students at the Metro Community College campus. If $\frac{7}{8}$ of the students work either full time or part time, find the total number of students who work.

CUMULATIVE REVIEW EXERCISES CHAPTERS 1–4

1. Write this number in words: 505,008,238

2. Write this number using digits: thirty-five billion, six hundred million, nine hundred sixteen.

3. **(a)** Round 60,719 to the nearest hundred.

(b) Round 99,505 to the nearest thousand.

(c) Round 3206 to the nearest ten.

4. Name the property illustrated by each example.

(a) $-3 \cdot 6 = 6 \cdot (-3)$

(b) $(7 + 18) + 2 = 7 + (18 + 2)$

(c) $5(-10 + 7) = 5 \cdot (-10) + 5 \cdot 7$

Simplify.

5. $9(-6)$

6. $-10 - 10$

7. $\frac{-14}{0}$

8. $6 + 3(2 - 7)^2$

9. $|-8| + |2|$

10. $-8 + 24 \div 2$

11. $(-4)^2 - 2^5$

12. $\frac{-45 \div (-5) \cdot 3}{-5 - 4(0 - 8)}$

13. $-3 \div \frac{3}{8}$

14. $-\frac{5}{6}(-42)$

15. $\frac{5a^2}{12} \cdot \frac{18}{a}$

16. $\frac{7}{x^2} \div \frac{7y^2}{3x}$

17. $\frac{3}{10} - \frac{5}{6}$

18. $-\frac{3}{8} + \frac{11}{16}$

19. $\frac{2}{3} - \frac{b}{7}$

20. $\frac{8}{5} + \frac{3}{n}$

21. $3\frac{1}{4} \div 2\frac{1}{4}$

22. $2\frac{2}{5} - 1\frac{3}{4}$

23. $\left(\frac{1}{2}\right)^3(-2)^3$

24. $\frac{\frac{7}{12}}{-\frac{14}{15}}$

Solve each equation. Show the steps you used.

25. $-5a + 2a = 12 - 15$

26. $y = -7y - 40$

27. $-10 = 6 + \frac{4}{9}k$

28. $20 + 5x = -2x - 8$

29. $3(-2m + 5) = -4m + 9 + m$

Find the perimeter and area of each figure.

30.

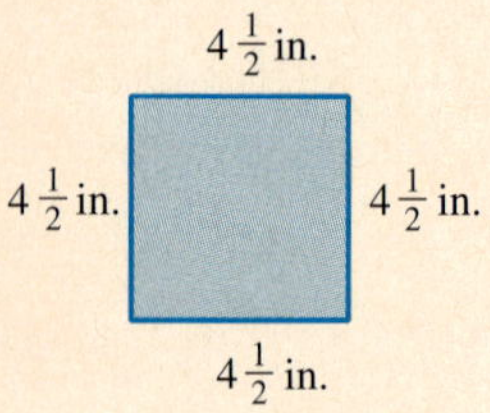

31.

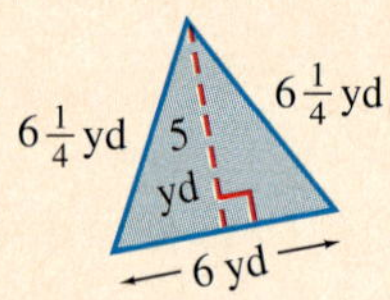

32.

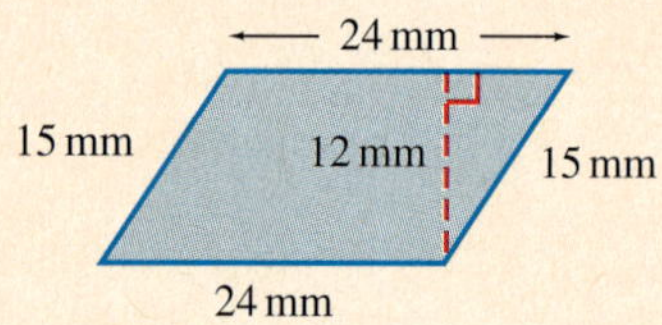

Solve each application problem by using the six problem solving steps.

33. Three new bags of birdseed each weighed the same amount. Sixteen pounds of seed were used from one bag and 25 pounds from another. There were still 79 pounds of seed left. How much did each bag weigh originally?

34. A rectangular parking lot is twice as long as it is wide. If the perimeter of the lot is 102 yd, find the length and width of the parking lot.

Positive and Negative Decimal Numbers

5

In Chapter 4 you used fractions to represent parts of a whole. In this chapter, **decimals** (DES-i-muls) are used as another way to show parts of a whole. For example, our money system is based on decimals. One dollar is divided into 100 equivalent parts. One cent ($0.01) is one of the parts, and a dime ($0.10) is 10 of the parts.

INTERNET
www.mathnotes.com

OBJECTIVES

1. Write parts of a whole as decimals.
2. Find the place value of a digit.
3. Read decimals.
4. Write decimals as fractions.

FOR EXTRA HELP

Tutorial | Tape 7 | SSM, Sec. 5.1

5.1 *Reading and Writing Decimal Numbers*

OBJECTIVE 1 Decimals are used when a whole is divided into 10 equivalent parts or into 100 or 1000 or 10,000 equivalent parts. In other words, decimals are fractions with denominators that are a power of 10. For example, the square below is cut into 10 equivalent parts. Written as a fraction, each part is $\frac{1}{10}$ of the whole. Written as a decimal, each part is 0.1. Both are read as "*one tenth.*"

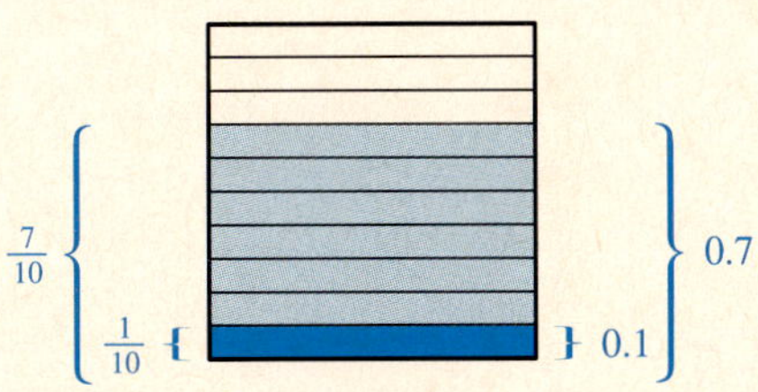

The dot in 0.1 is called the **decimal point.**

0.1
↑
Decimal point

The square above has 7 of its 10 parts shaded.

Written as a fraction, $\frac{7}{10}$ of the square is shaded.

Written as a decimal, **0.7** of the square is shaded.

Both are read as "*seven tenths.*"

1. There are 10 dimes in one dollar. Each dime is $\frac{1}{10}$ of a dollar. Write a fraction, a decimal, and the words that name the yellow shaded portion of each dollar.

(a)

(b)

(c)

2. Write the portion of each square that is shaded as a fraction, as a decimal, and in words.

(a)

(b)

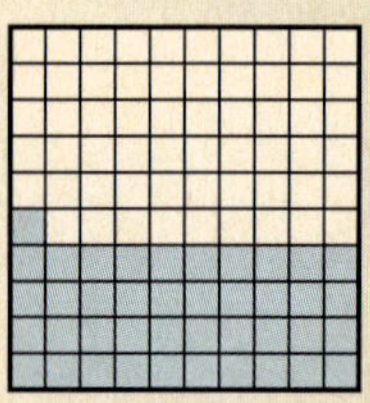

ANSWERS

1. (a) $\frac{1}{10}$; 0.1; one tenth **(b)** $\frac{3}{10}$; 0.3; three tenths
(c) $\frac{9}{10}$; 0.9; nine tenths

2. (a) $\frac{3}{10}$; 0.3; three tenths **(b)** $\frac{41}{100}$; 0.41; forty-one hundredths

◀◀ **WORK PROBLEM 1 AT THE SIDE.**

The square below is cut into 100 equivalent parts. Written as a fraction, each part is $\frac{1}{100}$ of the whole.

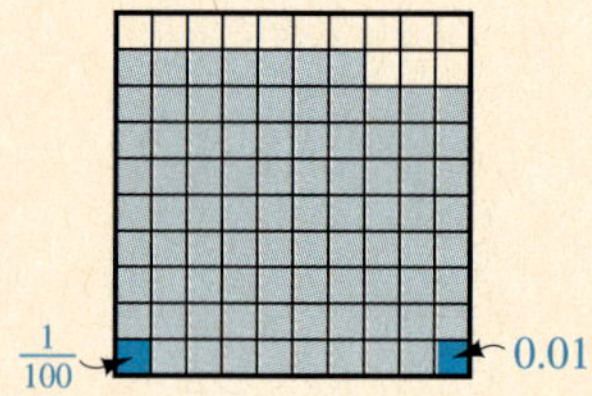

Written as a decimal, each part is

0.01 of the whole.
↑
Read "one hundredth."

The square has 87 parts shaded.

Written as a fraction, $\frac{87}{100}$ of the total area is shaded.

Written as a decimal, **0.87** of the total area is shaded.

Both are read as "*eighty-seven hundredths.*"

◀◀ **WORK PROBLEM 2 AT THE SIDE.**

The example below shows several numbers written as fractions, as decimals, and in words.

EXAMPLE 1 Using the Decimal Forms of Fractions

	Fraction	*Decimal*	*Read As*
(a)	$\frac{3}{10}$	0.3	three tenths
(b)	$-\frac{9}{100}$	−0.09	negative nine hundredths
(c)	$\frac{71}{100}$	0.71	seventy-one hundredths
(d)	$\frac{8}{1000}$	0.008	eight thousandths
(e)	$-\frac{45}{1000}$	−0.045	negative forty-five thousandths
(f)	$\frac{832}{1000}$	0.832	eight hundred thirty-two thousandths

WORK PROBLEM 3 AT THE SIDE.

OBJECTIVE 2 The decimal point separates the *whole number part* from the *fractional part* in a decimal number. In the chart below, you can see that the **place value names** for fractional parts are similar to those on the whole number side but end in "*ths.*"

Decimal Place Value Chart

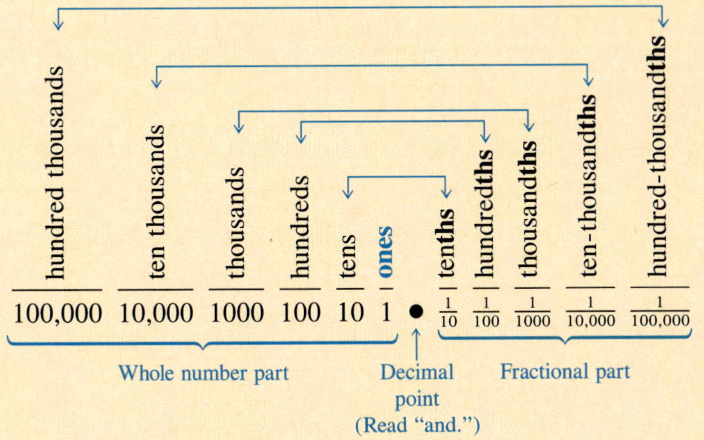

Notice that the ones place is at the center. (There is no "oneths" place.) Also notice that each place is 10 times the value of the place to its right.

Note

In this chapter, if a number does *not* have a decimal point, it is an integer. An integer has no fractional part. If you want to show the decimal point in an integer, it is just to the *right* of the digit in the ones place. For example:

$8 = 8.$ (Decimal point) $\qquad 306 = 306.$ (Decimal point) $\qquad -42 = -42.$ (Decimal point)

EXAMPLE 2 Identifying the Place Value of a Digit

Give the place values of the digits in each decimal.

(a) 178.36

hundreds	tens	ones		tenths	hundredths
1	7	8	.	3	6

(b) 0.0093

ones		tenths	hundredths	thousandths	ten-thousandths
0	.	0	0	9	3

Notice in Example 2(b) that we do *not* use commas on the right side of the decimal point.

WORK PROBLEM 4 AT THE SIDE.

3. Write each decimal as a fraction.

(a) -0.7

(b) 0.9

(c) -0.03

(d) 0.69

(e) 0.047

(f) -0.351

4. Identify the place value of each digit in these decimals.

(a) 971.54

(b) 0.4

(c) 5.60

(d) 0.0835

ANSWERS

3. (a) $-\frac{7}{10}$ **(b)** $\frac{9}{10}$ **(c)** $-\frac{3}{100}$ **(d)** $\frac{69}{100}$ **(e)** $\frac{47}{1000}$ **(f)** $-\frac{351}{1000}$

4. (a)

hundreds	tens	ones		tenths	hundredths
9	7	1	.	5	4

(b)

ones		tenths
0	.	4

(c)

ones		tenths	hundredths
5	.	6	0

(d)

ones		tenths	hundredths	thousandths	ten-thousandths
0	.	0	8	3	5

5. Write each decimal in words.

(a) 0.3

(b) 0.46

(c) 0.09

(d) 0.409

(e) 0.0003

(f) 0.0703

(g) 0.088

OBJECTIVE 3 A decimal is read according to its form as a fraction. We read 0.9 as "nine tenths" because 0.9 is the same as $\frac{9}{10}$. Notice that 0.9 ends in the tenths place.

ones	tenths
0.	9

We read 0.02 as "two hundredths" because 0.02 is the same as $\frac{2}{100}$. Notice that 0.02 ends in the hundredths place.

ones	tenths	hundredths
0.	0	2

EXAMPLE 3 Reading a Decimal Number

Write each decimal in words.

(a) 0.3

Because $0.3 = \frac{3}{10}$, write the decimal as three tenths.

(b) 0.49 Write it as forty-nine hundred**ths**.

(c) 0.08 Write it as eight hundred**ths**.

(d) 0.918 Write it as nine hundred eighteen thousand**ths**.

(e) 0.0106 Write it as one hundred six ten-thousand**ths**.

WORK PROBLEM 5 AT THE SIDE.

Reading a Decimal Number

Step 1 Read any whole number part to the *left* of the decimal point as you normally would.

Step 2 Read the decimal point as "*and*."

Step 3 Read the part of the number to the *right* of the decimal point as if it was an ordinary whole number.

Step 4 Finish with the place value name of the right-most digit; these names all end in "*ths*."

Note

If there is *no whole number part,* you will use only Steps 3 and 4.

EXAMPLE 4 Reading a Decimal

Read each decimal.

(a)

16.9 → 9 is in tenths place

sixteen **and** nine **tenths**

16.9 is read "sixteen and nine tenths."

(b)

482.35 → 5 is in hundredths place

four hundred eighty-two **and** thirty-five **hundredths**

482.35 is read "four hundred eighty-two and thirty-five hundredths."

CONTINUED ON NEXT PAGE

ANSWERS

5. (a) three tenths
(b) forty-six hundredths
(c) nine hundredths
(d) four hundred nine thousandths
(e) three ten-thousandths
(f) seven hundred three ten-thousandths
(g) eighty-eight thousandths

(3 is in thousandths place)

(c) 0.063 is sixty-three **thousandths**. (No whole number part.)

(d) 11.1085 is eleven **and** one thousand eighty-five **ten-thousandths**.

Note

Use "and" *only* when reading a decimal point. A common mistake is to read the whole number 405 as "four hundred *and* five." But there is no decimal point shown in 405, so it is read "four hundred five."

WORK PROBLEM 6 AT THE SIDE.

Objective 4 Knowing how to read decimals will help you when writing decimals as fractions.

Writing Decimals as Fractions or Mixed Numbers

Step 1 The digits to the right of the decimal point are the numerator of the fraction.

Step 2 The denominator is 10 for tenths, 100 for hundredths, 1000 for thousandths, 10,000 for ten-thousandths, and so on.

Step 3 If the decimal has a whole number part, the fraction will be a mixed number with the same whole number part.

EXAMPLE 5 Writing a Decimal as a Fraction or Mixed Number

Write each decimal as a fraction or mixed number.

(a) 0.19

The digits to the right of the decimal point, 19, are the numerator of the fraction. The denominator is 100 for hundredths because the right-most digit is in the hundredths place.

$$0.19 = \frac{19}{100} \leftarrow \text{100 for hundredths.}$$

(↑ Hundredths place)

(b) 0.863

$$0.863 = \frac{863}{1000} \leftarrow \text{1000 for thousandths.}$$

(↑ Thousandths place)

(c) 4.0099

The whole number part stays the same.

$$4.0099 = 4\frac{99}{10{,}000} \leftarrow \text{10,000 for ten-thousandths.}$$

(↑ Ten-thousandths place)

WORK PROBLEM 7 AT THE SIDE.

6. Write each decimal in words.

(a) 3.8

(b) 15.1

(c) 0.72

(d) 64.309

7. Write each decimal as a fraction or mixed number.

(a) 0.7

(b) 9.89

(c) 0.101

(d) 0.007

(e) 1.3717

Answers

6. **(a)** three and eight tenths **(b)** fifteen and one tenth **(c)** seventy-two hundredths **(d)** sixty-four and three hundred nine thousandths

7. **(a)** $\frac{7}{10}$ **(b)** $9\frac{89}{100}$ **(c)** $\frac{101}{1000}$ **(d)** $\frac{7}{1000}$ **(e)** $1\frac{3717}{10{,}000}$

8. Write each decimal as a fraction or mixed number in lowest terms.

(a) 0.2

(b) 12.6

(c) 0.85

(d) 3.05

(e) 0.225

(f) 420.0802

Note

After you write a decimal as a fraction or a mixed number, check to see if the fraction is in lowest terms.

EXAMPLE 6 Writing a Decimal as a Fraction or Mixed Number

Write each decimal as a fraction or mixed number in lowest terms.

(a) $0.4 = \frac{4}{10}$ ← 10 for tenths.

Write $\frac{4}{10}$ in lowest terms. $\frac{4}{10} = \frac{4 \div \mathbf{2}}{10 \div \mathbf{2}} = \frac{2}{5}$

(b) $0.75 = \frac{75}{100} = \frac{75 \div \mathbf{25}}{100 \div \mathbf{25}} = \frac{3}{4}$ Lowest terms

(c) $18.105 = 18\frac{105}{1000} = 18\frac{105 \div \mathbf{5}}{1000 \div \mathbf{5}} = 18\frac{21}{200}$ Lowest terms

(d) $42.8085 = 42\frac{8085}{10{,}000} = 42\frac{8085 \div \mathbf{5}}{10{,}000 \div \mathbf{5}} = 42\frac{1617}{2000}$ Lowest terms

◀◀ WORK PROBLEM 8 AT THE SIDE.

Calculator Tip: In this book you'll notice that we use a zero in the ones place for decimal fractions. We write **0**.45 instead of just .45, to emphasize that there is no whole number. Your calculator shows these zeros also. Enter [•] [4] [5]. Notice that the display automatically shows 0.45 even though you did not press 0. For comparison, enter the whole number 45 by pressing [4] [5] [+] and notice where the decimal point is shown in the display. (It automatically appears to the *right* of the 5.)

ANSWERS

8. **(a)** $\frac{1}{5}$ **(b)** $12\frac{3}{5}$ **(c)** $\frac{17}{20}$ **(d)** $3\frac{1}{20}$ **(e)** $\frac{9}{40}$ **(f)** $420\frac{401}{5000}$

5.1 Exercises

Name the digit that has the given place value.

Example: 3406.251
Solution: hundreds **4**
hundredths **5**
thousandths **1**

Example: 324.078
Solution: ones **4**
tenths **0**
tens **2**

1. 37.602
ones
tenths
tens

2. 135.296
ones
tenths
tens

3. 0.2518
hundredths
thousandths
ten-thousandths

4. 0.9347
hundredths
thousandths
ten-thousandths

5. 93.01472
thousandths
ten-thousandths
tenths

6. 0.51968
tenths
ten-thousandths
hundredths

7. 314.658
tens
tenths
hundreds

8. 51.325
tens
tenths
hundredths

9. 149.0832
hundreds
hundredths
ones

10. 3458.712
hundreds
hundredths
tenths

11. 6285.7125
thousands
thousandths
hundredths

12. 5417.6832
thousands
thousandths
ones

Write the decimal number that has the specified place values.

Example: 3 tenths, 5 ones, 9 thousandths, 0 hundredths
Solution: **5.309**

13. 0 ones, 5 hundredths, 1 ten, 4 hundreds, 2 tenths

14. 7 tens, 9 tenths, 3 ones, 6 hundredths, 8 hundreds

15. 3 thousandths, 4 hundredths, 6 ones, 2 ten-thousandths, 5 tenths

16. 8 ten-thousandths, 4 hundredths, 0 ones, 2 tenths, 6 thousandths

17. 4 hundredths, 4 hundreds, 0 tens, 0 tenths, 5 thousandths, 5 thousands, 6 ones

18. 7 tens, 7 tenths, 6 thousands, 6 thousandths, 3 hundreds, 3 hundredths, 2 ones

Write each decimal as a fraction or mixed number in lowest terms.

Example: 0.68

Solution: $0.68 = \frac{68}{100} = \mathbf{\frac{17}{25}}$ (Lowest terms)

Example: 4.005

Solution: $4.005 = 4\frac{5}{1000}$

$= \mathbf{4\frac{1}{200}}$ (Lowest terms)

19. 0.7 **20.** 0.1 **21.** 13.4 **22.** 9.8 **23.** 0.35

24. 0.85 **25.** 0.66 **26.** 0.33 **27.** 10.17 **28.** 31.99

29. 0.06 **30.** 0.08 **31.** 0.205 **32.** 0.805

33. 5.002 **34.** 4.008 **35.** 0.686 **36.** 0.492

Write each decimal in words.

Example:

16.028

Solution:

8 is in thousandths place.

16 . 028

sixteen **and** twenty-eight **thousandths**

37. 0.5 **38.** 0.9

39. 0.78 **40.** 0.55

41. 0.105 **42.** 0.609

43. 12.04

44. 86.09

45. 1.075

46. 4.025

Write each decimal in numbers.

47. six and seven tenths

48. eight and twelve hundredths

49. thirty-two hundredths

50. one hundred eleven thousandths

51. four hundred twenty and eight thousandths

52. two hundred and twenty-four thousandths

53. seven hundred three ten-thousandths

54. eight hundred and six hundredths

55. seventy-five and thirty thousandths

56. sixty and fifty hundredths

57. Anne read the number 4302 as "four thousand three hundred and two." Explain what is wrong with the way Anne read the number.

58. Jerry read the number 9.0106 as "nine and one hundred and six ten-thousandths." Explain the error he made.

Suppose your job is to take phone orders for precision parts. Use the table below. In Exercises 59–62, write the correct part number that matches what you hear the customer say over the phone. In Exercises 63–64, write the words you would say to the customer.

Part Number	*Size in Centimeters*
3-A	0.06
3-B	0.26
3-C	0.6
3-D	0.86
4-A	1.006
4-B	1.026
4-C	1.06
4-D	1.6
4-E	1.602

59. "Please send the six-tenths centimeter bolt."

Part number _______.

60. "The part missing from our order was the one and six hundredths size."

Part number _______.

61. "The size we need is one and six thousandths centimeters."

Part number _______.

62. "Do you still stock the twenty-six hundredths centimeter bolt?"

Part number _______.

63. "What size is part number 4-E?" Write your answer in words.

64. "What size is part number 4-B?" Write your answer in words.

65. Look back at the Decimal Place Value Chart in this section. What do you think would be the names of the next four places to the *right* of hundred-thousandths? What information did you use to come up with these names?

66. A common mistake is to think that the first place to the right of the decimal point is "oneths" and the second place is "tenths." Why might someone make that mistake? How would you explain why there is no "oneths" place?

67. Write 0.72436955 in words.

68. Write 0.000678554 in words.

69. Write 8006.500001 in words.

70. Write 20,060.000505 in words.

Review and Prepare

Round each of the following to the nearest ten, nearest hundred, and nearest thousand. (For help, see ***Section 1.5.****)*

	Ten	***Hundred***	***Thousand***
71. 8235	________	________	________
72. 3565	________	________	________
73. 19,705	________	________	________
74. 89,604	________	________	________

5.2 Rounding Decimal Numbers

Objectives

1. Learn the rules for rounding decimals.
2. Round decimals to any given place.
3. Round money amounts to the nearest cent or nearest dollar.

For Extra Help

Tutorial | Tape 7 | SSM, Sec. 5.2

Section 1.5 showed how to round integers. For example, 89 rounded to the nearest ten is 90, and 8512 rounded to the nearest hundred is 8500.

Objective 1 It is also important to be able to **round** decimals. For example, a store is selling 2 candy mints for \$0.75 but you want only one mint. The price of each mint is \$0.75 ÷ 2, which is \$0.375, but you cannot pay part of a cent. Is \$0.375 closer to \$0.37 or to \$0.38? Actually, it's exactly halfway between. When this happens in everyday situations, the rule is to round *up*. The store will charge you \$0.38 for the mint.

Rounding Decimals

Step 1 Find the place to which the rounding is being done. Draw a "cut-off" line **after** that place to show that you are cutting off and dropping the rest of the digits.

Step 2 Look **only** at the **first** digit you are cutting off.

Step 3A If this digit is **less than 5,** the part of the number you are keeping **stays the same.**

Step 3B If this digit is **5 or more,** you must **round up** the part of the number you are keeping.

Step 4 Use the "≈" sign to indicate that the rounded number is now an approximation (close, but not exact). "≈" means "is approximately equal to."

Note
Do **not** move the decimal point when rounding.

Objective 2 These examples show you how to round decimals.

EXAMPLE 1 Rounding a Decimal Number

Round 14.39652 to the nearest thousandth. Is it closer to 14.396 or to 14.397?

Step 1 Draw a "cut-off" line after the thousandths place.

14.396 | 52 (Thousandths ↑ at the 6) — You are cutting off the 5 and 2. They will be dropped.

Step 2 Look *only* at the *first* digit you are cutting off. Ignore the other digits you are cutting off.

14.396 | **5** 2 — Look only at the 5. Ignore the 2.

Step 3 If the first digit you are cutting off is 5 or more, round up the part of the number you are keeping.

```
  14.396 | 5 2
+  0.001
---------
  14.397
```

First digit cut is 5 or more, so round up by adding 1 thousandth to the part you are keeping.

So, 14.39652 rounded to the nearest thousandth is 14.397. Write it ≈14.397.

Note
When rounding integers in **Section 1.5,** you kept all the digits but changed some to zeros. With decimals, you cut off and *drop the extra digits.* In the example above, 14.39652 rounds to 14.397 **not** 14.39700.

1. Round to the nearest thousandth. Write your answers using the "≈" sign.

(a) 0.33492

(b) 8.00851

(c) 265.42068

(d) 10.70180

ANSWERS

1. (a) ≈0.335 **(b)** ≈8.009 **(c)** ≈265.421 **(d)** ≈10.702

◀◀ **WORK PROBLEM 1 AT THE SIDE.**

In Example 1, the rounded number 14.397 had *three* **decimal places.** Decimal places are the number of digits to the *right* of the decimal point. The first decimal place is tenths, the second is hundredths, the third is thousandths, and so on.

EXAMPLE 2 Rounding Decimals to Different Places

Round to the place indicated.

(a) 5.3496 to the nearest tenth (Is it closer to 5.3 or to 5.4?)

Step 1 Draw a cut-off line after the tenths place.

5 . 3 | 4 9 6 — You will be cutting off the 4, 9, and 6.
(Tenths → 3)

Step 2 5 . 3 | **4** 9 6 — Look only at the 4. (9 6: Ignore these digits.)

Step 3 5 . 3 | **4** 9 6 — First digit cut is less than 5 so the part you are keeping stays the same.
5 . 3 ← Stays the same

5.3496 rounded to the nearest tenth is 5.3 (one decimal place for tenths). Write it ≈5.3. Notice that it does **not** round to 5.3000 which would be ten-thousandths.

(b) 0.69738 to the nearest hundredth (Is it closer to 0.69 or to 0.70?)

Step 1 0 . 6 9 | 7 3 8 — Draw a cut-off line after the hundredths place.
(Hundredths → 9)

Step 2 0 . 6 9 | **7** 3 8 — Look only at the 7.

Step 3 0 . 6 9 | **7** 3 8 — First digit cut is 5 or more, so round up by adding 1 hundredth to the part you are keeping.

0 . 6 9	← Keep this part.
+ 0 . 0 1	← To round up, add 1 hundredth.
0 . 7 0	← 9 + 1 is 10; write 0 and carry 1 to the 6 in the tenths place.

0.69738 rounded to the nearest hundredth is 0.70. Hundredths is *two* decimal places so you *must* write the 0 in the hundredths place. Write it ≈0.70.

(c) 0.01806 to the nearest thousandth (Is it closer to 0.018 or to 0.019?)

0 . 0 1 8 | **0** 6 — First digit cut is less than 5 so the part you are keeping stays the same.
0 . 0 1 8

0.01806 rounded to the nearest thousandth is 0.018 (three decimal places for thousandths). Write it ≈0.018.

CONTINUED ON NEXT PAGE

(d) 57.976 to the nearest tenth (Is it closer to 57.9 or to 58.0?)

57.9 | 76 ← First digit cut is 5 or more so round up by adding 1 tenth to the part you are keeping.

$$\begin{array}{r} 57.9 \\ +\ 0.1 \\ \hline 58.0 \end{array}$$

← 9 + 1 is 10; write the 0 and carry 1 to the 7 in the ones place.

57.976 rounded to the nearest tenth is 58.0. Write it ≈58.0.
You *must* write the zero in the tenths place to show that the number was rounded to the nearest tenth.

Note

Check that your rounded answer shows **exactly** the number of decimal places called for, even if a zero is in that place. Be sure your answer shows one decimal place if you rounded to tenths, two decimal places for hundredths, or three decimal places for thousandths.

WORK PROBLEM 2 AT THE SIDE.

OBJECTIVE 3 When you are shopping in a store, money amounts are usually rounded to the nearest cent. There are 100 cents in a dollar.

$$\text{Each cent is } \frac{1}{100} \text{ of a dollar.}$$

Another way to write $\frac{1}{100}$ is 0.01. So rounding to the *nearest cent* is the same as rounding to the *nearest hundredth of a dollar*.

EXAMPLE 3 Rounding to the Nearest Cent

Round each of these money amounts to the nearest cent.

(a) \$2.4238 (Is it closer to \$2.42 or to \$2.43?)

\$2.42 | 38 ← Less than 5 so the part you are keeping stays the same.

\$2.42 ← You pay

(b) \$0.695 (Is it closer to \$0.69 or to \$0.70?)

\$0.69 | 5 ← 5 or more; round up

$$\begin{array}{r} \$0.69 \\ +\ \$0.01 \\ \hline \$0.70 \end{array}$$

← To round up, add 1 hundredth (1 cent)
← You pay

WORK PROBLEM 3 AT THE SIDE.

Note

Some stores round *all* money amounts up to the next higher cent, even if the next digit is *less* than 5. In Example 3(a) above, some stores would round \$2.4238 up to \$2.43, even though it is closer to \$2.42.

2. Round to the place indicated.

(a) 0.8988 to the nearest hundredth

(b) 5.8903 to the nearest hundredth

(c) 11.0299 to the nearest thousandth

(d) 0.545 to the nearest tenth

3. Round each of the following money amounts to the nearest cent.

(a) \$14.595
You pay ______

(b) \$578.0663
You pay ______

(c) \$0.849
You pay ______

(d) \$0.0548
You pay ______

ANSWERS

2. (a) ≈0.90 **(b)** ≈5.89 **(c)** ≈11.030 **(d)** ≈0.5

3. (a) \$14.60 **(b)** \$578.07 **(c)** \$0.85 **(d)** \$0.05

4. Round to the nearest dollar.

(a) \$29.10

(b) \$136.49

(c) \$990.91

(d) \$5949.88

(e) \$49.60

(f) \$0.55

(g) \$1.08

Answers

4. (a) ≈\$29 **(b)** ≈\$136 **(c)** ≈\$991 **(d)** ≈\$5950 **(e)** ≈\$50 **(f)** ≈\$1 **(g)** ≈\$1

It is also common to round money amounts to the nearest dollar. You can do that on your federal and state income tax, for example, to make the calculations easier.

EXAMPLE 4 Rounding to the Nearest Dollar

Round to the nearest dollar.

(a) \$48.69 (Is it closer to \$48 or to \$49?)

\$48.|69 ← First digit cut is 5 or more so round up by adding \$1.

```
  $48
+   1
  $49
```

Note

\$48.69 rounded to the nearest dollar is \$49. Write the answer as \$49 to show that the rounding is to the *nearest dollar*. Writing \$49.00 would show rounding to the nearest *cent*.

(b) \$594.36 (Is it closer to \$594 or to \$595?)

\$594.|36 ← Less than 5 so the part you keep stays the same.

\$594

\$594.36 rounded to the nearest dollar is \$594.

(c) \$349.88 (Is it closer to \$349 or to \$350?)

\$349.|88 ← 5 or more, so round up by adding \$1.

```
  $349
+    1
  $350
```

\$349.88 rounded to the nearest dollar is \$350.

(d) \$2689.50 rounded to the nearest dollar is \$2690.

(e) \$0.61 rounded to the nearest dollar is \$1.

Calculator Tip: Accountants and other people who work with money amounts often set their calculators to automatically round to 2 decimal places (nearest cent) or to round to 0 decimal places (nearest dollar). Your calculator may have this feature.

WORK PROBLEM 4 AT THE SIDE.

5.2 Exercises

Round each number to the place indicated. Write your answers using the "≈" sign.

Example:

5.7061 to the nearest hundredth

Solution:

Draw cut-off line after hundredths place.

First digit cut is 5 or more so round up the part you are keeping.

5.70 | 61

↑ Hundredths

$$\begin{array}{r} 5.70 \\ +\ 0.01 \\ \hline \approx \mathbf{5.71} \end{array}$$

5.70 ← Keep this part.

+ 0.01 ← Add 1 hundredth.

≈5.71 ← Check that answer has exactly 2 decimal places for hundredths.

1. 16.8974 to the nearest tenth

2. 193.845 to the nearest hundredth

3. 0.95647 to the nearest thousandth

4. 96.81584 to the nearest ten-thousandth

5. 0.799 to the nearest hundredth

6. 0.952 to the nearest tenth

7. 3.66062 to the nearest thousandth

8. 1.5074 to the nearest hundredth

9. 793.988 to the nearest tenth

10. 476.1196 to the nearest thousandth

11. 0.09804 to the nearest ten-thousandth

12. 176.004 to the nearest tenth

13. 48.512 to the nearest one

14. 3.385 to the nearest one

15. 9.0906 to the nearest hundredth

16. 30.1290 to the nearest thousandth

17. 82.000151 to the nearest ten-thousandth

18. 0.400594 to the nearest ten-thousandth

Nardos is grocery shopping. The store will round the amount she pays for each item to the nearest cent. Write the rounded amounts.

19. Soup is 3 cans for \$2.45, so one can is \$0.81666. Nardos pays ________

20. Orange juice is 2 cartons for \$2.69, so one carton is \$1.345. Nardos pays ________

21. Facial tissue is 4 boxes for \$4.89, so one box is \$1.2225. Nardos pays ________

22. Muffin mix is 3 packages for \$1.75, so one package is \$0.58333. Nardos pays ________

23. Candy bars are 6 for \$2.99, so one bar is \$0.4983. Nardos pays ________

24. Boxes of spaghetti are 4 for \$3.59, so one box is \$0.8975. Nardos pays ________

As she gets ready to do her income tax return, Ms. Chen rounds each amount to the nearest dollar. Write the rounded amounts.

25. Income from job, \$17,249.70

26. Income from interest on bank account, \$69.58

27. Union dues, \$310.08

28. Federal withholding, \$2150.49

29. Donations to charity, \$378.82

30. Medical expenses, \$609.38

31. Explain what happens when you round \$0.499 to the nearest dollar. Why does this happen?

32. Explain what happens when you round \$0.0015 to the nearest cent. Why does this happen?

33. Look again at Exercise 31. How else could you round \$0.499 that would be more helpful? What kind of guideline does this suggest about rounding to the nearest dollar?

34. Suppose you want to know which of these amounts is less, so you round them both to the nearest cent.

\$0.5968 \$0.6014

Explain what happens. Describe what you could do instead of rounding to the nearest cent.

Ⓖ *Round each of these money amounts.*

35. \$499.98 to the nearest dollar.

36. \$9899.59 to the nearest dollar.

37. \$0.996 to the nearest cent.

38. \$0.09929 to the nearest cent.

39. \$999.73 to the nearest dollar.

40. \$9999.80 to the nearest dollar.

Review and Prepare

First use front end rounding to ***estimate*** *each sum. Then find the* ***exact*** *answer. (For help, see* ***Section 1.5.****)*

41. *estimate* *exact*

estimate		exact
	← Rounds to	7929
	←	6076
+ ____	←	+ 8218

42. *estimate* *exact*

estimate		exact
	←	2078
	←	183
	←	231
+ ____	←	+ 7209

43. ____ + ____ + ____ = ____ *estimate*

81,976 + 98 + 785 = ____ *exact*

44. ____ + ____ + ____ = ____ *estimate*

1750 + 18,763 + 918 = ____ *exact*

5.3 Adding and Subtracting Decimal Numbers

Objectives

1. Add and subtract positive decimals.
2. Add and subtract negative decimals.
3. Estimate the answer when adding or subtracting decimals.

For Extra Help

Tutorial   Tape 7   SSM, Sec. 5.3

Objective 1 When adding or subtracting *whole* numbers, you line up the numbers in columns so that you are adding ones to ones, tens to tens, and so on. A similar idea applies to adding or subtracting *decimal* numbers. With decimals you line up the decimal points to be sure that you are adding tenths to tenths, hundredths to hundredths, and so on.

Adding and Subtracting Decimal Numbers

Step 1 Write the numbers in columns with the decimal points lined up.

Step 2 If necessary, write in zeros so both numbers have the same number of decimal places. Then add or subtract as if they were whole numbers.

Step 3 Line up the decimal point in the answer directly below the decimal points in the problem.

EXAMPLE 1 Adding Decimal Numbers

Add.

(a) 16.92 and 48.34

Step 1 Write the numbers in columns with the decimal points lined up.

tens	ones	.	tenths	hundredths
1	6	.	9	2
+ 4	8	.	3	4

Decimal points are lined up.

Step 2 Add as if these were whole numbers. Then line up the decimal point in the answer under the decimal points in the problem.

$$\begin{array}{r} {\scriptstyle 11} \\ 16.92 \\ +\ 48.34 \\ \hline 65.26 \end{array}$$

Step 3 Decimal point in answer is lined up under decimal points in problem.

(b) 5.897 + 4.632 + 12.174

Write the numbers vertically with decimal points lined up. Next, add.

$$\begin{array}{r} {\scriptstyle 11\ \ 21} \\ 5.897 \\ 4.632 \\ +\ 12.174 \\ \hline 22.703 \end{array}$$

Decimal points are lined up.

WORK PROBLEM 1 AT THE SIDE.

In Example 1(a), both numbers had *two* decimal places (two digits to the right of the decimal point). In Example 1(b), all the numbers had *three decimal places* (three digits to the right of the decimal point). That made it easy to add tenths to tenths, hundredths to hundredths, and so on.

If the number of decimal places does *not* match, you can write in zeros as placeholders to make them match. This is shown in Example 2.

1. Find each sum.

(a) 2.86 + 7.09

(b) 13.761 + 8.325

(c) 0.319 + 56.007 + 8.252

(d) 39.4 + 0.4 + 177.2

Answers

1. **(a)** 9.95 **(b)** 22.086 **(c)** 64.578 **(d)** 217.0

2. Find each of the following sums.

(a) 6.54 + 9.8

(b) 0.831 + 222.2 + 10

(c) 8.64 + 39.115 + 3.0076

(d) 5 + 429.823 + 0.76

3. Subtract.

(a) 22.7 from 72.9

(b) 6.425 from 11.813

(c) $20.15 − $19.67

ANSWERS

2. (a) 16.34 (b) 233.031 (c) 50.7626 (d) 435.583
3. (a) 50.2 (b) 5.388 (c) $0.48

EXAMPLE 2 Writing Zeros as Placeholders Before Adding

Add.

(a) 7.3 + 0.85

There are two decimal places in 0.85 (tenths and hundredths), so write a zero in the hundredths place in 7.3 so that it has two decimal places also.

$$\begin{array}{r} 7.3\mathbf{0} \\ +\ 0.85 \\ \hline 8.15 \end{array}$$
← One 0 is written in.

7.30 is equivalent to 7.3 because $7\frac{30}{100}$ in lowest terms is $7\frac{3}{10}$.

(b) 6.42 + 9 + 2.576

Write in zeros so that all the addends have three decimal places. Notice how the whole number 9 is written with the decimal point at the *far right* side. (If you put the decimal point on the *left* side of the 9, you would turn it into the decimal fraction 0.9.)

$$\begin{array}{r} 6.42\mathbf{0} \\ 9.\mathbf{000} \\ +\ 2.576 \\ \hline 17.996 \end{array}$$

6.420 ← One 0 is written in.
9.000 ← 9 is a whole number; decimal point and three 0's are written in.
2.576 ← No 0's are needed.

Note

Writing zeros to the right of a *decimal* number does *not* change the value of the number.

◀◀ WORK PROBLEM 2 AT THE SIDE.

EXAMPLE 3 Subtracting Decimal Numbers

Subtract.

(a) 15.82 from 28.93

Step 1
$$\begin{array}{r} 28.93 \\ -\ 15.82 \\ \hline \end{array}$$
Line up decimal points. Then you will be subtracting hundredths from hundredths and tenths from tenths.

Step 2
$$\begin{array}{r} 28.93 \\ -\ 15.82 \\ \hline 13\ 11 \end{array}$$
Both numbers have two decimal places; no need to write in zeros.

Subtract as if they were whole numbers.

Step 3
$$\begin{array}{r} 28.93 \\ -\ 15.82 \\ \hline 13.11 \end{array}$$
Decimal point in answer lined up.

(b) 146.35 minus 58.98

Borrowing is needed here.

$$\begin{array}{r} {\scriptstyle 0\ 13\ 15\ \ 12\ 15} \\ \not{1}\,\not{4}\,\not{6}.\not{3}\,\not{5} \\ -\ 58.98 \\ \hline 87.37 \end{array}$$
Line up decimal points.

◀◀ WORK PROBLEM 3 AT THE SIDE.

EXAMPLE 4 Writing Zeros as Placeholders Before Subtracting

Subtract.

(a) 16.5 from 28.362

Use the same steps as in the previous examples. Remember to write in zeros so both numbers have three decimal places.

$$\begin{array}{r} 28.362 \\ -\ 16.500 \\ \hline 11.862 \end{array}$$

28.362 ← Line up decimal points.
− 16.500 ← Write two 0's.
11.862 ← Subtract as usual.

(b) 59.7 − 38.914

$$\begin{array}{r} 59.700 \\ -\ 38.914 \\ \hline 20.786 \end{array}$$

59.700 ← Write two 0's.
20.786 ← Subtract as usual.

(c) 12 less 5.83

$$\begin{array}{r} 12.00 \\ -\ 5.83 \\ \hline 6.17 \end{array}$$

12.00 ← Write a decimal point and two 0's.
6.17 ← Subtract as usual.

WORK PROBLEM 4 AT THE SIDE.

OBJECTIVE 2 The rules that you used to add integers in **Section 1.3** will also work for positive and negative decimal numbers. To review:

To add two numbers with the *same* sign, add the absolute values of the numbers. Use the common sign as the sign of the sum.

To add two numbers with *unlike* signs, subtract the smaller absolute value from the larger absolute value. The sign of the sum is the sign of the number with the larger absolute value.

EXAMPLE 5 Adding Positive and Negative Decimal Numbers

Add.

(a) −3.7 + (−16)

Both addends are negative, so the sum will be negative. To begin, |−3.7| is 3.7 and |−16| is 16. Then add the absolute values.

$$\begin{array}{r} 3.7 \\ +\ 16.0 \\ \hline 19.7 \end{array}$$

16.0 ← Write a decimal point and one 0.

−3.7 + (−16) = −19.7

Both negative (−3.7, −16); Negative sum (−19.7)

Note

In Chapter 4 the negative sign was red to help you distinguish it from the subtraction symbol. From now on it will be black. We will continue to write parentheses around negative numbers when the negative sign might be confused with other symbols. Thus in part (a) above

−3.7 + (−16) means **negative** 3.7 **plus negative** 16.

CONTINUED ON NEXT PAGE

4. Subtract.

(a) 18.651 from 25.3

(b) 5.816 − 4.98

(c) 40 less 3.66

(d) 1 − 0.325

ANSWERS

4. (a) 6.649 **(b)** 0.836 **(c)** 36.34 **(d)** 0.675

5. Add.

(a) $13.245 + (-18)$

(b) $-0.7 + (-0.33)$

(c) $-6.02 + 100.5$

6. Subtract.

(a) $-0.37 - (-6)$

(b) $5.8 - 10.03$

(c) $-312.72 - 65.7$

(d) $0.8 - (6 - 7.2)$

Answers

5. **(a)** −4.755 **(b)** −1.03 **(c)** 94.48

6. **(a)** 5.63 **(b)** −4.23 **(c)** −378.42 **(d)** 2

(b) $-5.23 + 0.792$

The addends have different signs. To begin, $|-5.23|$ is 5.23 and $|0.792|$ is 0.792. Then subtract the smaller absolute value from the larger.

$$\begin{array}{r} 5.230 \\ -\ 0.792 \\ \hline 4.438 \end{array}$$ ← Write one 0.

$-5.23 + 0.792 = -4.438$

↑ Number with larger absolute value is negative. ↑ Answer is negative.

◀◀ WORK PROBLEM 5 AT THE SIDE.

In **Section 1.4** you rewrote subtraction of integers as addition of the first number to the opposite of the second number. This same strategy works with positive and negative decimal numbers.

EXAMPLE 6 Subtracting Positive and Negative Decimal Numbers

(a) $4.3 - 12.73$

Rewrite subtraction as adding the opposite.

$4.3 - 12.73$ → $4.3 + (-12.73)$ The opposite of 12.73 is −12.73

-12.73 has the larger absolute value, so the answer will be negative.

$4.3 + (-12.73) = -8.43$ ← Subtract the absolute values: $\begin{array}{r} 12.73 \\ -\ 4.30 \\ \hline 8.43 \end{array}$

↑ Answer is negative.

(b) $-3.65 - (-4.8)$

Rewrite subtraction as adding the opposite.

$-3.65 - (-4.8)$ → $-3.65 + 4.8$ The opposite of −4.8 is 4.8.

4.8 has the larger absolute value, so the answer will be positive.

$-3.65 + 4.8 = 1.15$ ← Subtract the absolute values: $\begin{array}{r} 4.80 \\ -\ 3.65 \\ \hline 1.15 \end{array}$

↑ Answer is positive.

(c) $14.2 - (1.69 + 0.48)$ Work inside parentheses first.

$14.2 - (2.17)$ Change subtraction to adding the opposite.

$14.2 + (-2.17)$ 14.2 has the larger absolute value, so the answer will be positive.

12.03

◀◀ WORK PROBLEM 6 AT THE SIDE.

OBJECTIVE 3 A common error in working decimal problems by hand is to misplace the decimal point in the answer. Or, when using a calculator, you may accidentally press the wrong key. Using front end rounding to estimate the answer will help you avoid these mistakes. Start by rounding each number so that there is only one nonzero digit (as you did in **Section 1.5**). Here are several examples. Notice that in the rounded numbers only the left-most digit is something other than zero.

3.25 rounds to 3	6.812 rounds to 7
532.6 rounds to 500	26.397 rounds to 30

EXAMPLE 7 Estimating a Decimal Answer

Round each number so there is only one nonzero digit. Then add or subtract the rounded numbers to get an estimated answer. Finally, find the exact answer.

(a) Add 194.2 and 6.825.

estimate		*exact*
200	← Rounds to	194.200
+ 7	← Rounds to	+ 6.825
207		201.025

The estimate goes out to the hundreds place (three places to the *left* of the decimal point), and so does the exact answer. Therefore, the decimal point is probably in the right place in the exact answer.

(b) \$69.42 − \$13.78

estimate		*exact*	
\$70	← Rounds to	\$69.42	
− 10	← Rounds to	− 13.78	Answer is close to estimate, so the
\$60		\$55.64	problem is probably set up correctly.

(c) −1.861 − 7.3

Rewrite subtraction as adding the opposite. Then round each number.

$$-1.861 \quad - \quad 7.3$$
$$\downarrow \qquad \downarrow$$
$$-1.861 \quad + \quad (-7.3)$$
$$\downarrow \quad \text{Rounded} \quad \downarrow$$
$$-2 \quad + \quad (-7) \quad = -9 \quad \text{Estimate}$$

To find the exact answer, add the absolute values. The answer will be negative because both numbers are negative.

$$-1.861 + (-7.3) = -9.161 \quad \text{Exact}$$

WORK PROBLEM 7 AT THE SIDE.

7. Use front end rounding to estimate the answer. Then find the exact answer.

(a) 2.83 + 5.009 + 76.1

estimate:

exact:

(b) \$19.28 less \$1.53

estimate:

exact:

(c) 11.365 − 38

estimate:

exact:

(d) −214.6 + 300.72

estimate:

exact:

ANSWERS

7. **(a)** 3 + 5 + 80 = 88; 83.939
(b) \$20 − \$2 = \$18; \$17.75
(c) 10 + (−40) = −30; −26.635
(d) −200 + 300 = 100; 86.12

Calculator Tip: If you are *adding* decimal numbers, you can enter them in any order on your calculator. Try these; jot down the answers.

9.82 [+] 1.86 [=] ______ 1.86 [+] 9.82 [=] ______

The answers are the same because addition is *commutative* (See **Section 1.3**). But subtraction is *not* commutative. It *does* matter which number you enter first. Try these:

9.82 [−] 1.86 [=] ______ 1.86 [−] 9.82 [=] ______

The answers are 7.96 and −7.96. As you know, positive numbers are *greater* than 0, but negative numbers are *less* than 0. So it is important to do subtraction in the correct order, particularly if it is in your checkbook!

5.3 Exercises

Find each sum.

1. $\begin{array}{r} 5.69 \\ 0.24 \\ +\ 11.79 \\ \hline \end{array}$

2. $\begin{array}{r} 372.1 \\ 33.7 \\ +\ 42.3 \\ \hline \end{array}$

3. $\begin{array}{r} 0.38 \\ 7 \\ +\ 4.6 \\ \hline \end{array}$

4. $\begin{array}{r} 3.7 \\ 0.812 \\ +\ 55 \\ \hline \end{array}$

5. 14.23 + 8 + 74.63 + 18.715 + 0.286

6. 197.4 + 0.72 + 17.43 + 25 + 1.4

7. 27.65 + 18.714 + 9.749 + 3.21

8. 58.546 + 19.2 + 8.735 + 14.58

9. Explain and correct the error that a student made when he added 0.72 + 6 + 39.5 this way:

$\begin{array}{r} 0.72 \\ 6 \\ +\ 39.50 \\ \hline 40.28 \end{array}$

10. Explain and correct the error that a student made when she added 7.21 + 65 + 13.15 this way:

$\begin{array}{r} 7.21 \\ .65 \\ +\ 13.15 \\ \hline 21.01 \end{array}$

11. Show why 0.3 is equivalent to 0.3000.

12. Explain why 7 could be written as 7.0 but not as 0.7.

Find each difference.

13. 90.5 − 0.8

14. 303.72 − 0.68

15. 0.4 − 0.291

16. 0.35 − 0.088

17. 6 − 5.09

18. 80 − 16.3

19. 15 − 8.339

20. 44 − 0.08

21. Explain and correct the error that a student made when he subtracted 7.45 from 15.32 this way:

$$\begin{array}{r} 7.45 \\ -\ 15.32 \\ \hline 12.13 \end{array}$$

22. Explain the difference between saying "subtract 2.9 from 8" and saying "2.9 minus 8."

Find each sum or difference.

23. $24.008 + (-0.995)$

24. $0.77 - 3.06$

25. $-6.05 + (-39.7)$

26. $-6.409 + 8.224$

27. $0.9 - 7.59$

28. $-489.7 - 38$

29. $-2 - 4.99$

30. $2.068 - (-32.7)$

31. $-5.009 + 0.73$

32. $-0.33 - 65$

33. $-1.7035 - (5 - 6.7)$

34. $60 + (-0.9345 + 1.4)$

35. $8000 - (8002.63 - 8)$

36. $-210 - (-0.7306 + 0.5)$

G *Use your estimation skills to pick the most reasonable answer for each example. Do* ***not*** *solve the problems. Circle your choice.*

37. 12 − 11.725

2.75 0.275 27.5

38. 20 − 1.37

0.1863 1.863 18.63

39. 6.5 + 0.007

6.507 0.6507 65.07

40. 9.67 + 0.09

0.976 9.76 0.00976

41. 456.71 − 454.9

18.1 181 1.81

42. 803.25 − 0.6

802.65 0.80265 8.0265

43. 6004.003 + 52.7172

605.67202 60,567.202 6056.7202

44. 128.35 + 97.0093

2253.593 225.3593 22.53593

Use front end rounding to estimate each sum or difference. Then find the exact answer to each application problem.

45. Tom has agreed to work 42.5 hours a week as a car wash attendant. So far this week he has worked 16.35 hours. How many more hours must he work?

estimate:
exact:

46. The U.S. population was 262.82 million in 1995. The Census Bureau estimates that it will be 393.93 million in 2050. The increase in population during that 55-year period is how many millions of people?

estimate:
exact:

47. Mrs. Little Owl put two checks in the deposit envelope at the automated teller machine. There was a \$310.14 paycheck and a \$0.95 refund check. How much did she deposit in her account?

estimate:
exact:

48. Rodney Green's paycheck stub showed wages of \$274.19 at the regular rate of pay and \$72.94 at the overtime rate. What were his total wages?

estimate:
exact:

49. The tallest known land mammal is a prehistoric ancestor of the rhino measuring 6.4 m. Find the combined heights of these NBA basketball stars: Charles Barkley at 1.98 m, Karl Malone at 2.06 m, and David Robinson at 2.16 m. Is their combined height greater or less than the prehistoric rhino?

estimate:
exact:

50. Sammy works in a veterinarian's office. He weighed two kittens. One was 3.9 ounces and the other was 4.05 ounces. What was the difference in the weight of the two kittens?

estimate:
exact:

51. Steven One Feather gave the cashier a $20 bill to pay for $9.12 worth of groceries. How much change did he get?

estimate:
exact:

52. The cost of Julie's tennis racket, with tax, is $41.09. She gave the clerk two $20 bills and a $10 bill. What amount of change did Julie receive?

estimate:
exact:

53. At the beginning of Lilia's trip to Dallas, her car odometer read 7942.1 miles. The distance to Dallas is 154.8 miles. What should the odometer read after driving to Dallas *and back?*

estimate:
exact:

54. Gonzalo runs his own package delivery service. On one trip he started from Atlanta and drove 226.6 miles to Charlotte, then 153.8 miles to Roanoke, and finally, 341.3 miles back to Atlanta. Find the total length of his trip.

estimate:
exact:

Yiangos works part-time at a factory. His time card for last week is shown below. Use the time card to solve Exercises 55–58. First use front end rounding to estimate the answer. Then find the exact answer.

Day	*Date*	*Hours*
Mon	6/1	4.5
Tue	6/2	0
Wed	6/3	0
Thr	6/4	6.2
Fri	6/5	5
Sat	6/6	9.5
Sun	6/7	4.8

55. Yiangos is paid a higher hourly wage for working on weekends. How many weekend hours did he work?

estimate:
exact:

56. How many hours did Yiangos work on weekdays?

estimate:
exact:

57. How many hours did Yiangos work in all?

estimate:
exact:

58. How many more hours did Yiangos work on weekdays than on the weekend?

estimate:
exact:

Find the perimeter of (distance around) these figures.

59.

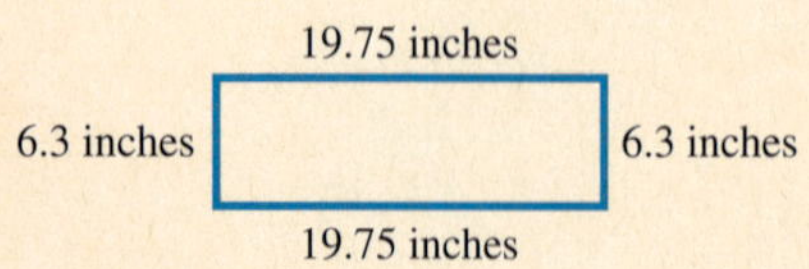

estimate:
exact:

60.

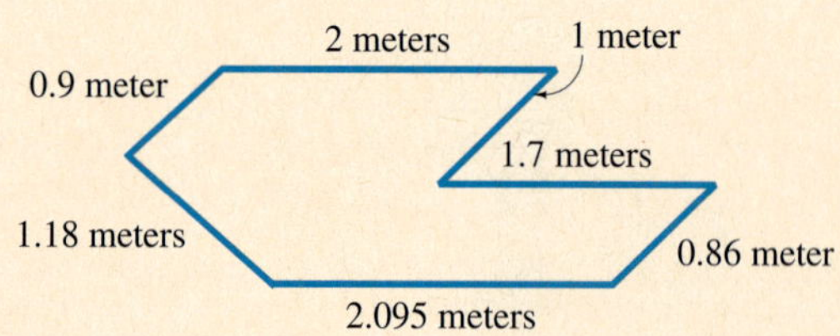

estimate:
exact:

Maria DeRisi kept track of her expenses for one month. Use her list to solve Exercises 61–66.

Monthly Expenses	
Rent	\$515
Car payment	\$190.78
Car repairs, gas	\$105
Cable TV	\$19.95
Electricity	\$42.10
Telephone	\$27.36
Groceries	\$95.81
Entertainment	\$57.75
Clothing, laundry	\$52

61. What were Maria's total expenses for the month?

62. How much did Maria pay for electricity, telephone, and cable TV?

63. What was the difference in the amounts spent for groceries and for the car payment?

64. Compare the amount Maria spent on entertainment to the amount spent on car repairs and gas. What is the difference?

65. How much more did Maria spend on rent than on all her car expenses?

66. How much less did Maria spend on clothing and laundry than on all her car expenses?

Find the length of the dashed line in each rectangle or circle.

67.

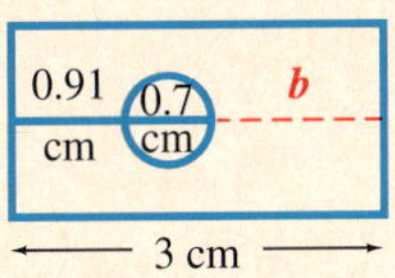

68.

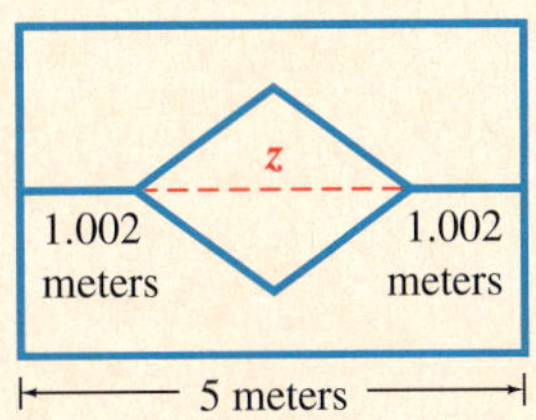

69.

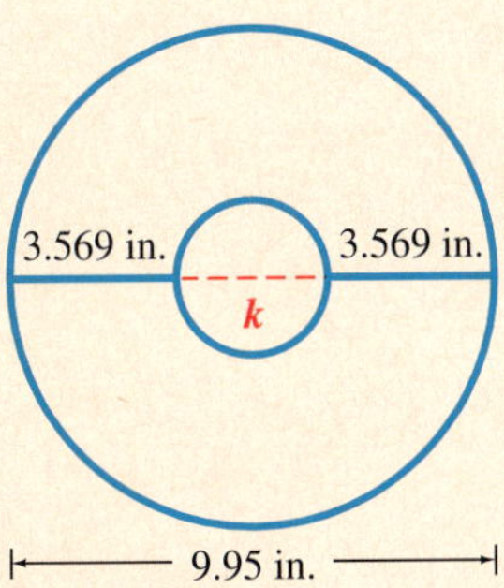

70.

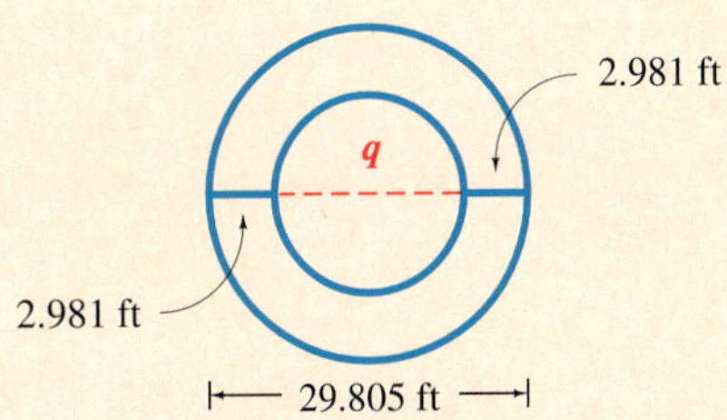

G *Solve each application problem. There may be extra information in the problem, or you may need to do several steps.*

71. Tameka keeps track of her business mileage so her company will pay for her travel. She is not paid for trips to lunch or for travel to and from home. Today she drove 12.6 miles to work, 35.4 miles to visit a client, 14.9 miles to visit another client, 8 miles to lunch, 40 miles to attend a business meeting, and 12.6 miles home. How many miles will her company pay for?

72. Tony wrote a lot of checks today. His tuition at the community college was \$476.44 and textbooks were \$80.06. He also paid \$17.99 for an oil change on his car, \$20.75 at the grocery store, and \$31.62 for brushes and paint for an art class he is taking at the college. What were his total school expenses?

73. The manual for Jason's car says the gas tank holds 16.6 gallons. Jason knows that the tank actually holds an extra 1.4 gallons. The gas station pump showed that Jason bought 8.628 gallons of gas to fill the tank. How much gas was in the tank before he filled it?

74. Tamara's rectangular garden plot is 4.75 meters on each long side and 2.9 meters on each short side. She has 20 meters of fencing. How much fencing will be left after Tamara puts fencing around all four sides of the garden?

75. In the 1996 Olympics, gymnast Dominique Dawes' score in the vault finals was 9.649. Her score in the floor exercise was 0.188 more than in the vault. What was her total score for the two events?

76. James jogged 3.25 kilometers this morning. His friend Anthony jogged with him and kept going another 1.4 kilometers after James stopped. What was the total distance run by the two men?

Review and Prepare

First use front end rounding to estimate each answer. Then find the exact answer. (For help, see ***Sections 1.5 and R.3.****)*

77. *estimate* (Rounds to ←) *exact*

$$\begin{array}{r} \\ \times\ \_\_\_ \end{array} \quad \longleftarrow \quad \begin{array}{r} 83 \\ \times\ 28 \\ \hline \end{array}$$

78. *estimate* *exact*

$$\begin{array}{r} \\ \times\ \_\_\_ \end{array} \qquad \begin{array}{r} 67 \\ \times\ 72 \\ \hline \end{array}$$

79. _____ × _____ = _____ *estimate*

3789 × 205 = _____ *exact*

80. _____ × _____ = _____ *estimate*

6381 × 709 = _____ *exact*

5.4 Multiplying Decimal Numbers

OBJECTIVE 1 The decimals 0.3 and 0.07 can be multiplied by writing them as fractions.

$$0.3 \times 0.07 = \frac{3}{10} \times \frac{7}{100} = \frac{21}{1000} = 0.021$$

1 decimal place + 2 decimal places → 3 decimal places

Can you see a way to multiply decimals without writing them as fractions? Try these steps. Remember that each number in a multiplication problem is called a *factor*.

Multiplying Two Decimal Numbers

Step 1 Multiply the numbers (the factors) as if they were whole numbers.

Step 2 Find the *total* number of decimal places in *both* factors.

Step 3 Write the decimal point in the answer (the product) so it has the same number of decimal places as the total from Step 2. You may need to write in extra zeros on the left side of the product in order to get the correct number of decimal places.

Step 4 If the factors have the same sign, the product is positive. If the factors have different signs, the product is negative.

Note

When multiplying decimals, you do **not** need to line up decimal points. (You **do** need to line up decimal points when adding or subtracting.)

EXAMPLE 1 Multiplying Decimal Numbers

Multiply 8.34 times (−4.2)

Step 1 Multiply the numbers as if they were whole numbers.

```
   8.3 4
 ×   4.2
 -------
   1 6 6 8
 3 3 3 6
 ---------
 3 5 0 2 8
```

Step 2 Count the total number of decimal places in both factors.

```
   8.3 4  ← 2 decimal places
 ×   4.2  ← 1 decimal place
 -------    ----------------------
   1 6 6 8  3 total decimal places
 3 3 3 6
 ---------
 3 5 0 2 8
```

Step 3 Count over 3 places and write the decimal point in the answer. Count from *right to left*.

```
   8.3 4  ← 2 decimal places
 ×   4.2  ← 1 decimal place
 -------    ----------------------
   1 6 6 8  3 total decimal places
 3 3 3 6
 ---------
 3 5.0 2 8  ← 3 decimal places in answer
```

Count over 3 places from right to left to position the decimal point.

Step 4 The factors have *different* signs, so the product is *negative:* 8.34 times (−4.2) = −35.028.

WORK PROBLEM 1 AT THE SIDE.

OBJECTIVES

1 Multiply positive and negative decimals.

2 Estimate the answer when multiplying decimals.

FOR EXTRA HELP

Tutorial | Tape 7 | SSM, Sec. 5.4

1. Multiply.

(a) −2.6(0.4)

(b) (45.2)(0.25)

(c)

```
   0.104  ← 3 decimal places
 ×     7  ← 0 decimal places
 -------  ← 3 decimal places
            in the answer
```

(d) $(-3.18)^2$

Hint: Recall that squaring a number means multiplying the number times itself, so this is (−3.18)(−3.18).

(e) 611(−3.7)

ANSWERS

1. (a) −1.04 (b) 11.300 or 11.3 (c) 0.728 (d) 10.1124 (e) −2260.7

2. Multiply.

(a) $0.04(-0.09)$

(b) $(0.2)(0.008)$

(c) $(-0.063)(-0.04)$

(d) $(0.003)^2$

(e) $(0.11)(0.0005)$

3. First use front end rounding to estimate the answer. Then multiply to find the exact answer.

(a) $(11.62)(4.01)$

(b) $(-5.986)(-33)$

(c) $8.31(4.2)$

(d) $58.6(-17.4)$

ANSWERS

2. (a) -0.0036 (b) 0.0016 (c) 0.00252 (d) 0.000009 (e) 0.000055
3. (a) $(10)(4) = 40$; 46.5962
(b) $(-6)(-30) = 180$; 197.538
(c) $8(4) = 32$; 34.902
(d) $60(-20) = -1200$; -1019.64

EXAMPLE 2 Writing Zeros as Placeholders in the Answer

Multiply $-0.042(-0.03)$

Start by multiplying and counting decimal places.

$$\begin{array}{r l} 0.042 & \leftarrow \text{3 decimal places} \\ \times \quad 0.03 & \leftarrow \text{2 decimal places} \\ \hline 126 & \leftarrow \text{5 decimal places needed in answer} \end{array}$$

The answer has only three decimal places, but five are needed. So write two zeros on the *left* side of the answer.

$$\begin{array}{r} 0.042 \\ \times \quad 0.03 \\ \hline 00126 \end{array} \qquad \begin{array}{r l} 0.042 & \leftarrow \text{3 decimal places} \\ \times \quad 0.03 & \leftarrow \text{2 decimal places} \\ \hline .00126 & \leftarrow \text{5 decimal places} \end{array}$$

Write two 0's on *left* side of answer.

Now count over 5 places and write in the decimal point.

The final answer is 0.00126, which has five decimal places. The product is positive because the factors have the same sign (both negative).

◀◀ WORK PROBLEM 2 AT THE SIDE.

OBJECTIVE 2 If you are doing multiplication problems by hand, estimating the answer helps you check that the decimal point is in the right place. When you are using a calculator, estimating helps you catch an error like pressing the [÷] key instead of the [×] key.

EXAMPLE 3 Estimating Before Multiplying

First estimate $(76.34)(12.5)$. Round each number so there is only one non-zero digit (front end rounding). Then find the exact answer.

estimate (Rounds to) *exact*

$$\begin{array}{r} 80 \\ \times \ 10 \\ \hline 800 \end{array} \qquad \begin{array}{r l} 76.34 & \leftarrow \text{2 decimal places} \\ \times \quad 12.5 & \leftarrow \text{1 decimal place} \\ \hline 38170 & \text{3 decimal places are in answer.} \\ 15268 & \\ 7634 & \\ \hline 954.250 & \end{array}$$

Both the estimate and the exact answer go out to the hundreds, so the decimal point in 954.250 is probably in the correct place.

◀◀ WORK PROBLEM 3 AT THE SIDE.

Calculator Tip: When working with money amounts, you need to write a zero in your answer. For example, try multiplying \$3.54 × 5 on your calculator. Write down the result.

3.54 [×] 5 [=] ______

Notice the result is 17.7 which is *not* the way to write a money amount. You have to add the zero in the hundredths place: \$17.70 is correct. The calculator does not show the "extra" zero because:

$$17.70 \text{ or } 17\frac{70}{100} \text{ reduces to } 17\frac{7}{10} \text{ or } 17.7.$$

So keep an eye on your calculator—it doesn't know when you're working with money amounts.

Name Date Class

5.4 Exercises

Multiply.

1. $\begin{array}{r} 0.042 \\ \times \quad 3.2 \\ \hline \end{array}$

2. $\begin{array}{r} 0.571 \\ \times \quad 2.9 \\ \hline \end{array}$

3. $-21.5(7.4)$

4. $-85.4(-3.5)$

5. $(-23.4)(-0.666)$

6. $0.896(-0.799)$

7. $\begin{array}{r} \$51.88 \\ \times \quad 665 \\ \hline \end{array}$

8. $\begin{array}{r} \$736.75 \\ \times \quad 118 \\ \hline \end{array}$

Use the fact that $72 \times 6 = 432$ to help you solve Exercises 9–16 by simply counting decimal places.

9. $72(-0.6)$

10. $7.2(-6)$

11. $(7.2)(0.06)$

12. $(0.72)(0.6)$

13. $-0.72(-0.06)$

14. $-72(-0.0006)$

15. $(0.0072)(0.6)$

16. $(0.072)(0.006)$

Multiply.

17. $(0.006)(0.0052)$

18. $(0.0052)(0.009)$

19. $(-0.003)^2$

20. $(0.0004)^2$

21. Do these multiplications.

(5.96)(10)	(3.2)(10)
(0.476)(10)	(80.35)(10)
(722.6)(10)	(0.9)(10)

What pattern do you see? Write a "rule" for multiplying by 10. What do you think the rule is for multiplying by 100? By 1000? Write the rules and try them out on the numbers above.

22. Do these multiplications.

(59.6)(0.1)	(3.2)(0.1)
(0.476)(0.1)	(80.35)(0.1)
(65)(0.1)	(523)(0.1)

What pattern do you see? Write a "rule" for multiplying by 0.1. What do you think the rule is for multiplying by 0.01? By 0.001? Write the rules and try them out on the numbers above.

Writing Calculator Small Group

First use front end rounding to estimate the answer. Then multiply to find the exact answer.

23. *estimate* — *exact* (Rounds to)

$\times$ ___ ← 39.6 × 4.8

24. *estimate* — *exact*

$\times$ ___ 18.7 × 2.3

25. *estimate* — *exact*

$\times$ ___ 37.1 × 42

26. *estimate* — *exact*

$\times$ ___ 5.08 × 71

27. *estimate* — *exact*

$\times$ ___ 6.53 × 4.6

28. *estimate* — *exact*

$\times$ ___ 7.51 × 8.2

29. *estimate* — *exact*

$\times$ ___ 2.809 × 6.85

30. *estimate* — *exact*

$\times$ ___ 73.52 × 22.34

G *Even with most of the problem missing, you can tell whether or not these answers are reasonable. Circle "reasonable" or "unreasonable." If the answer is unreasonable, move the decimal point or insert a decimal point to make the answer reasonable.*

31. How much was his car payment? $18.90
reasonable
unreasonable, should be ___________

32. How many hours did she work today? 25 hours
reasonable
unreasonable, should be ___________

33. How tall is her son? 60.5 inches
reasonable
unreasonable, should be ___________

34. How much does he pay for rent now? $4.92
reasonable
unreasonable, should be ___________

35. What is the price of one gallon of milk? $319
reasonable
unreasonable, should be ___________

36. How long is the living room? 16.8 feet
reasonable
unreasonable, should be ___________

37. How much did the baby weigh? 0.095 pounds
reasonable
unreasonable, should be ___________

38. What was the sale price of the jacket? $1.49
reasonable
unreasonable, should be ___________

Solve. If the problem involves money, round to the nearest cent, if necessary.

39. LaTasha worked 50.5 hours over the last two weeks. She earns $11.73 per hour. How much did she make?

40. Michael's time card shows 42.2 hours at $10.03 per hour. What are his gross earnings?

41. Sid needs 0.6 meter of canvas material to make a carry-all bag that fits on his wheelchair. If canvas is $4.09 per meter, how much will Sid spend? (*Note:* $4.09 *per* meter means $4.09 for *one* meter.)

42. How much will Mrs. Nguyen pay for 3.5 yards of lace trim that costs $0.87 per yard?

43. Michelle pumped 18.65 gallons of gas into her pickup truck. The price was $1.45 per gallon. How much did she pay for gas?

44. Spicy chicken wings were on sale for $0.98 per pound. Juma bought 1.7 pounds of wings. How much did the chicken wings cost?

45. Ms. Rolack is a real estate broker who helps people sell their homes. Her fee is 0.07 times the price of the home. What was her fee for selling a $125,300 home?

46. Alex Rodriguez, shortstop for the Seattle Mariners, has a 1996 batting average of 0.358. If he went to bat 601 times, how many hits did he get? (*Hint:* Multiply his batting average by the number of times at bat.) Round to the nearest whole number.

47. Judy Lewis pays $28.96 per month for cable TV. How much will she pay for cable over one year?

48. Chuck's car payment is $220.27 per month for three years. How much will he pay altogether?

49. Paper for the copy machine at the library costs $0.015 per sheet. How much will the library pay for 5100 sheets?

50. A student group collected 2200 pounds of plastic as a fund raiser. How much will they make if the recycling center pays $0.142 per pound?

Use the list of prices below from the Look Smart mail order catalog to solve Exercises 51 and 52.

Knit Shirt Ordering Information		
43–2A	short sleeve, solid colors	\$14.75 each
43–2B	short sleeve, stripes	\$16.75 each
43–3A	long sleeve, solid colors	\$18.95 each
43–3B	long sleeve, stripes	\$21.95 each
Extra-large size, add \$2 per shirt.		

51. Find the total cost of four long-sleeve, solid-color shirts and two short-sleeve, striped shirts, all in the extra-large size.

52. What is the total cost of eight long-sleeve shirts, five in solid colors and three striped?

53. Jack Burgess used 3.5 gallons of fertilizer on each of his 158.2 acres of corn. After he finished, how much fertilizer was left in a storage tank that originally contained 600 gallons?

54. Stan Johnson bought 7.8 yards of a Hawaiian print fabric at \$5.62 per yard. He paid for it with three \$20 bills. Find the amount of his change. (Ignore sales tax.)

55. Ms. Sanchez paid \$29.95 a day to rent a car, plus \$0.29 per mile. Find the cost of her rental for a four-day trip of 926 miles.

56. The Bell family rented a motor home for \$375 per week plus \$0.35 per mile. What was the rental cost for their three-week vacation trip of 2650 miles?

57. Barry bought 16.5 meters of rope at \$0.47 per meter and three meters of wire at \$1.05 per meter. How much change did he get from three \$5 bills?

58. Susan bought a VCR that cost \$229.88. She paid \$45 down and \$37.98 per month for six months. How much could she have saved by paying cash?

Review and Prepare

*First use front end rounding to estimate each answer. Then find the exact answer using long division and an **R** to express a remainder. (For help, see **Sections 1.5, R.4,** and **R.5.**)*

59. *estimate* $\overline{)\quad}$ *exact* $5\overline{)954}$

60. *estimate* $\overline{)\quad}$ *exact* $4\overline{)2223}$

61. *estimate* $\overline{)\quad}$ *exact* $21\overline{)19{,}020}$

62. *estimate* $\overline{)\quad}$ *exact* $28\overline{)93{,}621}$

5.5 Dividing Decimal Numbers

There are two kinds of decimal division problems; those in which a decimal is divided by an integer, and those in which a decimal is divided by a decimal. First recall the parts of a division problem.

$$\begin{array}{r} 8 \\ 4\overline{)33} \\ 32 \\ \hline 1 \end{array}$$

8 ← Quotient
Divisor → 4 ← 33 Dividend
1 ← Remainder

OBJECTIVE 1 When the divisor is an integer, use these steps.

Dividing a Decimal Number by an Integer

Step 1 Write the decimal point in the quotient (answer) directly above the decimal point in the dividend.

Step 2 Divide as if both numbers were whole numbers.

Step 3 If both numbers have the same sign, the quotient is positive. If they have different signs, the quotient is negative.

EXAMPLE 1 Dividing a Decimal by an Integer

Divide.

(a) $21.93 \div (-3)$ (21.93 is the Dividend; −3 is the Divisor)

First consider $21.93 \div 3$ $\qquad 3\overline{)21.93}$

Write the decimal point in the answer directly above the decimal point in the dividend. $\qquad$ $\begin{array}{r} . \\ 3\overline{)21.93} \end{array}$ (Decimal points lined up)

Divide as if the numbers were whole numbers. $\qquad$ $\begin{array}{r} 7.31 \\ 3\overline{)21.93} \end{array}$

Check by multiplying the quotient times the divisor.

Check

$$\begin{array}{r} 7.31 \\ \times \quad 3 \\ \hline 21.93 \end{array}$$

(Matches)

The quotient is −7.31 because the numbers had *different* signs.

$21.93 \div (-3) = -7.31$ (Different signs; Negative quotient)

(b) $9\overline{)470.7}$ (9 is the Divisor; 470.7 is the Dividend) Write the decimal point in the answer above the decimal point in the dividend. Then divide as if they were whole numbers.

Decimal points lined up

$$\begin{array}{r} 52.3 \\ 9\overline{)470.7} \\ 45 \\ \hline 20 \\ 18 \\ \hline 2\,7 \\ 2\,7 \\ \hline 0 \end{array}$$

Check

$$\begin{array}{r} 52.3 \\ \times \quad 9 \\ \hline 470.7 \end{array}$$

(Matches)

The quotient is positive because both numbers had the *same* sign.

WORK PROBLEM 1 AT THE SIDE.

OBJECTIVES

1. Divide a decimal by an integer.
2. Divide a decimal by a decimal.
3. Estimate the answer when dividing decimals.
4. Use the order of operations with decimals.

FOR EXTRA HELP

Tutorial — Tape 8 — SSM, Sec. 5.5

1. Divide. Check your answers by multiplying.

(a) $4\overline{)93.6}$

(b) $6\overline{)6.804}$

(c) $11\overline{)278.3}$

(d) $-0.51835 \div 5$

(e) $-213.45 \div (-15)$

ANSWERS

1. (a) 23.4; 23.4(4) = 93.6
(b) 1.134; 1.134(6) = 6.804
(c) 25.3; 25.3(11) = 278.3
(d) −0.10367; −0.10367(5) = −0.51835
(e) 14.23; 14.23(−15) = −213.45

2. Divide. Check your answers by multiplying.

(a) $5\overline{)6.4}$

(b) $30.87 \div (-14)$

(c) $\dfrac{-259.5}{-30}$

(d) $0.3 \div 8$

EXAMPLE 2 Writing Extra Zeros to Complete a Division

Divide 1.5 by 8.

Keep dividing until the remainder is zero, or until the digits in the answer begin to repeat in a pattern. In Example 1(b), you ended up with a remainder of 0. But sometimes you run out of digits in the dividend before that happens. If so, write extra zeros on the right side of the dividend so you can continue dividing.

$$\begin{array}{r l} 0.1 & \\ 8\overline{)1.5} & \leftarrow \text{All digits have been used.} \\ \underline{8} & \\ 7 & \leftarrow \text{Remainder is not yet 0.} \end{array}$$

Write a zero after the 5 so you can continue dividing. Keep writing more zeros in the dividend if needed. Recall that writing zeros to the right of a decimal number does **not** change its value.

$$\begin{array}{r l} 0.1875 & \\ 8\overline{)1.5000} & \leftarrow \text{Three 0's needed to complete the division.} \\ \underline{8} & \\ 70 & \\ \underline{64} & \\ 60 & \\ \underline{56} & \\ 40 & \\ \underline{40} & \\ 0 & \leftarrow \text{Stop dividing when the remainder is 0.} \end{array}$$

Check

$$\begin{array}{r} 0.1875 \\ \times \quad\quad 8 \\ \hline 1.5000 \end{array}$$

Matches dividend so 0.1875 is correct.

Calculator Tip: When *multiplying* numbers, you can enter them in any order because multiplication is commutative (see **Section 1.6**). But division is *not* commutative. It *does* matter which number you enter first. Try Example 2 both ways; jot down your answers.

1.5 [÷] 8 [=] ______ 8 [÷] 1.5 [=] ______

Notice that the first answer, 0.1875, matches the result from Example 2 above. But the second answer is much different: 5.333333333. Be careful to enter the dividend first.

Also notice that in decimals the dividend may *not* be the larger number, as it was in whole numbers. In Example 2 the dividend is 1.5, which is *smaller* than 8.

◀◀ WORK PROBLEM 2 AT THE SIDE.

The next example shows a quotient (answer) that must be rounded because you will never get a remainder of zero.

ANSWERS

2. (a) 1.28; 1.28(5) = 6.40 or 6.4
(b) −2.205; −2.205(−14) = 30.870 or 30.87
(c) 8.65; 8.65(−30) = −259.50 or −259.5
(d) 0.0375; 0.0375(8) = 0.3000 or 0.3

EXAMPLE 3 Rounding a Decimal Quotient

Divide 4.7 by 3. Round to the nearest thousandth.

Write extra zeros in the dividend so that you can continue dividing.

```
     1.5 6 6 6
  3)4.7 0 0 0  ← Three 0's added so far
    3
    1 7
    1 5
      2 0
      1 8
        2 0
        1 8
          2 0
          1 8
            2  ← Remainder is still not 0.
```

Notice that the digit 6 in the answer is repeating. It will continue to do so. The remainder will never be zero. There are two ways to show that the answer is a **repeating decimal** that goes on forever. Write three dots after the answer, or, write a bar above the digits that repeat (in this case, the 6).

$$1.5666\ldots \quad \text{or} \quad 1.5\overline{6}$$

Three dots ← Bar above repeating digit

When repeating decimals occur, round the answer according to the directions in the problem. In this example, to round to thousandths, divide out one *more* place, to ten-thousandths.

$$4.7 \div 3 = 1.5666\ldots \text{ rounds to } 1.567$$

Check the answer by multiplying 1.567 by 3. Because 1.567 is a rounded answer, the check will *not* give exactly 4.7, but it should be very close.

$$1.567 \cdot 3 = 4.701$$ ← Does not equal exactly 4.7 because 1.567 was rounded.

Note
When you're checking answers that you've rounded, the check will *not* match the dividend exactly, but it should be very close.

WORK PROBLEM 3 AT THE SIDE.

OBJECTIVE 2 To divide by a *decimal* divisor, first change the divisor to a whole number. Then divide as before. To see how this is done, write the problem in fraction form. For example:

$$1.2\overline{)6.36} \quad \text{can be written} \quad \frac{6.36}{1.2}.$$

In **Section 4.1** you learned that multiplying the numerator and denominator by the same number gives an equivalent fraction. We want the divisor (1.2) to be a whole number. Multiplying by 10 will accomplish that.

$$\frac{6.36}{1.2} = \frac{6.36 \cdot 10}{1.2 \cdot 10} = \frac{63.6}{12}$$

3. Divide. Round answers to the nearest thousandth. If it is a repeating decimal, also write the answer using a bar. Check your answers by multiplying.

(a) $13\overline{)267.01}$

(b) $6\overline{)20.5}$

(c) $\frac{10.22}{9}$

(d) $16.15 \div 3$

(e) $116.3 \div 11$

ANSWERS

3. (a) ≈20.539; no repeating digits visible on calculator; 20.539(13) = 267.007
(b) ≈3.417; $3.41\overline{6}$ 3.417(6) = 20.502
(c) ≈1.136; $1.13\overline{5}$ 1.136(9) = 10.224
(d) ≈5.383; $5.38\overline{3}$ 5.383(3) = 16.149
(e) ≈10.573; $10.5\overline{72}$ 10.573(11) = 116.303

4. Divide. If the quotient does not come out even, round to the nearest hundredth.

(a) $0.2\overline{)1.04}$

(b) $0.06\overline{)1.8072}$

(c) $0.005\overline{)32}$

(d) $-8.1 \div 0.025$

(e) $\dfrac{7}{1.3}$

(f) $-5.3091 \div (-6.2)$

Answers

4. **(a)** 5.2 **(b)** 30.12 **(c)** 6400 **(d)** −324 **(e)** ≈5.38 **(f)** ≈0.86

The short way to multiply by 10 is to move the decimal point one place to the right in both the divisor and the dividend.

$$1.2\overline{)6.36} \text{ is equivalent to } 12\overline{)63.6}$$

Note

Moving the decimal points the **same** number of places in **both** the divisor and dividend will **not** change the answer.

Dividing by a Decimal Number

Step 1 Count the number of decimal places in the divisor and move the decimal point that many places to the *right.* (This changes the divisor to a whole number.)

Step 2 Move the decimal point in the dividend the *same* number of places to the *right.* (Write in extra zeros if needed.)

Step 3 Write the decimal point in the answer directly above the decimal point in the dividend. Then divide as usual.

Step 4 If both numbers have the same sign, the quotient is positive. If they have different signs, the quotient is negative.

EXAMPLE 4 Dividing by a Decimal Number

(a) $0.003\overline{)27.69}$

Move the decimal point in the divisor *three* places to the *right* so 0.003 becomes the whole number 3. In order to move the decimal point in the dividend the same number of places, write in an extra zero.

$$0.003\overline{)27.690}$$

Move decimal points in divisor and dividend. Then line up decimal point in answer.

Moving decimal point three places is the same as multiplying by 1000.

$$\begin{array}{r} 9230. \\ 3\overline{)27690.} \end{array}$$

Divide as usual.

(b) Divide −5 by −4.2. Round to the nearest hundredth.

First consider 5 ÷ 4.2. Move the decimal point in the divisor one place to the right so that 4.2 becomes the whole number 42. The decimal point in the dividend starts on the right side of 5 and is also moved one place to the right.

$$\begin{array}{r} 1.190 \\ 4.2\overline{)5.0000} \\ \underline{42} \\ 80 \\ \underline{42} \\ 380 \\ \underline{378} \\ 20 \end{array}$$

← In order to round to hundredths, divide out one *more* place, to thousandths.

Round the quotient. It is ≈1.19 (nearest hundredth) and is *positive* because both the divisor and dividend were negative.

$$-5 \div (-4.2) \approx 1.19$$

Same sign — Positive quotient

◀◀ **WORK PROBLEM 4 AT THE SIDE.**

Objective 3 Estimating the answer to a division problem helps you catch errors. Compare the estimate to your exact answer. If they are very different, do the division again.

EXAMPLE 5 Estimating before Dividing

First use front end rounding to estimate the answer. Then divide to find the exact answer.

$$580.44 \div 2.8$$

Here is how one student solved this problem. She rounded 580.44 to 600 and 2.8 to 3 to estimate the answer.

estimate: $3\overline{)600}$ = **200**

exact: $2.8\overline{)580.44}$ = **27.3**

```
        2 7.3
2.8)5 8 0.4 4
    5 6
      2 0 4
      1 9 6
          8 4
          8 4
            0
```

Very different; need to rework the problem

Notice that the estimate, which is in the hundreds, is very different from the exact answer, which is only in the tens. This tells the student that she needs to rework the problem. Can you find the error? (The exact answer should be 207.3, which fits with the estimate of 200.)

WORK PROBLEM 5 AT THE SIDE.

Objective 4 Use the order of operations when a decimal problem involves more than one operation.

Order of Operations

1. Work inside parentheses or other grouping symbols.
2. Simplify expressions with exponents.
3. Do the remaining multiplications and divisions as they occur from left to right.
4. Do the remaining additions and subtractions as they occur from left to right.

EXAMPLE 6 Using the Order of Operations

Simplify by using the order of operations.

(a) $2.5 + (-6.3)^2 + 9.62$ Use the exponent: $(-6.3)(-6.3)$ is 39.69.

$2.5 + 39.69 + 9.62$ Add from left to right.

$42.19 + 9.62$

51.81

CONTINUED ON NEXT PAGE

5. Decide if each answer is reasonable by using front end rounding to estimate the answer. If the exact answer is *not* reasonable, find and correct the error.

(a) $42.75 \div 3.8 = 1.125$

estimate:

(b) $807.1 \div 1.76 = 458.580$ to nearest thousandth

estimate:

(c) $48.63 \div 52 = 93.519$ to nearest thousandth

estimate:

(d) $9.0584 \div 2.68 = 0.338$

estimate:

Answers

5. (a) Estimate is $40 \div 4 = 10$; exact answer not reasonable, should be 11.25
(b) Estimate is $800 \div 2 = 400$; exact answer is reasonable.
(c) Estimate is $50 \div 50 = 1$; exact answer is not reasonable, should be 0.935.
(d) Estimate is $9 \div 3 = 3$; exact answer is not reasonable, should be 3.38.

6. Simplify by using the order of operations.

(a) $-4.6 - 0.79 + 1.5^2$

(b) $3.64 \div 1.3(3.6)$

(c) $0.08 + 0.6(2.99 - 3)$

(d) $10.85 - 2.3(5.2) \div 3.2$

(b) $1.82 + (5.2 - 6.7)(5.8)$ Work inside parentheses.

$1.82 + (-1.5)\ (5.8)$ Multiply next.

$1.82 + (-8.7)$

-6.88 Add last.

(c) $3.7^2 - 1.8 \div 5(1.5)$ Do exponents first.

$13.69 - 1.8 \div 5(1.5)$ Multiply and divide from left to right.

$13.69 - 0.36\ (1.5)$

$13.69 - 0.54$ Subtract last.

13.15

WORK PROBLEM 6 AT THE SIDE.

Calculator Tip: Most scientific calculators that have parentheses keys [(] [)] can handle calculations like those in Example 6 just by entering the numbers in the order given. For example, the keystrokes for Example 6(b) are:

1.82 [+] [(] 5.2 [−] 6.7 [)] [×] 5.8 [=]

Standard, four-function calculators generally do *not* give the correct answer if you enter the numbers in the order given. Check the instruction manual that came with your calculator for information on "order of calculations" to see if your machine has the rules for order of operations built into it. For a quick check, try entering this problem:

2 [+] 2 [×] 2 [=]

If the result is 6, the calculator follows the order of operations. If the result is 8, it does *not* have the rules built into it. Use the space below to explain how this test works.

Answer: The test works because a calculator that follows the order of operations will automatically do the multiplication first. If the calculator does *not* have the rules built into it, it will work from left to right.

Following Order of Operations	Working from Left to Right
$2 + 2 \times 2$	$2 + 2 \times 2$
$2 + 4$	4×2
$6 \leftarrow$ Correct	$8 \leftarrow$ Incorrect

ANSWERS

6. (a) -3.14 **(b)** 10.08 **(c)** 0.074 **(d)** 7.1125

5.5 Exercises

Divide.

1. $27.3 \div (-7)$ **2.** $-50.4 \div 8$ **3.** $\dfrac{4.23}{9}$ **4.** $\dfrac{1.62}{6}$

5. $-20.01 \div (-0.05)$ **6.** $-16.04 \div (-0.08)$ **7.** $1.5\overline{)54}$ **8.** $2.4\overline{)132}$

Ⓖ *Use the fact that* $108 \div 18 = 6$ *to solve Exercises 9–16 simply by moving decimal points.*

9. $1.8\overline{)0.108}$ **10.** $18\overline{)10.8}$ **11.** $0.018\overline{)108}$ **12.** $0.18\overline{)1.08}$

13. $0.18\overline{)10.8}$ **14.** $0.18\overline{)108}$ **15.** $18\overline{)0.0108}$ **16.** $1.8\overline{)0.0108}$

In Exercises 17–20 round your answers to the nearest hundredth, if necessary.

17. $4.6\overline{)116.38}$ **18.** $2.6\overline{)4.992}$ **19.** $\dfrac{3.1}{0.006}$ **20.** $\dfrac{1.7}{0.09}$

In Exercises 21–24 round your answers to the nearest thousandth.

21. $-240 \div 9.88$ **22.** $-7643 \div (-5.36)$ **23.** $0.034\overline{)342.81}$ **24.** $0.043\overline{)1748.4}$

25. Do these division problems.
$3.77 \div 10$ $9.1 \div 10$
$0.886 \div 10$ $30.19 \div 10$
$406.5 \div 10$ $6625.7 \div 10$
What pattern do you see? Write a "rule" for dividing by 10. What do you think the rule is for dividing by 100? By 1000? Write the rules and try them out on the numbers above.

26. Do these division problems.
$40.2 \div 0.1$ $7.1 \div 0.1$
$0.339 \div 0.1$ $15.77 \div 0.1$
$46 \div 0.1$ $873 \div 0.1$
What pattern do you see? Write a "rule" for dividing by 0.1. What do you think the rule is for dividing by 0.01? By 0.001? Write the rules and try them out on the numbers above.

Writing Calculator Ⓖ Small Group

Decide if each answer is reasonable or unreasonable by using front end rounding to estimate the answer. If the exact answer is not reasonable, find the correct answer.

27. $37.8 \div 8 = 47.25$

estimate:

28. $345.6 \div 3 = 11.52$

estimate:

29. $54.6 \div 48.1 = 1.135$

estimate:

30. $2428.8 \div 4.8 = 50.6$

estimate:

31. $307.02 \div 5.1 = 6.2$

estimate:

32. $395.415 \div 5.05 = 78.3$

estimate:

33. $9.3 \div 1.25 = 0.744$

estimate:

34. $78 \div 14.2 = 0.182$

estimate:

Solve each application problem. Round money answers to the nearest cent, if necessary.

35. Alfred has discovered that Batman's favorite brand of superhero tights are on sale at six pairs for $23.98, but he's been told to buy only one pair for Robin. How much will he pay for one pair?

36. The bookstore is selling four notepads for $1.69. How much did Randall pay for one notepad?

37. It will take 21 months for Aimee to pay off her charge account balance of $408.66. How much is she paying each month?

38. Marcella Anderson bought 2.6 meters of suede fabric for $18.19. How much did she pay per meter?

39. Adrian Webb bought 619 bricks to build a barbecue pit, paying $185.70. Find the cost per brick. (*Hint:* Cost *per* brick means the cost for *one* brick.)

40. Lupe Wilson is a newspaper distributor. Last week she paid the newspaper $130.51 for 842 copies. Find the cost per copy.

41. Darren Jackson earned $235.60 for 40 hours of work. Find his earnings per hour.

42. At a record manufacturing company, 400 records cost $289. Find the cost per record.

43. It took 16.35 gallons of gas to fill Kim's car gas tank. She had driven 346.2 miles since her last fill-up. How many miles per gallon did she get? Round to the nearest tenth.

44. Mr. Rodriquez pays $53.19 each month to Household Finance. How many months will it take him to pay off a loan of $1436.13?

Use the table of 1996 Olympic long jumps to solve Exercises 45–50. To find an average, add the values that you're interested in and then divide the sum by the number of values. Round your answer to the nearest hundredth. Some of the other exercises may require subtraction or multiplication.

Country	Length
U.S.	8.50 meters
Jamaica	8.29 meters
U.S.	8.24 meters
France	8.19 meters
U.S.	8.17 meters
Slovenia	8.11 meters
Belarus	8.07 meters

45. Find the average length of the long jumps made by U.S. athletes.

46. Find the average length of all the long jumps listed in the table.

47. How much longer was the second place jump than the third place jump?

48. If the Olympic champion made six jumps of the same length, what was the total distance jumped?

49. What was the total length jumped by the top three athletes?

50. How much less was the last place jump than the next to last jump?

51. Soup is on sale at six cans for \$3.25, or you can purchase individual cans for \$0.57. How much will you save per can if you buy six cans? Round to the nearest cent.

52. Nadia's diet says she can eat 3.5 ounces of chicken nuggets. The package weighs 10.5 ounces and contains 15 nuggets. How many nuggets can Nadia eat?

53. The annual premium for Jenny's auto insurance policy is \$938. She can pay it in four quarterly installments, if she adds a \$2.75 service fee to each payment. Find the amount of each quarterly payment.

54. Lock and Store charges rent of \$936 per year for 200 square feet of storage space. To pay the rent monthly, \$1.25 must be added to each payment. Find the amount of each monthly payment.

Simplify by using the order of operations.

55. $7.2 - 5.2 + 3.5^2$

56. $6.2 + 4.3^2 - 9.72$

57. $38.6 + 11.6(10.4 - 13.4)$

58. $2.25 - 1.06(0.85 - 3.95)$

59. $-8.68 - 4.6(10.4) \div 6.4$

60. $25.1 + 11.4 \div 7.5(-3.75)$

61. $33 - 3.2(0.68 + 9) + (-1.3)^2$

62. $0.6 + (-1.89 + 0.11) \div 0.004(0.5)$

Review and Prepare

Write each fraction in lowest terms. (For help, see ***Section 4.2.****)*

63. $\frac{30}{60}$

64. $\frac{60}{80}$

65. $\frac{75}{125}$

66. $\frac{625}{1000}$

Write > or < between each pair of numbers. (For help, see ***Section 1.2.****)*

67. 0 _____ −2

68. −3 _____ 1

69. −15 _____ −10

70. 6 _____ 0

5.6 Fractions and Decimals

Writing fractions as equivalent decimals can help you do calculations more easily or compare the size of two numbers.

OBJECTIVE 1 Recall that a fraction is one way to show division (see **Section 1.7**). For example, $\frac{3}{4}$ means $3 \div 4$. If you are doing the division by hand, write it as $4\overline{)3}$. When you do the division, you will get the decimal equivalent of $\frac{3}{4}$.

Writing Fractions as Decimals

Step 1 Divide the numerator of the fraction by the denominator.

Step 2 If necessary, round the answer to the place indicated.

WORK PROBLEM 1 AT THE SIDE.

EXAMPLE 1 Writing a Fraction or Mixed Number as a Decimal

(a) Write $\frac{1}{8}$ as a decimal.

$\frac{1}{8}$ means $1 \div 8$. Write it as $8\overline{)1}$. The decimal point in the dividend is on the right side of the 1. Write extra zeros in the dividend so you can continue dividing until the remainder is 0.

$$\begin{array}{r} 0.125 \\ 8\overline{)1.000} \\ \underline{8} \\ 20 \\ \underline{16} \\ 40 \\ \underline{40} \\ 0 \end{array}$$

Decimal points lined up.
← Three extra 0's needed.
← Remainder is 0.

Therefore, $\frac{1}{8} = 0.125$. To check this, write 0.125 as a fraction, then change it to lowest terms.

$$0.125 = \frac{125}{1000} \quad \text{To write in lowest terms} \quad \frac{125 \div 125}{1000 \div 125} = \frac{1}{8}$$

Calculator Tip: When changing fractions to decimals on your calculator, enter the numbers from the top down. Remember that the order in which you enter the numbers *does* matter in division. Example 1(a) works like this:

$\frac{1}{8}$ ↓ Top down   Enter 1 [÷] 8 [=]

What happens if you enter 8 [÷] 1 [=] ? Do you see why that cannot possibly be the answer?

(b) Write $2\frac{3}{4}$ as a decimal.

One method is to divide 3 by 4 to get 0.75 for the fraction part. Then add the whole number part to 0.75.

$$\begin{array}{r} 0.75 \\ 4\overline{)3.00} \\ \underline{2\,8} \\ 20 \\ \underline{20} \\ 0 \end{array} \qquad \begin{array}{rr} \text{Whole number part} \rightarrow & 2.00 \\ \text{Fraction part} \rightarrow & \underline{+\ 0.75} \\ & 2.75 \end{array}$$

CONTINUED ON NEXT PAGE

OBJECTIVES

1. Write fractions as equivalent decimals.
2. Compare the size of fractions and decimals.

FOR EXTRA HELP

Tutorial   Tape 8   SSM, Sec. 5.6

1. Rewrite each fraction so you could do the division by hand. Do **not** complete the division.

(a) $\frac{1}{9}$ is written $9\overline{)}$

(b) $\frac{2}{3}$ is written $\overline{)}$

(c) $\frac{5}{4}$ is written $\overline{)}$

(d) $\frac{3}{10}$ is written $\overline{)}$

(e) $\frac{21}{16}$ is written $\overline{)}$

(f) $\frac{1}{50}$ is written $\overline{)}$

ANSWERS

1. (a) $9\overline{)1}$ (b) $3\overline{)2}$ (c) $4\overline{)5}$ (d) $10\overline{)3}$ (e) $16\overline{)21}$ (f) $50\overline{)1}$

2. Write each fraction or mixed number as a decimal.

(a) $\frac{1}{4}$

(b) $2\frac{1}{2}$

(c) $\frac{5}{8}$

(d) $4\frac{3}{5}$

(e) $\frac{7}{8}$

So, $2\frac{3}{4} = 2.75$ Check: $2.75 = 2\frac{75}{100} = 2\frac{3}{4}$ Lowest terms

Whole number parts match.

A second method is to write $2\frac{3}{4}$ as an improper fraction.

$$2\frac{3}{4} = \frac{11}{4} \leftarrow \frac{11}{4} \text{ means } 11 \div 4 \text{ or } 4\overline{)11}$$

$$\begin{array}{r} 2.75 \\ 4\overline{)11.00} \\ \underline{8} \\ 3\,0 \\ \underline{2\,8} \\ 20 \\ \underline{20} \\ 0 \end{array}$$

← Two extra 0's needed.

Whole number parts match.

So, $2\frac{3}{4} = 2.75$

$\frac{3}{4}$ is equivalent to $\frac{75}{100}$ or 0.75.

◀◀ WORK PROBLEM 2 AT THE SIDE.

EXAMPLE 2 Changing to a Decimal and Rounding

Write $\frac{2}{3}$ as a decimal and round to the nearest thousandth.

$\frac{2}{3}$ means $2 \div 3$. To round to thousandths, divide out one *more* place, to ten-thousandths.

$$\begin{array}{r} 0.6666 \\ 3\overline{)2.0000} \\ \underline{1\,8} \\ 20 \\ \underline{18} \\ 20 \\ \underline{18} \\ 20 \\ \underline{18} \\ 2 \end{array}$$

← Four 0's needed for ten-thousandths.

Written as a repeating decimal, $\frac{2}{3} = 0.\overline{6}$. Rounded to the nearest thousandth, $\frac{2}{3} \approx 0.667$.

Calculator Tip: Try Example 2 on your calculator. Enter 2 [÷] 3. Which answer do you get?

0.666666667 or **0.6666666**

Many scientific calculators will show a 7 as the last digit. Because the 6's keep on repeating forever, the calculator automatically rounds in the last decimal place it has room to show. If you have a 10-digit display space, the calculator is rounding like this:

0.6666666666 (11 digits) rounds to 0.666666667.

ANSWERS

2. **(a)** 0.25 **(b)** 2.5 **(c)** 0.625 **(d)** 4.6 **(e)** 0.875

Other calculators, especially standard, four-function ones, may *not* round. They just cut off, or truncate, the extra digits. Such a calculator would show 0.6666666 in the display.

Would this difference in calculators show up when changing $\frac{1}{3}$ to a decimal? Why not?

WORK PROBLEM 3 AT THE SIDE.

OBJECTIVE 2 You can use a number line to compare fractions and decimals. For example, the number line below shows the space between 0 and 1. The locations of some commonly used fractions are marked, along with their decimal equivalents.

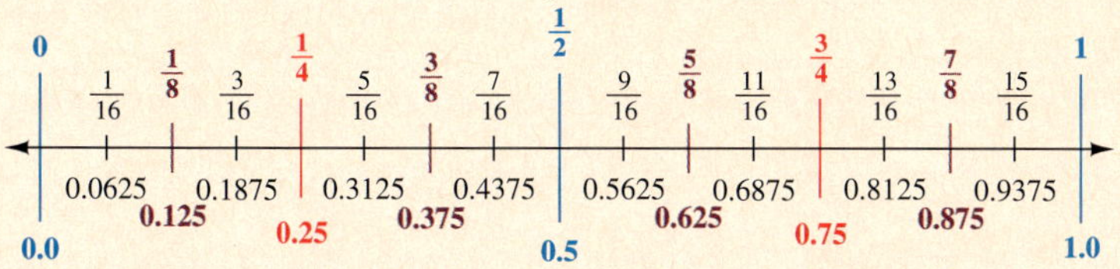

The next number line shows the locations of some commonly used fractions between 0 and 1 that are equivalent to repeating decimals. The decimal equivalents use a bar above repeating digits.

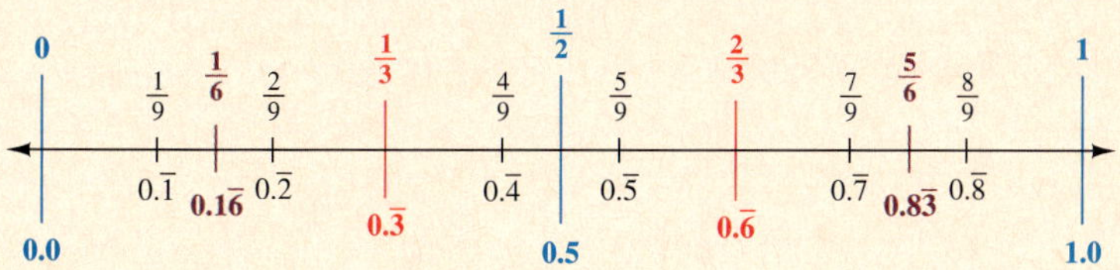

EXAMPLE 3 Using a Number Line to Compare Numbers

Use the number lines above to decide whether to write >, <, or = in the blank between each pair of numbers.

(a) 0.6875 ______ 0.625

You learned in **Section 1.2** that the number farther to the right on the number line is the greater number. On the first number line, 0.6875 is to the *right* of 0.625, so use the > symbol.

0.6875 is greater than 0.625   0.6875 > 0.625

(b) $\frac{3}{4}$ ______ 0.75

On the first number line, $\frac{3}{4}$ and 0.75 are at the same point on the number line. They are equivalent.

$$\frac{3}{4} = 0.75$$

CONTINUED ON NEXT PAGE

3. Write as decimals. Round to the nearest thousandth.

(a) $\frac{1}{3}$

(b) $2\frac{7}{9}$

(c) $\frac{10}{11}$

(d) $\frac{3}{7}$

(e) $3\frac{5}{6}$

ANSWERS

3. (a) ≈ 0.333 **(b)** ≈ 2.778 **(c)** ≈ 0.909 **(d)** ≈ 0.429 **(e)** ≈ 3.833

4. Use the number lines in the text to help you decide whether to write <, >, or = in each blank.

(a) 0.4375 ______ 0.5

(b) 0.75 ______ 0.6875

(c) 0.625 ______ 0.0625

(d) $\frac{2}{8}$ ______ 0.375

(e) $0.8\overline{3}$ ______ $\frac{5}{6}$

(f) $\frac{1}{2}$ ______ $0.\overline{5}$

(g) $0.\overline{1}$ ______ $0.1\overline{6}$

(h) $\frac{8}{9}$ ______ $0.\overline{8}$

(i) $0.\overline{7}$ ______ $\frac{4}{6}$

(j) $\frac{1}{4}$ ______ 0.25

5. Arrange in order from smallest to largest.

(a) 0.7, 0.703, 0.7029

(b) 6.39, 6.309, 6.4, 6.401

(c) 1.085, $1\frac{3}{4}$, 0.9

(d) $\frac{1}{4}, \frac{2}{5}, \frac{3}{7}$, 0.428

ANSWERS

4. **(a)** < **(b)** > **(c)** > **(d)** < **(e)** = **(f)** < **(g)** < **(h)** = **(i)** > **(j)** =

5. **(a)** 0.7, 0.7029, 0.703
(b) 6.309, 6.39, 6.4, 6.401
(c) 0.9, 1.085, $1\frac{3}{4}$
(d) $\frac{1}{4}, \frac{2}{5}$, 0.428, $\frac{3}{7}$

(c) 0.5 ______ $0.\overline{5}$

On the second number line, 0.5 is to the *left* of $0.\overline{5}$ (which is actually 0.555 . . .) so use the < symbol.

0.5 is less than $0.\overline{5}$ → $0.5 < 0.\overline{5}$

(d) $\frac{2}{6}$ ______ $0.\overline{3}$

Write $\frac{2}{6}$ in lowest terms as $\frac{1}{3}$.
On the second number line you can see that $\frac{1}{3} = 0.\overline{3}$.

WORK PROBLEM 4 AT THE SIDE.

You can also compare fractions by first writing each one as a decimal. You can then compare the decimals by writing each one with the same number of decimal places.

EXAMPLE 4 Arranging Numbers in Order

Write the following numbers in order, from smallest to largest.

(a) 0.49 0.487 0.4903.

It is easier to compare decimals if they are all tenths, or all hundredths, and so on. Because 0.4903 has four decimal places (ten-thousandths), write zeros to the right of 0.49 and 0.487 so they also have four decimal places. Writing zeros to the right of a decimal number does *not* change its value (see **Section 5.3**). Now find the smallest and largest number of ten-thousandths.

0.49 = 0.4900 = **4900** ten-thousandths ← 4900 is in the middle.
0.487 = 0.4870 = **4870** ten-thousandths ← 4870 is smallest.
0.4903 = **4903** ten-thousandths ← 4903 is largest.

From smallest to largest the correct order is:

0.487 0.49 0.4903.

(b) $2\frac{5}{8}$ 2.63 2.6

Write $2\frac{5}{8}$ as $\frac{21}{8}$ and divide $8\overline{)21}$ to get the decimal form, 2.625. Then, because 2.625 has three decimal places, write zeros so all the numbers have three decimal places.

$2\frac{5}{8}$ = 2.625 = 2 and **625** thousandths ← 625 is in the middle.
2.63 = 2.630 = 2 and **630** thousandths ← 630 is largest.
2.6 = 2.600 = 2 and **600** thousandths ← 600 is smallest.

From smallest to largest, the correct order is:

2.6 $2\frac{5}{8}$ 2.63.

WORK PROBLEM 5 AT THE SIDE.

5.6 Exercises

Write each fraction or mixed number as a decimal. Round to the nearest thousandth, if necessary.

1. $\frac{1}{2}$ **2.** $\frac{1}{4}$ **3.** $\frac{3}{4}$ **4.** $\frac{1}{10}$

5. $\frac{3}{10}$ **6.** $\frac{7}{10}$ **7.** $\frac{9}{10}$ **8.** $\frac{4}{5}$

9. $\frac{3}{5}$ **10.** $\frac{2}{5}$ **11.** $\frac{7}{8}$ **12.** $\frac{3}{8}$

13. $2\frac{1}{4}$ **14.** $1\frac{1}{2}$ **15.** $14\frac{7}{10}$ **16.** $23\frac{3}{5}$

17. $3\frac{5}{8}$ **18.** $2\frac{7}{8}$ **19.** $\frac{1}{3}$ **20.** $\frac{2}{3}$

21. $\frac{5}{6}$ **22.** $\frac{1}{6}$ **23.** $1\frac{8}{9}$ **24.** $5\frac{4}{7}$

25. Explain and correct the error that Keith made when changing a fraction to an equivalent decimal.

$$\frac{5}{9} = \begin{array}{r} 1.8 \\ 5\overline{)9.0} \\ \underline{5} \\ 4\,0 \\ \underline{4\,0} \\ 0 \end{array} \quad \text{so} \quad \frac{5}{9} = 1.8$$

26. Explain and correct the error Sandra made when writing $2\frac{7}{20}$ as a decimal.

$$2\frac{7}{20} = \begin{array}{r} 0.35 \\ 20\overline{)7.00} \\ \underline{6\,0} \\ 1\,00 \\ \underline{1\,00} \\ 0 \end{array} \quad \text{so} \quad 2\frac{7}{20} = 2.035$$

27. Ving knows that $\frac{3}{8} = 0.375$. How can he write $1\frac{3}{8}$ as a decimal *without* having to do a division? How can he write $3\frac{3}{8}$ as a decimal? $295\frac{3}{8}$? Explain your answer.

28. Iris has found a shortcut for writing mixed numbers as decimals:

$$2\frac{7}{10} = 2.7 \qquad 1\frac{13}{100} = 1.13$$

Does her shortcut work for all mixed numbers? Explain.

Find the decimal or fraction equivalent for each number. Write fractions in lowest terms.

	Fraction	Decimal		Fraction	Decimal
29.	________	0.4	30.	________	0.75
31.	________	0.625	32.	________	0.111
33.	________	0.35	34.	________	0.9
35.	$\frac{7}{20}$	________	36.	$\frac{1}{40}$	________
37.	________	0.04	38.	________	0.52
39.	________	0.15	40.	________	0.85
41.	$\frac{1}{5}$	________	42.	$\frac{1}{8}$	________
43.	________	0.09	44.	________	0.02

Ⓖ *Solve each application problem.*

45. The average length of a newborn baby is 20.8 inches. Charlene's baby is 20.08 inches long. Is her baby longer or shorter than the average? By how much?

46. The patient in room 830 is supposed to get 8.3 milligrams of medicine. She was actually given 8.03 milligrams. Did she get too much or too little medicine? What was the difference?

47. The label on the bottle of vitamins says that each capsule contains 0.5 gram of calcium. When checked, each capsule had 0.505 gram of calcium. Was there too much or too little calcium? What was the difference?

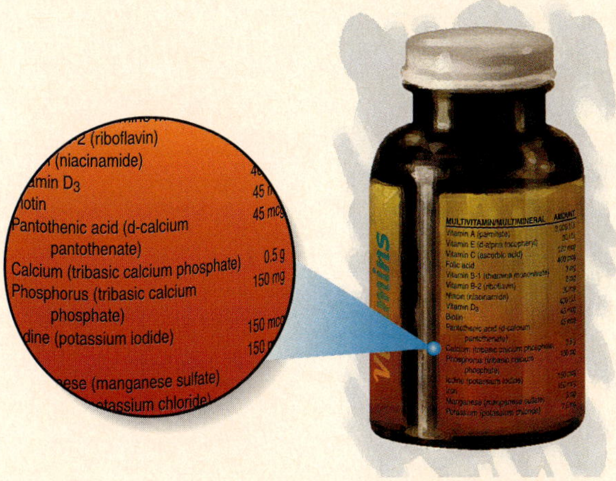

48. The glass mirror of the Hubble telescope had to be repaired in space in 1993 because it would not focus properly. The problem was that the mirror's outer edge had a thickness of 0.6248 cm when it was supposed to be 0.625 cm. Was the edge too thick or too thin? By how much?

49. Precision Medical Parts makes an artificial heart valve that must measure between 0.998 centimeter and 1.002 centimeters. Circle the lengths that are acceptable:
1.01 cm, 0.9991 cm, 1.0007 cm, 0.99 cm.

50. The mice in a medical experiment must start out weighing between 2.95 ounces and 3.05 ounces. Circle the weights that can be used: 3.0 ounces, 2.995 ounces, 3.055 ounces, 3.005 ounces.

51. Ginny Brown hoped her crops would get $3\frac{3}{4}$ inches of rain this month. The newspaper said the area received 3.8 inches of rain. Was that more or less than Ginny hoped for? By how much?

52. The mice in the experiment gained $\frac{3}{8}$ ounce. They were expected to gain 0.3 ounce. Was their actual gain more or less than expected? By how much?

Arrange in order from smallest to largest.

53. 0.54, 0.5455, 0.5399

54. 0.76, 0.7, 0.7006

55. 5.8, 5.79, 5.0079, 5.804

56. 12.99, 12.5, 13.0001, 12.77

57. 0.628, 0.62812, 0.609, 0.6009

58. 0.27, 0.281, 0.296, 0.3

59. 5.8751, 4.876, 2.8902, 3.88

60. 0.98, 0.89, 0.904, 0.9

61. 0.043, 0.051, 0.006, $\frac{1}{20}$

62. 0.629, $\frac{5}{8}$, 0.65, $\frac{7}{10}$

63. $\frac{3}{8}$, $\frac{2}{5}$, 0.37, 0.4001

64. 0.1501, 0.25, $\frac{1}{10}$, $\frac{1}{5}$

Some rulers show each inch divided into tenths. Use this scale drawing for Exercises 65–70. Change the measurements on the drawing to decimals and round them to the nearest tenth of an inch.

65. length (a) is ________

66. length (b) is ________

67. length (c) is ________

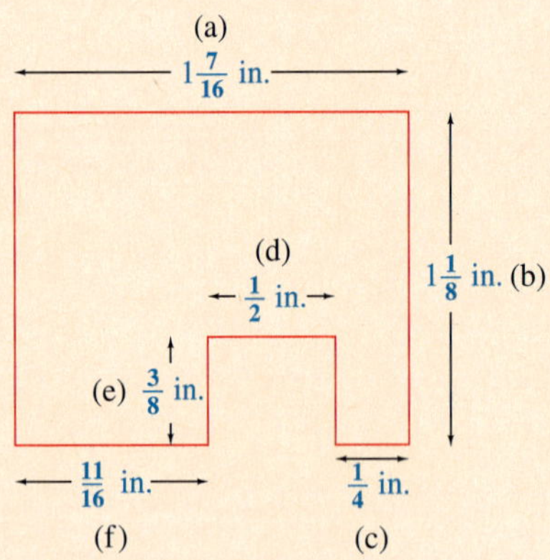

68. length (d) is ________

69. length (e) is ________

70. length (f) is ________

Review and Prepare

Simplify. Use the order of operations. (For help, see ***Section 1.8.****)*

71. $-6 + 18 \div 3$

72. $2 + 8(-2)$

73. $3 \cdot 4 + 6 \cdot 2$

74. $(-4 + 12) \div 2(3)$

75. $5 + 3(0 - 8)$

76. $6(-1) - 4(-3)$

5.7 Introduction to Statistics: Mean, Median, and Mode

OBJECTIVES

1. Find the mean of a list of numbers.
2. Find a weighted mean.
3. Find the median.
4. Find the mode.

FOR EXTRA HELP

Tutorial | Tape 8 | SSM, Sec. 5.7

The word *statistics* originally came from words that mean *state numbers.* State numbers refer to numerical information, or *data,* gathered by the government such as the number of births, deaths, or marriages in a population. Today the word *statistics* has a much broader meaning; data from the fields of economics, social science, science, and business can all be organized and studied under the branch of mathematics called *statistics.*

OBJECTIVE 1 Making sense of a long list of numbers can be hard. So when you analyze data, one of the first things to look for is a *measure of central tendency*—a single number that you can use to represent the entire list of numbers. One such measure is the *average* or **mean.** The mean can be found with the following formula.

Finding the Mean (Average)

$$\text{mean} = \frac{\text{sum of all values}}{\text{number of values}}$$

EXAMPLE 1 Finding the Mean

David had test scores of 84, 90, 95, 98, and 88. Find his average or mean score.

Use the formula for finding mean. Add up all the test scores and then divide by the number of tests.

$$\text{mean} = \frac{84 + 90 + 95 + 98 + 88}{5} \quad \begin{array}{l}\leftarrow \text{Sum of test scores}\\ \leftarrow \text{Number of tests}\end{array}$$

$$= \frac{455}{5} \quad \text{Divide.}$$

$$= 91$$

David has a mean score of 91.

WORK PROBLEM 1 AT THE SIDE.

EXAMPLE 2 Applying the Average or Mean

The sales of photo albums at Sarah's Card Shop for each day last week were

\$86, \$103, \$118, \$117, \$126, \$158, and \$149.

Find the mean daily sales of photo albums. Add all the daily sales amounts and then divide by the number of days (7).

$$\text{mean} = \frac{\$86 + \$103 + \$118 + \$117 + \$126 + \$158 + \$149}{7}$$

$$= \frac{\$857}{7}$$

$$\approx \$122.43 \quad \text{(rounded to nearest cent)}$$

WORK PROBLEM 2 AT THE SIDE.

1. Tanya had test scores of 96, 98, 84, 88, 82, and 92. Find her average or mean score.

2. Find the mean for each list of numbers.

   (a) Monthly phone bills of \$25.12, \$42.58, \$76.19, \$32, \$81.11, \$26.41, \$19.76, \$59.32, \$71.18, and \$21.03.

   (b) A list of the sales for one year at eight different office supply stores:
   \$749,820; \$765,480
   \$643,744; \$824,222
   \$485,886; \$668,178
   \$702,294; \$525,800

ANSWERS

1. 90
2. (a) $\frac{\$454.70}{10} = \$45.47$
   (b) $\frac{\$5,365,424}{8} = \$670,678$

3. Alison Nakano works downtown. Some days she can park in cheap lots that charge \$6 or \$7. Other days she has to park in lots that charge \$9 or \$10. Last month she kept track of the amount she spent each day for parking and the number of days she spent that amount. Find her average daily parking cost.

Value	*Frequency*
\$ 6	2
\$ 7	6
\$ 8	3
\$ 9	4
\$10	6

ANSWERS

3. ≈\$8.29 (to nearest cent)

OBJECTIVE 2 Some items in a list of data might appear more than once. In this case, we find a **weighted mean,** in which each value is "weighted" by multiplying it by the number of times it occurs.

EXAMPLE 3 Understanding the Weighted Mean

The following table shows the amount of contribution and the number of times the amount was given (frequency) to a food pantry. Find the weighted mean.

Contribution Value	*Frequency*
\$ 3	4
\$ 5	2
\$ 7	1
\$ 8	5
\$ 9	3
\$10	2
\$12	1
\$13	2

The same amount was given by more than one person: for example, \$5 was given twice and \$8 was given five times. Other amounts, such as \$12, were given once. To find the mean, multiply each contribution value by its frequency. Then add the products. Next, add the numbers in the *frequency* column to find the total number of values.

Value	*Frequency*	*Product*
\$ 3	4	(3 • 4) = \$12
\$ 5	2	(5 • 2) = \$10
\$ 7	1	(7 • 1) = \$ 7
\$ 8	5	(8 • 5) = \$40
\$ 9	3	(9 • 3) = \$27
\$10	2	(10 • 2) = \$20
\$12	1	(12 • 1) = \$12
\$13	2	(13 • 2) = \$26
Totals	**20**	**\$154**

Finally, divide the totals.

$$\text{mean} = \frac{\$154}{20} = \$7.70.$$

The mean contribution to the food pantry was \$7.70.

WORK PROBLEM 3 AT THE SIDE.

A common use of the weighted mean is to find a student's *grade point average,* as shown by the next example.

EXAMPLE 4 Applying the Weighted Mean

Find the grade point average for a student earning the following grades. Assume A = 4, B = 3, C = 2, D = 1, and F = 0. The number of credits determines how many times the grade is counted (the frequency).

Course	*Credits*	*Grade*	*Credits • Grade*
Mathematics	3	A (= 4)	3 • 4 = 12
Speech	3	C (= 2)	3 • 2 = 6
English	3	B (= 3)	3 • 3 = 9
Computer Science	3	A (= 4)	3 • 4 = 12
Lab for Computer Science	2	D (= 1)	2 • 1 = 2
Totals	**14**		**41**

It is common to round grade point averages to the nearest hundredth. So the grade point average for this student is shown below.

$$\frac{41}{14} \approx 2.93$$

WORK PROBLEM 4 AT THE SIDE.

Objective 3 Because it can be affected by extremely high or low numbers, the mean is often a poor indicator of central tendency for a list of numbers. In cases like this, another measure of central tendency, called the **median** (MEE-dee-un), can be used. The *median* divides a group of numbers in half; half the numbers lie above the median, and half lie below the median.

Find the median by listing the numbers *in order* from *smallest* to *largest*. If the list contains an *odd* number of items, the median is the *middle number*.

EXAMPLE 5 Using the Median

Find the median for the following list of prices.

$7, $23, $15, $6, $18, $12, $24

First arrange the numbers in numerical order from smallest to largest.

Smallest → 6, 7, 12, 15, 18, 23, 24 ← Largest

Next, find the middle number in the list.

6, 7, 12, 15, 18, 23, 24

Three are below — Middle number (15) — Three are above

The median price is $15.

WORK PROBLEM 5 AT THE SIDE.

If a list contains an *even* number of items, there is no single middle number. In this case, the median is defined as the mean (average) of the *middle two* numbers.

4. Find the grade point average for a student earning the following grades. Round to the nearest hundredth.

Course	*Credits*	*Grade*
Mathematics	3	A (= 4)
P.E.	1	C (= 2)
English	3	C (= 2)
Keyboarding	3	B (= 3)
History	3	B (= 3)

5. Find the median for the following number of customers helped each hour at the order desk.
35, 33, 27, 31, 39, 50, 59, 25, 30

Answers

4. ≈2.92

5. 33 customers (the middle number when the numbers are arranged from smallest to largest)

6. Find the median for the following list of measurements. 178 ft, 261 ft, 126 ft, 189 ft, 121 ft, 195 ft

7. Find the mode for each list of numbers.

(a) Ages of part-time employees (in years): 28, 16, 22, 28, 34

(b) Total points on a screening exam of 312, 219, 782, 312, 219, 426

(c) Monthly commissions of sales people: $1706, $1289, $1653, $1892, $1301, $1782

ANSWERS

6. 183.5 ft

7. (a) 28 years
(b) bimodal, 219 points and 312 points (this list has two modes)
(c) no mode (no number occurs more than once)

EXAMPLE 6 Finding the Median

Find the median for the following list of ages.

74, 7, 15, 13, 25, 28, 47, 59, 32, 68

First arrange the numbers in numerical order. Then find the middle two numbers.

Smallest → 7, 13, 15, 25, 28, 32, 47, 59, 68, 74 ← Largest

Middle two numbers (28, 32)

The median age is the mean of these two numbers.

$$\text{median} = \frac{28 + 32}{2} = \frac{60}{2} = 30 \text{ years}$$

◀◀ WORK PROBLEM 6 AT THE SIDE.

OBJECTIVE 4 The last important statistical measure is the **mode,** the number that occurs most often in a list of numbers. For example, if the test scores for 10 students were

74, 81, 39, 74, 82, 80, 100, 92, 74, and 85

then the mode is 74. Three students earned a score of 74, so 74 appears more times on the list than any other score.

A list can have two modes; such a list is sometimes called *bimodal.* If no number occurs more frequently than any other number in a list, the list has *no mode.*

EXAMPLE 7 Finding the Mode

Find the mode for each list of numbers.

(a) 51, 32, 49, 73, 49, 90

The number 49 occurs more often than any other number; therefore, 49 is the mode. (It is not necessary to place the numbers in numerical order when looking for the mode.)

(b) 482, 485, 483, 485, 487, 487, 489

Because both 485 and 487 occur twice, each is a mode. This list is *bimodal.*

(c) 10,708; 11,519; 10,972; 12,546; 13,905; 12,182

No number occurs more than once. This list has *no mode.*

Measures of Central Tendency

The **mean** is the sum of all the values divided by the number of values. It is the mathematical average.

The **median** is the middle number in a group of values that are listed from smallest to largest. It divides a group of numbers in half.

The **mode** is the value that occurs most often in a group of values.

◀◀ WORK PROBLEM 7 AT THE SIDE.

5.7 Exercises

Find the mean for each list of numbers. Round answers to the nearest tenth, if necessary.

1. Ages of infants at the child care center (in months) of 4, 9, 6, 4, 7, 10, 9

2. Monthly phone bills of \$53, \$77, \$38, \$29, \$49, \$48

3. Final exam scores of 92, 51, 59, 86, 68, 73, 49, 80

4. Quiz scores of 18, 25, 21, 8, 16, 13, 23, 19

5. Annual salaries of \$21,900, \$22,850, \$24,930, \$29,710, \$28,340, \$40,000

6. Numbers of people attending baseball games: 27,500; 18,250; 17,357; 14,298; 33,110

Solve the following application problems.

7. The Athletic Shoe Store sold shoes at the following prices: \$75.52, \$36.15, \$58.24, \$21.86, \$47.68, \$106.57, \$82.72, \$52.14, \$28.60, \$72.92. Find the average (mean) shoe sales amount.

8. In one evening, a waitress collected the following checks from her dinner customers: \$30.10, \$42.80, \$91.60, \$51.20, \$88.30, \$21.90, \$43.70, \$51.20. Find the average (mean) dinner check amount.

9. The following table shows the face value (policy amount) of life insurance policies sold and the number of policies sold for each amount by the New World Life Company during one week. Find the weighted mean amount for the policies sold.

Policy Amount	*Number of Policies Sold*
$10,000	6
$20,000	24
$25,000	12
$30,000	8
$50,000	5
$100,000	3
$250,000	2

10. Detroit Metro-Sales Company prepared the following table showing the gasoline mileage obtained by each of the cars in their automobile fleet. Find the weighted mean to determine the miles per gallon for the fleet of cars.

Miles per Gallon	*Number of Autos*
15	5
20	6
24	10
30	14
32	5
35	6
40	4

Find the weighted mean. Round answers to the nearest tenth, if necessary.

11.

Quiz Scores	*Frequency*
3	4
5	2
9	1
12	3

12.

Credits per Student	*Frequency*
9	3
12	5
15	1
18	1

13.

Hours Worked	*Frequency*
12	4
13	2
15	5
19	3
22	1
23	5

14.

Students per Class	*Frequency*
25	1
26	2
29	5
30	4
32	3
33	5

Find the median for the following lists of numbers.

15. Number of voice mail messages received:
9, 12, 14, 15, 23, 24, 28

16. Deliveries by a newspaper distributor: 99, 108, 109, 123, 126, 129, 146, 168, 170

17. Customers served each day:
328, 549, 420, 592, 715, 483

18. Number of cars in the parking lot each day:
520, 523, 513, 1283, 338, 509, 290, 420

19. The number of computer service calls taken each day:
51, 48, 96, 40, 47, 23, 95, 56, 34, 48

20. Number of gallons of paint sold per week:
1072, 1068, 1093, 1042, 1056, 205, 1009, 1081

Find the mode or modes for each list of numbers.

21. Number of samples taken each hour:
3, 8, 5, 1, 7, 6, 8, 4, 5, 8

22. Water bills of
\$21, \$32, \$46, \$32, \$49, \$32, \$49, \$25

23. Ages of retirees (in years) at the village:
74, 68, 68, 68, 75, 75, 74, 74, 70, 77

24. Tires balanced by different employees:
30, 19, 25, 78, 36, 20, 45, 85, 38

25. The number of boxes of candy sold by each child:
5, 9, 17, 3, 2, 8, 19, 1, 4, 20, 10, 6

26. The weights of soccer players (in pounds):
158, 161, 165, 162, 165, 157, 163, 162

27. Find the mean and the median for the home values listed below. Explain which measure is more appropriate and why.

\$82,000	\$115,000
\$64,000	\$ 97,000
\$91,000	\$892,000

28. Suppose you own a hat shop and can order a certain hat in only one size. You look at last year's sales to decide on the size to order. Should you find the mean, median, or mode for these sales? Explain your answer.

Ⓖ *Find the grade point average for students earning the following grades. Assume* A = 4, B = 3, C = 2, D = 1, *and* F = 0. *Round answers to the nearest hundredth.*

29.

Course	*Credits*	*Grade*
Biology	4	B
Biology Lab	2	A
Math	5	C
Health	1	F
Psychology	3	B

30.

Course	*Credits*	*Grade*
Chemistry	3	A
English	3	B
Math	4	B
Theater	2	C
Astronomy	3	C

Ⓖ **31.** Look again at the grades in Exercise 29. Find the student's grade point average in each of these situations.

(a) The student earned a B instead of an F in the 1-credit class.

(b) The student earned a B instead of a C in the 5-credit class.

(c) Both (a) and (b) happened.

Ⓖ **32.** List the credits for the courses you're taking at this time. List the lowest grades you think you will earn in each class and find your grade point average. Then list the highest grades you think you will earn and find your grade point average.

Review and Prepare

Simplify. (For help, see **Section 1.8.***)*

33. 8^2

34. 5^2

35. 15^2

36. 21^2

37. $3^2 + 9^2$

38. $10^2 + 4^2$

39. $12^2 - 11^2$

40. $16^2 - 14^2$

5.8 Geometry Applications: Pythagorean Theorem

In **Section 3.2** you used this formula for area of a square, $A = s^2$. The square on the left has an area of 25 cm^2 because 5 cm • 5 cm = 25 cm^2.

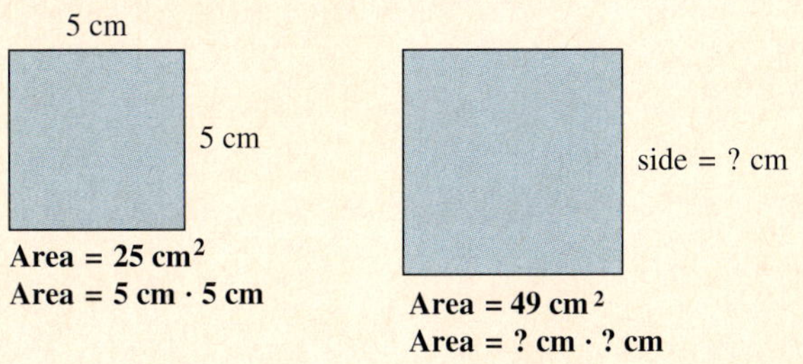

The square on the right has an area of 49 cm^2. To find the length of a side, ask yourself, "What number can be multiplied by itself to give 49?" Because 7 • 7 = 49, the length of each side is 7 cm. Also, because 7 • 7 = 49 we say that 7 is the *square root* of 49, or $\sqrt{49} = 7$.

Square Root

The **square root** of a positive number is one of two identical positive factors of that number. For example, $\sqrt{36} = 6$ because 6 • 6 = 36.

WORK PROBLEM 1 AT THE SIDE.

A number that has a whole number as its square root is called a *perfect square.* For example, 9 is a perfect square because $\sqrt{9} = 3$, and 3 is a whole number. The first few perfect squares are listed below.

$\sqrt{1} = 1$	$\sqrt{16} = 4$	$\sqrt{49} = 7$	$\sqrt{100} = 10$
$\sqrt{4} = 2$	$\sqrt{25} = 5$	$\sqrt{64} = 8$	$\sqrt{121} = 11$
$\sqrt{9} = 3$	$\sqrt{36} = 6$	$\sqrt{81} = 9$	$\sqrt{144} = 12$

OBJECTIVE 1 If a number is *not* a perfect square, then you can find its *approximate* square root by using a calculator with a square root key.

Calculator Tip: To find a square root, use the [√] key or the [√x] key. You do *not* need to use the [=] key. Try these. Jot down your answers.

To find $\sqrt{16}$ press: 16 [√x] Answer is 4.

To find $\sqrt{7}$ press: 7 [√x]

For $\sqrt{7}$, your calculator shows 2.645751311 which is an *approximate* answer. We will be rounding to the nearest thousandth so $\sqrt{7} \approx 2.646$. To check, multiply 2.646 times 2.646. Do you get 7 as the result? No, you get 7.001316 which is very close to 7. The difference is due to rounding.

EXAMPLE 1 Finding the Square Root of a Number

Use a calculator to find each square root. Round your answers to the nearest thousandth.

(a) $\sqrt{35}$ Calculator shows 5.916079783; round to 5.916

(b) $\sqrt{124}$ Calculator shows 11.13552873; round to 11.136

OBJECTIVES

1. Find square roots using the square root key on a calculator.
2. Find the unknown length in a right triangle.
3. Solve application problems involving right triangles.

FOR EXTRA HELP

Tutorial Tape 9 SSM, Sec. 5.8

1. Find each square root.

(a) $\sqrt{36}$

(b) $\sqrt{25}$

(c) $\sqrt{9}$

(d) $\sqrt{100}$

(e) $\sqrt{121}$

ANSWERS

1. (a) 6 **(b)** 5 **(c)** 3 **(d)** 10 **(e)** 11

2. Use a calculator with a square root key to find each square root. Round to the nearest thousandth if necessary.

(a) $\sqrt{11}$

(b) $\sqrt{40}$

(c) $\sqrt{56}$

(d) $\sqrt{196}$

(e) $\sqrt{147}$

ANSWERS

2. (a) ≈ 3.317 **(b)** ≈ 6.325 **(c)** ≈ 7.483 **(d)** 14 **(e)** ≈ 12.124

◀◀ WORK PROBLEM 2 AT THE SIDE.

OBJECTIVE 2 One place you will use square roots is when working with the *Pythagorean Theorem.* This theorem applies only to *right* triangles (triangles with a 90° angle). The longest side of a right triangle is called the **hypotenuse** (hy-POT-en-oos). It is opposite the right angle. The other two sides are called *legs.* The legs form the right angle.

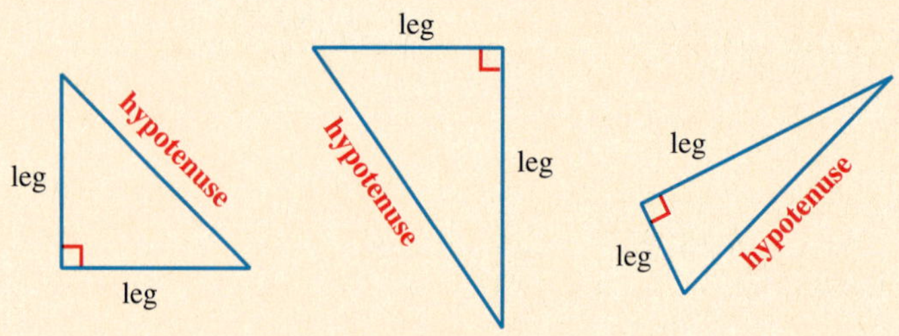

Examples of right triangles

Pythagorean Theorem

$$(\text{hypotenuse})^2 = (\text{leg})^2 + (\text{leg})^2$$

In other words, square the length of each side. After you have squared all the sides, the sum of the squares of the two legs will equal the square of the hypotenuse.

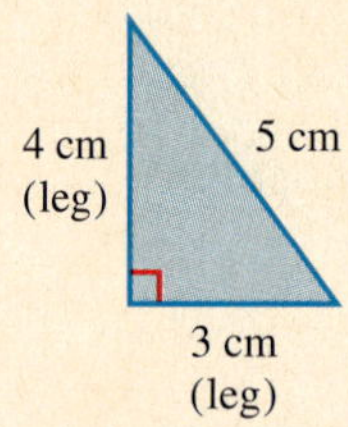

$$(\text{hypotenuse})^2 = (\text{leg})^2 + (\text{leg})^2$$
$$5^2 = 4^2 + 3^2$$
$$25 = 16 + 9$$
$$25 = 25$$

The theorem is named after Pythagoras, a Greek mathematician who lived about 2500 years ago. He and his followers may have used floor tiles to prove the theorem, as shown here.

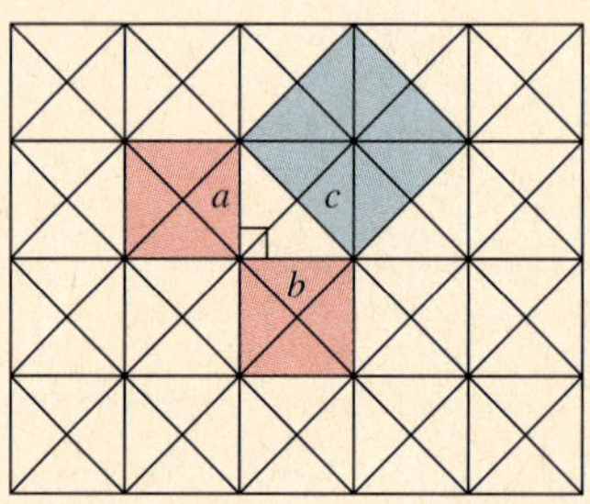

The right triangle in the center of the drawing has sides a, b, and c. The square drawn on side a contains four triangles. The square on side b contains four triangles. The square on side c contains eight triangles. The number of triangles in the square on side c equals the sum of the number of triangles in the squares on sides a and b, that is, 8 triangles = 4 triangles + 4 triangles. As a result, you often see the Pythagorean Theorem written as $c^2 = a^2 + b^2$.

If you know the lengths of any two sides in a right triangle, you can use the Pythagorean Theorem to find the length of the third side.

Using the Pythagorean Theorem

To find the hypotenuse, use this formula:

$$\textbf{hypotenuse} = \sqrt{(\textbf{leg})^2 + (\textbf{leg})^2}$$

To find a leg, use this formula:

$$\textbf{leg} = \sqrt{(\textbf{hypotenuse})^2 - (\textbf{leg})^2}$$

Note

Remember: A small square drawn in one angle of a triangle indicates a right angle. You can use the Pythagorean Theorem *only* on triangles that have a right angle.

EXAMPLE 2 Finding the Unknown Length in a Right Triangle

Find the unknown length in each right triangle.

(a)

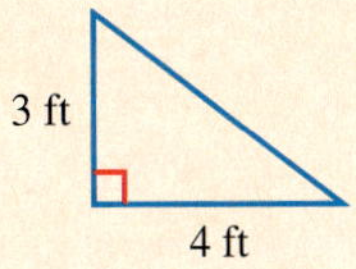

The length of the side opposite the right angle is unknown. That side is the hypotenuse, so use this formula.

$\text{hypotenuse} = \sqrt{(\text{leg})^2 + (\text{leg})^2}$	Find the hypotenuse.
$\text{hypotenuse} = \sqrt{(3)^2 + (4)^2}$	Legs are 3 and 4.
$= \sqrt{9 + 16}$	3 • 3 is 9 and 4 • 4 is 16
$= \sqrt{25}$	
$= 5$	

The hypotenuse is 5 ft long.

(b)

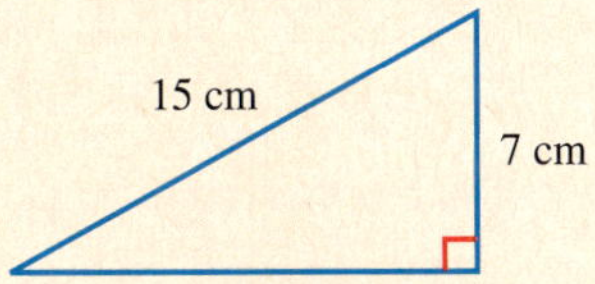

You *do* know the length of the hypotenuse (15 cm), so it is the length of one of the legs that is unknown. Use this formula.

$\text{leg} = \sqrt{(\text{hypotenuse})^2 - (\text{leg})^2}$	Find a leg.
$\text{leg} = \sqrt{(15)^2 - (7)^2}$	Hypotenuse is 15, one leg is 7.
$= \sqrt{225 - 49}$	15 • 15 is 225 and 7 • 7 is 49
$= \sqrt{176}$	Use calculator to find $\sqrt{176}$.
≈ 13.266	Round 13.26649916 to 13.266.

The length of the leg is approximately 13.266 cm.

Note

You use the Pythagorean Theorem to find the *length* of one side, *not* the area of the triangle. Your answer will be in linear units, such as ft, yd, cm, m, and so on (*not* ft^2, cm^2, m^2).

WORK PROBLEM 3 AT THE SIDE.

3. Find the unknown length in each right triangle. Round your answers to the nearest thousandth, if necessary.

(a)

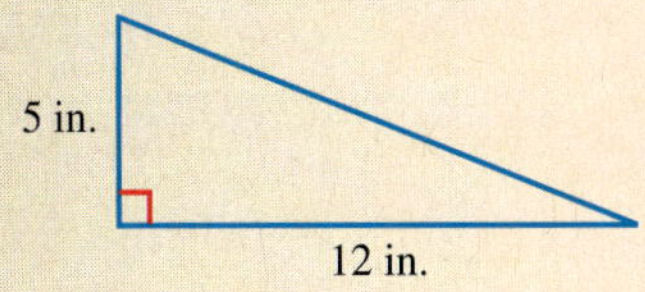

(b)

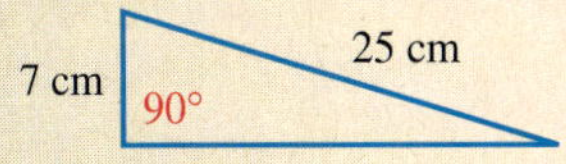

(c)

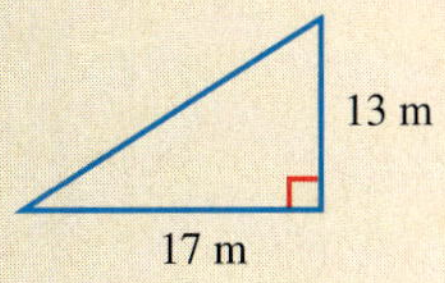

(d)

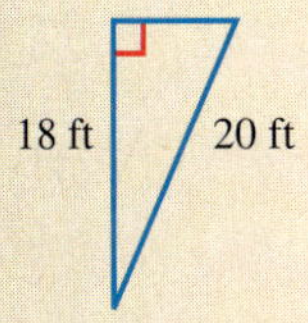

(e)

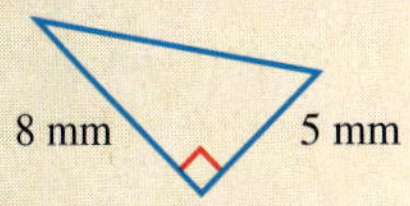

ANSWERS

3. **(a)** 13 in. **(b)** 24 cm **(c)** ≈21.401 m **(d)** ≈8.718 ft **(e)** ≈9.434 mm

4. These problems show ladders leaning against buildings. Find the unknown lengths. Round to the nearest thousandth of a foot, if necessary.

(a)

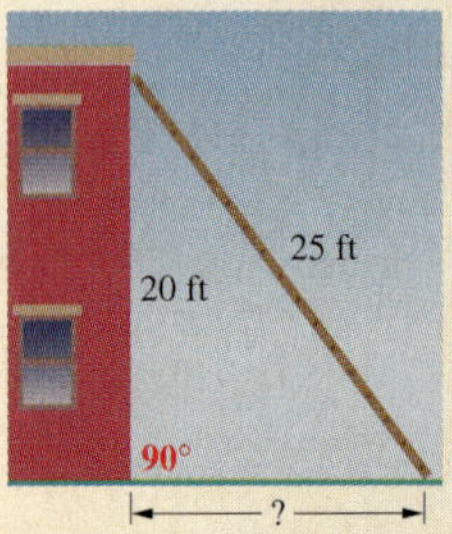

How far away from the building is the bottom of the ladder?

(b)

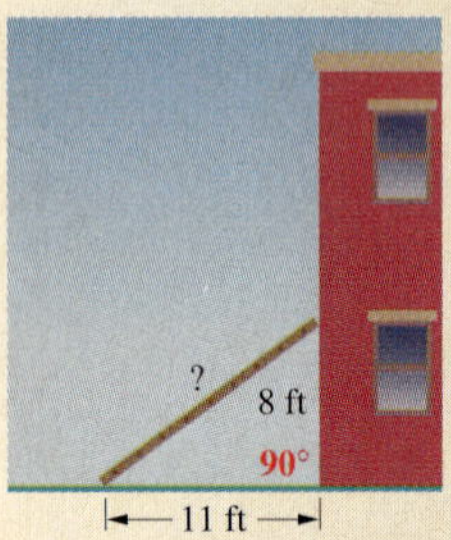

How long is the ladder?

(c) A 17-foot ladder is leaning against a building. The bottom of the ladder is 10 ft from the building. How high up on the building will the ladder reach? (*Hint:* Start by drawing the building and the ladder.)

ANSWERS

4. (a) $\sqrt{225} = 15$ ft
(b) $\sqrt{185} \approx 13.601$ ft
(c) $\sqrt{189} \approx 13.748$ ft

OBJECTIVE 3 The next example shows an application of the Pythagorean Theorem.

EXAMPLE 3 Using the Pythagorean Theorem

A television antenna is on the roof of a house, as shown. Find the length of the support wire.

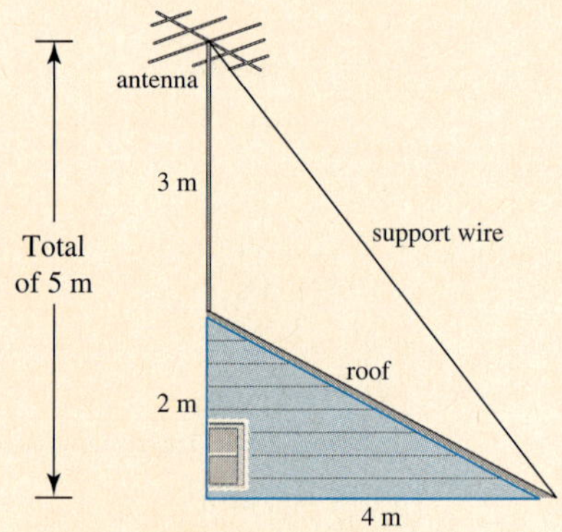

A right triangle is formed. The total length of the side at the left is 5 m.

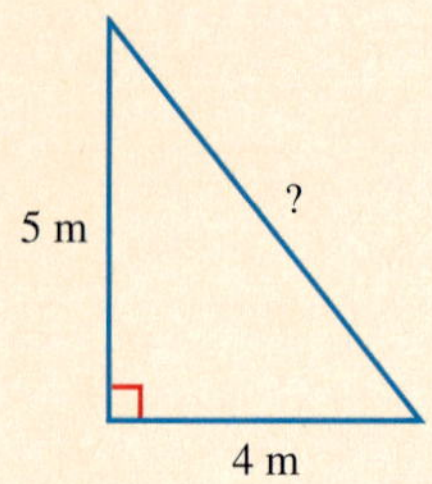

The support wire is the hypotenuse of the right triangle.

$$\text{hypotenuse} = \sqrt{(\text{leg})^2 + (\text{leg})^2} \quad \text{Find the hypotenuse.}$$
$$\text{hypotenuse} = \sqrt{(5)^2 + (4)^2} \quad \text{Legs are 5 and 4.}$$
$$= \sqrt{25 + 16} \quad 5^2 \text{ is 25 and } 4^2 \text{ is 16.}$$
$$= \sqrt{41} \quad \text{Use } \sqrt{x} \text{ key on a calculator.}$$
$$\approx 6.403 \quad \text{Round 6.403124237 to 6.403.}$$

The length of the support wire is ≈ 6.403 m.

◀◀ WORK PROBLEM 4 AT THE SIDE.

5.8 Exercises

Find each square root. Starting with Exercise 5, use the square root key on a calculator. Round your answers to the nearest thousandth, when necessary.

1. $\sqrt{16}$ **2.** $\sqrt{4}$ **3.** $\sqrt{64}$ **4.** $\sqrt{81}$

5. $\sqrt{11}$ **6.** $\sqrt{23}$ **7.** $\sqrt{5}$ **8.** $\sqrt{2}$

9. $\sqrt{73}$ **10.** $\sqrt{80}$ **11.** $\sqrt{101}$ **12.** $\sqrt{125}$

13. $\sqrt{190}$ **14.** $\sqrt{160}$ **15.** $\sqrt{1000}$ **16.** $\sqrt{2000}$

G **17.** You know that $\sqrt{25} = 5$ and $\sqrt{36} = 6$. Using just that information (no calculator), describe how you could estimate $\sqrt{30}$. How would you estimate $\sqrt{26}$ or $\sqrt{35}$? Now check your estimates using a calculator.

G **18.** Explain the relationship between squaring a number and finding the square root of a number.

Find the unknown length in each right triangle. Use a calculator to find square roots. Round your answers to the nearest thousandth, if necessary.

19.

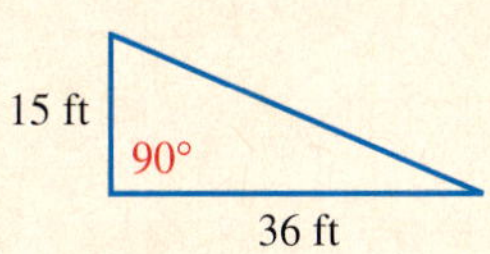

20.

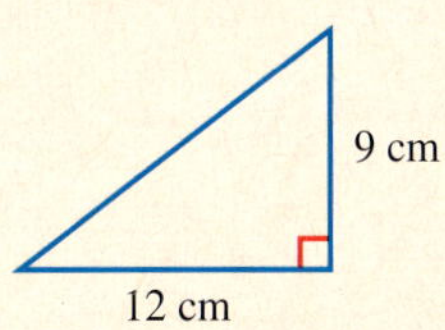

21.

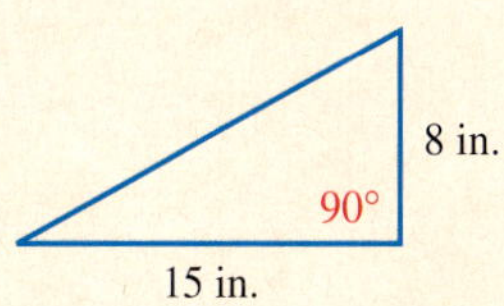

22.

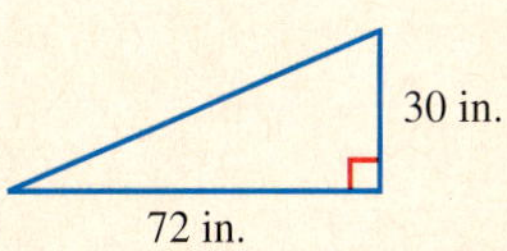

23.

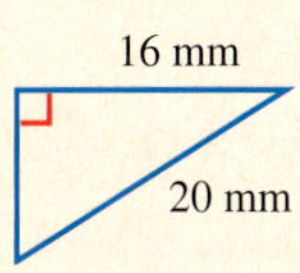

24.

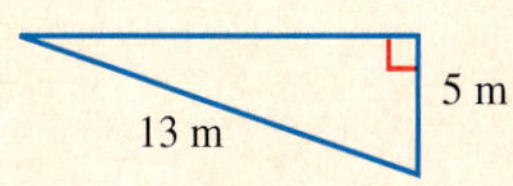

Writing Calculator G Small Group

25.

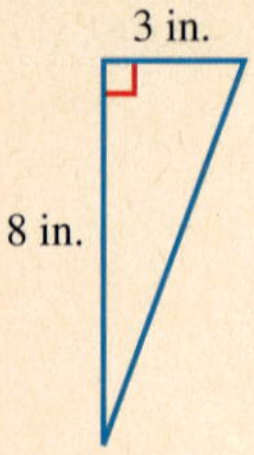

26.

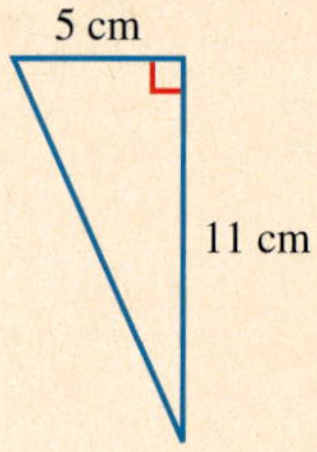

27.

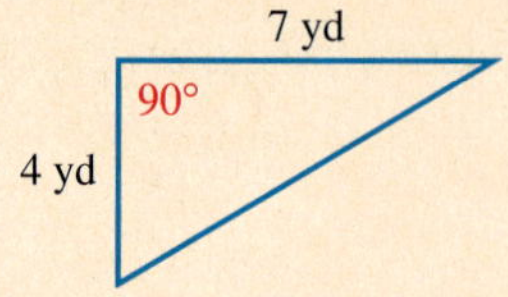

28.

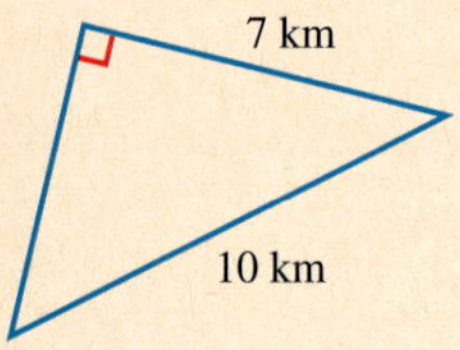

29.

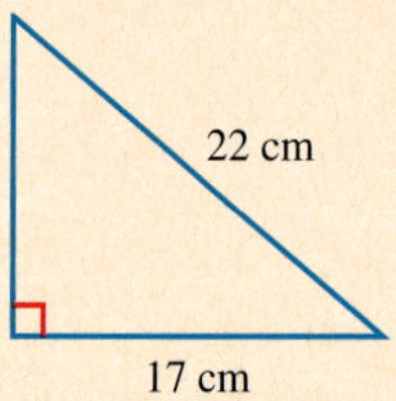

30.

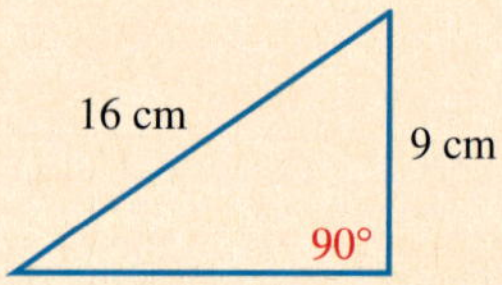

31.

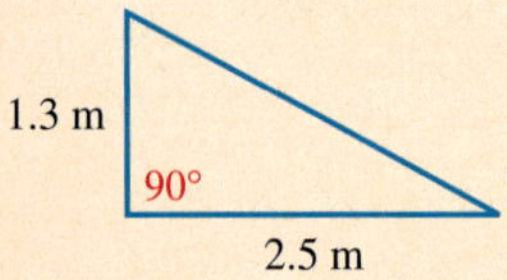

32.

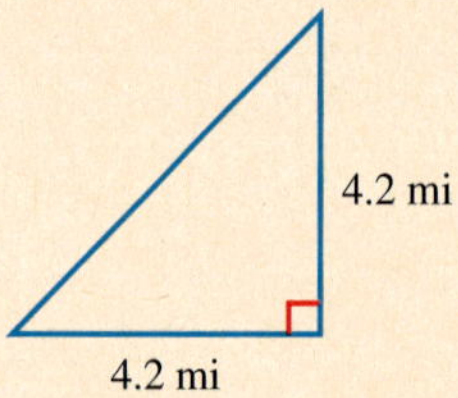

33.

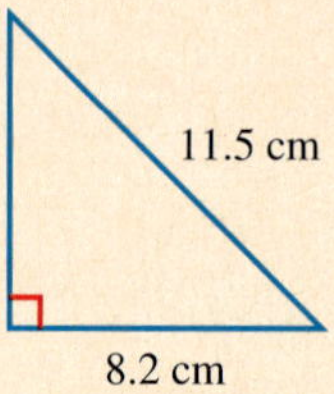

34.

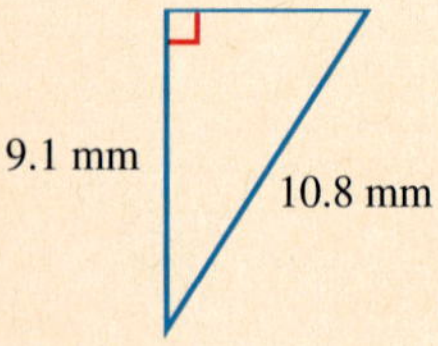

35.

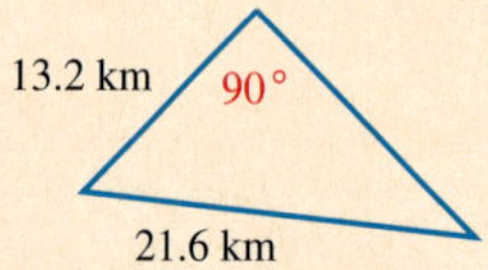

36.

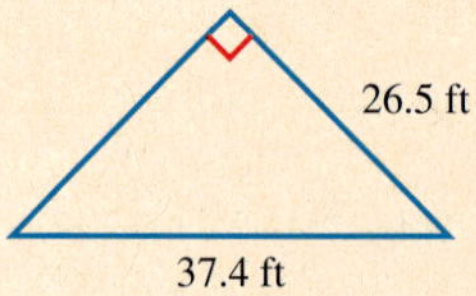

Solve each application problem. Round your answers to the nearest tenth when necessary.

37. Find the length of this loading ramp.

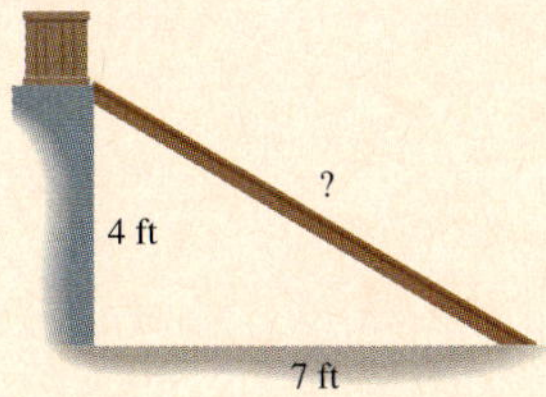

38. Find the unknown length in this roof plan.

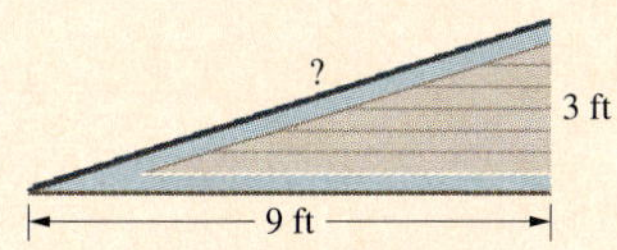

39. How high is the airplane above the ground?

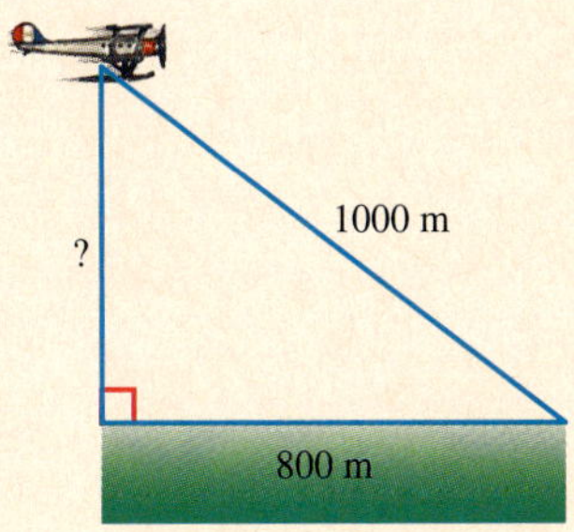

40. Find the height of this farm silo.

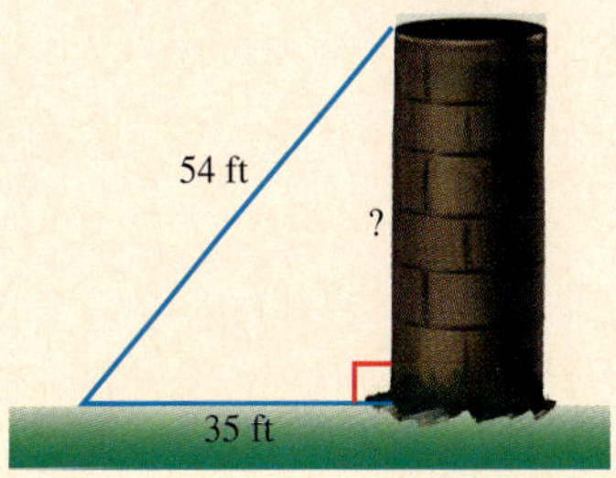

41. To reach his lady-love, a knight placed a 12-foot ladder against the castle wall. If the base of the ladder is 3 feet from the building, how high on the castle will the top of the ladder reach? Draw a sketch of the castle and ladder and solve the problem.

42. William drove his car 15 miles north, then made a right turn and drove 7 miles east. How far is he, in a straight line, from his starting point? Draw a sketch to illustrate the problem and solve it.

43. Explain the two errors made by a student in solving this problem. Also find the correct answer. Round to the nearest tenth.

$$? = \sqrt{(13)^2 + (20)^2}$$
$$= \sqrt{169 + 400}$$
$$= \sqrt{569} \approx 23.9 \text{ m}^2$$

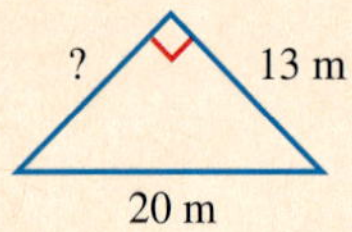

44. Explain the two errors made by a student in solving this problem. Also find the correct answer. Round to the nearest tenth.

$$? = \sqrt{(9)^2 + (7)^2}$$
$$= \sqrt{18 + 14}$$
$$= \sqrt{32} \approx 5.657 \text{ in.}$$

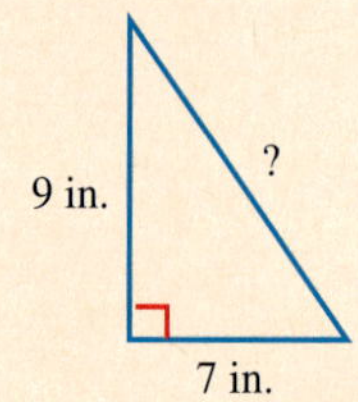

45. Find the lengths of $\overline{BC}$ and $\overline{BD}$.

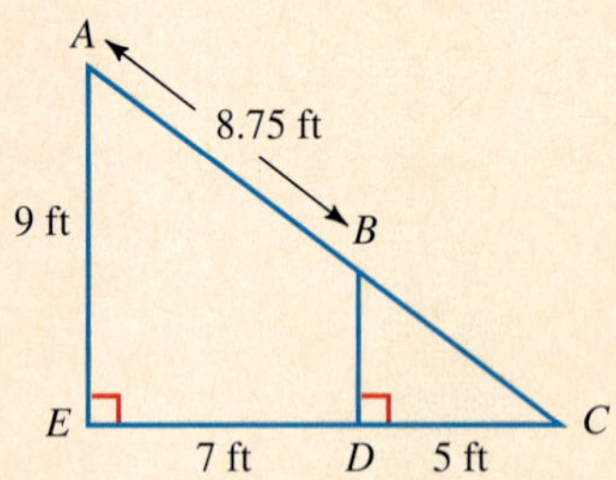

46. Find the lengths of $\overline{CD}$ and $\overline{DB}$. Round your answers to the nearest tenth.

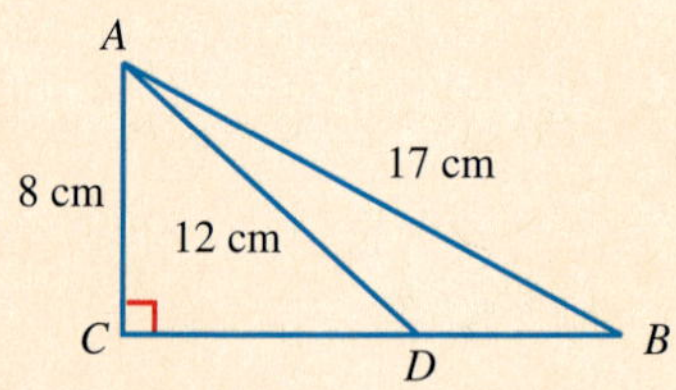

Review and Prepare

Solve each equation. Show the steps you used. (For help, see ***Sections 2.3, 2.4, and 2.5.****)*

47. $x - 18 = -30$

48. $10 = x + 15$

49. $48 = -8b + 16$

50. $3h = 33 - 6$

51. $3y - 5 = y + 3$

52. $-4k + 6 = 2k - 6$

5.9 Problem Solving: Equations Containing Decimals

OBJECTIVES

1. Solve equations containing decimals by using the addition property of equality.
2. Solve equations containing decimals by using the division property of equality.
3. Solve equations containing decimals by using both properties of equality.

FOR EXTRA HELP

Tutorial | Tape 9 | SSM, Sec. 5.9

OBJECTIVE 1 In **Section 2.3** you used the addition property of equality to solve an equation like $c + 5 = 30$. The addition property says that you can add the same number to both sides of an equation and still keep it balanced. You can also use this property when an equation contains decimal numbers.

EXAMPLE 1 Using the Addition Property of Equality

Solve each equation and check each solution.

(a) $w + 2.9 = -0.6$

The first step is to get the variable term (w) by itself on the left side of the equal sign. Use the addition property to "get rid of" the 2.9 on the left side by adding its opposite, -2.9, to both sides.

$$\begin{aligned} w + 2.9 &= -0.6 \\ -2.9 \quad &\quad -2.9 \qquad \text{Add } -2.9 \text{ to both sides.} \\ \underbrace{w + 0}_{} &= -3.5 \\ w &= -3.5 \end{aligned}$$

The solution is -3.5. To check the solution, go back to the original equation.

Check

$$\begin{aligned} w + 2.9 &= -0.6 \qquad \text{Replace } w \text{ with } -3.5. \\ \underbrace{-3.5 + 2.9}_{} &= -0.6 \\ -0.6 &= -0.6 \qquad \text{Balances, so } -3.5 \text{ is the correct solution.} \end{aligned}$$

(b) $7 = -4.3 + x$

To get x by itself on the right side of the equal sign, add 4.3 to both sides.

$$\begin{aligned} 7 &= -4.3 + x \\ +4.3 \quad &\quad +4.3 \qquad \text{Add 4.3 to both sides.} \\ 11.3 &= \underbrace{0 + x}_{} \\ 11.3 &= x \end{aligned}$$

The solution is 11.3. To check the solution, go back to the original equation.

Check

$$\begin{aligned} 7 &= -4.3 + x \qquad \text{Replace } x \text{ with } 11.3. \\ 7 &= \underbrace{-4.3 + 11.3}_{} \\ 7 &= 7 \qquad \text{Balances, so 11.3 is the correct solution.} \end{aligned}$$

WORK PROBLEM 1 AT THE SIDE. ▶▶

1. Solve each equation and check each solution.

(a) $8.1 = h + 9$ **Check**

(b) $-0.75 + y = 0$ **Check**

(c) $c - 6.8 = -4.8$ **Check**

ANSWERS

1. (a) $h = -0.9$ **Check** $8.1 = h + 9$; $8.1 = \underbrace{-0.9 + 9}$; Balances $8.1 = 8.1$

(b) $y = 0.75$ **Check** $-0.75 + y = 0$; $\underbrace{-0.75 + 0.75} = 0$; Balances $0 = 0$

(c) $c = 2$ **Check** $c - 6.8 = -4.8$; $\underbrace{2 - 6.8} = -4.8$; Balances $-4.8 = -4.8$

2. Solve each equation and check each solution.

(a) $-3y = -0.63$ **Check**

(b) $2.25r = -18$ **Check**

(c) $1.7 = 0.5n$ **Check**

OBJECTIVE 2 You can also use the division property of equality (from **Section 2.4**) when an equation contains decimals.

EXAMPLE 2 Using the Division Property of Equality

Solve each equation and check each solution.

(a) $5x = 12.4$

On the left side of the equation the variable is multiplied by 5. To undo the multiplication, divide both sides by 5.

$$5x = 12.4 \quad \text{Divide both sides by 5.}$$

5x means 5 • x.

$$\frac{5 \cdot x}{5} = \frac{12.4}{5} \quad \text{On the right side, } 12.4 \div 5 \text{ is } 2.48.$$

Divide out the common factor of 5.

$$\frac{\overset{1}{\cancel{5}} \cdot x}{\underset{1}{\cancel{5}}} = 2.48$$

$$x = 2.48$$

The solution is 2.48. To check the solution, go back to the original equation.

Check

$$5x = 12.4 \quad \text{Replace } x \text{ with } 2.48.$$

$$5(2.48) = 12.4$$

$$12.4 = 12.4 \quad \text{Balances, so 2.48 is the correct solution.}$$

(b) $-9.3 = 1.5t$

$$\frac{-9.3}{1.5} = \frac{\overset{1}{\cancel{1.5}}\,t}{\underset{1}{\cancel{1.5}}} \quad \text{Divide both sides by the coefficient of the variable term, 1.5.}$$

$$-6.2 = t$$

Signs are different, so quotient is negative.

The solution is -6.2. To check the solution, go back to the original equation.

Check

$$-9.3 = 1.5\,t \quad \text{Replace } t \text{ with } -6.2.$$

$$-9.3 = 1.5\,(-6.2)$$

$$-9.3 = -9.3 \quad \text{Balances, so } -6.2 \text{ is the correct solution.}$$

◀◀ **WORK PROBLEM 2 AT THE SIDE.**

OBJECTIVE 3 Sometimes you need to use both the addition and division properties to solve an equation.

ANSWERS

2. (a) $y = 0.21$ **Check** $-3y = -0.63$; $-3(0.21) = -0.63$; Balances $-0.63 = -0.63$

(b) $r = -8$ **Check** $2.25r = -18$; $2.25(-8) = -18$; Balances $-18 = -18$

(c) $n = 3.4$ **Check** $1.7 = 0.5n$; $1.7 = 0.5(3.4)$; Balances $1.7 = 1.7$

EXAMPLE 3 Solving Equations with Several Steps

(a) $2.5b + 0.35 = -2.65$

The first step is to get the variable term $2.5b$ by itself on the left side of the equal sign.

$$\begin{aligned} 2.5b + 0.35 &= -2.65 \\ -0.35 \quad & \quad -0.35 \quad \text{Add } -0.35 \text{ to both sides.} \\ 2.5b + 0 &= -3.00 \\ 2.5b &= -3 \end{aligned}$$

The next step is to divide both sides by the coefficient of the variable term. In $2.5b$, the coefficient is 2.5.

$$\frac{\overset{1}{\cancel{2.5}}b}{\underset{1}{\cancel{2.5}}} = \frac{-3}{2.5} \quad \text{On the right side, signs do not match, so the quotient is negative.}$$

$$b = -1.2$$

The solution is -1.2. To check the solution, go back to the original equation.

Check

$$\begin{aligned} 2.5b + 0.35 &= -2.65 \quad \text{Replace } b \text{ with } -1.2. \\ 2.5(-1.2) + 0.35 &= -2.65 \\ -3 + 0.35 &= -2.65 \\ -2.65 &= -2.65 \quad \text{Balances, so } -1.2 \text{ is the correct solution.} \end{aligned}$$

(b) $5x - 0.98 = 2x + 0.4$

There is a variable term on both sides of the equation. You can choose to keep the variable term on the left side or to keep the variable term on the right side. Either way will work. Just pick the left side or the right side.

Suppose that you decide to keep the variable term $5x$ on the left side. Use the addition property to "get rid off" $2x$ on the right side by adding its opposite, $-2x$, to both sides.

$$\begin{aligned} 5x - 0.98 &= 2x + 0.4 \\ -2x \quad & \quad -2x \quad \text{Add } -2x \text{ to both sides.} \\ 3x - 0.98 &= 0 + 0.4 \\ 3x + (-0.98) &= 0.4 \\ +0.98 \quad & \quad +0.98 \quad \text{Add 0.98 to both sides.} \\ 3x + 0 &= 1.38 \\ \frac{\overset{1}{\cancel{3}}x}{\underset{1}{\cancel{3}}} &= \frac{1.38}{3} \quad \text{Divide both sides by 3.} \\ x &= 0.46 \end{aligned}$$

The solution is 0.46. To check the solution, go back to the original equation.

Check

$$\begin{aligned} 5x - 0.98 &= 2x + 0.4 \quad \text{Replace } x \text{ with 0.46.} \\ 5(0.46) - 0.98 &= 2(0.46) + 0.4 \\ 2.3 - 0.98 &= 0.92 + 0.4 \\ 1.32 &= 1.32 \quad \text{Balances, so 0.46 is the correct solution.} \end{aligned}$$

WORK PROBLEM 3 AT THE SIDE.

3. Solve each equation and check each solution.

(a) $4 = 0.2c - 2.6$ **Check**

(b) $3.1k - 4 = 0.5k + 13.42$ **Check**

(c) $-2y + 3 = 3y - 6$ **Check**

ANSWERS

3. (a) $c = 33$ **Check** $4 = 0.2c - 2.6$; $4 = 0.2(33) - 2.6$; $4 = 6.6 - 2.6$; Balances $4 = 4$

(b) $k = 6.7$ **Check** $3.1k - 4 = 0.5k + 13.42$; $3.1(6.7) - 4 = 0.5(6.7) + 13.42$; $20.77 - 4 = 3.35 + 13.42$; Balances $16.77 = 16.77$

(c) $y = 1.8$ **Check** $-2y + 3 = 3y - 6$; $-2(1.8) + 3 = 3(1.8) - 6$; $-3.6 + 3 = 5.4 - 6$; Balances $-0.6 = -0.6$

NUMBERS IN THE *Real World*

collaborative investigations

Gotta Be Green

A lot's being said about personal responsibility these days, and the idea seems to be ending up on the front lawn—literally! Each spring, homeowners across the country gear up to green up their lawns, and the increased use of fertilizer has a lot of environmentalists concerned about the potential effects of chemical runoff into nearby rivers and streams.

Each year, according to a study conducted by the University of Minnesota's Department of Agriculture, each household in the Minneapolis/St. Paul metro area uses an average of 36 pounds of lawn fertilizer. That adds up to 25,529,295 pounds, or 12,765 tons. Add to that another 193,000 pounds of weed killer and you're looking at the total picture for keeping it green in the Twin Cities.

Source: Minneapolis Star Tribune

1. (a) According to the article, how many pounds of lawn fertilizer are used each year in the *entire metro area*?

(b) Use division to find the number of *households* in the metro area.

(c) Would it make sense to round your answer? If so, how would you round it?

2. (a) There are 2000 pounds in one ton. Find the number of tons equivalent to 25,529,295 pounds.

(b) Does your answer match the figure given in the article? If not, what did the writer of the article do to get 12,765 tons?

(c) Is the author's figure accurate? Why or why not?

3. (a) The article states that "each household in the Minneapolis/St. Paul metro area uses an average of 36 pounds of lawn fertilizer" each year. What mathematical operations do you do to find an average?

(b) When the calculations were done to find the average, the answer probably was not *exactly* 36 pounds. List six different answers that are *less than* 36 that would round to 36. List two answers that have one decimal place, two answers with two decimal places, and two answers with three decimal places.

(c) List six answers that are *greater than* 36 that would round to 36. Use one, two, and three decimal places.

(d) What is the *smallest* number that rounds to 36? The *largest* number?

4. (a) Use the memory keys on your calculator to find the average number of pounds of *weed killer* used by each household each year. Start by dividing to find the number of households, as you did in Exercise 1 above. If you have a scientific calculator, pressing the [STO] and [1] keys stores the answer in your calculator's first memory.

25,529,925 [÷] 36 [=] [STO] [1] (The calculator display will show 709147.0833 and a small M1 in the corner to show that the number is stored in the first memory.)

Now enter the number of pounds of weed killer. To divide by the number of households, use the [RCL] key to recall the number of households without having to enter the number again.

193,000 [÷] [RCL] [1] [=] What is your answer?

(b) How will you round your answer? Why?

Name Date Class

5.9 Exercises

Solve each equation and check each solution.

1. $h + 0.63 = 5.1$ **Check**

2. $-0.2 = k - 0.7$ **Check**

3. $-20.6 + n = -22$ **Check**

4. $g - 5 = 6.03$ **Check**

5. $0 = b - 0.008$ **Check**

6. $0.18 + m = -4.5$ **Check**

7. $2.03 = 7a$ **Check**

8. $-6.2c = 0$ **Check**

9. $0.8p = -96$ **Check**

10. $-10.16 = -4r$ **Check**

11. $-3.3t = -2.31$ **Check**

12. $8.3w = -49.8$ **Check**

13. $7.5x + 0.15 = -6$ **Check**

14. $0.8 = 0.2y + 3.4$ **Check**

15. $-7.38 = 2.05z - 7.38$ **Check**

16. $6.2h - 0.4 = 2.7$ **Check**

Solve each equation.

17. $2k + 2.1 = -1.5k$

18. $4n = 0.75 + 3.5n$

19. $3c + 10 = 6c + 8.65$

20. $2.1b + 5 = 1.6b + 10$

21. $8w - 6.4 = -6.4 + 5w$

22. $7r + 9.64 = -2.32 + 5r$

23. $-10.9 + 0.5p = 0.9p + 5.3$

24. $0.7x - 4.38 = x - 2.16$

25. $2(3y + 3.2) = 4y - 7.6$

26. $6a + 21 = 7(a + 8.6)$

27. Explain and correct the error in this student's work.

$$
\begin{array}{rcr}
-4a - 6.3 & = & 7.2 \\
+\ 6.3 & & +\ 6.3 \\
\hline
\underbrace{-4a + 0} & = & 13.5 \\
\dfrac{-4a}{4} & = & \dfrac{13.5}{4} \\
a & = & 3.375
\end{array}
$$

28. Explain and correct the error in this student's work.

$$
\begin{array}{rcr}
-0.3b - 6 & = & -2.4 \\
-6 & & -6 \\
\hline
\underbrace{-0.3b + 0} & = & -8.4 \\
\dfrac{-0.3b}{-0.3} & = & \dfrac{-8.4}{-0.3} \\
b & = & 28
\end{array}
$$

29. Write two different equations that have -0.5 as the solution and that require only the addition property of equality to solve them.

30. Write two different equations that have 1.2 as the solution and that require only the division property of equality to solve them.

5.10 Geometry Applications: Circles, Cylinders, and Cones

OBJECTIVE 1 Suppose you start with one dot on a piece of paper. Then you draw a bunch of dots that are each 2 cm away from the first dot. If you draw enough dots (points) you'll end up with a circle. Each point on the circle is exactly 2 cm away from the *center* of the circle. The 2 cm distance is called the radius (RAY-dee-us), r, of the circle. The distance across the circle (passing through the center) is called the diameter (dy-AH-meh-ter), d, of the circle.

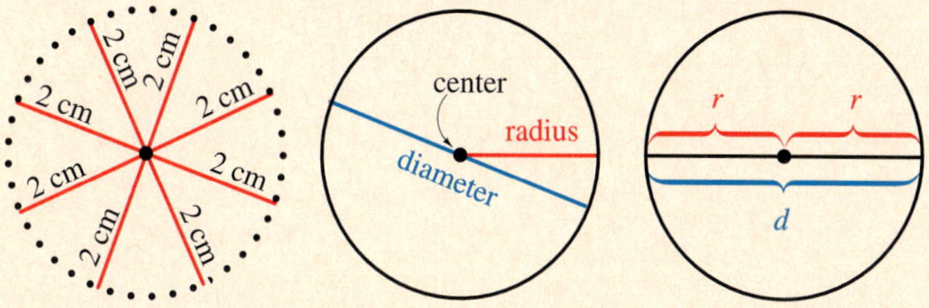

Circle, Radius, Diameter

A **circle** is a figure with all points the same distance from a fixed center point.

The **radius** (r) is the distance from the center of the circle to any point on the circle.

The **diameter** (d) is the distance across the circle passing through the center.

As the circle above on the right shows,

Finding the Diameter and Radius of a Circle

$$\text{diameter} = 2 \cdot \text{radius}$$

$$d = 2r$$

$$\text{and} \qquad r = \frac{d}{2}$$

EXAMPLE 1 Finding the Diameter and Radius of a Circle

Find the missing diameter or radius in each circle.

(a)

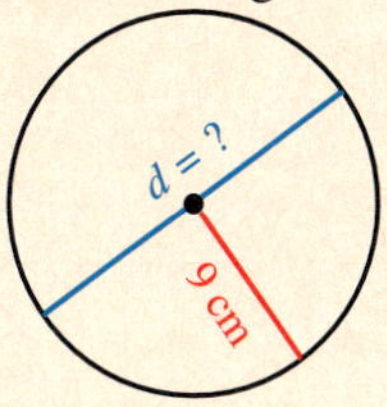

Because the radius is 9 cm, the diameter is twice as long.

$$d = 2 \cdot r$$

$$d = 2 \cdot 9 \text{ cm}$$

$$d = 18 \text{ cm}$$

(b)

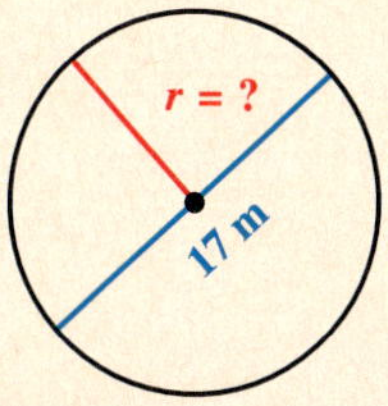

The radius is half the diameter.

$$r = \frac{d}{2} \quad \text{so} \quad r = \frac{17 \text{ m}}{2}$$

$$r = 8.5 \text{ m} \quad \text{or} \quad 8\frac{1}{2} \text{ m}$$

WORK PROBLEM 1 AT THE SIDE.

OBJECTIVES

1. Find the radius and diameter of a circle.
2. Find the circumference of a circle.
3. Find the area of a circle.
4. Find the volume of a cylinder and a cone.

FOR EXTRA HELP

Tutorial | Tape 9 | SSM, Sec. 5.10

1. Find the missing diameter or radius in each circle.

(a)

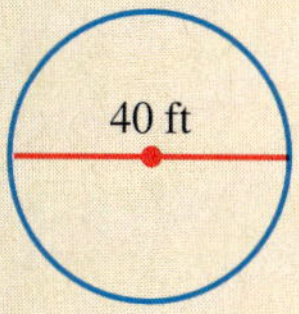

(b)

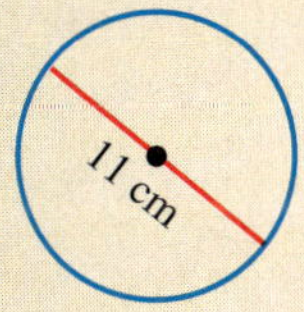

(c)

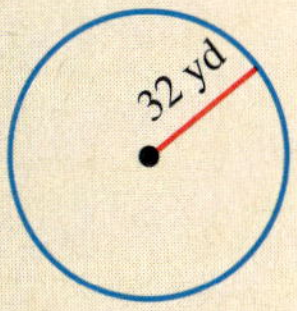

(d)

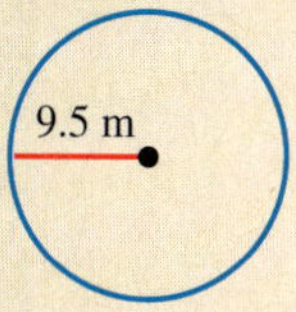

ANSWERS

1. (a) $r = 20$ ft
(b) $r = 5.5$ cm
(c) $d = 64$ yd
(d) $d = 19$ m

OBJECTIVE 2 The perimeter of a circle is called its **circumference** (sir-KUM-fer-ens). Circumference is the distance around the edge of a circle.

The diameter of the can in the drawing is about 10.6 cm, and the circumference of the can is about 33.3 cm. Dividing the circumference of the circle by the diameter gives an interesting result.

$$\frac{\text{Circumference}}{\text{diameter}} = \frac{33.3}{10.6} \approx 3.14 \quad \text{(Rounded)}$$

Dividing the circumference of *any* circle by its diameter *always* gives an answer close to 3.14. This means that going around the edge of any circle is a little more than 3 times as far as going straight across the circle.

This ratio of circumference to diameter is called π (the Greek letter **pi**, pronounced PIE). There is no decimal that is exactly equal to π, but approximately:

$$\pi \approx 3.14159265359.$$

Rounding the Value of *Pi* (π)

We usually round π to 3.14. Therefore, calculations involving π will give approximate answers and should be written using the $\approx$ symbol.

Use the following formulas to find the circumference of a circle.

Finding the Distance Around a Circle (Circumference)

$$\text{Circumference} = \pi \cdot \text{diameter}$$
$$C = \pi d$$

or, because $d = 2r$ then $C = \pi \cdot 2r$ usually written $C = 2\pi r$

EXAMPLE 2 Finding the Circumference of a Circle

Find the circumference of each circle. Use 3.14 as the approximate value for π. Round answers to the nearest tenth.

(a)

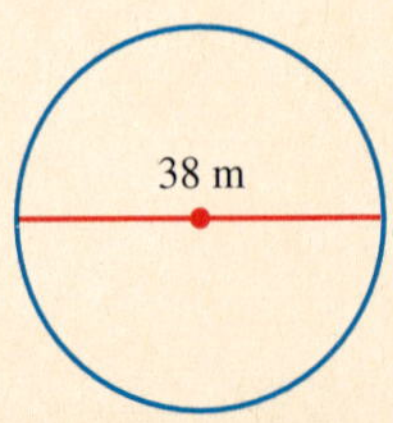

The diameter is 38 m, so use the formula with d in it.

$$C = \pi \cdot d$$
$$C \approx \mathbf{3.14} \cdot \mathbf{38\ m}$$
$$C \approx 119.3 \text{ m} \quad \text{(Rounded)}$$

CONTINUED ON NEXT PAGE

(b)

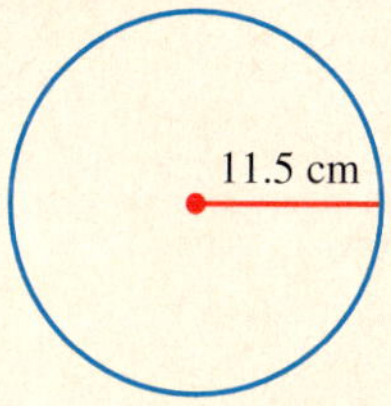

In this example, r is known, so it is easier to use the formula $C = 2\pi r$.

$$C = 2 \cdot \pi \cdot r$$
$$C \approx 2 \cdot 3.14 \cdot 11.5 \text{ cm}$$
$$C \approx 72.2 \text{ cm} \quad \text{(Rounded)}$$

Calculator Tip: Scientific calculators have a [π] key. Try pressing it. With a 10-digit display, you'll see the value of π to the nearest billionth.

3.141592654

But this is still an approximate value, although it is more precise than rounding π to 3.14. Try finding the circumference in Example 2(a) from the previous page using the [π] key.

[π] [×] 38 [=] 119.3805208

When you used 3.14 as the approximate value of π, the result was 119.32, so the answers are slightly different. In this book we will use 3.14 instead of the [π] key. Our measurements of radius and diameter are given as whole numbers or with tenths, so it is acceptable to round π to hundredths. And, you may be using a standard calculator without a [π] key or doing the calculations by hand.

WORK PROBLEM 2 AT THE SIDE.

OBJECTIVE 3 To find the formula for the area of a circle, start by cutting two circles into many pie-shaped pieces.

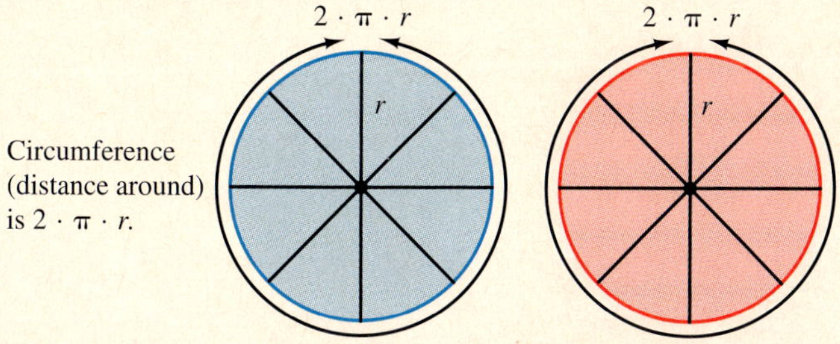

Unfold the circles, much as you might "unfold" a peeled orange, and put them together as shown here.

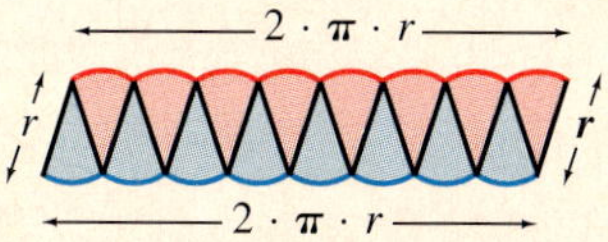

The figure is approximately a rectangle with width r (the radius of the original circle) and length $2 \cdot \pi \cdot r$ (the circumference of the original circle). The area of the "rectangle" is length times width.

$$\text{Area} = l \cdot w$$
$$\text{Area} = 2 \cdot \pi \cdot r \cdot r$$
$$\text{Area} = 2 \cdot \pi \cdot r^2$$

Because the "rectangle" was formed from *two* circles, the area of *one* circle is half as much.

$$\frac{1}{\cancel{2}_1} \cdot \cancel{2}^1 \cdot \pi \cdot r^2 = 1 \cdot \pi \cdot r^2 \quad \text{or simply} \quad \pi r^2$$

2. Find the circumference of each circle. Use 3.14 as the approximate value for π. Round answers to the nearest tenth.

(a)

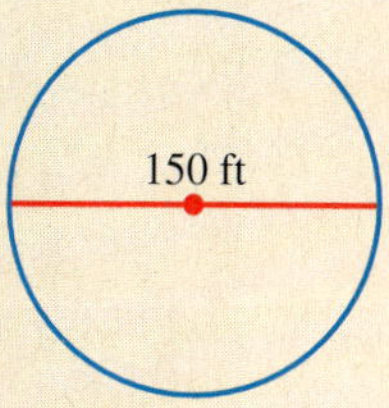

(b)

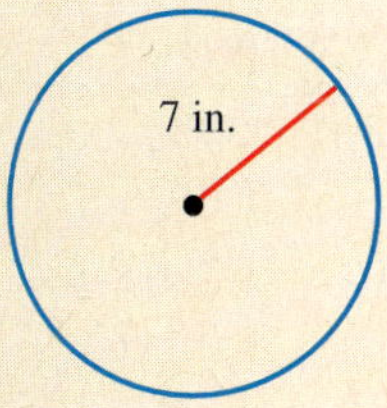

(c) diameter 0.9 km

(d) radius 4.6 m

ANSWERS

2. (a) $C \approx 471$ ft **(b)** $C \approx 44.0$ in. **(c)** $C \approx 2.8$ km **(d)** $C \approx 28.9$ m

Finding the Area of a Circle

Area of a circle = π • radius • radius

$$A = \pi r^2$$

Remember to use square units when measuring area.

EXAMPLE 3 Finding the Area of a Circle

Find the area of each circle. Use 3.14 for π. Round your answers to the nearest tenth.

(a) A circle with radius 8.2 cm.

Use the formula $A = \pi r^2$ which means $\pi \cdot r \cdot r$.

$$A = \pi \cdot r \cdot r$$

$$A \approx \mathbf{3.14} \cdot \mathbf{8.2\text{ cm}} \cdot \mathbf{8.2\text{ cm}}$$

$$A \approx 211.1\text{ cm}^2 \text{ (square units for area)}$$

(b)

To use the formula, you need to know the radius (r). In this circle, the diameter is 10 ft. First find the radius.

$$r = \frac{d}{2}$$

$$r = \frac{10\text{ ft}}{2} = 5\text{ ft}$$

Now find the area.

$$A \approx 3.14 \cdot 5\text{ ft} \cdot 5\text{ ft}$$

$$A \approx 78.5\text{ ft}^2$$

Note

When finding *circumference,* you can start with either the radius or the diameter. When finding *area,* you must use the *radius.* If you are given the diameter, divide it by 2 to find the radius. Then find the area.

Calculator Tip: You can find the area of the circle in Example 3(a) on your calculator. The first method works on both scientific and standard calculators:

3.14 [×] 8.2 [×] 8.2 [=] 211.1336

You round the answer to 211.1 (nearest tenth).

On a scientific calculator you can also use the [x^2] key, which automatically squares the number you enter (that is, multiplies the number times itself):

3.14 [×] 8.2 [x^2] 67.24 [=] 211.1336

Appears automatically; 8.2 × 8.2 is 67.24

3. Find the area of each circle. Use 3.14 for π. Round your answers to the nearest tenth.

(a)

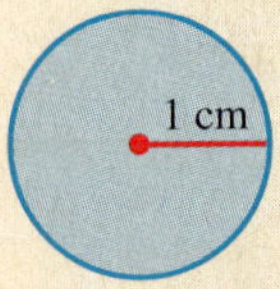

(b)

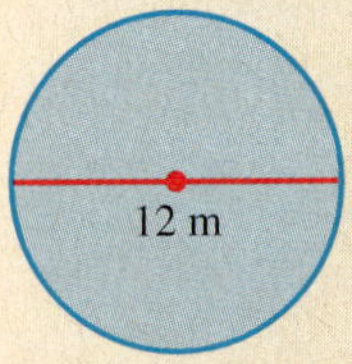

(*Hint:* The diameter is 12 m so r = ______ m)

(c)

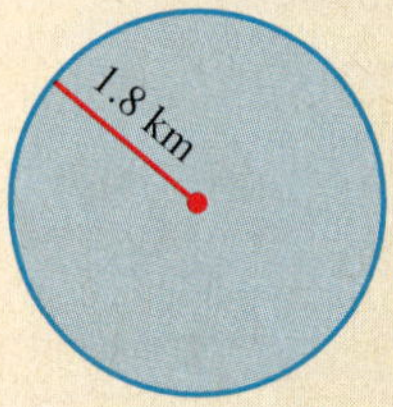

(d)

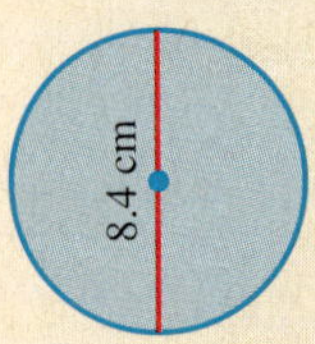

WORK PROBLEM 3 AT THE SIDE.

ANSWERS

3. **(a)** $\approx 3.1\text{ cm}^2$ **(b)** $\approx 113.0\text{ m}^2$ **(c)** $\approx 10.2\text{ km}^2$ **(d)** $\approx 55.4\text{ cm}^2$

EXAMPLE 4 Finding the Area of a Semicircle

Find the area of the semicircle. Use 3.14 for π. Round your answer to the nearest tenth.

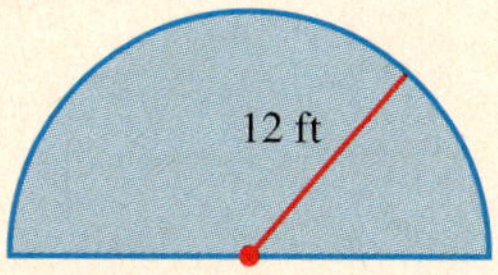

First, find the area of an entire circle with a radius of 12 ft.

$$A = \pi \cdot r \cdot r$$
$$A \approx 3.14 \cdot 12 \text{ ft} \cdot 12 \text{ ft}$$
$$A \approx 452.16 \text{ ft}^2 \quad \text{(Do not round yet.)}$$

Divide the area of the whole circle by 2 to find the area of the semicircle.

$$\frac{452.16 \text{ ft}^2}{2} = 226.08 \text{ ft}^2$$

The *last* step is rounding 226.08 to the nearest tenth.

$$\text{Area of semicircle} \approx 226.1 \text{ ft}^2$$

WORK PROBLEM 4 AT THE SIDE.

EXAMPLE 5 Applying the Concept of Circumference

A circular rug is 8 feet in diameter. The cost of fringe for the edge is \$2.25 per foot. What will it cost to add fringe to the rug? Use 3.14 for π.

$$\text{Circumference} = \pi \cdot d$$
$$C \approx 3.14 \cdot 8 \text{ ft}$$
$$C \approx 25.12 \text{ ft}$$
$$\text{Cost} = \text{Cost per foot} \cdot \text{Circumference}$$
$$\text{Cost} = \frac{\$2.25}{1 \not{\text{ft}}} \cdot \frac{25.12 \not{\text{ft}}}{1}$$
$$\text{Cost} = \$56.52$$

WORK PROBLEM 5 AT THE SIDE.

EXAMPLE 6 Applying the Concept of Area

Find the cost of covering the rug in Example 5 with a plastic cover. The material for the cover costs \$1.50 per square foot. Use 3.14 for π.

First find the radius.

$$r = \frac{d}{2} = \frac{8 \text{ ft}}{2} = 4 \text{ ft}$$

Then, $A = \pi \cdot r^2$

$$A \approx 3.14 \cdot 4 \text{ ft} \cdot 4 \text{ ft}$$
$$A \approx 50.24 \text{ ft}^2$$
$$\text{Cost} = \frac{\$1.50}{1 \not{\text{ft}^2}} \cdot \frac{50.24 \not{\text{ft}^2}}{1} = \$75.36$$

WORK PROBLEM 6 AT THE SIDE.

4. Find the area of each semicircle. Use 3.14 for π. Round your answers to the nearest tenth.

(a)

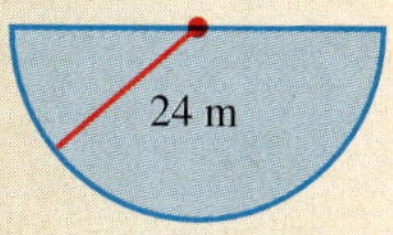

(b)

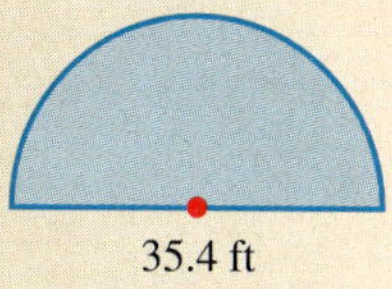

(c)

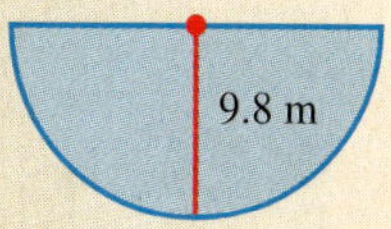

5. Find the cost of binding around the edge of a circular rug that is 3 meters in diameter. The binder charges \$4.50 per meter. Use 3.14 for π.

6. Find the cost of covering the rug in problem 5 above with a nonslip rubber backing. The rubber backing costs \$2 per square meter.

ANSWERS

4. (a) $\approx 904.3 \text{ m}^2$ (b) $\approx 491.9 \text{ ft}^2$ (c) $\approx 150.8 \text{ m}^2$
5. \$42.39
6. \$14.13

OBJECTIVE 4 Several *cylinders* are shown here.

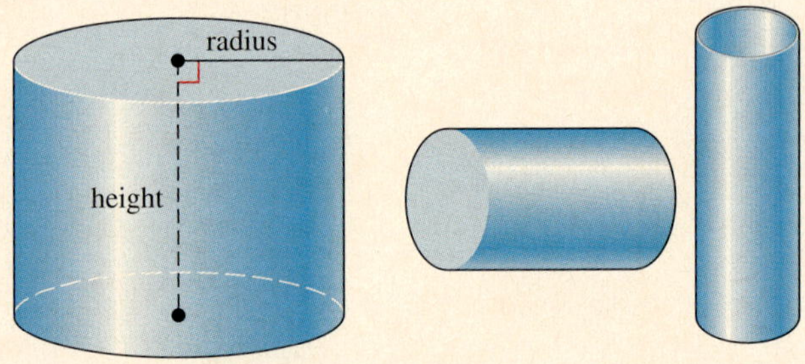

These are called *right circular cylinders* because the top and bottom are circles, and the side makes a right angle with the top and bottom. Examples of cylinders are a soup can, a home water heater, and a piece of pipe.

Use the following formula to find the *volume* of a cylinder. Notice that the first part of the formula, $\pi \cdot r \cdot r$, is the *area* of the circular base.

Finding the Volume of a Cylinder

$$\text{Volume of cylinder} = \pi \cdot r \cdot r \cdot h$$
$$V = \pi r^2 h$$

Remember to use cubic units when measuring volume.

EXAMPLE 7 Finding the Volume of a Cylinder

Find the volume of each cylinder. Use 3.14 as the approximate value of π. Round your answers to the nearest tenth, if necessary.

(a)

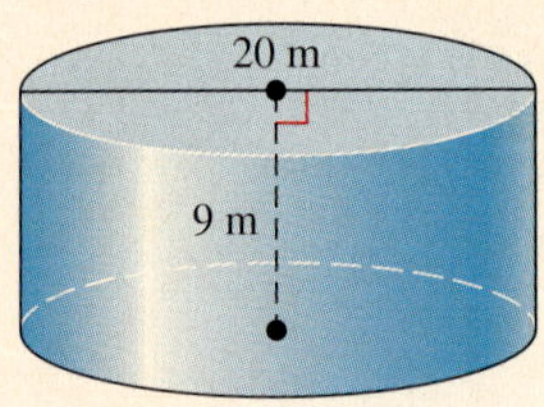

The diameter is 20 m so the radius is $\frac{20\text{ m}}{2} = 10$ m. The height is 9 m. Use the formula to find the volume.

$$V = \pi \cdot r \cdot r \cdot h$$
$$V \approx 3.14 \cdot 10\text{ m} \cdot 10\text{ m} \cdot 9\text{ m}$$
$$V \approx 2826\text{ m}^3$$

(b)

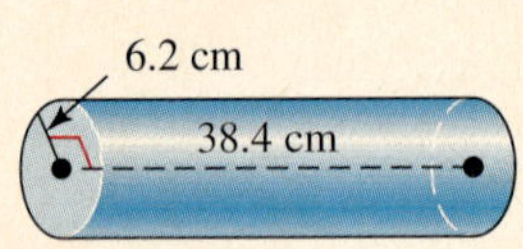

$$V \approx 3.14 \cdot 6.2\text{ cm} \cdot 6.2\text{ cm} \cdot 38.4\text{ cm}$$
$$V \approx 4634.94144 \quad \text{Now round to tenths.}$$
$$V \approx 4634.9\text{ cm}^3$$

WORK PROBLEM 7 AT THE SIDE.

A cone is a solid shape like the one shown below. The height is the perpendicular distance from the base to the highest point of the cone. The base of the cone is circular.

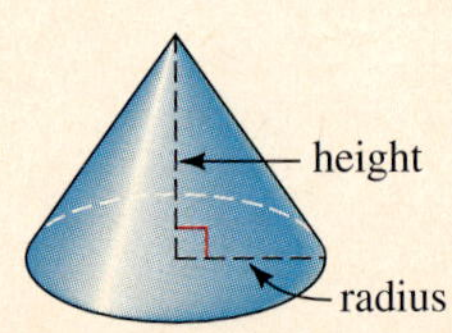

cone

7. Find the volume of each cylinder. Use 3.14 for π. Round your answers to the nearest tenth. (A calculator is helpful on these problems.)

(a)

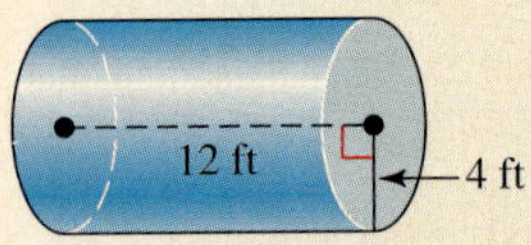

(b)

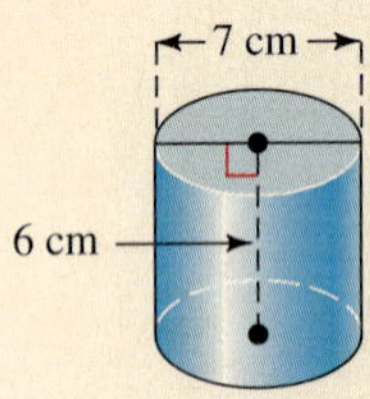

(c) radius 14.5 yd, height 3.2 yd

ANSWERS

7. (a) $\approx 602.9\text{ ft}^3$ **(b)** $\approx 230.8\text{ cm}^3$ **(c)** $\approx 2112.6\text{ yd}^3$

Use the following formula to find the volume of a cone. It is the same formula you used for the volume of a pyramid in **Section 4.8.**

Finding the Volume of a Cone

$$\text{Volume of cone} = \frac{1}{3} \cdot B \cdot h$$

$$\text{or} \quad V = \frac{Bh}{3}$$

where B is the area of the circular base of the cone and h is the height of the cone. Remember to use cubic units when measuring volume.

EXAMPLE 8 Finding the Volume of a Cone

Find the volume of the cone. Use 3.14 for π. Round your answer to the nearest tenth.

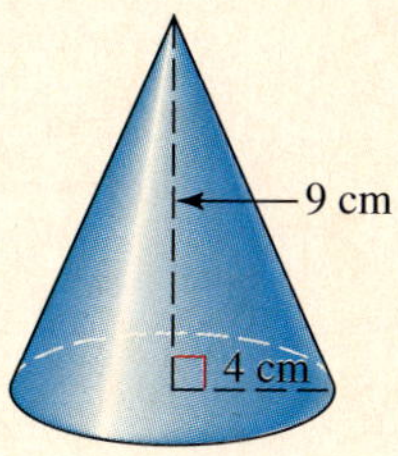

First find the area of the circular base. Recall that the formula for the area of a circle is πr^2.

$$B = \pi \cdot r \cdot r$$

$$B \approx 3.14 \cdot 4 \text{ cm} \cdot 4 \text{ cm}$$

$$\mathbf{B \approx 50.24\ cm^2} \quad \text{Do not round to tenths yet.}$$

Next, find the volume. The height is 9 cm.

$$V = \frac{B \cdot h}{3}$$

$$V \approx \frac{50.24 \text{ cm}^2 \cdot 9 \text{ cm}}{3}$$

$$V \approx 150.72 \text{ cm}^3 \quad \text{Now round to tenths.}$$

$$V \approx 150.7 \text{ cm}^3$$

WORK PROBLEM 8 AT THE SIDE.

8. Find the volume of a cone with base radius 2 ft and height 11 ft. Use 3.14 for π. Round your answer to the nearest tenth.

ANSWERS

8. $\approx 46.1 \text{ ft}^3$

NUMBERS IN THE *Real World*

collaborative investigations

A Census Bureau report gives the information at the right about family composition in 1970 and in 1995. There were approximately 99 million households in the United States in 1995.

Then and Now

	1970	1995
People per household	3.14	2.65
Households with five or more people	1 of 5	1 of 10
Households with people living alone	One-sixth	One-fourth
Families maintained by women with no husband present	5.6 million	12.2 million
Families maintained by men with no wife present	1.2 million	3.2 million
Households in metropolitan areas	2 of 3	4 of 5

Source: Commerce Department's Census Bureau

1. Write 99 million using digits.

2. In 1995, 1 of 10 households had five or more people living in it. 1 of 10 means the same as $\frac{1}{10}$. To find the number of households with five or more people, multiply $\frac{1}{10}$ times the number of households.

3. What shortcut involving division could you use to find $\frac{1}{10}$ of the number of households?

4. Why is multiplying by $\frac{1}{10}$ the same as dividing by 10?

5. Show how to use both multiplication and a division shortcut to find the number of households with people living alone in 1995.

6. Use a signed number to express the change from 1970 to 1995 in the number of people per household.

7. If $\frac{1}{4}$ of the households in 1995 have people living alone, what fraction of the households have more than one person?

8. Find the number of households in 1995 that have more than one person. Show *two* different ways to get the answer.

9. If $\frac{4}{5}$ of the households in 1995 are in metropolitan areas, what fraction of the households are in rural areas?

10. Use multiplication to find the number of households in metropolitan areas in 1995 and the number of households in rural areas. Can you also use a shortcut? If so, explain your shortcut.

5.10 Exercises

Find the unknown length in each circle.

1.

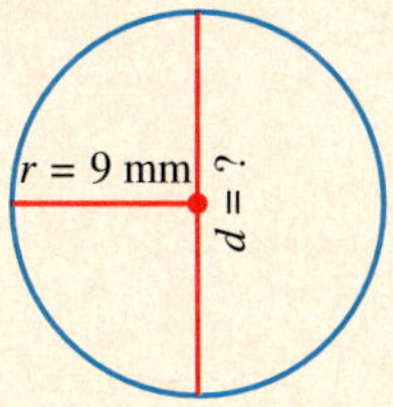

2.

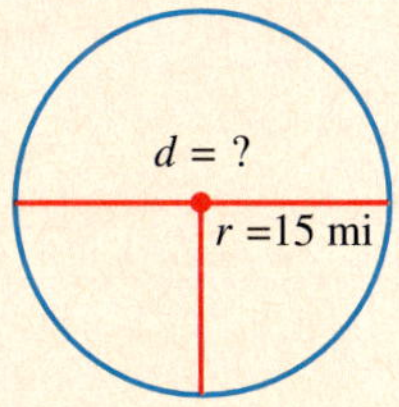

3.

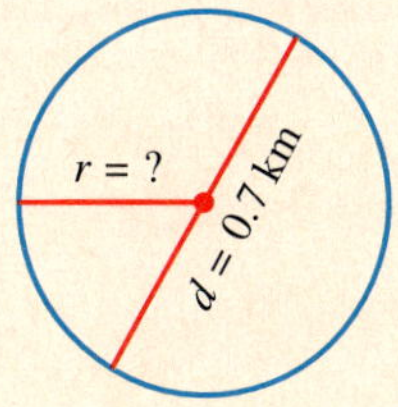

4.

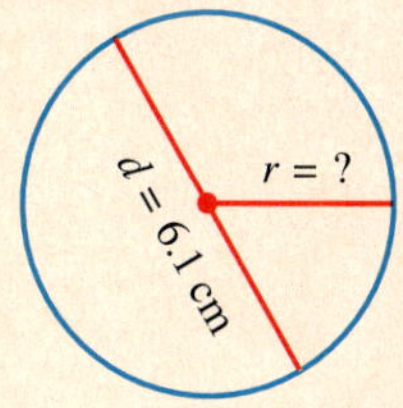

Find the circumference and area of each circle. Use 3.14 as the approximate value for π. Round your answers to the nearest tenth.

5.

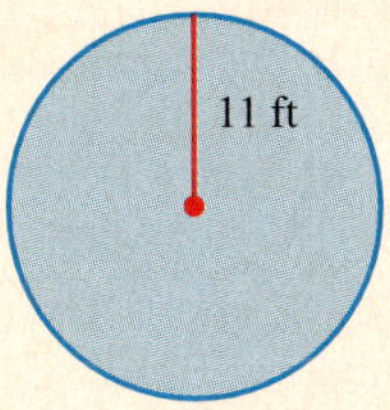

6.

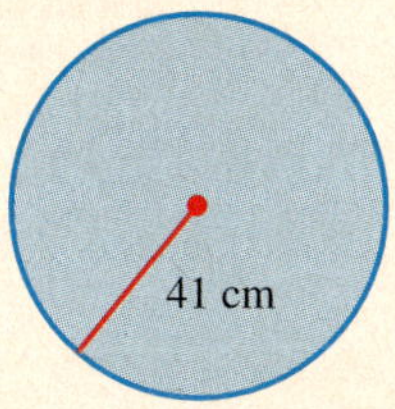

7.

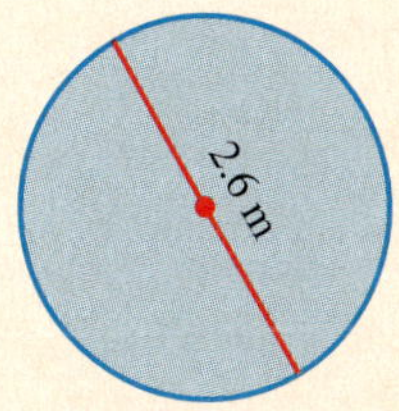

8.

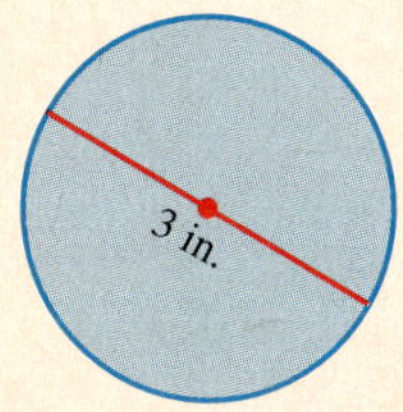

Find the circumference and area of circles having the following diameters. Use 3.14 for π. Round your answers to the nearest tenth.

9. $d = 15$ cm

10. $d = 39$ ft

11. $d = 7\frac{1}{2}$ ft

12. $d = 4\frac{1}{2}$ yd

13. $d = 8.65$ km

14. $d = 19.5$ mm

Solve each application problem. Use 3.14 for π. Round your answers to the nearest tenth, if necessary.

15. How far does a point on the tread of a tire move in one turn, if the diameter of the tire is 70 cm?

16. If you swing a ball held at the end of a string 2 m long, how far will the ball travel on each turn?

17. A wave energy extraction device is a huge undersea dome used to harness the power of ocean waves. The base of the dome is 250 ft in diameter. Find its circumference.

18. Find the area of the base of the dome in Exercise 17.

Writing Calculator Small Group

19. A radio station can be heard 150 miles in all directions during evening hours. How many square miles are in the station's broadcast area?

20. An earthquake was felt by people 900 km away in all directions from the epicenter (the source of the earthquake). How much area was affected by the quake?

21. The circumference of a circular swimming pool is 22 meters. What is the radius of the pool, to the nearest tenth of a meter?

22. A forest ranger measured 56 ft around the base of a giant sequoia tree. The ranger wanted to find the diameter of the tree without cutting it down. Find the diameter. Round your answer to the nearest tenth.

Use the table below to solve Exercises 23–28.

Find the "best buy" for each type of pizza. The "best buy" is the lowest cost per square inch of pizza. All the pizzas are circular in shape, and the measurement given on the menu board is the diameter of the pizza in inches. Use 3.14 as the approximate value of π. Round the area to the nearest tenth. Round cost per square inch to the nearest thousandth.

PIZZA MENU	Small $7\frac{1}{2}$"	Medium 13"	Large 16"
Cheese only	**$2.80**	**$6.50**	**$9.30**
"The Works"	**$3.70**	**$8.95**	**$14.30**
Deep-dish combo	**$4.35**	**$10.95**	**$15.65**

23. Find the area of a small pizza.

24. Find the area of a medium pizza.

25. Find the area of a large pizza.

26. What is the cost per square inch for each size of cheese pizza? Which size is the "best buy"?

27. What is the cost per square inch for each size of "the works" pizza? Which size is the "best buy"?

28. What is the cost per square inch for each size of deep dish combo pizza? Which size is the "best buy"?

29. How would you explain π to a friend who is not in your math class? Write an explanation. Then make up a test question which requires the use of π, and show how to solve it.

30. Explain how circumference and perimeter are alike. How are they different? Make up two problems, one involving perimeter, the other circumference. Show how to solve your problems.

Ⓖ *Find each shaded area in Exercises 31–34. Use 3.14 as the approximate value of π. Round your answers to the nearest tenth, if necessary.*

Example:

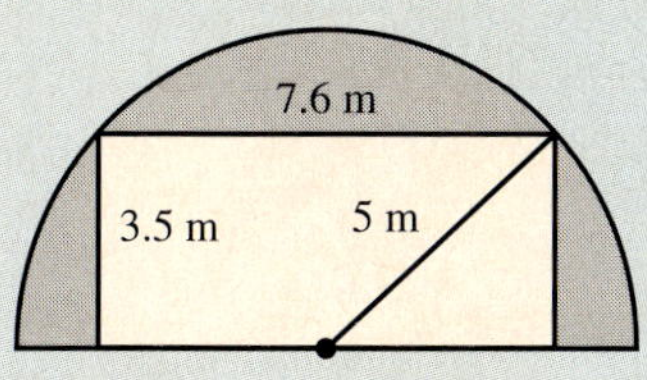

Solution:

First, find the area of the entire circle.

$$A = \pi \cdot r^2 \approx 3.14 \cdot 5 \text{ m} \cdot 5 \text{ m} \approx 78.5 \text{ m}^2$$

Next, find the area of the semicircle.

$$\frac{78.5 \text{ m}^2}{2} = 39.25 \text{ m}^2$$

Now, find the area of the white rectangle.

$$3.5 \text{ m} \cdot 7.6 \text{ m} = 26.6 \text{ m}^2$$

Finally, subtract to find the shaded area.

$$39.25 \text{ m}^2 - 26.6 \text{ m}^2 \approx \mathbf{12.7\ m^2} \quad \text{(Rounded)}$$

31.

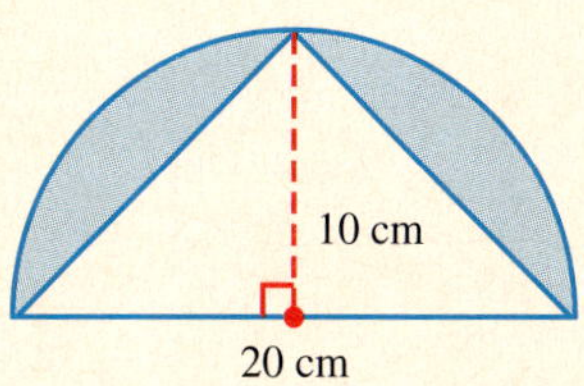

32.

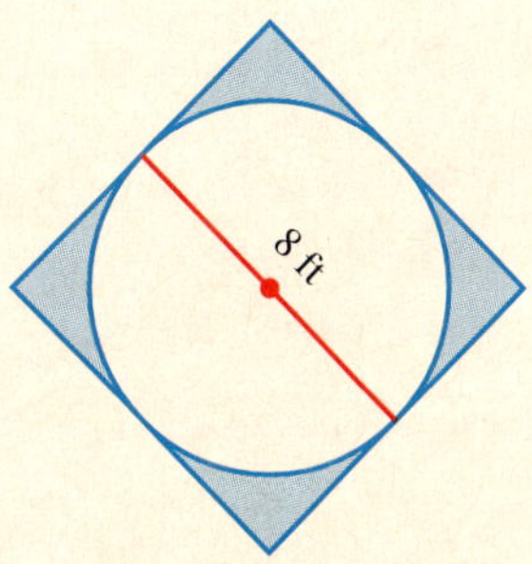

33.

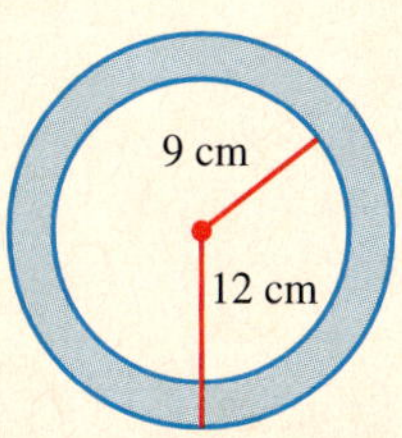

34.

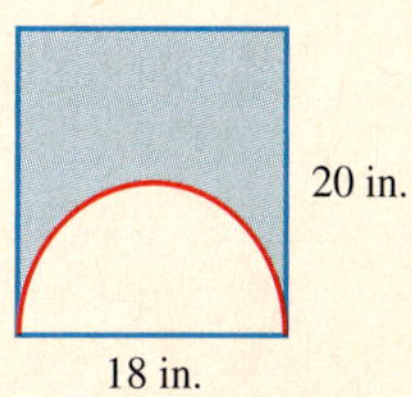

Find the volume of each figure. Use 3.14 as the approximate value of π. Round your answers to the nearest tenth, if necessary.

35.

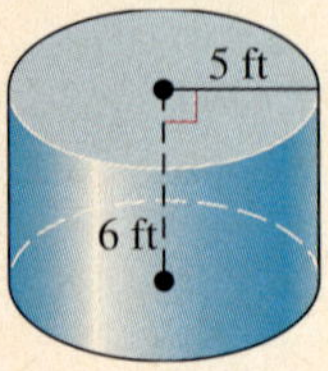

36.

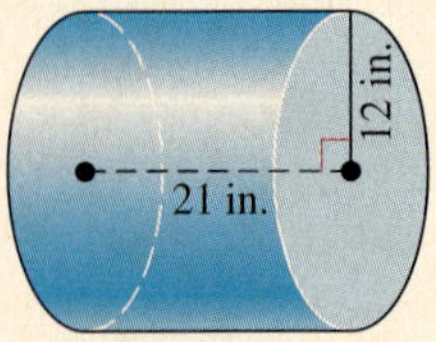

37.

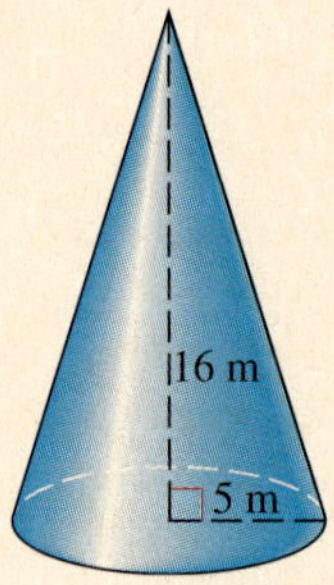

38.

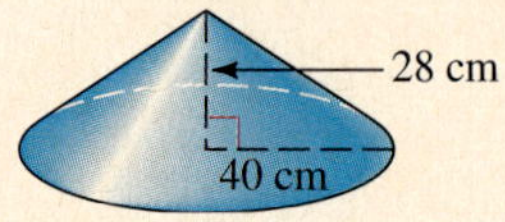

Solve each application problem. Use 3.14 as the approximate value of π. Round your answers to the nearest tenth, if necessary.

39. A city sewer pipe has a diameter of 5 ft and a length of 200 ft. Find the volume of the pipe.

40. A cylindrical woven basket made by a Northwest Coast tribe is 8 cm high and has a diameter of 11 cm. What is the volume of the basket?

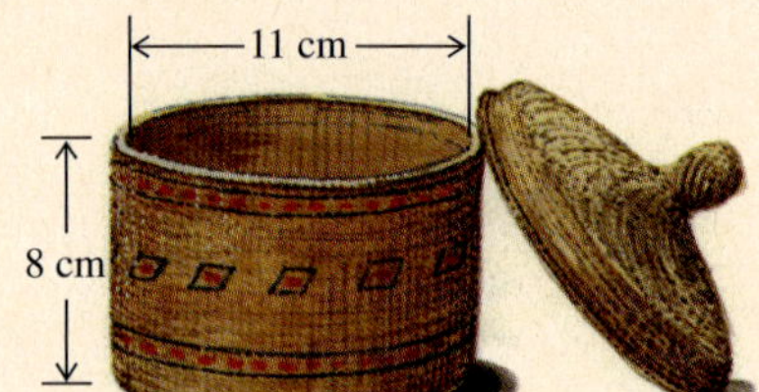

41. A paperweight in the shape of a cone is 4.5 in. tall. The base of the paperweight has a diameter of 3 in. Find its volume.

42. An ice cream cone has a diameter of 2 inches and a height of 4 inches. Find its volume.

43. Explain the two errors made by a student in finding the volume of a cylinder with a diameter of 7 cm and a height of 5 cm. Find the correct answer.

$$V \approx 3.14 \cdot 7 \cdot 7 \cdot 5$$
$$\approx 769.3 \text{ cm}^2$$

44. Compare the steps in finding the volume of a cylinder and a cone. How are they similar? Suppose you know the volume of a cylinder. Explain how can you find the volume of a cone with the same radius and height by doing just a one-step calculation.

Review and Prepare

Find each product or quotient. Write answers in simplest form. (For help, see ***Section 4.5.****)*

45. $3\frac{1}{3} \cdot 1\frac{3}{4}$

46. $1\frac{4}{5} \cdot 1\frac{2}{3}$

47. $(-5)\left(2\frac{3}{5}\right)$

48. $\left(-1\frac{1}{8}\right)(-6)$

49. $4\frac{1}{2} \div 1\frac{5}{6}$

50. $3\frac{1}{4} \div 2\frac{1}{4}$

51. $-1\frac{7}{10} \div (-5)$

52. $-3\frac{5}{9} \div 8$

KEY TERMS

5.1	**decimals**	Decimals, like fractions, are used to show parts of a whole.
	decimal point	The dot that is used to separate the whole number part from the fractional part of a decimal number.
	place value names	The value assigned to each place to the right or left of the decimal point. Whole numbers, such as ones and tens, are to the *left* of the decimal point. Fractional parts, such as tenths and hundredths, are to the *right* of the decimal point.
5.2	**round**	To "cut off" a number after a certain place, such as to round to the nearest hundredth. The rounded number is less accurate than the original number, so write the symbol "≈" in front of it to mean "approximately equal to."
	decimal places	The number of digits to the *right* of the decimal point; for example, 6.37 has two decimal places, 4.706 has three decimal places.
5.5	**repeating decimal**	A decimal with one or more digits that repeat forever, such as the 6 in 0.1666. . . . Use three dots to indicate that it is a repeating decimal; it never terminates (ends). You can also write it with a bar above the repeating digits, as in $0.1\overline{6}$.
5.7	**mean**	The mean is the sum of all the values divided by the number of values. It is often called the *average.*
	weighted mean	The weighted mean is a mean calculated so that each value is multiplied by its frequency.
	median	The median is the middle number in a group of values that are listed from smallest to largest. It divides a group of values in half. If there are an even number of values, the median is the mean (average) of the two middle values.
	mode	The mode is the value that occurs most often in a group of values.
5.8	**square root**	A square root of a positive number is one of two equal positive factors of the number.
	hypotenuse	The hypotenuse is the side of a right triangle opposite the 90° angle; it is the longest side.
5.10	**circle**	A circle is a figure with all points the same distance from a fixed center point.
	radius	Radius is the distance from the center of a circle to any point on the circle.
	diameter	Diameter is the distance across a circle, passing through the center.
	circumference	Circumference is the distance around a circle.
	π (pi)	π is the ratio of the circumference to the diameter of any circle. It is approximately equal to 3.14.

KEY FORMULAS

mean $= \dfrac{\text{sum of all values}}{\text{number of values}}$

hypotenuse $= \sqrt{(\text{leg})^2 + (\text{leg})^2}$

leg $= \sqrt{(\text{hypotenuse})^2 - (\text{leg})^2}$

diameter of a circle: $d = 2r$

radius of a circle: $r = \dfrac{d}{2}$

circumference of a circle: $C = \pi d$ or $C = 2\pi r$

area of a circle: $A = \pi r^2$

volume of a cylinder: $V = \pi r^2 h$

volume of a cone: $V = \dfrac{Bh}{3}$

Quick Review

Concepts	*Examples*

5.1 *Reading and Writing Decimals*

thousands	hundreds	tens	ones	decimal point ("**and**")	ten**ths**	hundred**ths**	thousand**ths**	ten-thousand**ths**
5	8	4	6	.	0	7	3	2

Whole number part (5 8 4 6) Fractional part (0 7 3 2)

Write each decimal in words.

15.38 — 8 is in hundredths place

fifteen **and** thirty-eight **hundredths**

0.0103 — 3 is in ten-thousandths place

one hundred three **ten-thousandths**

5.1 *Writing Decimals as Fractions*

The digits to the right of the decimal point are the numerator. The place value of the right-most digit determines the denominator. Reduce to lowest terms.

Write 0.45 as a fraction.

The numerator is 45. The right-most digit, 5, is in the hundredths place, so the denominator is 100. Then reduce.

$$\frac{45}{100} = \frac{45 \div 5}{100 \div 5} = \frac{9}{20} \quad \text{Lowest terms}$$

5.2 *Rounding Decimals*

Find the place to which you are rounding. Draw a cut-off line to the right of that place; the rest of the digits will be dropped. Look *only* at the first digit being cut. If it is *less than 5,* the part you are keeping stays the same. If it is *5 or more,* the part you are keeping rounds up. Do not move the decimal point when rounding. Write "≈" in front of the rounded number to mean "approximately equal to."

Round 0.17952 to the nearest thousandth. Is it closer to 0.179 or to 0.180?

0.179 | 52 — First digit cut is 5 or more so round up by adding 1 thousandth to the part you are keeping.

$$\begin{array}{r} 0.179 \\ +\ 0.001 \\ \hline 0.180 \end{array}$$

← To round up, add 1 thousandth.

0.17952 rounds to 0.180. Write it as ≈0.180.

5.3 *Adding Positive and Negative Decimals*

Estimate the answer by rounding each number so there is only one non-zero digit.

To find the exact answer, line up the decimal points. If needed, write in zeros as placeholders. Add or subtract the absolute values as if they were whole numbers. Line up the decimal point in the answer.

If the numbers have the same sign, use the common sign as the sign of the sum. If the numbers have different signs, the sign of the sum is the sign of the number with the larger absolute value.

Add 5.68 + 785.3 + 12 + 2.007.

estimate		*exact*
6	←	5.680
800	←	785.300
10	←	12.000
+ 2	←	+ 2.007
818		804.987

Use zeros as placeholders so all numbers have three decimal places.

Line up decimal points.

All addends are positive, so the sign of the sum is positive. The estimate and exact answer are both in hundreds, so the decimal point is probably in the correct place.

<table>
<tr><th>Concepts</th><th>Examples</th></tr>
<tr><td>
5.3 Subtracting Positive and Negative Decimals
Rewrite subtracting as adding the opposite of the second number. Then follow the rules for adding decimal numbers.
</td><td>
$4.2 - 12.91$
↓ ↓
$4.2 + (-12.91)$
$|4.2|$ is 4.2 and $|-12.91|$ is 12.91. Because −12.91 has the larger absolute value, the answer will be negative.
$4.2 + (-12.91) = -8.71$
</td></tr>
<tr><td>
5.4 Multiplying Positive and Negative Decimals
1. Multiply as you would for whole numbers.
2. Count the total number of decimal places in both factors.
3. Write the decimal point in the answer so it has the same number of decimal places as the total from Step 2. You may need to write extra zeros on the left side of the product in order to get enough decimal places in the answer.
4. If the factors have the same sign, the product is positive. If the factors have different signs, the product is negative.
</td><td>
Multiply 0.169(−0.21)
0.169 ← 3 decimal places
× 0.21 ← 2 decimal places
169   5 total decimal places
338
.03549 ← 5 decimal places in answer
Write in a zero so you can count over 5 decimal places.
The factors have different signs, so the product is negative:
$0.169(-0.21) = -0.03549$
</td></tr>
<tr><td>
5.5 Dividing by a Decimal
1. Change the divisor to a whole number by moving the decimal point to the right.
2. Move the decimal point in the dividend the same number of places to the right.
3. Write the decimal point in the answer directly above the decimal point in the dividend.
4. Divide as with whole numbers.
5. If the numbers have the same sign, the quotient is positive. If they have different signs, the quotient is negative.
</td><td>
Divide −52.8 by −0.75.
First consider 52.8 ÷ 0.75.
7 0.4
0.7 5)5 2.8 0 0
5 2 5
3 0 0
3 0 0
0
Move decimal point two places to the right in divisor and dividend.
Write zeros in the dividend so you can move the decimal point and continue dividing until the remainder is zero.
The quotient is positive because both the divisor and dividend were negative (same signs means positive quotient).
</td></tr>
<tr><td>
5.6 Writing Fractions as Decimals
Divide the numerator by the denominator. If necessary, round to the place indicated.
</td><td>
Write $\frac{1}{8}$ as a decimal.
$\frac{1}{8}$ means 1 ÷ 8. Write it as 8)1.
The decimal point is on the right side of 1.
0.125
8)1.000 ← Decimal point and three zeros written in so you can continue dividing.
8
20
16
40
40
0
Therefore, $\frac{1}{8}$ is equivalent to 0.125.
</td></tr>
</table>

Concepts	*Examples*
5.6 Comparing the Size of Fractions and Decimals **1.** Write any fractions as decimals. **2.** Write zeros so that all the numbers being compared have the same number of decimal places. **3.** Use < to mean "is less than," > to mean "is greater than," or list the numbers from smallest to largest.	Arrange in order from smallest to largest. $0.505 \quad \frac{1}{2} \quad 0.55$ $0.505 = 505$ thousandths $\frac{1}{2} = 0.5 = 0.500 = 500$ thousandths ← 500 is smallest. $0.55 = 0.550 = 550$ thousandths ← 550 is largest. (smallest) $\frac{1}{2}$ 0.505 0.55 (largest)
5.7 Finding the Mean (average) of a Set of Numbers **1.** Add all values to obtain a total. **2.** Divide the total by the number of values.	The test scores for Heather Hall in her math course were as follows: 85 76 93 91 78 82 87 85 Find Heather's test score average (mean) to the nearest tenth. Mean $= \frac{85 + 76 + 93 + 91 + 78 + 82 + 87 + 85}{8}$ $= \frac{677}{8} \approx 84.6$
5.7 Finding the Median of a Set of Numbers **1.** Arrange the data from smallest to largest. **2.** If there is an odd number of values, select the middle value. If there is an even number of values, find the average of the two middle values.	Find the median for Heather Hall's grades from the previous example. The data arranged from smallest to largest is as follows: 76 78 82 **85 85** 87 91 93 Middle values The middle two values are 85 and 85. The average of these two values is $\frac{85 + 85}{2} = 85$
5.7 Determining the Mode of a Set of Values **1.** Find the value that appears most often in the list of values. **2.** If no value appears more than once, there is no mode. **3.** If two different values appear the same number of times, the list is bimodal.	Find the mode for Heather's grades in the previous example. The most frequently occurring score is 85 (it occurs twice). Therefore, the mode is 85.

Concepts	*Examples*

5.8 Finding the Square Root of a Number

Use the square root key on a calculator, $\boxed{\sqrt{\ }}$ or $\boxed{\sqrt{x}}$. Round to the nearest thousandth, if necessary.

$\sqrt{64} = 8$

$\sqrt{43} \approx 6.557$ (6.557438524 is rounded to nearest thousandth)

5.8 Finding the Unknown Length in a Right Triangle

To find the **hypotenuse,** use:

$$\text{hypotenuse} = \sqrt{(\text{leg})^2 + (\text{leg})^2}$$

The hypotenuse is the side opposite the right angle.

Find the length of the hypotenuse. Round to thousandths.

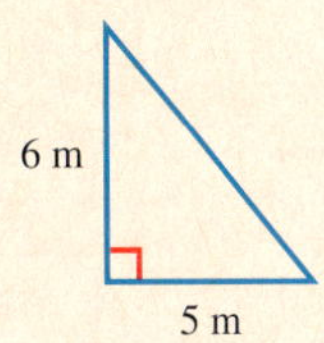

$$\begin{aligned}\text{hypotenuse} &= \sqrt{(6)^2 + (5)^2}\\ &= \sqrt{36 + 25}\\ &= \sqrt{61} \approx 7.810\end{aligned}$$

The hypotenuse is about 7.810 m long.

To find a **leg,** use:

$$\text{leg} = \sqrt{(\text{hypotenuse})^2 - (\text{leg})^2}$$

The legs are the sides that form the right angle.

Find the unknown measurement in this right triangle. Round to thousandths.

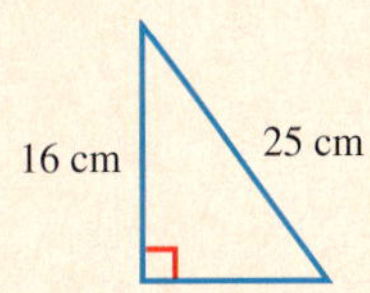

$$\begin{aligned}\text{leg} &= \sqrt{(25)^2 - (16)^2}\\ &= \sqrt{625 - 256}\\ &= \sqrt{369} \approx 19.209\end{aligned}$$

The leg is about 19.209 cm long.

5.9 Solving Equations Containing Decimals

1. Use the addition property of equality to get the variable term by itself on one side of the equal sign.
2. Divide both sides by the coefficient of the variable term to find the solution.
3. Check the solution by going back to the original equation. Replace the variable with the solution. If the equation balances, the solution is correct.

Solve and check the solution.

$$\begin{aligned}4.5x + 0.7 &= -5.15 \quad \text{Add } -0.7 \text{ to both sides.}\\ -0.7 \quad & \quad -0.7\\ \underbrace{4.5x + 0}_{} &= -5.85\end{aligned}$$

$$\frac{\overset{1}{\cancel{4.5}}x}{\underset{1}{\cancel{4.5}}} = \frac{-5.85}{4.5} \quad \text{Divide both sides by 4.5.}$$

$$x = -1.3$$

The solution is -1.3. To check the solution, go back to the original equation.

Check $4.5x + 0.7 = -5.15$ Replace x with -1.3.

$$\begin{aligned}\underbrace{4.5(-1.3)} + 0.7 &= -5.15\\ \underbrace{-5.85 \quad + 0.7} &= -5.15\\ -5.15 &= -5.15\end{aligned}$$

The equation balances, so -1.3 is the correct solution.

Concepts	*Examples*

5.10 Circles

Use this formula to find the **diameter** of a circle, given the radius.

$$\text{diameter} = 2 \cdot \text{radius}$$

Find the diameter of a circle if the radius is 3 meters.

$$d = 2 \cdot \mathbf{r} = 2 \cdot \mathbf{3\ m} = 6\text{ m}$$

Use this formula to find the **radius** of a circle, given the diameter.

$$\text{radius} = \frac{\text{diameter}}{2}$$

Find the radius of a circle if the diameter is 5 cm.

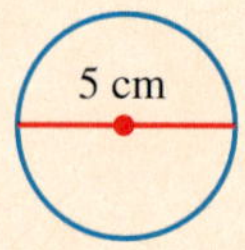

$$r = \frac{\mathbf{d}}{2} = \frac{\mathbf{5\ cm}}{2} = 2.5\text{ cm}$$

Use these formulas to find the **circumference** of a circle.

$$C = 2 \cdot \pi \cdot \text{radius}$$

$$\text{or} \quad C = \pi \cdot \text{diameter}$$

Use 3.14 as the approximate value for π.

Find the circumference of a circle with a radius of 3 cm.

$$\text{Circumference} = 2 \cdot \boldsymbol{\pi} \cdot \mathbf{r}$$

$$C \approx 2 \cdot \mathbf{3.14} \cdot \mathbf{3\ cm} \approx 18.84\text{ cm}$$

Use this formula to find the **area** of a circle.

$$A = \pi \cdot (\text{radius})^2$$

Area is measured in **square units**.

Find the area of the circle.

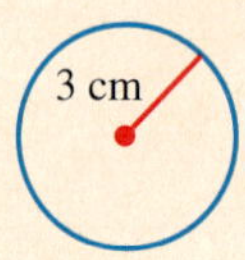

$$\text{Area} = \boldsymbol{\pi} \cdot \mathbf{r^2}$$

$$A \approx \mathbf{3.14} \cdot \mathbf{3\ cm} \cdot \mathbf{3\ cm}$$

$$A \approx 28.26\ \mathbf{cm^2}$$

5.10 Volume of a Cylinder

Use this formula to find the volume of a cylinder.

$$\text{Volume} = \pi \cdot r^2 \cdot h$$

where r is the radius of the circular base and h is the height of the cylinder.
Volume is measured in **cubic units**.

Find the volume of a cylinder that is 10 m high with a diameter of 8 m.

$$\text{Find the radius. } r = \frac{8\text{ m}}{2} = 4\text{ m}$$

$$V = \boldsymbol{\pi} \cdot \mathbf{r^2} \cdot \mathbf{h}$$

$$V \approx \mathbf{3.14} \cdot \mathbf{4\ m} \cdot \mathbf{4\ m} \cdot \mathbf{10\ m}$$

$$V \approx 502.4\ \mathbf{m^3}$$

5.10 Volume of a Cone

Use this formula to find the volume of a cone.

$$\text{Volume} = \frac{1}{3} \cdot B \cdot h$$

$$\text{or} \quad V = \frac{Bh}{3}$$

where B is the area of the circular base and h is the height of the cone.
Volume is measured in **cubic units**.

Find the volume of a cone, with a height of 9 inches and a base with a radius of 4 inches.

$$\text{Area of circular base} \approx 3.14 \cdot 4\text{ in.} \cdot 4\text{ in.}$$

$$\approx 50.24\text{ in.}^2$$

$$V = \frac{\mathbf{B} \cdot \mathbf{h}}{3}$$

$$V \approx \frac{\mathbf{50.24\ in.^2} \cdot \mathbf{9\ in.}}{3}$$

$$V \approx 150.72\ \mathbf{in.^3}$$

[5.1] *Name the digit that has the given place value.*

1. 243.059
tenths
hundredths

2. 0.6817
ones
tenths

3. $5824.39
hundreds
hundredths

4. 896.503
tenths
tens

5. 20.73861
tenths
ten-thousandths

Write each decimal as a fraction or mixed number in lowest terms.

6. 0.5

7. 0.75

8. 4.05

9. 0.875

10. 0.027

11. 27.8

Write each decimal in words.

12. 0.8

13. 400.29

14. 12.007

15. 0.0306

Write each decimal in numbers.

16. eight and three tenths

17. two hundred five thousandths

18. seventy and sixty-six ten-thousandths

19. thirty hundredths

[5.2] *Round to the place indicated. Write your answers using the "≈" sign.*

20. 275.635 to the nearest tenth

21. 72.789 to the nearest hundredth

22. 0.1604 to the nearest thousandth

23. 0.0905 to the nearest thousandth

24. 0.98 to the nearest tenth

Round to the nearest cent.

25. $15.8333

26. $0.698

27. $17,625.7906

Round each income or expense item to the nearest dollar.

28. Income from pancake breakfast was $350.48.

29. Members paid $129.50 in dues.

30. Refreshments cost $99.61.

31. Bank charges were $29.37.

Find each sum or difference.

32. $0.4 - 6.07$

33. $-20 + 19.97$

34. $-1.35 + 7.229$

35. $0.005 + (3 - 9.44)$

[5.3] *First use front end rounding to estimate each answer. Then find the exact answer.*

36. Tim agreed to donate 12.5 hours of work at his children's school. He has already worked 9.75 hours. How many more hours will he work?

estimate:
exact:

37. Today Jasmin wrote a check to the daycare center for \$215.53 and a check for \$44.47 at the grocery store. What was the total of the two checks?

estimate:
exact:

38. Joey spent \$1.59 for toothpaste, \$5.33 for a gift, and \$18.94 for a toaster. He gave the clerk three \$10 bills. How much change did he get?

estimate:

exact:

39. Roseanne is training for a wheelchair race. She raced 2.3 kilometers on Monday, 4 kilometers on Wednesday, and 5.25 kilometers on Friday. How far did she race altogether?

estimate:
exact:

[5.4] *First use front end rounding to estimate an answer. Then multiply to find the exact answer.*

40. *estimate* *exact*

$$\begin{array}{r} \\ \times \\ \hline \end{array} \qquad \begin{array}{r} 6.138 \\ \times \quad 3.7 \\ \hline \end{array}$$

41. *estimate* *exact*

$$\begin{array}{r} \\ \times \\ \hline \end{array} \qquad \begin{array}{r} 42.9 \\ \times \quad 3.3 \\ \hline \end{array}$$

Multiply.

42. $(-5.6)(-0.002)$

43. $(0.071)(-0.005)$

[5.5] *Decide if each answer is reasonable by rounding the numbers and estimating the answer. If the exact answer is not reasonable, find the correct answer.*

44. $706.2 \div 12 = 58.85$
estimate:

45. $26.6 \div 2.8 = 0.95$
estimate:

Divide. Round to the nearest thousandth, if necessary.

46. $3\overline{)43.4}$

47. $\dfrac{-72}{-0.06}$

48. $-0.00048 \div 0.0012$

[5.4–5.5] *Solve these application problems.*

49. Adrienne worked 36.5 hours this week. Her hourly wage is \$9.59. Find her total earnings to the nearest dollar.

50. A book of 12 tickets costs \$23.89 at the amusement park. What is the cost per ticket, to the nearest cent?

51. Stock in MathTronic sells for \$3.75 per share. Kenneth is thinking of investing \$500. How many whole shares could he buy?

52. Hamburger meat is on sale at \$0.89 per pound. How much will Ms. Lee pay for 3.5 pounds of hamburger, to the nearest cent?

Simplify by using the order of operations.

53. $3.5^2 + 8.7(-1.95)$

54. $11 - 3.06 \div (3.95 - 0.35)$

[5.6] *Write each fraction as a decimal. Round to the nearest thousandth, if necessary.*

55. $3\frac{4}{5}$

56. $\frac{16}{25}$

57. $1\frac{7}{8}$

58. $\frac{1}{9}$

Arrange in order from smallest to largest.

59. 3.68, 3.806, 3.6008

60. 0.215, 0.22, 0.209, 0.2102

61. 0.17, $\frac{3}{20}$, $\frac{1}{8}$, 0.159

[5.7] *Find the mean and the median for each set of data.*

62. Digital cameras sold:
18, 12, 15, 24, 9, 42, 54, 87, 21, 3

63. Number of insurance claims processed:
54, 28, 35, 43, 17, 37, 68, 75, 39

64. Find the weighted mean.

Dollar Value	*Frequency*
\$42	3
\$47	7
\$53	2
\$55	3
\$59	5

65. Find the mode or modes for each set of data.

(a) Hiking boots priced at \$80, \$72, \$64, \$64, \$72, \$53, \$64

(b) Boat launchings: 18, 25, 63, 32, 28, 37, 32, 26, 18

[5.8] *Find the unknown length in each right triangle. Use a calculator to find square roots. Round your answers to the nearest thousandth when necessary.*

66.

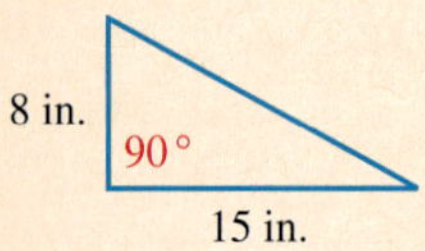

67.

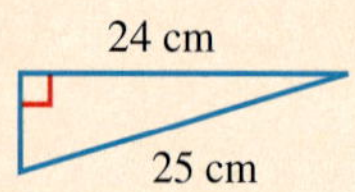

68.

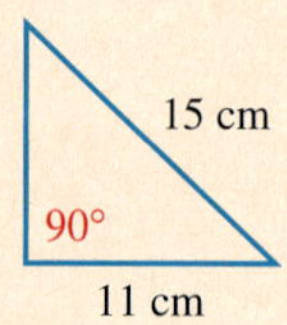

69.

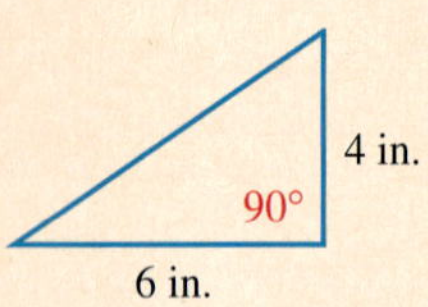

70.

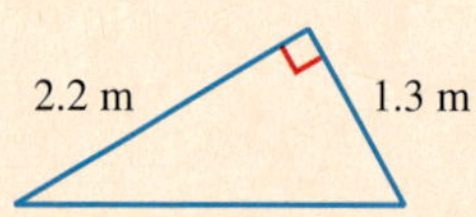

71.

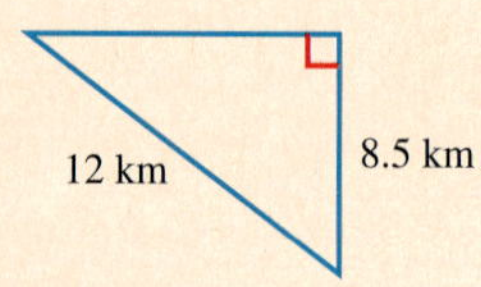

[5.9] *Solve each equation.*

72. $-0.1 = b - 0.35$

73. $-3.8x = 0$

74. $6.8 + 0.4n = 1.6$

75. $-0.375 + 1.75a = 2a$

76. $0.3y - 5.4 = 2.7 + 0.8y$

[5.10] *Find the missing value.*

77. The radius of a circular irrigation field is 68.9 m. What is the diameter of the field?

78. The diameter of a juice can is 3 inches. What is the radius of the can?

Find the circumference and area of each circle. Use 3.14 as the approximate value for π. Round your answers to the nearest tenth.

79.

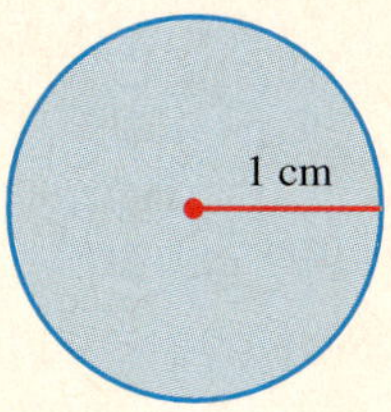

80.

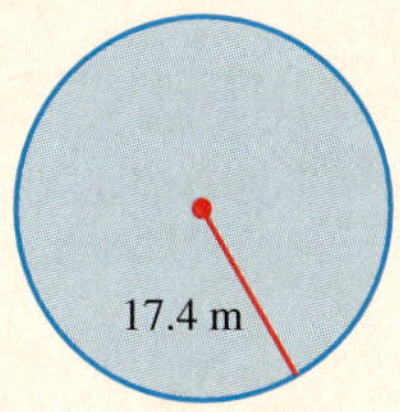

81.

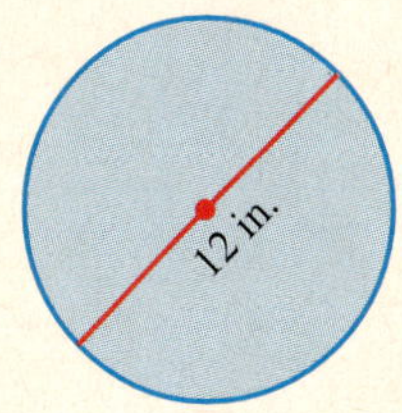

Find each volume. Use 3.14 as the approximate value for π. Round your answers to the nearest tenth, if necessary.

82.

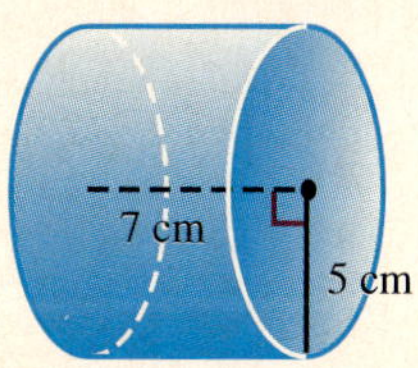

83.

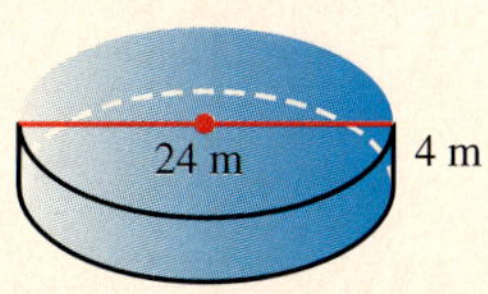

84.

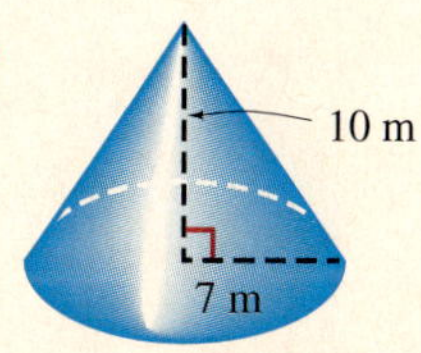

Mixed Review Exercises

Simplify.

85. $89.19 + 0.075 + 310.6 + 5$

86. $72.8(-3.5)$

87. $1648.3 \div 0.46$ Round to thousandths.

88. $30 - 0.9102$

89. $(4.38)(0.007)$

90. $0.005\overline{)0.047}$

91. $72.105 + 8.2 - 95.37$

92. $-81.36 \div (-9)$

93. $(0.6 - 1.22) + 4.8(-3.15)$

94. $0.455(18)$

95. $(-1.6)(-0.58)$

96. $0.218\overline{)7.63}$

97. $-21.059 - 20.8$

98. $18.3 - 3^2 \div 0.5$

Use the information in the ad to solve Exercises 99–103. Round money answers to the nearest cent. (Disregard any sales tax.)

99. How much would one pair of men's socks cost?

100. How much more would one pair of men's socks cost than one pair of children's socks?

101. How much would Fernando pay for a dozen pair of men's socks?

102. How much would Akiko pay for five pairs of teen jeans and four pairs of women's jeans?

103. What is the difference between the cheapest sale price for athletic shoes and the highest regular price?

Solve each equation.

104. $4.62 = -6.6y$

105. $1.05x - 2.5 = 0.8x + 5$

Solve these application problems. Round answers to the nearest tenth, if necessary.

106. A circular table has a diameter of 5 ft. How much rubber striping is needed to go around the edge of the table? What is the area of the table top?

107. Jerry missed one math test, so his test scores are 82, 0, 78, 93, 85. What is his average score?

108. LaRae drove 16 miles south, then made a 90° right turn and drove 12 miles west. How far is she, in a straight line, from her starting point?

109. A 7 in. tall juice can has a diameter of 4 in. What is the volume of the can?

Name ____________ Date ________ Class ________

CHAPTER 5 TEST

Write each decimal as a fraction or mixed number in lowest terms.

1. 18.4

2. 0.075

1. ____________

2. ____________

Write each decimal in words.

3. 60.007

4. 0.0208

3. ____________

4. ____________

First use front end rounding to estimate each answer. Then find the exact answer.

5. $7.6 + 82.0128 + 39.59$

6. $-5.79(1.2)$

7. $-79.1 - 3.602$

8. $-20.04 \div (-4.8)$

5. *estimate:* *exact:* ____________

6. *estimate:* *exact:* ____________

7. *estimate:* *exact:* ____________

8. *estimate:* *exact:* ____________

Find the exact answer.

9. $670 - 0.996$

10. $0.15\overline{)72}$

11. $(-0.006)(-0.007)$

9. ____________

10. ____________

11. ____________

12. Pat bought 3.4 meters of fabric. She paid \$15.47. What was the cost per meter?

12. ____________

13. Davida ran a race in 3.059 minutes. Angela ran the race in 3.5 minutes. Who won? By how much?

13. ____________

14. Mr. Yamamoto bought 1.85 pounds of cheese at \$2.89 per pound. How much did he pay for the cheese, to the nearest cent?

14. ____________

Solve each equation. Show the steps you used.

15. $-5.9 = y + 0.25$

16. $-4.2x = 1.47$

17. $3a - 22.7 = 10$

18. $-0.8n + 1.88 = 2n - 6.1$

15. ____________

16. ____________

17. ____________

18. ____________

19. ______________________

20. ______________________

21. ______________________

22. ______________________

23. ______________________

24. ______________________

25. ______________________

26. ______________________

27. ______________________

28. ______________________

29. ______________________

30. ______________________

31. ______________________

Arrange in order from smallest to largest

19. 0.44, 0.451, $\frac{9}{20}$, 0.4506

Use the order of operations to simplify.

20. $6.3^2 - 5.9 + 3.4(-0.5)$

21. Find the mean number of books loaned: 52, 61, 68, 69, 73, 75, 79, 84, 91, 98

22. Find the mode for hot tub temperatures (Fahrenheit) of 96°, 104°, 103°, 104°, 103°, 104°, 91°, 74°, 103°

23. Find the weighted mean.

Cost	*Frequency*
$ 6	7
$10	3
$11	4
$14	2
$19	3
$24	1

24. Find the median cost of a math textbook: $54.50, $48, $31.75, $89, $56.25, $49.30, $46.90, $51.80.

Find the unknown lengths. Round your answers to the nearest thousandth when necessary.

25.

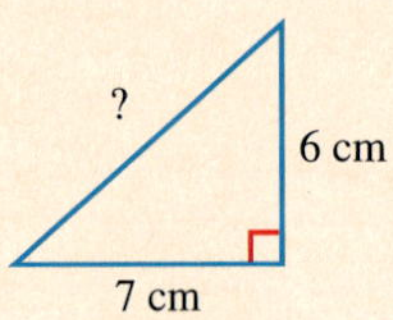

26.

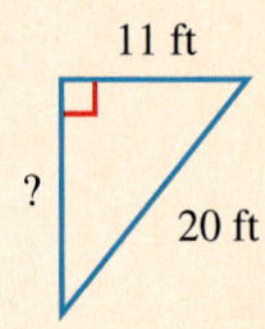

When necessary, use 3.14 as the approximate value for π and round your answers to the nearest tenth.

27. Find the radius.

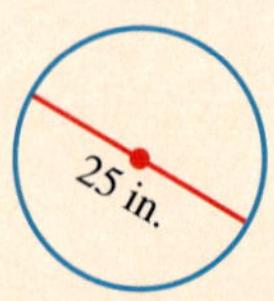

28. Find the circumference.

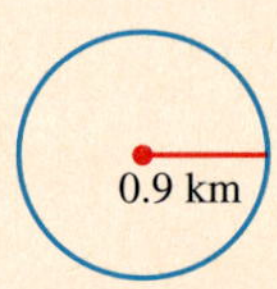

29. Find the area

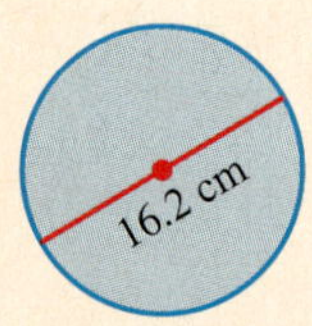

30. Find the volume.

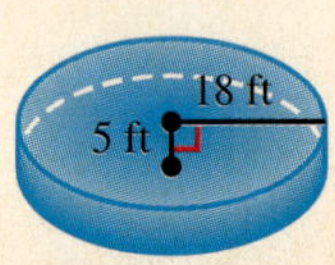

31. Find the volume.

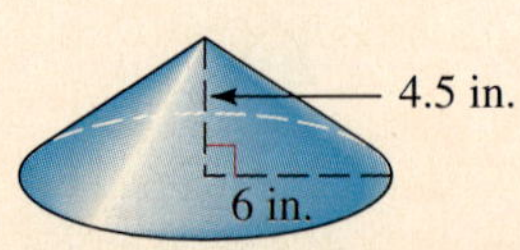

Name Date Class

CUMULATIVE REVIEW EXERCISES CHAPTERS 1–5

1. Write these numbers in words.

(a) 45.0203

(b) 30,000,650,008

2. Write these numbers using digits.

(a) one hundred sixty million, five hundred

(b) seventy-five thousandths

Round each number as indicated.

3. 46,908 to the nearest hundred

4. 6.197 to the nearest hundredth

5. 0.66148 to the nearest thousandth

6. 9951 to the nearest hundred

Simplify.

7. $-5 - 8$

8. $-0.003(0.02)$

9. $\frac{-7}{0}$

10. $8 + 4(2 - 5)^2$

11. $-\frac{3}{8}(-48)$

12. $|4| - |-10|$

13. $0.721 + 55.9$

14. $3\frac{1}{3} - 1\frac{5}{6}$

15. $\frac{3}{b^2} \cdot \frac{b}{8}$

16. $12 - 0.853$

17. $\frac{3}{10} - \frac{3}{4}$

18. $\frac{\frac{5}{16}}{-10}$

19. $-3.75 \div (-2.9)$ Round answer to nearest tenth.

20. $\frac{x}{2} + \frac{3}{5}$

21. $5 \div \left(-\frac{5}{8}\right)$

22. $2^5 - 4^3$

23. $\frac{-36 \div (-4) \cdot 2}{-6 - 4(0 - 6)}$

24. $(0.8)^2 - 3.2 + 4(-0.8)$

25. $\frac{3}{4} \div \frac{3}{10}\left(\frac{1}{4} + \frac{2}{3}\right)$

The ages (in years) of the students in a math class are: 19, 23, 24, 19, 20, 29, 26, 35, 20, 22, 26, 23, 25, 26, 20, 30. Use this data for Exercises 26–28.

26. Find the mean age, to the nearest tenth.

27. Find the median age.

28. Find the mode.

Solve each equation. Show the steps you used.

29. $3h - 4h = 16 - 12$

30. $-2x = x - 15$

31. $20 = 6r - 45.4$

32. $-3(y + 4) = 7y + 8$

33. $-0.8 + 1.4n = 2 + 0.7n$

Find the unknown length, perimeter, area, or volume. When necessary, use 3.14 as the approximate value of π and round answers to the nearest tenth.

34. Find the perimeter and the area.

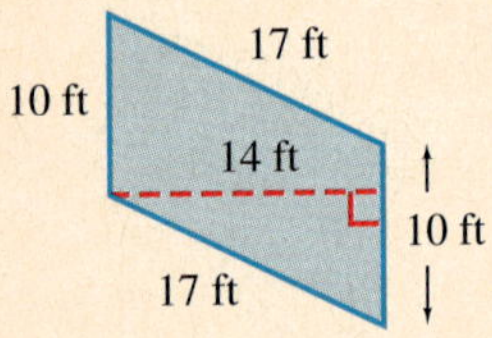

35. Find the circumference and the area.

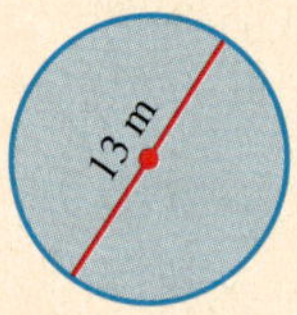

36. Find the length of the third side and the area.

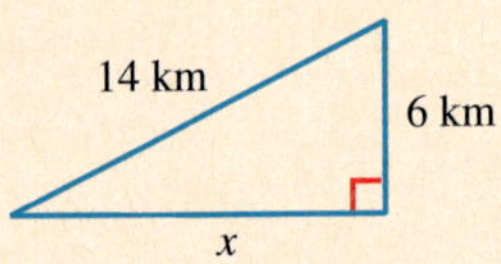

37. Find the volume.

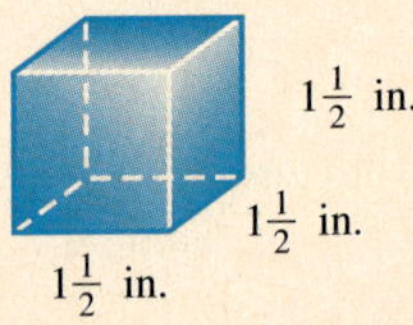

First use front end rounding to estimate each answer. Then find the exact answer.

38. Lameck had two \$10 bills. He spent \$7.96 on gasoline and \$0.87 for a candy bar at the convenience store. How much money does he have left?

estimate:
exact:

39. Toshihiro bought $2\frac{1}{3}$ yards of cotton fabric and $3\frac{7}{8}$ yards of wool fabric. How many yards did he buy in all?

estimate:
exact:

40. Paulette bought 2.7 pounds of grapes for \$2.56. What was the cost per pound, to the nearest cent?

estimate:
exact:

41. Carter Community College received a \$78,000 grant from a local computer company to help students pay tuition for computer classes. How much money could be given to each of 107 students? Round to the nearest dollar.

estimate:
exact:

6 Ratio and Proportion

A **ratio** (RAY-show) compares two quantities. You can compare two numbers, such as 8 and 4, or two measurements, such as 3 days and 12 days.

INTERNET
www.mathnotes.com

6.1 Ratios

OBJECTIVES

1. Write ratios as fractions.
2. Solve ratio problems involving decimals or mixed numbers.
3. Solve ratio problems after converting units.

FOR EXTRA HELP

Tutorial | Tape 10 | SSM, Sec. 6.1

OBJECTIVE 1 A ratio can be written in three ways.

Writing a Ratio

The ratio of \$7 **to** \$3 can be written:

$$7 \text{ to } 3 \quad \text{or} \quad 7:3 \quad \text{or} \quad \frac{7}{3} \leftarrow \text{Fraction bar indicates to}$$

":" indicates **to**

Writing a ratio as a fraction is the most common method, and the one we will use here. All three ways are read, "the ratio of 7 **to** 3." The word **to** separates the quantities being compared.

Writing a Ratio as a Fraction

Order is important when you're writing a ratio. The quantity mentioned **first** is the **numerator**. The quantity mentioned **second** is the **denominator**. For example:

The ratio of **5** to **12** is written $\frac{5}{12}$.

1. Shane spent $14 on meat, $5 on milk, and $7 on fresh fruit. Write these ratios as fractions.

(a) The ratio of amount spent on fruit to amount spent on milk.

(b) The ratio of amount spent on milk to amount spent on meat.

(c) The ratio of amount spent on meat to amount spent on milk.

EXAMPLE 1 Writing a Ratio

The Anasazi, ancestors of the Pueblo Indians, built multistory apartment towns in New Mexico about 1100 years ago. A room might measure 14 feet long, 11 feet wide, and 15 feet high.

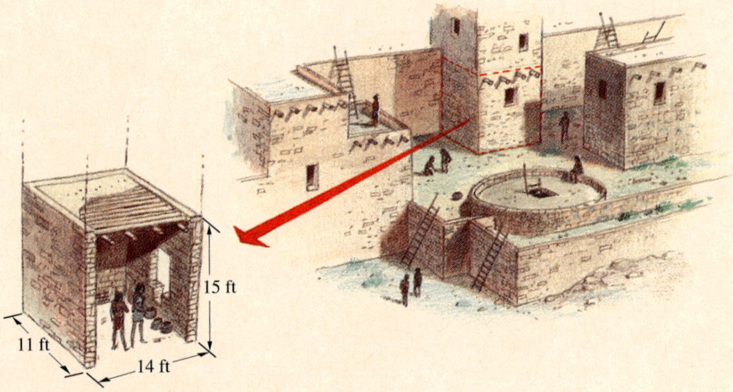

Write these ratios using the room measurements:

(a) ratio of length to width

$$\text{The ratio of } \textbf{length to width} \text{ is } \frac{14 \cancel{\text{ feet}}}{11 \cancel{\text{ feet}}} = \frac{14}{11}.$$

Numerator (mentioned first) — Denominator (mentioned second)

You can divide out common *units* just as you did with common *factors* when writing fractions in lowest terms. (See **Section 4.2.**) However, do *not* rewrite the fraction as a mixed number. Keep it as the ratio of 14 to 11.

(b) ratio of width to height

$$\text{The ratio of width to height is } \frac{11 \cancel{\text{ feet}}}{15 \cancel{\text{ feet}}} = \frac{11}{15}.$$

Note

Remember, the *order* of the numbers is important in a ratio. Look for the words "ratio of a to b." Write the ratio as $\frac{a}{b}$, **not** $\frac{b}{a}$. The quantity mentioned first is the numerator.

◀◀ WORK PROBLEM 1 AT THE SIDE.

Any ratio can be written as a fraction. Therefore, you can write a ratio in lowest terms, just as you do with any fraction.

EXAMPLE 2 Writing a Ratio in Lowest Terms

Write each ratio in lowest terms.

(a) 60 days to 20 days.

The ratio is $\frac{60}{20}$. Write this ratio in lowest terms by dividing numerator and denominator by 20.

$$\frac{60}{20} = \frac{60 \div 20}{20 \div 20} = \frac{3}{1} \quad \text{Ratio in lowest terms}$$

Note

In the fractions chapter you would have rewritten $\frac{3}{1}$ as 3. But a *ratio* compares *two* quantities, so you need to keep both parts of the ratio and write it as $\frac{3}{1}$.

CONTINUED ON NEXT PAGE

ANSWERS

1. (a) $\frac{7}{5}$ (b) $\frac{5}{14}$ (c) $\frac{14}{5}$

(b) 50 ounces of medicine to 120 ounces of medicine.
The ratio is $\frac{50}{120}$. Divide numerator and denominator by 10.

$$\frac{50}{120} = \frac{50 \div \mathbf{10}}{120 \div \mathbf{10}} = \frac{5}{12} \quad \left\{\text{Ratio in lowest terms}\right.$$

(c) 18 people in a large van to 8 people in a small van.

$$\text{The ratio is } \frac{18}{8} = \frac{18 \div \mathbf{2}}{8 \div \mathbf{2}} = \frac{9}{4} \quad \left\{\text{Ratio in lowest terms}\right.$$

Note

Although $\frac{9}{4} = 2\frac{1}{4}$, ratios are *not* written as mixed numbers. Nevertheless, in Example 2(c), the ratio $\frac{9}{4}$ does mean the large van holds $2\frac{1}{4}$ times as many people as the small van.

WORK PROBLEM 2 AT THE SIDE.

OBJECTIVE 2 Sometimes a ratio compares two decimal numbers or two fractions. It is easier to understand if we rewrite the ratio as a ratio of two whole numbers.

EXAMPLE 3 Using Decimal Numbers in a Ratio

The price of a Sunday newspaper increased from \$1.20 to \$1.50. Find the ratio of the increase in price **to** the original price.

Approach The words increase in price are mentioned first, so the increase will be the numerator. How much did the price go up? Use subtraction.

new price	−	original price	=	increase
\$1.50	−	\$1.20	=	\$0.30

The words original price are mentioned second, so the original price is the denominator.

Solution The ratio of increase in price **to** original price is

$$\frac{0.30}{1.20} \begin{matrix} \leftarrow \text{increase} \\ \leftarrow \text{original price} \end{matrix}$$

Now we rewrite the ratio as a ratio of whole numbers. Recall that if you multiply both the numerator and denominator of a fraction by the same number, you get an equivalent fraction. The decimals in this example are hundredths, so multiply by 100 to get whole numbers. (If the decimals are tenths, multiply by 10. If thousandths, multiply by 1000.) Then write the ratio in lowest terms.

$$\frac{0.30}{1.20} = \frac{0.30 \cdot \mathbf{100}}{1.20 \cdot \mathbf{100}} = \underbrace{\frac{30}{120}}_{\text{Ratio as two whole numbers}} = \frac{30 \div \mathbf{30}}{120 \div \mathbf{30}} = \frac{1}{4} \quad \left\{\text{Ratio in lowest terms}\right.$$

WORK PROBLEM 3 AT THE SIDE.

2. Write each ratio as a fraction in lowest terms.

(a) 9 hours to 12 hours

(b) 100 meters to 50 meters

(c) Write the ratio of width to length for this rectangle.

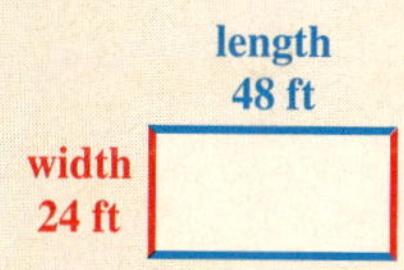

3. Write each ratio as a ratio of whole numbers in lowest terms.

(a) The price of Tamar's favorite brand of lipstick increased from \$3.75 to \$4.25. Find the ratio of the increase in price to the original price.

(b) Last week Lance worked 4.5 hours each day. This week he cut back to 3 hours each day. Find the ratio of the decrease in hours to the original number of hours.

ANSWERS

2. (a) $\frac{3}{4}$ **(b)** $\frac{2}{1}$ **(c)** $\frac{1}{2}$

3. (a) $\frac{0.50 \cdot 100}{3.75 \cdot 100} = \frac{50 \div 25}{375 \div 25} = \frac{2}{15}$

(b) $\frac{1.5 \cdot 10}{4.5 \cdot 10} = \frac{15 \div 15}{45 \div 15} = \frac{1}{3}$

EXAMPLE 4 Using Mixed Numbers in a Ratio

Write each ratio as a comparison of whole numbers in lowest terms.

(a) 2 days to $2\frac{1}{4}$ days

Write the ratio as follows. Divide out the common units.

$$\frac{2 \text{ days}}{2\frac{1}{4} \text{ days}} = \frac{2}{2\frac{1}{4}}$$

Next, write 2 as $\frac{2}{1}$ and $2\frac{1}{4}$ as the improper fraction $\frac{9}{4}$.

$$\frac{2}{2\frac{1}{4}} = \frac{\frac{2}{1}}{\frac{9}{4}}$$

Now rewrite the problem, using the "÷" symbol for division. Finally, multiply by the reciprocal of the divisor, as you did in **Section 4.3.**

$$\frac{\frac{2}{1}}{\frac{9}{4}} = \frac{2}{1} \div \frac{9}{4} = \frac{2}{1} \cdot \frac{4}{9} = \frac{8}{9}$$

Reciprocals

The ratio, in lowest terms, is $\frac{8}{9}$.

(b) $3\frac{1}{4}$ to $1\frac{1}{2}$

Write the ratio as $\frac{3\frac{1}{4}}{1\frac{1}{2}}$. Then write $3\frac{1}{4}$ and $1\frac{1}{2}$ as improper fractions.

$$3\frac{1}{4} = \frac{13}{4} \quad \text{and} \quad 1\frac{1}{2} = \frac{3}{2}$$

The ratio is

$$\frac{3\frac{1}{4}}{1\frac{1}{2}} = \frac{\frac{13}{4}}{\frac{3}{2}}.$$

Write as a division problem using the "÷" symbol. Then multiply by the reciprocal of the divisor.

$$\frac{13}{4} \div \frac{3}{2} = \frac{13}{4} \cdot \frac{2}{3} = \frac{13 \cdot \overset{1}{\cancel{2}}}{\underset{1}{\cancel{2}} \cdot 2 \cdot 3} = \frac{13}{6}$$

Reciprocals

Ratio in lowest terms

Note

We can also work Examples 4(a) and 4(b) by using decimals.

(a) $2\frac{1}{4}$ is equivalent to 2.25, so the ratio is

$$\frac{2}{2\frac{1}{4}} = \frac{2}{2.25} = \frac{2 \cdot 100}{2.25 \cdot 100} = \frac{200}{225} = \frac{200 \div 25}{225 \div 25} = \frac{8}{9}. \leftarrow \text{Same result}$$

(b) $3\frac{1}{4}$ is equivalent to 3.25 and $1\frac{1}{2}$ is equivalent to 1.5.

$$\frac{3\frac{1}{4}}{1\frac{1}{2}} = \frac{3.25}{1.5} = \frac{3.25 \cdot 100}{1.5 \cdot 100} = \frac{325}{150} = \frac{325 \div 25}{150 \div 25} = \frac{13}{6} \leftarrow \text{Same result}$$

This method would *not* work for fractions that are repeating decimals, such as $\frac{1}{3}$ or $\frac{5}{6}$.

WORK PROBLEM 4 AT THE SIDE.

OBJECTIVE 3 When a ratio compares measurements, both measurements must be in the *same* units. For example, *feet* must be compared to *feet, hours* to *hours, pints* to *pints,* and *inches* to *inches.*

EXAMPLE 5 Ratio Applications Using Measurement

(a) Write the ratio of the length of the board on the left to the board on the right. Compare in inches.

First, express 2 feet in inches. Because 1 foot has 12 inches, 2 feet is

$$2 \cdot \mathbf{12\ inches} = 24 \text{ inches.}$$

The length of the board on the left is 24 inches, so the ratio of the lengths is

$$\frac{24 \text{ inches}}{30 \text{ inches}} = \frac{24}{30}.$$

Write the ratio in lowest terms.

$$\frac{24}{30} = \frac{24 \div 6}{30 \div 6} = \frac{4}{5} \quad \text{Ratio in lowest terms}$$

The shorter board is $\frac{4}{5}$ the length of the longer board.

Note

Notice that we wrote the ratio using the smaller unit (inches are smaller than feet). Using the smaller unit will help you avoid working with fractions. If we wrote the ratio using feet, then:

$$30 \text{ inches} = 2\frac{1}{2} \text{ feet.}$$

So the ratio is:

$$\frac{2 \text{ feet}}{2\frac{1}{2} \text{ feet}} = \frac{2}{1} \div \frac{5}{2} = \frac{2}{1} \cdot \frac{2}{5} = \frac{4}{5}.$$

The ratio is the same, but it takes more steps to get the answer. Using the smaller unit is usually easier.

CONTINUED ON NEXT PAGE

4. Write each ratio as a ratio of whole numbers in lowest terms.

(a) $3\frac{1}{2}$ to 4

(b) $5\frac{5}{8}$ pounds to $3\frac{3}{4}$ pounds

(c) $3\frac{1}{2}$ inches to $\frac{7}{8}$ inch

ANSWERS

4. (a) $\frac{7}{8}$ **(b)** $\frac{3}{2}$ **(c)** $\frac{4}{1}$

5. Write each ratio as a fraction in lowest terms.

(a) 9 inches to 6 feet
Compare in inches.

(b) 2 days to 8 hours
Compare in hours.

(c) 7 yards to 14 feet
Compare in feet.

(d) 3 quarts to 3 gallons
Compare in quarts.

(e) 25 minutes to 2 hours
Compare in minutes.

(f) 4 pounds to 12 ounces
Compare in ounces.

(b) Write the ratio of 28 days to 3 weeks. Compare in days.
First express 3 weeks in days. Because 1 week has 7 days, 3 weeks is

$$3 \cdot \mathbf{7\ days} = 21 \text{ days.}$$

So the ratio in days is

$$\frac{28 \not{\text{days}}}{21 \not{\text{days}}} = \frac{28}{21} = \frac{28 \div 7}{21 \div 7} = \frac{4}{3}. \leftarrow \text{Lowest terms}$$

The following table will help you set up ratios that compare measurements. You will work with these measurements again in Chapter 8.

Length	Capacity (Volume)
1 foot = 12 inches	1 pint = 2 cups
1 yard = 3 feet	1 quart = 2 pints
1 mile = 5280 feet	1 gallon = 4 quarts

Weight	Time
1 pound = 16 ounces	1 week = 7 days
1 ton = 2000 pounds	1 day = 24 hours
	1 hour = 60 minutes
	1 minute = 60 seconds

◀◀ WORK PROBLEM 5 AT THE SIDE.

ANSWERS

5. (a) $\frac{1}{8}$ (b) $\frac{6}{1}$ (c) $\frac{3}{2}$ (d) $\frac{1}{4}$ (e) $\frac{5}{24}$ (f) $\frac{16}{3}$

6.1 Exercises

Write each ratio as a fraction in lowest terms.

1. 8 to 9

2. 11 to 15

3. \$100 to \$50

4. 35¢ to 7¢

5. 30 minutes to 90 minutes

6. 9 pounds to 36 pounds

7. 80 miles to 50 miles

8. 300 people to 450 people

9. 6 hours to 16 hours

10. 45 books to 35 books

Write each ratio as a ratio of whole numbers in lowest terms.

11. \$4.50 to \$3.50

12. \$0.08 to \$0.06

13. 15 to $2\frac{1}{2}$

14. 5 to $1\frac{1}{4}$

15. $1\frac{1}{4}$ to $1\frac{1}{2}$

16. $2\frac{1}{3}$ to $2\frac{2}{3}$

Write each ratio as a fraction in lowest terms. For help, use the table of measurement relationships in Example 5.

Example:
40 ounces to 2 pounds
Compare in ounces.

Solution:
1 pound has 16 ounces; 2 • 16 ounces = 32 ounces

$$\frac{40 \text{ ounces}}{32 \text{ ounces}} = \frac{40 \div 8}{32 \div 8} = \frac{5}{4} \leftarrow \text{Lowest terms}$$

17. 4 feet to 30 inches
Compare in inches.

18. 8 feet to 4 yards
Compare in feet.

19. 5 minutes to 1 hour
Compare in minutes.

20. 8 quarts to 5 pints
Compare in pints.

21. 15 hours to 2 days
Compare in hours.

22. 3 pounds to 6 ounces
Compare in ounces.

23. 5 gallons to 5 quarts
Compare in quarts.

24. 3 cups to 3 pints
Compare in cups.

Solve each application problem. Write each ratio as a fraction in lowest terms.

25. A 440-pound tiger may consume 60 pounds of food in one meal. Find the ratio of the tiger's weight to the weight of its meal.

26. When the Concorde jet lifts off the runway it is flying at 250 miles per hour. Fifty minutes later it is cruising at a supersonic speed of 1350 miles per hour. Write the ratio of the jet's takeoff speed to its cruising speed.

27. Our math class has 16 women and 20 men. What is the ratio of men to women?

28. Cherise sells souvenirs at baseball games. She sold 30 red hats and 40 blue hats. What is the ratio of blue hats to red hats?

29. The Sanchez Company made 400 washing machines. Four of them had defects. What is the ratio of defective washers to the total number of washers?

30. Andrew spends $500 per month on rent and $120 per month on utilities. Find the ratio of the amount spent on utilities to the amount spent on rent.

The table below shows the number of Americans who play various instruments. Use the information in the table to complete Exercises 31–34.

Instrument	Number Who Play
Piano	22 million
Guitar	20 million
Organ	6 million
Clarinet	4 million
Drums	3 million
Violin	2 million

31. Find the ratio of piano players to violin players.

32. Find the ratio of drum players to organ players.

33. Find the ratio of organ players to guitar players.

34. Find the ratio of guitar players to clarinet players.

35. Would you prefer that the ratio of your income to your friend's income be 1 to 3 or 3 to 1? Explain your answer.

36. Amelia said that the ratio of her age to her mother's age is 5 to 3. Is this possible? Explain your answer.

Use the circle graph below of one family's monthly budget to complete Exercises 37–40. Write each ratio as a fraction in lowest terms.

37. Find the ratio of taxes to transportation.

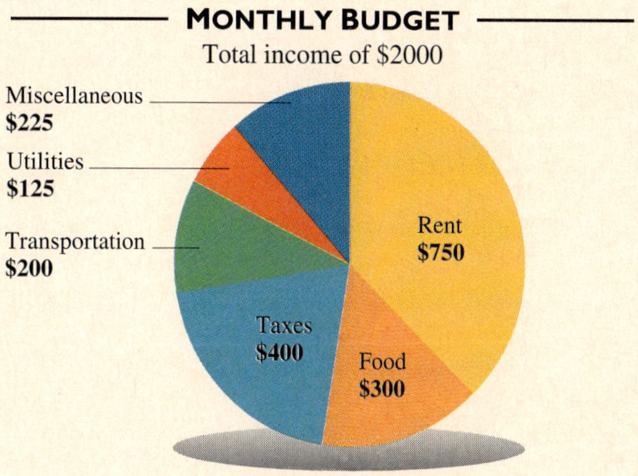

38. Find the ratio of rent to food.

39. Find the ratio of rent to total income.

40. Find the ratio of utilities to total income.

For each figure, find the ratio of the length of the longest side to the length of the shortest side. Write each ratio as a fraction in lowest terms.

41.

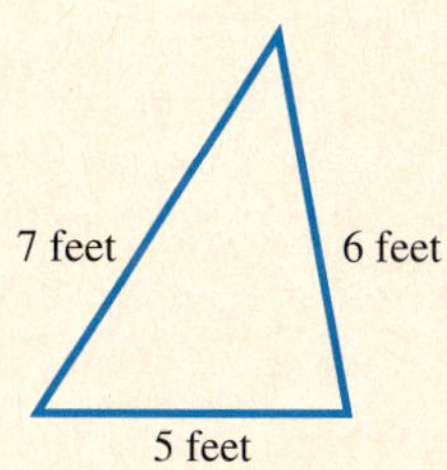

42.

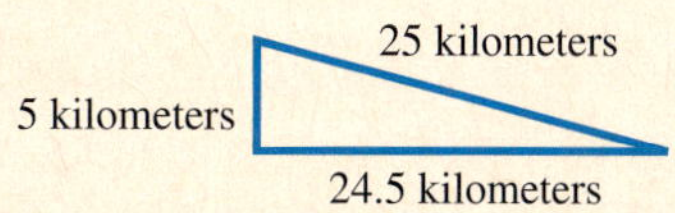

43.

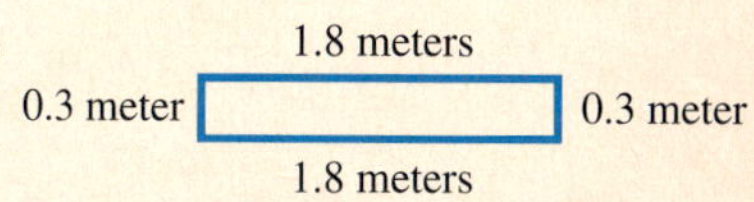

44.

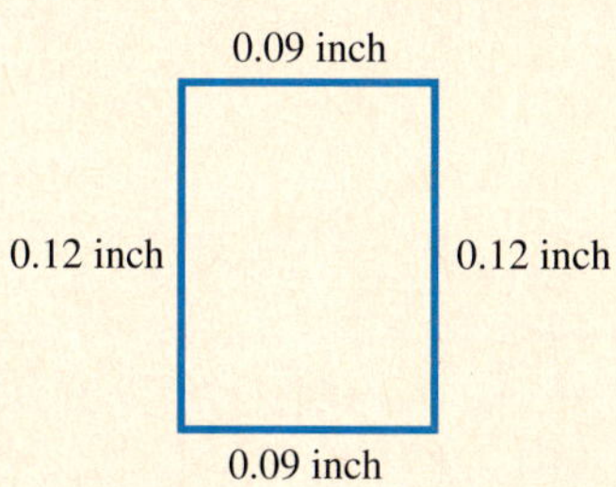

45.

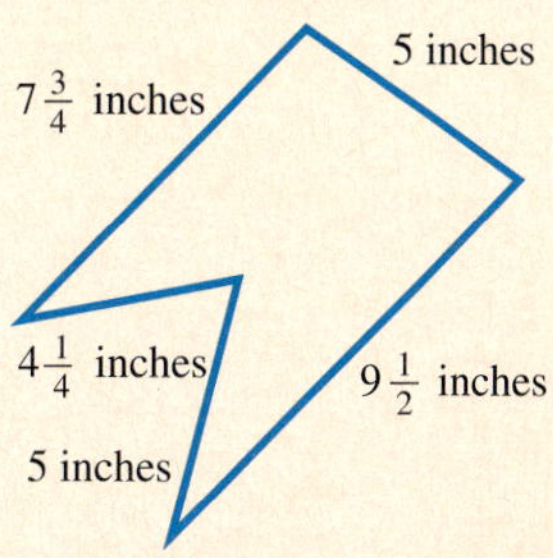

46.

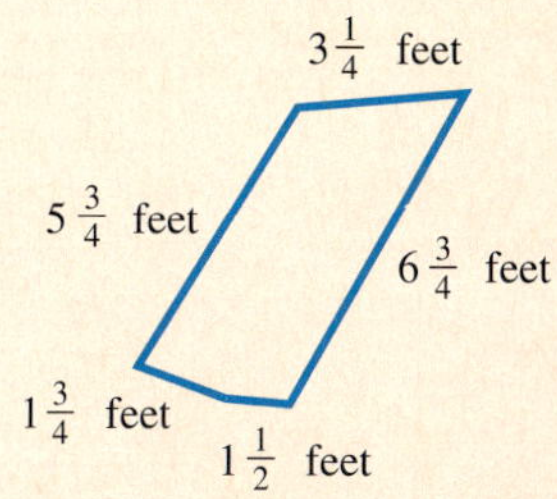

47. The price of oil recently went from \$6.60 to \$9.90 per case of 12 quarts. Find the ratio of the increase in price to the original price.

48. The price of an antibiotic decreased from \$8.80 to \$5.60 for a bottle of 100 tablets. Find the ratio of the decrease in price to the original price.

49. The first time a movie was made in Minnesota, the cast and crew spent $59\frac{1}{2}$ days filming winter scenes. The next year, another movie was filmed in $8\frac{3}{4}$ weeks. Find the ratio of the first movie's filming time to the second movie's time. Compare in weeks.

50. The percheron, a large draft horse, measures about $5\frac{3}{4}$ feet at the shoulder. The prehistoric ancestor of the horse measured only $15\frac{3}{4}$ inches at the shoulder. Find the ratio of the percheron's height to its prehistoric ancestor's height. Compare in inches.

51. The ratio of John's age to his sister's age is 4 to 5. One possibility is that John is 4 years old and his sister is 5 years old. Find six other possibilities that fit the 4 to 5 ratio.

52. In this painting, what is the ratio of the length of the longest side to the length of the shortest side? What other measurements could the painting have and still maintain the same ratio?

Review and Prepare

Divide. Round to the nearest thousandth, if necessary. (For help, see ***Section 5.5.****)*

53. $7\overline{)0.65}$ **54.** $3\overline{)7.33}$ **55.** $4\overline{)4.1}$ **56.** $0.95\overline{)41.8}$ **57.** $0.71\overline{)6.72}$ **58.** $4.6\overline{)116.38}$

6.2 *Rates*

A ratio compares two measurements with the same type of units, such as 9 feet **to** 12 feet (both length measurements). But many of the comparisons we make use measurements with different types of units, such as:

40 dollars **for** 8 hours **(money to time)**
450 miles **on** 18 gallons **(distance to capacity)**

This type of comparison is called a **rate.**

OBJECTIVE 1 For example, suppose you hiked 18 miles **in** 4 hours. The *rate* at which you hiked can be written as a fraction in lowest terms.

$$\frac{18 \text{ miles}}{4 \text{ hours}} = \frac{18 \text{ miles} \div 2}{4 \text{ hours} \div 2} = \frac{9 \text{ miles}}{2 \text{ hours}} \quad \text{Lowest terms}$$

In a rate, you often find these words separating the quantities you are comparing:

in **for** **on** **per** **from**

Note
When writing a rate, always include the units. Because the units in a rate are different, they do *not* divide out.

EXAMPLE 1 Write a Rate in Lowest Terms

Write each rate as a fraction in lowest terms.

(a) 5 gallons of chemical **for** $60.

$$\frac{5 \text{ gallons} \div 5}{60 \text{ dollars} \div 5} = \frac{1 \text{ gallon}}{12 \text{ dollars}}$$

(b) $1500 wages **in** 10 weeks

$$\frac{1500 \text{ dollars} \div 10}{10 \text{ weeks} \div 10} = \frac{150 \text{ dollars}}{1 \text{ week}}$$

(c) 2225 miles **on** 75 gallons of gas

$$\frac{2225 \text{ miles} \div 25}{75 \text{ gallons} \div 25} = \frac{89 \text{ miles}}{3 \text{ gallons}}$$

WORK PROBLEM 1 AT THE SIDE.

OBJECTIVE 2 When the *denominator* of a rate is 1, it is called a **unit rate.** We use unit rates frequently. For example, you earn $8.75 for *1 hour* of work. This unit rate is written:

$8.75 **per** hour or $8.75**/**hour.

Use **per** or a **/** mark when writing unit rates. You drive 28 miles on *1 gallon* of gas. This unit rate is written 28 miles **per** gallon, or 28 miles**/**gallon.

OBJECTIVES
1 Write rates as fractions.
2 Find unit rates.

FOR EXTRA HELP
Tutorial Tape 10 SSM, Sec. 6.2

1. Write each rate as a fraction in lowest terms.

(a) $6 for 30 packages

(b) 500 miles in 10 hours

(c) 4 teachers for 90 students

(d) 1270 bushels on 30 acres

ANSWERS

1. (a) $\frac{\$1}{5 \text{ packages}}$ **(b)** $\frac{50 \text{ miles}}{1 \text{ hour}}$ **(c)** $\frac{2 \text{ teachers}}{45 \text{ students}}$ **(d)** $\frac{127 \text{ bushels}}{3 \text{ acres}}$

2. Find each unit rate.

(a) $4.35 for 3 pounds of cheese

(b) 304 miles on 9.5 gallons of gas

(c) $850 in 5 days

(d) 24-pound turkey for 15 people

EXAMPLE 2 Finding a Unit Rate

Find each unit rate.

(a) 337.5 miles on 13.5 gallons of gas

Write the rate as a fraction.

$$\frac{337.5 \text{ miles}}{13.5 \text{ gallons}} \leftarrow \text{Fraction bar indicates division}$$

Divide 337.5 by 13.5 to find the unit rate.

$$13.5\overline{)337.5} = 25.$$

$$\frac{337.5 \text{ miles} \div \mathbf{13.5}}{13.5 \text{ gallons} \div \mathbf{13.5}} = \frac{25 \text{ miles}}{1 \text{ gallon}}$$

The unit rate is 25 miles/gallon.

(b) 549 miles in 18 hours

$$\frac{549 \text{ miles}}{18 \text{ hours}} \qquad \text{Divide } 18\overline{)549.0} = 30.5$$

The unit rate is 30.5 miles/hour.

(c) $810 in 6 days

$$\frac{810 \text{ dollars}}{6 \text{ days}} \qquad \text{Divide } 6\overline{)810} = 135$$

The unit rate is $135/day.

WORK PROBLEM 2 AT THE SIDE.

Cost per Unit

Cost per unit is a rate that tells how much you pay for *one* item or *one* unit. Examples are $1.25 per gallon, $47 per shirt, and $2.98 per pound. When shopping, you can save money by finding the lowest cost per unit.

EXAMPLE 3 Determining the Best Buy

The local store charges the following prices for pancake syrup. Find the best buy.

Size	Price
12 ounces	$1.28
24 ounces	$1.81
36 ounces	$2.73

CONTINUED ON NEXT PAGE

ANSWERS

2. **(a)** $1.45/pound **(b)** 32 miles/gallon **(c)** $170/day **(d)** 1.6 pounds/person

Approach The best buy is the container with the *lowest* cost per unit. All the containers are measured in *ounces,* so you first need to find the *cost per ounce* for each one. Divide the price of the container by the number of ounces in it. Round to the nearest thousandth, if necessary.

Solution

Size	*Cost per Unit (Rounded)*
12 ounces	$\dfrac{\$1.28}{12 \text{ ounces}} \approx \$0.107$ per ounce (highest)
24 ounces	$\dfrac{\$1.81}{24 \text{ ounces}} \approx \$0.075$ per ounce (lowest)
36 ounces	$\dfrac{\$2.73}{36 \text{ ounces}} \approx \$0.076$ per ounce

The lowest cost per ounce is \$0.075, so the 24-ounce container is the best buy.

Note

Earlier we rounded money amounts to the nearest hundredth (nearest cent). But when comparing unit costs, rounding to the nearest thousandth will help you see the difference between very similar unit costs. Notice that the 24-ounce and 36-ounce containers would both have rounded to \$0.08 if we had rounded to hundredths.

WORK PROBLEM 3 AT THE SIDE.

Calculator Tip: When using a calculator to find unit prices, remember that division is *not* commutative. In Example 3 you wanted to find cost per ounce. Let the *order* of the *words* help you enter the numbers in the correct order.

cost	per	**ounce**	
↓ Enter the cost ↓	↓ Per means divide ↓	↓ Enter number of ounces ↓	
\$2.73	÷	**36**	=

(If you entered 36 ÷ 2.73 = you'd get the number of *ounces* per *dollar*. How could you use that information to find the best buy?)

Finding the "best buy" is sometimes a complicated process. Things that affect the cost per unit can include "cents off" coupons and differences in how much use you'll get out of each unit.

3. Find the best buy (lowest cost per unit) for each purchase.

(a) 2 quarts for \$3.25
3 quarts for \$4.95
4 quarts for \$6.48

(b) 6 cans of cola for \$1.99
12 cans of cola for \$3.49
24 cans of cola for \$7

ANSWERS

3. **(a)** 4 quarts, at \$1.62 per quart
(b) 12 cans, at ≈\$0.291 per can

4. (a) Some batteries claim to last longer than others. If you believe these claims, which brand is the "best buy"?

Four-pack of AA size batteries for $2.79.

One AA size battery for $1.19. Lasts twice as long.

(b) Which tube of toothpaste is the better buy? You have a coupon for 85¢ off Brand C and a coupon for 20¢ off Brand D.

Brand C is $3.89 for 6 ounces

Brand D is $1.59 for 2.5 ounces

EXAMPLE 4 Solving Best Buy Applications

(a) There are many brands of liquid laundry detergent. If you feel they all do a good job of cleaning your clothes, you can base your purchase on cost per unit. But some brands are now "concentrated" so you use less detergent for each load of clothes. Which choice is the "best buy"?

To find Sudzy's unit cost, divide $3.99 by 64 ounces, not 50 ounces. You're getting as many clothes washed as if you bought 64 ounces. Similarly, to find White-O's unit cost, divide $9.89 by 256 ounces (twice 128 ounces, or $2 \cdot 128$ ounces = 256 ounces).

$$\text{Sudzy} \quad \frac{\$3.99}{64 \text{ ounces}} \approx \$0.062 \text{ per ounce}$$

$$\text{White-O} \quad \frac{\$9.89}{256 \text{ ounces}} \approx \$0.039 \text{ per ounce}$$

White-O has the lower cost per ounce and is the better buy. (However, if you try it and it really doesn't get out all the stains, Sudzy may be worth the extra cost.)

(b) "Cents-off" coupons also affect the best buy. Suppose you are looking at these choices for "extra strength" aspirin.

Brand X is $2.29 for 50 tablets

Brand Y is $10.75 for 200 tablets

You have a 40¢ coupon for Brand X and a 75¢ coupon for Brand Y. To find the better buy, first subtract the coupon amounts, then divide to find the lower cost per ounce.

$$\text{Brand X costs } \$2.29 - \$0.40 = \$1.89$$

$$\frac{\$1.89}{50 \text{ tablets}} \approx \$0.038 \text{ per tablet}$$

$$\text{Brand Y costs } \$10.75 - \$0.75 = \$10.00$$

$$\frac{\$10.00}{200 \text{ tablets}} = \$0.05 \text{ per tablet}$$

Brand X has the lower cost per tablet and is the better buy.

◀◀ WORK PROBLEM 4 AT THE SIDE.

ANSWERS

4. (a) One battery that lasts twice as long (like getting two) is the better buy. The cost per unit is $0.595 per battery. The four-pack is ≈$0.698 per battery.

(b) Brand C with the 85¢ coupon is the better buy at ≈$0.507 per ounce. Brand D with the 20¢ coupon is $0.556 per ounce.

6.2 Exercises

Write each rate as a fraction in lowest terms.

1. 10 cups for 6 people

2. \$12 for 30 pens

3. 15 feet in 35 seconds

4. 100 miles in 30 hours

5. 14 people for 28 dresses

6. 12 wagons for 48 horses

7. 25 letters in 5 minutes

8. 68 pills for 17 people

9. \$63 for 6 visits

10. 25 doctors for 310 patients

11. 72 miles on 4 gallons

12. 132 miles on 8 gallons

Find each unit rate.

Example: \$49.35 for 15 boxes

Solution: Divide $15\overline{)49.35}$ = 3.29
The unit rate is **\$3.29 per box**

13. \$60 in 5 hours

14. \$2500 in 20 days

15. 50 eggs from 10 chickens

16. 36 children from 12 families

17. 7.5 pounds for 6 people

18. 44 bushels from 8 trees

19. \$413.20 for 4 days

20. \$74.25 for 9 hours

Earl kept the record shown below of the gas he bought for his car. For each entry, find the number of miles he traveled and the unit rate. Round your answers to the nearest tenth.

	Date	Odometer at Start	Odometer at End	Miles Traveled	Gallons Purchased	Miles per Gallon
21.	2/4	27,432.3	27,758.2		15.5	
22.	2/9	27,758.2	28,058.1		13.4	
23.	2/16	28,058.1	28,396.7		16.2	
24.	2/20	28,396.7	28,704.5		13.3	

Find the best buy (based on the cost per unit) for each of the following.

25. black pepper
4 ounces for $0.89
8 ounces for $2.13

26. shampoo
8 ounces for $0.99
12 ounces for $1.47

27. cereal
15 ounces for $2.60
17 ounces for $2.89
21 ounces for $3.79

28. soup
2 cans for $0.75
3 cans for $1.17
5 cans for $1.79

29. chunky peanut butter
12 ounces for $1.09
18 ounces for $1.41
28 ounces for $2.29
40 ounces for $3.19

30. pork and beans
8 ounces for $0.37
16 ounces for $0.77
21 ounces for $0.99
31 ounces for $1.50

31. Suppose you are choosing between two brands of chicken noodle soup. Brand A is $0.38 per can and Brand B is $0.48 per can. But Brand B has more chunks of chicken in it. Which soup is the better buy? Explain your choice.

32. A small bag of potatoes costs $0.19 per pound. A large bag costs $0.15 per pound. But there are only two people in your family, so half the large bag would probably rot before you use it up. Which bag is the better buy? Explain.

Solve each application problem.

33. Makesha lost 10.5 pounds in six weeks. What was her rate of loss in pounds per week?

34. Enrique's taco recipe uses four pounds of meat to feed 10 people. Give the rate in pounds per person.

35. Russ works 7 hours to earn $85.82. What is his rate per hour?

36. Find the cost of 1 gallon of gas if 18 gallons cost $20.88.

37. Ms. Johnson bought 150 shares of stock for $1725. Find the cost of one share.

38. A company pays $6450 in dividends for the 2500 shares of its stock. Find the dividend per share.

39. In 1996, Michael Johnson ran the 200-meter sprint in a record time of approximately 20 seconds (actually 19.32 seconds). Give his rate in seconds per meter and in meters per second. Use 20 seconds as the time.

40. Sofia can clean and adjust five hearing aids in four hours. Give her rate in hearing aids per hour and in hours per hearing aid.

41. The 4.6 yards of fabric needed for a dress coat cost $51.75. Find the cost of 1 yard of fabric.

42. The cost to lay 42.4 square yards of carpet is $691.12. Find the cost of 1 square yard of carpet.

G **43.** If you believe the claims that some batteries last longer, which is the better buy: one AAA battery for \$1.79 that lasts three times as long, or an eight-pack of AAA batteries for \$4.99?

G **44.** Which is the better buy, assuming these laundry detergents both clean equally well: 64 fluid ounces for \$5.99, concentrated so you can wash twice as many loads as usual; or 150 fluid ounces for \$7.29 (not concentrated).

G **45.** Three brands of cornflakes are available. Brand G is priced at \$2.39 for 10 ounces. Brand K is \$3.99 for 20.3 ounces and Brand P is \$3.39 for 16.5 ounces. You have a coupon for 50¢ off Brand P and a coupon for 60¢ off Brand G. Which cereal is the best buy based on cost per unit?

G **46.** Two brands of facial tissue are available. Brand K is on special at three boxes of 175 tissues each for \$5. Brand S is priced at \$1.29 per box of 125 tissues. You have a coupon for 20¢ off one box of Brand S and a coupon for 45¢ off one box of Brand K. How can you get the best buy on one box of tissue?

Review and Prepare

Simplify. Write your answers as whole or mixed numbers when possible. (For help, see ***Section 4.6.)***

47. $4 + 2\frac{3}{4}\left(1 - \frac{1}{5}\right)$

48. $12 - 5\frac{2}{3}\left(\frac{2}{9} + \frac{4}{9}\right)$

49. $5\frac{2}{5} \div 18\left(\frac{5}{6}\right)$

50. $1\frac{5}{8} \div \left(2\frac{1}{4} - \frac{9}{16}\right)$

51. $\dfrac{\frac{15}{16}}{\frac{5}{32}}$

52. $\dfrac{6}{\frac{3}{4}}$

6.3 Proportions

OBJECTIVE 1 A **proportion** (proh-POR-shun) states that two ratios (or rates) are equivalent. For example,

$$\frac{\$20}{4 \text{ hours}} = \frac{\$40}{8 \text{ hours}}$$

is a proportion that says the rate $\frac{\$20}{4 \text{ hours}}$ is equivalent to the rate $\frac{\$40}{8 \text{ hours}}$. As the amount of money doubles, the number of hours also doubles. This proportion is read:

20 dollars **is to** 4 hours **as** 40 dollars **is to** 8 hours.

EXAMPLE 1 Writing a Proportion

Write each of the following proportions.

(a) 6 feet is to 11 feet as 18 feet is to 33 feet.

$$\frac{6 \,\cancel{\text{feet}}}{11 \,\cancel{\text{feet}}} = \frac{18 \,\cancel{\text{feet}}}{33 \,\cancel{\text{feet}}} \quad \text{so} \quad \frac{6}{11} = \frac{18}{33}$$

The common units (feet) divide out and are not written.

(b) \$9 is to 6 liters as \$3 is to 2 liters.

$$\frac{\$9}{6 \text{ liters}} = \frac{\$3}{2 \text{ liters}}$$

Units must be written.

WORK PROBLEM 1 AT THE SIDE.

OBJECTIVE 2 There are two ways to see whether a proportion is true. One way is to *write both of the ratios in lowest terms.*

EXAMPLE 2 Writing Both Ratios in Lowest Terms

Are the following proportions true?

(a) $\frac{5}{9} = \frac{18}{27}$ Write each ratio in lowest terms.

$\frac{5}{9}$ ← Already in lowest terms $\frac{18 \div \mathbf{9}}{27 \div \mathbf{9}} = \frac{2}{3}$ ← Lowest terms

Because $\frac{5}{9}$ is *not* equivalent to $\frac{2}{3}$, the proportion is *false.* The ratios are *not* proportional.

(b) $\frac{16}{12} = \frac{28}{21}$ Write each ratio in lowest terms.

$$\frac{16 \div \mathbf{4}}{12 \div \mathbf{4}} = \frac{4}{3} \quad \text{and} \quad \frac{28 \div \mathbf{7}}{21 \div \mathbf{7}} = \frac{4}{3}$$

Both ratios are equivalent to $\frac{4}{3}$, so the proportion is *true.* The ratios are proportional.

WORK PROBLEM 2 AT THE SIDE.

OBJECTIVES

1. Write proportions.
2. Decide whether proportions are true or false.
3. Find the unknown number in a proportion.

FOR EXTRA HELP

Tutorial Tape 10 SSM, Sec. 6.3

1. Write each proportion.

(a) \$7 is to 3 cans as \$28 is to 12 cans

(b) 9 meters is to 16 meters as 18 meters is to 32 meters

(c) 5 is to 7 as 35 is to 49

(d) 10 is to 30 as 60 is to 180

2. Are these proportions true or false?

(a) $\frac{6}{12} = \frac{15}{30}$

(b) $\frac{20}{24} = \frac{3}{4}$

(c) $\frac{25}{40} = \frac{30}{48}$

(d) $\frac{35}{45} = \frac{12}{18}$

(e) $\frac{21}{45} = \frac{56}{120}$

ANSWERS

1. (a) $\frac{\$7}{3 \text{ cans}} = \frac{\$28}{12 \text{ cans}}$ **(b)** $\frac{9}{16} = \frac{18}{32}$ **(c)** $\frac{5}{7} = \frac{35}{49}$ **(d)** $\frac{10}{30} = \frac{60}{180}$

2. (a) true **(b)** false **(c)** true **(d)** false **(e)** true

3. Cross multiply to see whether the following proportions are true or false.

(a) $\frac{5}{9} = \frac{10}{18}$

(b) $\frac{32}{15} = \frac{16}{8}$

(c) $\frac{10}{17} = \frac{20}{34}$

(d) $\frac{2.4}{6} = \frac{5}{12}$

$6 \cdot 5 =$

$2.4 \cdot 12 =$

(e) $\frac{3}{4.25} = \frac{24}{34}$

(f) $\frac{1\frac{1}{6}}{2\frac{1}{3}} = \frac{4}{8}$

Answers

3. **(a)** true **(b)** false **(c)** true **(d)** false **(e)** true **(f)** true

A second way to see if a proportion is true is to find *cross products.*

Using Cross Products to Test a Proportion

To see whether a proportion is true, first multiply along one diagonal, then multiply along the other diagonal, as shown here.

$$\frac{2}{5} = \frac{4}{10} \qquad \begin{aligned} 5 \cdot 4 &= \mathbf{20} \\ 2 \cdot 10 &= \mathbf{20} \end{aligned}$$

Cross products are equal

In this case the **cross products** are both 20. When cross products are *equal,* the proportion is *true.* If the cross products are *unequal,* the proportion is *false.*

Note

Why does the cross products test work? It is based on rewriting both fractions with the common denominator of $5 \cdot 10$ or 50. (We do not search for the *lowest* common denominator. We simply use the product of the two given denominators.)

$$\frac{2 \cdot \mathbf{10}}{5 \cdot \mathbf{10}} = \frac{20}{50} \qquad \frac{4 \cdot \mathbf{5}}{10 \cdot \mathbf{5}} = \frac{20}{50}$$

We take a shortcut by comparing only the two numerators (20 = 20).

EXAMPLE 3 Using Cross Products

Use cross multiplication to see whether the following proportions are true or false.

(a) $\frac{3}{5} = \frac{12}{20}$ Cross multiply one way and then the other way.

$$\frac{3}{5} = \frac{12}{20} \qquad \begin{aligned} 5 \cdot 12 &= \mathbf{60} \\ 3 \cdot 20 &= \mathbf{60} \end{aligned}$$

Equal

The cross products are equal, so the proportion is *true.*

(b) $\frac{2\frac{1}{3}}{3\frac{1}{3}} = \frac{9}{16}$ Cross multiply.

Changed to improper fractions

$$3\frac{1}{3} \cdot 9 = \frac{10}{\cancel{3}_{1}} \cdot \frac{\cancel{9}^{3}}{1} = \frac{30}{1} = \mathbf{30}$$

$$2\frac{1}{3} \cdot 16 = \frac{7}{3} \cdot \frac{16}{1} = \frac{112}{3} = \mathbf{37\frac{1}{3}}$$

Unequal

The cross products are unequal, so the proportion is *false.*

Note

The numbers in a proportion do *not* have to be whole numbers.

◀◀ WORK PROBLEM 3 AT THE SIDE.

OBJECTIVE 3 Four numbers are used in a proportion. If any three of these numbers are known, the fourth can be found. For example, find the unknown number that will make this proportion true.

$$\frac{3}{5} = \frac{x}{40}$$

The variable x represents the unknown number. Start by finding the cross products.

$$\frac{3}{5} = \frac{x}{40} \qquad \begin{matrix} 5 \cdot x \\ 3 \cdot 40 \end{matrix} \quad \text{Cross products}$$

To make the proportion true, the cross products must be equal. This gives us the following equation.

$$\underbrace{5 \cdot x}_{5x} = \underbrace{3 \cdot 40}_{120}$$

Recall from **Section 2.4** that we can solve an equation of this type by dividing both sides by the coefficient of the variable term. In this case, the coefficient of $5x$ is 5.

$$\frac{5x}{\mathbf{5}} = \frac{120}{\mathbf{5}} \quad \leftarrow \text{Divide both sides by 5.}$$

Divide out the common factor of 5.

$$\frac{\overset{1}{\cancel{5}} \cdot x}{\underset{1}{\cancel{5}}} = 24 \quad \text{Divide 120 by 5.}$$

Multiplying by 1 does *not* change a number, so $1 \cdot x$ is the same as x.

$$\frac{x}{1} = 24$$

Dividing by 1 does *not* change a number, so $\frac{x}{1}$ is the same as x.

$$x = 24$$

The missing number in the proportion is 24. The complete proportion is shown below.

$$\frac{3}{5} = \frac{24}{40} \quad \leftarrow x \text{ is } 24.$$

Check by finding the cross products. If they are equal, you solved the problem correctly. If they are unequal, rework the problem.

$$\frac{3}{5} = \frac{24}{40} \qquad \begin{matrix} 5 \cdot 24 = \mathbf{120} \\ 3 \cdot 40 = \mathbf{120} \end{matrix} \quad \text{Equal; proportion is true}$$

The cross products are equal, so the solution, $x = 24$, is correct.

Note

The solution is 24, which is the missing number in the proportion. 120 is **not** the solution; it is the cross product you get when checking the solution.

Solve a proportion for an unknown number with the following steps.

Finding an Unknown Number in a Proportion

Step 1 Find the cross products.

Step 2 Show that the cross products are equivalent.

Step 3 Divide both sides of the equation by the coefficient of the variable term.

EXAMPLE 4 Solving for an Unknown Number

Find the unknown number in each proportion. Round to hundredths, if necessary.

(a) $\frac{16}{x} = \frac{32}{20}$

Recall that ratios can be rewritten in lowest terms. If desired, you can do that *before* finding the cross products. In this example, write $\frac{32}{20}$ in lowest terms $\left(\frac{8}{5}\right)$ to get $\frac{16}{x} = \frac{8}{5}$.

Step 1 $\frac{16}{x} = \frac{8}{5}$ $x \cdot 8$, $16 \cdot 5$ ← Find cross products.

Step 2 $x \cdot 8 = 16 \cdot 5$ ← Show that cross products are equivalent.

$x \cdot 8 = 80$

Step 3 $\frac{x \cdot \cancel{8}^{1}}{\cancel{8}_{1}} = \frac{80}{8}$ ← Divide both sides by 8.

$x = 10$ ← Find x. (No rounding necessary.)

Write the complete proportion and check by finding the cross products.

$\frac{16}{10} = \frac{8}{5}$ $10 \cdot 8 = \mathbf{80}$, $16 \cdot 5 = \mathbf{80}$ ← Equal; proportion is true.

The cross products are equal, so 10 is the solution.

Note

It is not necessary to write the ratios in lowest terms before solving. However, if you do, you will have smaller numbers to work with.

(b) $\frac{7}{12} = \frac{15}{x}$

$\frac{7}{12} = \frac{15}{x}$ $12 \cdot 15 = 180$, $7 \cdot x$ ← Find cross products.

Show that cross products are equivalent.

$$7 \cdot x = 180$$

CONTINUED ON NEXT PAGE

Divide both sides by 7.

$$\frac{\overset{1}{\cancel{7}} \cdot x}{\underset{1}{\cancel{7}}} = \frac{180}{7}$$

$$x \approx 25.71 \text{ (rounded to nearest hundredth)}$$

When the division does not come out even, check for directions on how to round your answer. Divide out one more place, then round.

$$7\overline{)180.000} = 25.714$$ ← Divide out to thousandths. Round to hundredths.

Write the complete proportion and check by finding cross products.

$$\frac{7}{12} \approx \frac{15}{25.71}$$

$12 \cdot 15 = \mathbf{180}$

$7 \cdot 25.71 = \mathbf{179.97}$

Very close but not equal

The cross products are slightly different because you rounded the value of x. However, they are close enough to see that the problem was done correctly and 25.71 is the approximate solution.

WORK PROBLEM 4 AT THE SIDE.

The next example shows how to solve for the unknown number in a proportion with fractions or decimals.

EXAMPLE 5 Using Mixed Numbers and Decimals

Find the missing number in each proportion.

(a) $\frac{2\frac{1}{5}}{6} = \frac{x}{10}$ Cross multiply.

$$\frac{2\frac{1}{5}}{6} = \frac{x}{10}$$

$6 \cdot x$

$2\frac{1}{5} \cdot 10$

Find $2\frac{1}{5} \cdot 10$.

$$2\frac{1}{5} \cdot 10 = \frac{11}{5} \cdot \frac{10}{1} = \frac{11 \cdot 2 \cdot \overset{1}{\cancel{5}}}{\underset{1}{\cancel{5}} \cdot 1} = \frac{22}{1} = 22$$

Changed to improper fraction

CONTINUED ON NEXT PAGE

4. Find the unknown numbers. Round to hundredths, if necessary. Check your answers by finding cross products.

(a) $\frac{1}{2} = \frac{x}{12}$

(b) $\frac{6}{10} = \frac{15}{x}$

(c) $\frac{28}{x} = \frac{21}{9}$

(d) $\frac{x}{8} = \frac{3}{5}$

(e) $\frac{14}{11} = \frac{x}{3}$

ANSWERS

4. **(a)** $x = 6$ **(b)** $x = 25$ **(c)** $x = 12$ **(d)** $x = 4.8$ **(e)** $x \approx 3.82$ (rounded to nearest hundredth)

5. Find the unknown numbers. Round to hundredths on the decimal problems, if necessary. Check your answers by finding cross products.

(a) $\dfrac{3\frac{1}{4}}{2} = \dfrac{x}{8}$

(b) $\dfrac{x}{3} = \dfrac{1\frac{2}{3}}{5}$

(c) $\dfrac{0.06}{x} = \dfrac{0.3}{0.4}$

(d) $\dfrac{2.2}{5} = \dfrac{13}{x}$

(e) $\dfrac{x}{6} = \dfrac{0.5}{1.2}$

(f) $\dfrac{0}{2} = \dfrac{x}{7.092}$

Show that the cross products are equivalent.

$$6 \cdot x = 22$$

Divide both sides by 6.

$$\frac{\overset{1}{\cancel{6}} \cdot x}{\underset{1}{\cancel{6}}} = \frac{22}{6}$$

Write answer as a mixed number in lowest terms.

$$x = \frac{22 \div 2}{6 \div 2} = \frac{11}{3} = 3\frac{2}{3}$$

Write the complete proportion and check by finding cross products.

$$\frac{2\frac{1}{5}}{6} = \frac{3\frac{2}{3}}{10}$$

$$6 \cdot 3\frac{2}{3} = \frac{2 \cdot \overset{1}{\cancel{3}}}{1} \cdot \frac{11}{\underset{1}{\cancel{3}}} = \frac{22}{1} = 22$$

$$2\frac{1}{5} \cdot 10 = \frac{11}{\underset{1}{\cancel{5}}} \cdot \frac{2 \cdot \overset{1}{\cancel{5}}}{1} = \frac{22}{1} = 22$$

Equal

The cross products are equal, so $3\frac{2}{3}$ is the correct solution.

Note

You can use decimal numbers and your calculator to solve Example 5(a). $2\frac{1}{5}$ is equivalent to 2.2, so the cross products are

$$6 \cdot x = 2.2 \cdot 10$$

$$\frac{\overset{1}{\cancel{6}} \cdot x}{\underset{1}{\cancel{6}}} = \frac{22}{6}$$

When you divide 22 by 6 on your calculator, it shows 3.666666667. Write the answer using a bar to show the repeating digit: $3.\overline{6}$. Or round the answer to 3.67 (nearest hundredth).

(b) $\dfrac{1.5}{0.6} = \dfrac{2}{x}$

Show that cross products are equivalent.

$$1.5 \cdot x = 0.6 \cdot 2$$
$$1.5 \cdot x = 1.2$$

Divide both sides by 1.5.

$$\frac{\overset{1}{\cancel{1.5}} \cdot x}{\underset{1}{\cancel{1.5}}} = \frac{1.2}{1.5}$$

$$x = \frac{1.2}{1.5}$$ Complete the division. $1.5\overline{)1.20}$ = .8

$$x = 0.8$$

So the unknown number is 0.8. Check by finding cross products.

$$\frac{1.5}{0.6} = \frac{2}{0.8}$$

$$0.6 \cdot 2 = 1.2$$
$$1.5 \cdot 0.8 = 1.2$$

Equal

The cross products are equal, so 0.8 is the correct solution.

Answers

5. **(a)** $x = 13$ **(b)** $x = 1$ **(c)** $x = 0.08$ **(d)** $x \approx 29.55$ (rounded to nearest hundredth) **(e)** $x = 2.5$ **(f)** $x = 0$

◀◀ WORK PROBLEM 5 AT THE SIDE.

6.3 Exercises

Write each proportion.

1. \$9 is to 12 cans as \$18 is to 24 cans

2. 28 people is to 7 cars as 16 people is to 4 cars

3. 200 adults is to 450 children as 4 adults is to 9 children

4. 150 trees is to 1 acre as 1500 trees is to 10 acres

5. 120 feet is to 150 feet as 8 feet is to 10 feet

6. \$6 is to \$9 as \$10 is to \$15

7. 2.2 hours is to 3.3 hours as 3.2 hours is to 4.8 hours

8. 4 meters is to 4.75 meters as 6 meters is to 7.125 meters

Write each ratio in lowest terms in order to decide whether the following proportions are true or false.

9. $\frac{6}{10} = \frac{3}{5}$

10. $\frac{1}{4} = \frac{9}{36}$

11. $\frac{5}{8} = \frac{25}{40}$

12. $\frac{2}{3} = \frac{20}{27}$

13. $\frac{150}{200} = \frac{200}{300}$

14. $\frac{100}{120} = \frac{75}{100}$

Use cross multiplication to decide whether the following proportions are true or false. Circle the correct answer.

15. $\frac{2}{9} = \frac{6}{27}$

True False

16. $\frac{20}{25} = \frac{4}{5}$

True False

17. $\frac{20}{28} = \frac{12}{16}$

True False

18. $\frac{16}{40} = \frac{22}{55}$

True False

19. $\frac{110}{18} = \frac{160}{27}$

True False

20. $\frac{600}{420} = \frac{20}{14}$

True False

21. $\frac{3.5}{4} = \frac{7}{8}$

True False

22. $\frac{36}{23} = \frac{9}{5.75}$

True False

23. $\frac{18}{16} = \frac{2.8}{2.5}$

True False

24. $\frac{0.26}{0.39} = \frac{1.3}{1.9}$

True False

25. $\frac{6}{3\frac{2}{3}} = \frac{18}{11}$

True False

26. $\frac{16}{13} = \frac{2}{1\frac{5}{8}}$

True False

27. Suppose Jerome Walton of the Atlanta Braves had 16 hits in 50 times at bat, and Mariano Duncan of the New York Yankees was at bat 400 times and got 128 hits. Paul is trying to convince Jamie that the two men hit equally well. Show how you could use a proportion and cross products to see if Paul is correct.

28. Jay worked 3.5 hours and packed 91 cartons. Craig packed 126 cartons in 5.25 hours. To see if the men worked equally fast, Barry set up this proportion:

$$\frac{3.5}{91} = \frac{126}{5.25}.$$

Explain what is wrong with Barry's proportion and write a correct one. Is the correct proportion true or false?

Find the unknown number in each proportion. Round your answers to hundredths, if necessary. Check your answers by finding cross products.

29. $\frac{1}{3} = \frac{x}{12}$

30. $\frac{x}{6} = \frac{15}{18}$

31. $\frac{15}{10} = \frac{3}{x}$

32. $\frac{5}{x} = \frac{20}{8}$

33. $\frac{x}{11} = \frac{32}{4}$

34. $\frac{12}{9} = \frac{8}{x}$

35. $\frac{42}{x} = \frac{18}{39}$

36. $\frac{49}{x} = \frac{14}{18}$

37. $\frac{x}{25} = \frac{4}{20}$

38. $\frac{6}{x} = \frac{4}{8}$

39. $\frac{8}{x} = \frac{24}{30}$

40. $\frac{32}{5} = \frac{x}{10}$

41. $\frac{99}{55} = \frac{44}{x}$

42. $\frac{x}{12} = \frac{101}{147}$

43. $\frac{0.7}{9.8} = \frac{3.6}{x}$

44. $\frac{x}{3.6} = \frac{4.5}{6}$

45. $\frac{250}{24.8} = \frac{x}{1.75}$

46. $\frac{4.75}{17} = \frac{43}{x}$

Ⓖ *These proportions are* **not** *true. Change any* one *of the numbers in each proportion to make them true.*

47. $\frac{10}{4} = \frac{5}{3}$

48. $\frac{6}{8} = \frac{24}{30}$

Find the unknown number in each proportion. Write your answers as whole or mixed numbers when possible.

49. $\frac{15}{1\frac{2}{3}} = \frac{9}{x}$

50. $\frac{x}{\frac{3}{10}} = \frac{2\frac{2}{9}}{1}$

51. $\frac{2\frac{1}{3}}{1\frac{1}{2}} = \frac{x}{2\frac{1}{4}}$

52. $\frac{1\frac{5}{6}}{x} = \frac{\frac{3}{14}}{\frac{6}{7}}$

Ⓖ *Solve these proportions two different ways. First change all the numbers to decimal form and solve. Then change all the numbers to fraction form and solve; write your answers in lowest terms.*

53. $\frac{\frac{1}{2}}{x} = \frac{2}{0.8}$

54. $\frac{\frac{3}{20}}{0.1} = \frac{0.03}{x}$

55. $\frac{x}{\frac{3}{50}} = \frac{0.15}{1\frac{4}{5}}$

56. $\frac{8\frac{4}{5}}{1\frac{1}{10}} = \frac{x}{0.4}$

Review and Prepare

Write each set of rates as a proportion and use cross multiplication to decide whether it is true or false. Circle the correct answer. (For help, see **Sections 6.2 and 6.3.***)*

57. 25 feet in 18 seconds
15 feet in 10 seconds

True False

58. 170 miles on 6.8 gallons
330 miles on 13.2 gallons

True False

59. $14.75 for 2 hours
$33.25 for 4.5 hours

True False

6.4 Problem Solving: Proportions

OBJECTIVE 1 Proportions can be used to solve a wide variety of problems. Watch for problems in which you are given a ratio or rate and then asked to find part of a corresponding ratio or rate. Remember that a ratio or rate compares two quantities and often includes one of these indicator words:

in for on per from to

When setting up the proportion, use a variable to represent the unknown number. We have used the letter x, but you may use any letter you like.

EXAMPLE 1 Using a Proportion to Solve Problems

Mike's car can go 163 **miles on** 6.4 **gallons** of gas. How far can it go on a full tank of 14 **gallons** of gas? Round to the nearest whole mile.

Approach Decide what is being compared. This example compares **miles** to **gallons**. Write the two rates described in the example. Be sure that *both* rates compare miles to gallons in the same order. In other words, miles is in both numerators and gallons is in both denominators. Use a variable to represent the unknown number.

compares **miles** to **gallons** $\left\{ \dfrac{163 \text{ miles}}{6.4 \text{ gallons}} = \dfrac{x \text{ miles}}{14 \text{ gallons}} \right\}$ compares **miles** to **gallons**

Solution Both rates compare **miles** to **gallons**, so you can set them up as a proportion.

Note

Do **not** mix up the units in the rates.

compares **miles** to **gallons** $\left\{ \dfrac{163 \text{ miles}}{6.4 \text{ gallons}} \quad \dfrac{14 \text{ gallons}}{x \text{ miles}} \right\}$ compares **gallons** to **miles**

These rates do **not** compare things in the same order and **cannot** be set up as a proportion.

With the proportion set up correctly, solve for the unknown number of miles.

$$\frac{163 \text{ miles}}{6.4 \text{ gallons}} = \frac{x \text{ miles}}{14 \text{ gallons}}$$

Matching units (miles); Matching units (gallons)

Ignore the units while finding the cross products and dividing both sides by 6.4.

$6.4 \cdot x = 163 \cdot 14$ Show that cross products are equivalent.

$6.4 \cdot x = 2282$

$\dfrac{\overset{1}{\cancel{6.4}} \cdot x}{\underset{1}{\cancel{6.4}}} = \dfrac{2282}{6.4}$ Divide both sides by 6.4.

$x = 356.5625$

Rounded to the nearest mile, the car can go $\approx$357 **miles** on a full tank of gas. Be sure to *include the units* in your answer.

WORK PROBLEM 1 AT THE SIDE.

OBJECTIVE

1 Use proportions to solve application problems.

FOR EXTRA HELP

Tutorial Tape 11 SSM, Sec. 6.4

1. Set up and solve a proportion for each problem.

(a) If 2 pounds of fertilizer will cover 50 square feet of garden, how many pounds are needed for 225 square feet?

(b) A U.S. map has a scale of 1 inch to 75 miles. Lake Superior is 4.75 inches long on the map. What is the lake's actual length in miles?

(c) Cough syrup is to be given at the rate of 30 milliliters for each 100 pounds of body weight. How much should be given to a 34-pound child? Round to the nearest whole milliliter.

ANSWERS

1. (a) $\dfrac{2 \text{ pounds}}{50 \text{ sq feet}} = \dfrac{x \text{ pounds}}{225 \text{ sq feet}}$
$x = 9$ pounds

(b) $\dfrac{1 \text{ inch}}{75 \text{ miles}} = \dfrac{4.75 \text{ inches}}{x \text{ miles}}$
$x = 356.25$ miles or $x \approx 356$ miles

(c) $\dfrac{30 \text{ milliliters}}{100 \text{ pounds}} = \dfrac{x \text{ milliliters}}{34 \text{ pounds}}$
$x \approx 10$ milliliters

2. Solve each problem to find a reasonable answer. Then flip one side of your proportion to see what answer you get with an incorrect setup. Explain why the second answer is unreasonable.

(a) A survey showed that 2 out of 3 people would like to lose weight. At this rate, how many people in a group of 150 want to lose weight?

(b) In one state, 3 out of 5 college students receive financial aid. At this rate, how many of the 4500 students at Central Community College receive financial aid?

(c) An advertisement says that 9 out of 10 dentists recommend sugarless gum. If the ad is true, how many of the 60 dentists in our city would recommend sugarless gum?

ANSWERS

2. **(a)** 100 people (reasonable); incorrect setup gives 225 people (only 150 people in the group).
(b) 2700 students (reasonable); incorrect setup gives 7500 students (only 4500 students at the college).
(c) 54 dentists (reasonable); incorrect setup gives ≈67 dentists (only 60 dentists in the city).

EXAMPLE 2 More Proportion Applications

A newspaper report says that 7 out of 10 people surveyed watch the news on TV. At that rate, how many of the 3200 people in town would you expect to watch the news?

Approach You are comparing people who watch the news to people surveyed. Write the two rates described in the example. Be sure that both rates make the same comparison. "People who watch the news" is mentioned first, so it should be in the numerator of *both* ratios.

Solution Set up the two rates as a proportion and solve for the unknown number of people who watch the news.

$$\frac{7}{10} = \frac{x}{3200}$$

People who watch news → 7; x ← People who watch news
Total group (people surveyed) → 10; 3200 ← Total group (people in town)

$$10 \cdot x = 7 \cdot 3200 \quad \text{Show that cross products are equivalent.}$$

$$10 \cdot x = 22{,}400$$

$$\frac{\overset{1}{\cancel{10}} \cdot x}{\underset{1}{\cancel{10}}} = \frac{22{,}400}{10} \quad \text{Divide both sides by 10.}$$

$$x = 2240$$

You would expect 2240 people in town to watch the news on TV.

Note
To check the answer to an application problem, do *two* things:
1. Check that the answer is reasonable.
2. Put the answer back into the proportion and make sure the cross products are equal.

For example, suppose you set up the last proportion *incorrectly,* as shown below.

$$\frac{7}{10} = \frac{3200}{x} \quad \leftarrow \text{Incorrect setup}$$

$$7 \cdot x = 10 \cdot 3200$$

$$\frac{\overset{1}{\cancel{7}} \cdot x}{\underset{1}{\cancel{7}}} = \frac{32{,}000}{7}$$

$$x \approx 4571 \text{ people} \leftarrow \text{Unreasonable answer}$$

This answer, 4571 people, is an **unreasonable** answer because there are only 3200 people in the town; it is *not* possible for 4571 people to watch the news.

Note
Always check that your answer is reasonable. If it isn't, look at the way your proportion is set up. Be sure you have matching units in the numerators and matching units in the denominators.

◀◀ WORK PROBLEM 2 AT THE SIDE.

Name Date Class

6.4 Exercises

Set up and solve a proportion for each problem.

1. Caroline can sketch 4 cartoon strips in five hours. How long will it take her to sketch 18 strips?

2. The Cosmic Toads recorded 8 songs on their first album in 26 hours. How long will it take them to record 14 songs for their second album?

3. 60 newspapers cost \$27. Find the cost of 16 newspapers.

4. 22 guitar lessons cost \$330. Find the cost of 12 lessons.

5. Five pounds of grass seed cover 3500 square feet of ground. How many pounds are needed for 4900 square feet?

6. Anna earns \$1242.08 in 14 days. How much does she earn in 260 days?

7. Tom makes \$255.75 in 5 days. How much does he make in 3 days?

8. If 5 ounces of a medicine must be mixed with 11 ounces of water, how many ounces of medicine would be mixed with 99 ounces of water?

Use the floor plan shown below to complete Exercises 9–12. On the plan, one inch represents four feet.

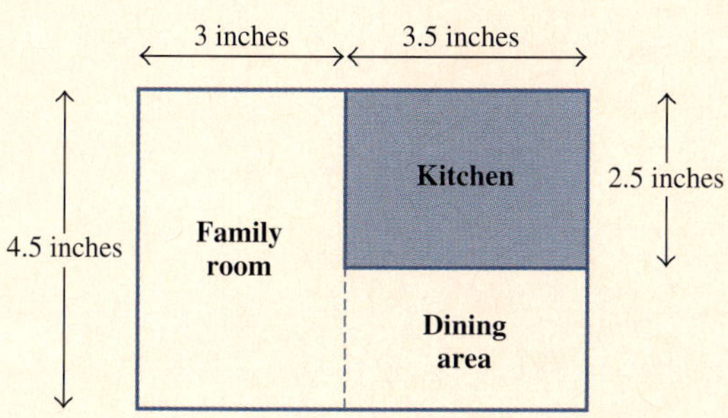

9. What is the actual length and width of the kitchen?

10. What is the actual length and width of the family room?

11. What is the actual length and width of the dining area?

12. What is the actual length and width of the entire floor plan?

Writing Calculator Small Group

13. The Cardinals' pitcher gave up 78 runs in 234 innings. At that rate, how many runs will he give up in a 9-inning game?

14. A quarterback completed 198 out of 318 passes last season. If he tries 30 passes in today's game, how many would you expect him to complete? Round to the nearest whole number of passes.

Ⓖ *Set up a proportion to solve each problem. Check to see if your answer is reasonable. Then flip one side of your proportion to see what answer you get with an incorrect setup. Explain why the second answer is unreasonable.*

15. About 7 out of 10 people entering our community college need to take a refresher math course. If we have 950 entering students, how many will probably need refresher math?

16. In a survey, only 3 out of 100 people like their eggs poached. At that rate, how many of the 60 customers at Soon-Won's restaurant ordered poached eggs this morning? Round to the nearest whole person.

17. Nearly 4 out of 5 people choose vanilla as their favorite ice cream flavor. If 238 people attend an ice cream social, how many would you expect to choose vanilla? Round to the nearest whole person.

18. In a test of 200 sewing machines, only one had a defect. At that rate, how many of the 5600 machines shipped from the factory have defects?

Solve each application problem by setting up a proportion.

19. The tax on a $20 item is $1. Find the tax on a $110 item.

20. A carpenter charges $195.50 to install a deck railing 10 feet long. How much would he charge to install a deck railing 18 feet long?

21. The stock market report says that 5 stocks went up for every 6 stocks that went down. If 750 stocks went down yesterday, how many went up?

22. Raoul paid $15 for 14 cans of oil. How much would 8 cans cost? Round to the nearest cent.

23. Terry's boat traveled 65 miles in 3 hours. At that rate, how long will it take her to travel 100 miles? Round to the nearest tenth.

24. The human body contains 90 pounds of water for every 100 pounds of body weight. How many pounds of water are in a child who weighs 80 pounds?

25. The ratio of the length of an airplane wing to its width is 8 to 1. If the length of a wing is 32.5 meters, how wide must it be? Round to the nearest hundredth.

26. The Rosebud School District wants a student-to-teacher ratio of 19 to 1. How many teachers are needed for 1850 students? Round to the nearest whole number.

27. At 3 P.M., Coretta's shadow is 1.05 meters long. Her height is 1.68 meters. At the same time, a tree's shadow is 6.58 meters long. How tall is the tree? Round to the nearest hundredth.

28. Refer to Exercise 27. Later in the day, the same woman had a shadow that was 2.95 meters long. How long a shadow did the tree have at that time? Round to the nearest hundredth.

29. Can you set up a proportion to solve this problem? Explain why or why not. Jim is 25 years old and weighs 180 pounds. How much will he weigh when he is 50 years old?

30. Write your own application problem that can be solved by setting up a proportion. Also show the proportion and the steps needed to solve your problem.

Ⓖ *A box of instant mashed potatoes has the list of ingredients shown below. Use this information to answer Exercises 31–34.*

Ingredient	*For 12 Servings*
Water	$3\frac{1}{2}$ cups
Margarine	6 tablespoons
Milk	$1\frac{1}{2}$ cups
Potato flakes	4 cups

31. Amount of each ingredient needed for 15 servings.

32. Amount of each ingredient needed for 18 servings.

33. Amount of each ingredient needed for 8 servings.

34. Amount of each ingredient needed for 4 servings.

35. A survey of college students shows that 4 out of 5 drink coffee. Of the students who drink coffee, 1 out of 8 adds cream to it. How many of the 38,000 students at the University of Minnesota would be expected to use cream in their coffee?

36. Nearly 9 out of 10 adults think it's a good idea to exercise regularly. But of the ones who think it is a good idea, only 1 in 6 actually exercises at least three times a week. At this rate, how many of the 300 employees in our company exercise regularly?

Review and Prepare

Find the perimeter and area of each figure. (For help, see ***Sections 3.1, 3.2, and 4.8.****)*

37. A rectangle that is 10 yd by 6 yd.

38. A square measuring 5 cm on each side.

39.

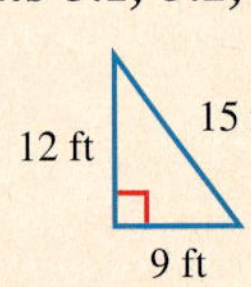

40.

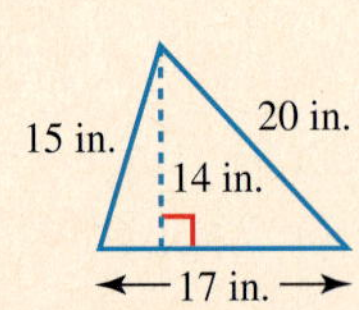

6.5 Geometry Applications: Similar Triangles

Two triangles with the same shape (but not necessarily the same size) are called **similar triangles.** Three pairs of similar triangles are shown here.

OBJECTIVE 1 The two triangles shown below are different sizes but have the same shape, so they are similar triangles. Angles A and P measure the same number of degrees and are called *corresponding angles.* They are marked on the triangles with a double red arc. Angles B and Q are corresponding angles. They are marked on the triangles with a single red arc. Angles C and R are also corresponding angles.

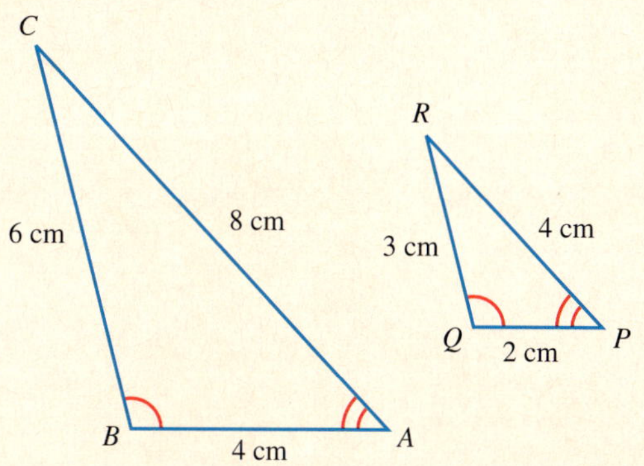

$\overline{PR}$ and $\overline{AC}$ are called *corresponding sides,* since they are *opposite* corresponding angles. Also, $\overline{QR}$ and $\overline{BC}$ are corresponding sides, as are $\overline{PQ}$ and $\overline{AB}$. Although corresponding angles measure the same number of degrees, corresponding sides do *not* need to be the same in length. In the triangles here, each side in the smaller triangle is half the length of the corresponding side in the larger triangle.

WORK PROBLEM 1 AT THE SIDE. ▶▶

OBJECTIVE 2 Similar triangles are useful because of the following property.

Similar Triangles

In similar triangles, the ratios of the lengths of corresponding sides are equal.

OBJECTIVES

1. Identify corresponding parts in similar triangles.
2. Find the unknown lengths of sides in similar triangles.
3. Solve problems with similar triangles.

FOR EXTRA HELP

Tutorial | Tape 11 | SSM, Sec. 6.5

1. Identify corresponding angles and sides in these similar triangles.

(a)

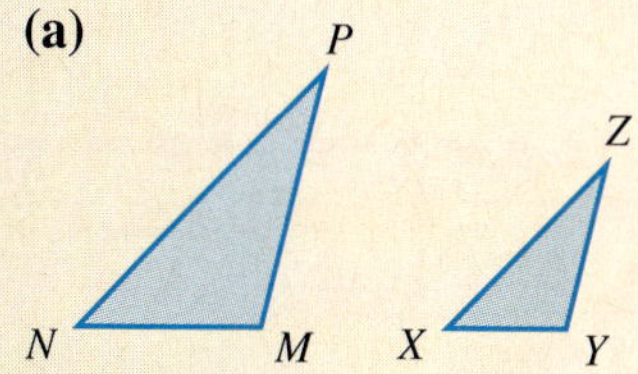

Angles:
P and ______
N and ______
M and ______
Sides:
$\overline{PN}$ and ______
$\overline{PM}$ and ______
$\overline{NM}$ and ______

(b)

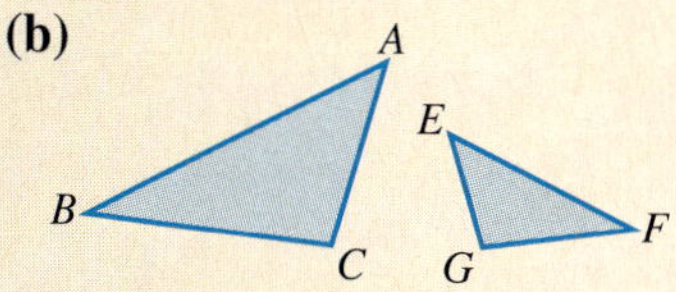

Angles:
A and ______
B and ______
C and ______
Sides:
$\overline{AB}$ and ______
$\overline{BC}$ and ______
$\overline{AC}$ and ______

ANSWERS

1. (a) Z; X; Y; $\overline{ZX}$; $\overline{ZY}$; $\overline{XY}$
(b) E; F; G; $\overline{EF}$; $\overline{GF}$; $\overline{EG}$

2. Find the length of $\overline{EF}$ in Example 1.

EXAMPLE 1 Finding the Unknown Lengths of Sides in Similar Triangles

Find the length of y in the smaller triangle. Assume the triangles are similar.

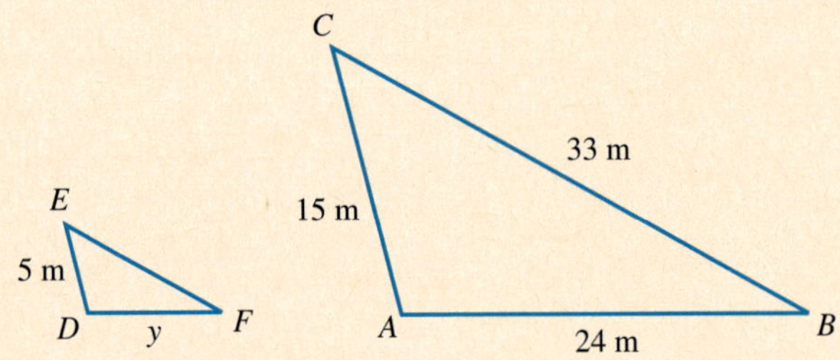

$\overline{ED}$ and $\overline{CA}$ are corresponding sides. The ratio of the lengths of these sides can be written as a fraction in lowest terms.

$$\frac{ED}{CA} = \frac{5\ \cancel{m}}{15\ \cancel{m}} = \frac{1}{3} \quad \text{Lowest terms}$$

As mentioned earlier, the ratios of the lengths of corresponding sides are equal. $\overline{DF}$ in the smaller triangle corresponds to $\overline{AB}$ in the larger triangle. Since the ratios of corresponding sides are equal,

$$\frac{DF}{AB} = \frac{1}{3}$$

Replace DF with y and AB with 24 to get the proportion

$$\frac{y}{24} = \frac{1}{3}.$$

Find cross products.

$$24 \cdot 1 = 24$$

$$\frac{y}{24} = \frac{1}{3}$$

$$y \cdot 3$$

Show that cross products are equivalent.

$$y \cdot 3 = 24$$

Divide both sides by 3.

$$\frac{y \cdot \cancel{3}^{1}}{\cancel{3}_{1}} = \frac{24}{3}$$

$$y = 8$$

$\overline{DF}$ has a length of 8 m.

◄◄ WORK PROBLEM 2 AT THE SIDE.

ANSWERS

2. 11 m

EXAMPLE 2 Finding an Unknown Length and the Perimeter

Find the perimeter of the smaller triangle. Assume the triangles are similar.

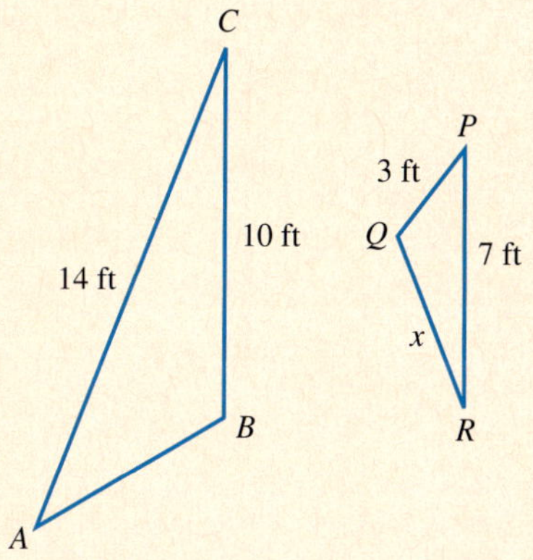

First find the length of $\overline{RQ}$, then add the sides to find the perimeter. The smaller triangle is turned "upside down" compared to the larger triangle, so be careful when identifying corresponding sides. $\overline{AC}$ is the longest side in the larger triangle, and $\overline{PR}$ is the longest side in the smaller triangle. So $\overline{PR}$ and $\overline{AC}$ are corresponding sides. The ratio of their lengths can be written as a fraction in lowest terms.

$$\frac{PR}{AC} = \frac{7\ \cancel{ft}}{14\ \cancel{ft}} = \frac{1}{2} \quad \text{Lowest terms}$$

The two triangles are similar, so the ratio of any pair of corresponding sides will also equal $\frac{1}{2}$. Because $\overline{RQ}$ and $\overline{CB}$ are corresponding sides,

$$\frac{RQ}{CB} = \frac{1}{2}.$$

Replace RQ with x and CB with 10 to make a proportion.

$$\frac{x}{10} = \frac{1}{2}$$

Find cross products.

$$\frac{x}{10} = \frac{1}{2} \qquad 10 \cdot 1 = 10 \qquad x \cdot 2$$

Show that cross products are equivalent.

$$x \cdot 2 = 10$$

Divide both sides by 2.

$$\frac{x \cdot \overset{1}{\cancel{2}}}{\underset{1}{\cancel{2}}} = \frac{10}{2}$$

$$x = 5$$

$\overline{RQ}$ has a length of 5 ft. Now add the lengths of all three sides to find the perimeter.

$$\text{Perimeter} = 5 \text{ ft} + 3 \text{ ft} + 7 \text{ ft} = 15 \text{ ft}$$

WORK PROBLEM 3 AT THE SIDE. ▶▶

3. (a) Find the perimeter of triangle ABC in Example 2.

(b) Find the perimeter of each triangle. Assume the triangles are similar.

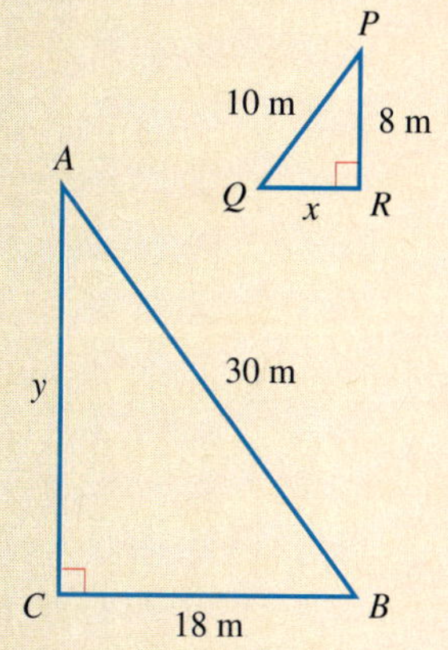

ANSWERS

3. (a) $\overline{AB}$ is 6 ft; perimeter = 14 ft + 10 ft + 6 ft = 30 ft
(b) x = 6 m, perimeter = 24 m; y = 24 m, perimeter = 72 m

4. Find the height of each flagpole.

(a)

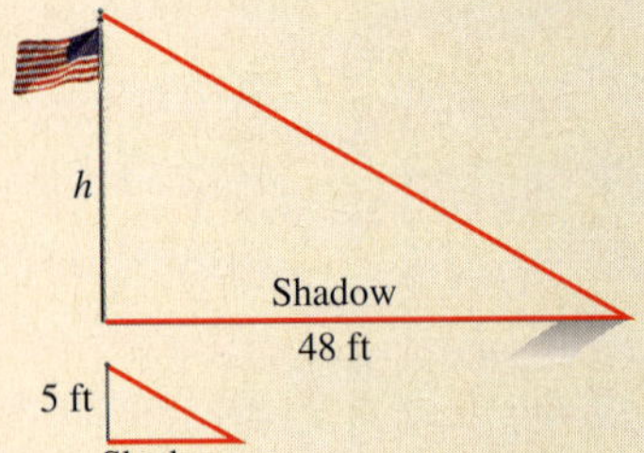

(b)

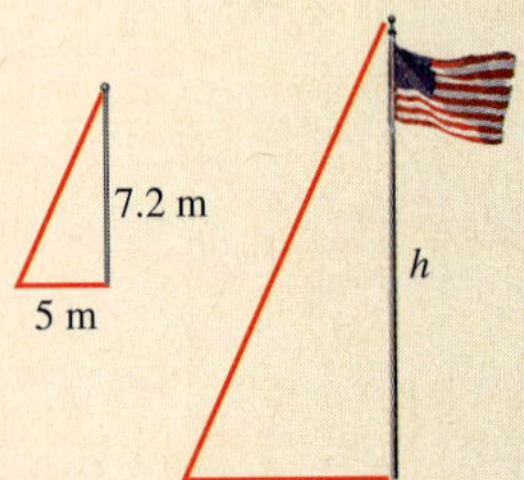

OBJECTIVE 3 The next example shows an application of similar triangles.

EXAMPLE 3 Using Similar Triangles in Applications

A flagpole casts a shadow 99 m long at the same time that a pole 10 m tall casts a shadow 18 m long. Find the height of the flagpole.

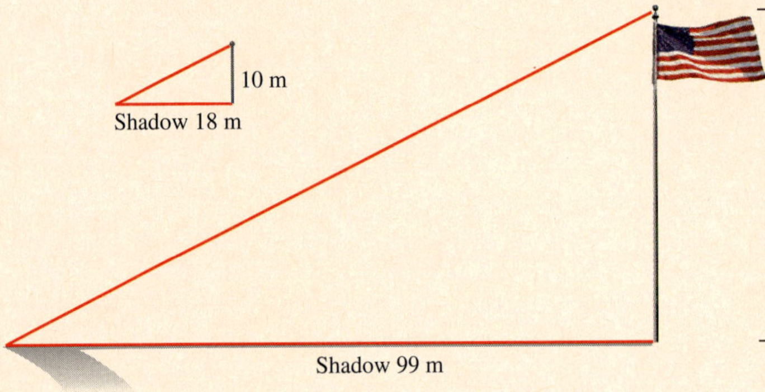

The triangles shown are similar, so write a proportion to find h.

Height in larger triangle → $\frac{h}{10} = \frac{99}{18}$ ← Shadow in larger triangle
Height in smaller triangle → ← Shadow in smaller triangle

Find cross products and show that they are equivalent.

$$h \cdot 18 = 10 \cdot 99$$
$$h \cdot 18 = 990$$

Divide both sides by 18.

$$\frac{h \cdot \overset{1}{\cancel{18}}}{\underset{1}{\cancel{18}}} = \frac{990}{18}$$
$$h = 55$$

The flagpole is 55 m high.

Note

There are several other correct ways to set up the proportion in Example 3. One is to simply flip the ratios on *both* sides of the equal sign:

$$\frac{10}{h} = \frac{18}{99}.$$

But there is another option, shown below.

Height in larger triangle → $\frac{h}{99} = \frac{10}{18}$ ← Height in smaller triangle
Shadow in larger triangle → ← Shadow in smaller triangle

Notice that both ratios compare *height* to *shadow* in the same order. The ratio on the left describes the larger triangle, and the ratio on the right describes the smaller triangle.

◀◀ WORK PROBLEM 4 AT THE SIDE.

ANSWERS

4. **(a)** 20 ft **(b)** 18 m

6.5 Exercises

Write similar *or* not similar *for each pair of triangles.*

1. 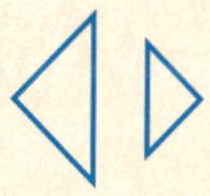

2. 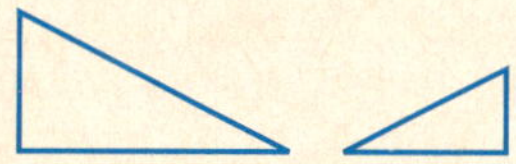

3. 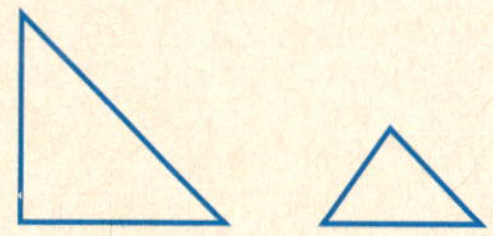

4. 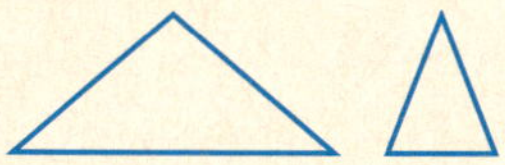

5. 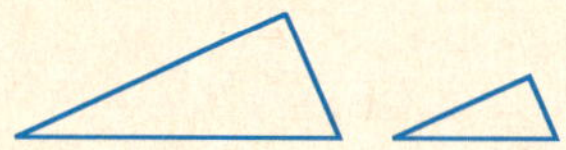

6. 

Name the corresponding angles and the corresponding sides in each pair of similar triangles.

7. 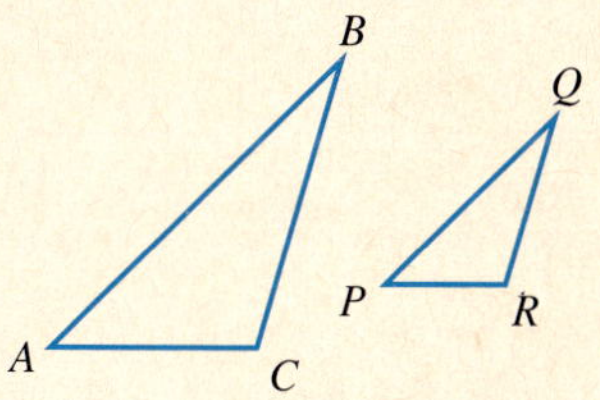

8. 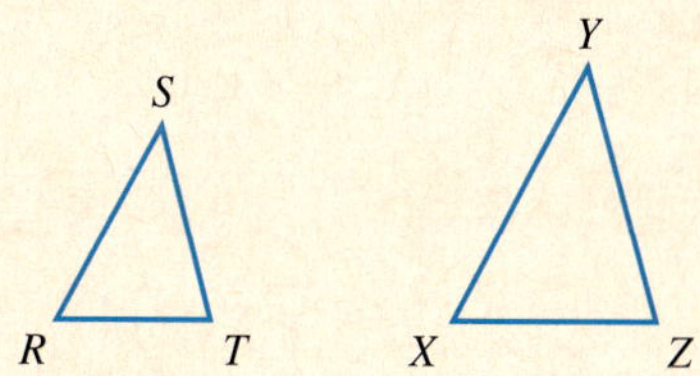

9. 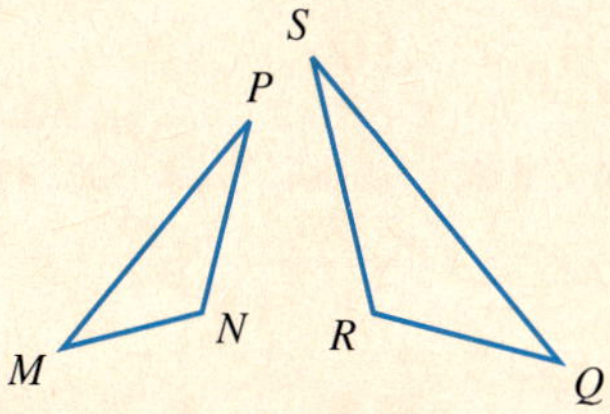

10. 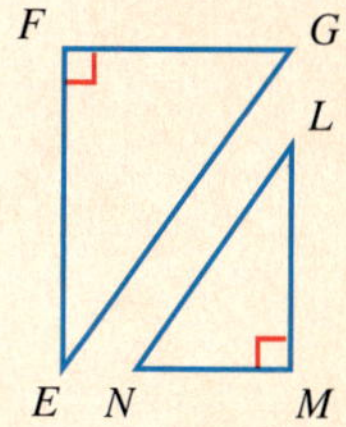

Writing Calculator Small Group

Find all the ratios for the triangles shown below. Write the ratios as fractions in lowest terms.

11. $\dfrac{AB}{PQ}; \dfrac{AC}{PR}; \dfrac{BC}{QR}$

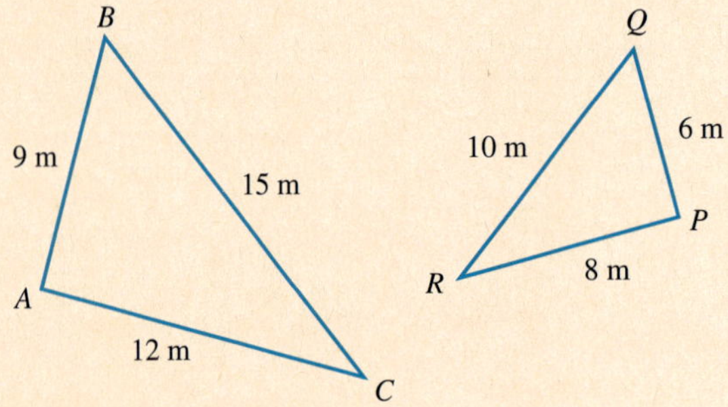

12. $\dfrac{AB}{PQ}; \dfrac{AC}{PR}; \dfrac{BC}{QR}$

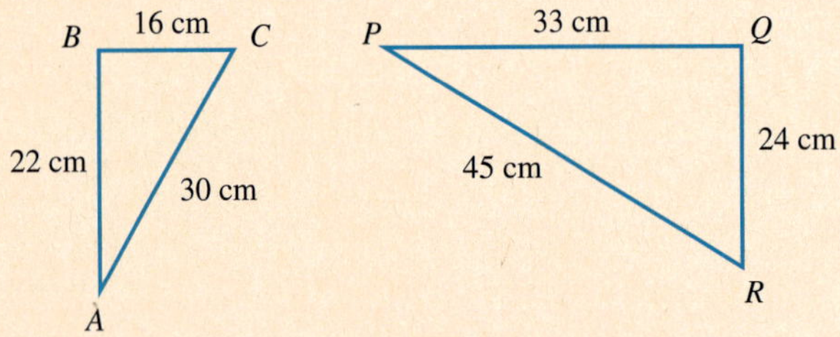

Find the unknown lengths in each pair of similar triangles.

13.

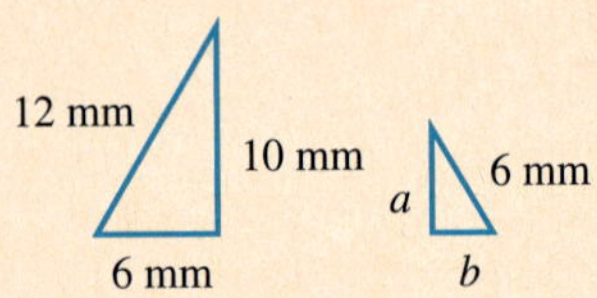

14.

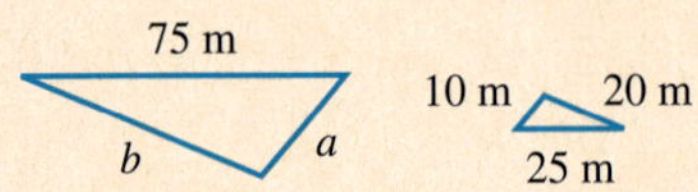

15.

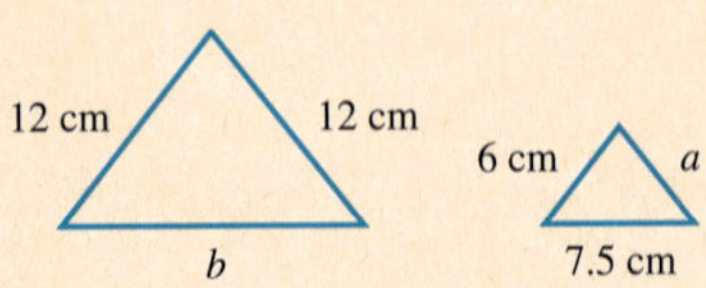

16.

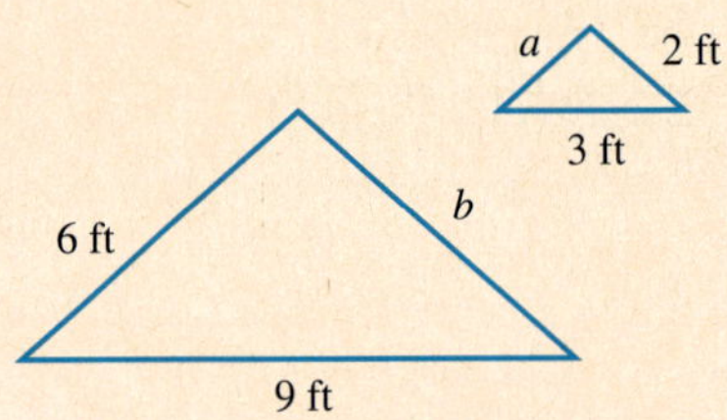

Find the perimeter of each triangle. Assume the triangles are similar.

17.

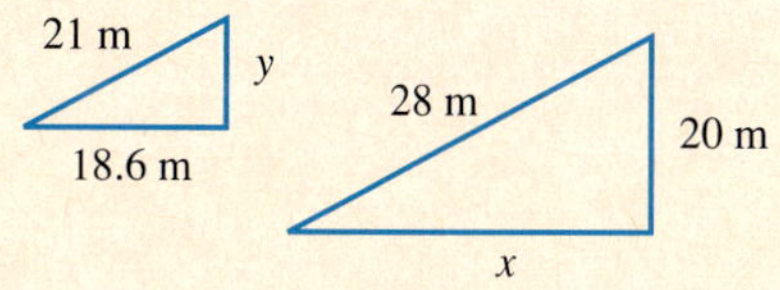

18.

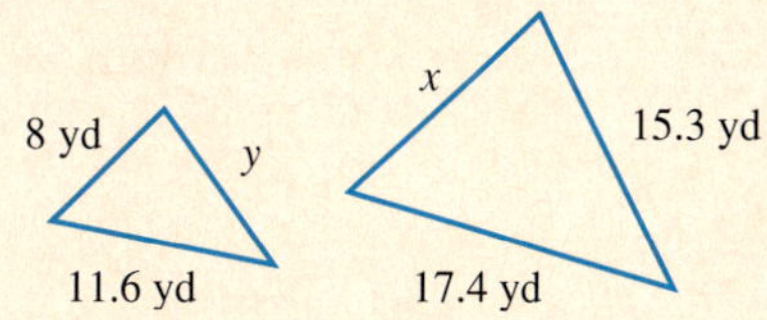

19. Triangles CDE and FGH are similar. Find the perimeter and area of triangle FGH.

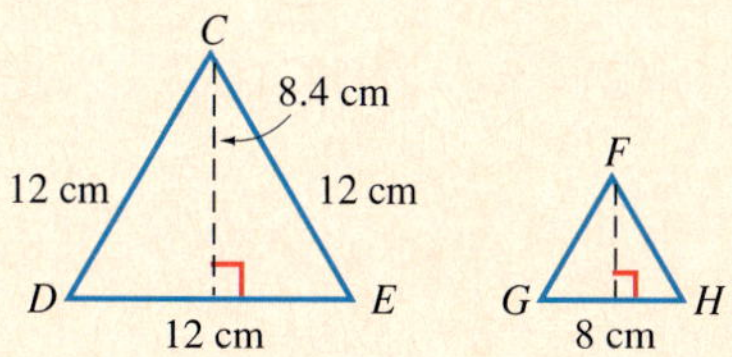

20. Triangles JKL and MNO are similar. Find the perimeter and area of triangle MNO.

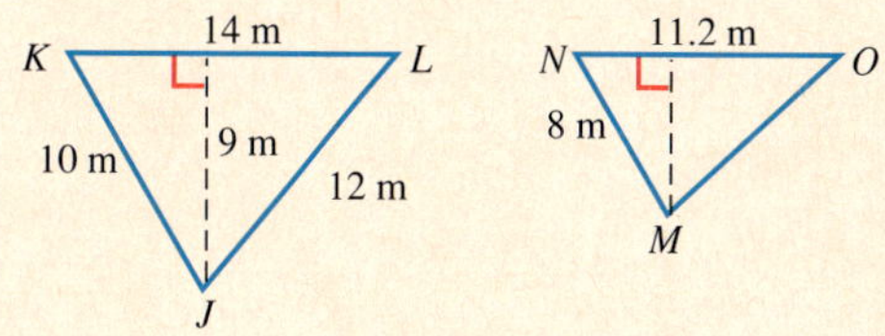

Solve the following application problems.

21. The height of the house shown here can be found by comparing its shadow to the shadow cast by a 3-foot stick. Find the height of the house by writing a proportion and solving it.

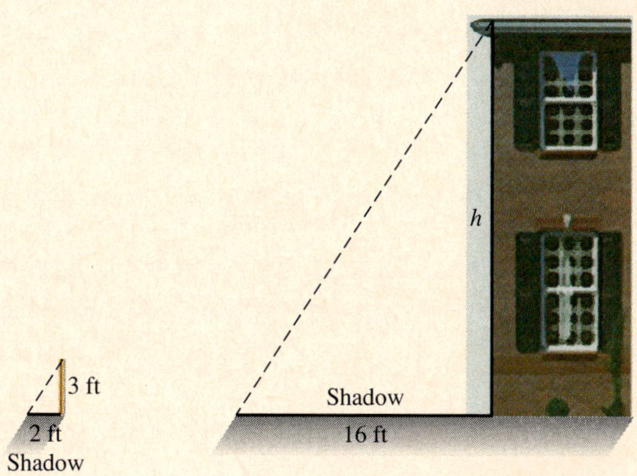

22. A fire lookout tower provides an excellent view of the surrounding countryside. The height of the tower can be found by lining up the top of the tower with the top of a 2-meter stick. Use similar triangles to find the height of the tower.

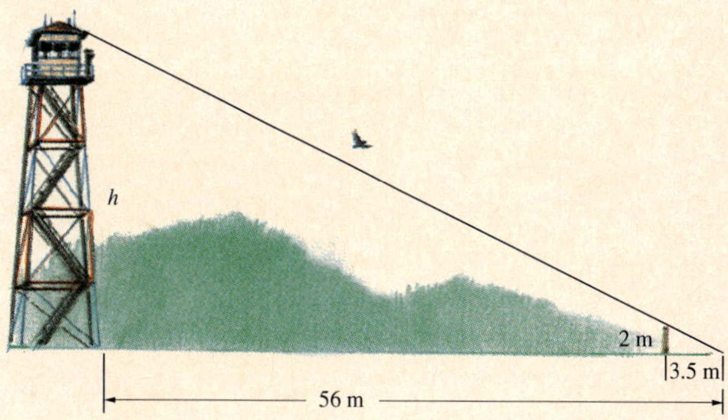

23. Look up the word *similar* in a dictionary. What is the nonmathematical definition of this word? Find two examples of similar objects at home or school.

24. *Congruent* objects have the same shape and the same size. Sketch a pair of congruent triangles. Find two examples of congruent objects at home or school.

Ⓖ *Find the unknown length. Round your answers to the nearest tenth.*
Note: When a line is drawn parallel to one side of a triangle, the smaller triangle that is formed will be similar to the original triangle.

25.

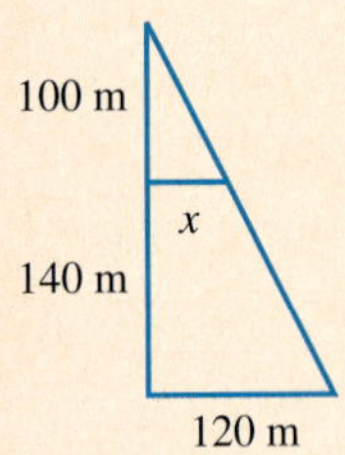

26.

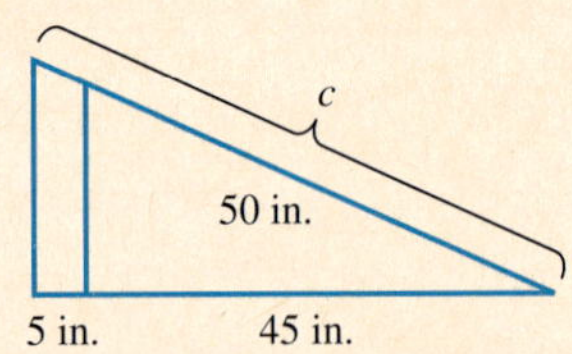

27. Use similar triangles and a proportion to find the length of the lake shown here. (*Hint:* The side 100 m long in the smaller triangle corresponds to a side of 100 + 120 = 220 m in the larger triangle.)

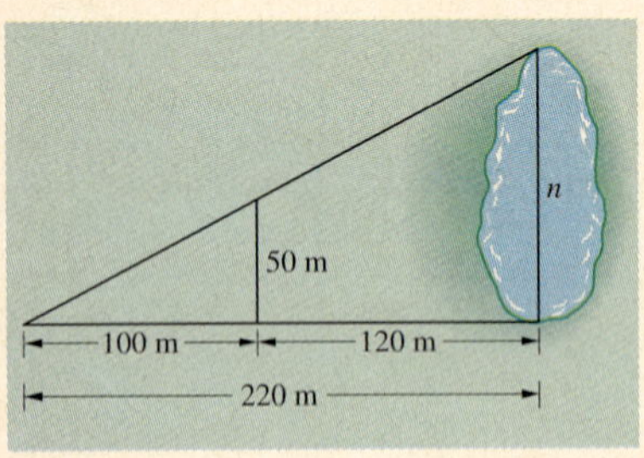

28. To find the height of the tree, find y and then add $5\frac{1}{2}$ ft for the distance from the ground to eye level.

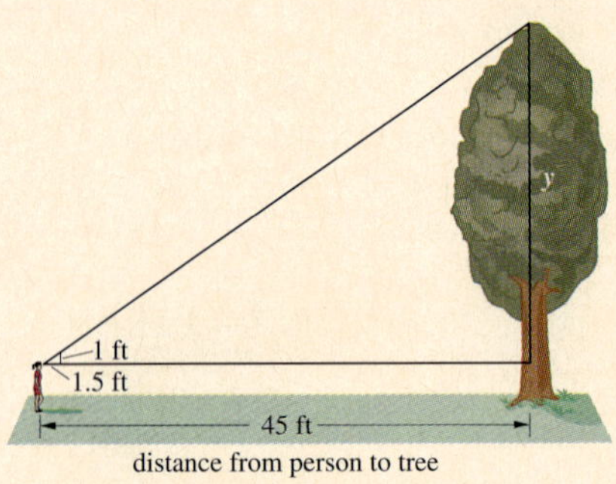

Review and Prepare

Multiply or divide as indicated. (For help, see ***Sections 5.4 and 5.5.****)*

29. $0.06 \cdot 100$

30. $6.1 \cdot 100$

31. $2.87(1000)$

32. $25.8 \div 100$

33. $1.93 \div 100$

34. $5 \div 1000$

KEY TERMS

6.1	**ratio**	A ratio compares two quantities. For example, the ratio of 6 apples to 11 apples is written in fraction form as $\frac{6}{11}$.
6.2	**rate**	A rate compares two measurements with different types of units. Examples are 96 dollars for 8 hours or 450 miles on 18 gallons.
	unit rate	A unit rate has 1 in the denominator.
	cost per unit	Cost per unit is a rate that tells how much you pay for one item or one unit. The lowest cost per unit is the best buy.
6.3	**proportion**	A proportion states that two ratios or rates are equivalent.
	cross products	Cross multiply to get the cross products of a proportion. If the cross products are equal, the proportion is true.
6.5	**similar triangles**	Similar triangles are triangles with the same shape but not necessarily the same size; corresponding angles measure the same number of degrees.

QUICK REVIEW

Concepts	*Examples*
6.1 Writing a Ratio A ratio compares two quantities. A ratio is usually written as a fraction with the number that is mentioned first in the numerator. The common units cancel. Check that the fraction is in lowest terms.	Write this ratio as a fraction in lowest terms. 60 ounces of medicine **to** 160 ounces of medicine $\frac{60 \text{ ounces}}{160 \text{ ounces}} = \frac{60 \div \mathbf{20}}{160 \div \mathbf{20}} = \frac{3}{8}$ ← Lowest terms ↑ Divide out common units.
6.1 Using Mixed Numbers in a Ratio If a ratio has mixed numbers, change the mixed numbers to improper fractions. Rewrite the problem using the "÷" symbol for division. Finally, multiply by the reciprocal of the divisor.	Write as a ratio of whole numbers in lowest terms. $2\frac{1}{2}$ to $3\frac{3}{4}$ $\frac{2\frac{1}{2}}{3\frac{3}{4}}$ Ratio in mixed numbers $= \frac{\frac{5}{2}}{\frac{15}{4}}$ Ratio in improper fractions $= \frac{5}{2} \div \frac{15}{4} = \frac{5}{2} \cdot \frac{4}{15} = \frac{\overset{1}{\cancel{5}} \cdot \overset{1}{\cancel{2}} \cdot 2}{\underset{1}{\cancel{2}} \cdot 3 \cdot \underset{1}{\cancel{5}}} = \frac{2}{3}$ Lowest terms Reciprocals

Concepts	*Examples*
6.1 Using Measurements in Ratios When a ratio compares measurements, both measurements must be in the *same* units. It is usually easier to compare the measurements using the smaller unit, for example, inches instead of feet.	Write as a ratio in lowest terms. Compare in inches. 8 inches to 6 feet Because 1 foot has 12 inches, 6 feet is $6 \cdot \mathbf{12\ inches} = 72$ inches. The ratio is $\frac{8 \text{ \sout{inches}}}{72 \text{ \sout{inches}}} = \frac{8 \div 8}{72 \div 8} = \frac{1}{9}.$
6.2 Writing Rates A rate compares two measurements with different types of units. The units do *not* divide out, so you must write them as part of the rate.	Write the rate as a fraction in lowest terms. 475 miles in 10 hours $\frac{475 \text{ miles} \div 5}{10 \text{ hours} \div 5} = \frac{95 \text{ miles}}{2 \text{ hours}}$ ← Must write units
6.2 Finding a Unit Rate A unit rate has 1 in the denominator. To find the unit rate, divide the numerator by the denominator. Write unit rates using the word "per" or a / mark.	Write as a unit rate: \$1278 in 9 days. $\frac{\$1278}{9 \text{ days}}$ ← Fraction bar indicates division $9\overline{)1278}$ = 142 so $\frac{\$1278 \div 9}{9 \text{ days} \div 9} = \frac{\$142}{1 \text{ day}}$ Write answer as \$142 per day or \$142/day.
6.2 Finding the Best Buy The best buy is the item with the lowest cost per unit. Divide the price by the number of units. Round to thousandths, if necessary. Then compare to find the lowest cost per unit.	Find the best buy on cheese. 2 pounds for \$2.25 3 pounds for \$3.40 Find cost per unit (cost per pound). $\frac{\$2.25}{2} = \$1.125$ per pound $\frac{\$3.40}{3} \approx \$1.133$ per pound The lower cost per pound is \$1.125, so 2 pounds for \$2.25 is the better buy.

Concepts	*Examples*

6.3 Writing Proportions

A proportion states that two ratios or rates are equivalent. This proportion,

$$\frac{5}{6} = \frac{25}{30},$$

is read as "5 is to 6 as 25 is to 30."

To see whether a proportion is true or false, cross multiply one way, then cross multiply the other way. If the two products are equal, the proportion is true. If the two products are unequal, the proportion is false.

Write as a proportion: 8 is to 40 as 32 is to 160.

$$\frac{8}{40} = \frac{32}{160}$$

Cross multiply to see whether the following proportion is true or false.

$$\frac{6}{8\frac{1}{2}} = \frac{24}{34}$$

Cross multiply.

$$8\frac{1}{2} \cdot 24 = \frac{17}{\cancel{2}_1} \cdot \frac{\overset{1}{\cancel{2}} \cdot 12}{1} = \mathbf{204}$$

$$\frac{6}{8\frac{1}{2}} = \frac{24}{34}$$

$$6 \cdot 34 = \mathbf{204}$$

Equal

Cross products are equal, so the proportion is true.

6.3 Solving Proportions

Solve for an unknown number in a proportion by using these steps.

Find the unknown number.

$$\frac{12}{x} = \frac{6}{8}$$

$$\frac{12}{x} = \frac{3}{4} \quad \text{Lowest terms}$$

Step 1 Find the cross products. (If desired, you can rewrite the ratios in lowest terms before finding the cross products.)

Step 1 $\frac{12}{x} = \frac{3}{4}$ $\quad x \cdot 3$, $12 \cdot 4$ Find cross products.

Step 2 Show that the cross products are equivalent.

Step 2 $x \cdot 3 = 12 \cdot 4$ Show that cross products are equivalent.

$$x \cdot 3 = 48$$

Step 3 Divide both sides of the equation by the coefficient of the variable term.

Step 3 $\frac{x \cdot \overset{1}{\cancel{3}}}{\cancel{3}_1} = \frac{48}{3}$ Divide both sides by 3.

$$x = 16$$

Check your answer by writing the complete proportion and finding the cross products.

Check

x is 16 → $\frac{12}{16} = \frac{6}{8}$

$$16 \cdot 6 = \mathbf{96}$$

$$12 \cdot 8 = \mathbf{96}$$

Equal

Cross products are equal, so 16 is the correct solution.

Concepts	*Examples*
6.4 Applications of Proportions	
Decide what is being compared, for example, pounds to square feet. Write the two rates described in the problem. Be sure that *both* rates compare things in the *same order*. Use a variable to represent the unknown number. Set up a proportion. Check that the numerators have matching units and the denominators have matching units. Solve for the unknown number.	If 3 pounds of grass seed cover 450 square feet of lawn, how much seed is needed for 1500 square feet of lawn? Matching units $\frac{3 \text{ pounds}}{450 \text{ square feet}} = \frac{x \text{ pounds}}{1500 \text{ square feet}}$ Matching units Both sides compare pounds to square feet. Ignore the units while finding cross products. $450 \cdot x = 3 \cdot 1500$ Show that cross products are equivalent. $450 \cdot x = 4500$ $\frac{\overset{1}{\cancel{450}} \cdot x}{\underset{1}{\cancel{450}}} = \frac{4500}{450}$ Divide both sides by 450. $x = 10$ 10 pounds of seed are needed.
6.5 Finding the Unknown Lengths in Similar Triangles	
Use the fact that in similar triangles, the ratios of the lengths of corresponding sides are equal. Write a proportion. Then find cross products and show that they are equivalent. Finish solving for the unknown length.	Find x and y if the triangles are similar. $\frac{x}{8} = \frac{5}{10}$ | $\frac{y}{12} = \frac{5}{10}$ $x \cdot 10 = 8 \cdot 5$ | $y \cdot 10 = 12 \cdot 5$ $\frac{x \cdot \overset{1}{\cancel{10}}}{\underset{1}{\cancel{10}}} = \frac{40}{10}$ | $\frac{y \cdot \overset{1}{\cancel{10}}}{\underset{1}{\cancel{10}}} = \frac{60}{10}$ $x = 4$ m | $y = 6$ m 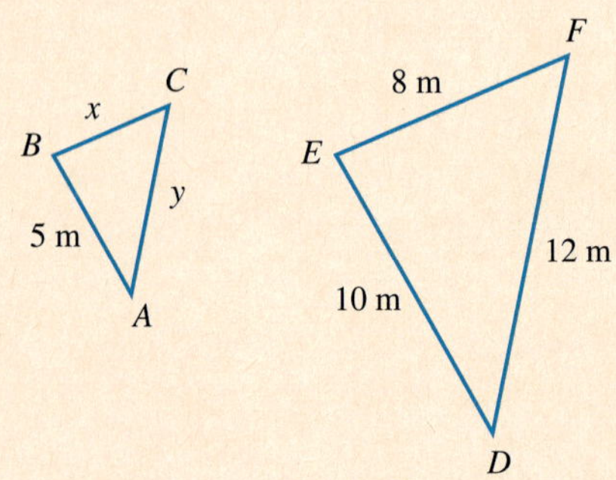

CHAPTER 6 REVIEW EXERCISES

[6.1] *Write each ratio as a fraction in lowest terms. Change to the same units when necessary, using the list of relationships in* ***Section 6.1.***

1. 3 oranges to 11 oranges

2. 19 miles to 7 miles

3. 9 doughnuts to 6 doughnuts

4. 90 feet to 50 feet

5. \$2.50 to \$1.25

6. \$0.30 to \$0.45

7. $1\frac{2}{3}$ cups to $\frac{2}{3}$ cup

8. $2\frac{3}{4}$ miles to $16\frac{1}{2}$ miles

9. 5 hours to 100 minutes
Compare in minutes.

10. 9 inches to 2 feet
Compare in inches.

11. 1 ton to 1500 pounds
Compare in pounds.

12. 8 hours to 3 days
Compare in hours.

13. Jake sold \$350 worth of kachina figures. Ramona sold \$500 worth of pottery. What is the ratio of her sales to his?

14. Ms. Wei's new car gets 35 miles per gallon. Her old car got 25 miles per gallon. Find the ratio of the new car's mileage to the old car's mileage.

15. This fall, 60 students are taking math and 72 students are taking English. Find the ratio of math students to English students.

16. There are 9 players on a baseball team and 5 players on a basketball team. What is the ratio of basketball players to baseball players?

[6.2] *Write each rate as a fraction in lowest terms.*

17. \$88 for 8 dozen

18. 96 children in 40 families

19. Explain the similarities and differences between a ratio and a rate. Give an example of each.

20. In your own words, explain the term "unit rate." Give three examples of unit rates.

21. In his keyboarding class, Patrick can type 4 pages in 20 minutes. Give his rate in pages per minute and minutes per page.

22. Elena made \$24 in 3 hours. Give her earnings in dollars per hour and hours per dollar.

Find the best buy.

23. minced onion
13 ounces for \$2.29
8 ounces for \$1.45
3 ounces for \$0.95

24. dog food; you have a coupon for \$1 off on 25 pounds or more.
50 pounds for \$19.95
25 pounds for \$10.40
8 pounds for \$3.40

[6.3] *Write each proportion.*

25. 5 is to 10 as 20 is to 40.

26. 7 is to 2 as 35 is to 10.

27. $1\frac{1}{2}$ is to 6 as $2\frac{1}{4}$ is to 9.

Use the method of writing in lowest terms or cross multiplication to decide whether the following proportions are true or false.

28. $\frac{6}{10} = \frac{9}{15}$

29. $\frac{16}{48} = \frac{9}{36}$

30. $\frac{47}{10} = \frac{98}{20}$

31. $\frac{64}{36} = \frac{96}{54}$

32. $\frac{1.5}{2.4} = \frac{2}{3.2}$

33. $\frac{3\frac{1}{2}}{2\frac{1}{3}} = \frac{6}{4}$

Find the missing number in each proportion. Round to hundredths, if necessary.

34. $\frac{4}{42} = \frac{150}{x}$

35. $\frac{16}{x} = \frac{12}{15}$

36. $\frac{100}{14} = \frac{x}{56}$

37. $\frac{5}{8} = \frac{x}{20}$

38. $\frac{x}{24} = \frac{11}{18}$

39. $\frac{7}{x} = \frac{18}{21}$

40. $\frac{x}{3.6} = \frac{9.8}{0.7}$

41. $\frac{13.5}{1.7} = \frac{4.5}{x}$

42. $\frac{0.82}{1.89} = \frac{x}{5.7}$

[6.4] *Set up and solve a proportion for each application problem.*

43. The ratio of cats to dogs at the animal shelter is 3 to 5. If there are 45 dogs, how many cats are there?

44. Danielle had 8 hits in 28 times at bat during last week's games. If she continues to hit at the same rate, how many hits will she get in 161 times at bat?

45. If 3.5 pounds of steak cost \$13.79, what will 5.6 pounds cost? Round to the nearest cent.

46. About 4 out of 10 students are expected to vote in campus elections. There are 8247 students. How many are expected to vote? Round to the nearest whole number.

47. The scale on Brian's model railroad is 1 inch to 16 feet. One of the scale model boxcars is 4.25 inches long. What is the length of a real boxcar in feet?

48. In the hospital pharmacy, Michiko sees that a certain medicine is to be given at the rate of 3.5 milligrams for every 50 pounds of body weight. How much medicine should be given to a patient who weighs 210 pounds?

49. Damien earns \$91 for 14 hours of part-time work at the convenience store. How long must he work to earn \$520 to pay his tuition?

50. Marvette makes necklaces to sell at a local gift shop. She made 2 dozen necklaces in $16\frac{1}{2}$ hours. How long will it take her to make 40 necklaces?

[6.5] *Find the unknown lengths in each pair of similar triangles.*

51.

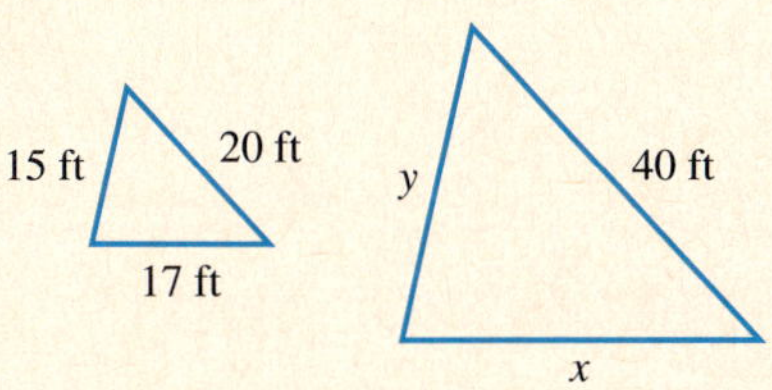

52.

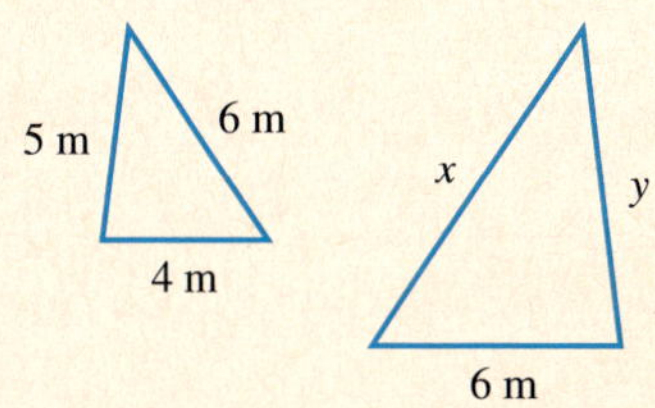

53.

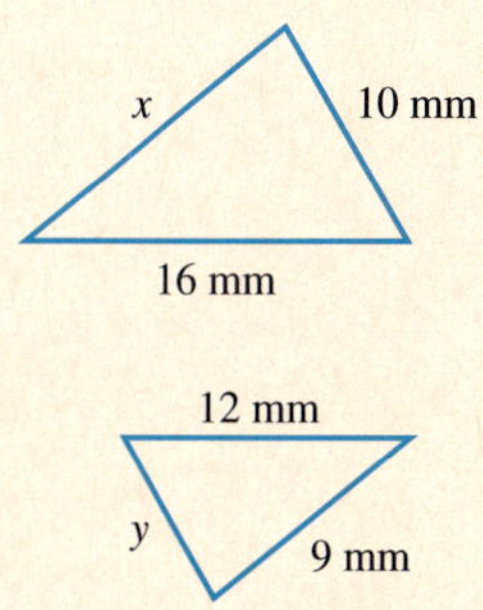

Mixed Review Exercises

Find the missing number in each proportion. Round to hundredths, if necessary.

54. $\frac{x}{45} = \frac{70}{30}$

55. $\frac{x}{52} = \frac{0}{20}$

56. $\frac{64}{10} = \frac{x}{20}$

57. $\frac{15}{x} = \frac{65}{100}$

58. $\frac{7.8}{3.9} = \frac{13}{x}$

59. $\frac{34.1}{x} = \frac{0.77}{2.65}$

Use cross multiplication to decide whether the following proportions are true or false. Circle the correct answer.

60. $\frac{55}{18} = \frac{80}{27}$

True False

61. $\frac{5.6}{0.6} = \frac{18}{1.94}$

True False

62. $\frac{\frac{1}{5}}{2} = \frac{1\frac{1}{6}}{11\frac{2}{3}}$

True False

Write each ratio as a fraction in lowest terms. Change to the same units when necessary.

63. 4 dollars to 10 quarters
Compare in quarters.

64. $4\frac{1}{8}$ inches to 10 inches

65. 10 yards to 8 feet
Compare in feet.

66. \$3.60 to \$0.90

67. 12 eggs to 15 eggs

68. 37 meters to 7 meters

69. 3 pints to 4 quarts
Compare in pints.

70. 15 minutes to 3 hours
Compare in minutes.

71. $4\frac{1}{2}$ miles to $1\frac{3}{10}$ miles

Set up and solve a proportion for each application problem.

72. Nearly 7 out of 8 fans buy something to drink at the ballpark. How many of the 28,500 fans at today's game would be expected to buy a beverage? Round to the nearest hundred fans.

73. Emily spent \$150 on car repairs and \$400 on car insurance. What is the ratio of amount spent on insurance to amount spent on repairs?

74. Antonio is choosing among three packages of plastic wrap. Is the best buy 25 feet for \$0.78; 75 feet for \$1.99; or 100 feet for \$2.59? He has a coupon for 50¢ off that is good for either of the larger two packages.

75. On this scale drawing of a backyard patio, 1 inch represents 6 feet. If the patio measures 2.75 inches long on the drawing, what will be the actual length of the patio when it is built?

76. An antibiotic is to be given at the rate of $1\frac{1}{2}$ teaspoons for every 24 pounds of body weight. How much should be given to an infant who weighs 8 pounds?

77. Charles made 251 points during 169 minutes of playing time last year. If he plays 14 minutes in tonight's game, how many points would you expect him to make? Round to the nearest whole number.

78. A lawn mower uses 0.8 gallon of gas every 3 hours. The gas tank holds 2 gallons. How long can the mower run on a full tank?

Name Date Class

CHAPTER 6 TEST

Write each rate or ratio as a fraction in lowest terms. Change to the same units when necessary.

1. 16 fish to 20 fish

2. 300 miles on 15 gallons

3. \$15 for 75 minutes

4. The little theater has 320 seats. The auditorium has 1200 seats. Find the ratio of auditorium seats to theater seats.

5. 3 quarts to 60 gallons
Compare in quarts.

6. 3 hours to 40 minutes
Compare in minutes.

7. Find the best buy on spaghetti sauce. You have a coupon for 75¢ off Brand X and a coupon for 25¢ off Brand Y.
28 ounces of Brand X for \$3.89
18 ounces of Brand Y for \$1.89
13 ounces of Brand Z for \$1.29

8. Suppose the ratio of your income last year to your income this year is 3 to 2. Explain what this means. Give an example of the dollars earned last year and this year that fits the 3 to 2 ratio.

Decide whether the following proportions are true or false.

9. $\frac{6}{14} = \frac{18}{45}$

10. $\frac{8.4}{2.8} = \frac{2.1}{0.7}$

Find the unknown number in each proportion. Round to hundredths, if necessary.

11. $\frac{5}{9} = \frac{x}{45}$

12. $\frac{3}{1} = \frac{8}{x}$

13. $\frac{x}{20} = \frac{6.5}{0.4}$

14. $\frac{2\frac{1}{3}}{x} = \frac{\frac{8}{9}}{4}$

1. ______
2. ______
3. ______
4. ______
5. ______
6. ______
7. ______
8. ______
9. ______
10. ______
11. ______
12. ______
13. ______
14. ______

15. ______________________

16. ______________________

17. ______________________

18. ______________________

19. ______________________

20. ______________________

21. ______________________

22. ______________________

Set up and solve a proportion for each application problem.

15. Pedro types 240 words in 5 minutes. How many words can he type in 12 minutes?

16. A boat travels 75 miles in 4 hours. At that rate, how long will it take to travel 120 miles?

17. About 2 out of every 15 people are left-handed. How many of the 650 students in our school would you expect to be left-handed? Round to the nearest whole number.

18. A student set up the proportion for Exercise 17 this way and arrived at an answer of 4875.

$$\frac{2}{15} = \frac{650}{x} \qquad \text{Check: } \frac{2}{15} = \frac{650}{4875} \qquad \begin{matrix} 15 \cdot 650 = 9750 \\ 2 \cdot 4875 = 9750 \end{matrix}$$

Because the cross products are equal, the student said the answer is correct. Is the student right? Explain why or why not.

19. A medication is given at the rate of 8.2 grams for every 50 pounds of body weight. How much should be given to a 145-pound person? Round to the nearest tenth.

20. On a scale model, 1 inch represents 8 feet. If a building in the model is 7.5 inches tall, what is the actual height of the building in feet?

21. Find the unknown lengths in these similar triangles.

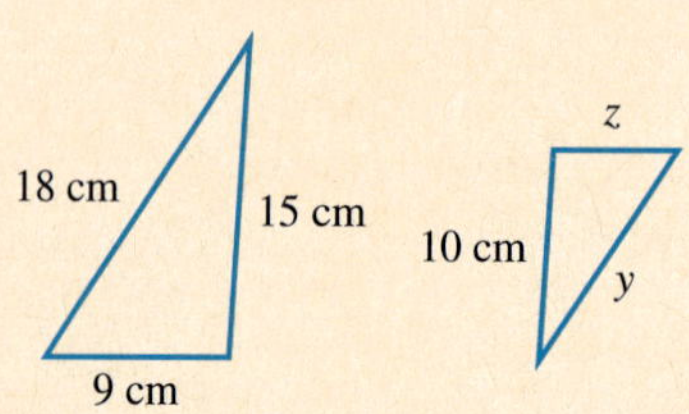

22. Find the perimeter of each of these similar triangles.

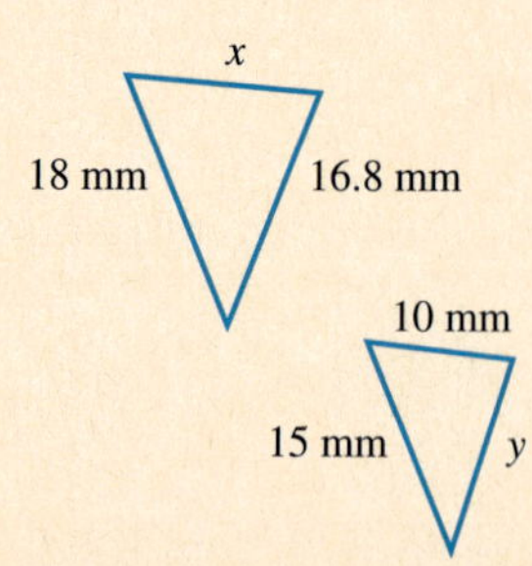

Name Date Class

CUMULATIVE REVIEW EXERCISES CHAPTERS 1–6

1. Write these numbers in words.
(a) 77,001,000,805

(b) 0.02

2. Write these numbers in digits.
(a) three and forty thousandths

(b) five hundred million, thirty-seven thousand

Simplify.

3. $\dfrac{0}{-16}$

4. $|0| + |-6 - 8|$

5. $4\dfrac{3}{4} - 1\dfrac{5}{6}$

6. $\dfrac{h}{5} - \dfrac{3}{10}$

7. $100 - 0.0095$

8. $\dfrac{-4 + 7}{9 - 3^2}$

9. $-6 + 3(0 - 4)$

10. $\dfrac{-8}{\frac{4}{7}}$

11. $\dfrac{5n}{6m^3} \div \dfrac{10}{3m^2}$

12. $(0.06)(-0.007)$

13. $\dfrac{x}{14y} \cdot \dfrac{7}{xy}$

14. $\dfrac{9}{n} + \dfrac{2}{3}$

15. $-40 + 8(-5) + 2^4$

16. $5.8 - (-0.6)^2 \div 0.9$

17. $\left(-\dfrac{1}{2}\right)^3\left(\dfrac{2}{3}\right)^2$

Solve each equation. Show the steps you used.

18. $7y + 5 = -3 - y$

19. $-2 + \dfrac{3}{5}x = 7$

20. $\dfrac{4}{x} = \dfrac{14}{35}$

Solve each application problem using the six problem-solving steps from Chapter 3.

21. Tuyen heard on the radio that the temperature had risen 15 degrees by noon, then dropped 23 degrees due to a storm, then risen 5 degrees and was now at 71 degrees. What was the starting temperature?

22. The width of a rectangular swimming pool is 14 ft less than the length. The perimeter of the pool is 100 ft. Find the length and the width.

Find the unknown length, perimeter, area, or volume. When necessary, use 3.14 as the approximate value of π and round answers to the nearest tenth.

23. Find the perimeter and the area.

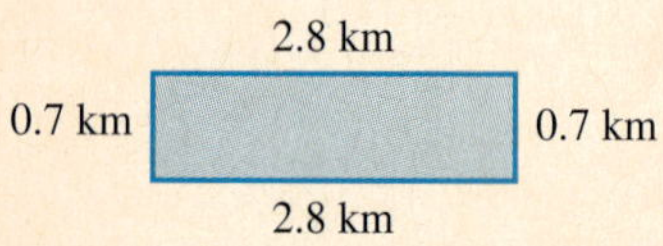

24. Find the diameter, circumference, and area.

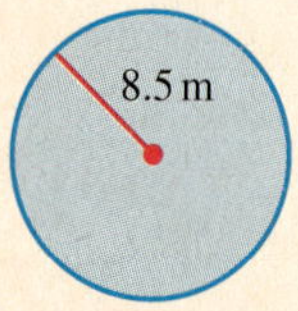

25. Find the volume.

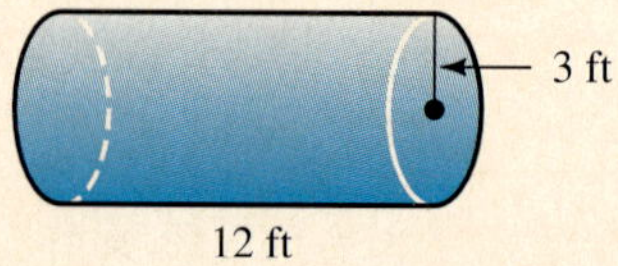

Use the circle graph below of one college's enrollment to complete Exercises 26–29. Where indicated, use front end rounding to estimate the answer. Then find the exact answer.

COLLEGE ENROLLMENT

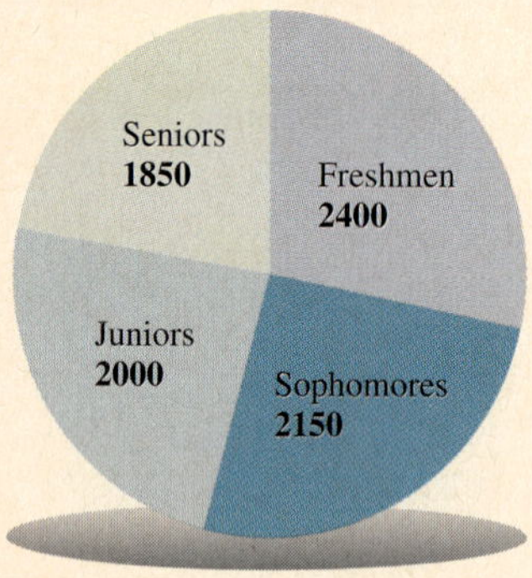

26. Find the total enrollment at the college.

estimate:

exact:

27. The college has budgeted \$186,400 for freshman orientation. How much is spent on each freshman, to the nearest dollar?

estimate:

exact:

28. **(a)** Write the ratio of freshmen to the total enrollment as a fraction in lowest terms.
(b) Write the ratio of freshmen and sophomores to juniors and seniors as a fraction in lowest terms.

29. The college collects a \$3.75 technology fee from each student to support the computer lab. What total amount is collected?

estimate:

exact:

30. The distance around Dunning Pond is $1\frac{1}{10}$ miles. Norma ran around the pond 4 times in the morning and $2\frac{1}{2}$ times in the afternoon. How far did she run in all?

estimate:

exact:

31. Rodney bought 49.8 gallons of gas for his truck while driving 896.5 miles on a vacation. How many miles per gallon did he get, rounded to the nearest tenth?

estimate:

exact:

32. The honor society has a goal of collecting 1500 pounds of food to fill Thanksgiving baskets. So far they've collected $\frac{5}{6}$ of their goal. How many more pounds do they need?

33. The directions on a can of plant food call for $\frac{1}{2}$ teaspoon in two quarts of water. How much plant food is needed for five quarts?

Percent

7

7.1 *The Basics of Percent*

OBJECTIVE 1 You have probably seen percents (per-SENTS) frequently in daily life. The symbol for percent is %. For example, in one day you may leave a 15% tip for the waitress at dinner, pay 7% sales tax on a CD player, and buy shoes at 25% off the regular price. The next day your score on a math test may be 89% correct.

The Meaning of Percent

A **percent** is a ratio with a denominator of 100. So percent means "per 100" or "how many out of 100." The symbol for percent is %. Read 15% as "fifteen percent."

EXAMPLE 1 Percent

(a) If you left a \$15 tip when the restaurant bill was \$100, then you left \$15 per \$100 or $\frac{15}{100}$ or 15%.

(b) If you pay \$7 in tax on a \$100 CD player, then the tax rate is \$7 per \$100 or $\frac{7}{100}$ or 7%.

(c) If you earn 89 points on a 100 point math test, then your score is 89 out of 100 or $\frac{89}{100}$ or 89%.

WORK PROBLEM 1 AT THE SIDE.

OBJECTIVE 2 In order to work with percents, you will need to write them as decimals or as fractions. We'll start by writing percents as equivalent decimal numbers. Twenty-five percent, or 25%, means 25 parts out of 100 parts, or $\frac{25}{100}$. Remember that the fraction bar indicates division. So we can write $\frac{25}{100}$ as 25 ÷ 100. When you do the division, 25 ÷ 100 is 0.25.

Indicates division → $\frac{25}{100}$ can be written as $25 \div 100 = 0.25$ (↑ Indicates division)

Another way to remember how to write a percent as a decimal is to use the meaning of the word *percent*. The first part of the word, *per*, is an indicator word for *division*. The last part of the word, *cent*, comes from the Latin word for *hundred*. (Recall that there are 100 *cents* in a dollar, and 100 years in a *cent*ury.)

25% is 25 **per** **cent**

↓ ↓ ↓

25 ÷ **100** = 0.25

---

www.mathnotes.com

OBJECTIVES

1. Learn the meaning of percent.
2. Write percents as decimals.
3. Write decimals as percents.
4. Write percents as fractions.
5. Write fractions as percents.
6. Use 100% and 50%.

FOR EXTRA HELP

Tutorial | Tape 12 | SSM, Sec. 7.1

1. Write a percent to describe each situation.

(a) You leave a \$20 tip for a restaurant bill of \$100. What percent tip did you leave?

(b) The tax on a \$100 graphing calculator is \$5. What is the tax rate?

(c) You earn 94 points on a 100 point test. What percent of the points did you earn?

ANSWERS

1. (a) 20% **(b)** 5% **(c)** 94%

2. Write each percent as a decimal.

(a) 68%

(b) 5%

(c) 40.6%

(d) 200%

(e) 350%

Writing a Percent as an Equivalent Decimal

To write a percent as a decimal, drop the % symbol and divide by 100.

EXAMPLE 2 Writing a Percent as a Decimal

Write each percent as a decimal.

(a) 47% $\quad 47\% = 47 \div 100 = 0.47$ Decimal form

(b) 3% $\quad 3\% = 3 \div 100 = 0.03$ Decimal form

(c) 28.2% $\quad 28.2\% = 28.2 \div 100 = 0.282$ Decimal form

(d) 100% $\quad 100\% = 100 \div 100 = 1.00$ Decimal form

(e) 135% $\quad 135\% = 135 \div 100 = 1.35$ Decimal form

Note

In Example 2(d) notice that 100% is 1.00, or 1, which is a whole number. Whenever you have a percent that is 100% or greater, the equivalent decimal number will be 1 or greater. Notice in Example 2(e) that 135% is 1.35 (greater than 1).

◀◀ WORK PROBLEM 2 AT THE SIDE.

In the exercise set for **Section 5.5**, you discovered a shortcut for *dividing* by 100: Move the decimal point *two* places to the *left*. You can use this shortcut when writing percents as decimals.

EXAMPLE 3 Changing Percents to Decimals by Moving the Decimal Point

Write each percent as a decimal by moving the decimal point two places to the left.

(a) 17%

17% = 17.% ← Decimal point starts at far right side.

.17 ← Percent symbol is dropped.

← Decimal point is moved two places to the left. This is a quick way to divide by 100.

17% = 0.17

(b) 160%

160% = 160.% = 1.60 or 1.6 Decimal point starts at far right side. 1.60 is equivalent to 1.6.

(c) 4.9%

.049% 0 is attached so the decimal point can be moved two places to the left.

4.9% = 0.049

(d) 0.6%

00.6% = 0.006 0 is attached so the decimal point can be moved.

Answers

2. **(a)** 0.68 **(b)** 0.05 **(c)** 0.406 **(d)** 2.00 or 2 **(e)** 3.50 or 3.5

Note

In Example 3(d) notice that 0.6% is less than 1%. Because 1% is equivalent to 0.01 or $\frac{1}{100}$, any fraction of a percent smaller than 1% is less than 0.01.

WORK PROBLEM 3 AT THE SIDE.

OBJECTIVE 3 You can write a decimal as a percent. For example, the decimal 0.25 is the same as the fraction

$$\frac{25}{100}.$$

This fraction means 25 out of 100 parts, or 25%. Notice that multiplying 0.25 by 100 gives the same result.

$$0.25 \cdot 100 = 25\%$$

This result makes sense because we are doing the opposite of what we did to change a percent to a decimal.

To change a percent to a decimal, *divide* by 100. So, to "go backwards" from a decimal to a percent, we multiply instead of divide.

To change a decimal to a percent, *multiply* by 100.

Writing a Decimal as a Percent

To write a decimal as a percent, multiply by 100 and attach a % symbol.

Note

A quick way to *multiply* a number by 100 is to move the decimal point *two* places to the *right*, which is the opposite of dividing by 100.

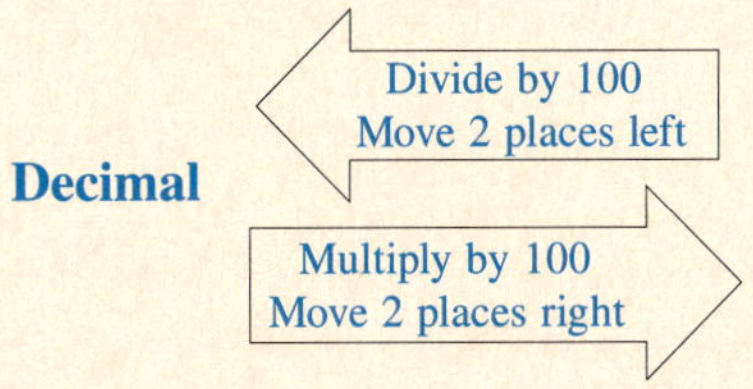

EXAMPLE 4 Changing Decimals to Percents by Moving the Decimal Point

Write each decimal as a percent.

(a) 0.21

0.21 ← Decimal point is moved two places to the right.

0.21 = 21% ← Percent symbol is attached after decimal point is moved.
← Decimal point is not written with whole number percents.

(b) 0.529 = 52.9% ← Percent symbol is attached after decimal point is moved.

(c) 1.92 = 192% ← Percent symbol is attached after decimal point is moved.

CONTINUED ON NEXT PAGE

3. Write each percent as a decimal.

(a) 90%

(b) 9%

(c) 900%

(d) 9.9%

(e) 0.9%

ANSWERS

3. **(a)** 0.90 or 0.9 **(b)** 0.09 **(c)** 9.00 or 9 **(d)** 0.099 **(e)** 0.009

4. Write each number as a percent.

(a) 0.95

(b) 0.18

(c) 0.09

(d) 0.617

(e) 0.834

(f) 5.34

(g) 2.8

(h) 4

Answers

4. (a) 95% (b) 18% (c) 9% (d) 61.7% (e) 83.4% (f) 534% (g) 280% (h) 400%

(d) 2.5

2.50% 0 is attached so the decimal point can be moved two places to the right.

2.5 = 250%

(e) 3

3. = 3.00% so 3 = 300%

Note

In Examples 4(c), 4(d), and 4(e), notice that 1.92, 2.5, and 3 are greater than 1. Because the number 1 is equivalent to 100%, all numbers greater than 1 will be equivalent to percents greater than 100%.

WORK PROBLEM 4 AT THE SIDE.

Objective 4 Percents can also be written as fractions. Recall that a percent is a ratio with a denominator of 100. For example, 89% is $\frac{89}{100}$. Because the fraction bar indicates division, we are dividing the percent by 100, just as we did when writing a percent as a decimal.

Writing a Percent as a Fraction

To write a percent as a fraction, drop the % symbol and write the number over 100. Then write the fraction in lowest terms.

EXAMPLE 5 Writing a Percent as a Fraction

Write each percent as a fraction or mixed number in lowest terms.

(a) 25% Drop the % symbol and write 25 over 100.

$$25\% = \frac{25}{100} \quad \text{(25 per 100)}$$

$$= \frac{25 \div 25}{100 \div 25} = \frac{1}{4} \quad \text{(Lowest terms)}$$

As a check, let's write 25% as a decimal.

$$25\% = 25 \div 100 = 0.25 \quad \text{(Percent sign dropped)}$$

Because 0.25 means 25 hundredths,

$$0.25 = \frac{25 \div 25}{100 \div 25} = \frac{1}{4}. \leftarrow \text{Same result as above}$$

(b) 76% Drop the % symbol and write 76 over 100.

The percent becomes the numerator.

$$76\% = \frac{76}{100}$$

The *denominator* is always 100 because percent means parts per 100.

CONTINUED ON NEXT PAGE

Write $\frac{76}{100}$ in lowest terms.

$$\frac{76 \div 4}{100 \div 4} = \frac{19}{25} \quad \text{(Lowest terms)}$$

(c) 150%

$$150\% = \frac{150}{100} = \frac{150 \div 50}{100 \div 50} = \frac{3}{2} = 1\frac{1}{2} \quad \text{(Mixed number)}$$

Note

Remember that percent means **per 100**.

WORK PROBLEM 5 AT THE SIDE.

Example 6 shows how to write decimal percents and fraction percents as fractions.

EXAMPLE 6 Writing a Decimal Percent or Fraction Percent as a Fraction

Write each percent as a fraction in lowest terms.

(a) 15.5%

Write 15.5 over 100.

$$15.5\% = \frac{15.5}{100}$$

To get a whole number in the numerator, we must multiply the numerator and denominator by 10. (Multiplying by $\frac{10}{10}$ is the same as multiplying by 1.)

$$\frac{15.5}{100} = \frac{15.5 \cdot 10}{100 \cdot 10} = \frac{155}{1000}$$

Write the fraction in lowest terms.

$$\frac{155 \div 5}{1000 \div 5} = \frac{31}{200}$$

(b) $33\frac{1}{3}\%$

Write $33\frac{1}{3}$ over 100.

$$33\frac{1}{3}\% = \frac{33\frac{1}{3}}{100}$$

When there is a mixed number in the numerator, write it as an improper fraction. So $33\frac{1}{3}$ is $\frac{100}{3}$.

$$\frac{33\frac{1}{3}}{100} = \frac{\frac{100}{3}}{100}$$

CONTINUED ON NEXT PAGE

5. Write each percent as a fraction or mixed number in lowest terms.

(a) 50%

(b) 19%

(c) 80%

(d) 6%

(e) 125%

(f) 210%

ANSWERS

5. (a) $\frac{1}{2}$ (b) $\frac{19}{100}$ (c) $\frac{4}{5}$ (d) $\frac{3}{50}$ (e) $1\frac{1}{4}$ (f) $2\frac{1}{10}$

6. Write as fractions in lowest terms.

(a) 18.5%

(b) 87.5%

(c) 6.5%

(d) $66\frac{2}{3}\%$

(e) $12\frac{1}{3}\%$

(f) $62\frac{1}{2}\%$

Answers

6. (a) $\frac{37}{200}$ **(b)** $\frac{7}{8}$ **(c)** $\frac{13}{200}$ **(d)** $\frac{2}{3}$ **(e)** $\frac{37}{300}$ **(f)** $\frac{5}{8}$

Now you have a complex fraction (see **Section 4.6**). Rewrite the complex fraction using the ÷ symbol for division. Then follow the steps for dividing fractions.

$$\frac{\frac{100}{3}}{100} = \frac{100}{3} \div 100 = \frac{100}{3} \div \frac{100}{1} = \frac{100}{3} \cdot \frac{1}{100} = \frac{\cancel{100}^{1} \cdot 1}{3 \cdot \cancel{100}_{1}} = \frac{1}{3}$$

reciprocals

Note
In Example 6(a) we could have changed 15.5% to $15\frac{1}{2}\%$ and then written it as the improper fraction $\frac{31}{2}$ over 100. But it is usually easier to work with decimal percents as they are.

WORK PROBLEM 6 AT THE SIDE.

Objective 5 Recall that when you wrote a *decimal* as a percent, you *multiplied* by 100. The same idea can be used to write a *fraction* as a percent. In Example 5, you saw that 25% is $\frac{1}{4}$. Now let's start with $\frac{1}{4}$ and change it to a percent. Multiplying by 100 will get us back to 25%.

Rewrite as improper fraction.

$$\frac{1}{4} \cdot 100 = \frac{1}{4} \cdot \frac{100}{1} = \frac{1}{4} \cdot \frac{4 \cdot 25}{1} = \frac{1 \cdot \cancel{4}^{1} \cdot 25}{\cancel{4}_{1} \cdot 1} = \frac{25}{1} = 25\%$$

Writing a Fraction as a Percent

To write a fraction as a percent, multiply by 100 and attach a % symbol.

EXAMPLE 7 Writing Fractions as Percents

Write each fraction as a percent. Round to the nearest tenth of a percent, if necessary.

(a) $\frac{2}{5}$ Multiply $\frac{2}{5}$ by 100 and attach a % symbol.

$$\frac{2}{5} \cdot 100 = \frac{2}{5} \cdot \frac{100}{1} = \frac{2}{5} \cdot \frac{5 \cdot 20}{1} = \frac{2 \cdot \cancel{5}^{1} \cdot 20}{\cancel{5}_{1} \cdot 1} = \frac{40}{1} = 40\%$$

Attach a % symbol.

To check the result, write 40% as $\frac{40}{100}$ and reduce to lowest terms.

$$40\% = \frac{40}{100} = \frac{40 \div 20}{100 \div 20} = \frac{2}{5} \leftarrow \text{Original fraction}$$

CONTINUED ON NEXT PAGE

(b) $\frac{5}{8}$ Multiply $\frac{5}{8}$ by 100 and attach a % symbol.

$$\frac{5}{8} \cdot 100 = \frac{5}{8} \cdot \frac{100}{1} = \frac{5}{2 \cdot 4} \cdot \frac{4 \cdot 25}{1} = \frac{5 \cdot \overset{1}{\cancel{4}} \cdot 25}{2 \cdot \underset{1}{\cancel{4}} \cdot 1} = \frac{125}{2} = 62\frac{1}{2}\%$$

Attach a % symbol.

You can also do the last step of simplifying $\frac{125}{2}$ on your calculator. Enter $\frac{125}{2}$ as 125 [÷] 2 [=]. The result is 62.5, so $\frac{5}{8} = 62\frac{1}{2}\%$ or $\frac{5}{8} = 62.5\%$.

(c) $\frac{1}{6}$ Multiply $\frac{1}{6}$ by 100 and attach a % symbol.

$$\frac{1}{6} \cdot 100 = \frac{1}{6} \cdot \frac{100}{1} = \frac{1}{2 \cdot 3} \cdot \frac{2 \cdot 50}{1} = \frac{1 \cdot \overset{1}{\cancel{2}} \cdot 50}{\underset{1}{\cancel{2}} \cdot 3 \cdot 1} = \frac{50}{3} = 16\frac{2}{3}\%$$

Attach a % symbol.

To simplify $\frac{50}{3}$ on your calculator, enter 50 [÷] 3 [=]. The result is 16.66666666, with the 6 continuing to repeat. The directions say to round to the nearest *tenth* of a percent.

16.66666666 rounds to 16.**7** %

Tenths place — Attach a % symbol.

So $\frac{1}{6} = 16\frac{2}{3}\%$ or $\frac{1}{6} \approx 16.7\%$.

Calculator Tip: In Example 7(a) you can use your calculator to write $\frac{2}{5}$ as a percent.

Step 1 Enter $\frac{2}{5}$ as 2 [÷] 5 [=]. The result is [0.4].

Decimal equivalent of $\frac{2}{5}$

Step 2 Change the decimal number 0.4 to a percent by moving the decimal point two places to the right (multiplying by 100).

0.40 = 40% ← Attach a % symbol.

Try this technique on Examples 7(b) and 7(c).

For $\frac{5}{8}$, enter 5 [÷] 8 [=]. The result is [0.625]. Move the decimal point in 0.625 two places to the right.

0.625 = 62.5% ← Attach a % symbol.

For $\frac{1}{6}$, enter 1 [÷] 6 [=]. The result is [0.166666666].

Some calculators show 7 in last place.

Move the decimal point two places to the right. Then round to the nearest tenth.

0.166666666 = 16.6666666% ≈ 16.7% ← Attach a % symbol.

With a fraction such as $\frac{1}{6}$, your calculator gives only an *approximate* answer. You can't get the exact answer of $16\frac{2}{3}\%$ by using your calculator.

WORK PROBLEM 7 AT THE SIDE.

7. Write each fraction as a percent. If you're using a calculator, first work each one by hand. Then use your calculator, and round to the nearest tenth of a percent, if necessary.

(a) $\frac{1}{2}$

(b) $\frac{3}{4}$

(c) $\frac{1}{10}$

(d) $\frac{7}{8}$

(e) $\frac{5}{6}$

(f) $\frac{2}{3}$

ANSWERS

7. (a) 50% **(b)** 75% **(c)** 10%

(d) $87\frac{1}{2}\%$ or 87.5% (both are exact answers)

(e) exactly $83\frac{1}{3}\%$ or ≈83.3%

(f) exactly $66\frac{2}{3}\%$ or ≈66.7%

8. Fill in the blanks.

(a) 100% of \$3.95 is ______.

(b) 100% of 3000 students is ______.

(c) 100% of 7 pages is ______.

(d) 100% of 305 miles is ______.

(e) 100% of $10\frac{1}{2}$ hours is ______.

9. Fill in the blanks.

(a) 50% of \$10 is ______.

(b) 50% of 36 cookies is ______.

(c) 50% of 6000 women is ______.

(d) 50% of 8 hours is ______.

(e) 50% of \$2.50 is ______.

Answers

8. **(a)** \$3.95 **(b)** 3000 students **(c)** 7 pages **(d)** 305 miles **(e)** $10\frac{1}{2}$ hours

9. **(a)** \$5 **(b)** 18 cookies **(c)** 3000 women **(d)** 4 hours **(e)** \$1.25

Objective 6 When working with percents, it is helpful to have several reference points. 100% and 50% are two helpful reference points.

100% means 100 parts out of 100 parts. That's *all* of the parts. If you pay 100% of a \$45 dentist bill, you pay \$45 (*all* of it).

EXAMPLE 8 Finding 100% of a Number

Fill in the blanks.

(a) 100% of \$34 is ______.
100% is *all* of the money.
So 100% of \$34 is <u>\$34</u>.

(b) 100% of 4 cats is ______.
100% is *all* of the cats.
So 100% of 4 cats is <u>4 cats</u>.

WORK PROBLEM 8 AT THE SIDE.

50% means 50 parts out of 100 parts, which is *half* of the parts ($\frac{50}{100} = \frac{1}{2}$). 50% of \$12 is \$6 (*half* of the money).

EXAMPLE 9 Finding 50% of a Number

Fill in the blanks.

(a) 50% of \$20 is ______.
50% is *half* of the money.
So 50% of \$20 is <u>\$10</u>.

(b) 50% of 280 miles is ______.
50% is *half* of the miles.
So 50% of 280 miles is <u>140 miles</u>.

WORK PROBLEM 9 AT THE SIDE.

Name Date Class

7.1 Exercises

Write each percent as a decimal.

1. 25%

2. 35%

3. 30%

4. 20%

5. 6%

6. 3%

7. 140%

8. 250%

9. 7.8%

10. 6.7%

11. 100%

12. 600%

13. 0.5%

14. 0.2%

15. 0.35%

16. 0.076%

Write each decimal as a percent.

17. 0.5

18. 0.6

19. 0.62

20. 0.18

21. 0.03

22. 0.09

23. 0.125

24. 0.875

25. 0.629

26. 0.494

27. 2

28. 5

29. 2.6

30. 1.8

31. 0.0312

32. 0.0625

Write each percent as a fraction or mixed number in lowest terms.

33. 20%

34. 40%

35. 50%

36. 75%

37. 55%

38. 35%

39. 37.5%

40. 87.5%

41. 6.25%

42. 43.75%

43. $16\frac{2}{3}\%$

44. $83\frac{1}{3}\%$

Writing Calculator Small Group

45. 130% **46.** 175% **47.** 250% **48.** 325%

Write each fraction as a percent. If you're using a calculator, first work each one by hand. Then use your calculator, and round to the nearest tenth of a percent, if necessary.

49. $\frac{1}{4}$ **50.** $\frac{1}{5}$ **51.** $\frac{3}{10}$ **52.** $\frac{9}{10}$

53. $\frac{3}{5}$ **54.** $\frac{3}{4}$ **55.** $\frac{37}{100}$ **56.** $\frac{63}{100}$

57. $\frac{3}{8}$ **58.** $\frac{1}{8}$ **59.** $\frac{1}{20}$ **60.** $\frac{1}{50}$

61. $\frac{5}{9}$ **62.** $\frac{7}{9}$ **63.** $\frac{1}{7}$ **64.** $\frac{5}{7}$

In each of the following, write percents as decimals and decimals as percents.

65. In Folsum, the sales tax rate is 8%.

66. At College of DuPage, 82% of the students work part time.

67. In one company, 65% of the salespeople are women.

68. Only 38.6% of those registered actually voted.

69. The property tax rate in Alpine County is 0.035.

70. A church building fund has 0.49 of the money needed.

71. The success rate in CPR training this session is 2 times that of the last session.

72. Attendance at the picnic this year is 3 times last year's attendance.

Write a fraction and a percent for the shaded part of each figure. Then write a fraction and a percent for the unshaded part of each figure.

73.

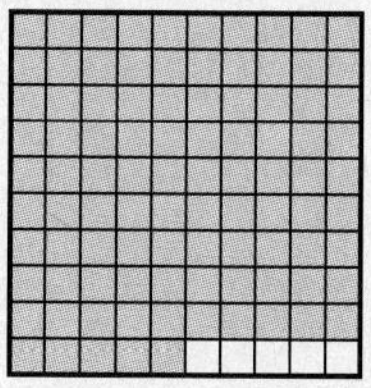

74.

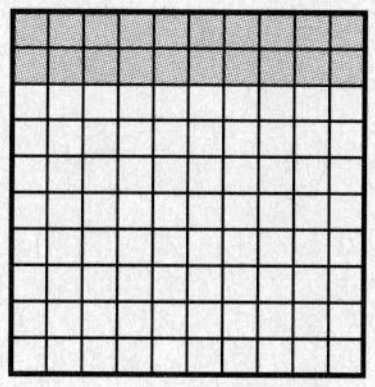

75.

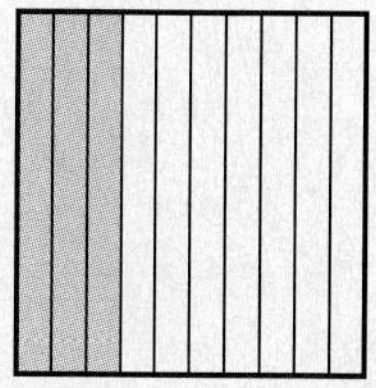

76.

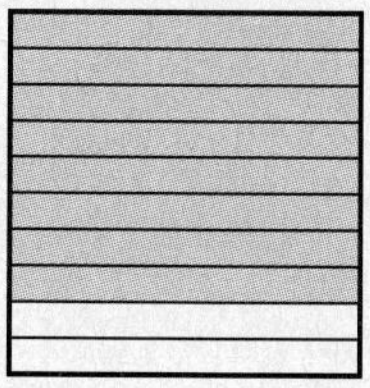

77.

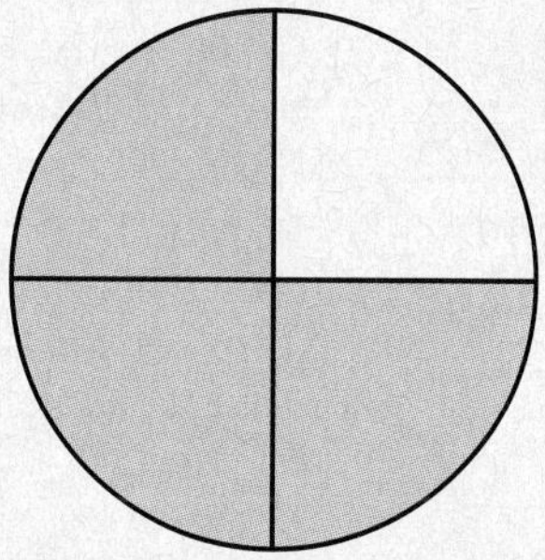

78.

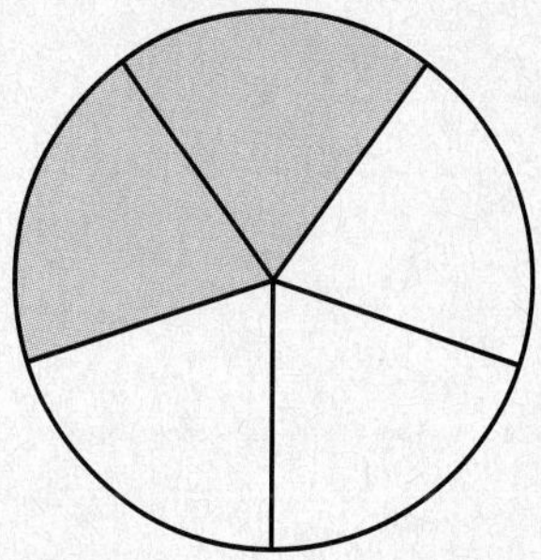

Complete this chart. Write fractions and mixed numbers in lowest terms.

	Fraction	Decimal	Percent
79.	$\frac{1}{100}$		
80.	$\frac{1}{10}$		
81.		0.2	
82.		0.25	
83.			30%
84.			40%
85.	$\frac{1}{2}$		
86.	$\frac{3}{4}$		
87.			90%
88.			100%
89.		1.5	
90.		2.25	

G *In the following application problems, write the answer as a fraction in lowest terms, as a decimal, and as a percent.*

91. The license applicant answered 76 of 100 questions correctly. What portion was correct?

92. Of 80 videos available, 20 are for children. What portion are for children?

93. The regular price of a laser jet printer was $750. It was reduced $150. By what portion was the price reduced?

94. Dick McIntosh must complete 200 hours of community service. So far, he has completed 75 hours of community service. What portion of the time has he completed?

95. Of the fifteen people at the office, nine are single parents. What portion are single parents?

96. A pizza shop has 25 employees. Of these, 14 are students. What portion are students?

97. An insurance office has 80 employees. If 64 of the employees have cellular phones, what portion of the employees do not have cellular phones?

98. A zoo has 125 animals, including 25 that are members of endangered species. What portion is not endangered?

99. An antibiotic is used to treat 380 people. If 342 people do not have any side effects from using the antibiotic, find the portion that do have side effects.

100. In apple grower's cooperative has 250 members. If 100 of the growers use a certain insecticide, find the portion that do not use the insecticide.

The circle graph shows the type of transportation used by 4200 students at Metro-Community College. It also shows the number of people using each type of transportation. Use this graph to answer Exercises 101–104, giving the answer as a fraction in lowest terms, as a decimal, and as a percent.

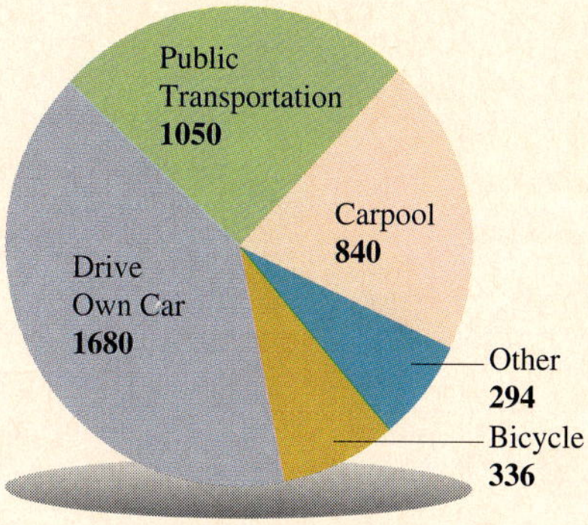

101. What portion of the students use public transportation?

102. What portion of the students use a bicycle?

103. Find the portion that drive their own cars.

104. Find the portion that carpool.

105. Explain and correct the errors made by students when they used their calculators on these problems.

(a) Write $\frac{7}{20}$ as a percent.

Student entered 7 [÷] 20 [=], and the result was 0.35. So $\frac{7}{20} = 0.35\%$

(b) Write $\frac{16}{25}$ as a percent.

Student entered 25 [÷] 16 [=], and the result was 1.5625. So $\frac{16}{25} = 156.25\%$

106. Explain and correct the errors made by students who moved decimal points to solve these problems.

(a) Write 3.2 as a percent.

03.2 = 0.032 so 3.2 = 0.032%

(b) Write 60% as a decimal.

00.60 = 0.0060 so 60% = 0.0060

Fill in the blanks. Remember that 100% is all *of something and 50% is* half *of it.*

107. 100% of \$78 is ______________.

108. 100% of 5 hours is ______________.

109. 50% of \$78 is ______________.

110. 50% of 5 hours is ______________.

111. There are 20 children in the preschool class. 100% of the children are served breakfast and lunch. How many children are served both meals? ______________.

112. The company owns 345 vans. 100% of the vans are painted white with blue lettering. How many vans are painted white with blue lettering? ______________.

113. 50% of 180 miles is ______________.

114. 50% of \$900 is ______________.

115. John owes \$285 for tuition. Financial aid will pay 50% of the cost. Financial aid will pay ______________.

116. The Animal Humane Society took in 20,000 animals last year. About 50% of them were dogs. The number of dogs taken in was about ______________.

117. 50% of 8200 college students is ______________.

118. 100% of 8200 college students is ______________.

119. Describe a shortcut way to find 100% of a number.

120. Describe a shortcut way to find 50% of a number.

Review and Prepare

Find the unknown number in each proportion. (For help, see **Section 6.3**.*)*

121. $\frac{8}{4} = \frac{x}{15}$

122. $\frac{n}{50} = \frac{8}{20}$

123. $\frac{4}{y} = \frac{12}{15}$

124. $\frac{5}{x} = \frac{12}{24}$

125. $\frac{42}{30} = \frac{14}{b}$

126. $\frac{6}{22} = \frac{a}{220}$

7.2 *The Percent Proportion*

We will show you two ways to solve percent problems. One is the proportion method, which we discuss in this section. The other is the percent equation method, which we explain in **Section 7.3**.

Objectives

1. Learn the percent proportion.
2. Solve percent problems using the percent proportion.

For Extra Help

Tutorial | Tape 12 | SSM, Sec. 7.2

Objective 1 You have learned that a statement of two equivalent ratios is called a proportion (see **Section 6.3**).

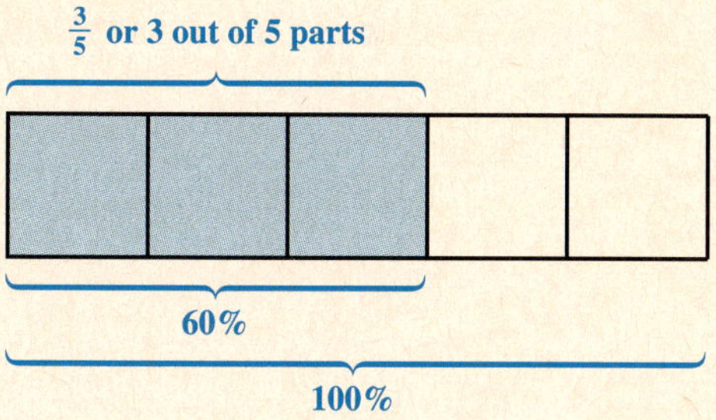

For example, the fraction $\frac{3}{5}$ is the same as the ratio 3 to 5, and 60% is the ratio 60 to 100. As the figure above shows, these two ratios are equivalent and make a proportion.

The **percent proportion** (per-SENT proh-POR-shun) can be used to solve percent problems.

The Percent Proportion

Percent is to *100* as *part* is to *whole.*

$$\frac{\text{percent}}{100} = \frac{\text{part}}{\text{whole}}$$

Always 100 because percent means "per 100." → 100

In some textbooks the percent proportion is written with the terms *amount* and *base.*

$$\frac{\text{percent}}{100} = \frac{\text{amount}}{\text{base}}$$

Here is the proportion for the figure at the top of the page.

$$\frac{60}{100} = \frac{3}{5}$$

60% means 60 parts out of 100 parts. 3 ← Shaded part (3 parts); 5 ← Whole (5 parts)

If we write $\frac{60}{100}$ in lowest terms, it is equal to $\frac{3}{5}$, so the proportion is true.

$$\frac{60}{100} = \frac{60 \div \mathbf{20}}{100 \div \mathbf{20}} = \frac{3}{5}$$

← Matches ratio on right side of proportion.

Objective 2 In order to use the percent proportion to solve problems, you must be able to pick out the *percent,* the *whole,* and the *part.* Look for the percent first, because it is the easiest to identify.

Finding the Percent

The **percent** is a ratio of a part to a whole, with 100 as the denominator. In a problem, the percent appears with the word "percent" or with the symbol **"%"** after it.

1. Identify the percent.

(a) Of the $2000, 15% will be spent on a washing machine.

(b) 60 employees is what percent of 750 employees?

(c) Find the sales tax by multiplying the $590 price by $6\frac{1}{2}$ percent.

(d) $30 is 150% of what number?

(e) 75 of the 110 rental cars were rented today. What percent were rented?

2. Identify the whole.

(a) Of the $2000, 15% will be spent on a washing machine.

(b) 60 employees is what percent of 750 employees?

(c) Find the amount of sales tax by multiplying the $590 price by $6\frac{1}{2}$ percent.

(d) $30 is 150% of what number?

(e) 75 of the 110 rental cars were rented today? What percent were rented?

ANSWERS

1. (a) 15 (b) unknown (c) $6\frac{1}{2}$ (d) 150 (e) unknown
2. (a) $2000 (b) 750 (c) $590 (d) unknown (e) 110

EXAMPLE 1 Finding the Percent in a Percent Problem

Find the percent in each of the following.

(a) 32% of the 900 women were retired. How many were retired?
(32 → percent)

The percent is 32. The number 32 appears with the symbol %.

(b) $150 is 25 percent of what number?
(25 → percent)

The percent is 25 because 25 appears with the word "percent."

(c) If 7 students failed, what percent of the 350 students failed?
(what percent → percent (unknown))

The word "percent" has no number with it, so the percent is the unknown part of the prblem.

WORK PROBLEM 1 AT THE SIDE.

The second thing to look for in a percent problem is the *whole* (sometimes called the *base*).

Finding the Whole

The **whole** is the entire quantity. In a problem the *whole* often appears after the word **of**.

EXAMPLE 2 Finding the Whole in a Percent Problem

These problems are the same as those in Example 1. Now find the whole.

(a) 32% **of** the 900 women were retired. How many were retired?
(32% → percent; 900 → whole)

The whole is 900. The number 900 appears after the word *of*.

(b) $150 is 25 percent **of** what number?
(25 → percent; what number → whole (unknown; follows *of*))

(c) If 7 students failed, what percent **of** the 350 students failed?
(what percent → percent (unknown); 350 → whole (follows *of*))

WORK PROBLEM 2 AT THE SIDE.

The third and final thing to find in a percent problem is the *part* (sometimes called the *amount*).

Finding the Part

The **part** is the number being compared to the whole.

Note
If you have trouble identifying the *part*, find the *whole* and the *percent* first. The remaining number is the *part*.

EXAMPLE 3 Finding the Part in a Percent Problem

Identify the part in each of the following. These problems are the same as those in Examples 1 and 2.

(a) 32% **of** the 900 women were retired. How many were retired?

32% → percent; 900 → whole; How many were retired? → part (unknown)

The part of the women who were retired is unknown. In other words, some part of 900 women were retired.

(b) $150 is 25 percent **of** what number?

$150 → part; 25 → percent; what number → whole (unknown)

150 is the remaining number, so 150 is the part.

(c) If 7 students failed, what percent **of** 350 students failed?

7 → part; what percent → percent (unknown); 350 → whole

WORK PROBLEM 3 AT THE SIDE.

Now you are ready to solve problems using the percent proportion.

EXAMPLE 4 Using the Percent Proportion to Find the Part

Use the percent proportion to answer this question.

15% of $165 is how much money?

First identify the percent by looking for the % symbol or the word *percent*. Then look for the whole (usually follows the word *of*). Finally, identify the part.

15% **of** $165 is how much money?

15% → percent; $165 → whole (follows *of*); how much money → part (unknown)

Set up the percent proportion. Here we use n as the variable representing the unknown part. You may use any letter you like.

$$\text{Percent} \rightarrow \frac{15}{100} = \frac{n}{165} \leftarrow \text{Part (unknown)}$$
Always 100 → 100; 165 ← Whole

Recall from **Section 6.3** that the first step in solving a proportion is to find the cross proucts.

Step 1 $\frac{15}{100} = \frac{n}{165}$

$100 \cdot n$
$15 \cdot 165$
Cross products

CONTINUED ON NEXT PAGE

3. Identify the part.

(a) Of the $2000, 15% will be spent on a washing machine.

(b) 60 employees is what percent of 750 employees?

(c) Find the sales tax by multiplying the $590 price by $6\frac{1}{2}$ percent.

(d) $30 is 150% of what number?

(e) 75 of the 110 rental cars were rented today. What percent were rented?

ANSWERS

3. (a) unknown **(b)** 60 **(c)** unknown **(d)** $30 **(e)** 75

4. Use the percent proportion to answer these questions.

(a) 9% of 3250 miles is how many miles?

(b) What is 20% of 180 calories?

(c) 78% of $5.50 is how much?

(d) What is $12\frac{1}{2}\%$ of 400 homes? (*Hint:* Write $12\frac{1}{2}\%$ as 12.5%.)

5. Use the percent proportion to answer these questions.

(a) 1200 books is what percent of 5000 books?

(b) What percent of $6.50 is $0.52?

(c) 20 athletes is what percent of 32 athletes?

Answers

4. (a) $\frac{9}{100} = \frac{n}{3250}$ The part is 292.5 miles.

(b) $\frac{20}{100} = \frac{n}{180}$ The part is 36 calories.

(c) $\frac{78}{100} = \frac{n}{5.50}$ The part is $4.29.

(d) $\frac{12.5}{100} = \frac{n}{400}$ The part is 50 homes.

5. (a) $\frac{p}{100} = \frac{1200}{5000}$; 24%

(b) $\frac{p}{100} = \frac{0.52}{6.50}$; 8%

(c) $\frac{p}{100} = \frac{20}{32}$; 62.5% or $62\frac{1}{2}\%$

Step 2 $100 \cdot n = \underbrace{15 \cdot 165}$ Show that cross products are equivalent.

$100 \cdot n = 2475$

Step 3 $\frac{\overset{1}{\cancel{100}} \cdot n}{\underset{1}{\cancel{100}}} = \frac{2475}{100}$ Divide both sides by the coefficient of the variable term. On the left side, divide out the common factor of 100.

$n = 24.75$ On the right side, $2475 \div 100$ is 24.75.

The part is **$24.75**, so 15% of $165 is $24.75.

Note

When you use the percent proportion, do **not** move the decimal point in the percent or in the answer.

WORK PROBLEM 4 AT THE SIDE.

EXAMPLE 5 Using the Percent Proportion to Find the Percent

Use the percent proportion to answer this question.

8 pounds is what percent **of** 160 pounds?

8 pounds → part; what percent → percent (unknown); 160 → whole (follows *of*)

Set up the percent proportion. Then find cross products.

$$\frac{p}{100} = \frac{8}{160}$$

Percent (unknown) → p; Always 100 → 100; 8 ← Part; 160 ← Whole

$$\frac{p}{100} = \frac{8}{160}$$

Cross products: $100 \cdot 8 = 800$; $p \cdot 160$

$p \cdot 160 = 800$ Show that cross products are equivalent.

$\frac{p \cdot \overset{1}{\cancel{160}}}{\underset{1}{\cancel{160}}} = \frac{800}{160}$ Divide both sides by 160.

$p = 5$

The percent is **5%**. So 8 pounds is 5% of 160 pounds.

Note

When you're finding an unknown percent, be careful to label your answer with the % symbol. Do **not** add a decimal point or move the decimal point in your answer.

WORK PROBLEM 5 AT THE SIDE.

EXAMPLE 6 Using the Percent Proportion to Find the Whole

Use the percent proportion to answer this question.

162 credits is 90% **of** how many credits?

162 → part; 90% → percent; how many credits → whole (unknown; follows *of*)

$$\frac{90}{100} = \frac{162}{n}$$

Percent → 90; Always 100 → 100; 162 ← Part; n ← Whole (unknown)

$$\frac{90}{100} = \frac{162}{n}$$

$100 \cdot 162 = 16{,}200$

$90 \cdot n$

Cross products

$90 \cdot n = 16{,}200$ Show that cross products are equivalent.

$$\frac{\overset{1}{\cancel{90}} \cdot n}{\underset{1}{\cancel{90}}} = \frac{16{,}200}{90}$$ Divide both sides by 90.

$$n = 180$$

The whole is **180 credits**. So 162 credits is 90% of 180 credits.

WORK PROBLEM 6 AT THE SIDE.

So far, in all the examples the part has been *smaller* than the whole. This is because all the percents have been less than 100%. Recall that 100% of something is *all* of it. When the percent is *less* than 100%, you have *less* than all of it.

Now let's look at percents *greater* than 100%. For example,

100% of \$20 is all of the money, or \$20.

150% of \$20 is *more* than \$20.

100%	+	**50%**	=	**150%**
of the money is		of the money is		of the money is
\$20	+	**\$10**	=	**\$30**

150% of \$20 is \$30.

When the percent is *greater* than 100%, the part is *larger* than the whole.

6. Use the percent proportion to answer these questions.

(a) 37 cars is 74% of how many cars?

(b) 45% of how much money is \$139.59?

(c) 1.2 tons is $2\frac{1}{2}$% of how many tons?

ANSWERS

6. (a) $\frac{74}{100} = \frac{37}{n}$; 50 cars

(b) $\frac{45}{100} = \frac{139.59}{n}$; \$310.20

(c) $\frac{2.5}{100} = \frac{1.2}{n}$; 48 tons

7. Use the percent proportion to answer each question.

(a) 350% of \$6 is how much?

(b) 23 hours is what percent of 20 hours?

(c) What percent of \$47.32 is \$106.47?

ANSWERS

7. (a) $\frac{350}{100} = \frac{n}{6}$; \$21

(b) $\frac{p}{100} = \frac{23}{20}$; 115%

(c) $\frac{p}{100} = \frac{106.47}{47.32}$; 225%

EXAMPLE 7 Working with Percents Greater Than 100%

Use the percent proportion to answer each question.

(a) How many students is 210% **of** 40 students?

How many students → part (unknown); 210% → percent; 40 → whole (follows *of*)

$$\frac{210}{100} = \frac{n}{40}$$

Percent → 210; Always 100 → 100; n ← Part (unknown); 40 ← Whole

Cross products: $100 \cdot n$ and $210 \cdot 40 = 8400$

$$100 \cdot n = 8400 \quad \text{Show that cross products are equivalent.}$$

$$\frac{\cancel{100}^{1} \cdot n}{\cancel{100}_{1}} = \frac{8400}{100} \quad \text{Divide both sides by 100.}$$

$$n = 84$$

The part is **84 students**, which is *more* than the whole of 40 students. This result makes sense because the percent is 210%. If it was exactly 200%, we would have *2 times the whole,* and 2 times 40 students is 80 students. So 210% should be even a little more than 80 students. Our answer of 84 students is reasonable.

(b) \$68 is what percent **of** \$50?

\$68 → part; what percent → percent (unknown); \$50 → whole (follows *of*)

$$\frac{p}{100} = \frac{68}{50}$$

Percent is unknown → p; Always 100 → 100; 68 ← Part; 50 ← Whole

Cross products: $100 \cdot 68 = 6800$ and $p \cdot 50$

$$p \cdot 50 = 6800 \quad \text{Show that cross products are equivalent.}$$

$$\frac{p \cdot \cancel{50}^{1}}{\cancel{50}_{1}} = \frac{6800}{50} \quad \text{Divide both sides by 50.}$$

$$p = 136$$

The percent is **136%**. This result makes sense because \$68 is *more* than \$50, so \$68 has to be *more than 100%* of \$50.

◀◀ **WORK PROBLEM 7 AT THE SIDE.**

7.2 Exercises

Use the percent proportion to answer these questions. If necessary, round money answers to the nearest cent and percent answers to the nearest tenth of a percent.

1. What is 10% of 3000 runners?

2. What is 35% of 2340 volunteers?

3. 4% of 120 feet is how many feet?

4. 9% of \$150 is how much money?

5. 16 pepperoni pizzas is what percent of 32 pizzas?

6. 35 hours is what percent of 140 hours?

7. What percent of 200 calories is 16 calories?

8. What percent of 350 parking spaces is 7 handicapped parking spaces?

9. 495 successful students is 90% of what number of students?

10. 84 letters is 28% of what number of letters?

11. $12\frac{1}{2}$% of what amount is \$3.50?

12. $5\frac{1}{2}$% of what amount is \$17.60?

13. 250% of 7 hours is how long?

14. What is 130% of 60 trees?

15. What percent of \$172 is \$32?

16. \$14 is what percent of \$398?

17. 748 books is 110% of what number of books?

18. 145% of what number of inches is 11.6 inches?

19. What is 14.7% of \$274?

20. 8.3% of \$43 is how much?

21. 105 employees is what percent of 54 employees?

22. What percent of 46 animals is 100 animals?

23. \$0.33 is 4% of what amount?

24. 6% of what amount is \$0.03?

25. A student turned in the following answers on a test. You can see that two of the answers are incorrect without working the problems. Find the incorrect answers and explain how you identified them (without actually solving the problems).

50% of \$84 is <u>\$42</u>
150% of \$30 is <u>\$20</u>
25% of \$16 is <u>\$32</u>
100% of \$217 is <u>\$217</u>

26. Identify the three parts in a percent problem. For each of these three parts, write a sentence telling how you will identify it.

27. Explain and correct the *two* errors that a student made when solving this problem: \$14 is what percent of \$8?

$$\frac{p}{100} = \frac{8}{14} \qquad p \cdot 14 = 100 \cdot 8$$

$$\frac{p \cdot \overset{1}{\cancel{14}}}{\underset{1}{\cancel{14}}} = \frac{800}{14}$$

The answer is ≈ 57.1

28. Explain and correct the *two* errors that a student made when solving this problem: 9 children is 30% of what number of children?

$$\frac{30}{100} = \frac{n}{9} \qquad 100 \cdot n = 30 \cdot 9$$

$$\frac{\overset{1}{\cancel{100}} \cdot n}{\underset{1}{\cancel{100}}} = \frac{270}{100}$$

$$n = 2.7$$

The answer is 2.7%.

Review and Prepare

Use shortcuts to answer these questions quickly. (For help, see ***Section 7.1****.)*

29. 100% of 15 miles is ______.

30. 100% of \$21 is ______.

31. 50% of 15 miles is ______.

32. 50% of \$21 is ______.

33. 50% of \$620 is ______.

34. 50% of 250 dogs is ______.

7.3 The Percent Equation

OBJECTIVES

1. Estimate answers to percent problems involving 25%, 10%, and 1%.
2. Solve basic percent problems by using the percent equation.

FOR EXTRA HELP

Tutorial | Tape 12 | SSM, Sec. 7.3

OBJECTIVE 1 Before showing you the percent equation, we need to do some more estimation. As you have learned when working with integers, fractions, and decimals, it is always a good idea to estimate the answer. Doing so helps you catch mistakes. Also, when you're out shopping or eating in a restaurant, you will be able to estimate the sales tax, discount, or tip.

In **Section 7.1** you developed shortcuts for 100% of a number (all of the number) and 50% of a number (divide the number by 2). Now let's look at quick ways to work with 25%, 10%, and 1%.

25% means 25 parts out of 100 parts, or $\frac{25}{100}$, which is the same as $\frac{1}{4}$.

25% of \$40 would be $\frac{1}{4}$ of \$40, or \$10.

A quick way to find $\frac{1}{4}$ of a number is to divide it by 4. Recall that the denominator, 4, tells you that the whole is divided into 4 equal parts.

EXAMPLE 1 Estimating 25% of a Number

Estimate the answer to each question.

(a) What is 25% of \$817?

Use front end rounding to round \$817 to \$800. Then divide \$800 by 4. The estimate is **\$200**.

(b) Find 25% of 19.7 miles.

Use front end rounding to round 19.7 miles to 20 miles. Then divide 20 miles by 4. The estimate is **5 miles**.

(c) 25% of 49 days is how long?

You could round 49 days to 50 days, using front end rounding. Then divide 50 by 4 to get an estimate of **12.5 days**. However, the division step is simpler if you notice that 48 is a multiple of 4. You can round 49 days to 48 days and divide by 4 to get an estimate of **12 days**. Either way gives you a fairly good idea of the correct answer.

WORK PROBLEM 1 AT THE SIDE.

1. Estimate the answer to each question.

(a) What is 25% of \$110.38?

(b) Find 25% of 7.6 hours.

(c) 25% of 34 pounds is how many pounds?

Ten percent, or 10%, means 10 parts out of 100 parts or $\frac{10}{100}$, which is the same as $\frac{1}{10}$. A quick way to find $\frac{1}{10}$ of a number is to divide by 10. The denominator, 10, tells you that the whole is divided into 10 equal parts. The shortcut for dividing by 10 is to move the decimal point *one* place to the *left*.

EXAMPLE 2 Finding 10% of a Number by Moving the Decimal Point

Find the exact answer to each question by moving the decimal point. Then round to get an estimated answer.

(a) What is 10% of \$817?

To find 10% of \$817, divide \$817 by 10. Do the division by moving the decimal point *one* place to the *left*. The decimal point starts at the far right side of \$817.

$$10\% \text{ of } \$817. = \$81.70$$

Write this zero because it's money.

The *exact* answer is **\$81.70**. For an *estimate,* you could round \$81.70 to **\$80**.

CONTINUED ON NEXT PAGE

ANSWERS

1. (a) \$100 ÷ 4 gives an estimate of \$25.
(b) 8 hours ÷ 4 gives an estimate of 2 hours.
(c) 30 pounds ÷ 4 gives an estimate of 7.5 pounds, or 32 pounds ÷ 4 gives an estimate of 8 pounds.

2. First, find the exact answer to each question by moving the decimal point. Then, round to get an estimated answer.

(a) What is 10% of \$110.38?

(b) Find 10% of 7.6 hours.

(c) 10% of 34 pounds is how many pounds?

3. First find the exact answer to each question by moving the decimal point. Then round to get an estimated answer.

(a) What is 1% of \$110.38?

(b) Find 1% of 7.6 hours.

(c) 1% of 34 pounds is how many pounds?

(b) Find 10% of 19.7 miles.

Move the decimal point *one* place to the *left.*

10% *of* 19.7 miles = 1.97 miles

The *exact* answer is **1.97 miles**. For an *estimate,* you could round 1.97 miles to **2 miles**.

WORK PROBLEM 2 AT THE SIDE.

One percent, or 1%, is 1 part out of 100 parts or $\frac{1}{100}$. This time the denominator of 100 tells you that the whole is divided into 100 parts. Recall that a quick way to divide by 100 is to move the decimal point *two* places to the *left*.

EXAMPLE 3 Finding 1% of a Number by Moving the Decimal Point

Find the exact answer to each question by moving the decimal point. Then round to get an estimated answer.

(a) What is 1% of \$817?

To find 1% of \$817, divide \$817 by 100. Do the division by moving the decimal point *two* places to the *left.*

1% of \$817. = \$8.17

The *exact* answer is **\$8.17**. For an *estimate,* you could round \$8.17 to **\$8**.

(b) Find 1% of 19.7 miles.

Move the decimal point *two* places to the *left.*

1% of 19.7 miles = 0.197 mile

The *exact* answer is **0.197 mile**. For an *estimate,* you could round 0.197 mile to **0.2 mile**.

WORK PROBLEM 3 AT THE SIDE.

Here is a summary of the shortcuts you can use with percents.

Percent Shortcuts

200% of a number is 2 times the number; 300% of a number is 3 times the number, and so on.

100% of a number is the entire number.

To find **50% of a number**, divide the number by 2.

To find **25% of a number**, divide the number by 4.

To find **10% of a number**, divide the number by 10. To do the division, move the decimal point in the number *one* place to the *left*.

To find **1% of a number**, divide the number by 100. To do the division, move the decimal point in the number *two* places to the *left*.

ANSWERS

2. (a) \$110.38 = \$11.038 (exact); estimate is \$11.
(b) 7.6 hours = 0.76 hour (exact); estimate is 0.8 hour.
(c) 34. pounds = 3.4 pounds (exact); estimate is 3 pounds.

3. (a) \$110.38 = \$1.1038 (exact); estimate is \$1.
(b) 07.6 hours = 0.076 hours (exact); estimate is 0.1 hour.
(c) 34. pounds = 0.34 pound (exact); estimate is 0.3 pound.

OBJECTIVE 2 In **Section 7.2** you used a proportion to solve percent problems. Now you will learn how to solve these problems by using the percent equation.

Percent Equation

percent **of** whole = part

The word **of** indicates multiplication, so the **percent equation** becomes

percent • whole = part

Be sure to write the percent as a decimal or fraction before using the equation.

The percent equation is just a rearrangement of the percent proportion. Recall that in the proportion you wrote the percent over 100. Because there is no 100 in the equation, you have to change the percent to a decimal or fraction by dividing by 100 *before* using the equation.

Note

Once you have set up a percent equation, we encourage you to use your calculator to do the multiplying or dividing needed to solve the equation. For this reason, we will always write the percent as a decimal. If you're doing the problems by hand, changing the percent to a fraction may be easier at times. Either method will work.

In Examples 4, 5, and 6, we ask the same percent questions that we did in the examples in **Section 7.2**. There we used a proportion to answer each question. Now we will use an equation to answer them. You can then compare the equation method with the proportion method.

EXAMPLE 4 Using the Percent Equation to Find the Part

Write and solve a percent equation to answer each question.

(a) 15% of \$165 is how much money?

Translate the sentence into an equation. Recall that *of* indicates multiplication and *is* translates to the equal sign. The percent must be written in decimal form. Use any letter you like to represent the unknown quantity. (We will use n for an unknown number and p for an unknown percent.)

15.% of \$165 is how much money?

$$0.15 \cdot 165 = n$$

To solve the equation, simplify the left side, multiplying 0.15 by 165.

$$\underbrace{0.15(165)} = n$$
$$24.75 = n$$

So 15% of \$165 is **\$24.75**, which matches the answer we got by using a proportion (see Example 4 in **Section 7.2**).

CONTINUED ON NEXT PAGE

4. Write and solve an equation to answer each question.

(a) 9% of 3250 miles is how many miles?

(b) 78% of \$5.50 is how much?

(c) What is $12\frac{1}{2}\%$ of 400 homes? (*Hint:* Write $12\frac{1}{2}\%$ as 12.5%. Then move the decimal point two places to the left.)

(d) How much is 350% of \$6?

Check We can use estimation to verify the solution. First we find 10% of \$165 by moving the decimal point.

10% of \$165. is \$16.50

5% of \$165 would be half as much, that is, half of \$16.50 or about \$8.

So our *estimate* for 15% of \$165 is \$16.50 + \$8 = \$24.50. The exact answer of \$24.75 is very close to this estimate.

(b) How many students is 210% of 40 students?

Translate the sentence into an equation. Write the percent in decimal form.

How many students is 210.% of 40 students?

$$n = 2.10 \cdot 40$$

This time the two sides of the percent equation are reversed: part = percent • whole. Recall that the variable may be on either side of the equal sign (see **Section 2.3**). To solve the equation, we simplify the right side, multiplying 2.10 by 40.

$$n = 2.10(40)$$
$$n = 84$$

So **84 students** is 210% of 40 students. This matches the answer we got using a proportion (see Example 7(a) in **Section 7.2**).

Check We can use estimation to verify the solution. 210% is close to 200%.

200% of 40 students is 2 times 40 students = 80 students ← Estimate

The exact answer of 84 students is close to the estimate.

WORK PROBLEM 4 AT THE SIDE.

EXAMPLE 5 Using the Percent Equation to Find the Percent

Write and solve a percent equation to answer each question.

(a) 8 pounds is what percent of 160 pounds?

Translate the sentence into an equation. This time the percent is unknown. Do **not** move the decimal point in the other numbers.

8 pounds is what percent of 160 pounds?

$$8 = p \cdot 160$$

To solve the equation, divide both sides by 160.

On the left side, divide 8 by 160. On the right side, divide out the common factor of 160.

$$\frac{8}{160} = \frac{p \cdot \cancel{160}^{1}}{\cancel{160}_{1}}$$

Solution in *decimal* form

$$0.05 = p$$

Now multiply the solution by 100 to change it from a *decimal* to a *percent*.

$$0.05 = 5\%$$

So 8 pounds is **5%** of 160 pounds. This matches the answer we got using a proportion (see Example 5 in **Section 7.2**).

CONTINUED ON NEXT PAGE

ANSWERS

4. (a) $0.09\,(3250) = n$; 292.5 miles
(b) $0.78(5.50) = n$; \$4.29
(c) $n = 0.125(400)$; 50 homes
(d) $n = 3.5(6)$; \$21

Check The answer makes sense because 10% of 160 pounds would be 16 pounds.

10% of 160. pounds is 16 pounds.

5% of 160 pounds is half as much, that is, half of 16 pounds, or 8 pounds.

8 pounds matches the number given in the original problem.

(b) What percent of \$50 is \$68?

Translate the sentence into an equation and solve it.

What percent of \$50 is \$68?

$$p \cdot 50 = 68$$

$$\frac{p \cdot \overset{1}{\cancel{50}}}{\underset{1}{\cancel{50}}} = \frac{68}{50} \quad \text{Divide both sides by 50.}$$

$$p = 1.36 \quad \text{Solution in } \textit{decimal} \text{ form.}$$

Now multiply the solution by 100 to change it from a *decimal* to a *percent*.

$$1.36 = 136\%$$

So **136%** of \$50 is \$68.

Check The answer makes sense because 100% of \$50 would be \$50 (all of it), and 200% of \$50 would be 2 times \$50, or \$100. So 135% of \$50 has to be between \$50 and \$100. The number given in the original problem, \$68, fits the estimate.

Note

When you use an equation to solve for an unknown percent, *the solution will be in decimal form.* Remember to **multiply the solution by 100** to change it from decimal form to a percent. The shortcut is to move the decimal point in the solution *two* places to the *right* and attach a % symbol.

WORK PROBLEM 5 AT THE SIDE.

EXAMPLE 6 Using the Percent Equation to Find the Whole

Write and solve a percent equation to answer each question.

(a) 162 credits is 90% of how many credits?

Translate the sentence into an equation. Write the percent in decimal form.

162 credits is 90.% of how many credits?

$$162 = 0.90 \cdot n$$

Recall that 0.90 is equivalent to 0.9, so use 0.9 in the equation.

$$\frac{162}{0.9} = \frac{\overset{1}{\cancel{0.9}} \cdot n}{\underset{1}{\cancel{0.9}}} \quad \text{Divide both sides by 0.9.}$$

$$180 = n$$

So 162 credits is 90% of **180 credits**.

CONTINUED ON NEXT PAGE

5. Write and solve an equation to answer each question.

(a) 1200 books is what percent of 5000 books?

(b) 23 hours is what percent of 20 hours?

(c) What percent of \$6.50 is \$0.52?

ANSWERS

5. (a) $1200 = p \cdot 5000$; $0.24 = 24\%$
(b) $23 = p \cdot 20$; $1.15 = 115\%$
(c) $p \cdot 6.50 = 0.52$; $0.08 = 8\%$

6. Write and solve an equation to answer each question.

(a) 74% of how many cars is 37 cars?

(b) 1.2 tons is $2\frac{1}{2}\%$ of how many tons?

(c) 216 calculators is 160% of how many calculators?

Answers

6. (a) $0.74 \cdot n = 37$; 50 cars
(b) $1.2 = 0.025 \cdot n$; 48 tons
(c) $216 = 1.6 \cdot n$; 135 calculators

Check The answer makes sense because 90% of 180 credits should be 10% less than 100% of the credits, and 10% of 180. credits is 18 credits.

100% of 180 credits	−	10% of 180 credits	=	90% of 180 credits	
↓		↓		↓	
180 credits	−	**18 credits**	=	**162 credits**	Matches the number given in the original problem

(b) 250% of what amount is $75?

Translate the sentence into an equation. Write the percent in decimal form.

250.% of what amount is $75?

$$2.5 \cdot n = 75$$

$$\frac{\overset{1}{\cancel{2.5}} \cdot n}{\underset{1}{\cancel{2.5}}} = \frac{75}{2.5} \quad \text{Divide both sides by 2.5.}$$

$$n = 30$$

So 250% of **$30** is $75.

Check The answer makes sense because 200% of $30 is 2 times $30 = $60, and 50% of $30 is $30 ÷ 2 = $15.

200% of $30	+	50% of $30	=	250% of $30	
↓		↓		↓	
$60	+	**$15**	=	**$75**	Matches the number given in the original problem

◀◀ WORK PROBLEM 6 AT THE SIDE.

7.3 Exercises

Ⓖ *Use your estimation skills and the percent shortcuts to select the most reasonable answers. Circle your choices. Do* ***not*** *write an equation and solve it.*

1. Find 50% of 3000 patients.

150 patients   1500 patients   300 patients

2. What is 50% of 192 pages?

48 pages   384 pages   96 pages

3. 25% of $60 is how much?

$15 $6 $30

4. Find 25% of $2840.

$28.40 $710 $284

5. What is 10% of 45 pounds?

0.45 pounds   22.5 pounds   4.5 pounds

6. 10% of 7 feet is how many feet?

0.7 feet   3.5 feet   14 feet

7. Find 200% of $3.50.

$0.35 $1.75 $7.00

8. What is 300% of $12?

$4 $36 $1.20

9. 1% of 5200 students is how many students?

520 students   52 students   2600 students

10. Find 1% of 460 miles

0.46 mile   46 miles   4.6 miles

Ⓖ **11.** **(a)** Describe a shortcut for finding 10% of a number and explain *why* your shortcut works.
(b) Once you know 10% of a certain number, explain how you could use that information to find 20% and 30% of the same number.

Ⓖ **12.** **(a)** Describe a shortcut for finding 1% of a number and explain *why* it works.
(b) Once you know 1% of a certain number, explain how you could use that information to find 2% and 3% of the same number.

Writing   Calculator   Ⓖ Small Group

Write and solve an equation to answer each question.

13. 35% of 660 programs is how many programs?

14. 55% of 740 cannisters is how many cannisters?

15. 70 truckloads is what percent of 140 truckloads?

16. 30 crew members is what percent of 75 crew members?

17. 476 circuits is 70% of what number of circuits?

18. 621 tons is 45% of what number of tons?

19. $12\frac{1}{2}\%$ of what number of people is 135 people?

20. $6\frac{1}{2}\%$ of what number of bottles is 130 bottles?

21. What is 65% of 1300 species?

22. What is 75% of 360 dosages?

23. 4% of $520 is how much?

24. 7% of $480 is how much?

25. 38 styles is what percent of 50 styles?

26. 75 offices is what percent of 125 offices?

27. What percent of \$264 is \$330?

28. What percent of \$480 is \$696?

29. 141 employees is 3% of what number of employees?

30. 16 books is 8% of what number of books?

31. 32% of 260 quarts is how many quarts?

32. 44% of 430 liters is how many liters?

33. \$1.48 is what percent of \$74?

34. \$0.51 is what percent of \$8.50?

35. How many tablets is 140% of 500 tablets?

36. How many patients is 175% of 540 patients?

37. 40% of what number of salads is 130 salads?

38. 75% of what number of wrenches is 675 wrenches?

39. What percent of 160 liters is 2.4 liters?

40. What percent of 600 miles is 7.5 miles?

41. 225% of what number of gallons is 11.25 gallons?

42. 180% of what number of ounces is 6.3 ounces?

43. What is 12.4% of 8300 meters?

44. What is 13.2% of 9400 acres?

45. Explain and correct the error in each of these solutions.

(a) 3 hours is what percent of 15 hours?

$$3 = p \cdot 15$$

$$\frac{3}{15} = \frac{p \cdot \overset{1}{\cancel{15}}}{\underset{1}{\cancel{15}}}$$

$$0.2 = p$$

The answer is 0.2%.

(b) \$50 is what percent of \$20?

$$50 \cdot p = 20$$

$$\frac{\overset{1}{\cancel{50}} \cdot p}{\underset{1}{\cancel{50}}} = \frac{20}{50}$$

$$p = 0.40 = 40\%$$

The answer is 40%.

46. Explain and correct the error in each of these solutions.

(a) 12 inches is 5% of what number of inches?

$$12 \cdot 0.05 = n$$

$$0.6 = n$$

The answer is 0.6 inch.

(b) What is 4% of 30 pounds?

$$n = 4 \cdot 30$$

$$n = 120$$

The answer is 120 pounds.

47. Suppose that you have this problem: $33\frac{1}{3}\%$ of \$162 is how much?

(a) First, change $33\frac{1}{3}\%$ to a fraction. (See **Section 7.1** for help.) Then, write an equation and solve it.

(b) Now solve the problem by changing $33\frac{1}{3}\%$ to a decimal. (*Hint:* Look at part (a) to see what fraction is equivalent to $33\frac{1}{3}\%$. Change the fraction to a decimal, using your calculator. Keep *all* the decimal places. Now write the equation and solve it, using your calculator.)

(c) Compare your answers from part **(a)** and part **(b)**. How different are they?

48. Do Exercise 47 first. Now suppose that you have this problem: 22 cans is $66\frac{2}{3}\%$ of what number of cans?

(a) First, change $66\frac{2}{3}\%$ to a fraction. (See **Section 7.1** for help.) Then, write an equation and solve it. (See **Section 4.7** for help.)

(b) Now solve the problem by changing $66\frac{2}{3}\%$ to a decimal. Use your calculator and keep all the decimal places. Now write the equation and solve it.

(c) Compare your answers from part **(a)** and part **(b)**. How different are they?

Review and Prepare

Write each percent as a decimal and each decimal as a percent. (For help, see ***Section 7.1****.)*

49. 3% **50.** 8% **51.** 12.5% **52.** 18.5%

53. 0.7 **54.** 0.3 **55.** 1.8 **56.** 2.4

7.4 *Problem Solving with Percent*

OBJECTIVE 1 Solving percent problems involves finding three items: the *percent,* the *whole,* and the *part*. Then you can write a percent equation and solve it to answer the question in the problem. As you read each problem, start by looking for the percent. Find a number with the % symbol or find the phrase "what percent."

EXAMPLE 1 Finding the Part

Write and solve an equation to answer each question.

(a) A new low-income housing project charges 30% of a family's income as rent. The Smith's family income is \$1260 per month. How much will the Smiths pay for rent?

Approach Notice that the percent is given in the problem: 30%. The key word **of** appears *right after* 30%, which means that you can use the phrase "30% **of** a family's income" to help you write one side of the equation. Write the percent as a decimal.

30.% **of** a family's income
↓ ↓
0.30 • \$1260 ← The family's income given in the problem

Solution You know the *percent* (30%) and the *whole* (\$1260 is the family's whole income). Write an equation to find the *part* of their income paid for rent.

30% **of** \$1260 is what amount?

$$0.30 \cdot 1260 = n$$

$$378 = n$$

The family will pay **\$378** for rent.

Check The answer makes sense because 10% of \$1260. is \$126, so 30% would be 3 times \$126, or \$378.

(b) When Britta received her first \$180 paycheck as a math tutor, $12\frac{1}{2}\%$ was withheld for federal income tax. How much was withheld?

Approach Use the percent equation. The *percent* is given: $12\frac{1}{2}\%$. Write $12\frac{1}{2}\%$ as 12.5% and then move the decimal point two places to the left. The key word *of* doesn't appear after $12\frac{1}{2}\%$. Instead, think about whether you know the *whole* or the *part*. You know Britta's *whole* paycheck is \$180, but you do *not* know what *part* of it was withheld.

Solution percent • whole = part

$$12.5\% \cdot \$180 = n$$

$$0.125 \cdot 180 = n$$

$$22.5 = n$$

\$22.50 was withheld from Britta's paycheck.

Check The answer makes sense because 10% of \$180. is \$18, so a little more than \$18 should be withheld.

WORK PROBLEM 1 AT THE SIDE.

OBJECTIVES

1 Solve percent application problems.

2 Solve problems involving percent increase or decrease.

FOR EXTRA HELP

Tutorial | Tape 12 | SSM, Sec. 7.4

1. Write and solve an equation to answer each question.

(a) About 65% of the students at City Center College receive some form of financial aid. How many of the 9280 students enrolled this year are receiving aid?

(b) There were 50 points on the first math test. Hue's score was 83% correct. How many points did Hue earn?

ANSWERS

1. (a) $0.65 \cdot 9280 = n$; 6032 students receive aid.
(b) $0.83 \cdot 50 = n$; Hue earned 41.5, or $41\frac{1}{2}$, points.

2. Write and solve an equation to answer each question.

(a) Valley College predicted that 1200 new students would enroll in the fall. It actually had 1620 new students enroll. The actual enrollment is what percent of the predicted number?

(b) The Los Angeles Lakers made 47 of 80 field goal attempts in one game. What percent is this, to the nearest whole percent?

EXAMPLE 2 Finding the Percent

Write and solve an equation to answer each question.

(a) On a 15 point quiz, Zenitia earned 13 points. What percent is this?

Approach Use the percent equation. There is no number with a % symbol in the problem. The question, "What percent is this?" tells you that the *percent* is unknown. The *whole* is all the points on the quiz (15 points), and 13 points is the *part* of the quiz that Zenitia did correctly.

Solution

$$\text{percent} \cdot \text{whole} = \text{part}$$

$$p \cdot 15 \text{ points} = 13 \text{ points}$$

$$\frac{p \cdot \cancel{15}^{1}}{\cancel{15}_{1}} = \frac{13}{15} \qquad \text{Divide both sides by 15.}$$

$$p = 0.8\overline{6} \qquad \text{The solution is a repeating decimal.}$$

Multiply the solution by 100 to change it from a *decimal* to a *percent*.

$$0.866666667 \approx 86.7\% \qquad \text{Rounded to nearest tenth of a percent}$$

The original problem didn't tell you how to round the answer, so you can choose to round Zenitia's score to **86.7%** (nearest tenth of a percent) or to **87%** (nearest whole percent).

Check The answer makes sense because she earned most of the possible points, so the percent should be fairly close to 100%.

(b) The rainfall in the Red River Valley was 33 inches this year. The average rainfall is 30 inches. This year's rainfall is what percent of the average rainfall?

Approach The percent is unknown. The key word **of** appears *right after* the word *percent*, so you can use that sentence to help you write the equation.

This year's rainfall is what percent **of** the average rainfall?

33 inches = p • 30 inches

Solution

$$33 = p \cdot 30$$

$$\frac{33}{30} = \frac{p \cdot \cancel{30}^{1}}{\cancel{30}_{1}} \qquad \text{Divide both sides by 30.}$$

Solution in *decimal* form $\quad 1.1 = p$

Multiply the solution by 100 to change it from a *decimal* to a *percent*.

$$1.10 = 110\%$$

This year's rainfall is **110%** of the average rainfall.

Check The answer makes sense because 33 inches is *more* than 30 inches (more than 100% of the average rainfall), so 33 inches must be *more* than 100% of 30 inches.

ANSWERS

2. (a) $p \cdot 1200 = 1620$; 135%
(b) $p \cdot 80 = 47$; ≈59%

◀◀ WORK PROBLEM 2 AT THE SIDE.

EXAMPLE 3 Finding the Whole

Write and solve an equation to answer the following question.

A newspaper article stated that 648 pints of blood were donated at the blood bank last month, which was only 72% of the number of pints needed. How many pints of blood were needed?

Approach The percent is given in the problem: 72%. The key word **of** appears *right after* 72%, so you can use the phrase "72% **of** the number of pints needed" to help you write one side of the equation.

72% **of** the number of pints needed
↓ ↓
$0.72 \cdot n$ ← The number of pints needed is unknown.

Solution You know the *percent* (72%) and the *part* (648 pints donated). Write an equation to find the *whole* (the number of pints needed).

72% **of** the number of pints needed is 648 pints
↓ ↓ ↓ ↓
$$0.72 \cdot n = 648$$

$$\frac{\overset{1}{\cancel{0.72}} \cdot n}{\underset{1}{\cancel{0.72}}} = \frac{648}{0.72} \quad \text{Divide both sides by 0.72.}$$

$$n = 900$$

900 pints of blood were needed.

Check The answer makes sense because 10% of 900 pints is 90 pints, so 70% would be 7 times 90 pints, or 630 pints, which is close to the number given in the problem (648 pints).

WORK PROBLEM 3 AT THE SIDE.

OBJECTIVE 2 We are often interested in looking at increases or decreases in prices, earnings, population, and many other numbers. This type of problem involves finding the percent of change. Use the following steps to find the **percent of increase**.

Finding the Percent of Increase

Step 1 Use subtraction to find the *amount* of increase.

Step 2 Use a form of the percent equation to find the *percent* of increase.

percent **of** whole = part
↓ ↓ ↓ ↓
percent **of** original value = amount of increase

---

3. Write and solve an equation to answer each question.

(a) Ezra did 15 problems correctly on a test, giving him a score of $62\frac{1}{2}\%$. How many problems were on the test?

(b) A frozen dinner advertises that only 18% of its calories are from fat. If the dinner contains 55 calories from fat, what is the total number of calories in the dinner? Round to the nearest whole number of calories.

ANSWERS

3. (a) $0.625 \cdot n = 15$; 24 problems
(b) $0.18 \cdot n = 55$; ≈306 calories

4. Write and solve an equation to answer each question.

(a) Over the last five years, Duyen's rent has increased from \$650 per month to \$767. What is the percent increase?

(b) A shopping mall increased the number of handicapped parking spaces from 8 to 20. What is the percent increase?

ANSWERS

4. (a) $767 - 650 = 117$; $p \cdot 650 = 117$; 18% increase
(b) $20 - 8 = 12$; $p \cdot 8 = 12$; 150% increase

EXAMPLE 4 Finding the Percent of Increase

Brad's hourly wage as assistant manager of a fast-food restaurant was raised from \$9.40 to \$9.87. What was the percent of increase?

Approach First subtract \$9.87 − \$9.40 to find how much Brad's wage went up. That is the *amount* of increase. Then write an equation to find the unknown *percent* of increase. Be sure to use his *original* wage (\$9.40) in the equation. Do **not** use the new wage of \$9.87 in the equation.

Solution

Step 1 $\$9.87 - 9.40 = \$0.47$ ← Amount of increase

Step 2 percent **of** original wage = amount of increase

$$p \cdot \$9.40 = \$0.47$$

$$\frac{p \cdot \cancel{9.40}^{1}}{\cancel{9.40}_{1}} = \frac{0.47}{9.40} \quad \text{Divide both sides by 9.40.}$$

$$p = 0.05 \quad \text{Solution in } \textit{decimal} \text{ form}$$

Multiply the solution by 100 to change it from a *decimal* to a *percent*.

$$0.05 = 5\%$$

Brad's hourly wage increased **5%**.

◀◀ WORK PROBLEM 4 AT THE SIDE.

Use a similar procedure to find the **percent of decrease**.

Finding the Percent of Decrease

Step 1 Use subtraction to find the *amount* of decrease.

Step 2 Use a form of the percent equation to find the *percent* of decrease.

percent **of** whole = part

percent **of** original value = amount of decrease

EXAMPLE 5 Finding the Percent of Decrease

Write and solve an equation to answer the following question.

Rozenia has been training for six months to run in a marathon race. Her weight has dropped from 137 pounds to 122 pounds. What is the percent of decrease? Round to the nearest whole percent.

Approach First subtract 137 pounds − 122 pounds to find how much Rozenia's weight went down. That is the *amount* of decrease. Then write an equation to find the unknown *percent* of decrease. Be sure to use her *original* weight (137 pounds) in the equation. Do **not** use the new weight of 122 pounds in the equation.

CONTINUED ON NEXT PAGE

Solution

Step 1 137 pounds − 122 pounds = 15 pounds ← Amount of decrease

Step 2 percent **of** original weight = amount of decrease

$$p \cdot 137 \text{ pounds} = 15 \text{ pounds}$$

$$\frac{p \cdot \cancel{137}^{1}}{\cancel{137}_{1}} = \frac{15}{137} \quad \text{Divide both sides by 137.}$$

$$p = 0.109489051 \quad \text{Solution in } \textit{decimal} \text{ form}$$

Multiply the solution by 100 to change it from a *decimal* to a *percent*.

$$0.109489051 = 10.9489051\% \approx 11\%$$

Rounded to nearest whole percent

Rozenia's weight decreased **≈11%**.

WORK PROBLEM 5 AT THE SIDE.

5. Write and solve an equation to answer each question. Round answers to the nearest whole percent.

(a) During a severe winter storm, average daily attendance at an elementary school fell from 425 students to 200 students. What was the percent decrease?

(b) The makers of a brand of spaghetti sauce claim that the number of calories from fat in each serving has been reduced by 20%. Is the claim correct if the number of calories from fat dropped from 70 calories to 60 calories per serving? Explain your answer.

Answers

5. (a) $425 - 200 = 225$; $p \cdot 425 = 225$; ≈53% decrease

(b) $70 - 60 = 10$; $p \cdot 70 = 10$; ≈14% decrease, so the claim is not true.

NUMBERS IN THE *Real World*

collaborative investigations

Ratios may be used in the directions for mixing two items. This allows you to make as much of the mixture as you need. One example is the recipe in the article below for a sugar mixture to be used in hummingbird feeders. Complete the tables below. Be sure you maintain the 1 to 4 ratio.

1.

Sugar	Water
1 cup	4 cups
	5 cups
	6 cups
	7 cups
2 cups	8 cups

2.

Sugar	Water
1 cup	4 cups
	3 cups
	2 cups
	1 cup

3. Suppose you wanted to use 3 cups of *sugar*. How much water would you need?

4. How much water for 4 cups of sugar?

5. If you had only $\frac{1}{3}$ cup of sugar, how much water could you use?

6. As you change the amounts of water and sugar, should you change the length of time that you boil the mixture? Explain your answer.

7. Will the length of time it takes to get the water hot enough to start boiling change? Explain your answer.

8. The article says that the nectar wildflowers visited by hummingbirds have an average sugar concentration of 21 percent. For mixtures, a percentage can be calculated using the mass (weight) of the ingredients. One cup of sugar weighs about 200 grams and one cup of water weighs about 235 grams. If you mix 1 cup of sugar and 4 cups of water, what is the weight of the resulting mixture?

9. What percent of the mixture's weight is sugar?

10. If you wanted the mixture to be 21% sugar by weight, how many grams of sugar should be mixed with 4 cups of water?

Feeding Hummingbirds

After getting a hummingbird feeder, the next step is to fill it! You have two choices at this point: you can either buy one of the commercial mixtures or you can make your own solution:

Recipe for Homemade Mixture:
1 part sugar (not honey)
4 parts water
Boil for 1 to 2 minutes. Cool.
Store extra in refrigerator.

The concentration of the sugar is important. A 1 to 4 ratio of sugar to water is recommended because it approximates the ratio of sugar to water found in the nectar of many hummingbird flowers. A recent study of 21 native California wildflowers visited by hummingbirds showed that their nectar had an average sugar concentration of 21 percent. This is sweet enough to attract the hummers without being too sweet. If you increase the concentration of sugar, it may be harder for the birds to digest; if you decrease the concentration, they may lose interest.

Boiling the solution helps retard fermentation. Sugar-and-water solutions are subject to rapid spoiling, especially in hot weather.

Source: The Hummingbird Book

7.4 Exercises

Write and solve an equation to answer each question. Round percent answers to the nearest tenth of a percent.

1. Robert Garrett, who works part-time, earns $110 per week and has 18% of this amount withheld for taxes, social security, and medicare. Find the amount withheld.

2. Most shampoos contain 75% to 90% water. If a 16-ounce bottle of shampoo contains 78% water, find the number of ounces of water in the 16-ounce bottle. Round to the nearest tenth of an ounce.

3. Sharon needs 64 credits to graduate from her community college. So far she has earned 40 credits. What percent is this?

4. There are about 55,000 words in Webster's Dictionary, but most educated people can identify only 20,000 of these words. What percent of the words in the dictionary can these people identify?

5. A survey at an intersection found that, of 2200 drivers, 38% were wearing seat belts. How many drivers in the survey were wearing seat belts?

6. For a tour of the eastern United States, a travel agent promised a trip of 3300 miles. Exactly 35% of the trip was by air. How many miles would be traveled by air?

7. Meadow Vista Bottled Water estimates that $117,000 will be spent this year on delivery costs alone. If total sales are estimated at $755,000, what percent of total sales will be spent on delivery?

8. According to industry figures there are 44,500 hotels and motels in America. Economy hotels and motels account for 16,910 of this total. What percent of the total are economy hotels and motels?

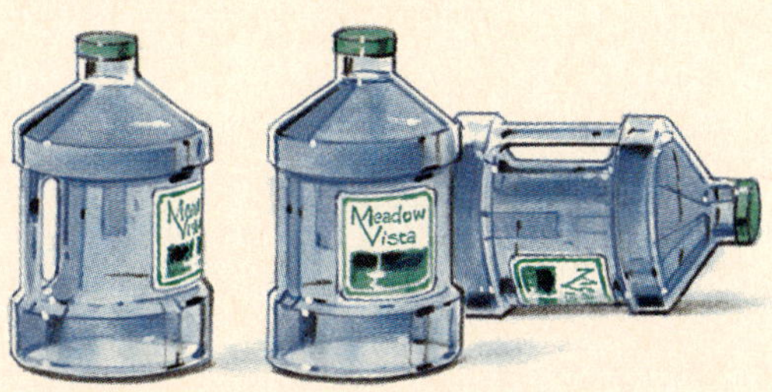

9. A newspaper article reported that Americans who are 65 years of age or older make up 12.7% of the total population. It said that there are 31.5 million Americans in this group. Find the total U.S. population. (Round to the nearest tenth of a million.)

10. Julie Ward has 8.5% of her earnings deposited into the credit union. If this amounts to $131.75 per month, find her monthly and annual earnings.

11. The campus honor society hoped to raise $50,000 in donations from businesses for scholarships. It actually raised $69,000. This amount was what percent of the goal?

12. Doug had budgeted $120 for textbooks but ended up spending $172.80. The amount he spent was what percent of his budget?

13. Alfonso earned a score of 87.5% on his test. He did 35 problems correctly. How many problems were on the test?

14. So far this season, Kevin Garnett has made 76% of his free throws. He has made 133 free throws. How many has he attempted?

15. A family of four with a monthly income of $2900 spends 95% of its income and saves the balance. Find **(a)** the monthly savings and **(b)** the annual savings of this family.

16. This year, there are 550 scholarship applicants. If 40% of the applicants will receive a scholarship, find **(a)** the number of students who will receive a scholarship, and **(b)** the percent of students who will not receive a scholarship.

17. An ad for steel-belted radial tires promises 15% better mileage. If Sheera's car has gotten 25.6 miles per gallon in the past, what mileage can she expect after the new tires are installed? (Round to the nearest tenth of a mile.)

18. The Saturn automobile dealers use a one-price, "no haggle" selling policy. Saturn dealers average 13% profit on new car sales. If a dealer pays $15,600 for a Saturn SC, find the selling price after adding the profit to the dealer's cost.

G *The graph (pictograph) below shows the percent of chicken noodle soup sold during the cold and flu season. Use this information to do Exercises 19–22.*

19. Which of the flu season months had the lowest sales of chicken noodle soup? How many cans were bought that month?

20. What percent of the chicken noodle soup sales take place during the flu months of October through March? What percent of sales take place in the *non-flu* season months?

21. Find the number of cans of soup sold in the highest sales month and in the second highest sales month.

22. How many more cans of soup were sold in October than in November? How many more were sold in November than December?

Write and solve an equation to find the percent increase or decrease. Round your answers to the nearest tenth of a percent.

23. The price per share of Toys "R" Us stock fell from \$35.50 to close at \$33.50. Find the percent of decrease in price.

24. In the past five years, the cost of generating electricity from the sun has been brought down from 24 cents per kilowatt hour to 8 cents (less than the newest nuclear power plants). Find the percent of decrease.

25. Students at Lane College were charged \$1449 as tuition this quarter. If the tuition was \$1228 last quarter, find the percent of increase.

26. Americans are eating more fish. This year the average American will eat $15\frac{1}{2}$ pounds compared to only $12\frac{1}{2}$ pounds per year a decade ago. Find the percent of increase. (*Hint:* $15\frac{1}{2} = 15.5$; $12\frac{1}{2} = 12.5$.)

27. Jordan's part-time work schedule has been reduced to 18 hours per week. He had been working 30 hours per week. What is the percent decrease?

28. Janis works as a hair stylist. She cut her price on haircuts from \$28 to \$25.50 to try to get more customers. By what percent did she decrease the price?

Explain why each answer given in Exercises 29–32 does **not** *make sense. Then find and correct the error.*

29. The recommended maximum daily amount of dietary fat is 65 grams. George ate 78 grams of fat today. He ate what percent of the recommended amount?

$$p \cdot 78 = 65$$

$$\frac{p \cdot \cancel{78}^{1}}{\cancel{78}_{1}} = \frac{65}{78}$$

$$p = 0.833 = 83.3\%$$

30. The Gobblers soccer team won 18 of its 25 games this season. What percent did the team win?

$$p \cdot 25 = 18$$

$$\frac{p \cdot \cancel{25}^{1}}{\cancel{25}_{1}} = \frac{18}{25}$$

$$p = 0.72\%$$

31. The human brain is $2\frac{1}{2}\%$ of total body weight. How much would the brain of a 150-pound person weigh?

$$2\tfrac{1}{2}\% \text{ of } 150 = n$$

$$\downarrow \quad \downarrow \quad \downarrow$$

$$\underbrace{2.5 \cdot 150}_{375} = n$$

$$375 = n$$

32. Yesterday, because of an ice storm, 80% of the students were absent. How many of the 800 students made it to class?

$$80\% \text{ of } 800 = n$$

$$\downarrow \quad \downarrow \quad \downarrow$$

$$\underbrace{0.80 \cdot 800}_{640} = n$$

$$640 = n$$

The answer is 640 students.

33. You can have an *increase* of 150% in the price of something. Could there be a 150% *decrease* in its price? Explain why or why not.

34. Show how to use a shortcut to find 25% of \$80. Then explain how to use the result to find 75% of \$80 and 125% of \$80 *without* solving a proportion or equation.

Review and Prepare

Find the exact answer by moving the decimal point (For help, see **Section 7.3**.*)*

35. 10% of \$432 **36.** 10% of \$6 **37.** 10% of 8 miles **38.** 10% of 35 ft

39. 1% of \$432 **40.** 1% of \$6 **41.** 1% of \$8 **42.** 1% of \$35

7.5 Consumer Applications: Sales Tax, Tips, and Discounts

Three of the more common uses of percent in daily life are sales taxes, tips, and discounts.

OBJECTIVE 1 Most states collect **sales taxes** on the purchases you make in stores. Your county or city may also add on a small amount of sales tax. For example, your state may charge $6\frac{1}{2}\%$ on purchases and your city may add on another $\frac{1}{2}\%$ for a total of 7%. The exact percent varies from place to place but is usually from 4% to 8%. The stores collect the tax and send it to the city or state government where it is used to pay for things like road repair, public schools, parks, police and fire protection, and so on.

You can use a form of the percent equation to calculate sales tax. The **tax rate** is the *percent*. The cost of the item(s) you are buying is the *whole*. The amount of tax you pay is the *part*.

Finding Sales Tax and Total Cost

Use a form of the percent equation to find sales tax.

percent • whole = part

tax rate • cost of item = amount you pay in sales tax

Then add to find how much you will pay in all.

cost of item + sales tax = total cost paid by you

EXAMPLE 1 Finding Sales Tax and Total Cost

Suppose that you buy a CD player for $289 from A-1 Electronics. The sales tax rate in your state is $6\frac{1}{2}\%$. How much is the tax? What is the total cost of the CD player?

Approach Use the sales tax equation. Write $6\frac{1}{2}\%$ as 6.5% and then move the decimal point two places to the left.

Solution

tax rate • cost of item = sales tax

6.5% • $289 = n

0.065 • 289 = n

18.785 = n

The store will round the tax to the nearest cent, so $18.785 rounds to **$18.79**.

Now add the sales tax to the cost of the CD player to find your total cost.

cost of item + sales tax = total cost

$289 + $18.79 = **$307.79**

Check Use estimation to check the amount of sales tax. Round $289 to $300. Then 1% of $300. is $3. Round $6\frac{1}{2}\%$ to 7%. Then 7% would be 7 times $3 or $21 for sales tax. This is close to our answer of $18.79.

WORK PROBLEM 1 AT THE SIDE.

OBJECTIVES

1. Find sales tax and total cost.
2. Calculate restaurant tips.
3. Find the discount and sale price.

FOR EXTRA HELP

Tutorial | Tape 13 | SSM, Sec. 7.5

1. Find the sales tax and the total cost. Round the sales tax to the nearest cent, if necessary. Then check your answer by estimating the sales tax.

(a) $495 camcorder; $5\frac{1}{2}\%$ sales tax

(b) $29.98 watch; 7% sales tax

(c) $1.19 candy bar; 4% sales tax

ANSWERS

1. (a) Sales tax = $27.23; total cost = $522.23. Estimate: 1% of $500. is $5; 6% is 6 times $5 = $30.
(b) Sales tax = $2.10; total cost = $32.08. Estimate: 1% of $30. is $0.30; 7% is 7 times $0.30 = $2.10.
(c) Sales tax = $0.05; total cost = $1.24 Estimate: 1% of $1 is $0.01; 4% is 4 times $0.01 = $0.04.

2. Find the sales tax rate on each purchase. Then use estimation to verify your solution.

(a) The tax on a \$57 textbook is \$3.42.

(b) The tax on a \$4 notebook is \$0.18.

(c) The tax on a \$998 sofa is \$49.90.

EXAMPLE 2 Finding the Sales Tax Rate

Ms. Ortiz bought a \$21,950 pickup truck. She paid an additional \$1646.25 in sales tax. What was the sales tax rate?

Approach Use the sales tax equation. This time the tax rate (the percent) is unknown.

Solution tax rate • cost of item = sales tax

$$p \cdot \$21{,}950 = \$1646.25$$

$$\frac{p \cdot \cancel{21{,}950}^{1}}{\cancel{21{,}950}_{1}} = \frac{1646.25}{21{,}950} \qquad \text{Divide both sides by 21,950.}$$

$$p = 0.075 \qquad \text{Solution in } \textit{decimal} \text{ form}$$

Multiply the solution by 100 to change it from a decimal to a percent: 0.075 = 7.5%. The sales tax rate is **7.5% (or $7\frac{1}{2}$%)**.

Check Use estimation to verify the solution.

If the tax rate was 1%, then 1% of \$21950. = \$219.50, or about \$200.

Round 7.5% to 8%. Then 8% would be 8 times \$200, or \$1600.

The tax amount given in the original problem, \$1646.25, is close to the estimate.

WORK PROBLEM 2 AT THE SIDE.

OBJECTIVE 2 Waiters and waitresses rely on tips as a major part of their income. The general rule of thumb is to leave 15% of your bill for food and beverages as a tip for the server. If you receive exceptional service or are eating in an upscale restaurant, consider leaving a 20% tip.

EXAMPLE 3 Estimating 15% and 20% Tips

First estimate each tip. Then calculate the exact amount.

(a) Kirby took his wife to dinner at a nice restaurant to celebrate her promotion at work. The bill came to \$77.85. How much should he leave for a 20% tip?

Estimate Round \$77.85 to \$80. Then 10% of \$80. is \$8.
20% would be 2 times \$8, or **\$16**.

Approach Use the percent equation. Write 20% as a decimal. The bill for food and beverages is the *whole* and the tip is the *part*.

Solution percent • whole = part

$$20.\% \cdot \$77.85 = n$$

$$0.20 \cdot 77.85 = n$$

$$15.57 = n$$

20% of \$77.85 is **\$15.57**, which is close to the estimate of \$16.
A tip is usually rounded off to a convenient amount, such as the nearest quarter or nearest dollar, so Kirby left \$16.

CONTINUED ON NEXT PAGE

ANSWERS

2. (a) 6% Estimate: 1% of \$60. is \$0.60; 6% would be 6 times \$0.60, or \$3.60 (which is close to \$3.42 given in the problem).
(b) 4.5% Estimate: 1% of \$4 is \$0.04; round 4.5% to 5%; 5% of \$4 is 5 times \$0.04, or \$0.20 (which is close to \$0.18 given in the problem).
(c) 5% Estimate: 10% of \$1000. is \$100; 5% is half of that, or \$50 (which is close to \$49.90 given in the problem).

(b) Linda, Peggy, and Mary ordered similarly priced lunches and agreed to split the bill plus a 15% tip. How much should each woman pay if the bill is $21.63?

Estimate Round $21.63 to $20. Then 10% of $20. is $2.

5% of $20 would be half as much, that is, half of $2, or $1.

So an estimate of the 15% tip is $2 + $1 = $3.

An estimate of the amount each woman should pay is ($20 + $3) ÷ 3 ≈ $8.

Approach Use the percent equation to calculate the 15% tip. Add the tip to the bill. Then divide the total by 3 to find the amount each woman should pay.

Solution

$$\text{percent} \cdot \text{whole} = \text{part}$$
$$15.\% \cdot \$21.63 = n$$
$$0.15 \cdot 21.63 = n$$
$$3.2445 = n$$

Round $3.2445 to **$3.24** (nearest cent), which is close to the estimate of $3 for the tip.

Add: $21.63 + $3.24 = $24.87 Total cost of lunch and tip

Divide: $24.87 ÷ 3 = **$8.29** Amount paid by each woman

WORK PROBLEM 3 AT THE SIDE.

OBJECTIVE 3 Most people prefer buying things when they are on sale. A store will reduce prices, or **discount**, to attract additional customers. You can use a form of the percent equation to calculate the discount. The rate of discount is the *percent*. The original price is the *whole*. The amount that will be discounted (subtracted from the original price) is the *part*.

Finding the Discount and Sale Price

Use a form of the percent equation to find the discount.

$$\text{percent} \cdot \text{whole} = \text{part}$$
$$\text{rate of discount} \cdot \text{original price} = \text{amount of discount}$$

Then subtract to find the sale price.

$$\text{original price} - \text{amount of discount} = \text{sale price}$$

3. First estimate each tip. Then calculate the exact tip.

(a) 20% tip on a bill of $58.37

(b) 15% tip on a bill of $11.93

(c) A bill of $89.02 plus a 15% tip shared equally by four friends. How much will each friend pay?

ANSWERS

3. (a) Estimate: 10% of $60 is $6; 20% is 2 times $6, or $12. Exact: $11.67 (rounded to nearest cent)

(b) Estimate: 10% of $12 is $1.20; 5% is half of $1.20, or $0.60; $1.20 + 0.60 = $1.80. Exact: $1.79 (rounded to nearest cent)

(c) Estimate of tip: 10% of $90 is $9; 5% is half of $9, or $4.50; $9 + $4.50 = $13.50. Exact tip: $13.35 (rounded to nearest cent). Each friend pays: ($89.02 + $13.35) ÷ 4 = $25.59 (rounded).

4. Find the amount of the discount and the sale price.

(a) An Easy-Boy leather recliner originally priced at \$950 is offered at a 35% discount. What is the sale price?

(b) Eastside Department Store has women's swimsuits on sale at 40% off. One swimsuit was originally priced at \$34. Another suit was originally priced at \$72. What is the sale price of each suit?

ANSWERS

4. (a) \$332.50; \$617.50
(b) \$13.60; \$20.40
\$28.80; \$43.20

EXAMPLE 4 Finding the Discount and Sale Price

The Oak Mill Furniture Store has an oak entertainment center with an original price of \$840 on sale at 15% off. Find the sale price of the entertainment center.

Approach This problem is solved in two steps. First, find the amount of the discount, that is, the amount that will be "taken off" (subtracted) by multiplying the original price (\$840) by the rate of discount (15%). The second step is to subtract the amount of discount from the original price. This gives you the sale price, or what you will actually pay for the entertainment center.

Solution First find the amount of the discount. Write 15% as 0.15.

$$\text{rate of discount} \cdot \text{original price} = \text{amount of discount}$$
$$\underbrace{0.15 \cdot 840}_{126} = n$$
$$126 = n$$

Find the sale price of the entertainment center by subtracting the amount of the discount (\$126) from the original price.

$$\text{original price} - \text{amount of discount} = \text{sale price}$$
$$\$840 - \$126 = \$714$$

During the sale, you can buy the entertainment center for \$714.

◀◀ **WORK PROBLEM 4 AT THE SIDE.**

Calculator Tip: In Example 4, you can use a calculator to find the amount of discount and subtract the discount from the original price.

840 [−] .15 [×] 840 [=] 714

(840: Original price; .15 × 840: Amount of discount; 714: Sale price)

Your scientific calculator observes the order of operations, so it will automatically do the multiplication before the subtraction.

7.5 Exercises

Find the amount of the sales tax or the tax rate and the total cost (cost of item + sales tax = total cost). Round money answers to the nearest cent.

	Cost of Item	*Tax Rate*	*Amount of Tax*	*Total Cost*
1.	\$100	6%	________	________
2.	\$200	4%	________	________
3.	\$68	________	\$2.04	________
4.	\$185	________	\$9.25	________
5.	\$365.98	6%	________	________
6.	\$28.49	7%	________	________
7.	\$2.10	$5\frac{1}{2}\%$	________	________
8.	\$7.00	$7\frac{1}{2}\%$	________	________
9.	\$12,600	________	\$567	________
10.	\$21,800	________	\$1417	________

For each restaurant bill, estimate a 15% tip and a 20% tip. Then find the exact amounts for a 15% tip and a 20% tip. Round exact amounts to the nearest cent, if necessary.

	Bill	*Estimate of 15% Tip*	*Exact 15% Tip*	*Estimate of 20% Tip*	*Exact 20% Tip*
11.	\$32.17	________	________	________	________
12.	\$21.94	________	________	________	________
13.	\$78.33	________	________	________	________
14.	\$67.85	________	________	________	________
15.	\$9.55	________	________	________	________
16.	\$52.61	________	________	________	________

Find the amount or rate of discount and the sale price. Round money answers to the nearest cent, if necessary.

	Original Price	*Rate of Discount*	*Amount of Discount*	*Sale Price*
17.	\$100	15%	________	________
18.	\$200	20%	________	________
19.	\$180	________	\$54	________
20.	\$38	________	\$9.50	________
21.	\$17.50	25%	________	________
22.	\$76	60%	________	________
23.	\$37.88	10%	________	________
24.	\$59.99	40%	________	________

Write an equation and solve it for each of these application problems. Round money answers to the nearest cent, if necessary.

25. The Diamond Center sells diamond jewelry at 40% off the regular price. Find the sale price of a diamond ring normally priced at $3850.

26. An Exer-Cycle Machine sells for $590 plus 7% sales tax. Find the amount of sales tax.

27. The sales tax rate is 5%, and the sales at Fort Bragg Gifts are $1050. Find the amount of sales tax.

28. Stephen Louis can purchase a new car at 8% below sticker price. Find his cost for a car with a window sticker price of $17,650.

29. An Anderson wood frame French door is priced at $1980 with a sales tax of $99. Find the sales tax rate.

30. Textbooks for three classes cost $135 plus sales tax of $8.10. Find the sales tax rate.

31. A "super 45% off sale" begins today. What is the sale price of a ski parka normally priced at $135?

32. What is the sale price of a $549 Maytag dishwasher with a discount of 35%?

33. Ricia and Seitu split a $43.70 dinner bill plus 15% tip. How much did each person pay?

34. Marvette took her brother out to dinner for his birthday. The bill for food was $58.36 and for wine was $10.44. How much was her 20% tip, rounded to the nearest dollar?

35. An 8-millimeter camcorder normally priced at $590 is on sale for 18% off. Find the discount and the sale price.

36. This week minivans are offered at 15% off manufacturers' suggested price. Find the discount and the sale price of a minivan originally priced at $23,500.

37. Vincente and Samuel ordered a large deep dish pizza for \$17.98. How much did they give the delivery person to pay for the pizza and a 15% tip, rounded to the nearest dollar?

38. Cher, Maya, and Adara shared a \$25.50 bill for a buffet lunch. Because the server only brought their beverages, they left a 10% tip instead of the usual 15%. How much did each person pay?

Use the information in the store ad to do Exercises 39–42. Round sale prices and sales tax to the nearest cent, if necessary.

STORE CLOSE-OUT!

All clothing is now 45% off

All jewelry is now 30% off

All electronics are now 65% off

6% sales tax added to jewelry and electronics purchases

CASH ONLY! ALL SALES ARE FINAL!

39. Danika bought a computer modem originally priced at \$129 and a \$60 pair of earrings. What was her bill for the two items?

40. Find David's total bill for a \$189 jacket and a \$75 graphing calculator.

41. Sergei purchased a television originally priced at \$287.95, two pairs of \$48 jeans, and a \$95 ring. Find his total bill.

42. Richard picked out three pairs of \$15 running shorts, two \$28 shirts, and a 35 mm camera originally priced at \$99.99. How much did he pay in all?

43. **(a)** College students are offered a 6% discount on a dictionary that sells for \$18.50. If the sales tax is 6%, find the cost of the dictionary, including the sales tax, to the nearest cent.

(b) In part (a) the rate of discount and the sales tax rate are the same percent. Explain why the answer did **not** end up back at \$18.50.

44. **(a)** A FAX machine priced at \$398 is marked down 7% to promote the new model. If the sales tax is also 7%, find the cost of the FAX machine, including sales tax, to the nearest cent.

(b) What rate of sales tax would have made the answer in part (a) end up back at \$398? Round your answer to the nearest hundredth of a percent.

Review and Prepare

Rewrite each fraction or mixed number as a decimal. (For help, see ***Section 5.6****.)*

45. $\frac{5}{8}$

46. $\frac{7}{20}$

47. $5\frac{1}{4}$

48. $1\frac{3}{4}$

49. $\frac{5}{12}$ Round to nearest hundredth.

50. $\frac{17}{18}$ Round to nearest hundredth.

7.6 Problem Solving: Simple and Compound Interest

OBJECTIVES

1. Calculate simple interest.
2. Find the total amount due on a loan.
3. Solve problems involving compound interest.

FOR EXTRA HELP

Tutorial | Tape 13 | SSM, Sec. 7.6

When you open a savings account you're actually lending money to the financial institution. It will, in turn, lend this money to individuals and businesses. These people then become borrowers. The financial institution pays a fee to you, the savings account holder, and charges a higher fee to its borrowers. This fee is called interest.

Interest is a fee paid, or a charge made, for lending or borrowing money. The amount of money borrowed is called the **principal**. The charge for interest is usually given as a percent, called the **interest rate**. The interest rate is assumed to be *per year* (for *one* year) unless stated otherwise.

OBJECTIVE 1 In most cases, interest is computed on the original principal and is called **simple interest**. Use the following **interest formula** to find simple interest.

Formula for Simple Interest

Interest = principal • rate • time

The formula is usually written using variables.

$$I = p \cdot r \cdot t \qquad \text{or} \qquad I = prt$$

Note

Simple interest calculations are used for most short-term business loans, most real estate loans, and many automobile and customer loans.

EXAMPLE 1 Finding Simple Interest for One Year

Find the simple interest on \$2000 at 6% for 1 year.

The amount borrowed, or principal (p), is \$2000. The interest rate (r) is 6%, which is 0.06 as a decimal, and the time of the loan (t) is 1 year. Use the interest formula.

$$I = p \cdot r \cdot t$$
$$I = (2000)(0.06)(1)$$
$$I = 120$$

The interest is **\$120**.

WORK PROBLEM 1 AT THE SIDE.

EXAMPLE 2 Finding Simple Interest for More Than One Year

Find the simple interest on \$4200 at 8% for three and a half years.

The principal (p) is \$4200. The rate (r) is 8% or 0.08 as a decimal, and the time (t) is $3\frac{1}{2}$ or 3.5 years. Use the formula.

$$I = p \cdot r \cdot t$$
$$I = (4200)(0.08)(3.5)$$
$$I = 1176$$

The interest is **\$1176**.

WORK PROBLEM 2 AT THE SIDE.

1. Find the simple interest.

(a) \$500 at 4% for 1 year

(b) \$1850 at $9\frac{1}{2}$% for 1 year (*Hint:* Write $9\frac{1}{2}$% as 9.5%. Then move the decimal point two places to the left to change 9.5% to a decimal.)

2. Find the simple interest.

(a) \$340 at 5% for $3\frac{1}{2}$ years

(b) \$2450 at 8% for $3\frac{1}{4}$ years (*Hint:* Write $3\frac{1}{4}$ years as 3.25 years.)

(c) \$14,200 at $7\frac{1}{2}$% for $2\frac{3}{4}$ years

ANSWERS

1. **(a)** \$20 **(b)** \$175.75

2. **(a)** \$59.50 **(b)** \$637 **(c)** \$2928.75

3. Find the simple interest.

(a) \$1500 at 7% for 4 months

(b) \$25,000 at $10\frac{1}{2}\%$ for 3 months

Interest rates are given *per year*. For loan periods of less than one year, be careful to express the time as a fraction of a year.

If the time is given in months, use a denominator of 12, because there are 12 months in a year. A loan of 9 months would be for $\frac{9}{12}$ of a year, a loan of 7 months would be for $\frac{7}{12}$ of a year, and so on.

EXAMPLE 3 Finding Simple Interest for Less Than 1 Year

Find the simple interest on \$840 at $8\frac{1}{2}\%$ for 7 months.

The principal is \$840. The rate is $8\frac{1}{2}\%$ or 0.085, and the time is $\frac{7}{12}$ of a year. Use the formula $I = prt$.

$$I = \underbrace{(840)(0.085)}\left(\frac{7}{12}\right) \quad \text{7 months} = \tfrac{7}{12} \text{ of a year.}$$

$$= (71.4)\left(\frac{7}{12}\right)$$

$$= \frac{71.4}{1} \cdot \frac{7}{12} \quad \text{Multiply numerators. Multiply denominators.}$$

$$= \frac{499.8}{12} = 41.65 \quad \text{Divide 499.8 by 12.}$$

The interest is **\$41.65**.

Calculator Tip: The calculator solution to Example 3 uses chain calculations.

840 [×] .085 [×] 7 [÷] 12 [=] 41.65

WORK PROBLEM 3 AT THE SIDE.

OBJECTIVE 2 When you repay a loan, the interest is added to the original principal to find the total amount due.

Finding the Total Amount Due

amount due = principal + interest

EXAMPLE 4 Calculating the Total Amount Due

Charlesetta borrowed \$1080 at 8% for three months to pay for tuition and books. Find the total amount due on her loan.

First find the interest.

$$I = (1080)(0.08)\left(\frac{3}{12}\right) \quad \text{3 months} = \tfrac{3}{12} \text{ of a year.}$$

$$I = 21.60$$

The interest is \$21.60.

CONTINUED ON NEXT PAGE

ANSWERS

3. (a) \$35 (b) \$656.25

Now add the principal and the interest to find the total amount due.

$$\begin{aligned}\textbf{amount due} &= \text{principal} + \text{interest}\\ &= \$1080 + \$21.60\\ &= \$1101.60\end{aligned}$$

The total amount due is **$1101.60**.

WORK PROBLEM 4 AT THE SIDE.

4. Find the total amount due on each of these loans.

(a) $2500 at $7\frac{1}{2}$% for 6 months

(b) $10,800 at 6% for 4 years

(c) $4300 at 10% for $2\frac{1}{2}$ years

OBJECTIVE 3 *Simple interest* is interest paid only on the original principal. Another common type of interest used with savings accounts and most investments is *compound interest.*

Compound Interest

Interest paid on principal plus past interest is **compound interest**.

Suppose that you make a single deposit of $1000 in a savings account that earns 5% per year. What will happen to your savings over three years if the interest is compounded yearly? At the end of the first year, one year's interest on the original deposit is found by using the simple interest formula. The *time* is 1 year because the interest is compounded yearly. We do the calculation *three* times because the money is in the account for *three* years.

$$\text{Interest} = \text{principal} \cdot \text{rate} \cdot \text{time}$$
$$I = prt$$

Year 1 ($1000)(0.05)(1) = **$50**
Add the interest to the $1000 to find the amount in your account at the end of the first year. $1000 + **$50** = **$1050**. The interest for the second year is found on $1050; that is, the interest is *compounded.*

Year 2 (**$1050**)(0.05)(1) = **$52.50**
Add this interest to the $1050 to find the amount in your account at the end of the second year. **$1050** + **$52.50** = **$1102.50**
The interest for the third year is found on $1102.50.

Year 3 (**$1102.50**)(0.05)(1) ≈ **$55.13**
Add this interest to the $1102.50 to find the amount in your account at the end of the third year. **$1102.50** + **$55.13** = **$1157.63**

At the end of three years, you will have **$1157.63** in your savings account. The $1157.63 is called the **compound amount.**

For comparison, let's see what would happen if you earned only *simple* interest for 3 years.

$$\begin{aligned}I &= (1000)(0.05)(3)\\ &= 150\end{aligned}$$

You would have $1000 + $150 = $1150 in your account. Compounding the interest *increased* your earnings by $7.63. ($1157.63 − $1150).

ANSWERS

4. **(a)** $2593.75 **(b)** $13,392 **(c)** $5375

5. First, find the compound amount for each of the following deposits. Then find how much more is earned because the interest is compounded.

(a) \$500 at 4% for 2 years

(b) \$1200 at 9% for 3 years

With *compound* interest, the interest earned during the second year is greater than the interest earned during the first year, and the interest earned during the third year is greater than the interest earned during the second year, and so on.

The reason is that the interest earned each year is *added* to the principal, and the new total is used to find the amount of interest in the next year.

EXAMPLE 5 Finding the Compound Amount

Nancy Wegener deposits \$3400 in an account that pays 6% interest compounded annually for 4 years. First, find the compound amount. Round to the nearest cent when necessary. Then compare the compound amount to what she would have had if she earned *simple* interest for 4 years.

Year	Interest	Compound Amount
1	(\$3400)(0.06)(1) = **\$204**	
	\$3400 + **\$204** =	**\$3604**
2	(\$3604)(0.06)(1) = **\$216.24**	
	\$3604 + **\$216.24** =	**\$3820.24**
3	(\$3820.24)(0.06)(1) ≈ **\$229.21**	
	\$3820.24 + **\$229.21** =	**\$4049.45**
4	(\$4049.45)(0.06)(1) ≈ **\$242.97**	
	\$4049.45 + **\$242.97** =	**\$4292.42**

The compound amount is **\$4292.42**.
The *simple* interest for 4 years is found by using $I = prt$.

$$I = (\$3400)(0.06)(4) = \$816$$

The amount in her account would be \$3400 + 816 = \$4216.

To compare, subtract \$4292.42 − \$4216 = \$76.42. Nancy earned **\$76.42** more because of compounding.

WORK PROBLEM 5 AT THE SIDE.

A more efficient way of finding the compound amount is to add the interest rate to 100% and then multiply by the original deposit. Notice that in Example 5, at the end of the first year, Nancy had \$3400 (100% of the original deposit) plus 6% (of the original deposit) or 106% (100% + 6% = 106%).

EXAMPLE 6 Finding the Compound Amount

Find the compound amount in Example 5 using multiplication.

Year 1 Year 2 Year 3 Year 4

$$(\$3400)(1.06)(1.06)(1.06)(1.06) \approx \$4292.42$$

Original deposit ↑ 100% + 6% = 106% = 1.06 ↑ Compound amount

The answer, **\$4292.42** is the same as in Example 5.

ANSWERS

5. (a) \$540.80; \$0.80 **(b)** ≈\$1554.03; \$30.03

Note

Adding the compound interest rate to 100% allows us to multiply by the original deposit. This method gives the compound amount at the end of each compound interest period.

Calculator Tip: You can use the $\boxed{y^x}$ key (exponent key) for Example 6.

$$3400 \ \boxed{\times}\ 1.06 \ \boxed{y^x}\ 4 \ \boxed{=}\ 4292.42 \text{ (rounded)}$$

↑ Number of years (points to the 4)

NUMBERS IN THE *Real World*

collaborative investigations

Calculating Compound Interest

The calculation of compound interest can be quite tedious. For this reason compound interest tables have been developed.

Suppose you deposit \$1 in a savings account today that earns 4% compounded annually and you allow the deposit to remain for three years. The diagram below shows the compound amount at the end of each of the 3 years.

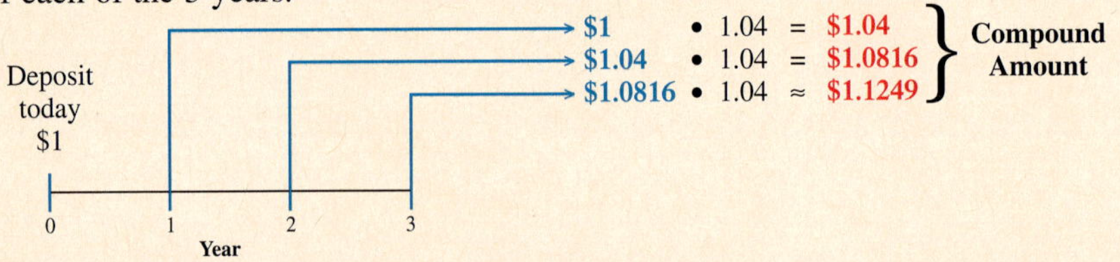

Using the compound amounts for \$1, a table can be formed. Look at the table below and find the column headed 4%. The first three numbers for years 1, 2, and 3 are the same as those we have calculated for \$1 at 4% for 3 years.

Compound Interest Table

Years	*3.00%*	*3.50%*	*4.00%*	*4.50%*	*5.00%*	*5.50%*	*6.00%*	*8.00%*
1	1.0300	1.0350	1.0400	1.0450	1.0500	1.0550	1.0600	1.0800
2	1.0609	1.0712	1.0816	1.0920	1.1025	1.1130	1.1236	1.1664
3	1.0927	1.1087	**1.1249**	1.1412	1.1576	1.1742	1.1910	1.2597
4	1.1255	1.1475	1.1699	1.1925	1.2155	1.2388	1.2625	1.3605
5	1.1593	1.1877	1.2167	1.2462	1.2763	1.3070	1.3382	1.4693
6	1.1941	1.2293	1.2653	1.3023	1.3401	1.3788	1.4185	1.5869
7	1.2299	1.2723	1.3159	1.3609	1.4071	1.4547	1.5036	1.7138
8	1.2668	1.3168	1.3686	1.4221	1.4775	1.5347	1.5938	1.8509
9	1.3048	1.3629	1.4233	1.4861	1.5513	1.6191	1.6895	1.9990
10	1.3439	1.4106	1.4802	1.5530	1.6289	1.7081	1.7908	2.1589

To find the compound amount for \$1000 deposited at 4% interest for 3 years, multiply the number from the table (1.1249) times the principal (\$1000).

(\$1000)(1.1249) = \$1124.90 ⟵ Amount in account after 3 years

1. **(a)** Use the table to find the compound amount for \$1000 deposited at 6% for 3 years, and the compound amount at 8% for 3 years.

(b) How much more do you earn in 3 years on \$1000 if the interest rate is 6% instead of 4%? How much more if the interest rate is 8% instead of 4%?

2. **(a)** Use the table to find the compound amount on \$15,000 invested for 10 years at 4%, at 6%, and at 8%.

(b) How much more do you earn at 6% than 4%? How much more at 8% than 4%?

(c) Make the calculations in part (a) using your calculator instead of the table. How much difference is there in the answers? Explain why this difference occurs.

3. Compound interest makes a tremendous difference when you invest money for retirement. Suppose a person plans to retire at age 65. Use your calculator to find the compound amount on \$5000 invested at 8% when the person is 45 years old, and left in the account until age 65. Then find the compound amount if the investment had been made at age 35, and if the investment had been made at age 25.

7.6 Exercises

Find the simple interest earned on each of these deposits. Round answers to the nearest cent, if necessary.

	Principal	Rate	Time	Interest
1.	\$200	4%	1 year	________
2.	\$100	5%	1 year	________
3.	\$600	6%	4 years	________
4.	\$800	7%	2 years	________
5.	\$2300	$8\frac{1}{2}\%$	$2\frac{1}{2}$ years	________
6.	\$4700	$5\frac{1}{2}\%$	$1\frac{1}{2}$ years	________
7.	\$9400	$6\frac{1}{2}\%$	$1\frac{1}{4}$ years	________
8.	\$10,000	$7\frac{1}{2}\%$	$3\frac{1}{4}$ years	________
9.	\$200	6%	6 months	________
10.	\$500	7%	9 months	________
11.	\$15,000	$7\frac{1}{4}\%$	7 months	________
12.	\$11,700	$4\frac{1}{2}\%$	5 months	________

Find the total amount due on each of these simple interest loans.

	Principal	*Rate*	*Time*	*Total Amount Due*
13.	\$300	14%	1 year	________
14.	\$600	11%	6 months	________
15.	\$740	6%	9 months	________
16.	\$1180	9%	2 years	________
17.	\$1500	10%	18 months	________
18.	\$3000	5%	5 months	________
19.	\$17,800	$7\frac{1}{2}$%	9 months	________
20.	\$20,500	$5\frac{1}{2}$%	6 months	________

21. The amount of interest paid on savings accounts and charged on loans can vary from one institution to another. However, when the amount of interest is calculated, three factors are used in the calculation. Name and describe these three factors.

22. Interest rates are usually given as a rate per year (annual rate). Explain what must be done when time is given in months. Write a problem giving time in months and then show how to solve it.

Find the answer to each of these simple interest application problems. Round your answers to the nearest cent, if necessary.

23. Gil Eckern deposits \$4850 at 6% for 1 year. How much interest will he earn?

24. The Jidobu family invests \$18,000 at 9% for 6 months. What amount of interest will the family earn?

25. A bank in New York City loans \$50,000 to a business at 9% for 18 months. How much interest will the bank earn?

26. Pat Martin, a retiree, deposits \$80,000 at 7% for 3 years. How much interest will she earn?

27. A student borrows \$1000 at 10% for 3 months to pay tuition. Find the total amount due.

28. A loan of \$1350 will be paid back with 12% interest at the end of 9 months. Find the total amount due.

29. Silvo Williams deposits \$7840 in a credit union account for 9 months. If the credit union pays $5\frac{1}{2}$% interest, find the amount of interest that he will earn, and the total amount in his account.

30. Norell Di Loreto, owner of Sunset Realtors, borrows \$27,000 to update her office computer system. If the loan is for 24 months at $7\frac{1}{4}$%, find the total amount due on the loan.

31. An investment fund pays $7\frac{1}{4}$% interest. If Beverly Habecker deposits \$8800 in her account for 3 months, find the amount of interest she will earn.

32. Ms. Henderson owes \$1900 in taxes. She is charged a penalty of $12\frac{1}{4}$% annual interest and pays the taxes and penalty after 6 months. Find the total amount she must pay.

33. A gift shop owner has additional profits of \$11,500 that are invested at $8\frac{3}{4}$% interest for $\frac{3}{4}$ of a year. Find the total amount in the account at the end of this time.

34. A pawn shop owner lends \$35,400 to another business for $\frac{1}{2}$ year at an interest rate of 14.9%. How much interest will be earned on the loan?

First, find the compound amount for each of the following deposits. Calculate the interest each year and then add it to the previous year's amount. Second, find how much more is earned because the interest is compounded. (See Example 5 for help.)

35. \$1000 at 5% for 2 years

36. \$700 at 3% for 2 years

37. \$3500 at 7% for 4 years

38. \$5500 at 6% for 4 years

Find each compound amount by multiplying the original amount deposited by 100% plus the compound rate. (See Example 6 for help.) Round answers to the nearest cent, if necessary.

39. \$800 at 4% for 3 years

40. \$600 at 6% for 3 years

41. \$1180 at 7% for 8 years

42. \$12,800 at 6% for 7 years

43. Glenda Wong deposits \$5280 in an account that pays 8% interest, compounded annually. Find the amount she will have (compound amount) at the end of 5 years.

44. John Hendrick borrows \$10,500 from his uncle to open Campus Bicycles. He will repay the loan at the end of 6 years with interest at 6% compounded annually. Find the amount he will repay.

45. Evelina Jones lends \$7500 to her son Rick, the owner of Rick's Limousine Service. He will repay the loan at the end of 6 years at 8% interest compounded annually. Find **(a)** the total amount that he will repay and **(b)** the amount of interest that Evelina will earn.

46. Sadie Simms has \$28,500 in an Individual Retirement Account (IRA) that pays 5% interest compounded annually. Find **(a)** the total amount that she will have at the end of 5 years and **(b)** the amount of interest that Sadie will earn.

47. Jennifer Boalt deposits \$10,000 at 6% compounded annually. Two years after she makes the first deposit, she adds another \$20,000, also at 6% compounded annually.

(a) What total amount will she have five years after her first deposit?

(b) What amount of interest will she have earned?

KEY TERMS

7.1	**percent**	Percent means "per one hundred." A percent is a ratio with a denominator of 100.
7.2	**percent proportion**	The proportion $\frac{\text{percent}}{100} = \frac{\text{part}}{\text{whole}}$ is used to solve percent problems.
	whole	The *whole* in a percent problem is the entire quantity or the total. It is sometimes called the *base.*
	part	The *part* in a percent problem is the number being compared to the *whole.*
7.3	**percent equation**	The percent equation is percent • whole = part. It is another way to solve percent problems.
7.4	**percent of increase or decrease**	Percent of increase or decrease is the amount of change (increase or decrease) expressed as a percent of the original value.
7.5	**sales tax**	Sales tax is a percent of the total sales charged as a tax.
	tax rate	The tax rate is the percent used when calculating the amount of tax.
	discount	Discount is often expressed as a percent of the original price; it is then deducted from the original price, resulting in the sale price.
7.6	**interest**	Interest is a fee paid or a charge made for lending or borrowing money.
	principal	Principal is the amount of money on which interest is earned.
	interest rate	Often referred to as *rate*, it is the charge for interest and is given as a percent.
	simple interest	When interest is computed on the original principal, it is called simple interest.
	interest formula	The interest formula is used to calculate interest. It is interest = principal • rate • time or $I = prt$.
	compound interest	Compound interest is interest paid on past interest as well as on the principal.
	compound amount	The total amount in an account, including compound interest and the original principal, is the compound amount.

KEY FORMULA

Finding simple interest: $I = prt$

QUICK REVIEW

Concepts	*Examples*
7.1 Basics of Percent	
Writing a Percent as a Decimal To write a percent as a decimal, move the decimal point *two* places to the *left* and drop the % sign.	50% = 50.% = 0.50 or 0.5 3% = 03.% = 0.03
Writing a Decimal as a Percent To write a decimal as a percent, move the decimal point *two* places to the *right* and attach a % sign.	0.75 = 0.75 = 75% 3.6 = 3.60 = 360%
Writing a Percent as a Fraction To write a percent as a fraction, drop the % symbol and write the number over 100. Then write the fraction in lowest terms.	$35\% = \frac{35}{100} = \frac{35 \div 5}{100 \div 5} = \frac{7}{20}$ ← Lowest terms $125\% = \frac{125}{100} = \frac{125 \div 25}{100 \div 25} = \frac{5}{4} = 1\frac{1}{4}$

Concepts	*Examples*

7.1 Basics of Percent (continued)

Writing a Fraction as a Percent

To write a fraction as a percent, multiply by 100 and attach a % symbol.

$$\frac{3}{5} = \frac{3}{5} \bullet \frac{100}{1} = \frac{3 \bullet \overset{1}{\cancel{5}} \bullet 20}{\underset{1}{\cancel{5}} \bullet 1} = \frac{60}{1} = 60\%$$

$$\frac{7}{8} = \frac{7}{8} \bullet \frac{100}{1} = \frac{7 \bullet \overset{1}{\cancel{4}} \bullet 25}{2 \bullet \underset{1}{\cancel{4}} \bullet 1} = \frac{175}{2} = 87\frac{1}{2}\%$$

7.2 Using the Percent Proportion

Percent is to 100 as part is to whole, or

$$\text{Always } 100 \rightarrow \frac{\text{percent}}{100} = \frac{\text{part}}{\text{whole}}.$$

Identify the percent first. It appears with the word *percent* or the % symbol.

The *whole* is the entire quantity or total. It often appears after the word **of**.

The *part* is the number being compared to the whole.

30 children is what percent **of** 75 children?

$$\text{Percent unknown} \rightarrow \frac{p}{100} = \frac{30}{75} \leftarrow \text{Part}$$
Always 100 → 100; 75 ← Whole (follows **of**)

To solve the proportion, find the cross products.

$$\frac{p}{100} = \frac{30}{75}$$

$100 \bullet 30 = 3000$ and $p \bullet 75$ — Cross products

$p \bullet 75 = 3000$ — Show that the cross products are equivalent.

$$\frac{p \bullet \overset{1}{\cancel{75}}}{\underset{1}{\cancel{75}}} = \frac{3000}{75}$$ Divide both sides by 75.

$p = 40$

30 children is **40%** of 75 children.

7.3 Percent Shortcuts

Concepts	Examples
200% of a number is 2 times the number.	200% of \$35 is 2 times \$35, or \$70.
100% of a number is the entire number.	100% of 600 women is *all* the women (600).
To find 50% of a number, divide the number by 2.	50% of \$8000 is \$8000 ÷ 2 = \$4000.
To find 25% of a number, divide the number by 4.	25% of 40 pens is 40 ÷ 4 = 10 pens.
To find 10% of a number, move the decimal point *one* place to the *left*.	10% of \$92.40 is \$9.24.
To find 1% of a number, move the decimal point *two* places to the *left*.	1% of 62 miles is 62. = 0.62 mile.

7.3–7.4 Using the Percent Equation

This is the percent equation.

$$\text{percent} \bullet \text{whole} = \text{part}$$

Write the percent as a decimal before using the equation.

If the percent is unknown, the solution will be in decimal form. Multiply by 100 to change it from a decimal to a percent.

Write and solve an equation to answer each question.

(a) Cherry's scholarship will pay 80% of her tuition. If tuition is \$2680, how much will the scholarship pay?

80.% **of** her tuition is how much?

$$\begin{aligned} 0.80 \bullet \$2680 &= n \\ 0.80(2680) &= n \\ 2144 &= n \end{aligned}$$

The scholarship will pay **\$2144**.

(Continued)

Concepts	Examples
7.3–7.4 Using the Percent Equation (continued)	**(b)** Todd's regular pay is \$540 per week. \$43.20 is taken out of each paycheck for medical insurance. What percent is that? percent • whole = part $p \cdot 540 = 43.20$ $\frac{p \cdot \overset{1}{\cancel{540}}}{\underset{1}{\cancel{540}}} = \frac{43.20}{540}$ $p = 0.08$ ← Decimal form Multiply 0.08 by 100 so that 0.08 = 8%. \$43.20 is **8%** of \$540.
7.4 Finding Percent of Increase or Decrease *Step 1* Use subtraction to find the *amount* of increase or decrease. *Step 2* Use this equation to find the *percent* of increase or decrease. percent **of** original value = amount of increase (or decrease)	Enrollment rose from 3820 students to 5157 students. Find the percent of increase. 5157 students − 3820 students = 1337 students ← Amount of increase $p \cdot 3820 = 1337$ $\frac{p \cdot \overset{1}{\cancel{3820}}}{\underset{1}{\cancel{3820}}} = \frac{1337}{3820}$ $p = 0.35$ ← Decimal form 0.35 = **35%** ← Percent increase
7.5 Consumer Applications To find **sales tax**, use the equation: tax rate • cost of item = sales tax. Write the percent as a decimal.	Find the sales tax on an \$89 pair of binoculars if the sales tax rate is 6%. tax rate • cost of item = sales tax 6% • \$89 = n $0.06(89) = n$ $5.34 = n$ The tax is **\$5.34**.
To estimate a 15% **restaurant tip**, first find 10% of the food bill by moving the decimal point *one* place to the *left*. Then add half of that amount for the other 5%.	Estimate a 15% tip on a restaurant bill of \$38.72. Then find the exact tip. Estimate: Round \$38.72 to \$40. 10% of \$40. is \$4. Half of \$4 is \$2. So an estimate of the 15% tip is \$4 + \$2 = \$6. (Continued)

Concepts	*Examples*
7.5 Consumer Applications (continued)	
To estimate a 20% tip, first find 10% by moving the decimal point one place to the left. Then double the amount.	Exact: percent • whole = part; 15.% • 38.72 = n; 0.15 • 38.72 = n; 5.808 = n. The exact tip is **\$5.81** (rounded to nearest cent).
To find a **discount** use this formula. rate of discount • original price = amount of discount. Then subtract to find the **sale price**. original price − amount of discount = sale price	All calculators are on sale at 20% off. Find the sale price of a calculator originally marked \$35. 20% • \$35 = n; 0.20 • 35 = n; 7 = n. Then \$35 − \$7 = **\$28** ← Sale price
7.6 Finding Simple Interest	
Use the formula ***I = prt*** **Interest = principal • rate • time**. Time (t) is in years. When the time is given in months, use a fraction with 12 in the denominator because there are 12 months in a year. Write the rate (the percent) as a decimal.	\$2800 is deposited at 8% for 5 months. Find the amount of interest. $I = p \cdot r \cdot t$ $= (2800)(0.08)\left(\frac{5}{12}\right)$ $= (224)\left(\frac{5}{12}\right) = \frac{224 \cdot 5}{12} \approx$ **\$93.33**
7.6 Finding Compound Amount and Compound Interest	
Two ways to find the compound amount are: **1.** Calculate the interest for each compound interest period and then add it to the principal. **2.** Multiply the original deposit by 100% plus the compound interest rate.	Use two methods to find the compound amount and interest if \$1500 is deposited at 5% interest for 3 years. (Interest / Compound Amount) 1. year 1 (\$1500)(0.05)(1) = **\$75**; \$1500 + **\$75** = **\$1575**. year 2 (\$1575)(0.05)(1) = **\$78.75**; \$1575 + **\$78.75** = **\$1653.75**. year 3 (\$1653.75)(0.05)(1) ≈ **\$82.69**; \$1653.75 + **\$82.69** = **\$1736.44**. 2. \$1500 • 1.05 • 1.05 • 1.05 ≈ \$1736.44 (\$1500 = Original deposit; \$1736.44 = Compound amount; 100% + 5% = 105% = 1.05)

[7.1] *Write each of the following percents as decimals and decimals as percents.*

1. 25% **2.** 180% **3.** 12.5% **4.** 0.085%

5. 2.65 **6.** 0.02 **7.** 0.875 **8.** 0.002

Write each percent as a fraction or mixed number in lowest terms and each fraction as percent.

9. 12% **10.** 37.5% **11.** 250% **12.** 5%

13. $\frac{3}{4}$ **14.** $\frac{5}{8}$ **15.** $3\frac{1}{4}$ **16.** $\frac{3}{50}$

Complete this chart.

Fraction	*Decimal*	*Percent*
$\frac{1}{8}$	**17.** ______	**18.** ______
19. ______	0.15	**20.** ______
21. ______	**22.** ______	180%

Fill in the blanks.

23. 100% of \$46 is ____________.

24. 50% of \$46 is ____________.

25. 100% of 9 hours is ____________.

26. 50% of 9 hours is ____________.

[7.2] *Write and solve a percent proportion to answer each question. If necessary, round percent answers to the nearest tenth of a percent.*

27. 338.8 meters is 140% of what number of meters?

28. 2.5% of what number of cases is 425 cases?

29. What is 6% of 450 telephones?

30. 60% of 1450 reference books is how many books?

31. What percent of 380 pairs is 36 pairs?

32. 1440 cans is what percent of 640 cans?

[7.3] *Write and solve a percent equation to answer each question. Round money answers to the nearest cent, if necessary.*

33. 11% of $23.60 is how much?

34. What is 125% of 64 days?

35. 1.28 ounces is what percent of 32 ounces?

36. $46 is 8% of what number of dollars?

37. 8 people is 40% of what number of people?

38. What percent of 174 ft is 304.5 ft?

[7.4] *Write and solve a percent equation to answer each question.*

39. A medical clinic found that 16.8% of the patients were late for their appointments last month. The number of patients who were late was 504. Find the total number of patients.

40. Coreen budgeted $280 for food on her vacation. She actually spent 130% of that amount. How much did she spend on food?

41. In a tree-planting project, 640 of the 800 trees planted were still living one year later. What percent of the trees planted was still living?

42. Scientists tell us that there are 9600 species of birds and that 1000 of these species are in danger of extinction. What percent of the bird species is in danger of extinction?

[7.5] *Find the amount of sales tax or the tax rate and the total cost.*

	Amount of Sale	Tax Rate	Amount of Tax	Total Cost
43.	\$210	4%	________	________
44.	\$780	________	\$58.50	________

For each restaurant bill, estimate a 15% tip and a 20% tip. Then find the exact amount for a 15% tip and a 20% tip. Round to the nearest cent.

	Bill	Estimated 15%	Exact 15%	Estimated 20%	Exact 20%
45.	\$42.73	________	________	________	________
46.	\$8.05	________	________	________	________

Find the amount or rate of discount and the amount paid after the discount.

	Original Price	Rate of Discount	Amount of Discount	Sale Price
47.	\$37.50	10%	________	________
48.	\$252	________	\$63	________

[7.6] *Find the simple interest and total amount due on the following loans.*

	Principal	Rate	Time	Interest	Total Amount Due
49.	\$350	$4\frac{1}{2}\%$	3 years	________	________
50.	\$1530	16%	9 months	________	________

51. Find the compound amount when \$3600 is invested for 3 years at 8% compounded annually.

52. Find the compound amount on a deposit of \$11,400 that earns 7% interest compounded annually for 4 years.

Mixed Review Exercises

The table shows the number of animals received during the first nine months of the year by the local Humane Society and the number placed in new homes. Use the table to do Exercises 53–55. Round percent answers to the nearest tenth of a percent.

Animals Received	Placed in New Homes
5371 dogs	2599 dogs
6447 cats	2346 cats
2223 other	406 other

53. **(a)** What percent of the dogs were placed in new homes?

(b) The number of dogs received so far this year is 75% of the total number expected. How many dogs are expected, to the nearest whole number?

54. **(a)** What percent of the cats were placed in new homes?

(b) During the first 9 months of last year, 2300 cats were received. What is the percent increase in cats received from last year to this year?

55. **(a)** What percent of all the animals received were placed in new homes?

(b) The Society's goal was to place 40% of all animals received. So far they have missed their goal by how many animals? Round to the nearest whole number.

Solve each of the following application problems. Round percent answers to the nearest tenth of a percent and money amounts to the nearest cent, if necessary.

56. Gladys Brooks, owner of Fair Oaks Hardware, deposited \$8520 at $5\frac{1}{2}\%$ for 9 months. Find the amount of interest earned.

57. Clarence Hanks borrowed \$1620 at 14% for 18 months to pay his hospital bill. Find the total amount due.

58. The mileage on a car dropped from 32.8 miles per gallon to 28.5 miles per gallon. Find the percent of decrease.

59. This month Dallas received 2.61 inches of rain. This is what percent of the normal rainfall of 1.8 inches?

Write each percent as a decimal and each decimal as a percent.

1. 75% **2.** 0.6

3. 1.8 **4.** 0.075

5. 300% **6.** 2%

1. ______
2. ______
3. ______
4. ______
5. ______
6. ______

Write as fractions or mixed numbers in lowest terms.

7. 62.5% **8.** 240%

7. ______
8. ______

Write each fraction or mixed number as a percent.

9. $\frac{1}{20}$ **10.** $\frac{7}{8}$ **11.** $1\frac{3}{4}$

9. ______
10. ______
11. ______

Write and solve a proportion or an equation to answer each question.

12. 16 files is 5% of what number of files? 12. ______

13. \$1250 is what percent of \$312.50? 13. ______

14. Erica Green has saved 75% of the amount needed for a down payment on a condominium. If she has saved \$14,625, find the total down payment needed. 14. ______

15. The price of a used car is \$5680 plus sales tax of $6\frac{1}{2}$%. Find the total cost of the car including sales tax. 15. ______

16. Enrollment in mathematics courses increased from 1440 students last semester to 1925 students this semester. Find the percent of increase, to the nearest whole percent. 16. ______

17. ____________________

18. ____________________

19. ____________________

20. ____________________

21. ____________________

22. ____________________

23. ____________________

24. ____________________

25. ____________________

17. Explain a shortcut for finding 50% of a number and a shortcut for finding 25% of a number. Show an example of how to use each shortcut.

18. Explain how you would *estimate* a 15% tip on a restaurant bill of \$31.94. Then explain how you would *estimate* a 20% tip on the same bill.

19. Find the exact 15% tip, to the nearest cent, for the restaurant bill in Exercise 18. If you and two friends are sharing the bill and the exact tip, how much will each person pay?

Find the amount of discount and the sale price. Round answers to the nearest cent, if necessary.

	Original Price	*Rate of Discount*
20.	\$48	8%
21.	\$179.95	45%

22. Jamal found a \$1089 computer on sale at 30% off because it was a "discontinued" model. The store will let him pay for it over 6 months with no interest charge. How much will each monthly payment be to cover the discounted price plus 7% sales tax?

Find the simple interest on each of the following.

	Principal	*Rate*	*Time*
23.	\$3500	5%	$1\frac{1}{2}$ years
24.	\$5200	4%	3 months

25. Kendra borrowed \$4600 to pay college expenses. The loan is for 6 months at 12% simple interest. Find the total amount due on the loan.

CUMULATIVE REVIEW EXERCISES Chapters 1–7

1. Write these numbers in words.
   (a) 90.105
   (b) 125,000,670

2. Write these numbers in digits.
   (a) thirty billion, five million
   (b) seventy-eight ten-thousandths

3. Round each number as indicated.
   (a) 49,617 to the nearest thousand
   (b) 0.7039 to the nearest hundredth
   (c) 8945 to the nearest hundred

4. Name the property illustrated by each example.
   (a) $-7(0.8) = 0.8(-7)$
   (b) $\left(\frac{2}{3} + \frac{3}{4}\right) + \frac{1}{2} = \frac{2}{3} + \left(\frac{3}{4} + \frac{1}{2}\right)$

5. Write > or < between each pair of numbers to make true statements.
   -18 ______ -8 $\qquad$ 0 ______ -5

6. Arrange in order from smallest to largest.
   $0.705 \quad 0.755 \quad \frac{3}{4} \quad 0.7005$

7. Find the mean and median for this set of data on rent prices: \$710, \$780, \$650, \$785, \$1125, \$695, \$740, \$685.

8. Find the best buy on cereal.
   17 ounces of Brand A for \$2.89
   21 ounces of Brand B for \$3.59
   15 ounces of Brand C for \$2.79

Simplify.

9. $50 - 1.099$

10. $(-3)^2 + 2^3$

11. $\frac{3b}{10a} \cdot \frac{15ab}{4}$

12. $3\frac{3}{10} - 2\frac{4}{5}$

13. $0 + 2(-6 + 1)$

14. $\frac{4}{5} + \frac{x}{4}$

15. $-20 - 20$

16. $(-0.5)(0.002)$

17. $\dfrac{-\frac{10}{11}}{-\frac{5}{6}}$

18. $\frac{3}{8}$ of 328

19. $\dfrac{-16 + 2^4}{-3 - 2}$

20. $\frac{7}{8} - \frac{2}{m}$

21. $\dfrac{4.8}{-0.16}$

22. $\dfrac{8}{9} \div 2n$

23. $1\dfrac{5}{6} + 1\dfrac{2}{3}$

24. $5 - 1\dfrac{7}{9}$

25. $|10 - 30| + (-4)^3$

26. $0.6 \div 12(3.6 - 4)$

27. $\dfrac{3}{10} - \left(\dfrac{1}{4} - \dfrac{3}{4}\right)^2 + \left(\dfrac{1}{2}\right)^2$

Evaluate each expression when w is −4, x is −2, and y is 3.

28. $-6w - 5$

29. $5x + 3y$

30. x^3y

31. $-5w^2x$

Simplify each expression.

32. $-2x^2 + 5x - 7x^2$

33. $ab - ab$

34. $-10(4w^3)$

35. $3(h - 4) + 2$

Solve each equation. Show the steps you used.

36. $2n - 3n = 0 - 5$

37. $12 - h = -3h$

38. $5a - 0.6 = 10.4$

39. $-\dfrac{7}{8} = \dfrac{3}{16}y$

40. $3 + \dfrac{1}{10}b = 5$

41. $\dfrac{0.2}{3.25} = \dfrac{10}{x}$

42. $32 - 3h = 5h + 8$

43. $3 + 2(x + 4) = -4x + 7 + 2x$

Translate each sentence into an equation and solve it.

44. If 5 is subtracted from four times a number, the result is −17. What is the number?

45. The sum of a number and 31 is three times the number plus 1. Find the number.

Solve each application problem using the six problem solving steps from Chapter 3.

46. Lawrence brought home three packages of disposable diapers. He used 17 diapers the first day and 19 the next day. There were 12 diapers left. How many diapers were originally in each package?

47. Susanna and Neoka are splitting $1620 for painting a house. Neoka worked twice as many hours, so she should earn twice as much. How much should each woman receive?

Find the unknown length, perimeter, circumference, area, or volume. When necessary, use 3.14 as the approximate value of π and round answers to the nearest tenth.

48. Find the perimeter and the area.

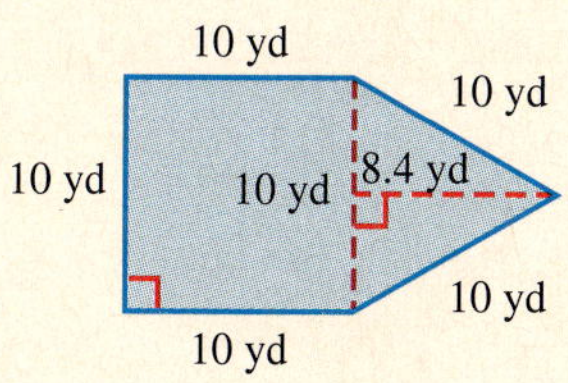

49. Find the circumference and the area.

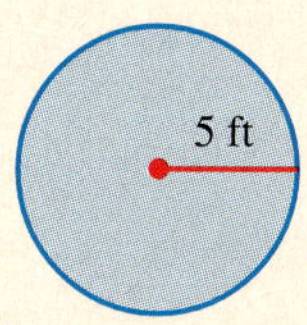

50. Find the unknown length in these similar triangles.

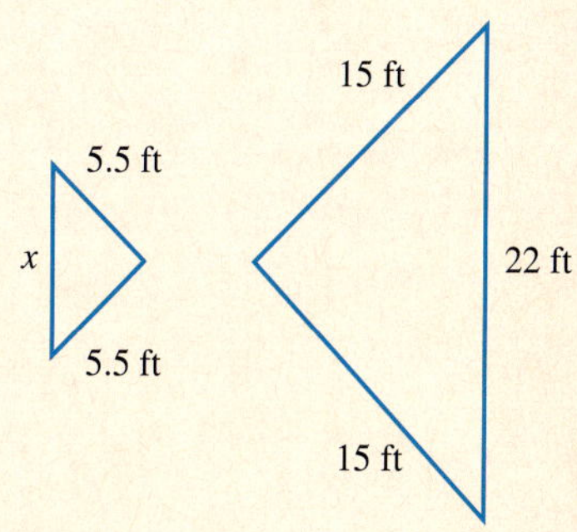

51. Find the volume.

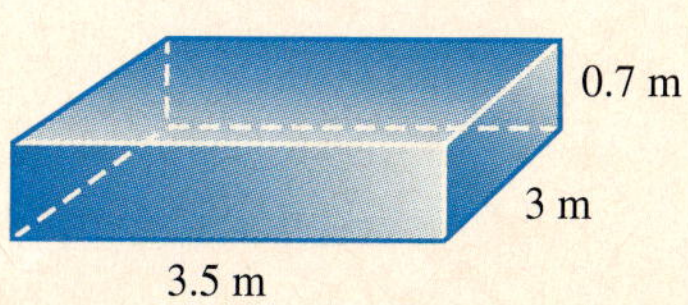

52. Find the volume.

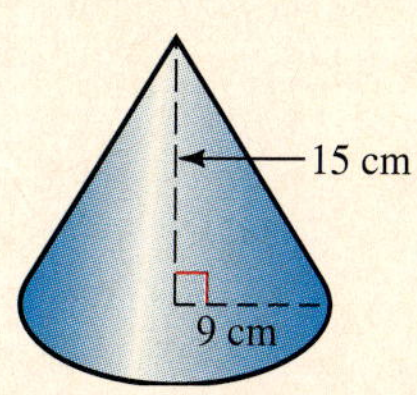

53. Find the unknown length.

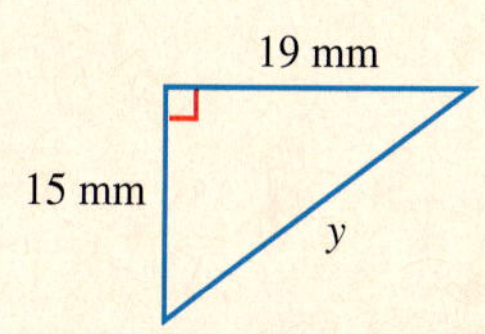

Solve each application problem.

54. Each bus holds 48 people. How many buses are needed to take 278 people to a baseball game?

55. The Americans With Disabilities Act provides the single parking space design below. Find the perimeter and area of this parking space, including the accessible aisle.

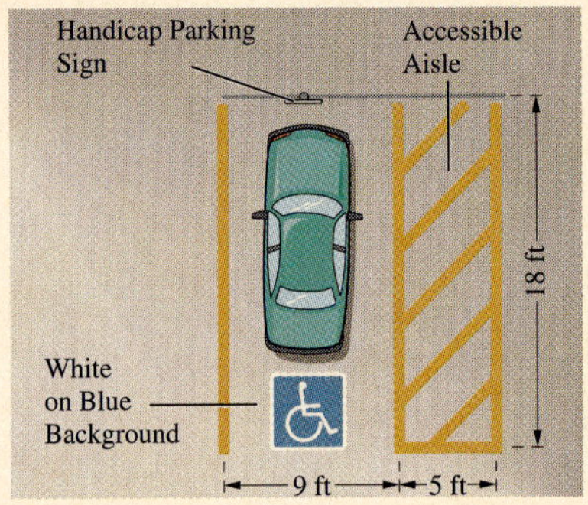

56. A survey of the 5600 students on our campus found that $\frac{3}{8}$ of the students work 20 hours or more per week. How many students work 20 hours or more?

57. An elevator has a weight limit of 1500 pounds. On it are twin boys weighing 88 pounds each and 8 adults weighing this number of pounds each: 240, 189, 127, 165, 143, 219, 116, and 124. The total weight of the people is how much above or below the limit.

58. The Jackson family is making three kinds of holiday cookies that require brown sugar. The recipes call for $2\frac{1}{4}$ cups, $1\frac{1}{2}$ cups, and $\frac{3}{4}$ cup, respectively. They bought two packages of brown sugar, each holding $2\frac{1}{3}$ cups. The amount bought is how much more or less than the amount needed?

59. Leather jackets are on sale at 30% off the regular price. Tracy likes a jacket with a regular price of $189. Find the amount of discount and the sale price she will pay.

60. On the Illinois map, one centimeter represents 12 kilometers. The center of Springfield is 7.8 cm from the center of Bloomington on the map. What is the actual distance in kilometers?

61. Dimitri took out a $3\frac{1}{2}$ year car loan for $8750 at 9% simple interest. Find the interest and the total amount due on the loan.

62. On a 35-problem math test, Juana solved 31 problems correctly. What percent of the problems were correct? Round to the nearest tenth of a percent.

63. Calbert bought an automatic focus camera for $64.95. He paid $7\frac{1}{2}$% sales tax. Find the amount of tax, to the nearest cent, and the total cost of the camera.

64. Danielle had a roll of 35 mm film developed. She received 24 prints for $10.25. What is the cost per print, to the nearest cent?

Measurement

8

We measure things all the time: the distance traveled on vacation, the floor area we want to cover with carpet, the amount of milk in a recipe, the weight of the bananas we buy at the store, the number of hours we work, and many more.

In the United States we still use the **English system** of measurement for many everyday activities. Examples of English units are inches, feet, quarts, ounces, and pounds. However, the fields of science, medicine, sports, and manufacturing increasingly use the **metric system** (meters, liters, and grams). And, because the rest of the world uses only the metric system, U.S. businesses are changing to the metric system in order to compete internationally.

INTERNET
www.mathnotes.com

OBJECTIVES

1. Know the basic units in the English system.
2. Convert among units.
3. Use unit fractions to convert among units.

FOR EXTRA HELP

Tutorial | Tape 14 | SSM, Sec. 8.1

8.1 *The English System*

OBJECTIVE 1 Until the switch to the metric system is complete, we still need to know how to use the English system of measurement. The table below lists the relationships you should memorize. The time relationships are used in both the English and metric systems.

LENGTH	WEIGHT
1 foot (ft) = 12 inches (in.)	1 pound (lb) = 16 ounces (oz)
1 yard (yd) = 3 feet (ft)	1 ton (T) = 2000 pounds (lb)
1 mile (mi) = 5280 feet (ft)	
CAPACITY	**TIME**
1 cup (c) = 8 fluid ounces	1 week (wk) = 7 days
1 pint (pt) = 2 cups (c)	1 day = 24 hours (h)
1 quart (qt) = 2 pints (pt)	1 hour (hr) = 60 minutes (min)
1 gallon (gal) = 4 quarts (qt)	1 minute (min) = 60 seconds (s)

As you can see, there is no simple or "natural" way to convert among these various measures. The units evolved over hundreds of years and were based on a variety of "standards." For example, one yard was the distance from the tip of a king's nose to his thumb when his arm was outstretched. An inch was three dried barleycorns laid end to end.

1. After memorizing the measurement conversions, answer these questions.

(a) 1 cup = ______ fluid ounces

(b) 1 gallon = ______ quarts

(c) 1 week = ______ days

(d) 1 yard = ______ feet

(e) 1 foot = ______ inches

(f) 1 pound = ______ ounces

(g) 1 ton = ______ pounds

(h) 1 hour = ______ minutes

(i) 1 pint = ______ cups

(j) 1 day = ______ hours

(k) 1 minute = ______ seconds

(l) 1 quart = ______ pints

(m) 1 mile = ______ feet

Answers

1. (a) 8 (b) 4 (c) 7 (d) 3 (e) 12 (f) 16 (g) 2000 (h) 60 (i) 2 (j) 24 (k) 60 (l) 2 (m) 5280

EXAMPLE 1 Knowing English Measurement Units

Memorize the English measurement conversions. Then answer these questions.

(a) 7 days = ______ week Answer: 1 week

(b) 1 yard = ______ feet Answer: 3 feet

WORK PROBLEM 1 AT THE SIDE.

Objective 2 You often need to convert from one unit of measure to another. Two methods of converting measurements are shown here. Study each way and use the method you prefer. You can make some conversions by deciding whether to multiply or divide.

Converting among Measurement Units

1. *Multiply* when converting from a larger unit to a smaller unit.
2. *Divide* when converting from a smaller unit to a larger unit.

EXAMPLE 2 Converting from One Unit of Measure to Another

Convert each measurement.

(a) 7 feet to inches

You are converting from a larger unit to a smaller unit (feet to inches), so multiply.

Because *1 foot = 12 inches,* multiply by 12.

$$7 \text{ feet} = 7 \cdot 12 = 84 \text{ inches}$$

(b) $3\frac{1}{2}$ pounds to ounces

You are converting from a larger unit to a smaller unit, pounds to ounces, so multiply.

Because *1 pound = 16 ounces,* multiply by 16.

$$3\frac{1}{2} \text{ pounds} = 3\frac{1}{2} \cdot 16 = \frac{7}{\cancel{2}_1} \cdot \frac{\cancel{2}^1 \cdot 8}{1} = \frac{56}{1} = 56 \text{ ounces}$$

(c) 20 quarts to gallons

You are converting from a smaller unit to a larger unit (quarts to gallons), so divide.

Because *4 quarts = 1 gallon,* divide by 4.

$$20 \text{ quarts} = \frac{20}{4} = 5 \text{ gallons}$$

(d) 45 minutes to hours

You are converting from a smaller unit to a larger unit (minutes to hours), so divide.

Because *60 minutes = 1 hour,* divide by 60 and write the fraction in lowest terms.

$$45 \text{ minutes} = \frac{45}{60} = \frac{45 \div 15}{60 \div 15} = \frac{3}{4} \text{ hour}$$

WORK PROBLEM 2 AT THE SIDE.

OBJECTIVE 3 If you have trouble deciding whether to multiply or divide when converting units, using **unit fractions** will solve the problem. You'll also find this method useful in science classes. A unit fraction is equivalent to 1. For example,

$$\frac{12 \text{ inches}}{12 \text{ inches}} = \frac{\overset{1}{\cancel{12}}\ \cancel{\text{inches}}}{\underset{1}{\cancel{12}}\ \cancel{\text{inches}}} = 1.$$

You know that 12 inches is the same as 1 foot. So you can substitute 1 foot for 12 inches in the numerator, or you can substitute 1 foot for 12 inches in the denominator. This makes two useful unit fractions.

$$\frac{\textbf{1 foot}}{12 \text{ inches}} = 1 \quad \text{or} \quad \frac{12 \text{ inches}}{\textbf{1 foot}} = 1$$

To convert from one measurement unit to another, just multiply by the appropriate unit fraction. Remember, a unit fraction is equivalent to 1. Multiplying something by 1 does *not* change its value.

Use these guidelines to choose the correct unit fraction.

Choosing a Unit Fraction

The *numerator* should use the measurement unit you want in the *answer*.
The *denominator* should use the measurement unit you want to *change*.

EXAMPLE 3 Using Unit Fractions with Length Measurement

(a) Convert 60 inches to feet.

Use a unit fraction with feet (the unit for your answer) in the numerator, and inches (the unit being changed) in the denominator. Because *1 foot = 12 inches,* the necessary unit fraction is

$$\frac{1 \text{ foot}}{12 \text{ inches}}. \quad \begin{matrix} \leftarrow \text{Unit for your answer} \\ \leftarrow \text{Unit being changed} \end{matrix}$$

Next, multiply 60 inches times this unit fraction. Write 60 inches as the fraction $\frac{60 \text{ inches}}{1}$. Then divide out common units and factors wherever possible.

$$\frac{60 \textbf{ inches}}{1} \cdot \frac{1 \text{ foot}}{12 \textbf{ inches}} = \frac{5 \cdot \overset{1}{\cancel{12}}\ \cancel{\text{inches}}}{1} \cdot \frac{1 \text{ foot}}{\underset{1}{\cancel{12}}\ \cancel{\text{inches}}} = \frac{5 \cdot 1 \text{ foot}}{1} = 5 \text{ ft}$$

Divide out common units.
Divide out common factor of 12.

CONTINUED ON NEXT PAGE

2. Convert each measurement.

(a) $5\frac{1}{2}$ feet to inches

(b) 64 ounces to pounds

(c) 6 yards to feet

(d) 2 tons to pounds

(e) 35 pints to quarts

(f) 20 minutes to hours

(g) 4 weeks to days

ANSWERS

2. **(a)** 66 inches **(b)** 4 pounds **(c)** 18 feet **(d)** 4000 pounds **(e)** $17\frac{1}{2}$ quarts **(f)** $\frac{1}{3}$ hour **(g)** 28 days

3. First write the unit fraction needed to make each conversion. Then complete the conversion.

(a) 36 inches to feet

unit fraction $\left.\right\}$ $\dfrac{1 \text{ foot}}{12 \text{ inches}}$

(b) 14 feet to inches

unit fraction $\left.\right\}$ $\dfrac{\underline{\quad} \text{ inches}}{\underline{\quad} \text{ foot}}$

(c) 60 inches to feet

unit fraction $\left.\right\}$ ______

(d) 4 yards to feet

unit fraction $\left.\right\}$ ______

(e) 39 feet to yards

unit fraction $\left.\right\}$ ______

(f) 2 miles to feet

unit fraction $\left.\right\}$ ______

ANSWERS

3. (a) 3 feet (b) $\dfrac{12 \text{ inches}}{1 \text{ foot}}$; 168 inches

(c) $\dfrac{1 \text{ foot}}{12 \text{ inches}}$; 5 feet

(d) $\dfrac{3 \text{ feet}}{1 \text{ yard}}$; 12 feet

(e) $\dfrac{1 \text{ yard}}{3 \text{ feet}}$; 13 yards

(f) $\dfrac{5280 \text{ feet}}{1 \text{ mile}}$; 10,560 feet

(b) Convert 9 feet to inches.

Select the correct unit fraction to change 9 feet to inches.

$$\frac{12 \text{ inches}}{1 \text{ foot}} \begin{array}{l} \leftarrow \text{Unit for your answer} \\ \leftarrow \text{Unit being changed} \end{array}$$

Multiply 9 feet times the unit fraction.

$$9 \text{ feet} \cdot \frac{12 \text{ inches}}{1 \text{ foot}} = \frac{9 \cancel{\text{ feet}}}{1} \cdot \frac{12 \text{ inches}}{1 \cancel{\text{ foot}}} = \frac{9 \cdot 12 \text{ inches}}{1} = 108 \text{ inches}$$

These units should match. Divide out feet

Note

If you cannot divide out any common units, you made a mistake in choosing the unit fraction.

◀◀ WORK PROBLEM 3 AT THE SIDE.

EXAMPLE 4 Using Unit Fractions with Capacity and Weight Measurement

(a) Convert 9 pints to quarts.

Select the correct unit fraction.

$$\frac{1 \text{ quart}}{2 \text{ pints}} \begin{array}{l} \leftarrow \text{Unit for your answer} \\ \leftarrow \text{Unit being changed} \end{array}$$

Next multiply.

Write as mixed number

$$9 \text{ pints} \cdot \frac{1 \text{ quart}}{2 \text{ pints}} = \frac{9 \cancel{\text{ pints}}}{1} \cdot \frac{1 \text{ quart}}{2 \cancel{\text{ pints}}} = \frac{9}{2} \text{ quarts} = 4\frac{1}{2} \text{ quarts}$$

These units should match. Divide out pints

(b) Convert $7\frac{1}{2}$ gallons to quarts.

Write as improper fraction

$$\frac{7\frac{1}{2} \cancel{\text{ gallons}}}{1} \cdot \frac{4 \text{ quarts}}{1 \cancel{\text{ gallon}}} = \frac{15}{2} \cdot \frac{4}{1} \text{ quarts}$$

$$= \frac{15}{\cancel{2}_1} \cdot \frac{\overset{1}{\cancel{2}} \cdot 2}{1} \text{ quarts}$$

$$= 30 \text{ quarts}$$

(c) Convert 36 ounces to pounds.

$$\frac{36 \text{ ounces}}{1} \cdot \frac{1 \text{ pound}}{16 \text{ ounces}} = \frac{\overset{1}{\cancel{4}} \cdot 9 \cancel{\text{ ounces}}}{1} \cdot \frac{1 \text{ pound}}{\underset{1}{\cancel{4}} \cdot 4 \cancel{\text{ ounces}}} = \frac{9}{4} \text{ pounds}$$

$$= 2\frac{1}{4} \text{ pounds}$$

Note

In Example 4(c) you get $\frac{9}{4}$ pounds. Recall that $\frac{9}{4}$ means $9 \div 4$. If you do $9 \div 4$ on your calculator, you get 2.25 pounds. English measurements usually use fractions or mixed numbers, like $2\frac{1}{4}$ pounds, or in Example 4(a), $4\frac{1}{2}$ quarts. However, 2.25 pounds is also correct and is the way grocery stores often show weights of produce, meat, and cheese.

WORK PROBLEM 4 AT THE SIDE.

EXAMPLE 5 Using Several Unit Fractions

Sometimes you may need to use two or three unit fractions in problems like these.

(a) Convert 63 inches to yards.

Use the unit fraction $\frac{1 \text{ foot}}{12 \text{ inches}}$ to change inches to feet and the unit fraction $\frac{1 \text{ yard}}{3 \text{ feet}}$ to change feet to yards. Notice how all the units divide out except yards, which is the unit you want in the answer.

$$\frac{63 \text{ \sout{inches}}}{1} \cdot \frac{1 \text{ \sout{foot}}}{12 \text{ \sout{inches}}} \cdot \frac{1 \text{ yard}}{3 \text{ \sout{feet}}} = \frac{63}{36} \text{ yards} = 1\frac{3}{4} \text{ yards}$$

You can also divide out common factors.

$$\frac{63}{1} \cdot \frac{1}{12} \cdot \frac{1}{3} = \frac{\overset{1}{\cancel{3}} \cdot \overset{1}{\cancel{3}} \cdot 7 \cdot 1 \cdot 1}{1 \cdot \underset{1}{\cancel{3}} \cdot 4 \cdot \underset{1}{\cancel{3}}} = \frac{7}{4} = 1\frac{3}{4} \text{ yards}$$

Instead of changing $\frac{7}{4}$ to $1\frac{3}{4}$, you can divide $7 \div 4$ on your calculator to get 1.75 yards. Both answers are correct because 1.75 is equivalent to $1\frac{3}{4}$.

(b) Convert 2 days to seconds.

Use three unit fractions. The first one changes days to hours, the second one changes hours to minutes, and the third one changes minutes to seconds. All the units divide out except seconds, which is what you want in your answer.

$$\frac{2 \text{ \sout{days}}}{1} \cdot \frac{24 \text{ \sout{hours}}}{1 \text{ \sout{day}}} \cdot \frac{60 \text{ \sout{minutes}}}{1 \text{ \sout{hour}}} \cdot \frac{60 \text{ seconds}}{1 \text{ \sout{minute}}} = 172{,}800 \text{ seconds}$$

WORK PROBLEM 5 AT THE SIDE.

4. Convert using unit fractions.

(a) 16 pints to quarts

(b) 16 quarts to gallons

(c) 3 cups to pints

(d) $2\frac{1}{4}$ gallons to quarts

(e) 48 ounces to pounds

(f) $3\frac{1}{2}$ tons to pounds

(g) $1\frac{3}{4}$ pounds to ounces

(h) 4 ounces to pounds

5. Convert using two or three unit fractions.

(a) 4 tons to ounces

(b) 90 inches to yards

(c) 3 miles to inches

(d) 36 pints to gallons

(e) 4 weeks to minutes

(f) $1\frac{1}{2}$ gallons to cups

ANSWERS

4. (a) 8 quarts **(b)** 4 gallons **(c)** $1\frac{1}{2}$ pints **(d)** 9 quarts **(e)** 3 pounds **(f)** 7000 pounds **(g)** 28 ounces **(h)** $\frac{1}{4}$ pound or 0.25 pound

5. (a) 128,000 ounces **(b)** $2\frac{1}{2}$ yards or 2.5 yards **(c)** 190,080 inches **(d)** $4\frac{1}{2}$ gallons or 4.5 gallons **(e)** 40,320 minutes **(f)** 24 cups

NUMBERS IN THE *Real World*

collaborative investigations

MAMMOTH RUSSIAN SUNFLOWER *Helianthus annuus*
Adults and children love this fast growing variety with enormous golden-yellow blossoms and edible seeds. Quickly grows to 8–10 feet with only minimal care. Great for a temporary hedge around the garden. Very easy to grow.

Type	Height	Planting Depth	Seed Spacing	Thinning		Days to Germinate
				Height	Spacing	
Annual	8–10' 2–3 m	1" 3 cm	1' 30 cm	3" 8 cm	2' 61 cm	10–14

PLANTING: Select a location with full sun to light shade and average soil. May be started indoors 4 weeks before last frost date. Keep young plants watered and weeded until established. Mature plants are quite drought tolerant.
NOTE: Save seeds for winter eating by hanging heads upside down in a cool, dry area for a few weeks. The seeds will brush off the flower head when they are dry.

The front and back of a seed packet for sunflowers are shown. Look at the front of the packet first.

There were 49 seeds in the packet and 46 of the seeds sprouted.
What was the cost per seed, to the nearest cent?
What percent of the seeds sprouted?
About how many seeds would weigh 1 gram?
How much does one seed weigh, rounded to the nearest tenth of a gram?

The back of the packet uses an apostrophe ('), which is a symbol for feet, and " which is a symbol for inches.

How tall will the plants grow in feet?
How tall will they grow in inches?
How tall will they grow in yards?

How many inches tall should the plants be when you thin them (remove less vigorous plants to give others room to grow)?
How tall is that in feet?

If you plant all 49 seeds in one long row, using the spacing given on the package, how long will your row be in feet?

If you thin the sprouts according to the directions, how many sprouts will you have to remove? Assume that all seeds sprout, and that you leave as many sprouts as possible.

Using the information in the article, how many gum wrappers are needed to make 1 foot of chain?
To make 1 inch of chain?

How many inches of chain are made from one wrapper?

Is the article correct in saying that 125 miles is 34 million gum wrappers? Find two different methods for calculating the comparison.

Chew on This!

Sometimes a person's life's work makes it into a museum. So it figures that Michael Knutson's 128-foot chain of gum wrappers now resides in the Yellow Medicine County Museum in Granite Falls, Minn.

According to the *Redwood Gazette*, the wrapper chain started in 1974 when Knutson, now a 37-year-old woodworker, became bored during study hall. "I've never been much of a gum chewer, but I found most of the wrappers on the streets of the city," he said. (FYI: It takes 6602 wrappers to make a 128-foot chain.)

Granite Falls is about 125 miles—or 34 million gum wrappers—west of the Twin Cities.

Source: Minneapolis Star Tribune

8.1 Exercises

Fill in the blanks with the measurement relationships you have memorized.

Example: 1 hour = ______ minutes **Solution:** 1 hour = **60** minutes

1. 1 yard = ______ feet
2. 1 foot = ______ inches
3. ______ fluid ounces = 1 cup
4. ______ quarts = 1 gallon
5. 1 mile = ______ feet
6. 1 week = ______ days
7. ______ pounds = 1 ton
8. ______ ounces = 1 pound
9. 1 minute = ______ seconds
10. 1 day = ______ hours

Convert these measurements using unit fractions.

Example: 3000 pounds = ______ tons

Solution: $3000 \text{ pounds} \cdot \frac{1 \text{ ton}}{2000 \text{ pounds}} = \frac{3 \cdot \overset{1}{\cancel{1000}}\ \cancel{\text{pounds}}}{1} \cdot \frac{1 \text{ ton}}{2 \cdot \underset{1}{\cancel{1000}}\ \cancel{\text{pounds}}} = \frac{3}{2} \text{ tons} = \mathbf{1\frac{1}{2} \text{ tons}}$

These units should match.

or **1.5 tons**

11. 120 seconds = ______ minutes
12. 180 minutes = ______ hours
13. 8 quarts = ______ gallons
14. 24 quarts = ______ gallons
15. An adult sperm whale may weigh 38 to 40 tons. How many pounds could it weigh?

16. Recent fossil finds indicate that the largest plant-eating dinosaur, Ultrasaurus, may have weighed 95 to 100 tons. How many pounds could it weigh?

17. 12 feet = __________ yards

18. 15 yards = __________ feet

19. 7 pounds = __________ ounces

20. 6 pounds = __________ ounces

21. 5 quarts = __________ pints

22. 13 quarts = __________ pints

23. 90 minutes = __________ hours

24. 45 seconds = __________ minutes

25. 3 inches = __________ foot

26. 30 inches = __________ feet

27. 24 ounces = __________ pounds

28. 36 ounces = __________ pounds

29. 5 cups = __________ pints

30. 15 quarts = __________ gallons

31. Mr. Kashpaws worked for 12 hours doing traditional harvesting of wild rice. What part of the day did he work?

32. Michelle prepares 4-ounce hamburgers at a fast-food restaurant. Each hamburger is what part of a pound?

33. $2\frac{1}{2}$ tons = __________ pounds

34. $4\frac{1}{2}$ pints = __________ cups

35. $4\frac{1}{4}$ gallons = __________ quarts

36. $2\frac{1}{4}$ hours = __________ minutes

37. Our premature baby weighed $2\frac{3}{4}$ pounds at birth. How many ounces did our baby weigh?

38. The NBA basketball star is $7\frac{1}{4}$ feet tall. What is his height in inches?

Use two or three unit fractions to convert the following.

Example: 3 days = ______ minutes

Solution: $\frac{3 \cancel{\text{days}}}{1} \cdot \frac{24 \cancel{\text{hours}}}{1 \cancel{\text{day}}} \cdot \frac{60 \text{ minutes}}{1 \cancel{\text{hour}}} = \textbf{4320 minutes}$

39. 6 yards = __________ inches

40. 2 tons = __________ ounces

41. 112 cups = __________ quarts

42. 336 hours = __________ weeks

43. 6 days = __________ seconds

44. 5 gallons = __________ cups

45. $1\frac{1}{2}$ tons = __________ ounces

46. $3\frac{1}{3}$ yards = __________ inches

47. The statement 8 = 2 is *not* true. But with appropriate measurement units, it *is* true.

8 *quarts* = 2 *gallons*

Attach measurement units to these numbers to make the statements true.

(a) 1 ________ = 16 ________
(b) 10 ________ = 20 ________
(c) 120 ________ = 2 ________
(d) 2 ________ = 24 ________
(e) 6000 ________ = 3 ________
(f) 35 ________ = 5 ________

48. Explain in your own words why you can add 2 feet + 12 inches to get 3 feet, but you cannot add 2 feet + 12 pounds.

Convert the following.

49. $2\frac{3}{4}$ miles = ________________ inches

50. $5\frac{3}{4}$ tons = ________________ ounces

51. $6\frac{1}{4}$ gallons = ________________ fluid ounces

52. $3\frac{1}{2}$ days = ________________ seconds

53. 24,000 ounces = ________________ ton

54. 57,024 inches = ________________ mile

55. 129,600 seconds = ________________ days

56. 1952 fluid ounces = ________________ gallons

Review and Prepare

Place < *or* > *in each blank to make a true statement. (For help, see* ***Section 1.2.****)*

57. 2 weeks ________ 15 days

58. 72 hours ________ 4 days

59. 4 hours ________ 185 minutes

60. 2 years ________ 28 months

61. 32 days ________ 4 weeks

62. 14 minutes ________ 780 seconds

8.2 *The Metric System—Length*

Around 1790, a group of French scientists developed the metric system of measurement. It is an organized system based on multiples of 10, like our number system and our money. After you are familiar with metric units, you will see that they are easier to use than the hodgepodge of English measurement relationships you used in **Section 8.1.**

OBJECTIVE 1 The basic unit of length in the metric system is the **meter** (also spelled *metre*). Use the symbol **m** for meter; do not put a period after it. If you put five of the pages from this textbook side by side, they would measure about 1 meter. Or, look at a yardstick—a meter is just a little longer. (A meter is about 39 inches long.)

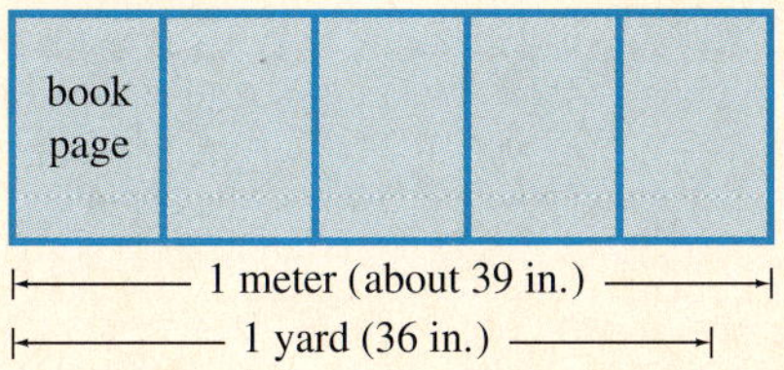

In the metric system you use meters for things like buying fabric for sewing projects, measuring the length of your living room, talking about heights of buildings, or describing track and field athletic events.

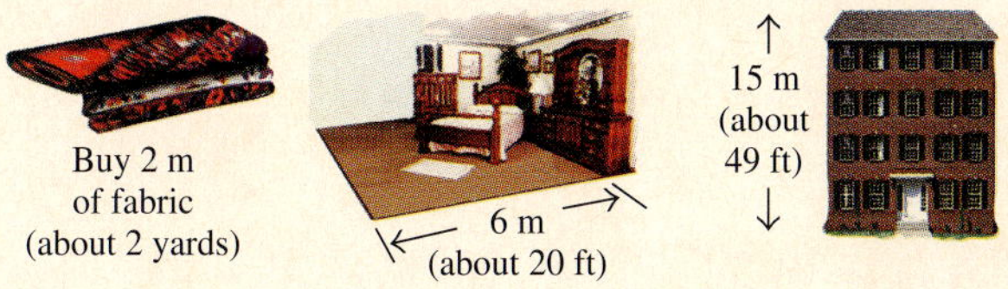

WORK PROBLEM 1 AT THE SIDE.

To make longer or shorter length units in the metric system, **prefixes** are written in front of the word meter. For example, the prefix ***kilo*** means 1000, so a ***kilo***meter is 1000 meters. The table below shows how to use the prefixes for length measurements. It is helpful to memorize the prefixes because they are also used with weight and capacity measurements. The colored boxes are the units you will use most often in daily life.

Prefix	*kilo-* meter	*hecto-* meter	*deka-* meter	meter	*deci-* meter	*centi-* meter	*milli-* meter
Meaning	1000 meters	100 meters	10 meters	1 meter	$\frac{1}{10}$ of a meter	$\frac{1}{100}$ of a meter	$\frac{1}{1000}$ of a meter
Symbol	km	hm	dam	m	dm	cm	mm

Here are some comparisons to help you get acquainted with the commonly used length units: km, m, cm, mm.

Kilometers are used instead of miles. A kilometer is 1000 meters. It is about 0.6 mile (a little more than half a mile) or about 6 city blocks. If you participate in a 10 km run, you'll go about 6 miles.

OBJECTIVES

1. Know the basic metric units of length.
2. Use unit fractions to convert among units.
3. Move the decimal point to convert among units.

FOR EXTRA HELP

Tutorial | Tape 14 | SSM, Sec. 8.2

1. Circle the items that measure about 1 meter.

length of a pencil

length of a baseball bat

height of doorknob from the floor

height of a house

basketball player's arm length

length of a paper clip

ANSWERS

1. baseball bat, height of doorknob, arm length

2. Write the most reasonable metric unit in each blank: km, m, cm, or mm.

(a) The woman's height is 168 ______.

(b) The man's waist is 90 ______ around.

(c) Louise ran the 100 ______ dash in the track meet.

(d) A postage stamp is 22 ______ wide.

(e) Michael paddled his canoe 2 ______ down the river.

(f) The pencil lead is 1 ______ thick.

(g) A stick of gum is 7 ______ long.

(h) The highway speed limit is 90 ______ per hour.

(i) The classroom was 12 ______ long.

(j) A penny is about 18 ______ across.

A meter is divided into 100 smaller pieces called ***centi***meters. Each centimeter is $\frac{1}{100}$ of a meter. Centimeters are used instead of inches. A centimeter is a little shorter than $\frac{1}{2}$ inch. The cover of this textbook is 21 cm wide. A nickel is about 2 cm across. Measure the width and length of your little finger on this centimeter ruler. The width of your little finger is probably about 1 centimeter.

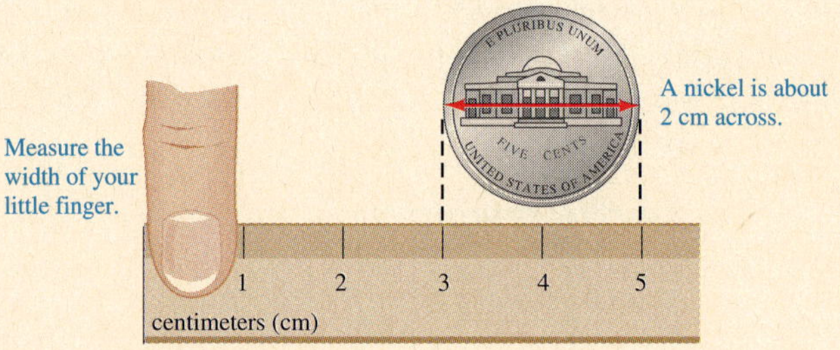

A meter is divided into 1000 smaller pieces called ***milli***meters. Each millimeter is $\frac{1}{1000}$ of a meter. It takes 10 mm to equal 1 cm, so it is a very small length. The thickness of a dime is about 1 mm. Measure the width of your pen or pencil and the width of your little finger on this millimeter ruler.

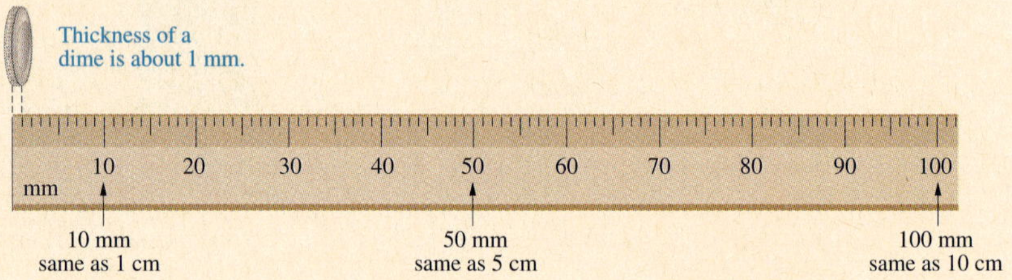

EXAMPLE 1 Using Metric Length Units

Write the most reasonable metric unit in each blank. Choose from km, m, cm, and mm.

(a) The distance from home to work is 20 ______.

20 km because kilometers are used instead of miles. 20 km is about 12 miles.

(b) My wedding ring is 4 ______ wide.

4 mm because the width of a ring is very small.

(c) The newborn baby is 50 ______ long.

50 cm, which is half of a meter; a meter is about 39 inches so half a meter is around 20 inches.

◀◀ WORK PROBLEM 2 AT THE SIDE.

ANSWERS

2. **(a)** cm **(b)** cm **(c)** m **(d)** mm **(e)** km **(f)** mm **(g)** cm **(h)** km **(i)** m **(j)** mm

OBJECTIVE 2 You can convert among metric length units by using unit fractions. Keep these relationships in mind when setting up the unit fractions.

Metric Length Relationships

1 km = 1000 m so the unit fractions are: $\frac{1\text{ km}}{1000\text{ m}}$ or $\frac{1000\text{ m}}{1\text{ km}}$	1 m = 1000 mm so the unit fractions are: $\frac{1\text{ m}}{1000\text{ mm}}$ or $\frac{1000\text{ mm}}{1\text{ m}}$
1 m = 100 cm so the unit fractions are: $\frac{1\text{ m}}{100\text{ cm}}$ or $\frac{100\text{ cm}}{1\text{ m}}$	1 cm = 10 mm so the unit fractions are: $\frac{1\text{ cm}}{10\text{ mm}}$ or $\frac{10\text{ mm}}{1\text{ cm}}$

EXAMPLE 2 Using Unit Fractions with Length Measurement

Convert the following.

(a) 5 km to m

Put the unit for the answer (meters) in the numerator of the unit fraction; the unit you want to change (km) in the denominator.

Unit fraction equivalent to 1. $\left\{\frac{1000\text{ m}}{1\text{ km}}\right.$ ← Unit for answer ← Unit being changed

Multiply. Divide out common units where possible.

$$5\text{ km} \cdot \frac{1000\text{ m}}{1\text{ km}} = \frac{5\text{ km}}{1} \cdot \frac{1000\text{ m}}{1\text{ km}} = \frac{5 \cdot 1000\text{ m}}{1} = 5000\text{ m}$$

These units should match.

The answer makes sense because a kilometer is much longer than a meter, so 5 km will contain many meters.

(b) 18.6 cm to m

Multiply by a unit fraction that allows you to divide out centimeters.

$$\frac{18.6\text{ cm}}{1} \cdot \underbrace{\frac{1\text{ m}}{100\text{ cm}}}_{\text{Unit fraction}} = \frac{18.6}{100}\text{ m} = 0.186\text{ m}$$

There are 100 cm in a meter, so 18.6 cm will be a small part of a meter. The answer makes sense.

WORK PROBLEM 3 AT THE SIDE.

3. First write the unit fraction needed to make each conversion. Then complete the conversion.

(a) 3.67 m to cm

unit fraction $\left.\right\} \frac{100\text{ cm}}{1\text{ m}}$

(b) 92 cm to m

unit fraction $\left.\right\} \frac{\_\_\_\text{ m}}{\_\_\_\text{ cm}}$

(c) 432.7 cm to m

unit fraction } ______

(d) 65 mm to cm

unit fraction } ______

(e) 0.9 m to mm

unit fraction } ______

(f) 2.5 cm to mm

unit fraction } ______

ANSWERS

3. (a) 367 cm **(b)** $\frac{1\text{ m}}{100\text{ cm}}$; 0.92 m
(c) $\frac{1\text{ m}}{100\text{ cm}}$; 4.327 m
(d) $\frac{1\text{ cm}}{10\text{ mm}}$; 6.5 cm
(e) $\frac{1000\text{ mm}}{1\text{ m}}$; 900 mm
(f) $\frac{10\text{ mm}}{1\text{ cm}}$; 25 mm

4. Do each multiplication or division by hand or on a calculator. Compare your answer to the one obtained by moving the decimal point.

(a) 43.5 • 10 = ______

43.5 gives 435.

(b) 43.5 ÷ 10 = ______

43.5 gives ______

(c) 28 • 100 = ______

28.00 gives ______

(d) 28 ÷ 100 = ______

28. gives ______

(e) 0.7 • 1000 = ______

0.700 gives ______

(f) 0.7 ÷ 1000 = ______

000.7 gives ______

Answers

4. **(a)** 435 **(b)** 4.35 **(c)** 2800 **(d)** 0.28 **(e)** 700 **(f)** 0.0007

Objective 3 By now you have probably noticed that conversions among metric units are made by multiplying or dividing by 10, by 100, or by 1000. A quick way to multiply by 10 is to move the decimal point one place to the *right*. Move it two places to multiply by 100, three places to multiply by 1000. Division is done by moving the decimal point to the *left*.

◀◀ WORK PROBLEM 4 AT THE SIDE.

An alternate conversion method to unit fractions is moving the decimal point using this **metric conversion line.**

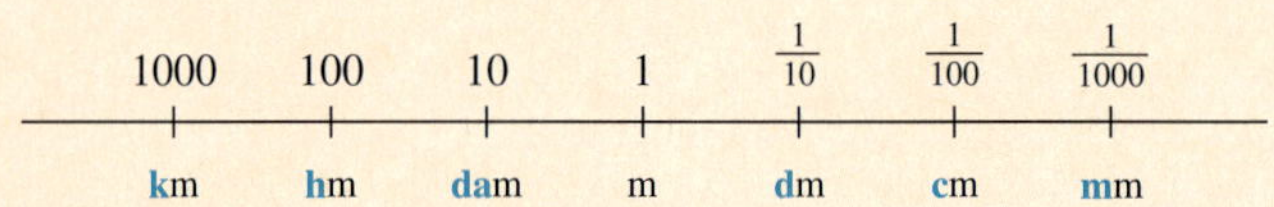

Here are the steps for using the conversion line.

Using the Metric Conversion Line

1. Find the unit you are given on the metric conversion line.
2. Count the number of places to get from the unit you are given to the unit you want in the answer.
3. Move the decimal point the *same number of places.* Move in the *same direction* as you did on the conversion line.

EXAMPLE 3 Using a Metric Conversion

Use the metric conversion line to make the following conversions.

(a) 5.702 km to m

Find **km** on the metric conversion line. To get to **m**, you move *three places* to the *right.* So move the decimal point in 5.702 *three places* to the *right.*

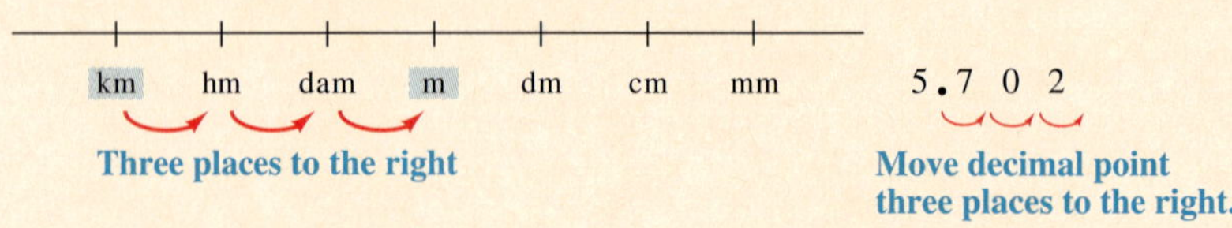

So 5.702 km = 5702 m.

(b) 69.5 cm to m

Find **cm** on the conversion line. To get to **m**, move *two places* to the *left.*

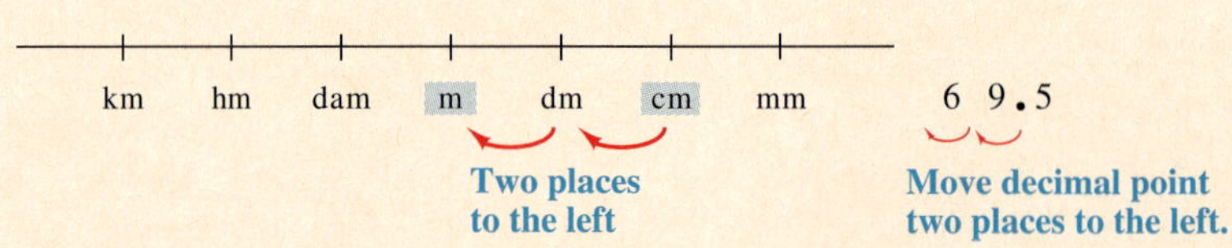

So 69.5 cm = 0.695 m.

CONTINUED ON NEXT PAGE

(c) 8.1 cm to mm

From **cm** to **mm** is *one place* to the *right*.

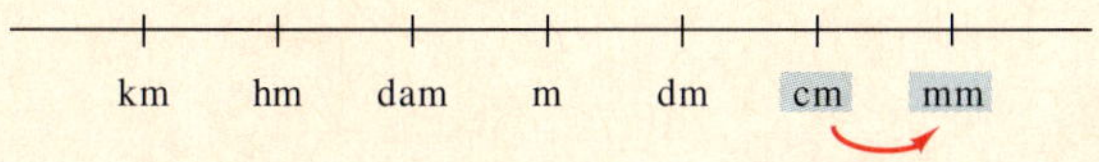

8.1

One place to the right

Move decimal point one place to the right.

So 8.1 cm = 81 mm.

WORK PROBLEM 5 AT THE SIDE.

EXAMPLE 4 Practicing Length Conversions

Convert using the metric conversion line.

(a) 1.28 m to mm

Moving from **m** to **mm** is going three places to the right. In order to move the decimal point in 1.28 three places to the right, you must write a zero as a placeholder.

1.28**0** Zero is written in as placeholder.

Move decimal point three places to the right.

1.28 m = 1280 mm

(b) 60 cm to m

From **cm** to **m** is two places to the left. The decimal point in 60 starts at the *far right side* because 60 is a whole number. Then move it two places to the left.

60. — Decimal point starts here.

60. — Move decimal point two places to the left.

60 cm = 0.60 m, which is equivalent to 0.6 m.

(c) 8 m to km

From **m** to **km** is three places to the left. The decimal point in 8 starts at the far right side. In order to move it three places to the left, you must write two zeros as placeholders.

8. — Decimal point starts here.

008. — Two zeros written in as placeholders. Move decimal point three places to the left.

8 m = 0.008 km.

WORK PROBLEM 6 AT THE SIDE.

5. Convert using the metric conversion line.

(a) 12.008 km to m

(b) 561.4 m to km

(c) 20.7 cm to m

(d) 20.7 cm to mm

(e) 4.66 m to cm

(f) 85.6 mm to cm

6. Convert using the metric conversion line.

(a) 9 m to mm

(b) 3 cm to m

(c) 14.6 km to m

(d) 5 mm to cm

(e) 70 m to km

(f) 0.8 m to cm

ANSWERS

5. **(a)** 12,008 m **(b)** 0.5614 km **(c)** 0.207 m **(d)** 207 mm **(e)** 466 cm **(f)** 8.56 cm

6. **(a)** 9000 mm **(b)** 0.03 m **(c)** 14,600 m **(d)** 0.5 cm **(e)** 0.07 km **(f)** 80 cm

NUMBERS IN THE *Real World*

collaborative investigations

HAIR AND NAIL GROWTH

Q How fast do hair and nails grow? Do they grow faster in the summer?

A Fingernails grow, on average, about one-tenth of a millimeter per day, although there is considerable variation among individuals. Fingernails grow faster than toenails, and nails on the longest fingers appear to grow the fastest.

Fingernails, as well as hair and skin, grow faster in the summer, presumably under the influence of sunlight, which expands blood vessels, bringing more oxygen and nutrients to the area and allowing for faster growth.

The rate the scalp hair grows is 0.3 to 0.4 millimeter per day, or about 6 inches a year.

Source: Minneapolis Star Tribune

- How much do nails grown in one week? One month? One year?
- How much does scalp hair grow in one week? One month? One year? (Use metric units.)
- When you have finished Section 8.5, come back to this article. Is the statement about hair growing 6 inches a year accurate? Explain your answer.

Measuring Up

New device measures distances within billionths of an inch

U.S. officials have unveiled "the ultimate ruler," a measuring device that can gauge distances to within billionths of an inch—the length of five individual atoms—and may help revolutionize high-tech manufacturing.

Developed at a cost of $8 million by the National Institute of Standards and Technology with other agencies, the Molecular Measuring Machine can measure distances to within 40 nanometers, or billionths of a meter. After further refinement it is expected to measure within 1 nanometer.

By way of comparison, the period at the end of this sentence is about 300,000 nanometers wide.

The machine is expected to prove a boon to U.S. manufacturers of computer chips and other tiny high-tech items that must meet exacting specifications. It can also help manufacturers better calibrate their own super-accurate equipment so that, for example, even more devices could be put onto silicon chips to increase their computing power.

"This is a key project where the United States is a long way in front of any other country," said Trevor Howe, a University of Connecticut professor of metallurgy and director of its Precision Manufacturing Center.

Source: Boston Globe

- The article states, "the Molecular Measuring Machine can measure distances to within 40 nanometers, or billionths of a meter." This suggests that the prefix *nano* means what?
- Write the two unit fractions you would use to convert between meters and nanometers. Use your unit fractions to change 40 nanometers to meters, and to change 300,000 nanometers to meters.
- Why do you suppose the headline and first paragraph talk about "billionths of an inch" when the rest of the article specifies nanometers, which are billionths of a meter? Would one–billionth of an inch be the same as one–billionth of a meter? How different would they be?
- Use the information in the article to find the length of one atom. Will you use inches or meters?

Name ______________________ Date ______________ Class ______________

8.2 Exercises

Use your knowledge of the meaning of metric prefixes to fill in the blanks.

Example: *hecto* means ______________ so 1 hm = ______________ m

Solution: *hecto* means ____**100**____ so 1 hm = ____**100**____ m

1. *kilo* means ______________ so 1 km = ______________ m

2. *deka* means ______________ so 1 dam = ______________ m

3. *milli* means ______________ so 1 mm = ______________ m

4. *deci* means ______________ so 1 dm = ______________ m

5. *centi* means ______________ so 1 cm = ______________ m

6. *hecto* means ______________ so 1 hm = ______________ m

Use this ruler to measure the following.

7. the width of your hand in centimeters

8. the width of your hand in millimeters

9. the width of your thumb in millimeters

10. the width of your thumb in centimeters

Write the most reasonable metric length unit in each blank. Choose from km, m, cm, and mm.

11. The child was 91 ________ tall.

12. The cardboard was 3 ________ thick.

13. Ming-Na swam in the 200 ________ backstroke race.

14. The bookcase is 75 ________ wide.

15. Adriana drove 400 ________ on her vacation.

16. The door is 2 ________ high.

17. An aspirin tablet is 10 ________ across.

18. Lamard jogs 4 ________ every morning.

19. A paper clip is about 3 ________ long.

20. My pen is 145 ________ long.

21. Dave's truck is 5 ________ long.

22. Wheelchairs need doorways that are at least 80 ________ wide.

23. Describe at least three examples of metric length units that you have come across in your daily life.

24. Explain one reason why the metric system would be easier for a child to learn than the English system.

Writing Calculator Small Group

Convert each measurement. Use unit fractions or the metric conversion line.

Example: 16 mm to m

Solution: $\frac{16\ \cancel{mm}}{1} \cdot \frac{1\ m}{1000\ \cancel{mm}} = \frac{16}{1000}\ m$ **= 0.016 m**

or: From mm to m is three places to the left 016. mm = **0.016 m**

25. 7 m to cm

26. 18 m to cm

27. 40 mm to m

28. 6 mm to m

29. 9.4 km to m

30. 0.7 km to m

31. 509 cm to m

32. 30 cm to m

33. 400 mm to cm

34. 25 mm to cm

35. 0.91 m to mm

36. 4 m to mm

37. Is 82 cm greater than or less than 1 m? What is the difference in the lengths?

38. Is 1022 m greater than or less than 1 km? What is the difference in the lengths?

39. Many cameras use film that is 35 mm wide. Movie film may be 70 mm wide. Using the ruler on the previous page, draw a line that is 35 mm long and a line 70 mm long. Then convert each measurement to centimeters.

40. Gold wedding bands may be very narrow or quite wide. Common widths are 3 mm, 5 mm, and 10 mm. Using the ruler on the previous page, draw lines that are 3 mm, 5 mm, and 10 mm long. Then convert each measurement to centimeters.

Ⓖ **41.** Use two unit fractions to convert 5.6 mm to km.

Ⓖ **42.** Use two unit fractions to convert 16.5 km to mm.

Review and Prepare

Write each decimal as a fraction in lowest terms. (For help, see ***Section 5.1.****)*

43. 0.875

44. 0.6

45. 0.08

46. 0.075

8.3 *The Metric System—Capacity and Weight (Mass)*

OBJECTIVES

1. Know the basic metric units of capacity.
2. Convert among metric capacity units.
3. Know the basic metric units of weight (mass).
4. Convert among metric weight (mass) units.
5. Distinguish among basic metric units of length, capacity, and weight (mass).

FOR EXTRA HELP

We use capacity units to measure liquids, such as the amount of milk in a recipe, the gasoline in our car tank, and the water in an aquarium. (The English capacity units we've been using are cups, pints, quarts, and gallons.) The basic metric unit for capacity is the **liter** (LEE-ter) (also spelled *litre*). The capital letter **L** is the symbol for liter, to avoid confusion with the numeral 1.

OBJECTIVE 1 The liter is related to metric length in this way: A box that measures 10 cm on every side holds exactly one liter. (The volume of the box is 1000 cubic centimeters. Volume was discussed in **Section 4.8.**) A liter is just a little more than 1 quart.

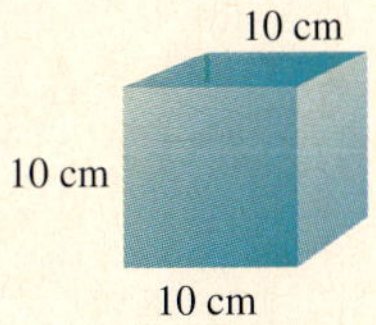

Holds exactly 1 liter (L)

A liter is a little more than one quart (just $\frac{1}{4}$ cup more).

In the metric system you use liters for things like buying milk at the store, filling a pail with water, and describing the size of your home aquarium.

Buy a 4 L jug of milk

Use a 12 L pail to wash floors

Watch the fish in your 40 L aquarium

WORK PROBLEM 1 AT THE SIDE.

1. Which things would you measure in liters?

amount of water in the bathtub

length of the bathtub

width of your car

amount of gasoline you buy for your car

weight of your car

height of a pail

amount of water in a pail

To make larger or smaller capacity units we use the same *prefixes* as we did with length units. For example, ***kilo*** means 1000 so a ***kilo***meter is 1000 meters. In the same way, a ***kilo***liter is 1000 liters.

Prefix	*kilo-* liter	*hecto-* liter	*deka-* liter	liter	*deci-* liter	*centi-* liter	*milli-* liter
Meaning	1000 liters	100 liters	10 liters	1 liter	$\frac{1}{10}$ of a liter	$\frac{1}{100}$ of a liter	$\frac{1}{1000}$ of a liter
Symbol	**k**L	**h**L	**da**L	L	**d**L	**c**L	**m**L

ANSWERS

1. water in bathtub, gasoline, water in a pail

2. Write the most reasonable metric unit in each blank. Choose from L and mL.

(a) I bought 8 _______ of milk at the store.

(b) The nurse gave me 10 _______ of cough syrup.

(c) This is a 100 _______ garbage can.

(d) It took 10 _______ of paint to cover the bedroom walls.

(e) My car's gas tank holds 50 _______.

(f) I added 15 _______ of oil to the pancake mix.

(g) The can of orange soda holds 350 _______.

(h) My friend gave me a 30 _______ bottle of expensive perfume.

Answers

2. (a) L **(b)** mL **(c)** L **(d)** L **(e)** L **(f)** mL **(g)** mL **(h)** mL

The capacity units you will use most often in daily life are liters (L) and ***milli***liters (mL). A tiny box that measures 1 cm on every side holds exactly one milliliter. (In medicine, this small amount is also called 1 cubic centimeter, or 1 cc for short.) It takes 1000 mL to make 1 L. Here are some other useful comparisons.

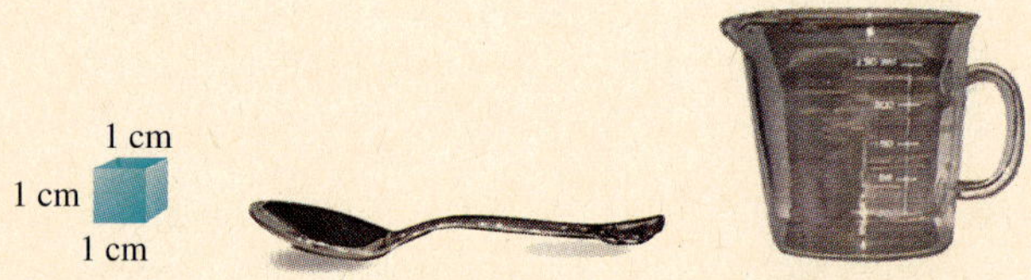

Holds exactly 1 milliliter (mL) — Teaspoon holds 5 mL — One cup holds about 250 mL

EXAMPLE 1 Using Metric Capacity Units

Write the most reasonable metric unit in each blank. Choose from L and mL.

(a) The bottle of shampoo held 500 _______.

500 mL because 500 L would be about 500 quarts, which is too much.

(b) I bought a 2 _______ carton of orange juice.

2 L because 2 mL would be less than a teaspoon.

WORK PROBLEM 2 AT THE SIDE.

Objective 2 Just as with length units, you can convert between milliliters and liters using unit fractions. The units fractions you need are:

$$\frac{1000 \text{ mL}}{1 \text{ L}} \qquad \frac{1 \text{ L}}{1000 \text{ mL}}$$

Or you can use a metric conversion line to decide how to move the decimal point.

1000	100	10	1	$\frac{1}{10}$	$\frac{1}{100}$	$\frac{1}{1000}$
kL	hL	daL	L	dL	cL	mL

EXAMPLE 2 Conversions among Metric Capacity Units

Convert using the metric conversion line or unit fractions.

(a) 2.5 L to mL

Using the metric conversion line:

From **L** to **mL** is *three places* to the *right*.

2.500 — Write two zeros as placeholders.

2.5 L = 2500 mL

Using unit fractions:

Multiply by a unit fraction that allows you to divide out liters.

$$\frac{2.5 \not{L}}{1} \cdot \frac{1000 \text{ mL}}{1 \not{L}} = 2500 \text{ mL}$$

CONTINUED ON NEXT PAGE

(b) 80 mL to L

Using the metric conversion line:
From **mL** to **L** is *three places* to the *left*.

80. → 080.

Decimal point starts here. Move three places left.

80 mL = 0.080 L or 0.08 L

Using unit fractions:

Multiply by a unit fraction that allows you to divide out mL.

$$\frac{80\ \cancel{\text{mL}}}{1} \cdot \frac{1\ \text{L}}{1000\ \cancel{\text{mL}}} = \frac{80}{1000}\text{L} = 0.08\ \text{L}$$

WORK PROBLEM 3 AT THE SIDE.

Objective 3 The **gram** is the basic metric unit for mass. Although we often call it "weight," there is a difference. Weight is a measure of the pull of gravity; the farther you are from the center of the earth, the less you weigh. In outer space you become weightless, but your mass, the amount of matter in your body, stays the same regardless of where you are. We will use the word "weight" for everyday purposes.

The gram is related to metric length in this way: the weight of the water in a box measuring 1 cm on every side is 1 gram. This is a very tiny amount of water (1 mL) and a very small weight. One gram is also the weight of a dollar bill or a single raisin. A nickel weighs 5 grams. A regular hamburger weighs from 175 to 200 grams.

The 1 mL of water in this box weighs 1 gram.

A nickel weighs 5 grams.

A dollar bill weighs 1 gram.

A hamburger weighs 175 to 200 grams.

WORK PROBLEM 4 AT THE SIDE.

3. Convert.

(a) 9 L to mL

(b) 0.75 L to mL

(c) 500 mL to L

(d) 5 mL to L

(e) 2.07 L to mL

(f) 3275 mL to L

4. Which things would weigh about 1 gram?

a small paperclip

a pair of scissors

one playing card from a deck of cards

a calculator

an average-size apple

the check you wrote at the grocery store

Answers

3. (a) 9000 mL **(b)** 750 mL **(c)** 0.5 L **(d)** 0.005 L **(e)** 2070 mL **(f)** 3.275 L

4. paperclip, playing card, check

To make larger or smaller weight units, we use the same *prefixes* as we did with length and capacity units. For example, ***kilo*** means 1000 so a ***kilo***meter is 1000 meters, a ***kilo***liter is 1000 liters, and a ***kilo***gram is 1000 grams.

Prefix	*kilo-gram*	*hecto-gram*	*deka-gram*	*gram*	*deci-gram*	*centi-gram*	*milli-gram*
Meaning	1000 grams	100 grams	10 grams	1 gram	$\frac{1}{10}$ of a gram	$\frac{1}{100}$ of a gram	$\frac{1}{1000}$ of a gram
Symbol	**kg**	**hg**	**dag**	g	**dg**	**cg**	**mg**

The units you will use most often in daily life are kilograms (kg), grams (g), and milligrams (mg). ***Kilo***grams are used instead of pounds. A kilogram is 1000 grams. It is about 2.2 pounds. This textbook weighs about 1.7 kg. An average newborn baby weighs 3 to 4 kg; a college football player might weigh 100 to 110 kg.

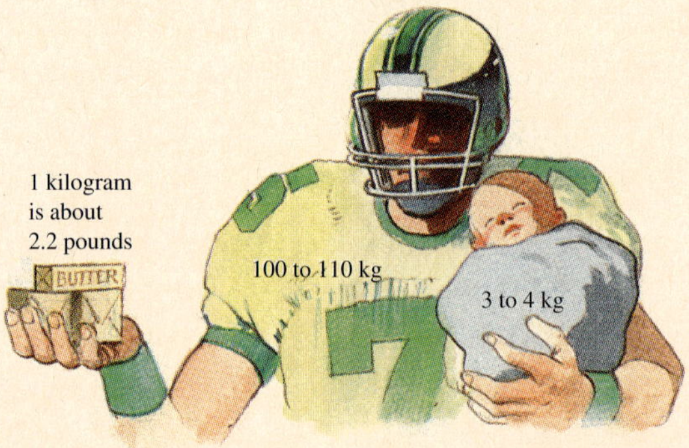

Extremely small weights are measured in ***milli***grams. It takes 1000 mg to make 1 g. Recall that a dollar bill weighs about 1 g. Imagine cutting it into 1000 pieces; the weight of one tiny piece would be 1 mg. Dosages of medicine and vitamins are given in milligrams. You will also use milligrams in science classes.

Cut a dollar bill into 1000 pieces.
One tiny piece weighs 1 milligram.

EXAMPLE 3 Using Metric Weight Units

Write the most reasonable metric unit in each blank. Choose from kg, g, and mg.

(a) Ramon's suitcase weighed 20 ______.

20 kg because kilograms are used instead of pounds. 20 kg is about 44 pounds.

(b) LeTia took a 350 ______ aspirin tablet.

350 mg because 350 g would be more than the weight of a hamburger, which is too much.

(c) Jenny mailed a letter that weighed 30 ______.

30 g because 30 kg would be much too heavy and 30 mg is less than the weight of a dollar bill.

WORK PROBLEM 5 AT THE SIDE.

5. Write the most reasonable metric unit in each blank. Choose from kg, g, and mg.

(a) A thumbtack weighs 800 ______.

(b) A teenager weighs 50 ______.

(c) This large cast-iron frying pan weighs 1 ______.

(d) Jerry's basketball weighed 600 ______.

(e) Tamlyn takes a 500 ______ calcium tablet every morning.

(f) On his diet, Greg can eat 90 ______ of meat for lunch.

(g) One strand of hair weighs 2 ______.

(h) One banana might weigh 150 ______.

ANSWERS

5. **(a)** mg **(b)** kg **(c)** kg **(d)** g **(e)** mg **(f)** g **(g)** mg **(h)** g

OBJECTIVE 4 As with length and capacity, you can convert among metric weight units by using unit fractions. The unit fractions you need are shown here.

Converting between grams and kilograms	Converting between milligrams and grams
$\frac{1000\text{ g}}{1\text{ kg}}$ or $\frac{1\text{ kg}}{1000\text{ g}}$	$\frac{1000\text{ mg}}{1\text{ g}}$ or $\frac{1\text{ g}}{1000\text{ mg}}$

Or you can use a metric conversion line to decide how to move the decimal point.

1000	100	10	1	$\frac{1}{10}$	$\frac{1}{100}$	$\frac{1}{1000}$
kg	hg	dag	g	dg	cg	mg

EXAMPLE 4 Conversions among Metric Weight Units

Convert, using the metric conversion line or unit fractions.

(a) 7 mg to g

Using the metric conversion line:
From **mg** to **g** is *three places* to the *left*.

7. ↑ Decimal point starts here.

007. Move three places left.

7 mg = 0.007 g

Using unit fractions:

Multiply by a unit fraction that allows you to divide out mg.

$$\frac{7\ \cancel{\text{mg}}}{1} \cdot \frac{1\text{ g}}{1000\ \cancel{\text{mg}}} = \frac{7}{1000}\text{ g}$$

$$= 0.007\text{ g}$$

(b) 13.72 kg to g

Using the metric conversion line:
From **kg** to **g** is *three* places to the *right*.

13.720 Decimal point moves three places to the right.

13.72 kg = 13,720 g
↑ A comma (not a decimal point)

Using unit fractions:

Multiply by a unit fraction that allows you to divide out kg.

$$\frac{13.72\ \cancel{\text{kg}}}{1} \cdot \frac{1000\text{ g}}{1\ \cancel{\text{kg}}} = 13{,}720\text{ g}$$

↑ A comma (not a decimal point)

WORK PROBLEM 6 AT THE SIDE.

6. Convert.

(a) 10 kg to g

(b) 45 mg to g

(c) 6.3 kg to g

(d) 0.077 g to mg

(e) 5630 g to kg

(f) 90 g to kg

ANSWERS

6. (a) 10,000 g **(b)** 0.045 g **(c)** 6300 g **(d)** 77 mg **(e)** 5.63 kg **(f)** 0.09 kg

7. First decide which type of unit is needed: length, capacity, or weight. Then write the most appropriate unit in the blank. Choose from km, m, cm, mm, L, mL, kg, g, and mg.

(a) Gail bought a 4 ________ can of paint.

Use ________ units.

(b) The bag of chips weighed 450 ________.

Use ________ units.

(c) Give the child 5 ________ of liquid aspirin.

Use ________ units.

(d) The width of the window is 55 ________.

Use ________ units.

(e) Akbar drives 18 ________ to work.

Use ________ units.

(f) Each computer weighs 5 ________.

Use ________ units.

(g) A credit card is 55 ________ wide.

Use ________ units.

Answers

7. **(a)** L; capacity **(b)** g; weight **(c)** mL; capacity **(d)** cm; length **(e)** km; length **(f)** kg; weight **(g)** mm; length

Objective 5 As you encounter things to be measured at home, on the job, or in your classes at school, be careful to use the correct type of measurement unit.

Use *length units* (kilometers, meters, centimeters, millimeters) to measure:

how long	how high	how far away
how wide	how tall	how far around (perimeter)
how deep	distance	

Use *capacity units* (liters, milliliters) to measure liquids (things that can be poured) such as:

water	shampoo	gasoline
milk	perfume	oil
soft drinks	cough syrup	paint

Also use liters and milliliters to describe how much liquid something can hold, such as an eyedropper, measuring cup, pail, or bathtub.

Use *weight units* (kilogram, grams, milligrams) to measure:

the weight of something   how heavy something is

In Chapters 3 and 4 you used square units (such as square meters) to measure area, and cubic units (such as cubic centimeters) to measure volume.

EXAMPLE 5 Using a Variety of Metric Units

First decide which type of unit is needed: length, capacity, or weight. Then write the most appropriate metric unit in the blank. Choose from km, m, cm, mm, L, mL, kg, g, and mg.

(a) The letter needs another stamp because it weighs 40 ________.
Use ________ units.

The letter weighs 40 **grams** because 40 mg is less than the weight of a dollar bill and 40 kg would be about 88 pounds.
Use **weight** units because of the word "weighs."

(b) The swimming pool is 3 ________ deep at the deep end.
Use ________ units.

The pool is 3 **meters** deep because 3 cm is only about an inch and 3 km is about 1.8 miles.
Use **length** units because of the word "deep."

(c) This is a 340 ________ can of juice.
Use ________ units.

It is a 340 **milliliter** can because 340 liters would be more than 340 quarts.
Use **capacity** units because juice is a liquid.

◀◀ WORK PROBLEM 7 AT THE SIDE.

8.3 Exercises

Write the most reasonable metric unit in each blank. Choose from L, mL, kg, g, and mg.

1. The glass held 250 _______ of water.

2. Hiromi used 20 _______ of water to wash the kitchen floor.

3. Dolores can make 10 _______ of soup in that pot.

4. Jay gave 2 _______ of vitamin drops to the baby.

5. Our labrador dog grew up to weigh 40 _______.

6. One dime weighs 2 _______.

7. Lori caught a small sunfish weighing 150 _______.

8. A small safety pin weighs 750 _______.

9. Andre donated 500 _______ of blood today.

10. Barbara bought the large 2 _______ bottle of cola.

11. The patient received a 250 _______ tablet of medication each hour.

12. The 8 people on the elevator weighed a total of 500 _______.

13. The gas can for the lawn mower holds 4 _______.

14. Kevin poured 10 _______ of vanilla into the bowl.

15. Pam's backpack weighs 5 _______ when it is full of books.

16. One grain of salt weighs 2 _______.

Ⓖ *Today, medical measurements are usually given in the metric system. Since we convert among metric units of measure by moving the decimal point, it is possible that mistakes can be made. Examine the following dosages and indicate whether they are reasonable or unreasonable.*

17. Drink 4.1 liters of Kaopectate after each meal.

18. Drop 1 mL of solution into the eye twice a day.

19. Soak your feet in 5 kilograms of Epsom salts per liter of water.

20. Inject 0.5 liter of insulin each morning.

21. Take 15 milliliters of cough syrup every four hours.

22. Take 200 milligrams of vitamin C each day.

23. Take 350 milligrams of aspirin three times a day.

24. Buy a tube of ointment weighing 0.002 gram.

Ⓖ **25.** Describe at least two examples of metric capacity units and two examples of metric weight units that you have come across in your daily life.

Ⓖ **26.** Explain in your own words how the meter, liter, and gram are related.

Ⓖ **27.** Describe how you decide which unit fraction to use when converting 6.5 kg to grams.

Ⓖ **28.** Write out an explanation of each step you would use to convert 20 mg to grams using the metric conversion line.

Convert each measurement. Use unit fractions or the metric conversion line.

Example: 9 g to kg

Solution: $\frac{9\ \not{g}}{1} \cdot \frac{1\ \text{kg}}{1000\ \not{g}} = \frac{9}{1000}\text{kg} = \mathbf{0.009\ kg}$

or: From g to kg is three places to the left. $009.\ \text{g} = \mathbf{0.009\ kg}$

29. 15 L to mL

30. 6 L to mL

31. 3000 mL to L

32. 18,000 mL to L

33. 925 mL to L

34. 200 mL to L

35. 8 mL to L

36. 25 mL to L

37. 4.15 L to mL

38. 11.7 L to mL

39. 8000 g to kg

40. 25,000 g to kg

41. 5.2 kg to g

42. 12.42 kg to g

43. 0.85 g to mg

44. 0.2 g to mg

45. 30,000 mg to g

46. 7500 mg to g

47. 598 mg to g

48. 900 mg to g

49. 60 mL to L

50. 6.007 kg to g

51. 3 g to kg

52. 12 mg to g

53. 0.99 L to mL

54. 13,700 mL to L

Write the most appropriate metric unit in each blank. Choose from km, m, cm, mm, L, mL, kg, g, and mg.

55. The masking tape is 19 ______ wide.

56. The roll has 55 ______ of tape on it.

57. Buy a 60 ______ jar of acrylic paint for art class.

58. One onion weighs 200 ______.

59. My waist measurement is 65 ______.

60. Add 2 ______ of windshield washer fluid to your car.

61. A single postage stamp weighs 90 ______.

62. The hallway is 10 ______ long.

Solve the following application problems.

63. The doctor told Sara to drink two liters of water each day. How many milliliters is that?

64. A juice can holds 1500 mL. How many liters of juice does it hold?

65. The premature infant weighed only 950 grams. How many kilograms did he weigh?

66. Bill bought three kilograms of potatoes. How many grams did he buy?

67. A healthy human heart pumps about 70 mL of blood per beat. How many liters of blood does it pump per beat?

68. In one sip, an elephant can suck 7.6 L of water into its trunk. How many milliliters does it suck into its trunk?

69. A small adult cat weighs about 3 kg. How many grams does it weigh?

70. If the letter you are mailing weighs 29 g, you must put additional postage on it. How many kilograms does the letter weigh?

71. Is 1005 mg greater than or less than 1 g? What is the difference in the weights?

72. Is 990 mL greater than or less than 1 L? What is the difference in the amounts?

Ⓖ **73.** One nickel weighs 5 grams. How many nickels are in 1 kilogram of nickels?

Ⓖ **74.** Seawater contains about 3.5 grams of salt per 1000 milliliters of water. How many grams of salt would be in 1 liter of seawater?

Ⓖ **75.** Helium weighs about 0.0002 grams per milliliter. How much would 1 liter of helium weigh?

Ⓖ **76.** About 1500 grams of sugar can be dissolved in a liter of warm water. How much sugar could be dissolved in 1 milliliter of warm water?

Review and Prepare

Name the digit that has the given place value in each of the following. (For help, see ***Section 5.1.****)*

77. 7250.6183
thousands
hundredths
thousandths
hundreds

78. 1358.0256
ten-thousandths
tenths
tens
ones

8.4 Problem Solving: Metric Measurement

OBJECTIVE 1 One advantage of the metric system is the ease of comparing measurements in application situations. Just be sure that you are comparing similar units: mg to mg, km to km, and so on.

EXAMPLE 1 Solving Metric Applications

(a) Cheddar cheese is on sale at \$8.99 per kg. Jake bought 350 grams of the cheese. How much did he pay, to the nearest cent?

The price is \$8.99 per *kilogram,* but the amount purchased is in *grams.* Convert grams to kilograms (the unit in the price). Then multiply the weight times the cost per kilogram.

$$350\text{ g} = 0.35\text{ kg}$$

$$\frac{\$8.99}{1\ \cancel{\text{kg}}} \cdot \frac{0.35\ \cancel{\text{kg}}}{1} = \$3.1465$$

Jake paid \$3.15, to the nearest cent.

(b) Olivia has 2.5 meters of lace. How many centimeters of lace can she use to trim each of six hair ornaments? Round to the nearest tenth of a centimeter.

The given amount is in *meters,* but the answer must be in *centimeters,* so convert meters to centimeters. Then divide by the number of hair ornaments.

$$2.5\text{ m} = 250\text{ cm}$$

$$\frac{250\text{ cm}}{6\text{ ornaments}} \approx 41.6666\text{ cm/ornament}$$

Olivia can use ≈41.7 cm of lace on each ornament (rounded to the nearest tenth).

WORK PROBLEM 1 AT THE SIDE.

EXAMPLE 2 Measurement Applications

(a) Rubin measured a board and found that the length was 3 meters plus an additional 5 centimeters. He cut off a piece measuring 1 meter 40 centimeters for a shelf. Find the length of the remaining piece in meters.

The lengths involve two units, m and cm. To make the calculations easier, write each length in terms of meters (the unit called for in the answer). Then subtract to find the leftover length.

Board		Shelf	
3m →	3.00 m	1 m →	1.0 m
plus 5 cm →	+ 0.05 m	plus 40 cm →	+ 0.4 m
	3.05 m		1.4 m

Subtract to find leftover length.

$$\begin{array}{r} 3.05\text{ m} \leftarrow \text{Board} \\ -\ 1.40\text{ m} \leftarrow \text{Shelf} \\ \hline 1.65\text{ m} \leftarrow \text{Leftover piece} \end{array}$$

The length of the leftover piece is 1.65 m.

CONTINUED ON NEXT PAGE

OBJECTIVE

1 Solve application problems involving metric measurements.

FOR EXTRA HELP

Tutorial | Tape 15 | SSM, Sec. 8.4

1. Solve each problem.

(a) Satin ribbon is on sale at \$0.89 per meter. How much will 75 cm cost, to the nearest cent?

(b) Lucinda's doctor wants her to take 1.2 grams of medication each day in three equal doses. How many milligrams should be in each dose?

ANSWERS

1. (a) \$0.67 (rounded to the nearest cent)
(b) 400 mg/dose

(b) Amy put a basket of nuts on her scale and saw that they weighed 4 kilograms plus 140 grams. She plans to put the nuts into three gift packs of equal size. Find the number of kilograms in each pack.

Write 4 kg 140 g in terms of kg (the unit for the answer). Then divide by the number of gift packs.

$$\begin{array}{rr} 4 \text{ kg} \rightarrow & 4.00 \text{ kg} \\ \text{plus } 140 \text{ g} \rightarrow & +\ 0.14 \text{ kg} \\ \hline & 4.14 \text{ kg} \end{array}$$

$$\frac{4.14 \text{ kg}}{3 \text{ packs}} = 1.38 \text{ kg/pack}$$

Each gift pack will have 1.38 kg of nuts.

◀◀ WORK PROBLEM 2 AT THE SIDE.

2. (a) Andrea has two pieces of fabric. One measures 2 meters 35 centimeters and the other measures 1 meter 85 centimeters. How many meters of fabric does she have in all?

(b) Mr. Green has 9 m 20 cm of rope. He is cutting it into eight pieces so his Scout troop can practice knot tying. How many meters of rope will each Scout get?

ANSWERS

2. (a) 4.2 m **(b)** 1.15 m/Scout

8.4 Exercises

Solve each application problem.

Example: A basket of strawberries weighed 1 kg 80 g. Find the cost of the strawberries if they are priced at \$2.19 per kg.

Solution: Write 1 kg 80 g in terms of kg (the unit in the price).

$$\begin{array}{lr} 1 \text{ kg} \rightarrow & 1.00 \text{ kg} \\ \text{plus } 80 \text{ g} \rightarrow & +\ 0.08 \text{ kg} \\ \hline & 1.08 \text{ kg} \end{array} \qquad \frac{\$2.19}{1\ \cancel{\text{kg}}} \cdot \frac{1.08\ \cancel{\text{kg}}}{1} = \$2.3652$$

In consumer situations, prices are rounded to the nearest cent, so the strawberries cost **\$2.37.**

1. Bulk rice is on special at \$0.65 per kilogram. Pam scooped some rice into a bag and put it on the scale. How much will she pay for 2 kg 50 g of rice?

2. Lanh is buying a piece of plastic tubing for the science lab that measures 3 m 15 cm. The price is \$4.75 per meter. How much will Lanh pay?

3. Kendal works for a garden store. He put 15 grams of fertilizer on each of 650 tomato plants. How many kilograms of fertilizer did he use?

4. The garden store ordered a 50 L drum of liquid plant food. They repackaged the plant food into bottles that hold 125 mL. How many bottles were filled?

5. An adult human body contains about 5 L of blood. If each beat of the heart pumps 70 mL of blood, how many times must the heart beat to pass all the blood through the heart? Round to the nearest whole number of beats.

6. A floor tile measures 30 cm by 30 cm and weighs 185 g. How many kilograms would a stack of 24 tiles weigh?

7. Rosa is building a bookcase. She has one board that is 2 m 8 cm long and another that is 2 m 95 cm long. How long are the two boards together in meters?

8. Eric's Scottie dog weighs 8 kg 600 g. Rob's Great Dane weighs 50 kg 50 g. The Great Dane weighs how much more than the Scottie, in kilograms?

9. The apartment building caretaker puts 750 mL of chlorine into the swimming pool every day. How many liters should he order to have a one-month (30-day) supply on hand?

10. Janet has 10 m 30 cm of fabric. She wants to make curtains for three windows that are all the same size. How much fabric is available for each window, to the nearest tenth of a meter?

Ⓖ *It is difficult to weigh very light objects, such as a single sheet of paper or a single staple (unless you have a very expensive scientific scale). One way around this problem is to weigh a large number of the items and then divide to find the weight of one item. Of course, before dividing, you must subtract the weight of the box or wrapper that the items are packaged in to find the net weight. Complete this table.*

Item	***Total weight***	***Weight of packaging***	***Net weight***	***Weight of one item in grams***	***Weight of one item in milligrams***
11. Box of 50 envelopes	255 g	40 g	______	______	______
12. Box of 1000 staples	350 g	20 g	______	______	______
13. Ream of paper (500 sheets)	______ kg	50 g	______	______	3000 mg
14. Box of 100 small paper clips	______	5 g	______	______	500 mg

Ⓖ **15.** As a fund raiser, the PTA bought 40 kg of nuts for $113.50. They sold the nuts in 250 g bags for $2.95 each. Find the amount of profit.

Ⓖ **16.** In chemistry class, each of the 45 students needs 85 mL of acid. How many one-liter bottles of acid need to be ordered?

Ⓖ **17.** Which case of shampoo is the better buy: a $16 case that holds 12 1-liter bottles or an $18 case that holds 36 400-mL bottles?

Ⓖ **18.** James needs 3 m 80 cm of wood molding to frame a picture. The price is $5.89 per meter plus a 7% sales tax. How much will James pay?

Review and Prepare

Multiply. (For help, see **Section 5.4.***)*

19. $\begin{array}{r} 0.035 \\ \times\ 18 \\ \hline \end{array}$

20. $\begin{array}{r} 28.35 \\ \times\ 12 \\ \hline \end{array}$

21. (6.3)(0.91)

22. (14.7)(2.2)

8.5 Metric–English Conversions and Temperature Formulas

OBJECTIVE 1 Until the United States has switched completely from the English system to the metric system, it will be necessary to make conversions from one system to the other. *Approximate* conversions can be made with the help of the following table, in which the values have been rounded to the nearest hundredth or thousandth.

Metric to English		*English to Metric*	
1 kilometer	≈ 0.62 mile	1 mile	≈ 1.61 kilometers
1 meter	≈ 1.09 yards	1 yard	≈ 0.91 meter
1 meter	≈ 3.28 feet	1 foot	≈ 0.30 meter
1 centimeter	≈ 0.39 inch	1 inch	≈ 2.54 centimeters
1 liter	≈ 0.26 gallon	1 gallon	≈ 3.78 liters
1 liter	≈ 1.06 quarts	1 quart	≈ 0.95 liter
1 kilogram	≈ 2.20 pounds	1 pound	≈ 0.45 kilogram
1 gram	≈ 0.035 ounce	1 ounce	≈ 28.35 grams

EXAMPLE 1 Converting Metric and English (Length)

Convert using unit fractions. Round your answers to the nearest tenth, if necessary.

10 meters to yards

We're changing from a metric unit to an English unit. In the "Metric to English" part of the table, you can see that 1 meter ≈ 1.09 yards. Two unit fractions can be written using that information:

$$\frac{1 \text{ meter}}{1.09 \text{ yards}} \quad \text{or} \quad \frac{1.09 \text{ yards}}{1 \text{ meter}}.$$

Multiply by the unit fraction that allows you to divide out meters (that is, meters is in the denominator).

$$10 \textbf{ meters} \cdot \frac{1.09 \text{ yards}}{1 \textbf{ meter}} = \frac{10 \cancel{\text{meters}}}{1} \cdot \frac{1.09 \text{ yards}}{1 \cancel{\text{meter}}} = 10.9 \text{ yards}$$

These units should match.

10 meters ≈ 10.9 yards

Note

You could also use the other numbers from the table involving meters and yards: 1 yard ≈ 0.91 meter.

$$\frac{10 \cancel{\text{meters}}}{1} \cdot \frac{1 \text{ yard}}{0.91 \cancel{\text{meter}}} = \frac{10}{0.91} \text{ yards} \approx 10.99 \text{ yards}$$

The answer is slightly different because all the values in the table are approximate. Also, you have to divide instead of multiply, which is usually more difficult to do without a calculator. We will use the first method in this chapter.

WORK PROBLEM 1 AT THE SIDE.

OBJECTIVES

1. Use unit fractions to convert from metric to English or English to metric units.
2. Know common temperatures on the Celsius scale.
3. Convert temperatures by using the order of operations.

FOR EXTRA HELP

Tutorial | Tape 15 | SSM, Sec. 8.5

1. Convert using unit fractions. Round your answers to the nearest tenth.

(a) 23 meters to yards

(b) 40 centimeters to inches

(c) 5 miles to kilometers (Look at the "English to Metric" side of the table.)

(d) 12 inches to centimeters

ANSWERS

1. (a) ≈25.1 yd
(b) ≈15.6 in.
(c) ≈8.1 km
(d) ≈30.5 cm

2. Convert. Use the values from the table on the previous page to make unit fractions. Round answers to the nearest tenth.

(a) 17 kilograms to pounds

(b) 5 liters to quarts

(c) 90 grams to ounces

(d) 3.5 gallons to liters

(e) 145 pounds to kilograms

(f) 8 ounces to grams

ANSWERS

2. (a) ≈37.4 pounds (b) ≈5.3 quarts (c) ≈3.2 ounces (d) ≈13.2 L (e) ≈65.3 kg (f) ≈226.8 g

EXAMPLE 2 Converting Metric and English (Weight and Capacity)

Convert, using unit fractions. Round your answers to the nearest tenth.

(a) 3.5 kilograms to pounds

Look at the "Metric to English" side of the table on the previous page to see that 1 kilogram ≈ 2.20 pounds. Use this information to write a unit fraction that allows you to divide out kilograms.

$$\frac{3.5 \text{ \sout{kilograms}}}{1} \cdot \frac{2.20 \text{ pounds}}{1 \text{ \sout{kilogram}}} = \frac{3.5 \cdot 2.20 \text{ pounds}}{1} = 7.7 \text{ pounds}$$

3.5 kilograms ≈ 7.7 pounds

(b) 18 gallons to liters

Look at the "English to Metric" side of the same table to see that 1 gallon ≈ 3.78 liters. Write a unit fraction that will allow you to divide out gallons.

$$\frac{18 \text{ \sout{gallons}}}{1} \cdot \frac{3.78 \text{ liters}}{1 \text{ \sout{gallon}}} = \frac{18 \cdot 3.78 \text{ liters}}{1} = 68.04 \text{ liters}$$

68.04 rounded to the nearest tenth is 68.0 so 18 gallons ≈ 68.0 liters.

(c) 300 grams to ounces

In the "Metric to English" side of the same table, 1 gram ≈ 0.035 ounce.

$$\frac{300 \text{ \sout{grams}}}{1} \cdot \frac{0.035 \text{ ounce}}{1 \text{ \sout{gram}}} = 10.5 \text{ ounces}$$

300 grams ≈ 10.5 ounces

Note

Because the metric and English systems were developed independently, there are no exact comparisons. Your answers should be written with the "≈" symbol to show they are approximate.

◀◀ WORK PROBLEM 2 AT THE SIDE.

OBJECTIVE 2 In the metric system, temperature is measured on the **Celsius** (SELL-see-us) **scale.** On the Celsius scale, water freezes at 0 °C and boils at 100 °C. The small raised circle stands for "degrees," and the capital **C** is for Celsius. Read the temperatures like this:

Water freezes at 0 degrees Celsius (0 °C).

Water boils at 100 degrees Celsius (100 °C).

The English temperature system that we now use is measured on the **Fahrenheit** (FAIR-en-hite) **scale.** On this scale:

Water freezes at 32 degrees Fahrenheit (32 °F).

Water boils at 212 degrees Fahrenheit (212 °F).

The thermometer below shows some typical temperatures in both Celsius and Fahrenheit. For example, comfortable room temperature is about 20 °C or 68 °F, and normal body temperature is about 37 °C or 98.6 °F.

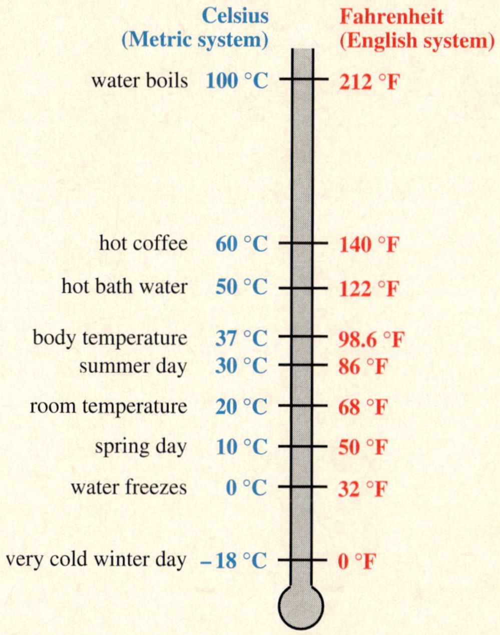

Note

The freezing and boiling temperatures are exact. The other temperatures are approximate. Even normal body temperature varies slightly from person to person.

EXAMPLE 3 Using Celsius Temperatures

Circle the Celsius temperature that is most reasonable for each situation.

(a) warm summer day   29 °C   64 °C   90 °C

29 °C is reasonable. 64 °C and 90 °C are too hot; they're both above the temperature of hot bath water (above 122 °F).

(b) inside a freezer   −10 °C   3 °C   25 °C

−10 °C is the reasonable temperature because it is the only one below the freezing point of water (0 °C). Your frozen foods would start thawing at 3 °C or 25 °C.

WORK PROBLEM 3 AT THE SIDE.

OBJECTIVE 3 You can use these formulas to convert temperatures.

Celsius–Fahrenheit Conversion Formulas

Converting from Fahrenheit (F) to Celsius (C)	Converting from Celsius (C) to Fahrenheit (F)
$C = \dfrac{5(F - 32)}{9}$	$F = \dfrac{9C}{5} + 32$

3. Circle the Celsius temperature that is most reasonable for each situation.

(a) Set the living room thermostat at:
11 °C   21 °C   71 °C

(b) The baby has a fever of:
29 °C   39 °C   49 °C

(c) Wear a sweater outside because it's:
15 °C   25 °C   50 °C

(d) My iced tea is:
−5 °C   5 °C   30 °C

(e) Time to go swimming! It's:
95 °C   65 °C   35 °C

(f) Inside a refrigerator (not the freezer) it's:
−15 °C   0 °C   3 °C

(g) There's a blizzard outside. It's:
10 °C   0 °C   −20 °C

(h) I need hot water to get these clothes clean. It should be:
55 °C   105 °C   200 °C

ANSWERS

3. (a) 21 °C **(b)** 39 °C **(c)** 15 °C **(d)** 5 °C **(e)** 35 °C **(f)** 3 °C **(g)** −20 °C **(h)** 55 °C

4. Convert to Celsius. Round your answers to the nearest degree, if necessary.

(a) 72 °F

(b) 20 °F

(c) 212 °F

(d) 98.6 °F

5. Convert to Fahrenheit. Round your answers to the nearest degree, if necessary.

(a) 100 °C

(b) −25 °C

(c) 32 °C

(d) 5 °C

Answers

4. (a) ≈22 °C **(b)** ≈−7 °C **(c)** 100 °C **(d)** 37 °C

5. (a) 212 °F **(b)** −13 °F **(c)** ≈90 °F **(d)** 41 °F

As you use these formulas, be sure to follow the order of operations.

1. Work inside parentheses or other grouping symbols.
2. Simplify expressions with exponents.
3. Do the remaining multiplications and divisions as they occur from left to right.
4. Do the remaining additions and subtractions as they occur from left to right.

EXAMPLE 4 Converting Fahrenheit to Celsius

Convert 10 °F to Celsius. Round your answer to the nearest degree.

Use the formula and the order of operations.

$$C = \frac{5(F - 32)}{9}$$ Replace F with 10.

$$= \frac{5(10 - 32)}{9}$$ Work inside parentheses first. 10 − 32 becomes 10 + (−32).

$$= \frac{5(-22)}{9}$$ Multiply; positive times negative gives a negative product.

$$= \frac{-110}{9}$$ Divide; negative divided by positive gives a negative quotient.

$$= -12.\overline{2}$$ Round to −12.

So 10 °C ≈ −12 °F.

◀◀ WORK PROBLEM 4 AT THE SIDE.

EXAMPLE 5 Converting Celsius to Fahrenheit

Convert 15 °C to Fahrenheit.

Use the formula and the order of operations.

$$F = \frac{9C}{5} + 32$$ Replace C with 15.

$$= \frac{9 \cdot 15}{5} + 32$$

$$= \frac{9 \cdot 3 \cdot \overset{1}{\cancel{5}}}{\underset{1}{\cancel{5}}} + 32$$ Divide out common factors if possible. Multiply.

$$= 27 + 32$$ Add.

$$= 59$$

Thus 15 °C = 59 °F.

◀◀ WORK PROBLEM 5 AT THE SIDE.

8.5 Exercises

Use the table on the first page of this section and unit fractions to make approximate conversions from metric to English or English to metric. Round your answers to the nearest tenth.

Example: 36 meters to yards

Solution: In the "Metric to English" part of the table, 1 meter ≈ 1.09 yards.

$$\frac{36 \cancel{\text{meters}}}{1} \cdot \frac{1.09 \text{ yards}}{1 \cancel{\text{meter}}} = \frac{36 \cdot 1.09 \text{ yards}}{1} = 39.24 \text{ yards}$$

39.24 rounds to 39.2 so **36 meters ≈ 39.2 yards**

1. 20 meters to yards

2. 8 kilometers to miles

3. 80 meters to feet

4. 85 centimeters to inches

5. 16 feet to meters

6. 3.2 yards to meters

7. 150 grams to ounces

8. 2.5 ounces to grams

9. 248 pounds to kilograms

10. 7.68 kilograms to pounds

11. 28.6 liters to quarts

12. 15.75 liters to gallons

13. Manuela's new sports car has a 16-gallon gas tank. How many liters does the tank hold?

14. Jamal's foot is 11 inches long. How long is it in centimeters?

Circle the more reasonable temperature for each of the following.

15. A snowy day

28 °C 28 °F

16. Brewing coffee

80 °C 80 °F

17. A high fever

40 °C 40 °F

18. Swimming pool water

78 °C 78 °F

19. Oven temperature

150 °C 150 °F

20. Light jacket weather

10 °C 10 °F

21. Would a drop of 20 Celsius degrees be more or less than a drop of 20 Fahrenheit degrees? Explain your answer.

22. Describe one advantage of switching from the Fahrenheit temperature scale to the Celsius scale. Describe one disadvantage.

Use the temperature conversion formulas and the order of operations to convert Fahrenheit temperatures to Celsius and Celsius temperatures to Fahrenheit. Round your answers to the nearest degree, if necessary.

23. 60 °F

24. 80 °F

25. -4 °F

26. 15 °F

27. 8 °C

28. 18 °C

29. -5 °C

30. 0 °C

Solve the following application problems. Round your answers to the nearest degree, if necessary.

31. The highest temperature ever recorded on earth was 136 °F at Aziza, Libya. The lowest was -129 °F in Antarctica. Convert these temperatures to Celsius.

32. The greatest temperature change during a 24-hour period in the United States occurred in Billings, Montana. The temperature fell from 8 °C to -48 °C. Convert these temperatures to Fahrenheit.

33. Here is the tag on a pair of boots. In what kind of weather would you wear these boots?

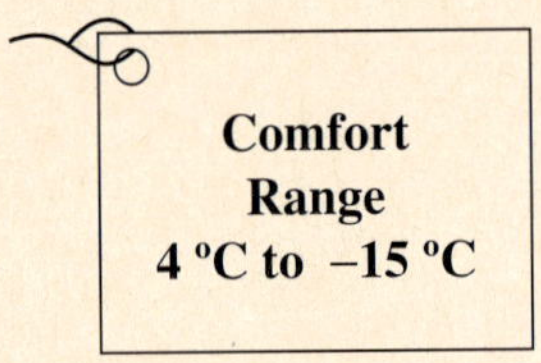

For what Fahrenheit temperatures are the boots designed?

What range of metric temperatures would you have in January where you live?

34. Explain what the picture directions on this tea bag package are telling you.

What Fahrenheit temperature would give the same result?

How many seconds should the tea bag be left in the water?

35. Paint sells for \$9.20 per gallon. Find the cost of 4 liters.

36. A 3-liter bottle of beverage sells for \$2.80. A 1-gallon bottle of the same beverage sells for \$3.50. What is the better value?

Review and Prepare

Simplify. (For help, see ***Section 1.8.****)*

37. $2(-8) + 2(-8)$

38. $2(12.2) + 2(5.6)$

39. $(-9)^2$

40. 2^7

41. $(5^2) + (-4)^3$

42. $(-12)^2 + (-3)^3$

KEY TERMS

8.1	**English system**	The English system of measurement (American system of units) is the system used for many daily activities in the United States. Common units in this system include quarts, pounds, feet, miles, and degrees Fahrenheit.
	metric system	The metric system of measurement is an international system of measurement used in manufacturing, science, medicine, sports, and other fields. The system uses meters, liters, grams, and degrees Celsius.
	unit fraction	A unit fraction involves measurement units and is equivalent to 1. Unit fractions are used to convert among different measurements.
8.2	**meter**	The meter is the basic unit of length in the metric system. The symbol **m** is used for meter. One meter is a little longer than a yard.
	prefixes	Attaching a prefix to meter, liter, or gram produces larger or smaller units. For example, the prefix *kilo* means 1000, so a *kilo*meter is 1000 meters.
	metric conversion line	The metric conversion line is a line showing the various metric measurement prefixes and their size relationship to each other.
8.3	**liter**	The liter is the basic unit of capacity in the metric system. The symbol **L** is used for liter. One liter is a little more than one quart.
	gram	The gram is the basic unit of weight (mass) in the metric system. The symbol **g** is used for gram. One gram is the weight of 1 milliliter of water or one dollar bill.
8.5	**Celsius**	The Celsius scale is the scale used to measure temperature in the metric system. Water boils at 100 °C and freezes at 0 °C.
	Fahrenheit	The Fahrenheit scale is the scale used to measure temperature in the English system. Water boils at 212 °F and freezes at 32 °F.

KEY FORMULAS

Converting from Celsius to Fahrenheit: $F = \frac{9C}{5} + 32$

Converting from Fahrenheit to Celsius: $C = \frac{5(F - 32)}{9}$

QUICK REVIEW

Concepts	*Examples*
8.1 The English System of Measurement Memorize the basic measurement relationships. Then, to convert units, multiply when changing from a larger unit to a smaller unit; divide when changing from a smaller unit to a larger unit.	Convert each measurement. **(a)** 5 feet to inches 5 feet $= 5 \cdot \mathbf{12} = 60$ inches **(b)** 3 pounds to ounces 3 pounds $= 3 \cdot \mathbf{16} = 48$ ounces **(c)** 15 quarts to gallons 15 quarts $= \frac{15}{\mathbf{4}} = 3\frac{3}{4}$ gallons

Concepts	*Examples*
8.1 Using Unit Fractions Another, more useful, conversion method is multiplying by a unit fraction. The unit you want in the answer should be in the numerator. The unit you want to change should be in the denominator. Then divide out the common units and any common factors.	Convert 32 ounces to pounds. $\frac{32 \text{ ounces}}{1} \cdot \underbrace{\frac{1 \text{ pound}}{16 \text{ ounces}}}_{\text{Unit fraction}}$ $= \frac{2 \cdot \overset{1}{\cancel{16}} \cancel{\text{ounces}}}{1} \cdot \frac{1 \text{ pound}}{\underset{1}{\cancel{16}} \cancel{\text{ounces}}}$ $= 2$ pounds
8.2 Knowing Basic Metric Length Units Use approximate comparisons to judge which units are appropriate: 1 mm is the thickness of a dime. 1 cm is about $\frac{1}{2}$ inch. 1 m is a little more than 1 yard. 1 km is about 0.6 mile.	Write the most reasonable metric unit in each blank. Choose from km, m, cm, mm. The room is 6 __m__ long. A paper clip is 30 __mm__ long. He drove 20 __km__ to work.
8.2 and 8.3 Converting Within the Metric System **Using Unit Fractions** One conversion method is to multiply by a unit fraction. Use a fraction with the unit you want in the answer in the numerator and the unit you want to change in the denominator.	Convert 9 g to kg $\frac{9 \cancel{g}}{1} \cdot \frac{1 \text{ kg}}{1000 \cancel{g}} = \frac{9}{1000} \text{ kg} = 0.009 \text{ kg}$ Convert 3.6 m to cm $\frac{3.6 \cancel{m}}{1} \cdot \frac{100 \text{ cm}}{1 \cancel{m}} = 360 \text{ cm}$
Using the Metric Conversion Line Another conversion method is to find the unit you are given on the metric conversion line. Count the number of places to get from the unit you are given to the unit you want. Move the decimal point the same number of places and in the same direction.	Convert each of the following. **(a)** 68.2 kg to g From kg to g is three places to the right. 6 8.2 0 0 — Decimal point is moved three places to the right. 68.2 kg = 68,200 g **(b)** 300 mL to L From mL to L is three places to the left. 3 0 0. — Decimal point is moved three places to the left. 300 mL = 0.3 L **(c)** 825 cm to m From cm to m is two places to the left. 8 2 5. — Decimal point is moved two places to the left. 825 cm = 8.25 m

Concepts	*Examples*
8.3 Knowing Basic Metric Capacity Units Use approximate comparisons to judge which units are appropriate: 1 L is a little more than 1 quart. 1 mL is the amount of water in a cube 1 cm on each side. 5 mL is about one teaspoon. 250 mL is about one cup.	Write the most appropriate metric unit in each blank. Choose from L or mL. The pail holds 12 <u>L</u>. The milk carton from the vending machine holds 250 <u>mL</u>.
8.3 Knowing Basic Metric Weight (Mass) Units Use approximate comparisons to judge which units are appropriate: 1 kg is about 2.2 pounds. 1 g is the weight of 1 mL of water or one dollar bill. 1 mg is $\frac{1}{1000}$ of a gram; very tiny!	Write the most appropriate metric unit in each blank. Choose from kg, g, and mg. The wrestler weighed 95 <u>kg</u>. She took a 500 <u>mg</u> aspirin tablet. One banana weighs 150 <u>g</u>.
8.4 Solving Metric Application Problems Convert units so you are comparing kg to kg, cm to cm, and so on. When a measurement involves two units, such as 6 m 20 cm, write it in terms of the unit called for in the answer (6.2 m or 620 cm).	**(a)** Grapes are \$3.95 per kg. How much will 400 g of grapes cost? 400 g = 0.4 kg $\quad \frac{0.4\ \cancel{kg}}{1} \cdot \frac{\$3.95}{1\ \cancel{kg}} = \$1.58$ **(b)** How many meters are left if 1 m 35 cm is cut off a board measuring 3 m? 1 m → 1.00 m plus 35 cm → + 0.35 m 1.35 m 3.00 m ← Board − 1.35 m ← Cut off 1.65 m ← Left
8.5 Converting from Metric to English and English to Metric Use the values in the table of conversion factors to write a unit fraction. Because the values in the table are rounded, your answers will be approximate and should be written with the ≈ symbol.	Convert. Round answers to the nearest tenth. **(a)** 23 meters to yards From the table, 1 meter ≈ 1.09 yards. $\frac{23\ \cancel{\text{meters}}}{1} \cdot \frac{1.09\ \text{yards}}{1\ \cancel{\text{meter}}} = 25.07\ \text{yards}$ 25.07 rounds to 25.1 so 23 meters ≈ 25.1 yards **(b)** 4 ounces to grams $\frac{4\ \cancel{\text{ounces}}}{1} \cdot \frac{28.35\ \text{grams}}{1\ \cancel{\text{ounce}}} = 113.4\ \text{grams}$ So 4 ounces ≈ 113.4 grams.

Concepts	*Examples*
8.5 Knowing Common Celsius Temperatures	
Use approximate and exact comparisons to judge which temperatures are appropriate.	Circle the Celsius temperature that is most reasonable.
Exact comparisons: 0 °C is freezing point (32 °F) 100 °C is boiling point (212 °F)	**(a)** Hot summer day: (35 °C) 90 °C 110 °C **(b)** The first snowy day in winter. −20 °C (0 °C) 15 °C
Approximate comparisons: 10 °C for a spring day (50 °F) 20 °C for room temperature (68 °F) 30 °C for summer day (86 °F) 37 °C for body temperature (98.6 °F)	
8.5 Converting between Fahrenheit and Celsius Temperatures	
Use these formulas. $C = \dfrac{5(F - 32)}{9}$	Convert 100 °F to Celsius. Round your answer to the nearest degree, if necessary. $C = \dfrac{5(100 - 32)}{9} = \dfrac{5(68)}{9} = \dfrac{340}{9} = 37.\overline{7}$ Round to 38. 100 °F ≈ 38 °C
$F = \dfrac{9C}{5} + 32$	Convert −8 °C to Fahrenheit. Round your answer to the nearest degree, if necessary. $F = \dfrac{9(-8)}{5} + 32 = \dfrac{-72}{5} + 32 = -14.4 + 32 = 17.6$ Round to 18. −8 °C ≈ 18 °F

Name ______ Date ______ Class ______

CHAPTER 8 REVIEW EXERCISES

[8.1] *Fill in the blanks with the measurement relationships you have memorized.*

1. 1 pound = ______ ounces

2. ______ feet = 1 yard

3. 1 ton = ______ pounds

4. ______ hours = 1 day

5. 1 hour = ______ minutes

6. 1 cup = ______ fluid ounces

7. ______ quarts = 1 gallon

8. ______ feet = 1 mile

9. ______ inches = 1 foot

10. 1 week = ______ days

11. ______ seconds = 1 minute

12. 1 pint = ______ cups

Convert using unit fractions.

13. 4 feet = ______ inches

14. 15 yards = ______ feet

15. 64 ounces = ______ pounds

16. 6000 pounds = ______ tons

17. 150 minutes = ______ hours

18. 11 cups = ______ pints

19. 18 hours = ______ day

20. 9 quarts = ______ gallons

21. $6\frac{1}{2}$ feet = ______ inches

22. $1\frac{3}{4}$ pounds = ______ ounces

23. 7 gallons = ______ cups

24. 4 days = ______ seconds

[8.2] *Write the most reasonable metric length unit in each blank. Choose from km, m, cm, mm.*

25. My thumb is 20 ______ wide.

26. Her waist measurement is 66 ______.

27. The two towns are 40 ______ apart.

28. A basketball court is 30 ______ long.

29. The height of the picnic bench is 45 ______.

30. The eraser on the end of my pencil is 5 ______ long.

Convert using unit fractions or the metric conversion line.

31. 5 m to cm

32. 8.5 km to m

33. 85 mm to cm

34. 370 cm to m

35. 70 m to km

36. 0.93 m to mm

[8.3] *Write the most reasonable metric unit in each blank. Choose from L, mL, kg, g, and mg.*

37. The eyedropper holds 1 ______.

38. I can heat 3 ______ of water in this pan.

39. Loretta's hammer weighed 650 ______.

40. Yongshu's suitcase weighed 20 ______ when it was packed.

41. My fish tank holds 80 ______ of water.

42. I'll buy the 500 ______ bottle of mouthwash.

43. Mara took a 200 ______ antibiotic pill.

44. This piece of chicken weighs 100 ______.

Convert using unit fractions or the metric conversion line.

45. 5000 mL to L

46. 8 L to mL

47. 4.58 g to mg

48. 0.7 kg to g

49. 6 mg to g

50. 35 mL to L

[8.4] *Solve each application problem.*

51. Each serving of punch at the wedding reception will be 180 mL. How many liters of punch are needed for 175 guests?

52. Jason is serving a 10 kg turkey to 28 people. How many grams of meat is he allowing for each person? Round to the nearest whole gram.

53. Yerald weighed 92 kg. Then he lost 4 kg 750 g. What is his weight now in kilograms?

54. Young-Mi bought 2 kg 20 g of onions. The price was $1.49 per kilogram. How much did she pay, to the nearest cent?

[8.5] *Use the table at the beginning of Section 8.5 and unit fractions to make approximate conversions. Round your answers to the nearest tenth, if necessary.*

55. 6 m to yards

56. 30 cm to inches

57. 108 km to miles

58. 800 miles to km

59. 23 quarts to L

60. 41.5 L to quarts

Write the appropriate metric (Celsius) temperature in each blank.

61. Water freezes at ______.

62. Water boils at ______.

63. Norm al body temperature is about ______.

64. Comfortable room temperature is about ______.

Use the temperature conversion formulas to convert each temperature to Fahrenheit or Celsius. Round to the nearest degree, if necessary.

65. 77 °F

66. 5 °F

67. −2 °C

68. 40 °C

Mixed Review Exercises

Write the most reasonable metric unit in each blank. Choose from km, m, cm, mm, L, mL, kg, g, and mg.

69. I added 1 ______ of oil to my car.

70. The box of books weighed 15 ______.

71. Larry's shoe is 30 ______ long.

72. Jan used 15 ______ of shampoo on her hair.

73. My fingernail is 10 ______ wide.

74. I walked 2 ______ to school.

75. The tiny bird weighed 15 ______.

76. The new library building is 18 ______ wide.

77. The cookie recipe uses 250 ______ of milk.

78. Renee's pet mouse weighs 30 ______.

79. One postage stamp weighs 90 ______.

80. I bought 30 ______ of gas for my car.

Convert the following using unit fractions, the metric conversion line, or the temperature conversion formulas.

81. 10.5 cm to mm

82. 45 minutes to hours

83. 90 inches to feet

84. 1.3 m to cm

85. 25 °C to Fahrenheit

86. $3\frac{1}{2}$ gallons to quarts

87. 700 mg to g

88. 0.81 L to mL

89. 5 pounds to ounces

90. 60 kg to g

91. 1.8 L to mL

92. 30 °F to Celsius; round to the nearest degree

93. 0.36 m to cm

94. 55 mL to L

Solve the following application problems.

95. Peggy had a board measuring 2 m 4 cm. She cut off 78 cm. How long is the board now, in meters?

96. Imported wool fabric is $12.99 per meter. What is the cost, to the nearest cent, of a piece that measures 3 m 70 cm?

97. Olivia is sending a recipe to her mother in Mexico. Among other things, the recipe calls for 4 ounces of rice and a baking temperature of 350 °F. Convert these measurements to metric, rounding to the nearest gram and nearest degree.

98. While on vacation in Canada, Jalo became ill and went to a health clinic. They said he weighed 80.9 kilograms and was 1.83 meters tall. Find his weight in pounds and height in feet. Round to the nearest tenth.

Name ____ Date ____ Class ____

CHAPTER 8 TEST

Convert the following measurements.

1. 9 gallons = ______ quarts
2. 45 feet = ______ yards
3. 135 minutes = ______ hours
4. 9 inches = ______ foot
5. $3\frac{1}{2}$ pounds = ______ ounces
6. 5 days = ______ minutes

Write the most reasonable metric unit in each blank. Choose from km, m, cm, mm, L, mL, kg, g, and mg.

7. My husband weighs 75 ______.
8. I hiked 5 ______ this morning.
9. She bought 125 ______ of cough syrup.
10. This apple weighs 180 ______.
11. This page is 21 ______ wide.
12. My watch band is 10 ______ wide.
13. I bought 10 ______ of soda for the picnic.
14. The bracelet is 16 ______ long.

Convert these measurements.

15. 250 cm to meters
16. 4.6 km to meters
17. 5 mm to centimeters
18. 325 mg to grams
19. 16 L to milliliters
20. 0.4 kg to grams
21. 10.55 m to centimeters
22. 95 mL to liters

1. ______
2. ______
3. ______
4. ______
5. ______
6. ______
7. ______
8. ______
9. ______
10. ______
11. ______
12. ______
13. ______
14. ______
15. ______
16. ______
17. ______
18. ______
19. ______
20. ______
21. ______
22. ______

23. ______________________

23. Stan's cat weighed 3 kg 740 g. His dog weighed 10 kg 60 g. How much heavier is the dog in kilograms?

24. ______________________

24. Denise is making five matching pillows. She needs 1 m 35 cm of braid to trim each pillow. How many meters of braid should she buy?

Pick the metric temperature that is most appropriate in each situation.

25. ______________________

25. The water is almost boiling.
210 °C 155 °C 95 °C

26. ______________________

26. The tomato plants may freeze tonight.
30 °C 20 °C 0 °C

Use the table in **Section 8.5** *and unit fractions to convert the following measurements. Round your answers to the nearest tenth, if necessary.*

27. ______________________

28. ______________________

27. 6 feet to meters

28. 125 pounds to kilograms

29. ______________________

30. ______________________

29. 50 liters to gallons

30. 8.1 kilometers to miles

Use the conversion formulas to convert each temperature. Round your answers to the nearest degree, if necessary.

31. ______________________

32. ______________________

31. 74 °F to Celsius

32. −12 °C to Fahrenheit

33. ______________________

33. Describe two benefits the United States would achieve by switching entirely to the metric system.

34. ______________________

34. Give two reasons why you think the United States has resisted changing to the metric system.

Name Date Class

CUMULATIVE REVIEW EXERCISES CHAPTERS 1–8

1. Write these numbers in words.
(a) 603,005,040,000

(b) 9.040

2. Write these numbers in digits.
(a) eighty and eight hundredths

(b) two hundred million, sixty-five thousand, four

3. Round each number as indicated.
(a) 0.9802 to the nearest tenth
(b) 495 to the nearest ten
(c) 306,472,000 to the nearest million

4. Find the mean and median, to the nearest tenth, for this set of data on hours worked: 22, 18, 40, 18, 20, 21, 45, 25.

Simplify.

5. $(-2)^3 - 3^2$

6. $2\frac{2}{5} - \frac{3}{4}$

7. $\frac{-4}{0.16}$

8. $\frac{7}{6x^2} \bullet \frac{9x}{14y}$

9. $(-0.003)(-0.05)$

10. $0.083 - 42$

11. $\frac{3}{c} - \frac{5}{6}$

12. $-15 - 15$

13. $10 - 3(6 - 7)$

14. $\frac{-3(-4)}{27 - 3^3}$

15. $\frac{3}{5}$ of (-400)

16. $3\frac{7}{12} - 4$

17. $\frac{9y^2}{8x} \div \frac{y}{6x^2}$

18. $\frac{2}{3} + \frac{n}{m}$

19. $1\frac{1}{6} + 1\frac{2}{3}$

20. $\frac{\frac{14}{15}}{-6}$

21. $\frac{12 \div (2 - 5) + 12(-1)}{2^3 - (-4)^2}$

22. $(-0.8)^2 \div (0.8 - 1)$

23. $\left(-\frac{1}{3}\right)^2 - \frac{1}{4}\left(\frac{4}{9}\right)$

Writing Calculator Small Group

Evaluate each expression when $k = -6$, $m = -1$, and $n = 2$.

24. $-3k + 4$

25. $4k - 5n$

26. $3m^3n$

Simplify each expression.

27. $3p - 3p^2 - 4p$

28. $-5(x + 2) - 4$

29. $7(-2r^3)$

Solve each equation. Show the steps you used.

30. $2b + 2 = -5 + 5$

31. $12.92 - a = 4.87$

32. $7(t + 6) = 42$

33. $-5n = n - 12$

34. $2 = \frac{1}{4}w - 3$

35. $\frac{1.5}{45} = \frac{x}{12}$

36. $4y - 3 = 7y + 12$

37. $3(k - 6) - 4 = -2(k + 1)$

Translate each sentence into an equation and solve it.

38. If eight times a number is subtracted from eleven times the number, the result is -9. Find the number.

39. When twice a number is decreased by 8, the result is the number increased by 7. Find the number.

Solve each application problem using the six problem solving steps from Chapter 3.

40. The perimeter of a rectangle is 124 cm. The width is 25 cm. Find the length.

41. A 90 ft pipe is cut into two pieces so that one piece is 6 ft shorter than the other. Find the length of each piece.

Find the unknown length, perimeter, circumference, area, or volume. When necessary, use 3.14 as the approximate value of π and round answers to the nearest tenth.

42. Find the perimeter and area.

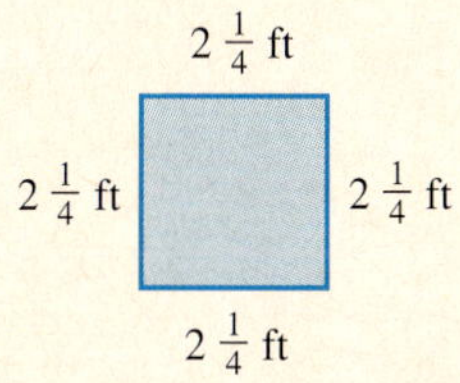

43. Find the circumference and the area.

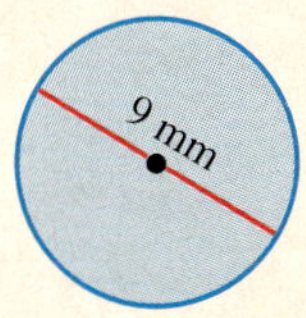

44. Find the perimeter and the area.

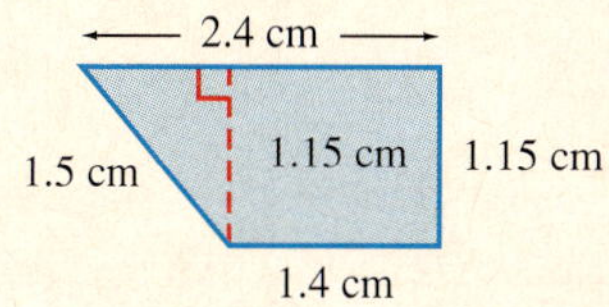

45. Find the unknown length.

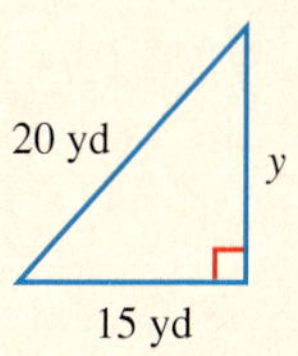

46. Find the volume.

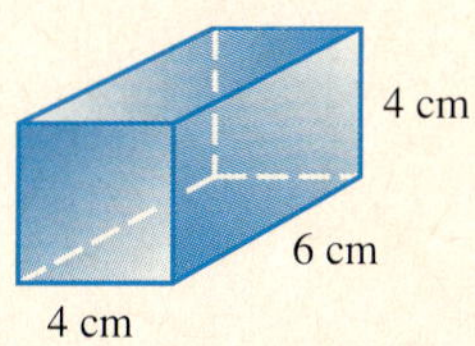

47. Find the unknown length in these similar triangles.

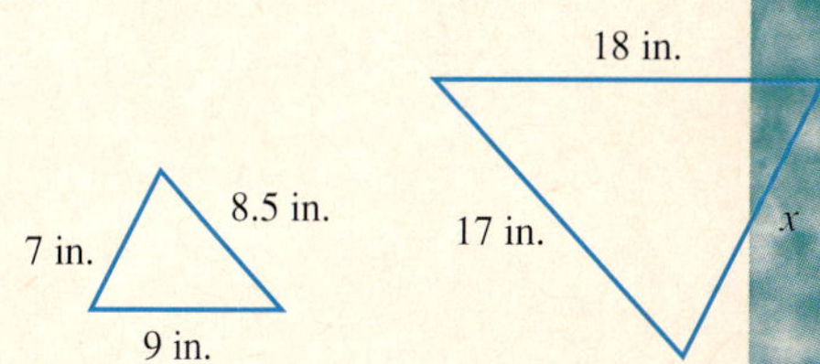

Convert the following using unit fractions, the metric conversion line, or the temperature conversion formulas.

48. $4\frac{1}{2}$ ft to inches

49. 72 hours to days

50. 3.7 kg to grams

51. 60 cm to meters

52. 7 mL to liters

53. -20 °C to Fahrenheit

Solve the following application problems.

54. Mei Ling must earn 90 credits to receive an associate of arts degree. She has 53 credits. What percent of the necessary credits does she have? Round to the nearest whole percent.

55. Which bag of chips is the best buy: Brand T is $15\frac{1}{2}$ ounces for \$2.99, Brand F is 14 ounces for \$2.49, and Brand H is 18 ounces for \$3.89. You have a coupon for 40¢ off Brand H and another for 30¢ off Brand T.

56. A coffee can has a diameter of 13 cm and a height of 17 cm. Find the volume of the can. Use 3.14 for π and round your answer to the nearest tenth.

57. Swimsuits are on sale in August at 65% off the regular price. How much will Lanece pay for a suit that has a regular price of \$44?

58. Steven bought $4\frac{1}{2}$ yards of canvas material to repair the tents used by the scout troop. He used $1\frac{2}{3}$ yards on one tent and $1\frac{3}{4}$ yards on another. How much material is left?

59. Mark bought 650 grams of maple sugar candy on his vacation in Montreal. The candy is priced at \$14.98 per kilogram. How much did Mark pay, to the nearest cent?

60. Bags of slivered almonds weigh 115 g each. They are packed in a carton that weighs 450 g. How many kilograms would a carton containing 48 bags weigh?

61. Akuba is knitting a scarf. Six rows of knitting result in 5 cm of scarf. At that rate, how many rows will she knit to make a 100 cm scarf?

62. A loan of \$3500 will be paid back with $7\frac{1}{2}$% interest at the end of 6 months. Find the total amount due.

63. A shelter for homeless people received a \$400 donation to buy blankets. If the blankets cost \$17 each, how many blankets can be purchased? How much money will be left over?

9 Graphs

9.1 *Reading and Constructing Circle Graphs*

Throughout this book you have used numbers, expressions, formulas, and equations to communicate rules or information. Graphs are another way to communicate. They are visual; that is, they show a *picture* of the information. This picture can often be understood faster and more easily than a formula or a list of numbers (data).

OBJECTIVE 1 The graphs described in this section and the next can help you make sense of a collection of data. The **circle graph** is used to show how a total amount is divided into parts. The following circle graph shows you how 24 hours in the life of a college student are divided among different activities.

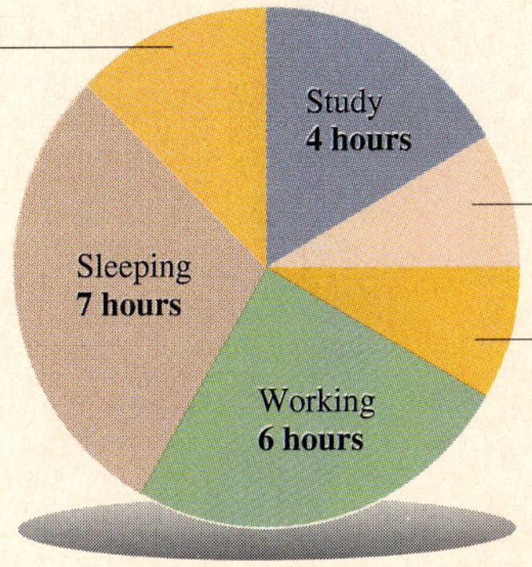

WORK PROBLEM 1 AT THE SIDE.

OBJECTIVE 2 This circle graph uses pie-shaped pieces called *sectors* to show the amount of time spent on each activity (the total must be 24 hours); a circle graph can therefore be used to compare the time spent on one activity to the total number of hours in the day.

EXAMPLE 1 Using a Circle Graph

Find the ratio of time spent in college classes to the total number of hours in a day. Write the ratio as a fraction in lowest terms. (See **Section 6.1**.)

The circle graph shows that 3 of the 24 hours in a day are spent in class. The ratio of class time to the hours in a day is

$$\frac{3 \text{ hours } \textbf{(college classes)}}{24 \text{ hours } \textbf{(whole day)}} = \frac{3 \text{ ~~hours~~}}{24 \text{ ~~hours~~}} = \frac{\overset{1}{\cancel{3}}}{\underset{1}{\cancel{3}} \cdot 8} = \frac{1}{8} \leftarrow \text{Lowest terms}$$

www.mathnotes.com

OBJECTIVES

1. Read and understand a circle graph.
2. Use a circle graph.
3. Draw a circle graph by using a protractor.

FOR EXTRA HELP

Tutorial | Tape 16 | SSM, Sec. 9.1

1. Use the circle graph to answer each of the following.

(a) The greatest number of hours is spent in which activity?

(b) How many more hours are spent working than studying?

(c) Find the total number of hours spent studying, working, and attending classes.

ANSWERS

1. (a) sleeping **(b)** 2 hours **(c)** 13 hours

2. Use the circle graph to find the following ratios. Write the ratios as fractions in lowest terms.

(a) hours spent driving to whole day

(b) hours spent studying to whole day

(c) hours spent sleeping and doing other to all day

(d) hours spent working and studying to whole day

3. Use the circle graph to find the following ratios. Write the ratios as fractions in lowest terms.

(a) hours spent studying to hours spent working

(b) hours spent working to hours spent sleeping

(c) hours spent studying to hours spent driving

(d) hours spent in class to hours spent for other

Answers

2. (a) $\frac{1}{12}$ (b) $\frac{1}{6}$ (c) $\frac{3}{8}$ (d) $\frac{5}{12}$

3. (a) $\frac{2}{3}$ (b) $\frac{6}{7}$ (c) $\frac{2}{1}$ (d) $\frac{3}{2}$

◀◀ WORK PROBLEM 2 AT THE SIDE.

This circle graph can also be used to find the ratio of the time spent on one activity to the time spent on any other activity.

EXAMPLE 2 Finding a Ratio from a Circle Graph

Find the ratio of working time to class time.

The circle graph shows 6 hours spent working and 3 hours spent in class. The ratio of working time to class time is

$$\frac{6 \text{ hours } \mathbf{(working)}}{3 \text{ hours } \mathbf{(class)}} = \frac{6 \text{ ~~hours~~}}{3 \text{ ~~hours~~}} = \frac{\overset{1}{\cancel{3}} \cdot 2}{\underset{1}{\cancel{3}}} = \frac{2}{1} \quad \text{Ratio in lowest terms}$$

◀◀ WORK PROBLEM 3 AT THE SIDE.

A circle graph often shows data as percents. For example, the recorded music industry had total annual sales of \$12 billion in 1995. The next circle graph shows how the sales were divided among the various types of recording formats. The entire circle represents the \$12 billion in sales. Each sector represents the sales of one format as a percent of the total sales. The sectors must add up to 100%.

SMALL PACKAGE—BIG SALES

Source: RIAA

EXAMPLE 3 Calculating Amounts by Using a Circle Graph

Use the circle graph on recorded music sales to find the amount spent on CD albums for the year.

Recall the percent equation:

$$\text{percent} \cdot \text{whole} = \text{part}$$

The percent for CD albums is 65%. Rewrite 65% as a decimal. The *whole* is the total sales of \$12 billion (the entire circle).

$$\begin{aligned} \text{percent} \cdot \text{whole} &= \text{part} \\ 65.\% \cdot \$12 \text{ billion} &= n \\ 0.65 \cdot \$12 \text{ billion} &= n \\ \$7.8 \text{ billion} &= n \end{aligned}$$

The amount spent on CD albums was \$7.8 billion, or \$7,800,000,000.

WORK PROBLEM 4 AT THE SIDE. ▶▶

OBJECTIVE 3 The coordinator of the Fair Oaks Youth Soccer League organizes teams in five age groups. She places the players in various age groups as follows.

Age Group	*Percent of Total*
Under 8 years old	20%
8–9 years old	15%
10–11 years old	25%
12–13 years old	25%
14–15 years old	15%
Total	100%

You can show these percents by using a circle graph. A circle has 360 degrees (written 360°). The 360° represents the entire league, or 100% of the soccer league.

EXAMPLE 4 Drawing a Circle Graph

Using the data on *age groups*, find the number of degrees in the sector that would represent the "Under 8" group, and begin constructing a circle graph.

A complete circle has 360°. Because the "Under 8" group makes up 20% of the total number of players, the number of degrees needed for the "Under 8" sector of the circle graph is 20% of 360°.

$$20.\% \text{ of } 360^\circ = n$$
$$0.20 \cdot 360^\circ = n$$
$$72^\circ = n$$

Use a tool called a **protractor** (PRO-trak-ter) to make a circle graph. First, using a ruler or straight edge, draw a line from the center of a circle to the left edge. Place the hole in the protractor over the center of the circle, making sure that zero on the protractor lines up with the line that was drawn. Find 72° and make a mark as shown in the illustration. Then remove the protractor and use the straight edge to draw a line from the center of the circle to the 72° mark at the edge of the circle.

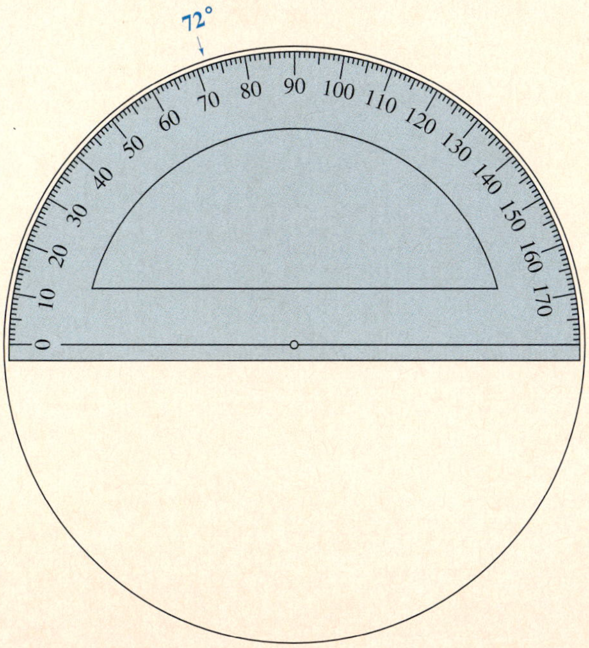

CONTINUED ON NEXT PAGE

4. Use the circle graph on recorded music sales to find the following.

(a) the amount spent on CD singles

(b) the amount spent on LP albums, 7–12 inch singles

(c) the amount spent on cassette albums

(d) the amount spent on cassette singles

ANSWERS

4. **(a)** \$0.36 billion, or \$360,000,000
(b) \$0.12 billion, or \$120,000,000
(c) \$3 billion, or \$3,000,000,000
(d) \$0.6 billion, or \$600,000,000

5. Using the information on the soccer age groups in the table, find the number of degrees needed for each of the following sectors. Then complete the circle graph at the bottom right on this page.

(a) "10–11 years" sector

(b) "12–13 years" sector

(c) "14–15 years" sector

To draw the "8–9 years" sector, begin by finding the number of degrees in the sector. The "8–9 years" group is 15% of the total circle.

$$15.\% \text{ of } 360° = n$$
$$0.15 \cdot 360° = n$$
$$54° = n$$

Again, place the hole of the protractor at the center of the circle, but this time align zero on the second line that was drawn. Make a mark at 54° and draw a line as before. This sector is 54° and represents the "8–9 years" group.

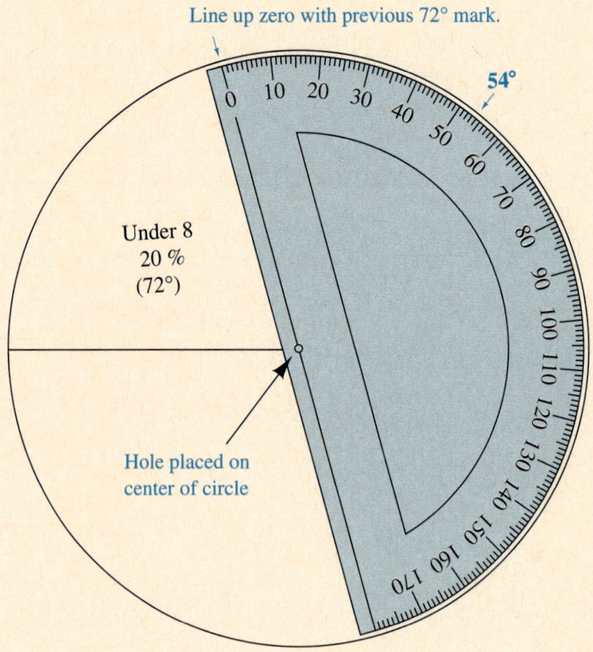

Note
You must be certain that the hole in the protractor is placed on the exact center of the circle each time you measure the size of a sector.

◀◀ **WORK PROBLEM 5 AT THE SIDE.**

Use this circle for Problem 5 at the side.

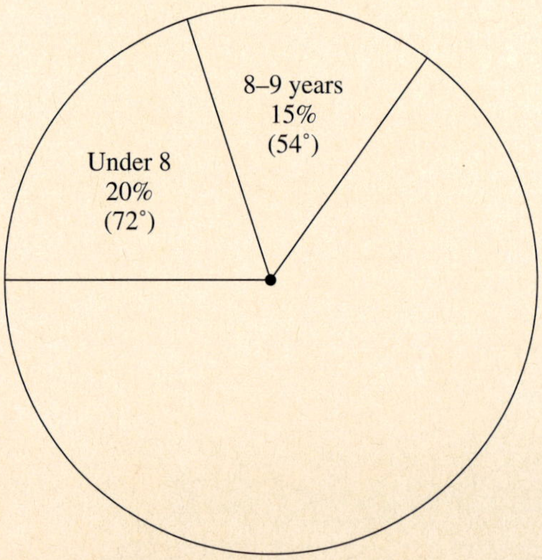

ANSWERS

5. **(a)** 90° **(b)** 90° **(c)** 54°

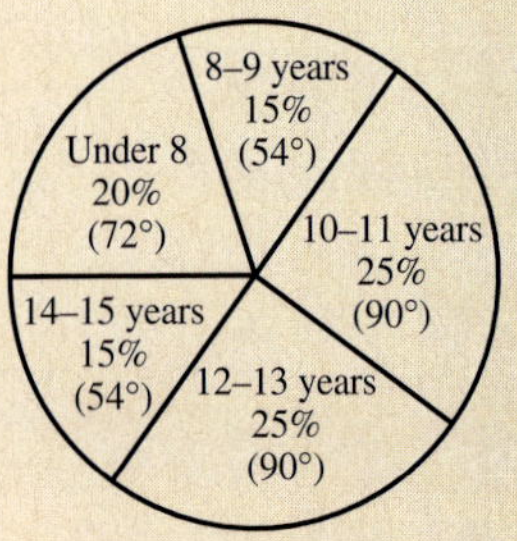

9.1 Exercises

Use the following circle graph for Exercises 1–6. The circle graph shows the cost of adding an art studio to an existing building. Write ratios as fractions in lowest terms.

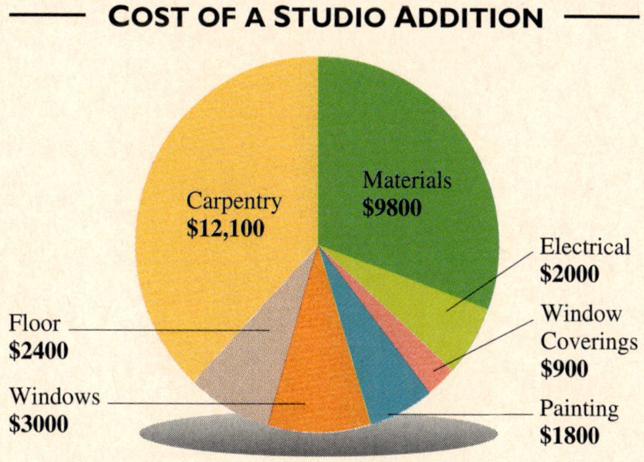

1. **(a)** Find the total cost of adding the art studio. **(b)** What is the largest single expense?

2. **(a)** What is the second largest expense in adding the studio? **(b)** What is the smallest expense?

3. **(a)** Find the ratio of the cost of materials to the total remodeling cost. **(b)** Find the ratio of the cost of windows to the cost of electrical.

4. **(a)** Find the ratio of the cost of painting to the total remodeling cost. **(b)** Find the ratio of the cost of windows to the cost of window coverings.

5. Find the ratio of the cost of carpentry, windows, and window coverings to the total remodeling cost.

6. Find the ratio of the cost of windows and electrical to the cost of the floor and painting.

7. The circle graph at the right shows the number of students at Rockfield College who are enrolled in various majors. **(a)** In which major are the fewest students enrolled? **(b)** In which major are the second fewest students enrolled?

8. Use the circle graph at the right. **(a)** Which major has the most students enrolled? **(b)** Which major has the second most students enrolled?

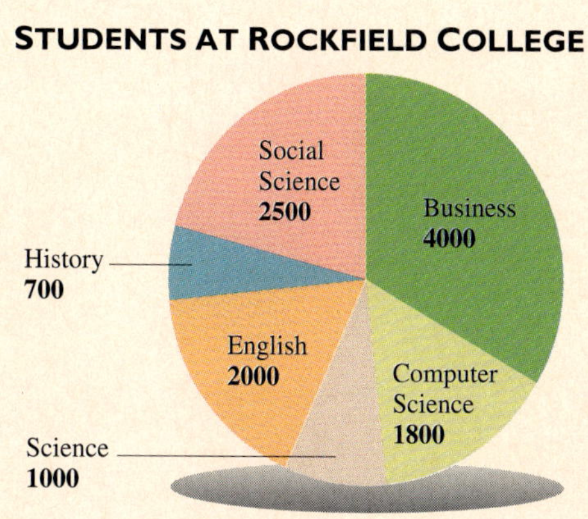

Use the circle graph from Exercises 7 and 8 to find each ratio. Write the ratios as fractions in lowest terms.

9. business majors to the total number of students

10. English majors to the total number of students

11. computer science majors to the number of English majors

12. history majors to the number of social science majors

13. business majors to the number of science majors

14. science majors to the number of history majors

The following circle graph shows the costs necessary to comply with the Americans with Disabilities Act (ADA) at the Dos Pueblos College. Each cost item is expressed as a percent of the total cost of $1,740,000. Use the graph to find the dollar amount spent for each of the following.

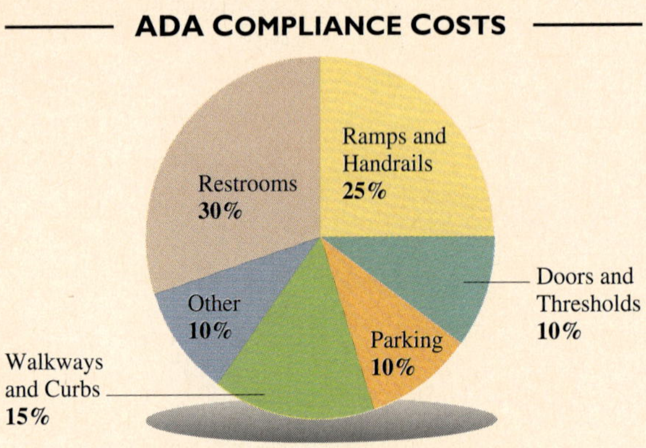

15. restrooms

16. ramps and handrails

17. doors and thresholds

18. parking

19. walkways and curbs

20. other

The circle graph below is adapted from USA TODAY *and shows the percent of employees who take time off from work during one year because of colds. If Folsum Electronics has 8740 employees, use the graph to find the number of employees who miss work for various lengths of time because of colds. Round to the nearest whole number.*

Source: Ketchum Public Relations for SmithKline Beecham

21. 1–2 days

22. don't know

23. 5–6 days

24. 3–4 days

25. 7 or more days

26. never take off for colds

27. Describe the procedure for determining how large each sector must be to represent each of the items in a circle graph.

28. A protractor is the tool used to draw a circle graph. Give a brief explanation of what the protractor does and how you would use it to measure and draw each sector in the circle graph.

During one semester Zoë Werner spent $4200 for school expenses as shown in the following chart. Find all numbers missing from the chart.

Item	*Dollar Amount*	*Percent of Total*	*Degrees of a Circle*
29. rent	$1050	25%	________
30. food	$840	________	72°
31. clothing	$420	________	________
32. books	$420	10%	________
33. entertainment	$630	________	54°
34. savings	$210	________	________
35. other	________	________	54°

36. Draw a circle graph by using the above information.

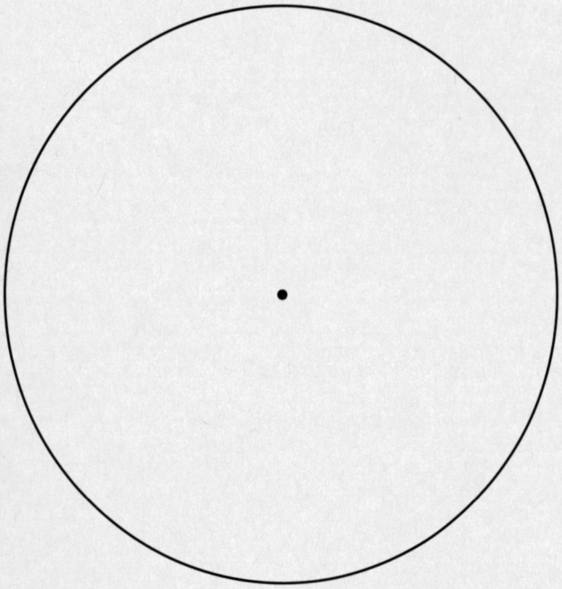

37. White Water Rafting Company divides its annual sales into five categories, as follows.

Category	*Annual Sales*
Adventure classes	\$12,500
Grocery and provision sales	\$40,000
Equipment rentals	\$60,000
Rafting tours	\$50,000
Equipment sales	\$37,500

(a) Find the total sales for the year.

(b) Find the number of degrees in a circle graph for each item.

(c) Draw a circle graph showing this information.

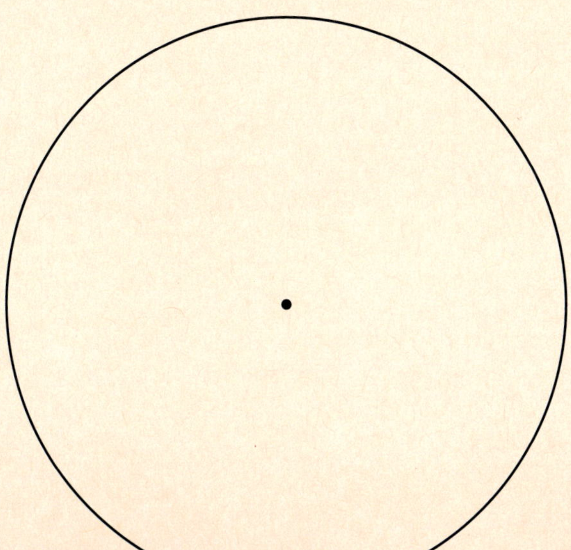

38. A book publisher had 25% of total sales in mysteries, 10% in biographies, 15% in cookbooks, 15% in romantic novels, 20% in science, and the rest in business books.

(a) Find the number of degrees in a circle graph for each type of book.

(b) Draw a circle graph, using the information given.

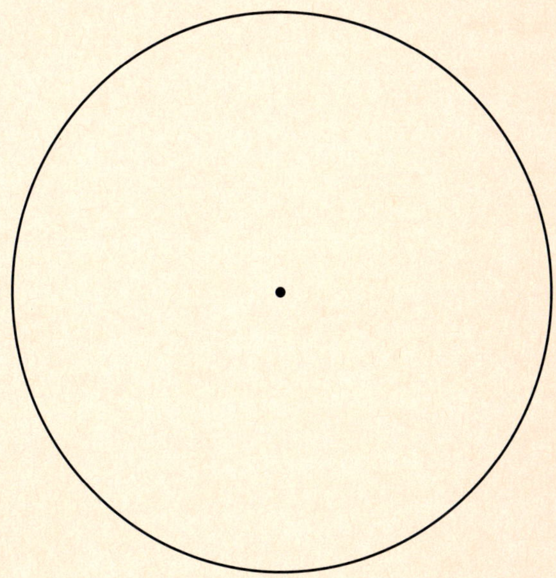

39. A family kept track of its expenses for a year and recorded the following results. Complete the chart and draw a circle graph.

Item	*Amount*	*Percent of Total*	*Number of Degrees*
Housing	\$9600	____	____
Food	\$6400	____	____
Automobile	\$4800	____	____
Clothing	\$3200	____	____
Medical	\$1600	____	____
Savings	\$1600	____	____
Other	\$4800	____	____
Total	______		

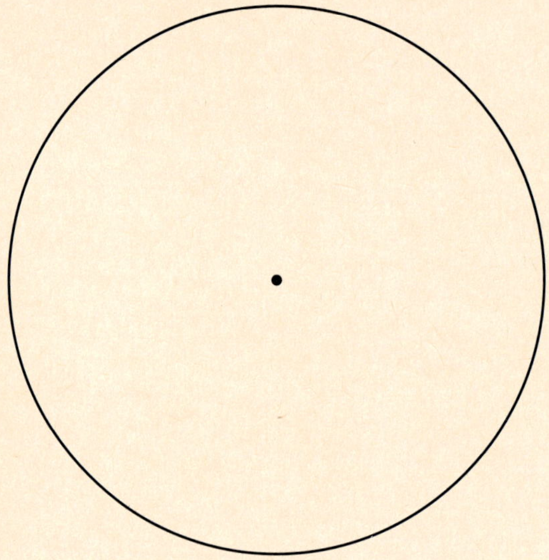

Review and Prepare

Write < or > to make a true statement. (For help, see ***Sections 1.2*** *and* ***5.6****.)*

40. $\frac{1}{4}$ ______ $\frac{3}{8}$

41. $\frac{4}{5}$ ______ $\frac{7}{8}$

42. 0.4219 ______ 0.422

43. 0.0118 ______ 0.01

44. 38.25% ______ 38.29%

45. 25.9% ______ 26.01%

46. 60,500 ______ 60,498

47. 44,272.68 ______ 44,272.098

48. 799,802 ______ 799,899

9.2 *Bar Graphs and Line Graphs*

OBJECTIVE 1 A **bar graph** is useful for showing comparisons. For example, the following bar graph compares the number of college graduates who continued taking advanced courses in their major field during each of five years.

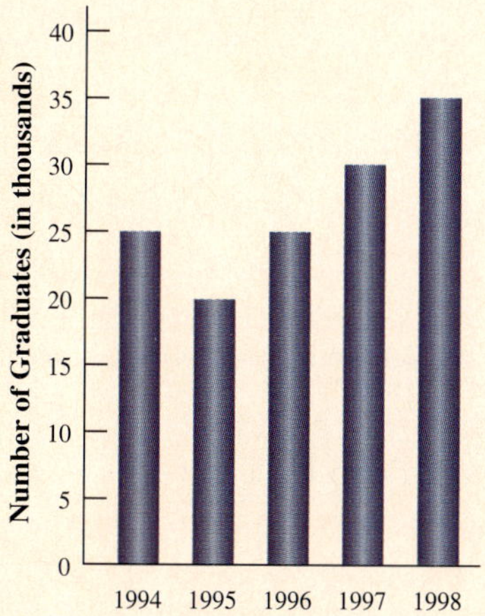

EXAMPLE 1 Using a Bar Graph

How many college graduates took advanced classes in their major field in 1996?

The bar for 1996 rises to 25. Notice the label along the left side of the graph that says "Number of Graduates (in thousands)." The phrase *in thousands* means you have to multiply 25 by 1000 to get 25,000. So, 25,000 (not 25) graduates took advanced classes in their major field in 1996.

WORK PROBLEM 1 AT THE SIDE.

OBJECTIVE 2 A **double-bar graph** can be used to compare two sets of data. This graph shows the number of new cable television installations each quarter for two different years.

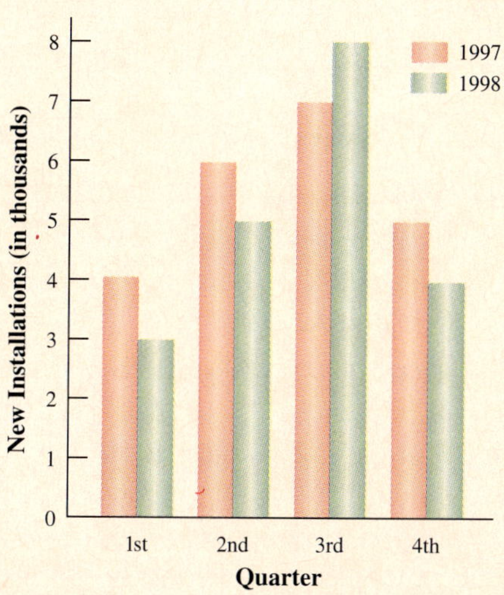

OBJECTIVES

Read and understand

1. a bar graph;
2. a double-bar graph;
3. a line graph;
4. a comparison line graph.

FOR EXTRA HELP

Tutorial — Tape 16 — SSM, Sec. 9.2

1. Use the bar graph in the text to find the number of college graduates who took advanced classes in their major field in each of these years.

(a) 1994

(b) 1995

(c) 1997

(d) 1998

ANSWERS

1. (a) 25,000 **(b)** 20,000 **(c)** 30,000 **(d)** 35,000

2. Use the double-bar graph to find the number of new cable television installations in 1997 and 1998 for each of the following quarters.

(a) 1st quarter

(b) 3rd quarter

(c) 4th quarter

(d) Find the greatest number of installations. Identify the quarter and the year in which they occurred.

EXAMPLE 2 Reading a Double-Bar Graph

Use the double-bar graph on the previous page to find each of the following.

(a) the number of new cable television installations in the second quarter of 1997

There are two bars for the second quarter. The color code in the upper right-hand corner of the graph tells you that the **red bars** represent 1997. So the **red bar** on the *left* is for the 2nd quarter of 1997. It rises to 6. Multiply 6 by 1000 because the label on the left side of the graph says *in thousands.* So there were 6000 new installations for the second quarter in 1997.

(b) the number of new cable television installations in the second quarter of 1998

The **green bar** for the second quarter rises to 5 and 5 times 1000 is 5000. So, in the second quarter of 1998, there were 5000 new installations.

Note

Use a ruler or straight edge to line up the top of the bar with the number on the left side of the graph.

WORK PROBLEM 2 AT THE SIDE.

3. Use the line graph at the right to find the number of trout stocked in each of the following months.

(a) June

(b) May

(c) April

(d) July

OBJECTIVE 3 A **line graph** is often useful for showing a trend. The line graph that follows shows the number of trout stocked along the Feather River over a five-month period. Each dot indicates the number of trout stocked during the month directly below that dot.

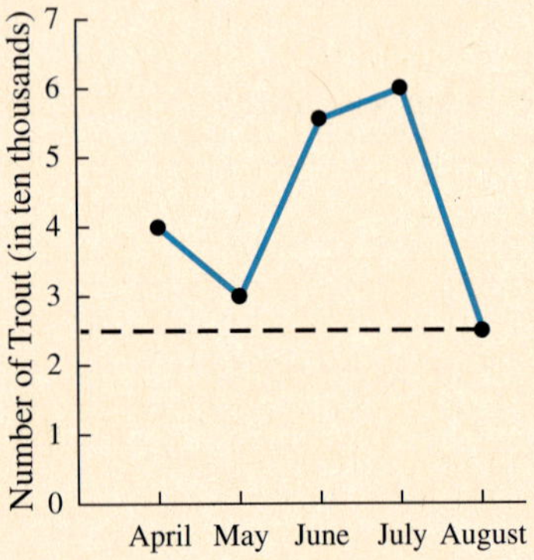

EXAMPLE 3 Understanding a Line Graph

Use the line graph to find the following.

(a) In which month were the least number of trout stocked?

The lowest point on the graph is the dot directly over August, so the least number of trout were stocked in August.

(b) How many trout were stocked in August?

Use a ruler or straight edge to line up the August dot with the numbers along the left edge of the graph. The August dot is halfway between the 2 and 3. Notice the label on the left side says "in ten thousands." So August is halfway between 2 • 10,000 and 3 • 10,000. It is halfway between 20,000 and 30,000. That means 25,000 trout were stocked in August.

WORK PROBLEM 3 AT THE SIDE.

ANSWERS

2. **(a)** 4000; 3000 **(b)** 7000; 8000 **(c)** 5000; 4000 **(d)** 8000; 3rd quarter of 1998

3. **(a)** 55,000 **(b)** 30,000 **(c)** 40,000 **(d)** 60,000

OBJECTIVE 4 You can also compare two sets of data by drawing two line graphs together as a **comparison line graph.** For example, the following line graph compares the number of thermal-paper fax machines and the number of plain-paper fax machines sold during each of five years.

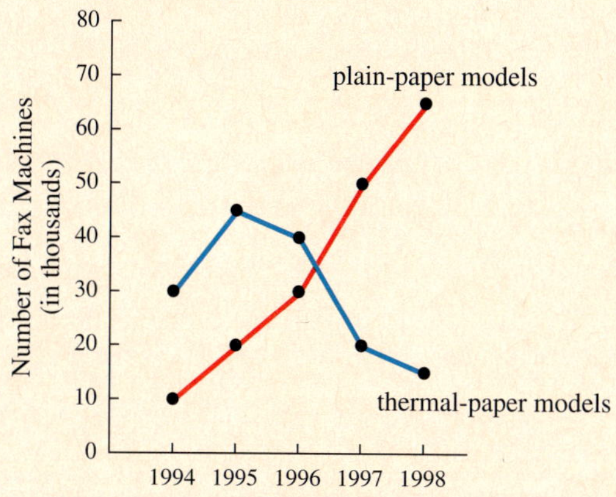

EXAMPLE 4 Interpreting a Comparison Line Graph

Use the comparison line graph above to find the following.

(a) the number of thermal-paper machines sold in 1995

Find the dot on the **blue line** above 1995. Use a ruler or straight edge to line up the dot with the numbers along the left edge. The dot is halfway between 40 and 50, which is 45. Then, 45 times 1000 is 45,000 thermal-paper machines that were sold in 1995.

(b) the number of plain-paper machines sold in 1998

The **red line** on the graph shows that 65,000 plain-paper machines were sold in 1998.

Note

Both the double-bar graph and the comparison line graph are used to compare two or more sets of data.

WORK PROBLEM 4 AT THE SIDE.

4. Use the comparison line graph at the left to find the following.

(a) the number of thermal-paper machines sold in 1994, 1996, 1997, and 1998

(b) the number of plain-paper machines sold in 1994, 1995, 1996, and 1997

(c) the first full year in which the number of plain-paper machines sold was greater than the number of thermal-paper machines sold

ANSWERS

4. **(a)** 30,000; 40,000; 20,000; 15,000
(b) 10,000; 20,000; 30,000; 50,000
(c) 1997

NUMBERS IN THE *Real World* *collaborative investigations*

1. The graph at the right is a double-bar graph. What is it about? Write several sentences describing the general purpose of the graph.

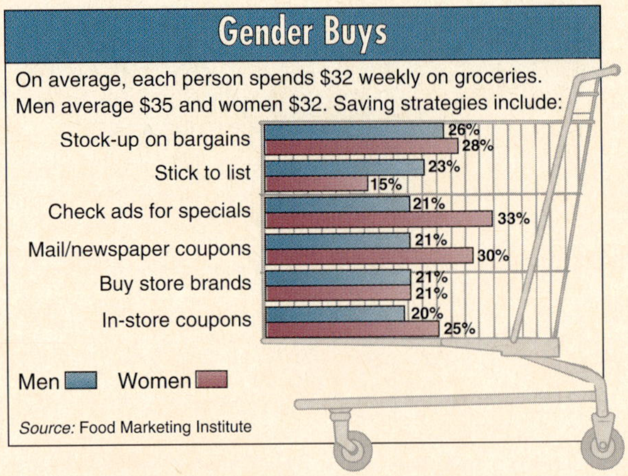

2. Which two savings strategies do women use most often? Which two do men use most often?

3. Which two strategies show the greatest difference in use between men and women? Which two strategies show the least difference?

4. Sometimes people "jump to conclusions" without enough evidence. Which of these conclusions are appropriate, *based on the information in the graph*?

(a) Men and women use a variety of savings strategies when grocery shopping.
(b) Men spend more for groceries than women.
(c) Men eat more groceries than women do.
(d) There are some differences in the grocery shopping strategies used by men and women.
(e) Women are better grocery shoppers than men.

5. Write two additional conclusions that you can make, based on the information in the graph. Share your conclusions with other class members. Be prepared to defend your conclusions.

6. Conduct a survey of your class members. Find out how many of them regularly use each of the savings strategies shown in the graph. Complete the table below.

Total number of women in survey ______			Total number of men in survey ______	
	Number of Women Using Strategy	Percent of Women Using Strategy	Number of Men Using Strategy	Percent of Men Using Strategy
Stock up on bargains				
Stick to list				
Check ads for specials				
Mail/Newspaper coupons				
Buy store Brands				
In-store coupons				

Now make a double-bar graph showing your survey data. How is your data similar to the graph above? How is it different?

9.2 Exercises

The National Insurance Crime Bureau says that "thieves are stealing fewer cars but more motorcycles." There has been a 46% increase in motorcycle thefts since 1992. The bar graph below shows the number of motorcycles stolen in a large eastern city over a five-year period.

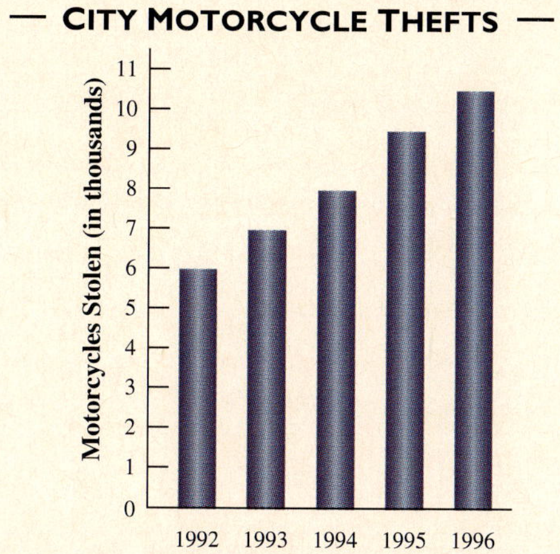

1. Find the number of motorcycles stolen in 1994.

2. Find the number of motorcycles stolen in 1993.

3. Which year had the greatest number of motorcycles stolen? How many were stolen that year?

4. Which year had the least number of motorcycles stolen? How many were stolen that year?

5. How many more motorcycles were stolen in 1995 than in 1993?

6. How many more motorcycles were stolen in 1996 than in 1992?

This double-bar graph shows the number of workers who were unemployed in a city during the first six months of 1997 and 1998.

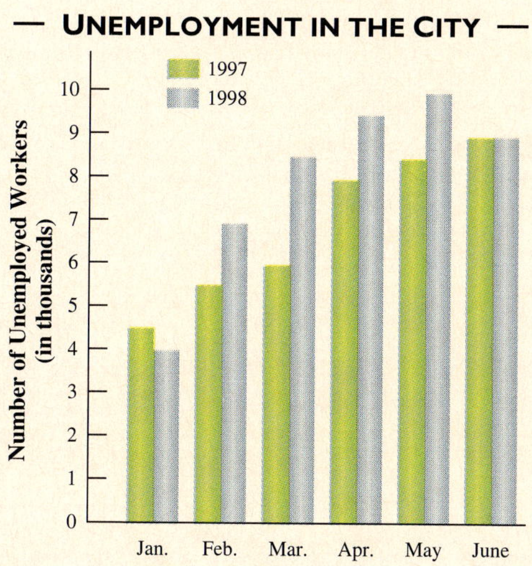

7. In which month in 1998 were the greatest number of workers unemployed? What was the total number unemployed in that month?

8. How many workers were unemployed in January of 1997?

9. How many more workers were unemployed in February of 1998 than in February of 1997?

10. How many fewer workers were unemployed in March of 1997 than in March of 1998?

11. Find the increase in the number of unemployed workers from February 1997 to April 1997.

12. Find the increase in the number of unemployed workers from January 1998 to June 1998.

This double-bar graph shows sales of super unleaded and supreme unleaded gasoline at a service station for each of five years.

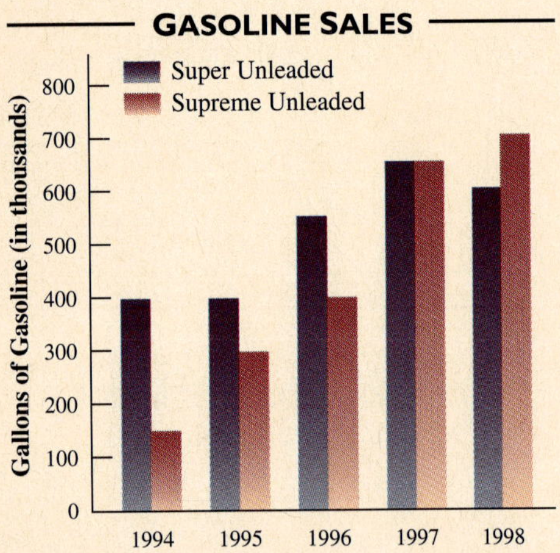

13. How many gallons of supreme unleaded gasoline were sold in 1994?

14. How many gallons of super unleaded gasoline were sold in 1997?

15. In which year did the greatest difference in sales between super unleaded and supreme unleaded gasoline occur? Find the difference.

16. In which year did the sales of supreme unleaded gasoline surpass the sales of super unleaded gasoline?

17. Find the increase in supreme unleaded gasoline sales from 1994 to 1998.

18. Find the increase in super unleaded gasoline sales from 1994 to 1998.

This line graph shows the number of burglaries in a community during the first six months of last year.

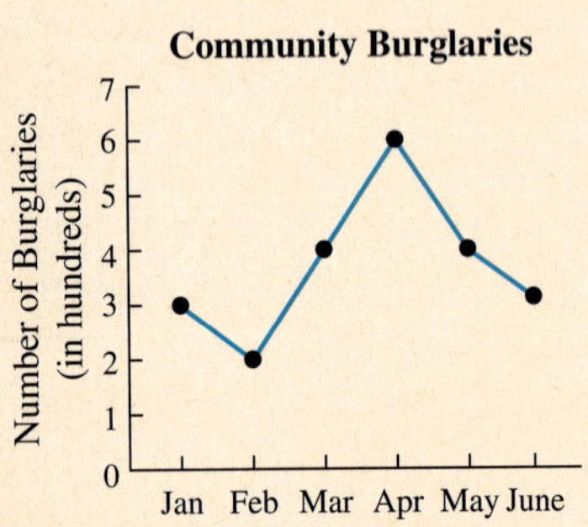

19. In which month did the greatest number of burglaries occur? How many were there?

20. In which month did the least number of burglaries occur? How many were there?

21. Find the increase in the number of burglaries from March to April.

22. Find the decrease in the number of burglaries from April to May.

23. Give two possible explanations for the decrease in burglaries during February.

24. Give two possible explanations for the increase in burglaries in March and April.

This comparison line graph shows the number of compact discs (CDs) sold by two different chain stores during each of five years. Find the number of CDs sold in each of the following years.

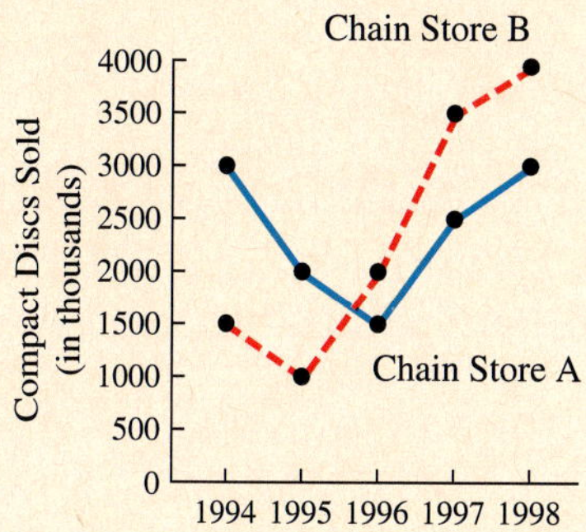

25. Chain Store A in 1998

26. Chain Store A in 1997

27. Chain Store A in 1996

28. Chain Store B in 1998

29. Chain Store B in 1997

30. Chain Store B in 1996

31. Looking at the comparison line graph above, which store would you like to own? Explain why. Based on the graph, what amount of sales would you predict for your store in 1999?

32. In the comparison line graph above, Store B used to have lower sales than Store A. What might have happened to cause this change? Give two possible explanations.

33. Explain in your own words why a bar graph or a line graph (not a double-bar graph or comparison line graph) can be used to show only one set of data.

34. The double-bar graph and the comparison line graph are both useful for comparing two sets of data. Explain how this works and give your own example.

This comparison line graph shows the sales and profits of Tacos-To-Go for each of four years. Use the graph to answer the following questions.

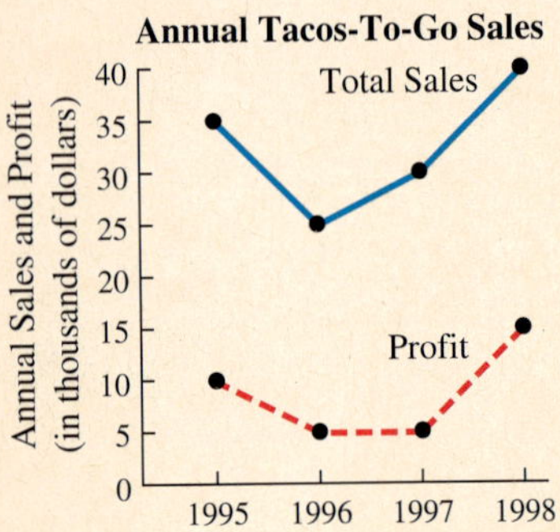

35. total sales in 1998

36. total sales in 1997

37. total sales in 1996

38. profit in 1998

39. profit in 1997

40. profit in 1996

41. Give two possible explanations for the decrease in sales from 1995 to 1996 and two possible explanations for the increase in sales from 1996 to 1998.

42. Based on the graph, what conclusion can you make about the relationship between sales and profits?

Review and Prepare

The circle graph shows how the city of Santa Barbara spent its annual budget. Use the graph to find the amount spent on each budget item. Round to the nearest thousand dollars. (For help, see **Section 9.1**.*)*

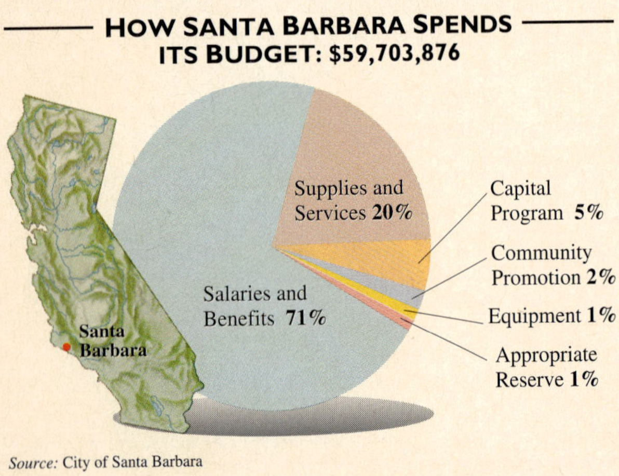

43. equipment

44. capital program

45. salaries and benefits

46. community promotion

47. supplies and services

48. appropriate reserve

9.3 *The Rectangular Coordinate System*

OBJECTIVES

1. Plot a point, given the coordinates, and find the coordinates, given a point.
2. Identify the four quadrants and determine which points lie within each one.
3. Determine whether a set of points lies on a straight line and, if so, find the coordinates of other points on the line.

FOR EXTRA HELP

Tutorial Tape 16 SSM, Sec. 9.3

OBJECTIVE 1 A bar graph or line graph shows the relationship between two things. The following is the line graph used for Exercises 19–24 in Section 9.2. It shows the relationship between the month of the year and the number of burglaries in a community.

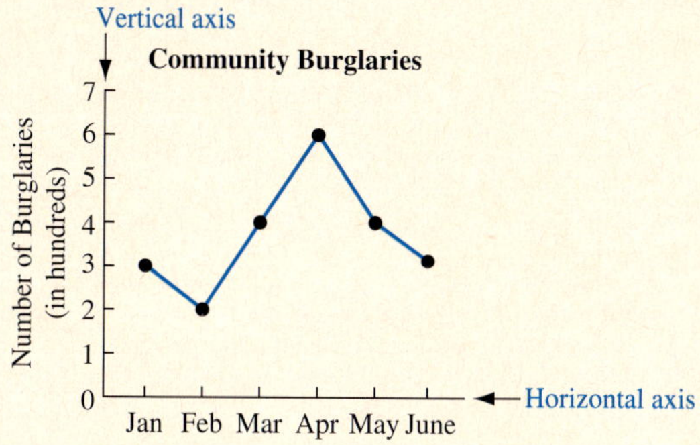

Each black dot on the graph represents a particular month paired with a particular number of burglaries. This is an example of **paired data.** We write each pair inside parentheses, with a comma separating the two items. To be consistent, we will always list the item on the *horizontal axis* (AKS-iss) first. In this case, the months are shown on the **horizontal axis** (the line that goes "left and right"), and the number of burglaries is shown along the **vertical axis** (the line that goes "up and down").

Paired Data from Line Graph on Community Burglaries

(Jan, 300) (Feb, 200) (Mar, 400) (Apr, 600) (May, 400) (June, 300)

Each data pair gives you the location of a particular spot on the graph, and that spot is marked with a dot. This idea of paired data can be used to locate particular places on any flat surface. Think of a small town laid out in a grid of square blocks, as shown below. To tell a taxi driver where to go, you could say, "the corner of 4th Avenue and 2nd Street" or just "4th and 2nd." As an ordered pair, it would be (4, 2). Of course, both you and the taxi driver need to know that the avenue is mentioned first (the number on the horizontal axis) and that the street is mentioned second (the number on the vertical axis). If the driver goes to (2, 4) instead, you'll be at the wrong corner.

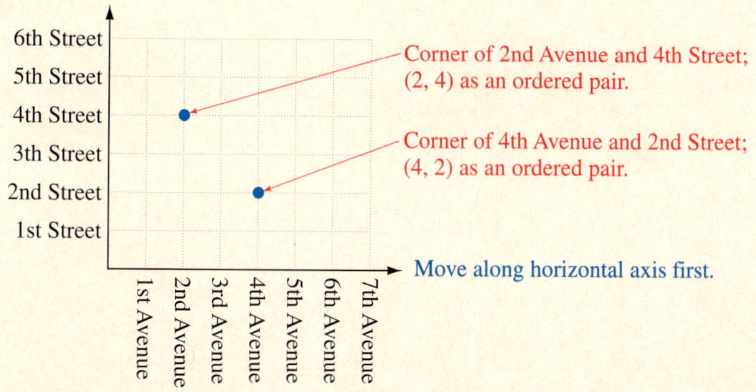

1. Plot each point on the grid. Write the ordered pair next to each point.

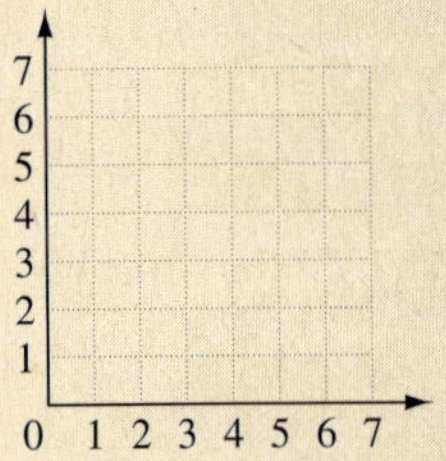

(a) (1, 4)

(b) (5, 2)

(c) (4, 1)

(d) (3, 3)

EXAMPLE 1 Plotting Points on a Grid

Use the grid at the right to plot each of these points.

(a) (3, 5)

Move *to the right* along the horizontal axis until you reach 3. Then move *up* 5 units so that you are opposite 5 on the vertical axis. Make a dot. This is the plot, or graph, of the point (3, 5).

(b) (5, 3)

Move *to the right* along the horizontal axis until you reach 5. Then move *up* 3 units so that you are opposite 3 on the vertical axis. Make a dot. This is the plot, or graph, of the point (5, 3).

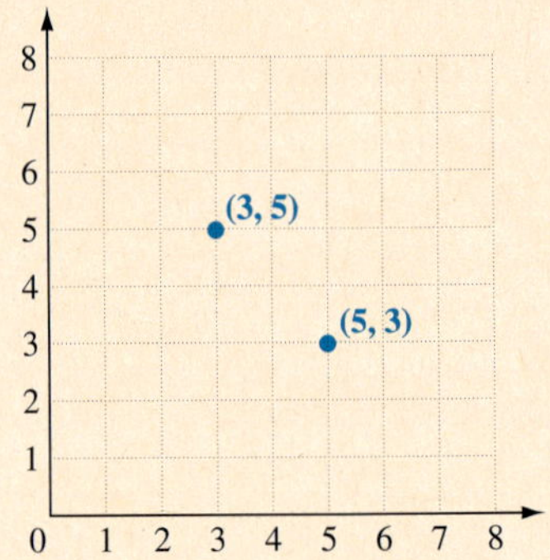

Move along horizontal axis first.

Note

In Example 1, the points (3, 5) and (5, 3) are **not** the same. The order within the pair is important. Always move along the horizontal axis first.

WORK PROBLEM 1 AT THE SIDE.

You have been using both positive and negative numbers throughout this book. We can extend our grid system to include negative numbers, as shown below. The horizontal axis is now a number line with 0 at the center, positive numbers extending to the right and negative numbers to the left. This horizontal number line is called the ***x*-axis.** The vertical axis is also a number line, with positive numbers extending upward and negative numbers extending downward. The vertical axis is called the ***y*-axis.** Together, the x-axis and the y-axis form a rectangular **coordinate system** (koh-OR-din-et SISS-tem).

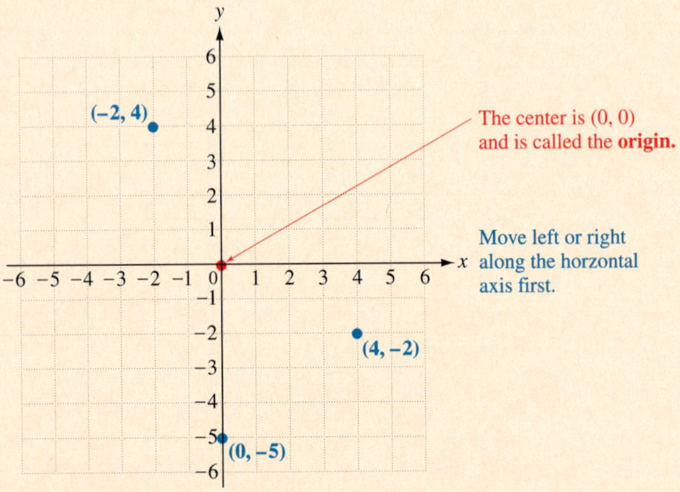

ANSWERS

1.

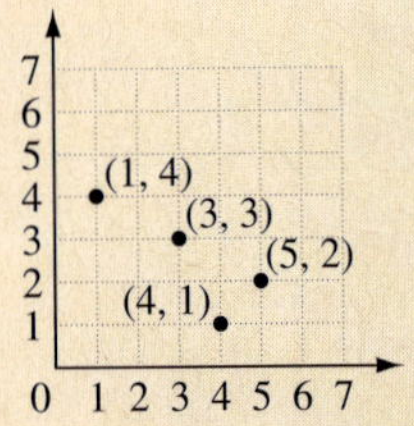

EXAMPLE 2 Plotting Points on a Rectangular Coordinate System

Plot each point on the rectangular coordinate system shown at the bottom of the previous page.

(a) $(4, -2)$

Move left or right along the horizontal x-axis first. Because 4 is *positive,* move *to the right* until you reach 4. Now, because the 2 is *negative,* move *down* 2 units so that you are opposite -2 on the y-axis. Make a dot and label it $(4, -2)$.

(b) $(-2, 4)$

Move left or right along the horizontal x-axis first. In this case, move *to the left* until you reach -2. Then move *up* 4 units. Make a dot and label it $(-2, 4)$. Notice that $(-2, 4)$ is **not** the same as $(4, -2)$.

(c) $(0, -5)$

Move left or right along the horizontal x-axis first. However, because the first number is 0, stay right at the center of the coordinate system. Then move *down* 5 units. Make a dot and label it $(0, -5)$.

Note

When the *first* number in an ordered pair is 0, the point is on the y-axis, as in Example 2(c) above. When the *second* number in an ordered pair is 0, the point is on the x-axis—for example, $(4, 0)$.

WORK PROBLEM 2 AT THE SIDE.

Once we have drawn a coordinate system, we can show the location of any point with an ordered pair. The numbers in the ordered pair are called the **coordinates** of the point.

EXAMPLE 3 Finding the Coordinates of Points

Find the coordinates of points A, B, C, and D.

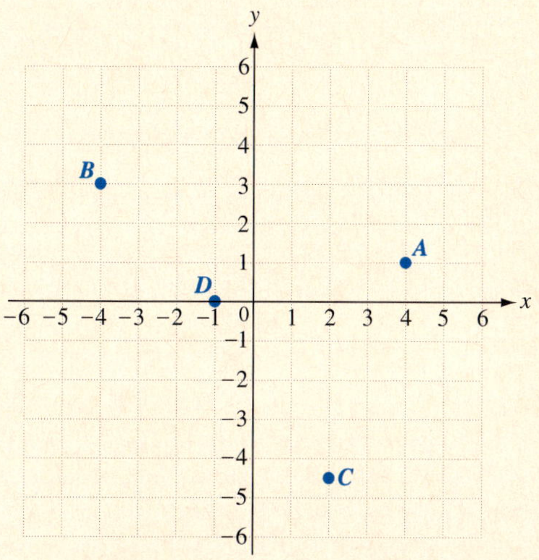

To reach point A from the origin, move 4 units *to the right;* then move *up* 1 unit. The coordinates are $(4, 1)$.

To reach point B from the origin, move 4 units *to the left;* then move *up* 3 units. The coordinates are $(-4, 3)$.

CONTINUED ON NEXT PAGE

2. Plot each point on the coordinate system shown. Write the ordered pair next to each point.

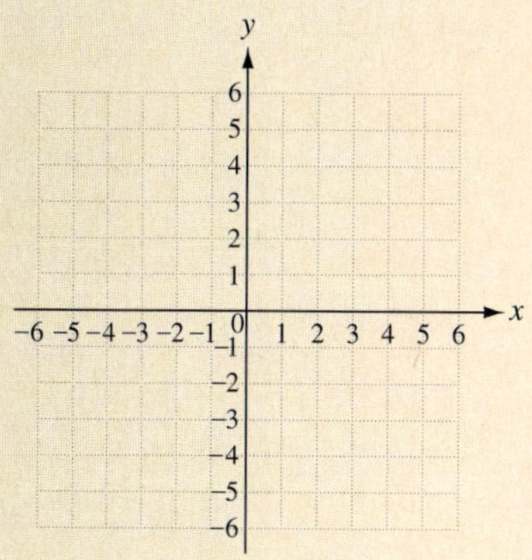

(a) $(5, -3)$

(b) $(-5, 3)$

(c) $(0, 3)$

(d) $(-4, -4)$

(e) $(-2, 0)$

ANSWERS

2.

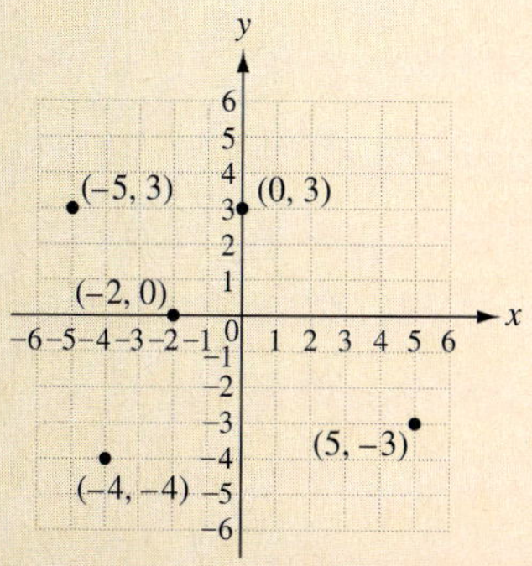

3. Find the coordinates of points A, B, C, D, and E.

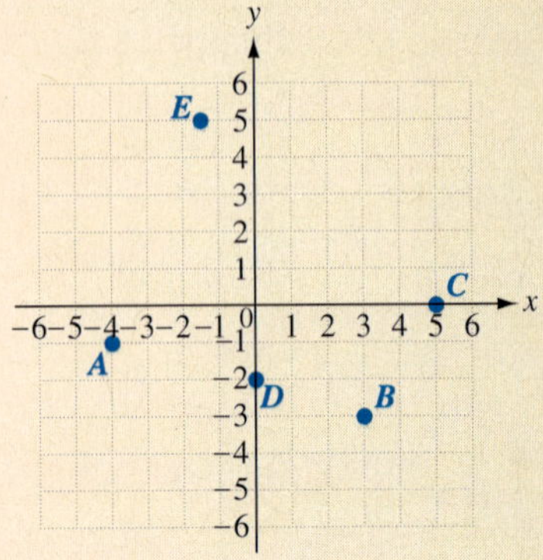

To reach point C from the origin, move 2 units *to the right;* then move *down* approximately $4\frac{1}{2}$ units. The approximate coordinates are $(2, -4\frac{1}{2})$.

To reach point D from the origin, move 1 unit *to the left;* then do *not* move either up or down. The coordinates are $(-1, 0)$.

Note

If a point is between the lines on the coordinate system, you can use fractions to give the approximate coordinates. For example, the approximate coordinates of point C, above, are $(2, -4\frac{1}{2})$.

WORK PROBLEM 3 AT THE SIDE.

Objective 2 The x-axis and y-axis divide the coordinate system into four regions, called **quadrants** (KWAD-runts). These quadrants are numbered with Roman numerals, as shown below. Points on the axes themselves are not in any quadrant.

4. (a) All points in the fourth quadrant are similar in what way? Give two examples of points in the fourth quadrant.

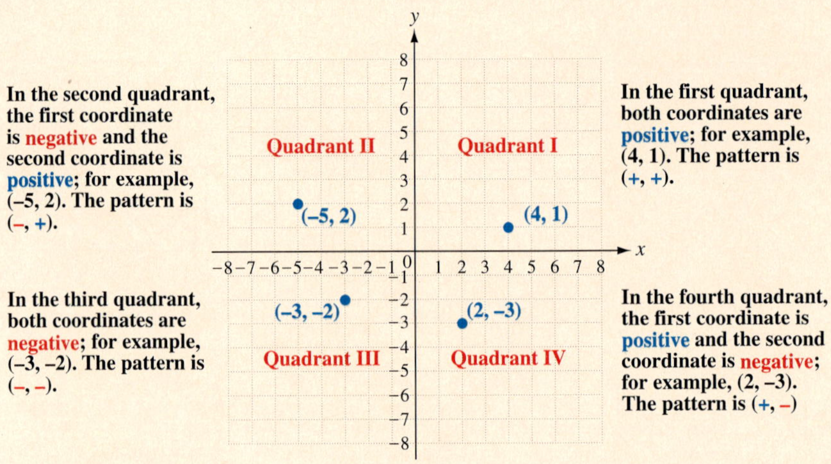

(b) In which quadrant is each point located: $(-2, -6)$; $(0, 5)$; $(-3, 1)$; $(4, -1)$?

EXAMPLE 4 Working with Quadrants

(a) All points in the third quadrant are similar in what way? Give two examples of points in the third quadrant.

For all points in quadrant III, both coordinates are negative. The pattern is $(-, -)$. There are many possible examples, such as $(-2, -5)$ and $(-4, -4)$. Just be sure that both numbers are negative.

(b) In which quadrant is each point located: $(3, 5)$; $(1, -6)$; $(-4, 0)$?

For $(3, 5)$ the pattern is $(+, +)$, so the point is in **quadrant I**.

For $(1, -6)$ the pattern is $(+, -)$, so the point is in **quadrant IV**.

The point corresponding to $(-4, 0)$ is on the x-axis, so it isn't in any quadrant.

WORK PROBLEM 4 AT THE SIDE.

Answers

3. A is $(-4, -1)$; B is $(3, -3)$; C is $(5, 0)$; D is $(0, -2)$; E is approximately $(-1\frac{1}{2}, 5)$

4. (a) The pattern for all points in quadrant IV is $(+, -)$. Examples will vary; just be sure that they fit the $(+, -)$ pattern.
(b) III; no quadrant; II; IV.

OBJECTIVE 3 To get ready for the next section, we will plot groups of points that lie on a straight line.

EXAMPLE 5 Plotting Points on a Straight Line

(a) Plot each set of points. Which set lies on a straight line? Draw the line.

(4, 1) (5, 2) (6, 2) or **(1, 2) (2, 4) (3, 6)**

(b) Which of these points are also on the straight line?

(0, 3) (4, 8) (−2, −1) (−1, −2)

(a) The first set of points is shown in red on the coordinate system below. It is *not* possible to draw a straight line that passes through all three points. The second set of points is shown in blue. These points do lie on a straight line, also shown in blue. The line can be extended in both directions, as shown.

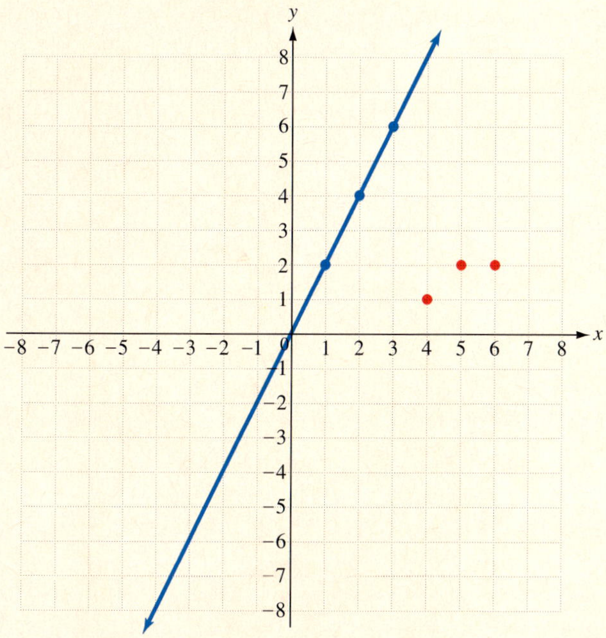

(b) The line passes through (4, 8) and (−1, −2). The other two points are *not* on the line.

WORK PROBLEM 5 AT THE SIDE.

5. (a) Plot each set of points. Which set lies on a straight line? (0, 4) (1, 3) (2, 2) or (−4, 2) (−3, 1) (−2, 3)

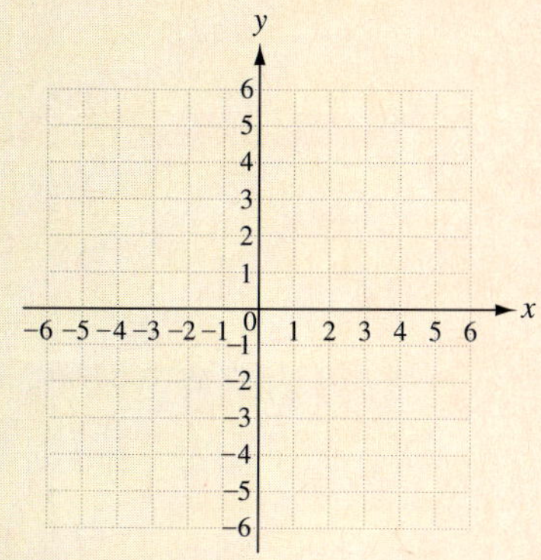

(b) Which of these points are also on the straight line: (−1, 2) (5, 1) (−1, 5) (4, 0)?

ANSWERS

5. (a)

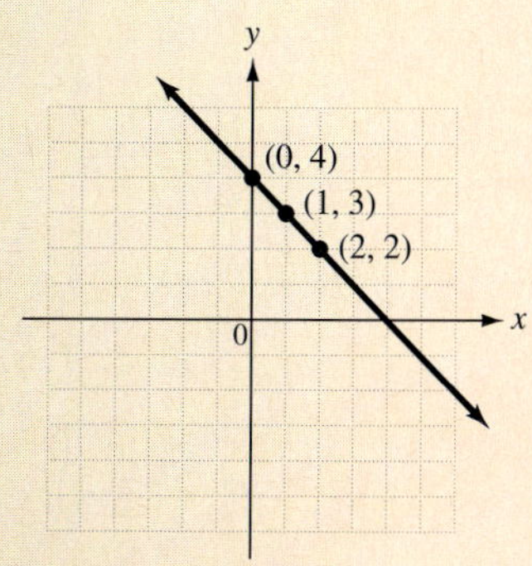

(b) Points (−1, 5) and (4, 0) are on the line.

NUMBERS IN THE *Real World*

collaborative investigations

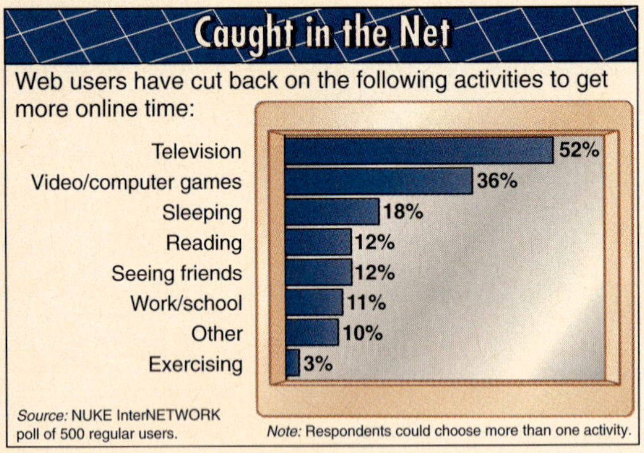

1. Look at the "Source" information at the bottom of the graph. How were the numbers in the graph obtained?

2. What *percent* of people in the poll said they cut back on television in order to find time to use the World Wide Web? How *many* people is that?

3. Find the *number* of people in the poll who cut back on each of the other activities shown in the graph.

4. Add up all the numbers of people you just calculated for all the different activities (Questions 2 and 3). Why is the total more than the 500 people that were in the poll?

5. Suppose you took a similar poll of 100 students at your school who regularly use the Web. Using the information in the graph, how many students would you expect to choose each of the activities?

6. Why might your poll give different results than the ones shown in the graph? Give at least *two* possible explanations.

7. Suppose you were writing a paper about the Web and wanted to include some of the information from the graph, but without showing the whole graph and *without using percents*. You also decide that it would be all right to round the data up or down a little bit in order to make it easier for people to understand. Write a sentence that expresses the information about people who cut back on television. Also write sentences about the people who cut back on video/computer games, those who cut back on sleep, and those who cut back on work/school.

Name ______________________ Date ____________ Class ____________

9.3 Exercises

Plot each point on the rectangular coordinate system. Label each point with its coordinates.

1. (3, 7) (−2, 2) (−3, −7) (2, −2)
(0, 6) (6, 0) (0, −4) (−4, 0)

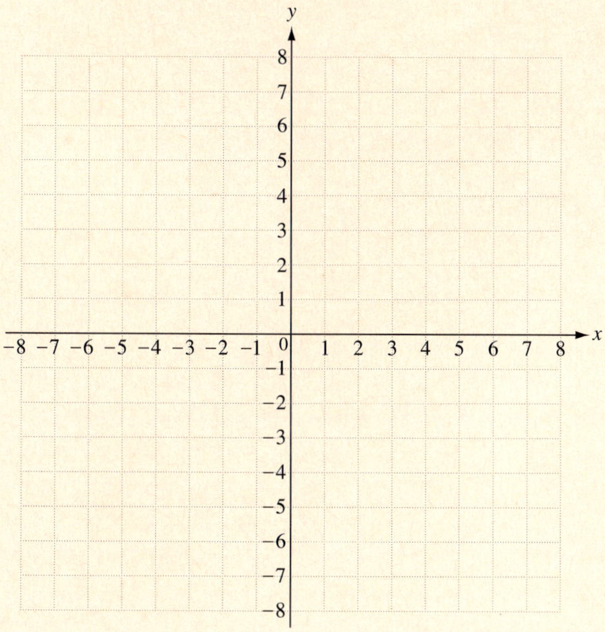

2. (5, 2) (−3, −3) (4, −1) (−4, 1)
(−1, 0) (0, 3) (2, 0) (0, −5)

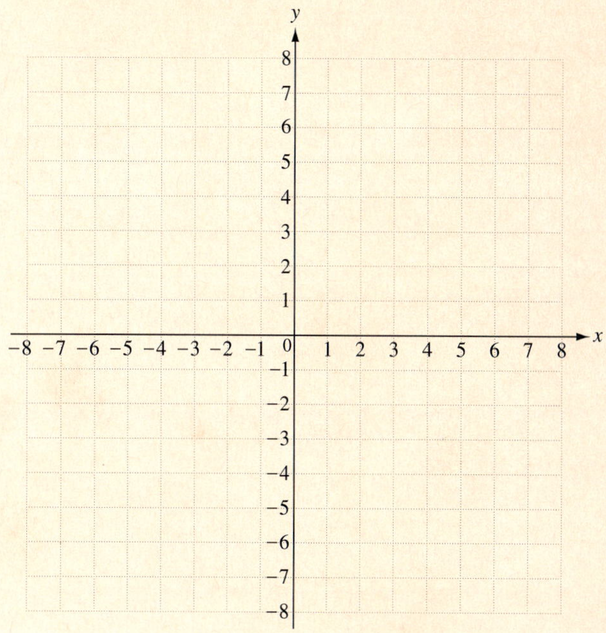

G **3.** (−5, 3) (4, 4) $(-2\frac{1}{2}, 0)$ (3, −5)
(0, 0) $(2, \frac{1}{2})$ (−7, −5) (−1, −6)

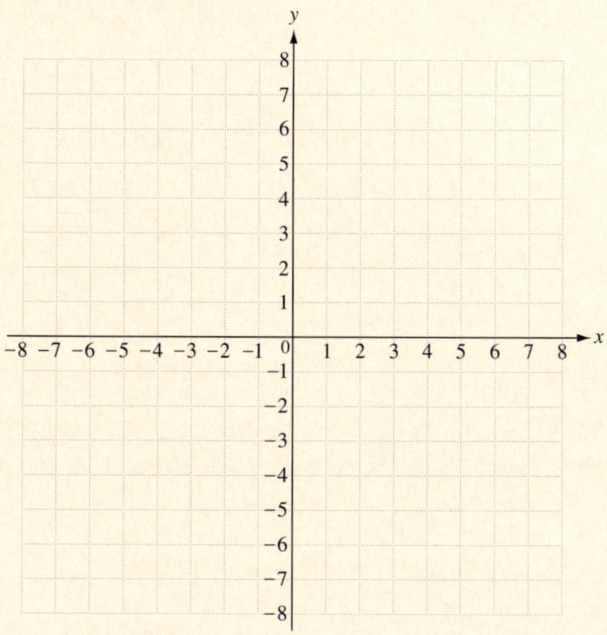

G **4.** (1, 7) $(0, 3\frac{1}{2})$ (−5, −1) (6, −2)
(−2, 6) (0, 0) (−3, 3) $(-\frac{1}{2}, -2)$

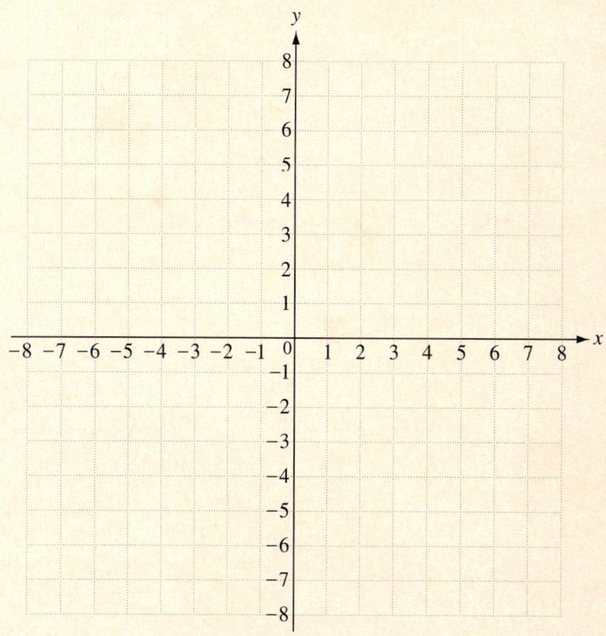

5. In which quadrant is each point located?
(−3, −7) (0, 4) (10, −16) (−9, 5)

6. In which quadrant is each point located?
(1, 12) (20, −8) (−5, 0) (−14, 14)

Give the coordinates of each point.

7.

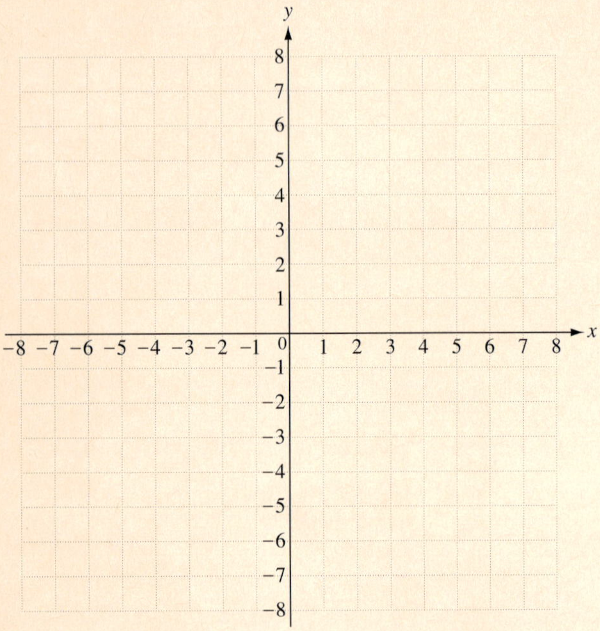

8.

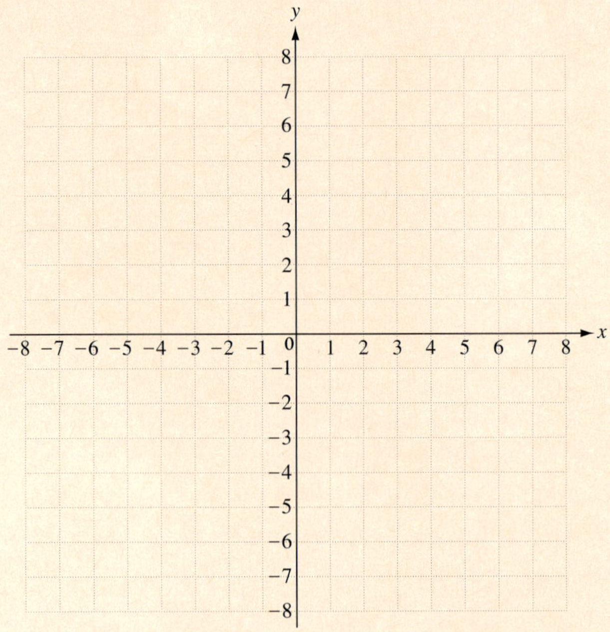

9. Explain how to graph the ordered pair (a, b), where a and b are positive or negative integers.

10. Explain how to graph the ordered pair (a, b) when a is zero and b is an integer. Explain how to graph (a, b) when a is an integer and b is zero.

11. **(a)** Plot each set of points. Which set lies on a straight line? Draw the line.
(3, 3) (2, 4) (0, 3) or
(6, 3) (4, 2) (2, 1)
(b) Which of these points are also on the straight line:
(−2, −1) (−4, 2) (4, 8) (0, 0)

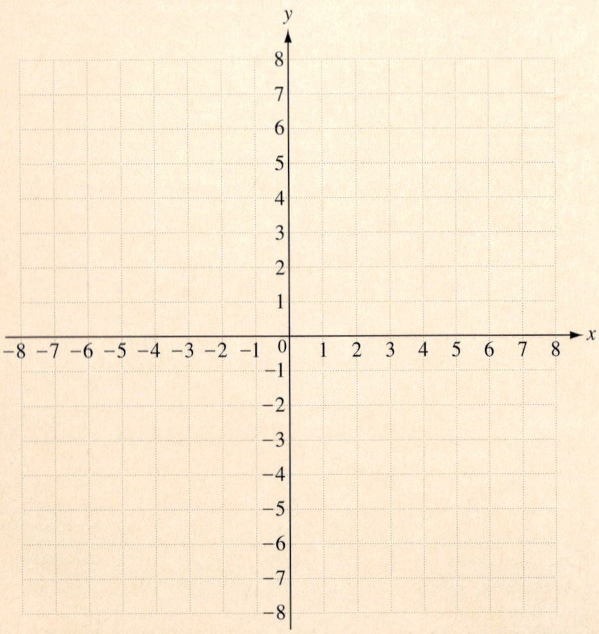

12. **(a)** Plot each set of points. Which set lies on a straight line? Draw the line.
(0, 5) (2, 4) (4, 3) or
(1, 1) (2, 0) (3, 2)
(b) Which of these points are also on the straight line:
(−2, −6) (−4, 7) (6, 2) (8, 0)

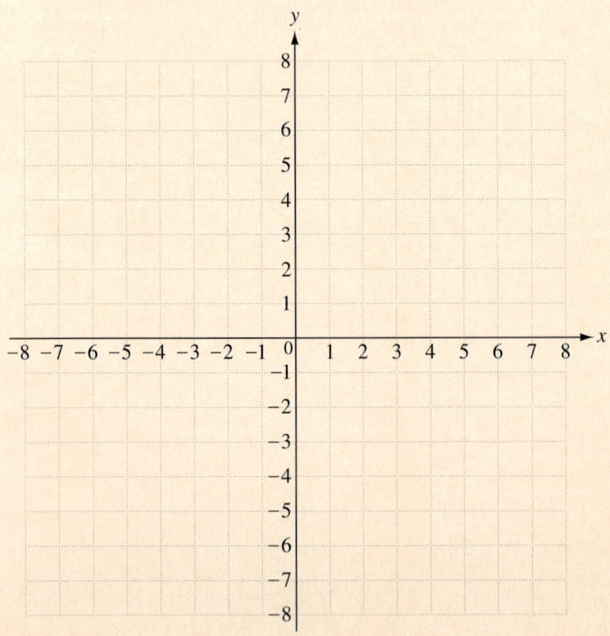

9.4 *Graphing Linear Equations*

OBJECTIVES

1. Graph linear equations in two variables.
2. Identify the slope of a line as positive or negative.

FOR EXTRA HELP

Tutorial — Tape 16 — SSM, Sec. 9.4

In Chapters 2–7 you solved equations that had only one variable, such as $2n - 3 = 7$ or $\frac{1}{3}x = 10$. Each of these equations had exactly one solution: n is 5 in the first equation, and x is 30 in the second equation. In other words, there was only *one* number that could replace the variable and make the equation balance. As you take more algebra courses, you will work with equations that have two variables and many different numbers that will make the equation balance. This section will get you started.

OBJECTIVE 1 Suppose that you have 6 hours of study time planned during a weekend. You plan to study math and psychology. For example, you could spend 4 hours on math and then 2 hours on psychology, for a total of 6 hours. Or you could spend $1\frac{1}{2}$ hours on math and then $4\frac{1}{2}$ hours on psychology, for a total of 6 hours. Here is a list of *some* of the possible combinations.

Hours on Math	**+**	**Hours on Psychology**	**=**	**Total Hours Studying**
0	+	**6**	=	6
1	+	**5**	=	6
$1\frac{1}{2}$	+	$4\frac{1}{2}$	=	6
3	+	**3**	=	6
4	+	**2**	=	6
$5\frac{1}{2}$	+	$\frac{1}{2}$	=	6
6	+	**0**	=	6

We can write an equation to represent this situation.

Hours studying math	+	**Hours studying psychology**	=	Total of 6 hours
m	+	p	=	6

This equation has two variables. The hours spent on math (m) can vary, and the hours spent on psychology (p) can vary.

As you can see, there is more than one solution for this equation. We can list possible solutions as ordered pairs. The first number in the pair is the value of m, and the second number is the corresponding value of p.

$$(m, p) \to (0, 6) \quad (m, p) \to (1, 5) \quad (m, p) \to \left(1\frac{1}{2}, 4\frac{1}{2}\right) \quad (m, p) \to (3, 3) \quad (m, p) \to (4, 2) \quad (m, p) \to \left(5\frac{1}{2}, \frac{1}{2}\right) \quad (m, p) \to (6, 0)$$

Another way to show the solutions is to plot the ordered pairs. This method will give us a "picture" of the solutions that we listed on the previous page.

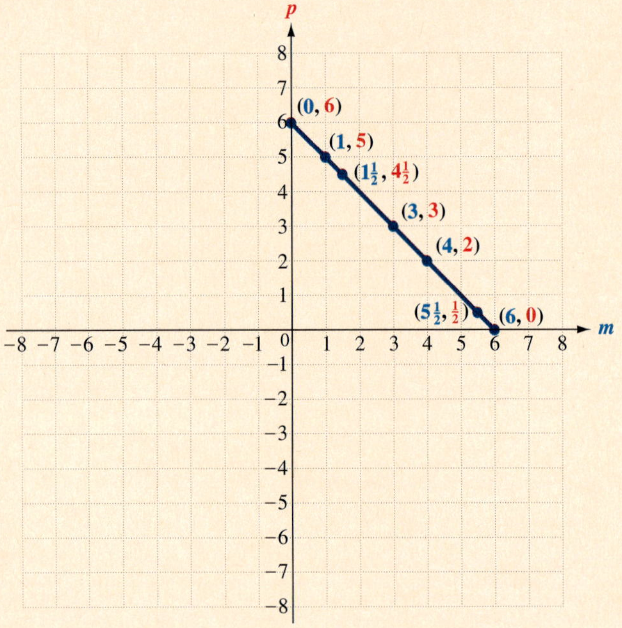

Notice that all the solutions (all the ordered pairs) lie on a straight line. When you draw a line connecting the ordered pairs, you have graphed the solutions. *Every* point on the line is a solution. You can use the line to find additional solutions besides the ones that we listed. For example, the point (5, 1) is on the line. This point tells you that another solution is 5 hours on math and 1 hour on psychology. The fact that the line is a *straight* line tells you that $m + p = 6$ is a *linear equation.* (The word ***line*** is part of the word ***line***ar.) Later on in algebra you will work with equations whose solutions form a curved line when you graph them.

To draw the line for $m + p = 6$, we really needed only two solutions (two ordered pairs). But it's a good idea to use a third ordered pair as a check. If the three ordered pairs are *not* in a straight line, there is an error in your work.

Graphing a Linear Equation

To **graph a linear equation**, find at least three ordered pairs that satisfy the equation. Then plot the ordered pairs on a coordinate system and connect them with a straight line. *Every* point on the line is a solution to the equation.

EXAMPLE 1 Graphing Linear Equations

Graph $x + y = 3$ by finding three solutions and plotting the ordered pairs. Then use the graph to find a fourth solution to the equation.

There are many possible solutions. Start by picking three different values for x. You can choose any numbers you like, but 0 and small numbers usually are easy to use. Then find the value of y that will make the sum equal to 3. Set up a table to organize the information.

CONTINUED ON NEXT PAGE

x	y	Check that $x + y = 3$	Ordered pair (x, y)
0	**3**	$\mathbf{0} + \mathbf{3} = 3$	$(\mathbf{0}, \mathbf{3})$
1	**2**	$\mathbf{1} + \mathbf{2} = 3$	$(\mathbf{1}, \mathbf{2})$
2	**1**	$\mathbf{2} + \mathbf{1} = 3$	$(\mathbf{2}, \mathbf{1})$

Plot the ordered pairs and draw a line through the points, extending it in both directions as shown below.

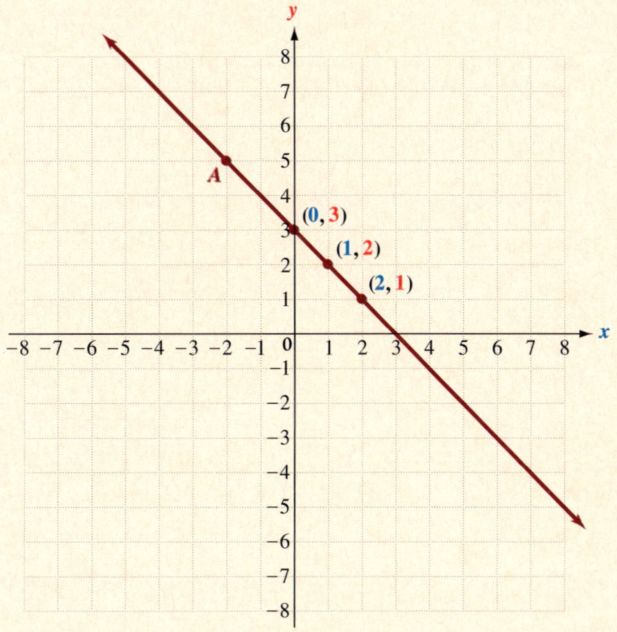

Now you can use the graph to find more solutions to $x + y = 3$. *Every* point on the line is a solution. Suppose that you pick **point A**. The coordinates are $(-2, 5)$.

To check that $(-2, 5)$ is a solution, substitute -2 for x and 5 for y in the original equation.

$$x + y = 3$$
$$\downarrow \quad \downarrow$$
$$\underbrace{-2 + 5}_{} = 3$$
$$3 = 3 \quad \text{Balances}$$

The equation balances, so $(-2, 5)$ is another solution for $x + y = 3$.

Note

The line in Example 1 was extended in both directions because *every* point on the line is a solution to $x + y = 3$. However, when we graphed the line for the hours spent studying, $m + p = 6$, we did *not* extend the line. That is because the variables m and p represented hours, and hours can only be positive numbers; all the solutions had to be in the first quadrant.

WORK PROBLEM 1 AT THE SIDE. ▶▶

1. Graph $x + y = 5$ by finding three solutions and plotting the ordered pairs. Then use the graph to find *two* other solutions to the equation.

x	y	Check that $x + y = 5$	Ordered pair (x, y)
0			
1			
2			

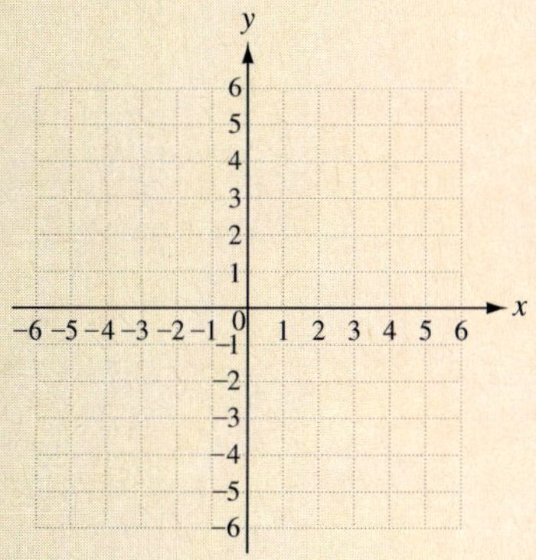

Two other solutions are (___, ___) and (___, ___).

Answers

1. Plot (0, 5), (1, 4), and (2, 3).

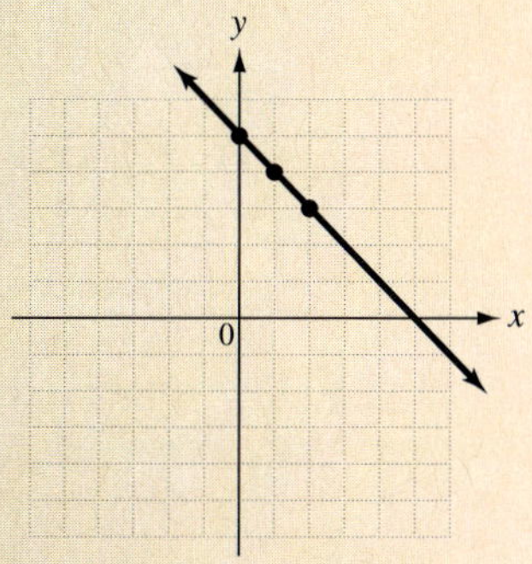

Some other solutions are $(-1, 6)$; $(3, 2)$; $(4, 1)$; $(5, 0)$; $(6, -1)$.

EXAMPLE 2(a) Graphing Linear Equations

Graph $y = -3x$ by finding three solutions and plotting the ordered pairs. Then use the graph to find a fourth solution to the equation.

You can choose any three values for x, but small numbers such as 0, 1, and 2 are easy to use. Then $y = -3x$ tells you that y is -3 times the value of x.

$$y = -3x \qquad -3x \text{ means } -3 \text{ times } x.$$

$$y \text{ is } -3 \text{ times } x$$

First set up a table.

x	y (must be -3 times x)	Ordered pair (x, y)
0	$-3 \cdot 0$ is 0	$(0, 0)$
1	$-3 \cdot 1$ is -3	$(1, -3)$
2	$-3 \cdot 2$ is -6	$(2, -6)$

Plot the ordered pairs and draw a line through the points.

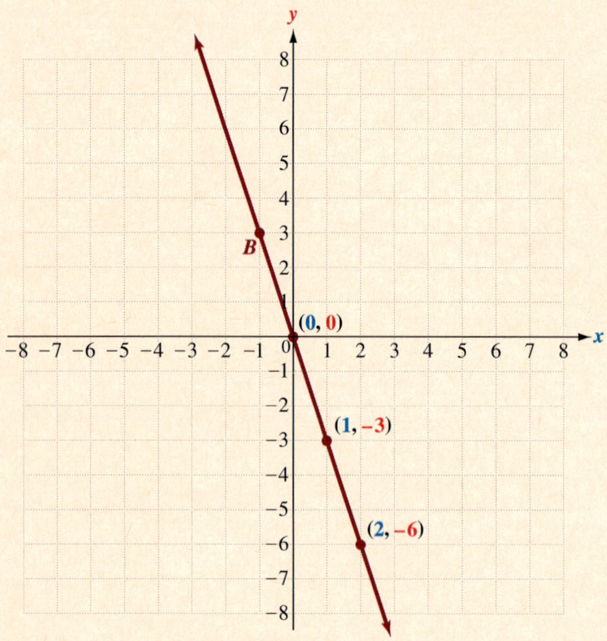

Now use the graph to find more solutions. *Every* point on the line is a solution.

Suppose that you pick **point *B***. The coordinates are $(-1, 3)$. To check that $(-1, 3)$ is a solution, substitute -1 for x and 3 for y in the original equation.

$$y = -3x$$

$$3 = -3(-1)$$

$$3 = 3 \qquad \text{Balances}$$

The equation balances, so $(-1, 3)$ is another solution for $y = -3x$.

WORK PROBLEM 2(a) AT THE SIDE.

2. (a) Graph $y = 2x$ by finding three solutions and plotting the ordered pairs. Then use the graph to find *two* other solutions to the equation.

x	y (must be $2 \cdot x$)	Ordered pair
0		
1		
2		

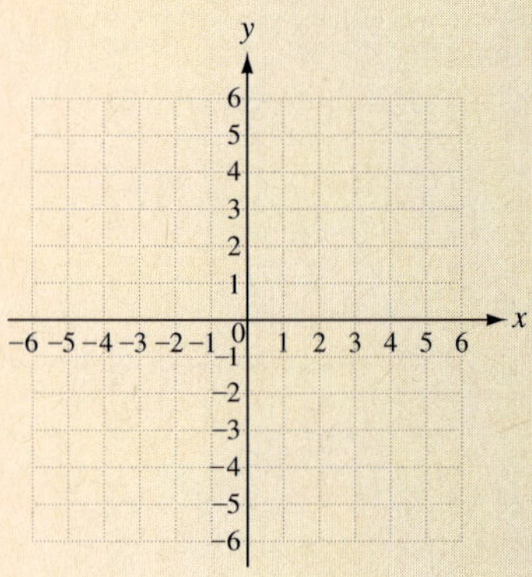

Two other solutions are (__, __) and (__, __).

ANSWER

2. (a) Plot (0, 0), (1, 2), and (2, 4).

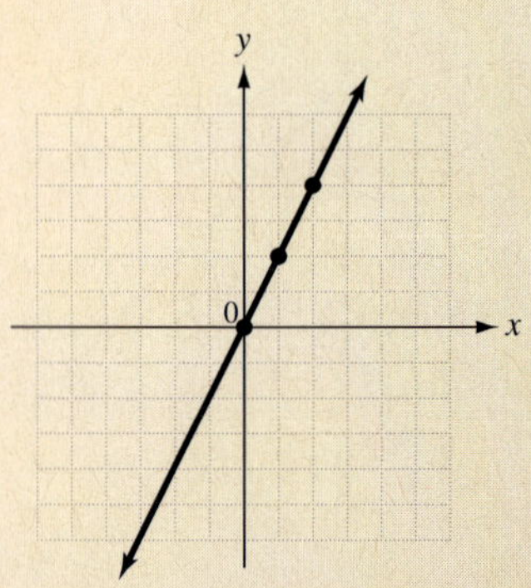

Some other solutions are (3, 6); (−1, −2); (−2, −4); (−3, −6).

EXAMPLE 2(b) Graphing Linear Equations

Graph $y = \frac{1}{2}x$.

Complete the table. The coefficient of x is $\frac{1}{2}$, so we chose 2, 4, and 6 as values for x because they are easy to divide in half. The equation $y = \frac{1}{2}x$ tells you that y is $\frac{1}{2}$ times the value of x, or $\frac{1}{2}$ of x.

x	y (must be $\frac{1}{2}$ of x)	Ordered pair
2	$\frac{1}{2}$ of 2 is 1	(2, 1)
4	$\frac{1}{2}$ of 4 is 2	(4, 2)
6	$\frac{1}{2}$ of 6 is 3	(6, 3)

Plot the ordered pairs and draw a line through the points.

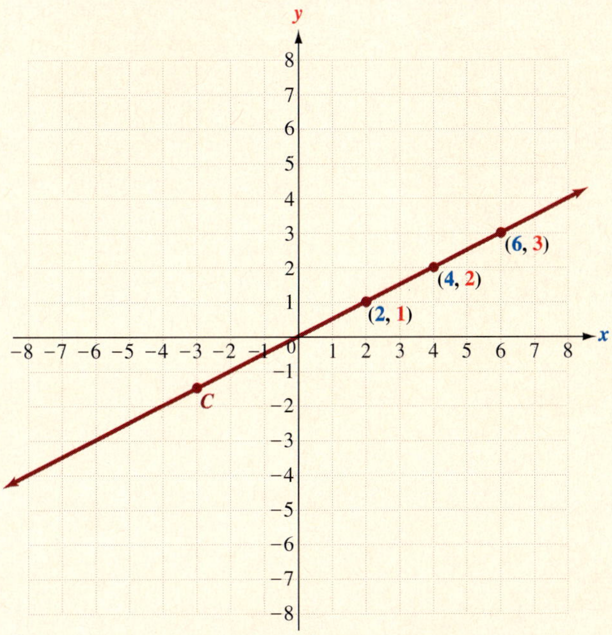

Now use the graph to find more solutions. Every point on the line is a solution. Suppose that you pick **point *C***. The coordinates are $(-3, -1\frac{1}{2})$. Check that $(-3, -1\frac{1}{2})$ is a solution by substituting -3 for x and $-1\frac{1}{2}$ for y.

$$y = \frac{1}{2}x$$

$$-1\frac{1}{2} = \frac{1}{2}(-3)$$

$$-1\frac{1}{2} = -\frac{3}{2}$$ Balances; $-1\frac{1}{2}$ is equivalent to $-\frac{3}{2}$.

The equation balances, so $(-3, -1\frac{1}{2})$ is another solution for $y = \frac{1}{2}x$.

WORK PROBLEM 2(b) AT THE SIDE.

2. (b) Graph $y = -\frac{1}{2}x$.

x	y (must be $-\frac{1}{2} \cdot x$)	Ordered pair
2		
4		
6		

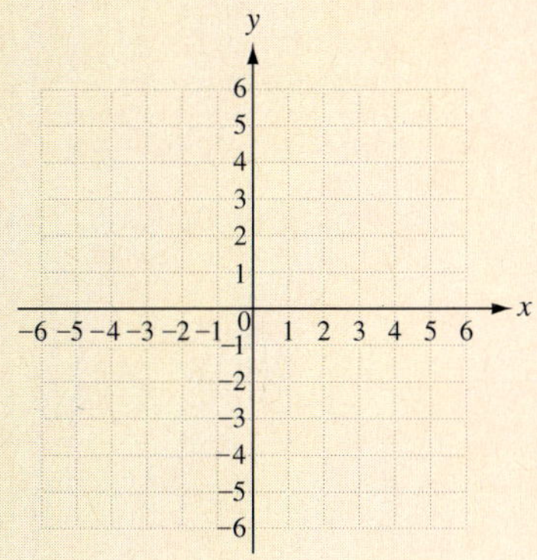

Two other solutions are (__, __) and (__, __).

ANSWER

2. (b) Plot (2, −1), (4, −2), and (6, −3).

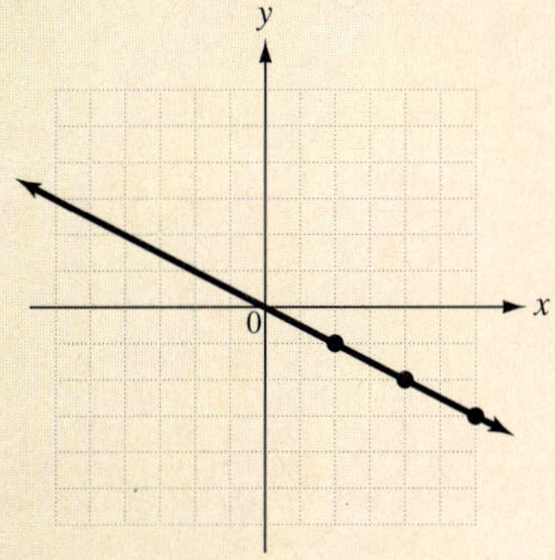

Some other solutions are (0, 0); (−2, 1); (−4, 2); (−6, 3); $(1, -\frac{1}{2})$; $(3, -1\frac{1}{2})$; $(5, -2\frac{1}{2})$.

3. Graph the equation $y = x - 5$ by finding three solutions and plotting the ordered pairs. Then use the graph to find *two* more solutions.

x	y (must be 5 less than x)	Ordered pair
1		
2		
3		

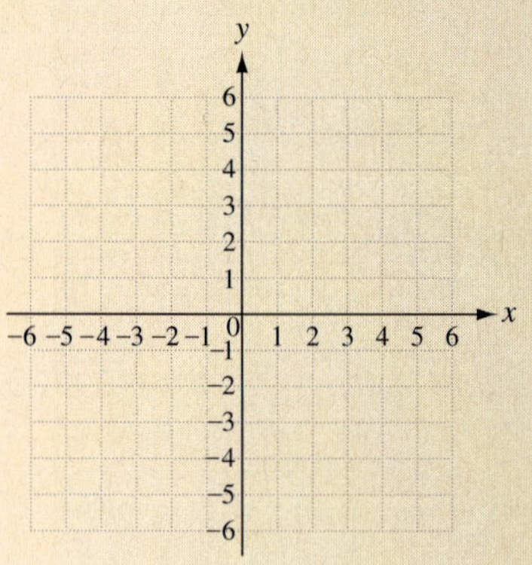

Two other solutions are (___, ___) and (___, ___).

Answer

3. Plot $(1, -4)$, $(2, -3)$, and $(3, -2)$.

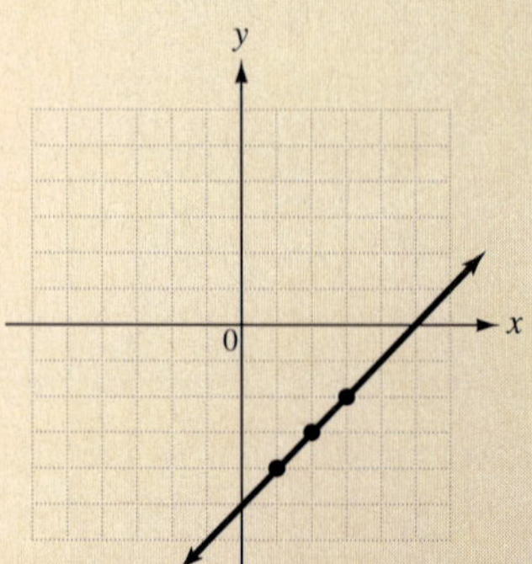

Some other solutions are $(-1, -6)$; $(0, -5)$; $(4, -1)$; $(5, 0)$; $(6, 1)$.

EXAMPLE 3 Graphing Linear Equations

Graph the equation $y = x + 4$ by finding three solutions and plotting the ordered pairs. Then use the graph to find two more solutions to the equation.

First set up a table. The equation $y = x + 4$ tells you that y must be 4 more than the value of x.

x	y (must be 4 more than x)	Ordered pair
0	0 + 4 is 4	(0, 4)
1	1 + 4 is 5	(1, 5)
2	2 + 4 is 6	(2, 6)

Plot the ordered pairs and draw a line through the points.

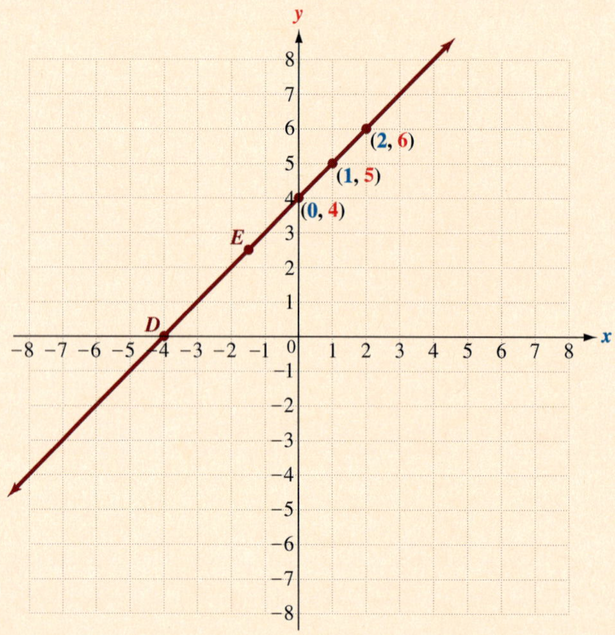

Now use the graph to find two more solutions. Every point on the line is a solution. Suppose that you pick **point *D*** at $(-4, 0)$ and **point *E*** at $(-1\frac{1}{2}, 2\frac{1}{2})$. Check that both ordered pairs are solutions.

Check $(-4, 0)$

$$y = x + 4$$
$$0 = -4 + 4$$
$$0 = 0 \quad \text{Balances}$$

Check $\left(-1\frac{1}{2}, 2\frac{1}{2}\right)$

$$y = x + 4$$
$$2\frac{1}{2} = -1\frac{1}{2} + 4$$
$$\frac{5}{2} = -\frac{3}{2} + \frac{8}{2}$$
$$\frac{5}{2} = \frac{5}{2} \quad \text{Balances}$$

Both equations balance, so $(-4, 0)$ and $(-1\frac{1}{2}, 2\frac{1}{2})$ are both solutions.

◀◀ WORK PROBLEM 3 AT THE SIDE.

OBJECTIVE 2 Let's look again at some of the lines that we graphed for various equations. All are straight lines, but some are almost flat and some tilt steeply upward or downward.

Graph from Example 1: $x + y = 3$.

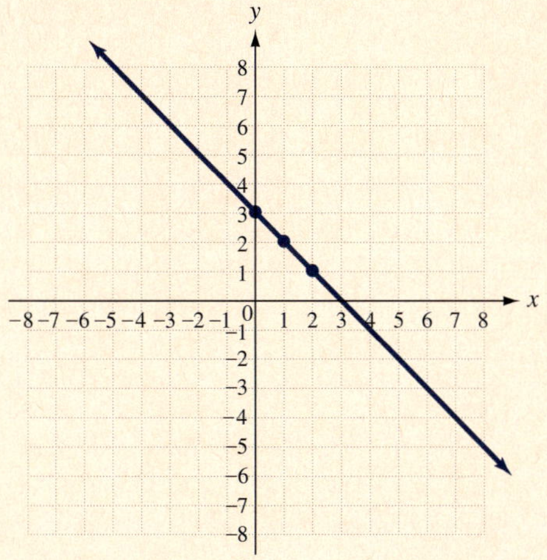

As you move from left to right, the line slopes downward, as if you were going down a hill. When a line tilts downward, we say that it has a *negative slope*.

Now look at the table of solutions we used to draw the line.

x	y
0	3
1	2
2	1

The value of x is *increasing* from 0 to 1 to 2.

The value of y is *decreasing* from 3 to 2 to 1.

As the value of x *increases,* the value of y does the *opposite*—it *decreases*. Whenever one variable increases while the other variable decreases, the line will have a negative slope.

Graph from Example 2(b): $y = \frac{1}{2}x$.

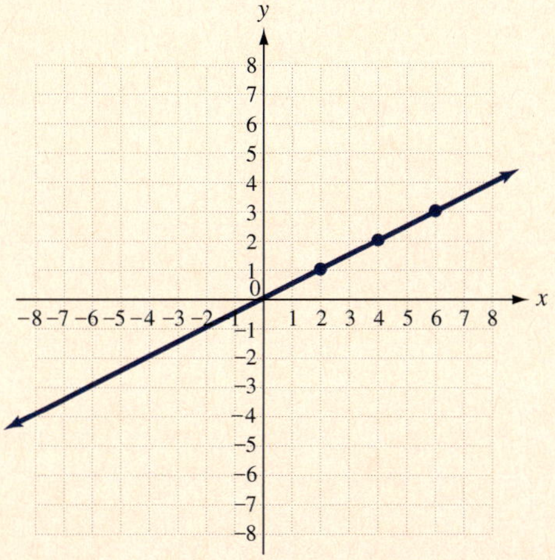

As you move from left to right, this line slopes upward, as if you were walking up a hill. When a line tilts upward, we say that it has a *positive slope*.

4. Look back at the graphs in Margin Exercises 2(a) and 2(b). Then complete these sentences.

(a) The graph of $y = 2x$ has a ____________ slope. As the value of x increases, the value of y ____________.

(b) The graph of $y = -\frac{1}{2}x$ has a ____________ slope. As the value of x increases, the value of y ____________.

Now look at the table of solutions we used to draw the line.

The value of x is *increasing* from 2 to 4 to 6.

x	y
2	1
4	2
6	3

The value of y is *increasing* from 1 to 2 to 3.

As the value of x *increases,* the value of y does the *same* thing—it also *increases.* Whenever both variables do the same thing (both increase or both decrease) the line will have a positive slope.

Positive and Negative Slopes

As you move from left to right, a line with a *positive* slope tilts *upward* or rises. As you move from left to right, a line with a *negative* slope tilts *downward* or falls.

EXAMPLE 4 Identifying Lines with a Positive or Negative Slope

Look back at the graph of $y = -3x$ in Example 2(a). Then complete these sentences.

The graph of $y = -3x$ has a ____________ slope. As the value of x increases, the value of y ____________.

The line has a negative slope because it tilts downward. As the value of x increases, the value of y decreases (does the opposite).

◀◀ **WORK PROBLEM 4 AT THE SIDE.**

ANSWERS

4. (a) positive; increases
(b) negative; decreases

9.4 Exercises

Graph each equation by completing the table to find three solutions and plotting the ordered pairs. Then use the graph to find two *other solutions.*

1. $x + y = 4$

x	y	Ordered pair
0		
1		
2		

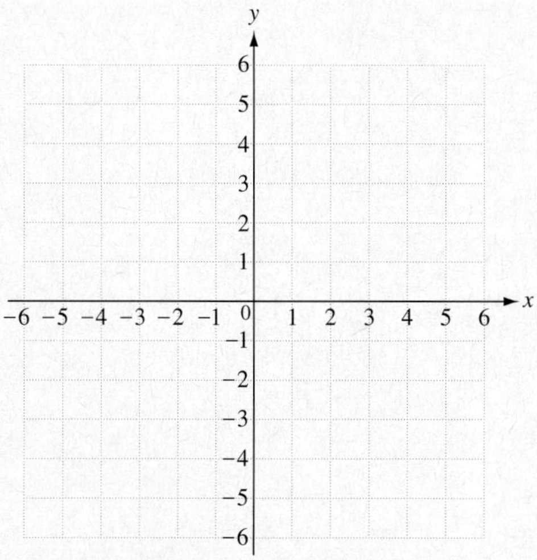

Two other solutions are (__, __) and (__, __).

2. $x + y = -4$

x	y	Ordered pair
0		
1		
2		

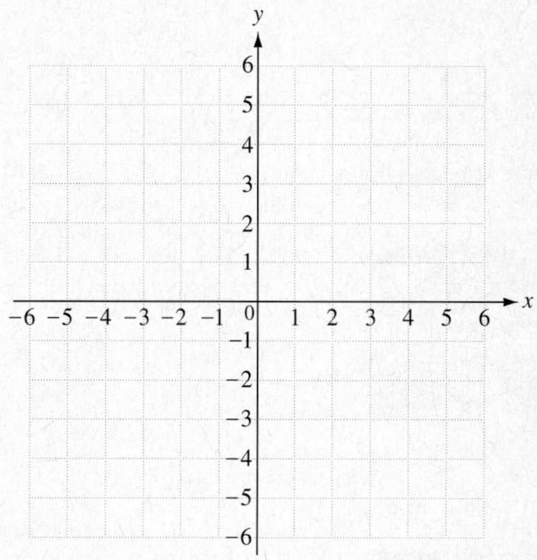

Two other solutions are (__, __) and (__, __).

3. $x + y = -1$

x	y	Ordered pair
0		
1		
2		

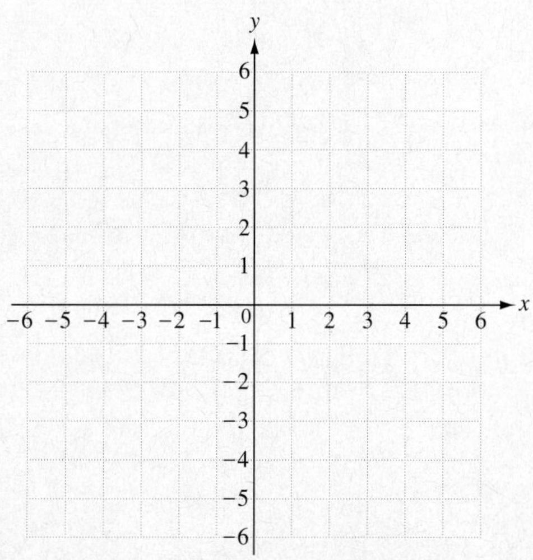

Two other solutions are (__, __) and (__, __).

4. $x + y = 1$

x	y	Ordered pair
0		
1		
2		

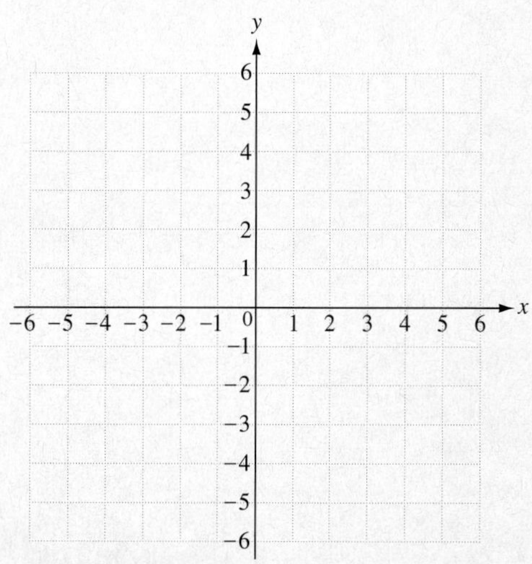

Two other solutions are (__, __) and (__, __).

5. The line in Exercise 1 crosses the y-axis at what point? ______. The line in Exercise 3 crosses the y-axis at what point? ______. Based on these examples, where would the graph of $x + y = -6$ cross the y-axis? Where would the graph of $x + y = 99$ cross the y-axis?

6. Look at where the line crosses the x-axis and where it crosses the y-axis in Exercises 2 and 4. What pattern do you see?

Graph each equation. Make your own table using the listed values of x.

7. $y = x - 2$
Use 1, 2, and 3 as the values of x.

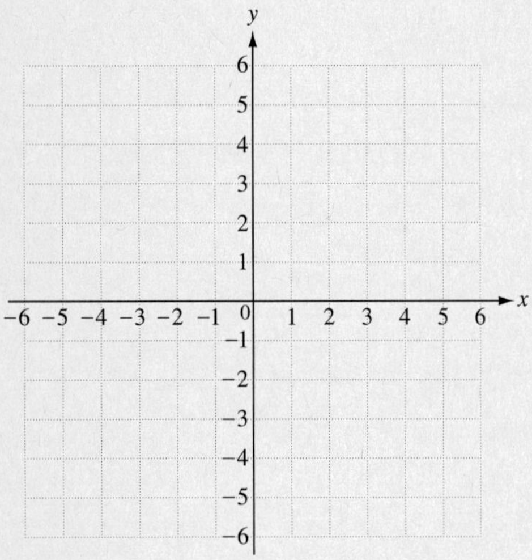

8. $y = x + 1$
Use 1, 2, and 3 as the values of x.

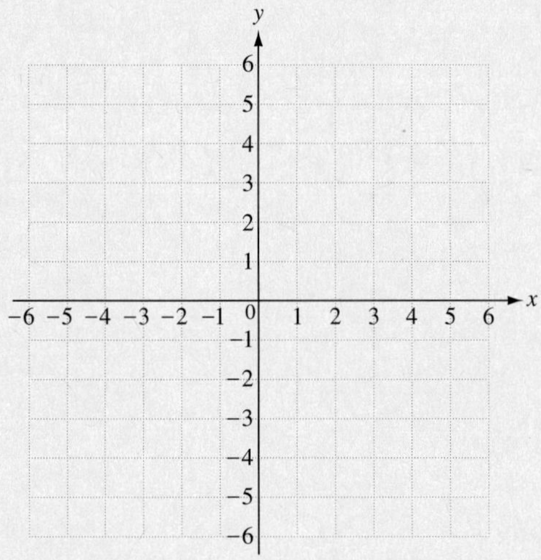

9. $y = x + 2$
Use 0, -1, and -2 as the values of x.

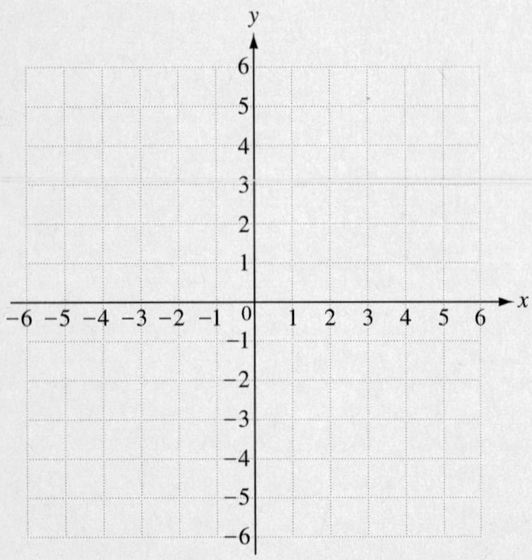

10. $y = x - 1$
Use 0, -1, and -2 as the values of x.

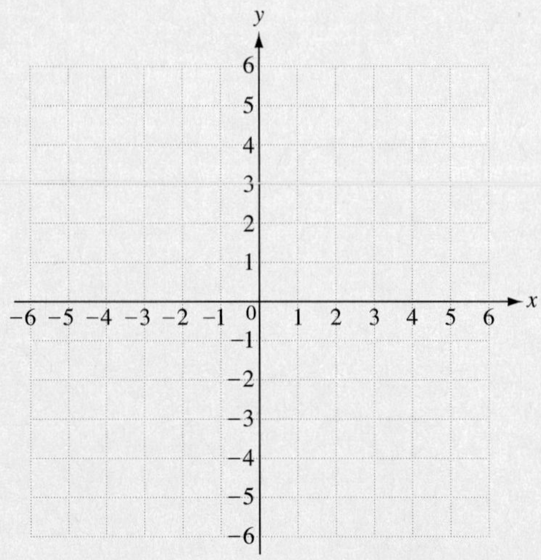

11. $y = -3x$
Use 0, 1, and 2 as the values of x.

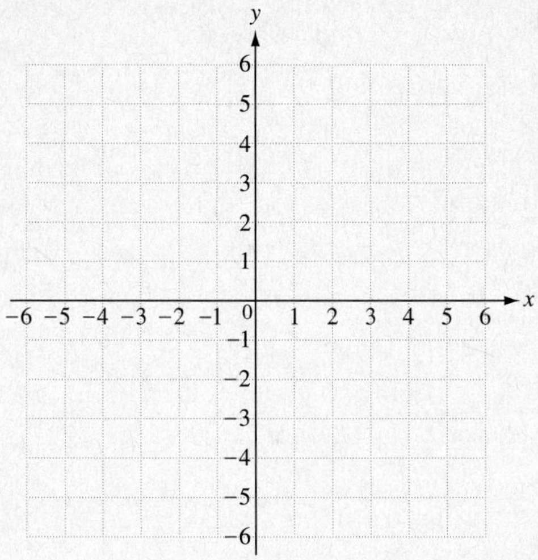

12. $y = -2x$
Use 0, 1, and 2 as the values of x.

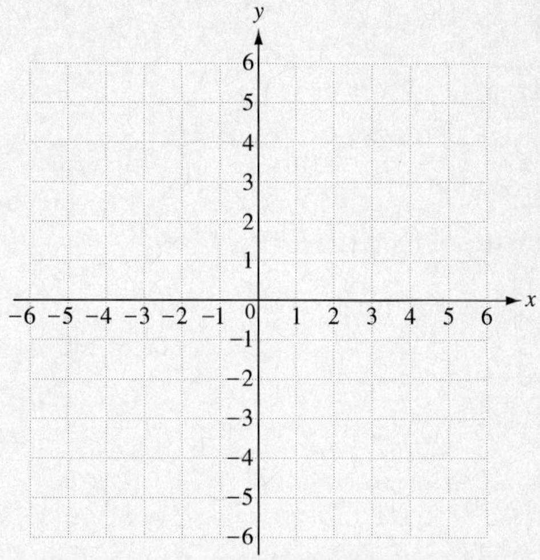

13. Look back at the graphs in Exercises 1, 3, 7, 9, and 11. Which lines have a positive slope? Which lines have a negative slope?

14. Look back at the graphs in Exercises 2, 4, 8, 10, and 12. Which lines have a positive slope? Which lines have a negative slope?

Graph each equation. Make your own table using the listed values of x.

15. $y = \frac{1}{3}x$
Use 0, 3, and 6 as the values of x.

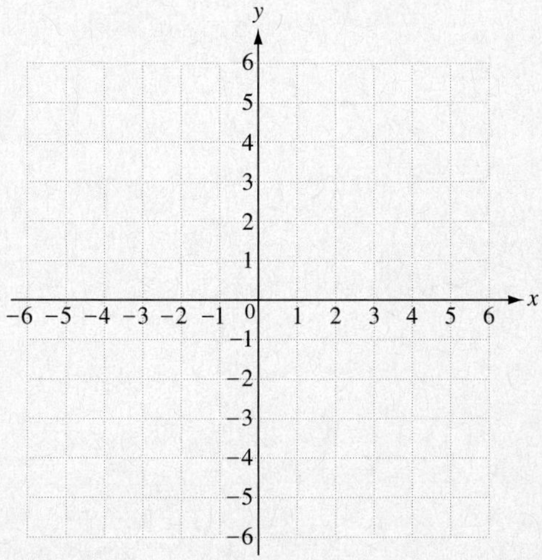

16. $y = \frac{1}{2}x$
Use 0, 2, and 4 as the values of x.

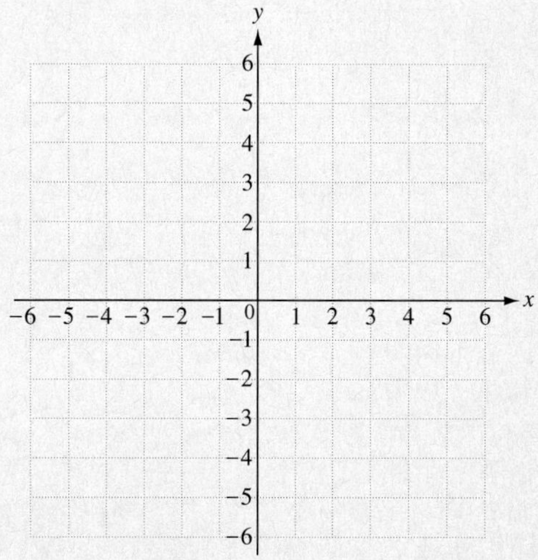

17. $y = x$
Use -1, -2, and -3 as the values of x.

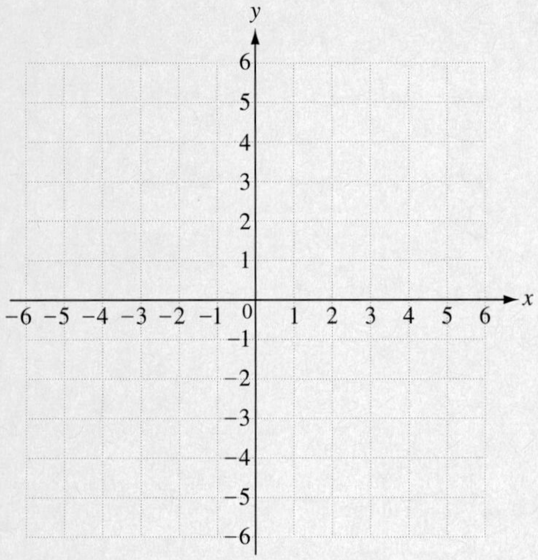

18. $x + y = 0$
Use 1, 2, and 3 as the values of x.

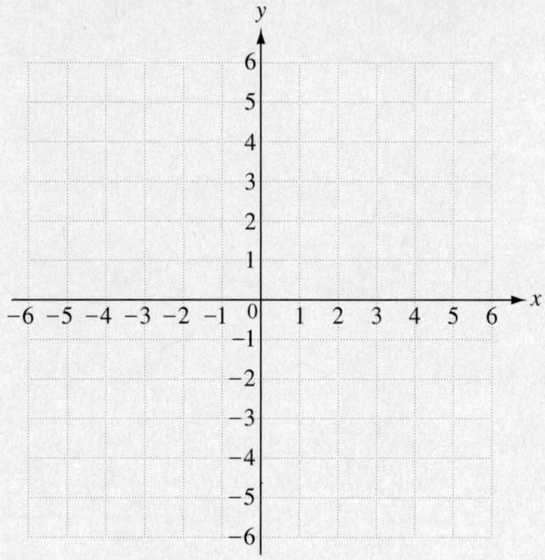

19. $y = -2x + 3$
Use 0, 1, and 2 as the values of x.

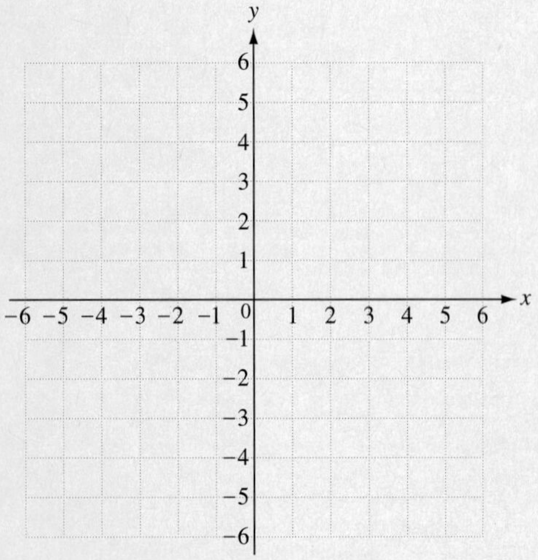

20. $y = 3x - 4$
Use 0, 1, and 2 as the values of x.

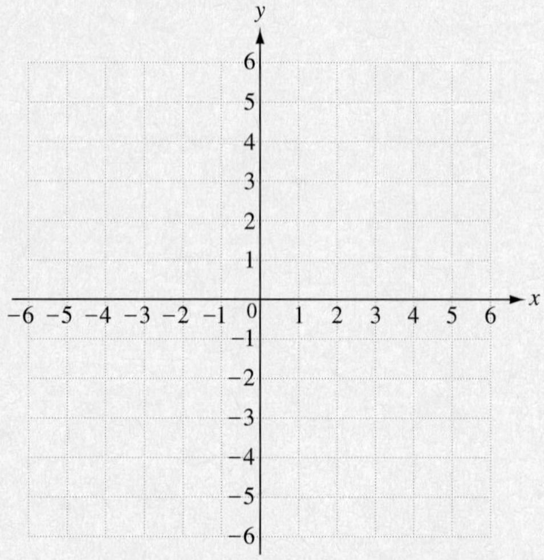

Graph each equation. Choose three values for x. Make a table showing your x values and the corresponding y values. After graphing the equation, state whether the line has a positive or negative slope.

21. $x + y = -3$

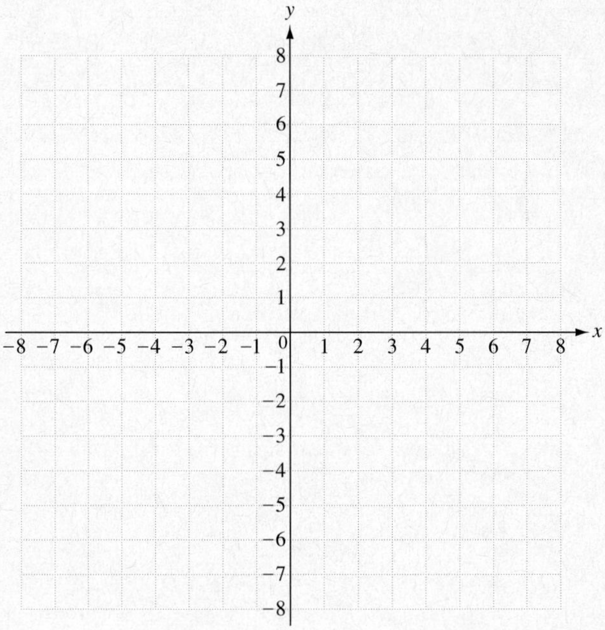

22. $x + y = 2$

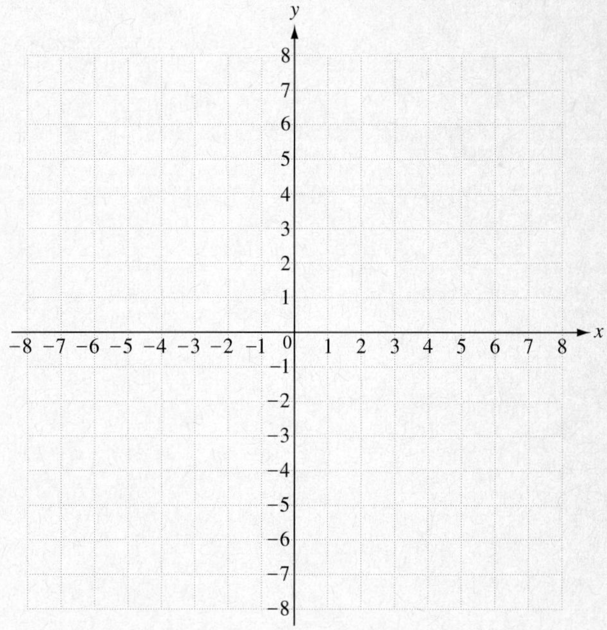

23. $y = \frac{1}{4}x$

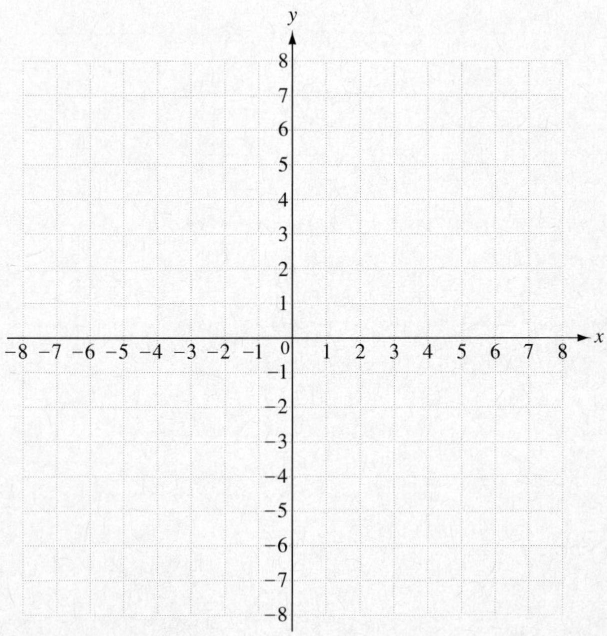

24. $y = -\frac{1}{3}x$

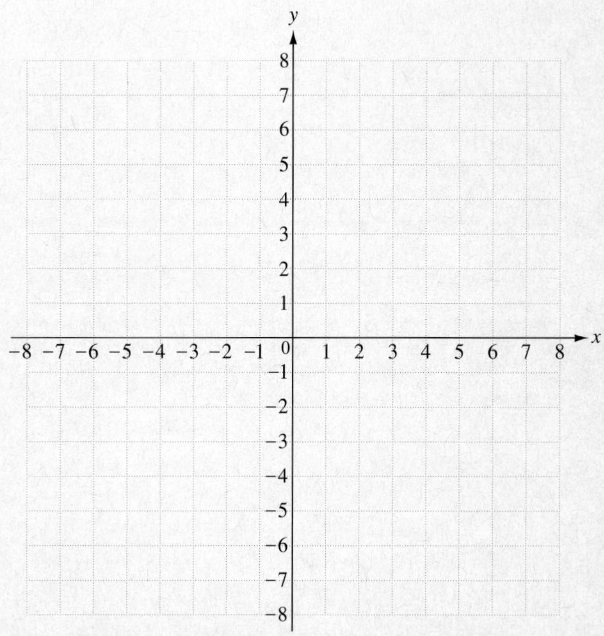

25. $y = x - 5$

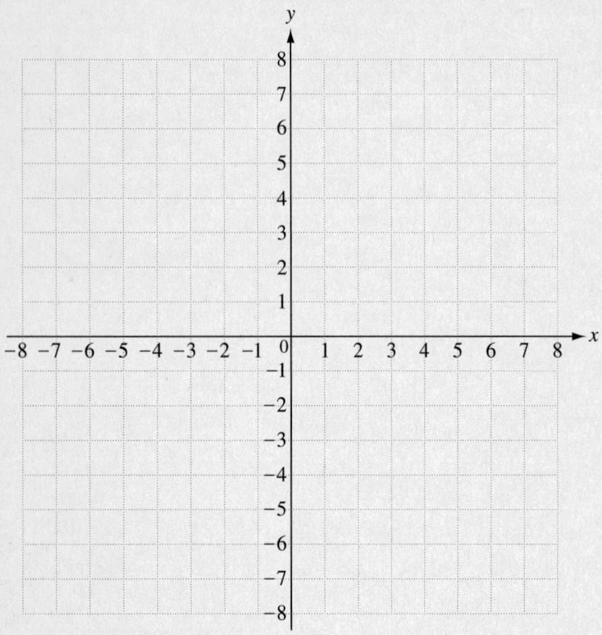

26. $y = x + 4$

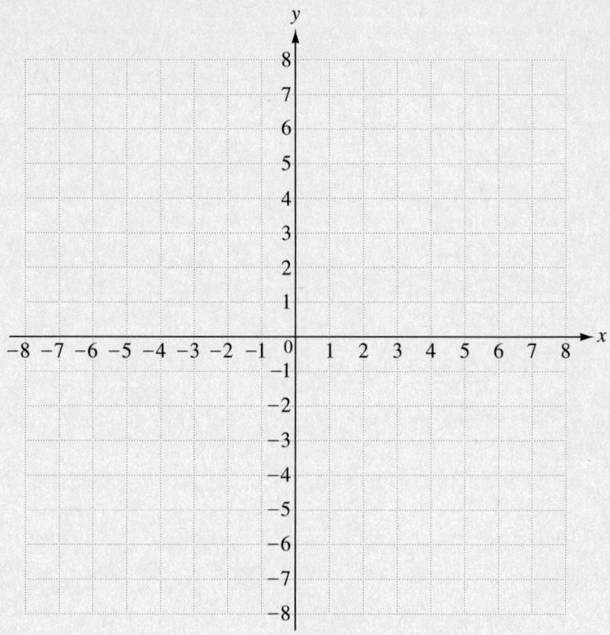

27. $y = -3x + 1$

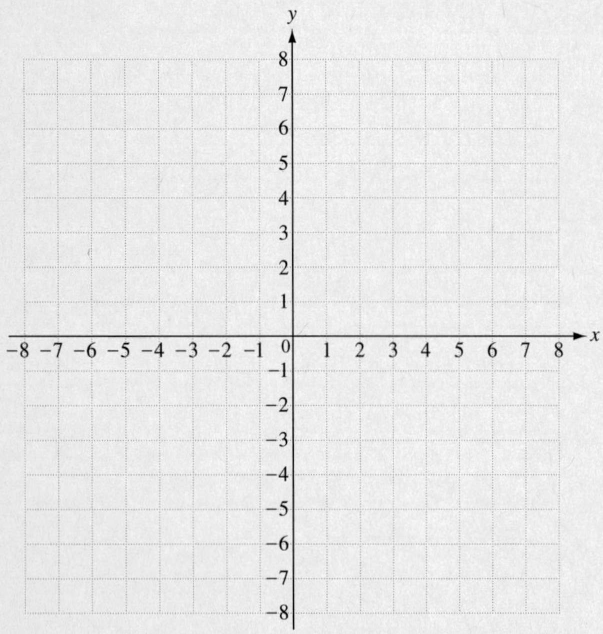

28. $y = 2x - 2$

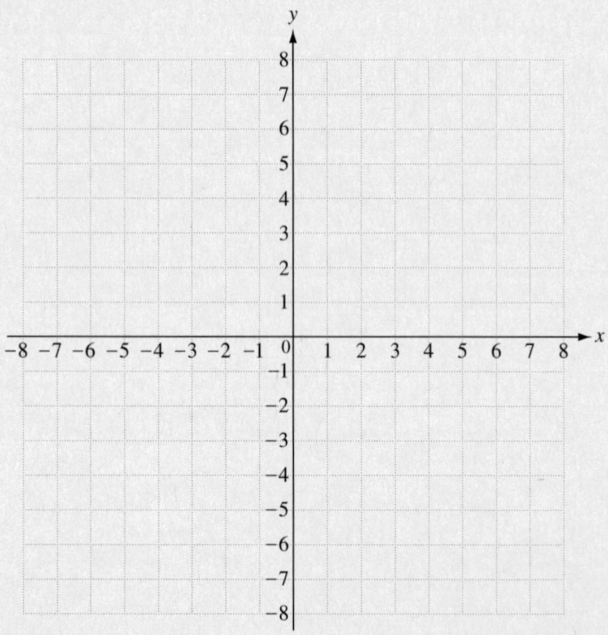

KEY TERMS

9.1	**circle graph**	A circle graph shows how a total amount is divided into parts or sectors. It is based on percents of 360°.
	protractor	A protractor is a device (usually in the shape of a half-circle) used to measure the number of degrees in an angle or parts of a circle.
9.2	**bar graph**	A bar graph uses bars of various heights to show quantity or frequency.
	double-bar graph	A double-bar graph compares two sets of data by showing two sets of bars.
	line graph	A line graph uses dots connected by lines to show trends.
	comparison line graph	A comparison line graph shows how several different items relate to each other by showing a line graph for each item.
9.3	**paired data**	When each number in a set of data is matched with another number by some rule of association, we call it paired data.
	horizontal axis	The horizontal axis is the number line in a coordinate system that goes "left and right."
	vertical axis	The vertical axis is the number line in a coordinate system that goes "up and down."
	***x*-axis**	The horizontal axis is called the x-axis.
	***y*-axis**	The vertical axis is called the y-axis.
	coordinate system	Together, the x-axis and the y-axis form a rectangular coordinate system.
	coordinates	Coordinates are the numbers in the ordered pair that specify the location of a point on a rectangular coordinate system.
	quadrants	The x-axis and the y-axis divide the coordinate system into four regions called quadrants; they are designated with Roman numerals.
9.4	**graph a linear equation**	All the solutions of a linear equation (all the ordered pairs that satisfy the equation) lie along a straight line. When you draw the line you have graphed the equation.

QUICK REVIEW

Concepts

9.1 Constructing a Circle Graph

1. Determine the percent of the total for each item.
2. Find the number of degrees out of 360° that each percent represents.
3. Use a protractor to measure the number of degrees for each item in the circle.

Examples

Construct a circle graph for the following table, which lists expenses for a business trip.

Item	***Amount***
Transportation	\$200
Lodging	\$300
Food	\$250
Entertainment	\$150
Other	\$100
Total	**\$1000**

(Continued)

Concepts	*Examples*

9.1 Constructing a Circle Graph (continued)

Item	***Amount***	***Percent of Total***	***Sector Size***
Transportation	\$200	$\frac{\$200}{\$1000} = \frac{1}{5} = \mathbf{20\%}$ so $20\% \cdot 360^\circ = 0.20 \cdot 360$	$= 72^\circ$
Lodging	\$300	$\frac{\$300}{\$1000} = \frac{3}{10} = \mathbf{30\%}$ so $30\% \cdot 360^\circ = 0.30 \cdot 360$	$= 108^\circ$
Food	\$250	$\frac{\$250}{\$1000} = \frac{1}{4} = \mathbf{25\%}$ so $25\% \cdot 360^\circ = 0.25 \cdot 360$	$= 90^\circ$
Entertainment	\$150	$\frac{\$150}{\$1000} = \frac{3}{20} = \mathbf{15\%}$ so $15\% \cdot 360^\circ = 0.15 \cdot 360$	$= 54^\circ$
Other	\$100	$\frac{\$100}{\$1000} = \frac{1}{10} = \mathbf{10\%}$ so $10\% \cdot 360^\circ = 0.10 \cdot 360$	$= 36^\circ$

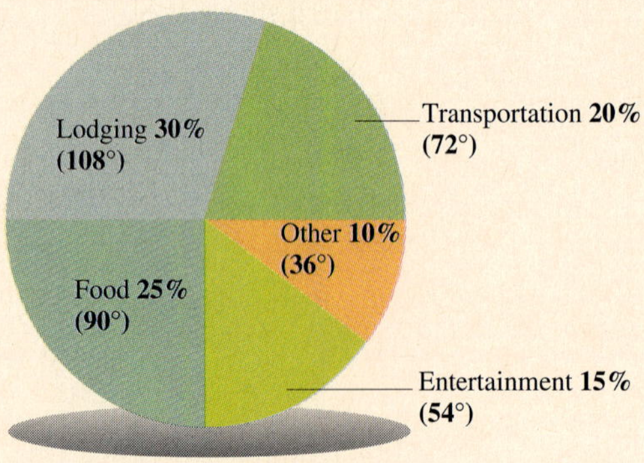

9.2 Reading a Bar Graph

The height of the bar is used to show the quantity or frequency (number) in a specific category. Use a ruler or straight edge to line up the top of each bar with the numbers on the left side of the graph.

Use the bar graph below to determine the number of students who earned each letter grade.

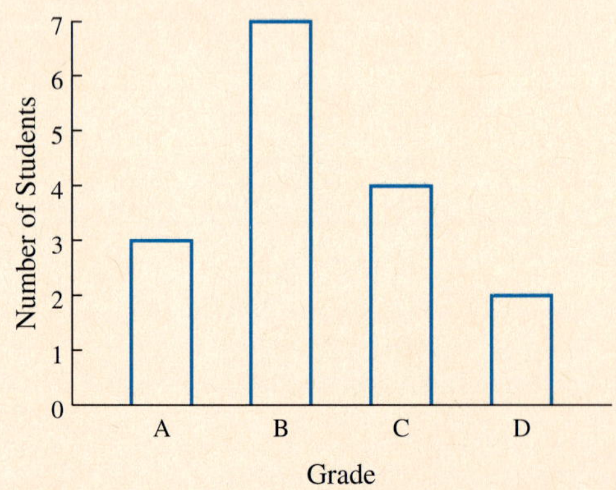

Grade	***Number of Students***
A	3
B	7
C	4
D	2

Concepts | *Examples*

9.2 Reading a Line Graph

A dot is used to show the number or quantity in a specific class. The dots are connected with lines. This kind of graph is used to show a trend.

The line graph below shows the sales volume for each of four years.

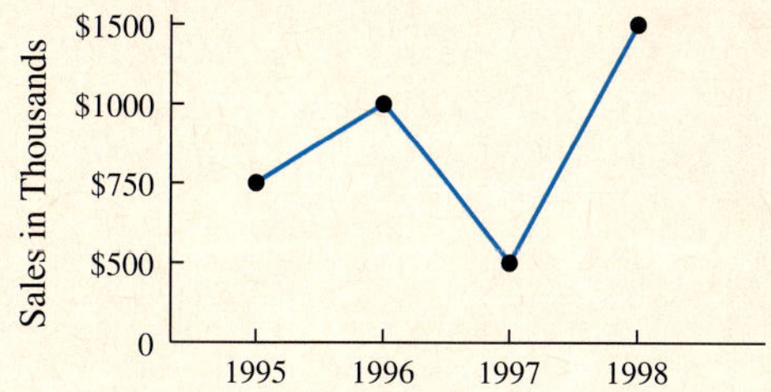

Find the sales in each year.

Year	***Total Sales***
1995	$750 • 1000 = $750,000
1996	$1000 • 1000 = $1,000,000
1997	$500 • 1000 = $500,000
1998	$1500 • 1000 = $1,500,000

9.3 Plotting Points

Start at the center of the coordinate system (the origin). The first number in an ordered pair tells you how far to move *left* or *right* along the horizontal axis; *positive* numbers are to the *right, negative* numbers to the *left.* The second number in an ordered pair tells you how far to move *up* or *down; positive* numbers are *up, negative* numbers *down.*

To plot **(3, −2)**, move to the *right* 3 units and then move *down* 2 units.

To plot **(−2, 3)**, move to the *left* 2 units and then move *up* 3 units.

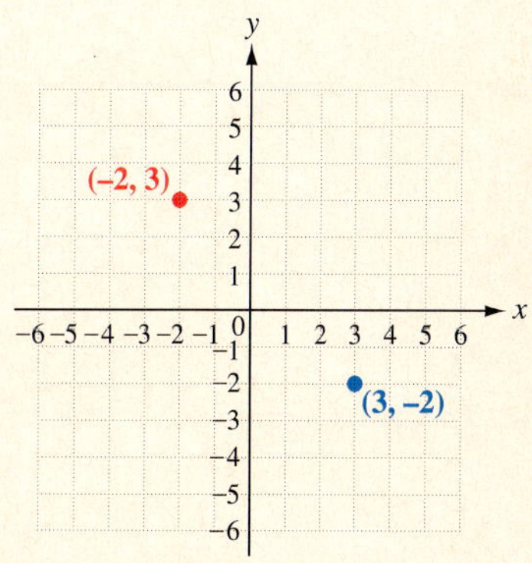

9.3 Identifying Quadrants

The x-axis and y-axis divide the coordinate system into four regions called quadrants. The quadrants are designated with Roman numerals.

Points in the first quadrant fit the pattern $(+, +)$.

Points in the second quadrant fit the pattern $(-, +)$.

Points in the third quadrant fit the pattern $(-, -)$.

Points in the fourth quadrant fit the pattern $(+, -)$.

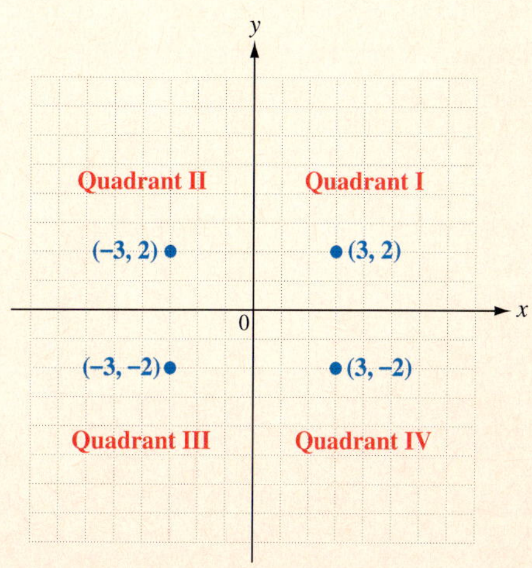

Concepts	*Examples*
9.4 Graphing Linear Equations	
Choose any three values for x. Then find the corresponding values of y. Plot the three ordered pairs on a coordinate system. Draw a line through the points, extending it in both directions. If the three points do *not* lie on a straight line, there is an error in your work. Every point on the line is a solution of the given equation.	Graph $y = -2x$ (see table and graph below)

x	y (must be -2 times x)	Ordered pair (x, y)
1	$-2 \cdot 1$ is -2.	$(1, -2)$
2	$-2 \cdot 2$ is -4.	$(2, -4)$
3	$-2 \cdot 3$ is -6.	$(3, -6)$

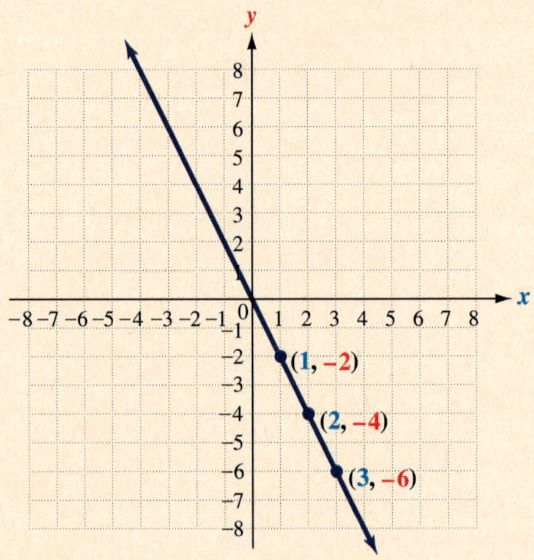

9.4 Identifying the Slope of a Line

As you move from left to right, a line with *positive* slope tilts *upward* or rises. A positive slope means that, as the value of x increases, the value of y also increases.

A line with *negative* slope tilts *downward* or falls. A negative slope means that, as the value of x increases, the value of y decreases.

Look at the graph of $y = -2x$ in the preceding example. Then complete these sentences.

The graph of $y = -2x$ has a ____________ slope. As the value of x increases, the value of y ____________.

The line has a negative slope because it tilts downward from left to right. As the value of x increases, the value of y decreases (does the opposite).

Name Date Class

CHAPTER 9 REVIEW EXERCISES

[9.1] *Use this circle graph for Exercises 1–5.*

1. (a) What is the largest single expense of the vacation? How much is that item?

(b) What is the second most expensive item? How much is that item?

COST OF A FAMILY VACATION

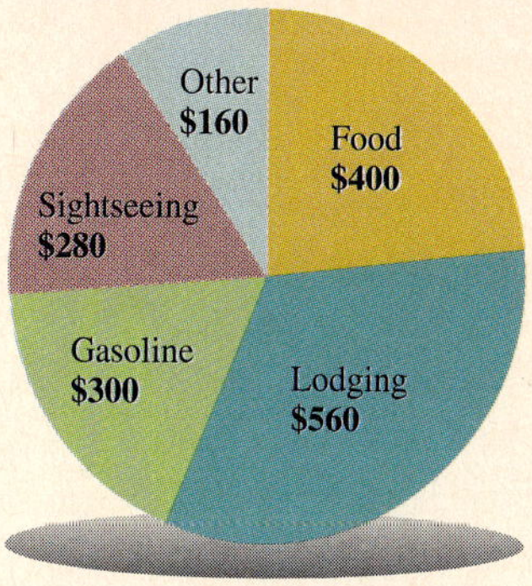

Find each ratio. Write the ratio as a fraction in lowest terms.

2. cost of the food to the total cost of the vacation

3. cost of the gasoline to the total cost of the vacation

4. cost of sightseeing to the total cost of the vacation

5. cost of gasoline to the cost of the *other* category

[9.2] *The bar graph shown is adapted from* USA TODAY. *If there were 16,500,000 cellular phones sold last year, find the number of cellular phones sold in each of the following categories.*

TEENS AND CELLULAR PHONES

28.5 million teenagers in the U.S. have nearly $100 billion to spend each year — making them a major market for cellular phones.

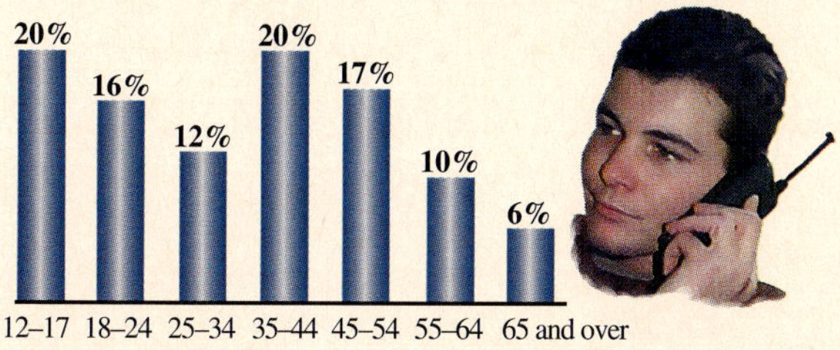

Note: Adds up to 101% due to rounding.
Source: Interep Radio Store

6. ages 12–17

7. ages 45–54

8. ages 55–64

9. ages 25–34

10. What two age groups buy the most cellular phones? Give one possible explanation why these age groups buy more phones than others.

11. What two age groups buy the fewest cellular phones? Give one possible explanation why these age groups buy fewer phones than others.

This double-bar graph shows the number of acre-feet of water in Lake Natoma for each of the first six months of 1997 and 1998.

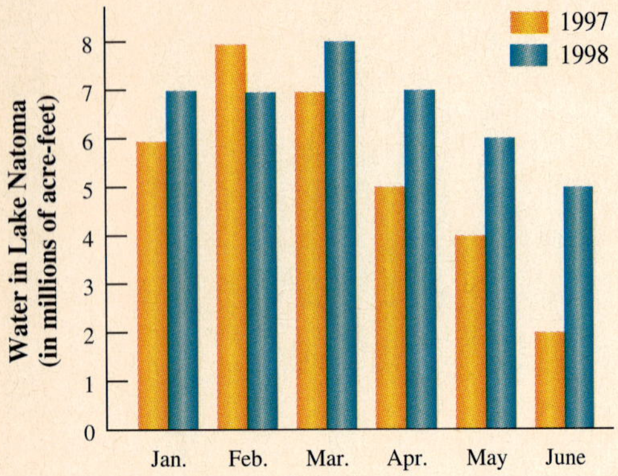

12. (a) During which month in 1998 was the greatest amount of water in the lake? How much was there?

(b) During which month in 1997 was the least amount of water in the lake? How much was there?

13. (a) How many acre-feet of water were in the lake in June 1998?

(b) How many acre-feet of water were in the lake in May 1997?

14. Find the decrease in the amount of water in the lake from March 1997 to June 1997.

15. Find the decrease in the amount of water in the lake from April 1998 to June 1998.

This comparison line graph shows the annual grocery purchases of two different childcare centers during each of five years. Find the amount of annual grocery purchases in each of the following years.

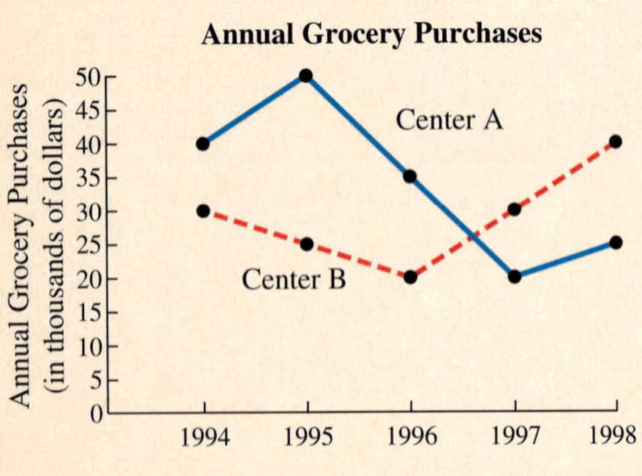

16. Center A in 1995

17. Center A in 1997

18. Center B in 1996

19. Center B in 1998

20. What trend do you see in Center A's purchases? Why might this have happened?

21. What trend do you see in Center B's purchases since 1996? Why might this have happened?

G [9.1] *The Broadway Hair Salon spent a total of $22,400 to open a new shop. The breakdown of expenditures for various items is shown. Find all the missing numbers in Exercises 22–26.*

Item	*Dollar Amount*	*Percent of Total*	*Degrees of Circle*
22. plumbing and electrical changes	$2240	10%	______
23. work stations	$7840	______	126°
24. small appliances	$4480	______	72°
25. interior decoration	$5600	______	90°
26. supplies	$2240	10%	______

27. Draw a circle graph by using the information in Exercises 22–26.

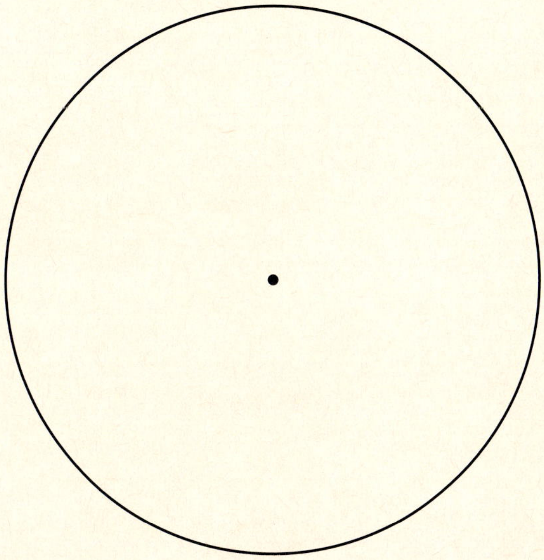

[9.3] *In Exercise 28, plot each point on the rectangular coordinate system and label it with its coordinates. In Exercise 29, give the coordinates of each point.*

28. $(1, 7)$ $(-1\frac{1}{2}, 0)$ $(-4, -2)$ $(0, 3)$
$(2, -\frac{1}{2})$ $(-7, 1)$ $(5, -5)$ $(0, -4)$

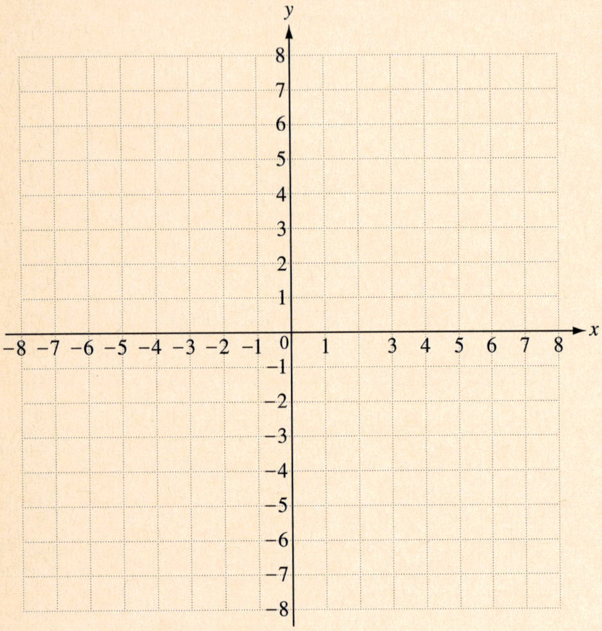

29.

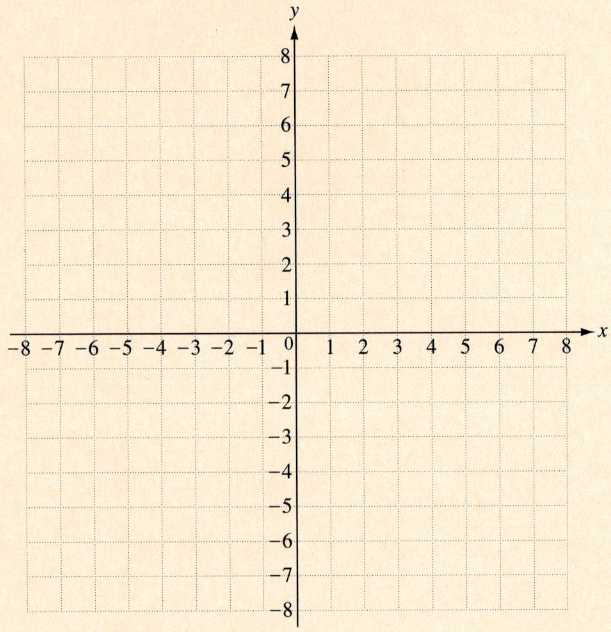

G **[9.4]** *Graph each equation. Make your own table using the listed values of x. State whether each line has a positive or negative slope.*

30. $x + y = -2$
Use 0, 1, and 2 as the values of x.

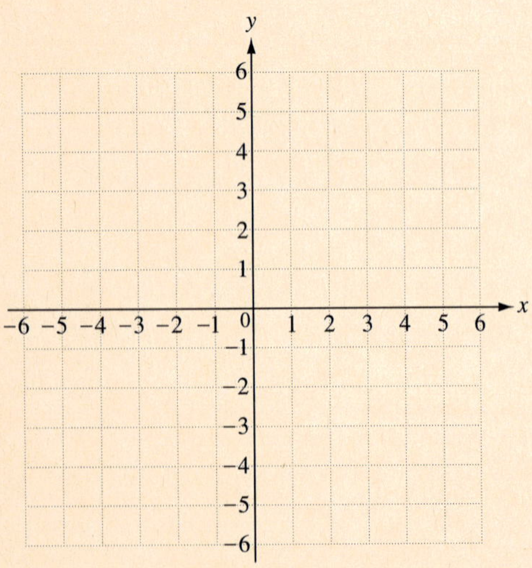

31. $y = x + 3$
Use 0, 1, and 2 as the values of x.

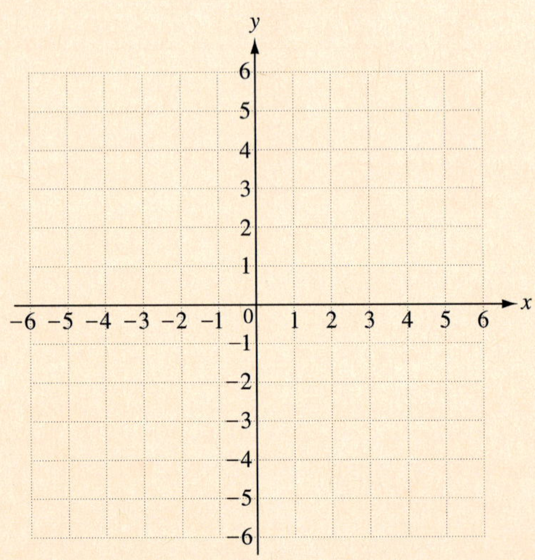

Name Date Class

CHAPTER 9 TEST

This circle graph shows the development costs for the Shady Brook Public Housing Subdivision. The total budget of the subdivision is $5,600,000.

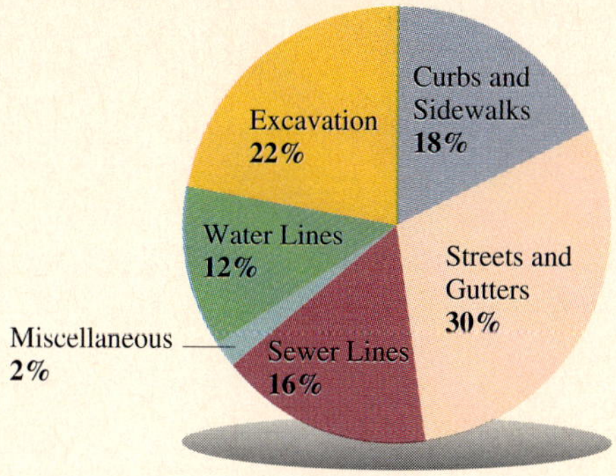

1. How much is budgeted for excavation?

2. How much is budgeted for curbs and sidewalks?

3. Which item costs the most? How much will it cost?

4. Which item costs the least? How much will it cost?

1. ________
2. ________
3. ________
4. ________

This graph shows one student's income and expenses for four years.

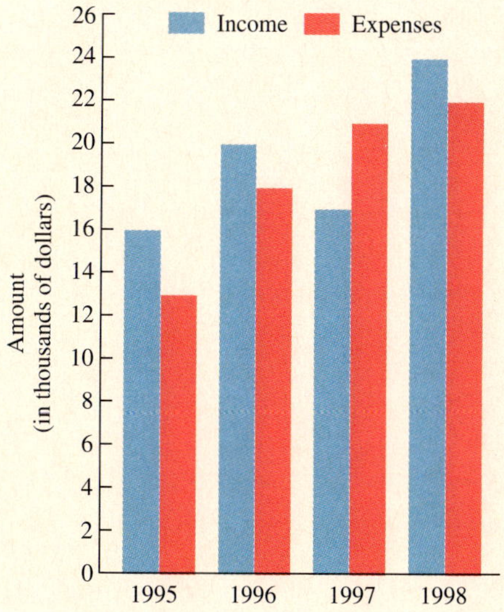

5. In what year did the student's expenses exceed income? By how much?

5. ________

6. How much did the student's expenses increase from 1995 to 1998?

6. ________

7. In what year did the student's income decline? Give two possible explanations for the decline.

7. ________

This graph shows enrollment at two community colleges.

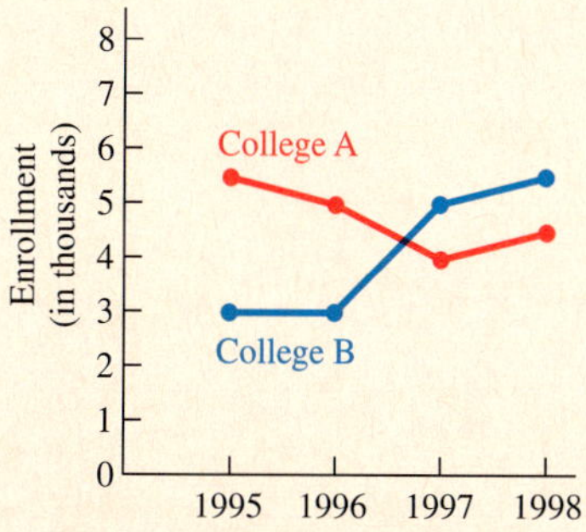

8. What was college A's enrollment in 1995? What was college B's enrollment in 1996?

8. ________

9. Which college had higher enrollment in 1998? How much higher?

9. ________

10. Give two possible explanations for the fact that college B's enrollment passed college A's.

10. ________

During a one-year period, Oak Mill Furniture Sales had the following expenses. Find all the missing numbers.

11. ____________

12. ____________

13. ____________

14. ____________

15. ____________

Item	***Dollar Amount***	***Percent of Total***	***Degrees of a Circle***
11. salaries	$36,000	30%	________
12. delivery expense	$12,000	10%	________
13. advertising	$24,000	20%	________
14. rent	$36,000	30%	________
15. other	$12,000	________	36°

16. Draw a circle graph using the information in Exercises 11–15 and a protractor.

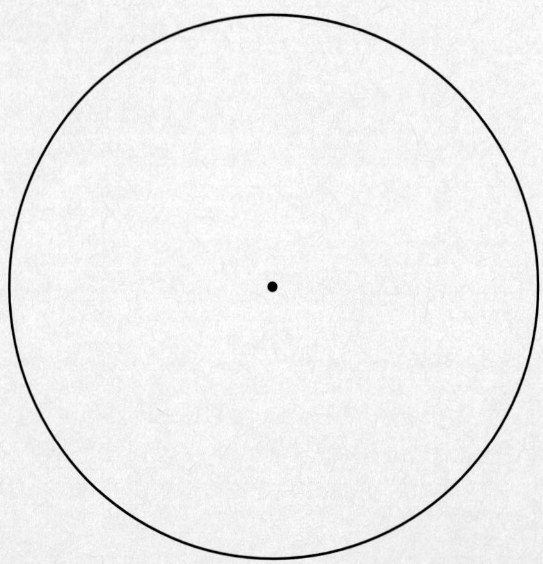

Plot each point on the coordinate system below. Label each point with its coordinates.

17. (−5, 3) **18.** (1, −4) **19.** (0, 6) **20.** (7, 0)

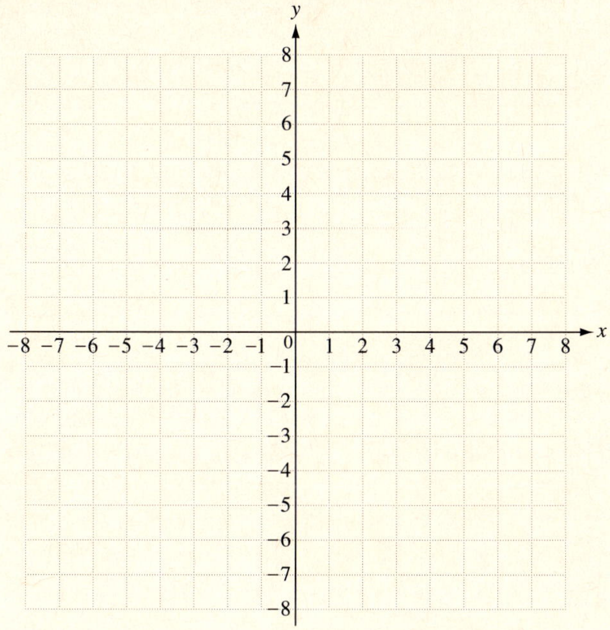

Give the coordinates of each lettered point shown below.

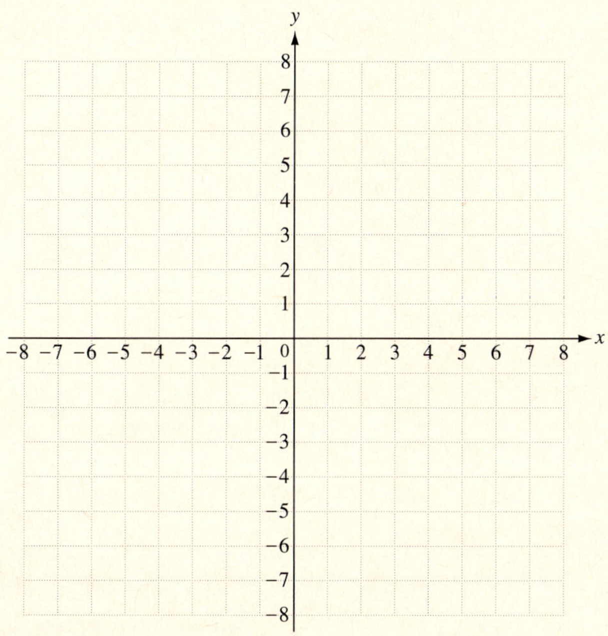

21. Point *A* **22.** Point *B* **23.** Point *C* **24.** Point *D*

Name the quadrant in which each lettered point is located.

25. Point *A* **26.** Point *B* **27.** Point *C* **28.** Point *D*

29. Explain why (−4, 3) and (3, −4) do *not* indicate the same point.

21. ______________

22. ______________

23. ______________

24. ______________

25. ______________

26. ______________

27. ______________

28. ______________

29. ______________

30. ____________________

31. ____________________

Graph each equation. Make your own table using the given values of x.

30. $x + y = -5$ Use 0, 1, and 2 as the values of x.

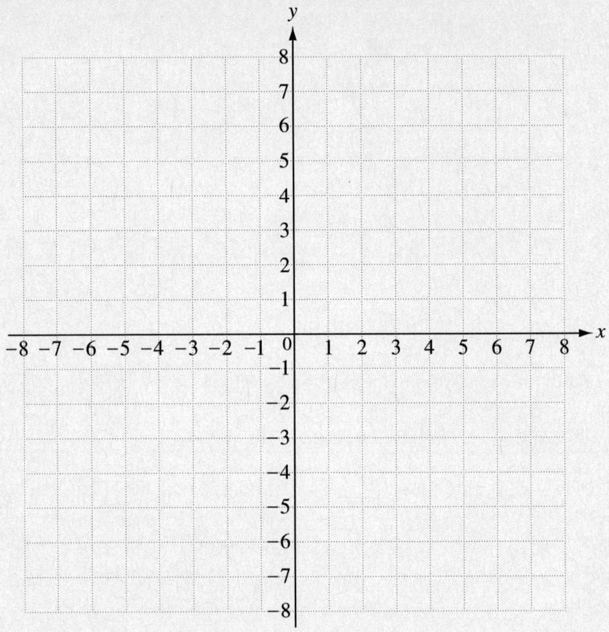

31. $y = -4x$ Use 0, 1, and 2 as the values of x.

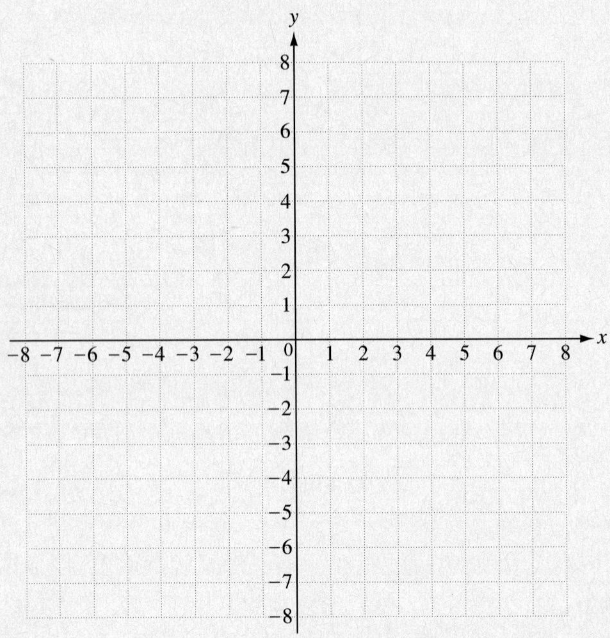

CUMULATIVE REVIEW EXERCISES CHAPTERS 1–9

1. Write these numbers in words.
   (a) 0.0602

   (b) 300,000,560

2. Write these numbers in digits.
   (a) Seventy billion, five million, forty-three

   (b) Eighteen and nine hundredths

3. Round each number as indicated.
   (a) 3.049 to the nearest tenth.

   (b) 0.7982 to the nearest hundredth.

   (c) 68,592,000 to the nearest million.

4. Find the mean, median, and mode for this set of data on student ages in years: 23, 29, 18, 23, 36, 62, 23, 19, 27, 30.

Simplify.

5. $\frac{4}{5} + 2\frac{1}{3}$

6. $\frac{0.8}{-3.2}$

7. $5^2 + (-4)^3$

8. $(0.002)(-0.05)$

9. $\frac{4a}{9} \cdot \frac{6b}{2a^3}$

10. $1\frac{1}{4} - 3\frac{5}{6}$

11. $-13 + 2.993$

12. $\frac{2}{7} - \frac{8}{x}$

13. $-3 - 33$

14. $\frac{10m}{3n^2} \div \frac{2m^2}{5n}$

15. $\dfrac{-3}{-\frac{9}{10}}$

16. $\frac{3(-7)}{2^4 - 16}$

17. $10 - 2\frac{5}{8}$

18. $\frac{3}{w} + \frac{x}{6}$

19. $8 + 4(2 - 5)$

20. $\frac{7}{8}$ of 960

21. $\dfrac{(-4)^2 + 8(0 - 2)}{8 \div 2(-3 + 5) - 10}$

22. $0.5 - 0.25(3.2)^2$

23. $6\left(-\frac{1}{2}\right)^3 + \frac{2}{3}\left(\frac{3}{5}\right)$

Evaluate each expression when $a = 2$, $b = 3$, and $c = -5$.

24. $20 + 4c$

25. $7b - 4c$

26. $-2ac^2$

Simplify each expression.

27. $-4x + x^2 - x$

28. $3(y - 4) + 2y$

29. $-5(8h^2)$

Solve each equation. Show the steps you used.

30. $2x - 3 = -20 + 3$

31. $-12 = 3(y + 2)$

32. $6x = 14 - x$

33. $-8 = \frac{2}{3}m + 2$

34. $\frac{2}{13.5} = \frac{2.4}{n}$

35. $-20 = -w$

36. $3.4x - 6 = 8 + 1.4x$

37. $2(h - 1) = -3(h + 12) - 11$

Translate each sentence into an equation and solve it.

38. If five times a number is subtracted from 12, the result is the number. Find the number.

39. When -8 is added to twice a number, the result is -28. What is the number?

Solve each application problem using the six problem solving steps from Chapter 3.

40. An $1800 lottery prize is to be split between two people so that one person gets $500 more than the other. How much will each person receive?

41. The length of a rectangular movie theater is three times the width. The perimeter is 280 ft. Find the width and the length.

Find the unknown length, perimeter, circumference, area, or volume. When necessary, use 3.14 as the approximate value of π, and round answers to the nearest tenth.

42. Find the perimeter and area.

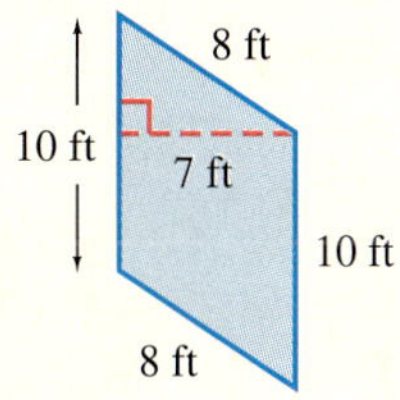

43. Find the perimeter and area.

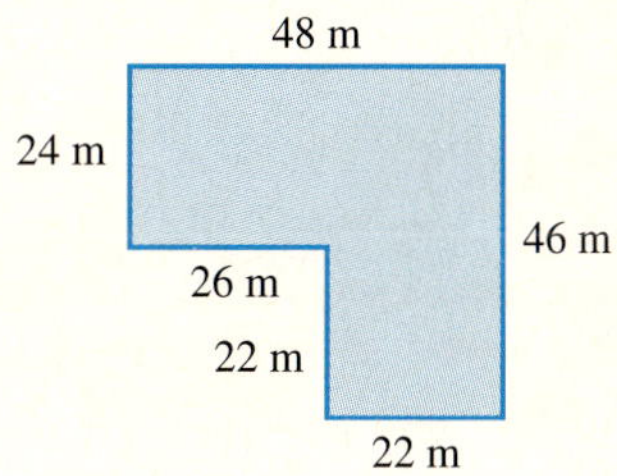

44. Find the unknown length, the perimeter, and the area.

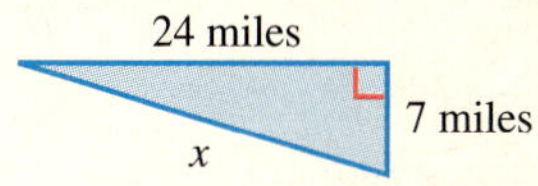

45. A circular hot tub has a diameter of 6 ft. Find its circumference and the area of the tub's floor.

46. A cylindrical oil tank is 12 m tall and has a diameter of 20 m. Find its volume.

47. Find the volume of a storage shed that is 4 yd long, 3 yd wide, and $2\frac{1}{2}$ yd high.

Convert the following using unit fractions, the metric conversion line, or the temperature conversion formulas.

48. $2\frac{1}{4}$ hours to minutes

49. 54 in. to feet

50. 1.85 L to milliliters

51. 35 mm to centimeters

52. 10 g to kilograms

53. 25 °F to Celsius; round to the nearest whole degree.

Solve these application problems.

54. A survey found that 19 out of 25 adults are nonsmokers. If Mathtronic has 732 employees, how many would be expected to be nonsmokers? Round to the nearest whole number.

55. Esther bought a cellular phone at a 15% discount. The regular price was \$129. She also paid $6\frac{1}{2}\%$ sales tax. What was her total cost for the phone, to the nearest cent?

56. Abiola earned 167 points out of 180 on the prealgebra final exam. What percent of the points did she earn, to the nearest tenth?

57. Century College received a \$30,000 technology grant. **(a)** How many \$957 computers were they able to buy with the grant money? How much was left over? **(b)** How many \$19 calculators could they buy with the money left over from the computer purchase? How much was left then?

58. Wayne bought $2\frac{1}{4}$ pounds of lunch meat. He used $\frac{1}{6}$ pound of meat on each of five sandwiches. How much meat is left?

59. Plot the following points and label each one with its coordinates. Which of the points are in the second quadrant? Which are in the third quadrant?

$(3, -5)$ $(0, 4)$ $(-2, 3\frac{1}{2})$ $(1, 0)$ $(4, 2)$

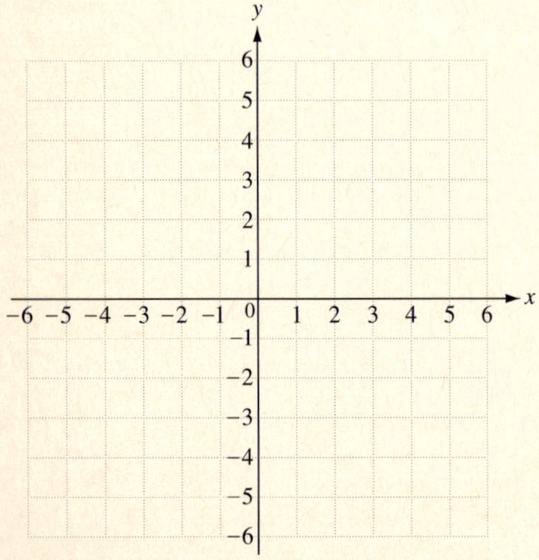

60. Graph $y = x - 4$. Use 0, 1, and 2 as the values of x. After graphing the equation, state whether the line has a positive or negative slope.

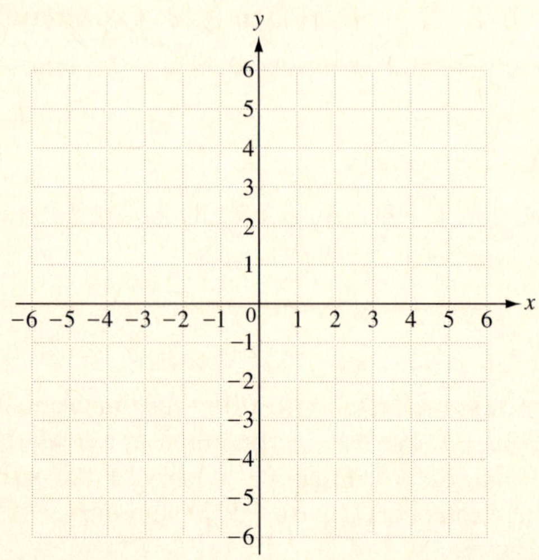

Exponents and Polynomials 10

1 The Product Rule and Power Rules for Exponents

OBJECTIVES

1. Use exponents.
2. Use the product rule for exponents.
3. Use the rule $(a^m)^n = a^{mn}$.
4. Use the rule $(ab)^m = a^m b^m$.
5. Use the rule $\left(\frac{a}{b}\right)^m = \frac{a^m}{b^m}$.
6. Use combinations of the rules for exponents.

OBJECTIVE 1 We have seen how to use exponents to write repeated products. Recall that in the expression 5^2, the number 5 is called the *base* and 2 is called the *exponent* or *power*. The expression 5^2 is called an *exponential expression.* Usually we do not write a quantity with an exponent of 1; however, sometimes it is convenient to do so. In general, for any quantity a, $a = a^1$.

EXAMPLE 1 Using Exponents

Write $3 \cdot 3 \cdot 3 \cdot 3 \cdot 3$ in exponential form, and evaluate the exponential expression.

Since 3 occurs as a factor five times, the base is 3 and the exponent is 5. The exponential expression is 3^5, read "3 to the fifth power" or simply "3 to the fifth." The value is

$$3^5 = 3 \cdot 3 \cdot 3 \cdot 3 \cdot 3 = 243.$$

WORK PROBLEM 1 AT THE SIDE.

1. Write $2 \cdot 2 \cdot 2 \cdot 2$ in exponential form, and evaluate.

EXAMPLE 2 Evaluating an Exponential Expression

Evaluate each exponential expression. Name the base and the exponent.

	Base	*Exponent*
(a) $5^4 = 5 \cdot 5 \cdot 5 \cdot 5 = 625$	5	4
(b) $-5^4 = -1 \cdot 5^4 = -1 \cdot (5 \cdot 5 \cdot 5 \cdot 5) = -625$	5	4
(c) $(-5)^4 = (-5)(-5)(-5)(-5) = 625$	-5	4

Caution

It is important to understand the difference between parts (b) and (c) of Example 2. In -5^4 the lack of parentheses shows that the exponent 4 refers only to the base 5, and not -5; in $(-5)^4$ the parentheses show that the exponent 4 refers to the base -5. In summary, $-a^n$ and $(-a)^n$ are not always the same.

Expression	*Base*	*Exponent*	*Example*
$-a^n$	a	n	$-3^2 = -(3 \cdot 3) = -9$
$(-a)^n$	$-a$	n	$(-3)^2 = (-3)(-3) = 9$

ANSWERS

1. $2^4 = 16$

2. Evaluate each exponential expression. Name the base and the exponent.

(a) $(-2)^5$ **(b)** -2^5

(c) -4^2 **(d)** $(-4)^2$

3. Find each product by the product rule, if possible.

(a) $8^2 \cdot 8^5$

(b) $(-7)^5 \cdot (-7)^3$

(c) $y^3 \cdot y$

(d) $4^2 \cdot 3^5$

(e) $6^4 + 6^2$

ANSWERS

2. (a) −32; −2; 5 **(b)** −32; 2; 5 **(c)** −16; 4; 2 **(d)** 16; −4; 2

3. (a) 8^7 **(b)** $(-7)^8$ **(c)** y^4 **(d)** cannot use the product rule **(e)** cannot use the product rule

◀◀ **WORK PROBLEM 2 AT THE SIDE.**

OBJECTIVE 2 By the definition of exponential expressions,

$$2^4 \cdot 2^3 = \overbrace{(2 \cdot 2 \cdot 2 \cdot 2)}^{4 \text{ factors}}\overbrace{(2 \cdot 2 \cdot 2)}^{3 \text{ factors}}$$
$$= \underbrace{2 \cdot 2 \cdot 2 \cdot 2 \cdot 2 \cdot 2 \cdot 2}_{4 + 3 = 7 \text{ factors}}$$
$$= 2^7.$$

Also,

$$6^2 \cdot 6^3 = (6 \cdot 6)(6 \cdot 6 \cdot 6)$$
$$= 6 \cdot 6 \cdot 6 \cdot 6 \cdot 6$$
$$= 6^5.$$

Generalizing from these examples, $2^4 \cdot 2^3 = 2^{4+3} = 2^7$ and $6^2 \cdot 6^3 = 6^5$. In each case, adding the exponents gives the exponent of the product, suggesting the **product rule for exponents.**

Product Rule for Exponents

For any positive integers m and n, $a^m \cdot a^n = a^{m+n}$.
(Keep the same base and add the exponents.)
Example: $6^2 \cdot 6^5 = 6^{2+5} = 6^7$

EXAMPLE 3 Using the Product Rule

Use the product rule for exponents to find each product, if possible.

(a) $6^3 \cdot 6^5 = 6^{3+5} = 6^8$ by the product rule.

(b) $(-4)^7(-4)^2 = (-4)^{7+2} = (-4)^9$ by the product rule.

(c) $x^2 \cdot x = x^2 \cdot x^1 = x^{2+1} = x^3$

(d) $m^4 \cdot m^3 = m^{4+3} = m^7$

(e) The product rule does not apply to the product $2^3 \cdot 3^2$ because the bases are different.

$$2^3 \cdot 3^2 = 8 \cdot 9 = 72.$$

(f) The product rule does not apply to $2^3 + 2^4$ because it is a *sum,* not a *product.*

$$2^3 + 2^4 = 8 + 16 = 24$$

Caution

The bases must be the same before we can apply the product rule for exponents.

◀◀ **WORK PROBLEM 3 AT THE SIDE.**

EXAMPLE 4 Using the Product Rule

Multiply $2x^3$ and $3x^7$.

Since $2x^3$ means $2 \cdot x^3$ and $3x^7$ means $3 \cdot x^7$, we use the associative and commutative properties and the product rule to get

$$2x^3 \cdot 3x^7 = 2 \cdot 3 \cdot \boldsymbol{x^3} \cdot \boldsymbol{x^7} = 6\boldsymbol{x^{10}}.$$

Caution

Be sure you understand the difference between *adding* and *multiplying* exponential expressions. For example,

$$8\boldsymbol{x^3} + 5\boldsymbol{x^3} = 13\boldsymbol{x^3}$$

but $$(8\boldsymbol{x^3})(5\boldsymbol{x^3}) = 8 \cdot 5 \cdot \boldsymbol{x^{3+3}} = 40\boldsymbol{x^6}.$$

WORK PROBLEM 4 AT THE SIDE.

OBJECTIVE 3 We can simplify an expression such as $(8^3)^2$ with the product rule for exponents, as follows.

$$(8^3)^2 = (8^3)(8^3) = 8^{3+3} = 8^6$$

The product of the exponents, 3 and 2, in $(8^3)^2$ gives the exponent in 8^6. As another example,

$$\begin{aligned}(5^2)^3 &= 5^2 \cdot 5^2 \cdot 5^2 \\ &= 5^{2+2+2} \\ &= 5^6,\end{aligned}$$

and $2 \cdot 3 = 6$. These examples suggest power rule (a) for exponents.

Power Rule (a) for Exponents

For any positive integers m and n, $\qquad (a^m)^n = a^{mn}$.
(Raise a power to a power by multiplying exponents.)
Example: $(3^2)^4 = 3^{2 \cdot 4} = 3^8$

EXAMPLE 5 Using Power Rule (a)

Use power rule (a) for exponents to simplify each expression.

(a) $(2^5)^3 = 2^{5 \cdot 3} = 2^{15}$

(b) $(5^7)^2 = 5^{7 \cdot 2} = 5^{14}$

(c) $(x^2)^5 = x^{2 \cdot 5} = x^{10}$

(d) $(n^3)^2 = n^{3 \cdot 2} = n^6$

WORK PROBLEM 5 AT THE SIDE.

OBJECTIVE 4 Properties we studied previously can be used to develop two more rules for exponents. Using the definition of an exponential expression and the commutative and associative properties, we can evaluate the expression $(4x)^3$ as shown below.

$(4x)^3$	$= (4x)(4x)(4x)$	Definition of exponent
	$= 4 \cdot 4 \cdot 4 \cdot x \cdot x \cdot x$	Commutative and associative properties
	$= 4^3x^3$	Definition of exponent

This example suggests power rule (b) for exponents.

4. Multiply.

(a) $5m^2 \cdot 2m^6$

(b) $3p^5 \cdot 9p^4$

(c) $-7p^5 \cdot (3p^8)$

5. Simplify each expression.

(a) $(5^3)^4$

(b) $(6^2)^5$

(c) $(3^2)^4$

(d) $(a^6)^5$

ANSWERS

4. (a) $10m^8$ **(b)** $27p^9$ **(c)** $-21p^{13}$

5. (a) 5^{12} **(b)** 6^{10} **(c)** 3^8 **(d)** a^{30}

6. Simplify.

(a) $5(mn)^3$

(b) $(3a^2b^4)^5$

(c) $(5m^2)^3$

7. Simplify. Assume all variables represent nonzero real numbers.

(a) $\left(\frac{5}{2}\right)^4$

(b) $\left(\frac{p}{q}\right)^2$

(c) $\left(\frac{r}{t}\right)^3$

ANSWERS

6. (a) $5m^3n^3$ **(b)** $3^5a^{10}b^{20}$ **(c)** 5^3m^6

7. (a) $\frac{5^4}{2^4}$ **(b)** $\frac{p^2}{q^2}$ **(c)** $\frac{r^3}{t^3}$

Power Rule (b) for Exponents

For any positive integer m, $(ab)^m = a^m b^m$.
(Raise a product to a power by raising each factor to the power.)
Example: $(2p)^5 = 2^5p^5$

EXAMPLE 6 Using Power Rule (b)

Use power rule (b) to simplify each expression.

(a) $(3xy)^2 = 3^2x^2y^2$
$= 9x^2y^2$

(b) $9(pq)^2 = 9(p^2q^2)$ Power rule (b)
$= 9p^2q^2$ Multiply.

(c) $5(2m^2p^3)^4 = 5[2^4(m^2)^4(p^3)^4]$ Power rule (b)
$= 5(2^4m^8p^{12})$ Power rule (a)
$= 5 \cdot 2^4m^8p^{12}$
$= 80m^8p^{12}$ $5 \cdot 2^4 = 5 \cdot 16 = 80$

WORK PROBLEM 6 AT THE SIDE.

OBJECTIVE 5 Since the quotient $\frac{a}{b}$ can be written as $a \cdot \frac{1}{b}$, we can use power rule (b), together with some of the properties of real numbers, to get power rule (c) for exponents.

Power Rule (c) for Exponents

For any positive integer m, $\left(\frac{a}{b}\right)^m = \frac{a^m}{b^m}$ $(b \neq 0)$.
(Raise a quotient to a power by raising both the numerator and the denominator to the power.)
Example: $\left(\frac{5}{3}\right)^2 = \frac{5^2}{3^2}$

EXAMPLE 7 Using Power Rule (c)

Simplify each expression.

(a) $\left(\frac{2}{3}\right)^5 = \frac{2^5}{3^5}$

(b) $\left(\frac{m}{n}\right)^4 = \frac{m^4}{n^4}$, $n \neq 0$

WORK PROBLEM 7 AT THE SIDE.

The rules for exponents discussed in this section are now summarized. These rules are basic to the study of algebra and should be *memorized.*

Rules for Exponents

For positive integers m and n:

			Examples
Product rule		$a^m \cdot a^n = a^{m+n}$	$6^2 \cdot 6^5 = 6^{2+5} = 6^7$
Power rules	(a)	$(a^m)^n = a^{mn}$	$(3^2)^4 = 3^{2 \cdot 4} = 3^8$
	(b)	$(ab)^m = a^m b^m$	$(2p)^5 = 2^5 p^5$
	(c)	$\left(\frac{a}{b}\right)^m = \frac{a^m}{b^m}$ $(b \neq 0)$	$\left(\frac{5}{3}\right)^2 = \frac{5^2}{3^2}$

OBJECTIVE 6 As shown in the next example, more than one rule may be needed to simplify an expression.

EXAMPLE 8 Using Combinations of Rules

Simplify each expression.

(a) $\left(\frac{2}{3}\right)^2 \cdot 2^3$

$$\left(\frac{2}{3}\right)^2 \cdot 2^3 = \frac{2^2}{3^2} \cdot \frac{2^3}{1} \quad \text{Power rule (c)}$$

$$= \frac{2^2 \cdot 2^3}{3^2 \cdot 1} \quad \text{Multiply fractions.}$$

$$= \frac{2^5}{3^2} \quad \text{Product rule}$$

(b) $(5x)^3(5x)^4$

$$(5x)^3(5x)^4 = (5x)^7 \quad \text{Product rule}$$

$$= 5^7x^7 \quad \text{Power rule (b)}$$

(c) $(2x^2y^3)^4(3xy^2)^3$

$$(2x^2y^3)^4(3xy^2)^3 = 2^4(x^2)^4(y^3)^4 \cdot 3^3x^3(y^2)^3 \quad \text{Power rule (b)}$$

$$= 2^4 \cdot 3^3x^8y^{12}x^3y^6 \quad \text{Power rule (a)}$$

$$= 16 \cdot 27x^{11}y^{18} \quad \text{Product rule}$$

$$= 432x^{11}y^{18}$$

Caution

Refer to Example 8(c). Notice that

$$(2x^2y^3)^4 \neq (2 \cdot 4)x^{2 \cdot 4}y^{3 \cdot 4}.$$

Do not multiply the coefficient 2 and the exponent 4.

WORK PROBLEM 8 AT THE SIDE.

8. Simplify.

(a) $(2m)^3(2m)^4$

(b) $\left(\frac{5k^3}{3}\right)^2$

(c) $\left(\frac{1}{5}\right)^4(2x)^2$

(d) $(3xy^2)^3(x^2y)^4$

ANSWERS

8. (a) 2^7m^7 **(b)** $\frac{5^2k^6}{3^2}$ **(c)** $\frac{2^2x^2}{5^4}$ **(d)** $3^3x^{11}y^{10}$

1 Exercises

Write each expression using exponents. See Example 1.

1. $3 \cdot 3 \cdot 3 \cdot 3 \cdot 3 \cdot 3 \cdot 3$

2. $(-6)(-6)(-6)(-6)$

3. $(-2)(-2)(-2)(-2)(-2)$

4. $t \cdot t \cdot t \cdot t \cdot t \cdot t$

5. $w \cdot w \cdot w \cdot w \cdot w \cdot w$

6. $\left(\frac{1}{2}\right)\left(\frac{1}{2}\right)\left(\frac{1}{2}\right)\left(\frac{1}{2}\right)\left(\frac{1}{2}\right)\left(\frac{1}{2}\right)$

7. $\left(-\frac{1}{4}\right)\left(-\frac{1}{4}\right)\left(-\frac{1}{4}\right)\left(-\frac{1}{4}\right)\left(-\frac{1}{4}\right)$

8. $(-8p)(-8p)$

9. $(-7x)(-7x)(-7x)(-7x)$

10. Explain how the expressions $(-3)^4$ and -3^4 are different.

11. Explain how the expressions $(5x)^3$ and $5x^3$ are different.

Identify the base and the exponent for each exponential expression. In Exercises 12–15, also evaluate the expression. See Example 2.

12. 3^5

13. 2^7

14. $(-3)^5$

15. $(-2)^7$

16. $(-6x)^4$

17. $(-8x)^4$

18. $-6x^4$

19. $-8x^4$

20. Explain why the product rule does not apply to the expression $5^2 + 5^3$. Then evaluate the expression by finding the individual powers and then adding the results.

INTERPRETING TECHNOLOGY (EXERCISES 21–22)

21. Two graphing calculator screens are shown here. What do you think the calculator notation $a \wedge b$ means?

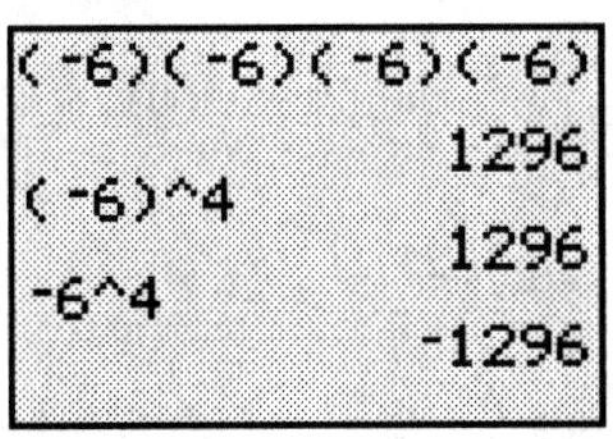

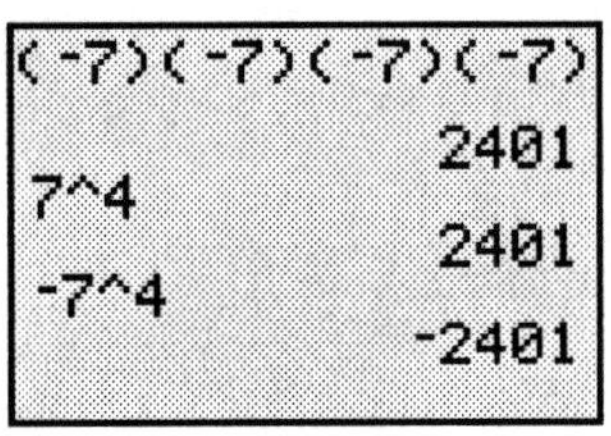

22. Use your answer to Exercise 21 to evaluate $4 \wedge 3$.

23. What exponent is understood on the base x in the expression xy^2?

24. How are the expressions 3^2, 5^3, and 7^4 read?

Use the product rule to simplify each expression. Write each answer in exponential form. See Examples 3 and 4.

25. $5^2 \cdot 5^6$ **26.** $3^6 \cdot 3^7$ **27.** $4^2 \cdot 4^7 \cdot 4^3$ **28.** $5^3 \cdot 5^8 \cdot 5^2$

29. $(-7)^3(-7)^6$ **30.** $(-9)^8(-9)^5$ **31.** $t^3 \cdot t^8 \cdot t^{13}$ **32.** $n^5 \cdot n^6 \cdot n^9$

33. $(-8r^4)(7r^3)$ **34.** $(10a^7)(-4a^3)$ **35.** $(-6p^5)(-7p^5)$ **36.** $(-5w^8)(-9w^8)$

37. Explain why the product rule does not apply to the expression $3^2 \cdot 4^3$. Then evaluate the expression by finding the individual powers and multiplying the results.

38. Repeat Exercise 37 for the expression $(-3)^3 \cdot (-2)^5$.

In each of the following exercises, first add the given terms. Then start over and multiply them.

39. $5x^4$, $9x^4$ **40.** $8t^5$, $3t^5$ **41.** $-7a^2$, $2a^2$, $10a^2$ **42.** $6x^3$, $9x^3$, $-2x^3$

Use the power rules for exponents to simplify each expression. Write each answer in exponential form. See Examples 5–7.

43. $(4^3)^2$ **44.** $(8^3)^6$ **45.** $(t^4)^5$ **46.** $(y^6)^5$

47. $(7r)^3$ **48.** $(11x)^4$ **49.** $(5xy)^5$ **50.** $(9pq)^6$

51. $(-5^2)^6$ **52.** $(-9^4)^8$ **53.** $(-8^3)^5$ **54.** $(-7^5)^7$

55. $8(qr)^3$ **56.** $4(vw)^5$ **57.** $\left(\frac{1}{2}\right)^3$ **58.** $\left(\frac{1}{3}\right)^5$

59. $\left(\frac{a}{b}\right)^3$ $(b \neq 0)$ **60.** $\left(\frac{r}{t}\right)^4$ $(t \neq 0)$ **61.** $\left(\frac{9}{5}\right)^8$ **62.** $\left(\frac{12}{7}\right)^3$

Use a combination of the rules of exponents introduced in this section to simplify each expression. See Example 8.

63. $\left(\frac{5}{2}\right)^3 \cdot \left(\frac{5}{2}\right)^2$ **64.** $\left(\frac{3}{4}\right)^5 \cdot \left(\frac{3}{4}\right)^6$ **65.** $\left(\frac{9}{8}\right)^3 \cdot 9^2$ **66.** $\left(\frac{8}{5}\right)^4 \cdot 8^3$

67. $(2x)^9(2x)^3$ **68.** $(6y)^5(6y)^8$ **69.** $(-6p)^4(-6p)$ **70.** $(-13q)^3(-13q)$

71. $(6x^2y^3)^5$ **72.** $(5r^5t^6)^7$ **73.** $(x^2)^3(x^3)^5$ **74.** $(y^4)^5(y^3)^5$

75. $(2w^2x^3y)^2(x^4y)^5$ **76.** $(3x^4y^2z)^3(yz^4)^5$ **77.** $(-r^4s)^2(-r^2s^3)^5$

78. $(-ts^6)^4(-t^3s^5)^3$ **79.** $\left(\frac{5a^2b^5}{c^6}\right)^3$ $(c \neq 0)$ **80.** $\left(\frac{6x^3y^9}{z^5}\right)^4$ $(z \neq 0)$

81. A student tried to simplify $(10^2)^3$ as 1000^6. Is this correct? If not, how is it simplified using the product rule for exponents?

82. Explain why $(3x^2y^3)^4$ is *not* equivalent to $(3 \cdot 4)x^8y^{12}$.

2 *Integer Exponents and the Quotient Rule*

OBJECTIVES

1. Use zero as an exponent.
2. Use negative numbers as exponents.
3. Use the quotient rule for exponents.
4. Use combinations of rules.

OBJECTIVE 1 In the previous section we studied the product rule for exponents. In all of our work, exponents were positive integers. To develop meanings for exponents other than positive integers (such as 0 and negative integers), we want to define them in such a way that rules for exponents are the same, regardless of the kind of number used for the exponents.

Suppose we want to find a meaning for an expression such as

$$6^0,$$

where 0 is used as an exponent. If we were to multiply this by 6^2, for example, we would want the product rule to still be valid. Therefore, we would have

$$6^0 \cdot 6^2 = 6^{0+2} = 6^2.$$

So multiplying 6^2 by 6^0 should give 6^2. Because 6^0 is acting as if it were 1 here, we should define 6^0 to equal 1. This is the definition for 0 used as an exponent with any nonzero base.

Definition of Zero Exponent

For any nonzero real number a, $\quad a^0 = 1.$

Example: $17^0 = 1$

EXAMPLE 1 Using Zero Exponents

Evaluate each exponential expression.

(a) $60^0 = 1$

(b) $(-60)^0 = 1$

(c) $-60^0 = -(1) = -1$

(d) $y^0 = 1$, if $y \neq 0$

(e) $-r^0 = -1$, if $r \neq 0$

Caution

Notice the difference between parts (b) and (c) of Example 1. In Example 1(b) the base is -60 and the exponent is 0. Any nonzero base raised to a zero exponent is 1. But in Example 1(c), the base is 60. Then $60^0 = 1$, and $-60^0 = -1$.

WORK PROBLEM 1 AT THE SIDE.

1. Evaluate.

(a) 28^0

(b) $(-16)^0$

(c) -7^0

(d) m^0, $m \neq 0$

(e) $-p^0$, $p \neq 0$

OBJECTIVE 2 Now let us consider how we can define negative integers as exponents. Suppose that we want to give a meaning to

$$6^{-2}$$

so that the product rule is still valid. If we multiply 6^{-2} by 6^2, we get

$$6^{-2} \cdot 6^2 = 6^{-2+2} = 6^0 = 1.$$

The expression 6^{-2} is acting as if it were the reciprocal of 6^2, because their product is 1. The reciprocal of 6^2 may be written $\frac{1}{6^2}$, leading us to define 6^{-2} as $\frac{1}{6^2}$. This is a particular case of the definition of negative exponents.

ANSWERS

1. (a) 1 (b) 1 (c) -1 (d) 1 (e) -1

Definition of Negative Exponents

For any nonzero real number a and any integer n, $\quad a^{-n} = \dfrac{1}{a^n}$.

Example: $3^{-2} = \dfrac{1}{3^2}$

By definition, a^{-n} and a^n are reciprocals, since

$$a^n \cdot a^{-n} = a^n \cdot \frac{1}{a^n} = 1.$$

Since $1^n = 1$, the definition of a^{-n} also can be written

$$a^{-n} = \frac{1}{a^n} = \frac{1^n}{a^n} = \left(\frac{1}{a}\right)^n.$$

For example,

$$6^{-3} = \left(\frac{1}{6}\right)^3 \quad \text{and} \quad \left(\frac{1}{3}\right)^{-2} = 3^2.$$

EXAMPLE 2 Using Negative Exponents

Simplify by writing with positive exponents.

(a) $3^{-2} = \dfrac{1}{3^2} = \dfrac{1}{9}$

(b) $5^{-3} = \dfrac{1}{5^3} = \dfrac{1}{125}$

(c) $\left(\dfrac{1}{2}\right)^{-3} = 2^3 = 8$ $\quad$ $\frac{1}{2}$ and 2 are reciprocals.

As shown above, we can change the base to its reciprocal if we also change the sign of the exponent.

(d) $\left(\dfrac{2}{5}\right)^{-4} = \left(\dfrac{5}{2}\right)^4$ $\quad$ $\frac{2}{5}$ and $\frac{5}{2}$ are reciprocals.

(e) $4^{-1} - 2^{-1} = \dfrac{1}{4} - \dfrac{1}{2} = \dfrac{1}{4} - \dfrac{2}{4} = -\dfrac{1}{4}$

Apply the exponents first, then subtract.

(f) $p^{-2} = \dfrac{1}{p^2}, p \neq 0$

(g) $\dfrac{1}{x^{-4}}, x \neq 0$

$$\begin{aligned}\frac{1}{x^{-4}} &= \frac{1^{-4}}{x^{-4}} && 1^{-4} = 1\\ &= \left(\frac{1}{x}\right)^{-4} && \text{Power rule (c)}\\ &= x^4 && \tfrac{1}{x} \text{ and } x \text{ are reciprocals.}\end{aligned}$$

Caution

A negative exponent does not indicate a negative number; negative exponents lead to reciprocals.

Expression	*Example*	
a^{-n}	$3^{-2} = \frac{1}{3^2} = \frac{1}{9}$	Not negative
$-a^{-n}$	$-3^{-2} = -\frac{1}{3^2} = -\frac{1}{9}$	Negative

WORK PROBLEM 2 AT THE SIDE.

The definition of negative exponents allows us to move factors in a fraction if we also change the signs of the exponents. For example,

$$\frac{2^{-3}}{3^{-4}} = \frac{\frac{1}{2^3}}{\frac{1}{3^4}} = \frac{1}{2^3} \cdot \frac{3^4}{1} = \frac{3^4}{2^3}$$

so that

$$\frac{2^{-3}}{3^{-4}} = \frac{3^4}{2^3}$$

Changing from Negative to Positive Exponents

For any nonzero numbers a and b, and any integers m and n,

$$\frac{a^{-m}}{b^{-n}} = \frac{b^n}{a^m}.$$

Example: $\frac{3^{-5}}{2^{-4}} = \frac{2^4}{3^5}$

EXAMPLE 3 Changing from Negative to Positive Exponents

Write with only positive exponents. Assume all variables represent nonzero real numbers.

(a) $\frac{4^{-2}}{5^{-3}} = \frac{5^3}{4^2}$

(b) $\frac{m^{-5}}{p^{-1}} = \frac{p^1}{m^5} = \frac{p}{m^5}$

(c) $\frac{a^{-2}b}{3d^{-3}} = \frac{bd^3}{3a^2}$

(d) $x^3y^{-4} = \frac{x^3y^{-4}}{1} = \frac{x^3}{y^4}$

WORK PROBLEM 3 AT THE SIDE.

2. Write with positive exponents.

(a) 4^{-3}

(b) 6^{-2}

(c) $\left(\frac{2}{3}\right)^{-2}$

(d) $2^{-1} + 5^{-1}$

(e) $m^{-5}, m \neq 0$

(f) $\frac{1}{z^{-4}}, z \neq 0$

3. Write with only positive exponents. Assume all variables represent nonzero real numbers.

(a) $\frac{7^{-1}}{5^{-4}}$

(b) $\frac{x^{-3}}{y^{-2}}$

(c) $\frac{4h^{-5}}{m^{-2}k}$

(d) p^2q^{-5}

ANSWERS

2. **(a)** $\frac{1}{4^3}$ **(b)** $\frac{1}{6^2}$ **(c)** $\left(\frac{3}{2}\right)^2$
(d) $\frac{1}{2} + \frac{1}{5} = \frac{7}{10}$ **(e)** $\frac{1}{m^5}$ **(f)** z^4

3. **(a)** $\frac{5^4}{7}$ **(b)** $\frac{y^2}{x^3}$ **(c)** $\frac{4m^2}{h^5k}$ **(d)** $\frac{p^2}{q^5}$

4. Simplify. Write answers with positive exponents.

(a) $\dfrac{5^{11}}{5^8}$

(b) $\dfrac{4^7}{4^{10}}$

(c) $\dfrac{6^{-5}}{6^{-2}}$

(d) $\dfrac{8^4 \cdot m^9}{8^5 \cdot m^{10}}, m \neq 0$

ANSWERS

4. (a) 5^3 (b) $\dfrac{1}{4^3}$ (c) $\dfrac{1}{6^3}$ (d) $\dfrac{1}{8m}$

OBJECTIVE 3 What about the quotient of two exponential expressions with the same base? We know that

$$\frac{6^5}{6^3} = \frac{6 \cdot 6 \cdot 6 \cdot 6 \cdot 6}{6 \cdot 6 \cdot 6} = 6^2.$$

Notice that **5 − 3 = 2**. Also,

$$\frac{6^2}{6^4} = \frac{6 \cdot 6}{6 \cdot 6 \cdot 6 \cdot 6} = \frac{1}{6^2} = 6^{-2}.$$

Here, **2 − 4 = −2**. These examples suggest the quotient rule for exponents.

Quotient Rule for Exponents

For any nonzero real number a and any integers m and n,

$$\frac{a^m}{a^n} = a^{m-n}.$$

(Keep the base and subtract the exponents.)

Example: $\dfrac{5^8}{5^4} = 5^{8-4} = 5^4$

EXAMPLE 4 Using the Quotient Rule for Exponents

Simplify, using the quotient rule for exponents. Write answers with positive exponents.

(a) $\dfrac{5^8}{5^6} = 5^{8-6} = 5^2$

(b) $\dfrac{4^2}{4^9} = 4^{2-9} = 4^{-7} = \dfrac{1}{4^7}$

(c) $\dfrac{5^{-3}}{5^{-7}} = 5^{-3-(-7)} = 5^4$

(d) $\dfrac{q^5}{q^{-3}} = q^{5-(-3)} = q^8, q \neq 0$

(e) $\dfrac{3^2x^5}{3^4x^3} = \dfrac{3^2}{3^4} \cdot \dfrac{x^5}{x^3} = 3^{2-4} \cdot x^{5-3}$

$= 3^{-2}x^2 = \dfrac{x^2}{3^2}, x \neq 0$

Sometimes numerical expressions with small exponents, such as 3^2, are evaluated. Doing that would give the result as $\frac{x^2}{9}$.

◀ WORK PROBLEM 4 AT THE SIDE.

Since exponential expressions with negative exponents can be written with positive exponents, the rules for exponents also are true for negative exponents.

The definitions and rules for exponents given in this section and the previous one are summarized below.

Definitions and Rules for Exponents

For any integers m and n:		*Examples*
Product rule	$a^m \cdot a^n = a^{m+n}$	$7^4 \cdot 7^5 = 7^9$
Zero exponent	$a^0 = 1 \quad (a \neq 0)$	$(-3)^0 = 1$
Negative exponent	$a^{-n} = \frac{1}{a^n} \quad (a \neq 0)$	$5^{-3} = \frac{1}{5^3}$
Quotient rule	$\frac{a^m}{a^n} = a^{m-n} \ (a \neq 0)$	$\frac{2^2}{2^5} = 2^{2-5} = 2^{-3} = \frac{1}{2^3}$
Power rules (a)	$(a^m)^n = a^{mn}$	$(4^2)^3 = 4^6$
(b)	$(ab)^m = a^m b^m$	$(3k)^4 = 3^4 k^4$
(c)	$\left(\frac{a}{b}\right)^m = \frac{a^m}{b^m} \ (b \neq 0)$	$\left(\frac{2}{3}\right)^{-2} = \frac{2^{-2}}{3^{-2}} = \frac{3^2}{2^2}$
Negative to positive exponent	$\frac{a^{-m}}{b^{-n}} = \frac{b^n}{a^m}$	$\frac{2^{-4}}{5^{-3}} = \frac{5^3}{2^4}$

OBJECTIVE 4 As shown in the next example, we may sometimes need to use more than one rule to simplify an expression.

EXAMPLE 5 Using a Combination of Rules

Use a combination of the rules for exponents to simplify each expression. Assume all variables represent nonzero real numbers.

(a) $\frac{(4^2)^3}{4^5}$

$$\frac{(4^2)^3}{4^5} = \frac{4^6}{4^5} \qquad \text{Power rule (a)}$$

$$= 4^{6-5} \qquad \text{Quotient rule}$$

$$= 4^1 = 4$$

(b) $(2x)^3(2x)^2$

$$(2x)^3(2x)^2 = (2x)^5 \qquad \text{Product rule}$$

$$= 2^5x^5 \text{ or } 32x^5 \qquad \text{Power rule (b)}$$

(c) $\left(\frac{2x^3}{5}\right)^{-4}$

By the definition of a negative exponent and the power rules,

$$\left(\frac{2x^3}{5}\right)^{-4} = \left(\frac{5}{2x^3}\right)^4 \qquad \text{Change the base to its reciprocal, and change the sign of the exponent.}$$

$$= \frac{5^4}{2^4x^{12}}. \qquad \text{Power rules}$$

(d) $\left(\frac{3x^{-2}}{4^{-1}y^3}\right)^{-3} = \frac{3^{-3}x^6}{4^3y^{-9}}$ Power rules

$$= \frac{x^6y^9}{4^3 \cdot 3^3} \qquad \text{Negative to positive exponent rule}$$

CONTINUED ON NEXT PAGE

5. Simplify. Assume all variables represent nonzero real numbers.

(a) $12^5 \cdot 12^{-7} \cdot 12^6$

(b) $y^{-2} \cdot y^5 \cdot y^{-8}$

(c) $\dfrac{(6x)^{-1}}{(3x^2)^{-2}}$

(d) $\dfrac{3^9 \cdot (x^2y)^{-2}}{3^3 \cdot x^{-4}y}$

Answers

5. (a) 12^4 **(b)** $\dfrac{1}{y^5}$ **(c)** $\dfrac{3x^3}{2}$ **(d)** $\dfrac{3^6}{y^3}$

(e) $\dfrac{(4m)^{-3}}{(3m)^{-4}} = \dfrac{4^{-3}m^{-3}}{3^{-4}m^{-4}}$ Power rule (b)

$= \dfrac{3^4m^4}{4^3m^3}$ Negative to positive exponent rule

$= \dfrac{3^4m^{4-3}}{4^3}$ Quotient rule

$= \dfrac{3^4m}{4^3}$ ■

Note

Since the steps can be done in several different orders, there are many equally good ways to simplify problems like Examples 5(d) and 5(e).

◀◀ WORK PROBLEM 5 AT THE SIDE.

2 Exercises

Each of the following expressions is equal to either 0, 1, *or* −1. *Decide which is correct. See Example 1.*

1. $(-4)^0$ **2.** $(-10)^0$ **3.** -9^0 **4.** -5^0

5. $(-2)^0 - 2^0$ **6.** $(-8)^0 - 8^0$ **7.** $(5x)^0 \quad (x \neq 0)$ **8.** $\left(\frac{3}{y}\right)^0 \quad (y \neq 0)$

Evaluate each expression. See Examples 1 and 2.

9. $7^0 + 9^0$ **10.** $8^0 + 6^0$ **11.** 4^{-3} **12.** 5^{-4}

13. $\left(\frac{1}{2}\right)^{-4}$ **14.** $\left(\frac{1}{3}\right)^{-3}$ **15.** $\left(\frac{6}{7}\right)^{-2}$ **16.** $\left(\frac{2}{3}\right)^{-3}$

17. $(-3)^{-4}$ **18.** $(-4)^{-3}$ **19.** $5^{-1} + 3^{-1}$ **20.** $6^{-1} + 2^{-1}$

Decide whether each of the following is positive, negative, or zero.

21. $(-2)^{-3}$ **22.** $(-3)^{-2}$ **23.** -2^4 **24.** -3^6

25. $\left(\frac{1}{4}\right)^{-2}$ **26.** $\left(\frac{1}{5}\right)^{-2}$ **27.** $1 - 5^0$ **28.** $1 - 7^0$

Mathematical Connections (Exercises 29–32)

In Objective 1, we used the product rule to motivate the definition of a zero exponent. We can also use the quotient rule. To see this, work Exercises 29–32 in order.

29. Consider the expression $\frac{25}{25}$. What is its simplest form?

30. Write the quotient in Exercise 29 using the fact that $25 = 5^2$.

31. Apply the quotient rule for exponents to your answer for Exercise 30. Give the answer as a power of 5.

32. Because your answers for Exercises 29 and 31 both represent $\frac{25}{25}$, they must be equal. Write this equality. What definition does it support?

Use the quotient rule to simplify each expression. Write each expression with positive exponents. Assume that all variables represent nonzero real numbers. See Examples 2, 3, and 4.

33. $\frac{9^4}{9^5}$ **34.** $\frac{7^3}{7^4}$ **35.** $\frac{6^{-3}}{6^2}$ **36.** $\frac{4^{-2}}{4^3}$

37. $\frac{1}{6^{-3}}$ **38.** $\frac{1}{5^{-2}}$ **39.** $\frac{2}{r^{-4}}$ **40.** $\frac{3}{s^{-8}}$

41. $\frac{4^{-3}}{5^{-2}}$ **42.** $\frac{6^{-2}}{5^{-4}}$ **43.** p^5q^{-8} **44.** $x^{-8}y^4$

45. $\frac{r^5}{r^{-4}}$ **46.** $\frac{a^6}{a^{-4}}$ **47.** $\frac{6^4x^8}{6^5x^3}$ **48.** $\frac{3^8y^5}{3^{10}y^2}$

Use a combination of the rules for exponents to simplify each expression. Write answers with only positive exponents. Assume that all variables represent nonzero real numbers. See Example 5.

49. $\frac{(7^4)^3}{7^9}$ **50.** $\frac{(5^3)^2}{5^2}$ **51.** $x^{-3} \cdot x^5 \cdot x^{-4}$ **52.** $y^{-8} \cdot y^5 \cdot y^{-2}$

53. $\frac{(3x)^{-2}}{(4x)^{-3}}$ **54.** $\frac{(2y)^{-3}}{(5y)^{-4}}$ **55.** $\left(\frac{x^{-1}y}{z^2}\right)^{-2}$ **56.** $\left(\frac{p^{-4}q}{r^{-3}}\right)^{-3}$

57. $(6x)^4(6x)^{-3}$ **58.** $(10y)^9(10y)^{-8}$ **59.** $\frac{(m^7n)^{-2}}{m^{-4}n^3}$ **60.** $\frac{(m^8n^{-4})^2}{m^{-2}n^5}$

61. $\frac{5x^{-3}}{(4x)^2}$ **62.** $\frac{-3k^5}{(2k)^2}$ **63.** $\left(\frac{2p^{-1}q}{3^{-1}m^2}\right)^2$ **64.** $\left(\frac{4xy^2}{x^{-1}y}\right)^{-2}$

3 An Application of Exponents: Scientific Notation

OBJECTIVES

1 Express numbers in scientific notation.

2 Convert numbers in scientific notation to numbers without exponents.

3 Use scientific notation in calculations.

OBJECTIVE 1 One example of the use of exponents comes from science. The numbers occurring in science are often extremely large (such as the distance from the earth to the sun, 93,000,000 miles) or extremely small (the wavelength of yellow-green light is approximately .0000006 meter). Because of the difficulty of working with many zeros, scientists often express such numbers with exponents. Each number is written as $a \times 10^n$, where $1 \le |a| < 10$ and n is an integer. This form is called **scientific notation.** There is always one nonzero digit before the decimal point. For example, 35 is written 3.5×10^1, or 3.5×10; 56,200 is written 5.62×10^4, since

$$56{,}200 = 5.62 \times \mathbf{10{,}000} = 5.62 \times \mathbf{10^4},$$

and .09 is written as 9×10^{-2}.

The steps involved in writing a number in scientific notation are given below. For negative numbers, follow these steps using the absolute value of the number; then make the result negative.

Writing a Number in Scientific Notation

Step 1 Move the decimal point to the right of the first nonzero digit.

Step 2 Count the number of places you moved the decimal point.

Step 3 The number of places in Step 2 is the absolute value of the exponent on 10.

Step 4 The exponent on 10 is positive if you made the number smaller in Step 1; the exponent is negative if you made the number larger in Step 1.

EXAMPLE 1 Using Scientific Notation

Write each number in scientific notation.

(a) 93,000,000

The number will be written in scientific notation as 9.3×10^n. To find the value of n, first compare 9.3 with 93,000,000. Since 9.3 is *smaller* than 93,000,000, we must multiply by a *positive* power of 10 so the product 9.3×10^n will equal the larger number.

Move the decimal point to follow the first nonzero digit. Count the number of places the decimal point was moved.

9.3 000 000 7 places

Since the decimal point was moved 7 places, and since n is positive, $93{,}000{,}000 = 9.3 \times 10^7$.

(b) $463{,}000{,}000{,}000{,}000 = 4.63\ 000\ 000\ 000\ 000$ 14 places

$= 4.63 \times 10^{14}$

(c) $-302{,}100 = -3.021 \times 10^5$

(d) .00462

Move the decimal point to the right of the first nonzero digit and count the number of places the decimal point was moved.

004.62 3 places

CONTINUED ON NEXT PAGE

Because 4.62 is *larger* than .00462, the exponent must be *negative*.

$$.00462 = 4.62 \times \mathbf{10^{-3}}$$

(e) $.0000762 = 7.62 \times 10^{-5}$

◀◀ WORK PROBLEM 1 AT THE SIDE.

OBJECTIVE 2 To convert a number written in scientific notation to a number without exponents, work in reverse. Multiplying a number by a positive power of 10 will make the number larger; multiplying by a negative power of 10 will make the number smaller.

EXAMPLE 2 Writing Numbers without Exponents

Write each number without exponents.

(a) 6.2×10^3

Since the exponent is positive, make 6.2 larger by moving the decimal point 3 places to the right.

$$6.2 \times \mathbf{10^3} = 6.200 = 6200$$

(b) $4.283 \times 10^5 = 4.28300 = 428{,}300$ Move 5 places to the right.

(c) $-9.73 \times 10^{-2} = -09.73 = -.0973$ Move 2 places to the left.

As these examples show, the exponent tells the number of places and the direction that the decimal point is moved.

◀◀ WORK PROBLEM 2 AT THE SIDE.

OBJECTIVE 3 The next example shows how scientific notation can be used with products and quotients.

EXAMPLE 3 Multiplying and Dividing with Scientific Notation

Write each product or quotient without exponents.

(a) $(6 \times 10^3)(5 \times 10^{-4})$

$(6 \times 10^3)(5 \times 10^{-4})$

$= (6 \times 5)(10^3 \times 10^{-4})$ Commutative and associative properties

$= 30 \times 10^{-1}$ Product rule for exponents

$= 30. = 3$ Write without exponents.

(b) $\dfrac{6 \times 10^{-5}}{2 \times 10^3} = \dfrac{6}{2} \times \dfrac{10^{-5}}{10^3} = 3 \times 10^{-8} = .00000003$

◀◀ WORK PROBLEM 3 AT THE SIDE.

Calculators usually have a key labeled EE or EXP for scientific notation. See An Introduction to Scientific Calculators at the front of this book for more information.

---

1. Write each number in scientific notation.

(a) 63,000

(b) 5,870,000

(c) .0571

(d) .000062

2. Write without exponents.

(a) 4.2×10^3

(b) 8.7×10^5

(c) 6.42×10^{-3}

3. Simplify, and write without exponents.

(a) $(2.6 \times 10^4)(2 \times 10^{-6})$

(b) $\dfrac{4.8 \times 10^2}{2.4 \times 10^{-3}}$

ANSWERS

1. (a) 6.3×10^4 (b) 5.87×10^6 (c) 5.71×10^{-2} (d) 6.2×10^{-5}
2. (a) 4200 (b) 870,000 (c) .00642
3. (a) .052 (b) 200,000

3 Exercises

Write the numbers (other than dates) mentioned in the following statements in scientific notation.

1. How is it that the average CEO in Japan receives an income of \$300,000, while the average CEO in the United States earns \$2.8 million? (Andrew Zimbalist, *Baseball and Billions*. New York: Basic Books, 1992, p. 78.)

2. In 1995, women made up about 22 percent or 5.4 million, of all golfers. (From "Driving Force," an article in the *Sacramento Bee,* October 6, 1996, Sec. D.)

3. In 1995 sales of imported stemmed roses increased to 752 million, while sales of U.S.-produced roses decreased to 394 million. (Knight-Ridder Tribune graphic in the *Sacramento Bee,* Sept. 22, 1996, Sec. E, p. 3.)

4. The number of engineers employed in the U.S. during 1993 was 1.67 million. (Engineering Workforce Commission)

Determine whether or not the given number is written in scientific notation as defined in Objective 1. If it is not, write it as such.

5. 4.56×10^3 **6.** 7.34×10^5 **7.** 5,600,000 **8.** 34,000

9. $.8 \times 10^2$ **10.** $.9 \times 10^3$ **11.** .004 **12.** .0007

13. Explain in your own words what it means for a number to be written in scientific notation.

14. Explain how to multiply a number by a positive power of ten. Then explain how to multiply a number by a negative power of ten.

Write each number in scientific notation. See Example 1.

15. 5,876,000,000 **16.** 9,994,000,000 **17.** 82,350 **18.** 78,330

19. .000007 **20.** .0000004 **21.** $-.00203$ **22.** $-.0000578$

Write each number without exponents. See Example 2.

23. 7.5×10^5 **24.** 8.8×10^6 **25.** 5.677×10^{12} **26.** 8.766×10^9

27. -6.21×10^0 **28.** -8.56×10^0 **29.** 7.8×10^{-4}

30. 8.9×10^{-5} **31.** 5.134×10^{-9} **32.** 7.123×10^{-10}

Perform the indicated operations, and write the answers in scientific notation and then without exponents. See Example 3.

33. $(2 \times 10^8) \times (3 \times 10^3)$

34. $(4 \times 10^7) \times (3 \times 10^3)$

35. $(5 \times 10^4) \times (3 \times 10^2)$

36. $(8 \times 10^5) \times (2 \times 10^3)$

37. $(3.15 \times 10^{-4}) \times (2.04 \times 10^8)$

38. $(4.92 \times 10^{-3}) \times (2.25 \times 10^7)$

Perform the indicated operations, and write the answers in scientific notation. See Example 3.

39. $\dfrac{9 \times 10^{-5}}{3 \times 10^{-1}}$

40. $\dfrac{12 \times 10^{-4}}{4 \times 10^{-3}}$

41. $\dfrac{8 \times 10^3}{2 \times 10^2}$

42. $\dfrac{5 \times 10^4}{1 \times 10^3}$

43. $\dfrac{2.6 \times 10^{-3} \times 7.0 \times 10^{-1}}{2 \times 10^2 \times 3.5 \times 10^{-3}}$

44. $\dfrac{9.5 \times 10^{-1} \times 2.4 \times 10^4}{5 \times 10^3 \times 1.2 \times 10^{-2}}$

If the number in the statement is written in scientific notation, write it without exponents. If it is written without exponents, write it in scientific notation. See Examples 1 and 2.

45. The number of possible hands in contract bridge is about 6.35×10^{11}.

46. If there are forty numbers to choose from in a lottery, and a player must choose six different ones, the player has about 3.84×10^6 ways to make a choice.

47. Quentin Tarantino's black comedy, *Pulp Fiction,* boosted John Travolta's movie career by grossing a little over $45,500,000.

48. Whitney Houston's starring debut film, *The Bodyguard,* grossed $415,000,000 worldwide.

49. The body of a 150-pound person contains about 2.3×10^{-4} pounds of copper and about 6×10^{-3} pounds of iron.

50. The mean distance from Venus to the sun is about 6.7×10^6 miles.

4 Addition and Subtraction of Polynomials

OBJECTIVE 1 Recall that in an expression such as

$$4x^3 + 6x^2 + 5x + 8,$$

the quantities that are added, $4x^3$, $6x^2$, $5x$, and 8 are called *terms.* In the term $4x^3$, the number 4 is called the *numerical coefficient,* or simply the *coefficient* (koh-uh-FISH-ent), of x^3. In the same way, 6 is the coefficient of x^2 in the term $6x^2$, 5 is the coefficient of x in the term $5x$, and 8 is the coefficient in the term 8. A constant term, like 8 in the polynomial above, can be thought of as $8x^0$, where x^0 is defined to equal 1.

EXAMPLE 1 Identifying Coefficients

Name the coefficient of each term in these expressions.

(a) $4x^3$

The coefficient is 4.

(b) $x - 6x^4$

The coefficient of x is 1 because $x = 1 \cdot x$. The coefficient of x^4 is -6, since $x - 6x^4$ can be written as the sum $x + (-6x^4)$.

(c) $5 - v^3$

The coefficient of the term 5 is 5 since $5 = 5v^0$. By writing $5 - v^3$ as a sum, $5 + (-v^3)$, or $5 + (-1v^3)$, the coefficient of v^3 can be identified as -1.

WORK PROBLEM 1 AT THE SIDE.

OBJECTIVE 2 You will recall that *like terms* have exactly the same combination of variables, with the same exponents on the variables. Only the coefficients may be different. Examples of like terms are

$$19m^5 \quad \text{and} \quad 14m^5;$$
$$6y^9, \quad -37y^9, \quad \text{and} \quad y^9;$$
$$3pq^2, \quad -2pq^2, \quad \text{and} \quad 4pq^2.$$

By the distributive property, we add like terms by adding their coefficients.

EXAMPLE 2 Adding Like Terms

Simplify each expression by adding like terms.

(a) $-4x^3 + 6x^3 = (-4 + 6)x^3$ Distributive property

$= 2x^3$

(b) $9x^6 - 14x^6 + x^6 = (9 - 14 + 1)x^6 = -4x^6$

(c) $12m^2 + 5m + 4m^2 = (12 + 4)m^2 + 5m$

$= 16m^2 + 5m$

(d) $3x^2y + 4x^2y - x^2y = (3 + 4 - 1)x^2y = 6x^2y$

OBJECTIVES

1. Identify terms and coefficients.
2. Add like terms.
3. Know the vocabulary for polynomials.
4. Add polynomials.
5. Subtract polynomials.
6. Apply the rules and definitions for polynomials to multivariable polynomials.

1. Name the coefficient of each term in these expressions.

(a) $3m^2$

(b) $2x^2 - x$

(c) $x + 8$

ANSWERS

1. (a) 3 **(b)** 2; -1 **(c)** 1; 8

2. Add like terms.

(a) $5x^4 + 7x^4$

(b) $9pq + 3pq - 2pq$

(c) $r^2 + 3r + 5r^2$

(d) $8t + 6w$

3. Choose all descriptions that apply for each of the expressions in parts (a)–(d).

(1) Polynomial

(2) Polynomial written in descending order

(3) Not a polynomial

(a) $3m^5 + 5m^2 - 2m + 1$

(b) $2p^4 + p^6$

(c) $\frac{1}{x} + 2x^2 + 3$

(d) $x - 3$

ANSWERS

2. (a) $12x^4$ (b) $10pq$ (c) $6r^2 + 3r$ (d) cannot be added—unlike terms

3. (a) 1 and 2 (b) 1 (c) 3 (d) 1 and 2

In Example 2(c), it is not possible to add $16m^2$ and $5m$. These two terms are unlike because the exponents on the variables are different. *Unlike terms* have different variables or different exponents on the same variables.

WORK PROBLEM 2 AT THE SIDE.

OBJECTIVE 3 Polynomials are basic to algebra. A **polynomial** (pah-luh-NOH-mee-ul) **in x** is a term or the sum of a finite number of terms of the form ax^n, for any real number a and any whole number n. For example,

$$16x^8 - 7x^6 + 5x^4 - 3x^2 + 4$$

is a polynomial in x (here 4 can be written as $4x^0$). This polynomial is written in **descending** (dee-SEND-ing) **powers** of the variable, because the exponents on x decrease from left to right. On the other hand,

$$2x^3 - x^2 + \frac{4}{x}, \quad 5\sqrt{x} - 2x, \quad \text{and} \quad x^{-2} + 4 - x^2$$

are not polynomials in x. The first expression is not a polynomial because $\frac{4}{x} = 4x^{-1}$ is not a *product, ax^n*, for a whole number n. The second expression is not a polynomial because of the term $5\sqrt{x}$, and the third expression is not a polynomial because of the negative exponent on x. (Of course, a polynomial could be defined using any variable, or variables, and not just x.)

WORK PROBLEM 3 AT THE SIDE.

The **degree of a term with one variable** is the exponent on the variable. For example, $3x^4$ has degree 4, $6x^{17}$ has degree 17, $5x$ has degree 1, and -7 has degree 0 (since -7 can be written as $-7x^0$). The **degree of a polynomial** is the highest degree in any nonzero term of the polynomial. For example, $3x^4 - 5x^2 + 6$ is degree 4, the polynomial $5x + 7$ is degree 1, and 3 (or $3x^0$) is degree 0.

Three types of polynomials are very common and are given special names. A polynomial with exactly three terms is called a **trinomial** (TRY-noh-mee-ul). (*Tri-* means "three," as in *tri*angle.) Examples are

$$9m^3 - 4m^2 + 6, \quad \frac{19}{3}y^2 + \frac{8}{3}y + 5, \quad \text{and} \quad -3m^5 - 9m^2 + 2.$$

A polynomial with exactly two terms is called a **binomial** (BY-noh-mee-ul). (*Bi-* means "two," as in *bi*cycle.) Examples are

$$-9x^4 + 9x^3, \quad 8m^2 + 6m, \quad \text{and} \quad 3m^5 - 9m^2.$$

A polynomial with only one term is called a **monomial** (MAH-noh-mee-ul). (*Mon(o)-* means "one," as in *mono*rail.) Examples are

$$9m, \quad -6y^5, \quad a^2, \quad \text{and} \quad 6.$$

EXAMPLE 3 Classifying Polynomials

For each polynomial, first simplify if possible by combining like terms. Then give the degree and tell whether the polynomial is a monomial, a binomial, a trinomial, or none of these.

(a) $2x^3 + 5$

The polynomial cannot be simplified. The degree is 3. The polynomial is a binomial.

CONTINUED ON NEXT PAGE

(b) $4x - 5x + 2x$

Add like terms to simplify: $4x - 5x + 2x = x$. The degree is 1. The simplified polynomial is a monomial.

WORK PROBLEM 4 AT THE SIDE.

OBJECTIVE 4 Polynomials may be added, subtracted, multiplied, and divided. Polynomial addition and subtraction are explained in the rest of this section.

Adding Polynomials

To add two polynomials, add like terms.

EXAMPLE 4 Adding Polynomials Horizontally

(a) Add $6x^3 - 4x^2 + 3$ and $-2x^3 + 7x^2 - 5$.

Write the sum.

$$(6x^3 - 4x^2 + 3) + (-2x^3 + 7x^2 - 5)$$

Rewrite this sum with the parentheses removed.

$$6x^3 + (-4x^2) + 3 + (-2x^3) + 7x^2 + (-5)$$

Place like terms together.

$$6x^3 + (-2x^3) + (-4x^2) + 7x^2 + 3 + (-5)$$

Combine like terms to get

$$4x^3 + 3x^2 + (-2) \quad \text{or} \quad 4x^3 + 3x^2 - 2.$$

(b) Add $(2x^2 - 4x + 3)$ and $(x^3 + 5x)$.

$$(2x^2 - 4x + 3) + (x^3 + 5x) = 2x^2 - 4x + 3 + x^3 + 5x$$
$$= x^3 + 2x^2 + x + 3 \quad \text{Combine like terms.}$$

WORK PROBLEM 5 AT THE SIDE.

The polynomials in Example 4 also could be added vertically, as shown in the next example. This form of addition is used in multiplication of polynomials.

EXAMPLE 5 Adding Polynomials Vertically

Add $6x^3 - 4x^2 + 3$ and $-2x^3 + 7x^2 - 5$.

Write like terms in columns.

$$\begin{array}{r} 6x^3 - 4x^2 + 3 \\ \underline{-2x^3 + 7x^2 - 5} \end{array}$$

Now add, column by column.

$$\begin{array}{r} 6x^3 \\ \underline{-2x^3} \\ \mathbf{4x^3} \end{array} \qquad \begin{array}{r} -4x^2 \\ \underline{7x^2} \\ \mathbf{3x^2} \end{array} \qquad \begin{array}{r} 3 \\ \underline{-5} \\ \mathbf{-2} \end{array}$$

Add the three sums together.

$$\mathbf{4x^3 + 3x^2 + (-2)} = 4x^3 + 3x^2 - 2$$

This is the same answer found in Example 4(a).

4. For each polynomial, first simplify if possible. Then give the degree and tell whether the polynomial is a monomial, binomial, trinomial, or none of these.

(a) $3x^2 + 2x - 4$

(b) $x^3 + 4x^3$

(c) $x^8 - x^7 + 2x^8$

5. Find each sum.

(a) $(2x^4 - 6x^2 + 7) + (-3x^4 + 5x^2 + 2)$

(b) $(3x^2 + 4x + 2) + (6x^3 - 5x - 7)$

ANSWERS

4. **(a)** degree 2; trinomial **(b)** degree 3; monomial (simplify to $5x^3$) **(c)** degree 8; binomial (simplify to $3x^8 - x^7$)

5. **(a)** $-x^4 - x^2 + 9$
(b) $6x^3 + 3x^2 - x - 5$

6. Add each pair of polynomials.

(a) $$\begin{array}{r} 4x^3 - 3x^2 + 2x \\ \underline{6x^3 + 2x^2 - 3x} \end{array}$$

(b) $$\begin{array}{r} x^2 - 2x + 5 \\ \underline{4x^2 + 3x - 2} \end{array}$$

7. Subtract, and check your answers by addition.

(a) $(14y^3 - 6y^2 + 2y - 5)$
$\quad - (2y^3 - 7y^2 - 4y + 6)$

(b) $\left(\frac{7}{2}y^2 - \frac{11}{3}y + 8\right)$
$\quad - \left(-\frac{3}{2}y^2 + \frac{4}{3}y + 6\right)$

Answers

6. (a) $10x^3 - x^2 - x$ (b) $5x^2 + x + 3$

7. (a) $12y^3 + y^2 + 6y - 11$
(b) $5y^2 - 5y + 2$

◀◀ WORK PROBLEM 6 AT THE SIDE.

Objective 5 Earlier, the difference $x - y$ was defined as $x + (-y)$. (We find the difference $x - y$ by adding x and the opposite of y.) For example,

$$7 - 2 = 7 + (-2) = 5$$

and

$$-8 - (-2) = -8 + 2 = -6.$$

A similar method is used to subtract polynomials.

Subtracting Polynomials

To subtract polynomials, change all the signs of the second polynomial and add the result to the first polynomial.

EXAMPLE 6 Subtracting Polynomials

Subtract $(5x - 2) - (3x - 8)$.

By the definition of subtraction, we should change the signs of the second polynomial and then add the two polynomials.

$$\begin{aligned}(5x - 2) - (3x - 8) &= (5x - 2) + (-3x + 8) \\ &= 5x - 2 - 3x + 8 \\ &= 2x + 6\end{aligned}$$

EXAMPLE 7 Subtracting Polynomials

Subtract $6x^3 - 4x^2 + 2$ from $11x^3 + 2x^2 - 8$.

Start with

$$(11x^3 + 2x^2 - 8) - (6x^3 - 4x^2 + 2).$$

Change all the signs on the second polynomial and add like terms.

$$\begin{aligned}&(11x^3 + 2x^2 - 8) + (-6x^3 + 4x^2 - 2) \\ &= 11x^3 + 2x^2 - 8 - 6x^3 + 4x^2 - 2 \\ &= 5x^3 + 6x^2 - 10\end{aligned}$$

We can check a subtraction problem by using the fact that if $a - b = c$, then $a = b + c$. For example, $6 - 2 = 4$. Check by writing $6 = 2 + 4$, which is correct. Check the polynomial subtraction above by adding $6x^3 - 4x^2 + 2$ and $5x^3 + 6x^2 - 10$. Since the sum is $11x^3 + 2x^2 - 8$, the subtraction was performed correctly.

◀◀ WORK PROBLEM 7 AT THE SIDE.

Subtraction also can be done in columns. We will use vertical subtraction in polynomial division later in this chapter.

EXAMPLE 8 Subtracting Polynomials Vertically

Use the method of subtracting by columns to find

$$(14y^3 - 6y^2 + 2y - 5) - (2y^3 - 7y^2 - 4y + 6).$$

Step 1 Arrange like terms in columns.

$$\begin{array}{r} 14y^3 - 6y^2 + 2y - 5 \\ \underline{2y^3 - 7y^2 - 4y + 6} \end{array}$$

Step 2 Change all signs in the second row, and then add (*Step 3*).

$$\begin{array}{rl} 14y^3 - 6y^2 + 2y - 5 & \\ \underline{-2y^3 + 7y^2 + 4y - 6} & \text{All signs changed} \\ 12y^3 + y^2 + 6y - 11 & \text{Add.} \end{array}$$

Either the horizontal or the vertical method may be used for adding and subtracting polynomials.

WORK PROBLEM 8 AT THE SIDE.

EXAMPLE 9 Adding and Subtracting More than Two Polynomials

Perform the indicated operations to simplify the expression

$$(4 - x + 3x^2) - (2 - 3x + 5x^2) + (8 + 2x - 4x^2).$$

Since additions and subtractions are performed from left to right, we subtract first to get

$$\begin{aligned} &(4 - x + 3x^2) - (2 - 3x + 5x^2) + (8 + 2x - 4x^2) \\ &= (4 - x + 3x^2) + (-2 + 3x - 5x^2) + (8 + 2x - 4x^2). \end{aligned}$$

Now add the first two polynomials, then add the sum and the third polynomial.

$$\begin{aligned} &= (2 + 2x - 2x^2) + (8 + 2x - 4x^2) \\ &= 10 + 4x - 6x^2 \end{aligned}$$

WORK PROBLEM 9 AT THE SIDE.

OBJECTIVE 6 A **multivariable** (muhl-tih-VAIR-ee-uh-bul) **polynomial** has more than one variable. Some examples are $a + ab + b$ and $6x^2y + xy - 5x$. Most of the definitions and rules in this section also apply to multivariable polynomials. The **degree of a term with more than one variable** is the sum of the exponents on the variables. Thus, the term $2x^2y$ has degree $2 + 1 = 3$, and the polynomial $6x^2y + xy - 5$ has degree 3.

EXAMPLE 10 Determining the Degree of a Multivariable Polynomial

Give the degree of each polynomial.

(a) $7m^3n + 5m^2n - 8mn$

The degree of each term is $3 + 1 = \mathbf{4}$, $2 + 1 = 3$, and $1 + 1 = 2$, respectively. Thus, the degree of the polynomial is **4**, the largest of these three numbers.

CONTINUED ON NEXT PAGE

8. Subtract, using the method of subtracting by columns.

$$\begin{aligned} &(14y^3 - 6y^2 + 2y) \\ &\quad - (2y^3 - 7y^2 + 6) \end{aligned}$$

9. Perform the indicated operations.

$$\begin{aligned} &(6p^4 - 8p^3 + 2p - 1) \\ &\quad - (-7p^4 + 6p^2 - 12) \\ &\quad + (p^4 - 3p + 8) \end{aligned}$$

ANSWERS

8. $12y^3 + y^2 + 2y - 6$

9. $14p^4 - 8p^3 - 6p^2 - p + 19$

10. Add or subtract. Give the degree of the answer.

(a) $(3mn + 2m - 4n)$
$+ (-mn + 4m + n)$

(b) $(5p^2q^2 - 4p^2 + 2q)$
$- (2p^2q^2 - p^2 - 3q)$

(b) $-x^2y + 5xy^3 - 4y + x^3y^2$

The four terms have the following degrees, respectively: $2 + 1 = 3$, $1 + 3 = 4$, 1, and $3 + 2 = 5$. The polynomial has degree 5.

Multivariable polynomials are added and subtracted by combining like terms, just as we did with single variable polynomials.

EXAMPLE 11 Adding and Subtracting Multivariable Polynomials

Add or subtract as indicated, and give the degree of the answer.

(a) $(4a + 2ab - b) + (3a - ab + b)$

$$(4a + 2ab - b) + (3a - ab + b)$$
$$= 4a + 2ab - b + 3a - ab + b$$
$$= 7a + ab$$

The polynomial $7a + ab$ has degree 2.

(b) $(2x^2y + 3xy + y^2) - (3x^2y - xy - 2y^2)$

$$(2x^2y + 3xy + y^2) - (3x^2y - xy - 2y^2)$$
$$= 2x^2y + 3xy + y^2 - 3x^2y + xy + 2y^2$$
$$= -x^2y + 4xy + 3y^2$$

The polynomial $-x^2y + 4xy + 3y^2$ has degree 3.

WORK PROBLEM 10 AT THE SIDE.

ANSWERS

10. **(a)** $2mn + 6m - 3n$; degree 2
(b) $3p^2q^2 - 3p^2 + 5q$; degree 4

4 Exercises

Tell whether each statement is true always, sometimes, *or* never.

1. A polynomial is a binomial.

2. A polynomial is a trinomial.

3. A trinomial is a polynomial.

4. A binomial is a polynomial.

5. A trinomial is a binomial.

6. A binomial is a trinomial.

7. A monomial is a polynomial.

8. A polynomial is a monomial.

For each of the following, determine the number of terms in the polynomial, and name the coefficients of the terms. See Example 1.

9. $6x^4$

10. $-9y^5$

11. t^4

12. s^7

13. $-19r^2 - r$

14. $2y^3 - y$

15. $x + 8x^2$

16. $v - 2v^3$

In each polynomial, add like terms whenever possible. Write the result in descending powers of the variable. See Example 2.

17. $-3m^5 + 5m^5$

18. $-4y^3 + 3y^3$

19. $2r^5 + (-3r^5)$

20. $-19y^2 + 9y^2$

21. $.2m^5 - .5m^2$

22. $-.9y + .9y^2$

23. $-3x^5 + 2x^5 - 4x^5$

24. $6x^3 - 8x^3 + 9x^3$

25. $-4p^7 + 8p^7 + 5p^9$

26. $-3a^8 + 4a^8 - 3a^2$

27. $-4y^2 + 3y^2 - 2y^2 + y^2$

28. $3r^5 - 8r^5 + r^5 + 2r^5$

For each polynomial, first simplify, if possible, and write it in descending powers of the variable. Then give the degree of the resulting polynomial, and tell whether it is a monomial, a binomial, a trinomial, or none of these. See Example 3.

29. $6x^4 - 9x$

30. $7t^3 - 3t$

31. $5m^4 - 3m^2 + 6m^5 - 7m^3$

32. $6p^5 + 4p^3 - 8p^4 + 10p^2$

33. $\frac{5}{3}x^4 - \frac{2}{3}x^4 + \frac{1}{3}x^2 - 4$

34. $\frac{4}{5}r^6 + \frac{1}{5}r^6 - r^4 + \frac{2}{5}r$

35. $.8x^4 - .3x^4 - .5x^4 + 7$

36. $1.2t^3 - .9t^3 - .3t^3 + 9$

MATHEMATICAL CONNECTIONS (EXERCISES 37–40)

If an object is thrown upward under certain conditions, its height in feet is given by the trinomial $-16x^2 + 60x + 80$, *where x is in seconds. If we evaluate the polynomial for a specific value of x, we will get one and only one value as a result. This idea is basic to the study of* functions, *one of the most important concepts in mathematics. Work Exercises 37–40 in order.*

37. If gasoline costs \$1.25 per gallon, then the monomial $1.25x$ gives the cost, in dollars, of x gallons. Evaluate this monomial for $x = 4$, then use the result to fill in the blanks. If ______ gallons are purchased, the cost is ______.

38. If it costs \$15 to rent a chain saw plus \$2 per day, the binomial $2x + 15$ gives the cost, in dollars, to rent the chain saw for x days. Evaluate this polynomial for $x = 6$ and use the result to fill in the blanks. If the saw is rented for ______ days, the cost is ______.

39. Evaluate the polynomial $-16x^2 + 60x + 80$, given in the directions above, for $x = 2.5$ and use the result to fill in the blanks. If ______ seconds have elapsed, the height of the object is ______ feet.

40. A certain viral infection causes a fever that typically lasts 6 days. A patient's temperature in °F on day x is given by the trinomial

$$-\frac{2}{3}x^2 + \frac{14}{3}x + 96.$$

Evaluate the trinomial for $x = 3$ and fill in the blanks. On day ______, the patient has a temperature of ______ °F.

Add or subtract as indicated. See Examples 5 and 8.

41. Add.

$$\begin{array}{r} 3m^2 + 5m \\ \underline{2m^2 - 2m} \end{array}$$

42. Add.

$$\begin{array}{r} 4a^3 - 4a^2 \\ \underline{6a^3 + 5a^2} \end{array}$$

43. Subtract.

$$\begin{array}{r} 12x^4 - x^2 \\ \underline{8x^4 + 3x^2} \end{array}$$

44. Subtract.

$$\begin{array}{r} 13y^5 - y^3 \\ \underline{7y^5 + 5y^3} \end{array}$$

45. Add.

$$\begin{array}{r} \frac{2}{3}x^2 + \frac{1}{5}x + \frac{1}{6} \\ \underline{\frac{1}{2}x^2 - \frac{1}{3}x + \frac{2}{3}} \end{array}$$

46. Add.

$$\begin{array}{r} \frac{4}{7}y^2 - \frac{1}{5}y + \frac{7}{9} \\ \underline{\frac{1}{3}y^2 - \frac{1}{3}y + \frac{2}{5}} \end{array}$$

47. Subtract.

$$\begin{array}{r} 12m^3 - 8m^2 + 6m + 7 \\ \underline{-3m^3 + 5m^2 - 2m - 4} \end{array}$$

48. Subtract.

$$\begin{array}{r} 5a^4 - 3a^3 + 2a^2 - a + 6 \\ \underline{-6a^4 + a^3 - a^2 + a - 1} \end{array}$$

Perform the indicated operations. See Examples 4, 6, 7, and 9.

49. $(2r^2 + 3r - 12) + (6r^2 + 2r)$

50. $(3r^2 + 5r - 6) + (2r - 5r^2)$

51. $(8m^2 - 7m) - (3m^2 + 7m - 6)$

52. $(x^2 + x) - (3x^2 + 2x - 1)$

53. $(16x^3 - x^2 + 3x) + (-12x^3 + 3x^2 + 2x)$

54. $(-2b^6 + 3b^4 - b^2) + (b^6 + 2b^4 + 2b^2)$

55. $(7y^4 + 3y^2 + 2y) - (18y^5 - 5y^3 + y)$

56. $(8t^5 + 3t^3 + 5t) - (19t^4 - 6t^2 + t)$

57. $[(8m^2 + 4m - 7) - (2m^3 - 5m + 2)] - (m^2 + m + 1)$

58. $[(9b^3 - 4b^2 + 3b + 2) - (-2b^3 - 3b^2 + b)] - (8b^3 + 6b + 4)$

Find the perimeter of each of the following geometric figures.

59.

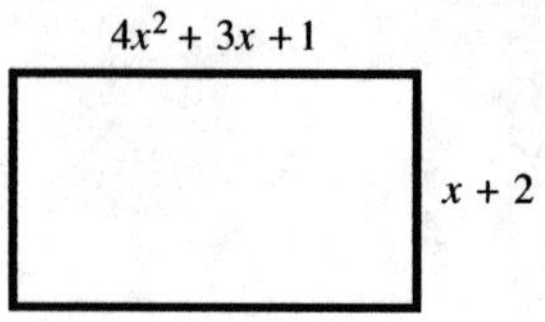

60.

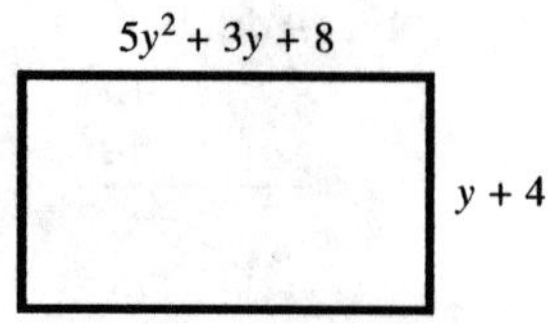

61.

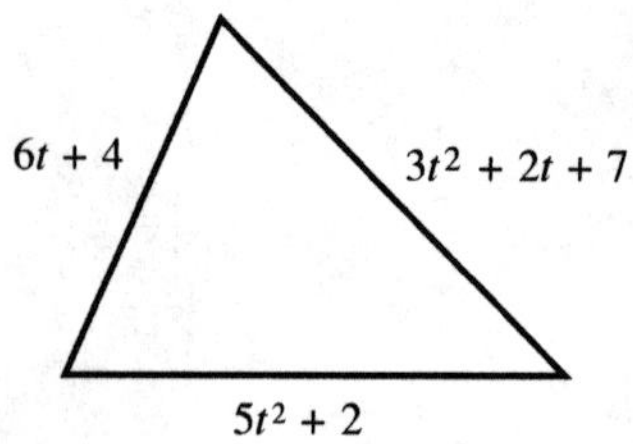

62.

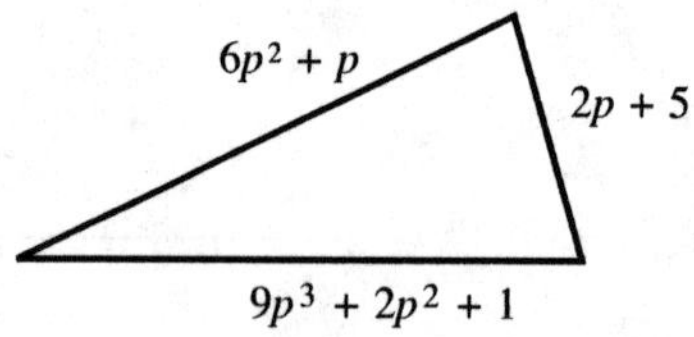

63. Subtract $9x^2 - 3x + 7$ from $-2x^2 - 6x + 4$.

64. Subtract $-5w^3 + 5w^2 - 7$ from $6w^3 + 8w + 5$.

65. Explain why the degree of the term 3^4 is not 4. What is its degree?

66. Can the sum of two polynomials in x, both of degree 3, be of degree 2? If so, give an example.

Add or subtract as indicated. Give the degree of the answer. See Examples 10 and 11.

67. $(9a^2b - 3a^2 + 2b) + (4a^2b - 4a^2 - 3b)$

68. $(4xy^3 - 3x + y) + (5xy^3 + 13x - 4y)$

69. $(2c^4d + 3c^2d^2 - 4d^2) - (c^4d + 8c^2d^2 - 5d^2)$

70. $(3k^2h^3 + 5kh + 6k^3h^2) - (2k^2h^3 - 9kh + k^3h^2)$

71. Subtract. $\begin{array}{r} 9m^3n - 5m^2n^2 + 4mn^2 \\ \underline{-3m^3n + 6m^2n^2 + 8mn^2} \end{array}$

72. Subtract. $\begin{array}{r} 12r^5t + 11r^4t^2 - 7r^3t^3 \\ \underline{-8r^5t + 10r^4t^2 + 3r^3t^3} \end{array}$

5 *Multiplication of Polynomials*

OBJECTIVE 1 As shown earlier, we find the product of two monomials by using the rules for exponents and the commutative and associative properties. For example,

$$(6x^3)(4x^4) = 6 \cdot 4 \cdot x^3 \cdot x^4 = 24x^7.$$

Also,

$$(-8m^6)(-9n^6) = (-8)(-9)(m^6)(n^6) = 72m^6n^6.$$

To find the product of a monomial and a polynomial with more than one term, we use the distributive property and then the method shown above.

EXAMPLE 1 Multiplying a Monomial and a Polynomial

Use the distributive property to find each product.

(a) $4x^2(3x + 5)$

$$\mathbf{4x^2}(3x + 5) = (\mathbf{4x^2})(3x) + (\mathbf{4x^2})(5) \qquad \text{Distributive property}$$
$$= 12x^3 + 20x^2 \qquad \text{Multiply monomials.}$$

(b) $-8m^3(4m^3 + 3m^2 + 2m - 1)$

$$= (\mathbf{-8m^3})(4m^3) + (\mathbf{-8m^3})(3m^2)$$
$$+ (\mathbf{-8m^3})(2m) + (\mathbf{-8m^3})(-1) \qquad \text{Distributive property}$$
$$= -32m^6 - 24m^5 - 16m^4 + 8m^3 \qquad \text{Multiply monomials.}$$

WORK PROBLEM 1 AT THE SIDE.

OBJECTIVE 2 The distributive property is used also to find the product of any two polynomials. For example, to find the product of the polynomials $x + 1$ and $x - 4$, think of $x + 1$ as a single quantity and use the distributive property as follows.

$$(\mathbf{x + 1})(x - 4) = (\mathbf{x + 1})\,x + (\mathbf{x + 1})(-4)$$

Now use the distributive property to find $(x + 1)x$ and $(x + 1)(-4)$.

$$(x + 1)\mathbf{x} + (x + 1)(\mathbf{-4}) = x(\mathbf{x}) + 1(\mathbf{x}) + x(\mathbf{-4}) + 1(\mathbf{-4})$$
$$= x^2 + x + (-4x) + (-4)$$
$$= x^2 - 3x - 4$$

EXAMPLE 2 Multiplying Two Binomials

Find the product $(2x + 1)(3x + 5)$.

$$(2x + 1)(3x + 5) = (2x + 1)(3x) + (2x + 1)(5)$$
$$= (2x)(3x) + (1)(3x) + (2x)(5) + (1)(5)$$
$$= 6x^2 + 3x + 10x + 5$$
$$= 6x^2 + 13x + 5$$

WORK PROBLEM 2 AT THE SIDE.

OBJECTIVES

1. Multiply a monomial and a polynomial.
2. Multiply two polynomials.
3. Multiply vertically.
4. Find higher powers of binomials.

1. Find each product.

(a) $5m^3(2m + 7)$

(b) $2x^4(3x^2 + 2x - 5)$

(c) $-4y^2(3y^3 + 2y^2 - 4y + 8)$

2. Find each product.

(a) $(4x + 3)(2x + 1)$

(b) $(3k - 2)(2k + 1)$

(c) $(m + 5)(3m - 4)$

ANSWERS

1. **(a)** $10m^4 + 35m^3$
(b) $6x^6 + 4x^5 - 10x^4$
(c) $-12y^5 - 8y^4 + 16y^3 - 32y^2$

2. **(a)** $8x^2 + 10x + 3$
(b) $6k^2 - k - 2$
(c) $3m^2 + 11m - 20$

3. Multiply.

(a) $(m^3 - 2m + 1) \cdot (2m^2 + 4m + 3)$

(b) $(6p^2 + 2p - 4)(3p^2 - 5)$

Answers

3. **(a)** $2m^5 + 4m^4 - m^3 - 6m^2 - 2m + 3$
(b) $18p^4 + 6p^3 - 42p^2 - 10p + 20$

A rule for multiplying any two polynomials is given below.

Multiplying Polynomials

Multiply two polynomials by multiplying each term of the second polynomial by each term of the first polynomial and adding the products.

EXAMPLE 3 Multiplying Any Two Polynomials

Find the product $(3y^2 + 2y - 1)(y^3 + y^2 - 5)$.

Multiply each term in the first polynomial by each term in the second polynomial, then add any like terms.

$$(\mathbf{3y^2} + \mathbf{2y} - \mathbf{1})(y^3 + y^2 - 5)$$
$$= \mathbf{3y^2}(y^3) + \mathbf{3y^2}(y^2) + \mathbf{3y^2}(-5) + \mathbf{2y}(y^3) + \mathbf{2y}(y^2) + \mathbf{2y}(-5) - \mathbf{1}(y^3) - \mathbf{1}(y^2) - \mathbf{1}(-5)$$
$$= 3y^5 + 3y^4 - 15y^2 + 2y^4 + 2y^3 - 10y - y^3 - y^2 + 5$$
$$= 3y^5 + 5y^4 + y^3 - 16y^2 - 10y + 5$$

WORK PROBLEM 3 AT THE SIDE.

Objective 3 We can also multiply two polynomials by writing one polynomial above the other.

EXAMPLE 4 Multiplying Polynomials Vertically

Multiply $2x^2 + 4x + 1$ by $3x + 5$.

Start with

$$\begin{array}{r} 2x^2 + 4x + 1 \\ 3x + 5. \\ \hline \end{array}$$

It is not necessary to line up like terms in columns, because any terms may be multiplied (not just like terms). Begin by multiplying each of the terms in the top row by 5.

Step 1

$$\begin{array}{rl} \mathbf{2x^2 + 4x + 1} & \\ 3x + \mathbf{5} & \\ \hline 10x^2 + 20x + 5 & \leftarrow \text{Product of 5 and } 2x^2 + 4x + 1 \end{array}$$

Notice how this process is similar to multiplication of whole numbers. Now multiply each term in the top row by $3x$. Be careful to place the like terms in columns, since the final step will involve addition (as in multiplying two whole numbers).

Step 2

$$\begin{array}{rl} \mathbf{2x^2 + 4x + 1} & \\ \mathbf{3x} + 5 & \\ \hline 10x^2 + 20x + 5 & \\ 6x^3 + 12x^2 + 3x \qquad\qquad & \leftarrow \text{Product of } 3x \text{ and } 2x^2 + 4x + 1 \end{array}$$

CONTINUED ON NEXT PAGE

Step 3 Add like terms.

$$\begin{array}{r} 2x^2 + 4x + 1 \\ 3x + 5 \\ \hline 10x^2 + 20x + 5 \\ 6x^3 + 12x^2 + 3x \phantom{{}+5} \\ \hline \mathbf{6x^3 + 22x^2 + 23x + 5} \end{array}$$

The product is $6x^3 + 22x^2 + 23x + 5$.

WORK PROBLEM 4 AT THE SIDE.

EXAMPLE 5 Multiplying Polynomials Vertically

Find the product of $3p - 5q$ and $2p + 7q$.

$$\begin{array}{rl} 3p - 5q & \\ 2p + 7q & \\ \hline 21pq - 35q^2 & \leftarrow 7q(3p - 5q) \\ 6p^2 - 10pq \phantom{{}-35q^2} & \leftarrow 2p(3p - 5q) \\ \hline 6p^2 + 11pq - 35q^2 & \end{array}$$

EXAMPLE 6 Multiplying Polynomials Vertically

Find the product of $4m^3 - 2m^2 + 4m$ and $\frac{1}{2}m^2 + \frac{5}{2}$.

$$\begin{array}{rl} 4m^3 - 2m^2 + 4m & \\ \frac{1}{2}m^2 + \frac{5}{2} & \\ \hline 10m^3 - 5m^2 + 10m & \text{Terms of top row multiplied by } \frac{5}{2} \\ 2m^5 - m^4 + 2m^3 \phantom{{}- 5m^2 + 10m} & \text{Terms of top row multiplied by } \frac{1}{2}m^2 \\ \hline 2m^5 - m^4 + 12m^3 - 5m^2 + 10m & \end{array}$$

WORK PROBLEM 5 AT THE SIDE.

OBJECTIVE 4 To find higher powers of binomials, such as $(x + 5)^3$, we use the definition of exponent and perform repeated multiplication.

EXAMPLE 7 Squaring a Polynomial

Find $(3n^2 + 5n - 1)^2$.

By definition, $(3n^2 + 5n - 1)^2 = (3n^2 + 5n - 1)(3n^2 + 5n - 1)$. Use the vertical method of multiplication.

$$\begin{array}{r} 3n^2 + 5n - 1 \\ 3n^2 + 5n - 1 \\ \hline -3n^2 - 5n + 1 \\ 15n^3 + 25n^2 - 5n \phantom{{}+1} \\ 9n^4 + 15n^3 - 3n^2 \phantom{{}-5n+1} \\ \hline 9n^4 + 30n^3 + 19n^2 - 10n + 1 \end{array}$$

4. Find each product.

(a) $\begin{array}{r} 4k - 6 \\ 2k + 5 \\ \hline \end{array}$

(b) $\begin{array}{r} 3x^2 + 4x - 5 \\ x + 4 \\ \hline \end{array}$

5. Find each product.

(a) $\begin{array}{r} 2m + 3p \\ 5m - 4p \\ \hline \end{array}$

(b) $\begin{array}{r} k^3 - k^2 + k + 1 \\ \frac{2}{3}k - \frac{1}{3} \\ \hline \end{array}$

(c) $\begin{array}{r} a^3 + 3a - 4 \\ 2a^2 + 6a + 5 \\ \hline \end{array}$

ANSWERS

4. (a) $8k^2 + 8k - 30$
(b) $3x^3 + 16x^2 + 11x - 20$

5. (a) $10m^2 + 7mp - 12p^2$
(b) $\frac{2}{3}k^4 - k^3 + k^2 + \frac{1}{3}k - \frac{1}{3}$
(c) $2a^5 + 6a^4 + 11a^3 + 10a^2 - 9a - 20$

6. Find each product.

(a) $(2x^2 - 3x + 4)^2$

(b) $(m + 1)^3$

(c) $(3k - 2)^4$

ANSWERS

6. (a) $4x^4 - 12x^3 + 25x^2 - 24x + 16$
(b) $m^3 + 3m^2 + 3m + 1$
(c) $81k^4 - 216k^3 + 216k^2 - 96k + 16$

EXAMPLE 8 Cubing a Binomial

Find $(x + 5)^3$.

Since $(x + 5)^3 = (x + 5)(x + 5)(x + 5)$, the first step is to find the product $(x + 5)(x + 5)$.

$$\begin{aligned}(x + 5)(x + 5) &= x^2 + 5x + 5x + 25\\ &= x^2 + 10x + 25\end{aligned}$$

Now multiply this result by $x + 5$.

$$\begin{aligned}(x + 5)(x^2 + 10x + 25) &= x^3 + 10x^2 + 25x + 5x^2 + 50x + 125\\ &= x^3 + 15x^2 + 75x + 125\end{aligned}$$

EXAMPLE 9 Finding the Fourth Power of a Binomial

Find $(2y - 3)^4$.

One way to proceed is to note that $(2y - 3)^4 = (2y - 3)^2(2y - 3)^2$.

$$\begin{aligned}(2y - 3)^4 &= (2y - 3)^2(2y - 3)^2\\ &= (4y^2 - 6y - 6y + 9)(4y^2 - 6y - 6y + 9)\\ &= (4y^2 - 12y + 9)(4y^2 - 12y + 9)\end{aligned}$$

Multiplying vertically will help to keep like terms together so they can be combined.

$$\begin{array}{rrrrr} & & 4y^2 & -\ 12y & +\ 9\\ & & 4y^2 & -\ 12y & +\ 9\\ \hline & & 36y^2 & -\ 108y & +\ 81\\ & -\ 48y^3 & +\ 144y^2 & -\ 108y & \\ 16y^4 & -\ 48y^3 & +\ 36y^2 & & \\ \hline 16y^4 & -\ 96y^3 & +\ 216y^2 & -\ 216y & +\ 81\end{array}$$

WORK PROBLEM 6 AT THE SIDE.

5 Exercises

1. In multiplying a monomial by a polynomial, such as in $4x(3x^2 + 7x^3) = 4x(3x^2) + 4x(7x^3)$, the first property that is used is the ________________ property.

2. In multiplying two monomials, we use (but do not always show) the commutative and associative properties. In each item below, name the property that has been used.

$$(5x^3)(6x^5) = (5x^3 \cdot 6)x^5 \quad \_\_\_\_\_\_\_\_\_\_$$
$$= (5 \cdot 6x^3)x^5 \quad \_\_\_\_\_\_\_\_\_\_$$
$$= (5 \cdot 6)(x^3 \cdot x^5) \quad \_\_\_\_\_\_\_\_\_\_$$
$$= 30x^8$$

Find each product. See Example 1.

3. $-2m(3m + 2)$

4. $-5p(6 + 3p)$

5. $\frac{3}{4}p(8 - 6p + 12p^3)$

6. $\frac{4}{3}x(3 + 2x + 5x^3)$

7. $2y^5(3 + 2y + 5y^4)$

8. $2m^4(3m^2 + 5m + 6)$

Find each binomial product. See Examples 2 and 5.

9. $(n - 2)(n + 3)$

10. $(r - 6)(r + 8)$

11. $(4r + 1)(2r - 3)$

12. $(5x + 2)(2x - 7)$

13. $(3x + 2)(3x - 2)$

14. $(7x + 3)(7x - 3)$

15. $(3q + 1)(3q + 1)$

16. $(4w + 7)(4w + 7)$

17. $(3t + 4s)(2t + 5s)$

18. $(8v + 5w)(2v + 3w)$

19. $(-.3t + .4)(t + .6)$

20. $(-.5x + .9)(x - .2)$

21. Perform the following multiplications: $(x + 4)(x - 4)$; $(y + 2)(y - 2)$; $(r + 7)(r - 7)$. Observe your answers, and explain the pattern that can be found in the answers.

22. Repeat Exercise 21 for the following: $(x + 4)(x + 4)$; $(y - 2)(y - 2)$; $(r + 7)(r + 7)$.

Find each product. See Examples 3, 4, and 6.

23. $(6x + 1)(2x^2 + 4x + 1)$

24. $(9y - 2)(8y^2 - 6y + 1)$

25. $(4m + 3)(5m^3 - 4m^2 + m - 5)$

26. $(y + 4)(3y^3 - 2y^2 + y + 3)$

27. $(5x^2 + 2x + 1)(x^2 - 3x + 5)$

28. $(2m^2 + m - 3)(m^2 - 4m + 5)$

Find each product. See Examples 7, 8, and 9.

29. $(x + 12)^2$

30. $(y + 11)^2$

31. $(3t + 1)^2$

32. $(4y + 3)^2$

33. $(5x - 2y)^2$

34. $(6p - 5s)^2$

35. $(h - 5)^3$

36. $(n - 3)^3$

37. $(r + 1)^4$

38. $(k - 1)^4$

39. $(3x^2 + x - 4)^2$

40. $(4y^2 - y + 3)^2$

MATHEMATICAL CONNECTIONS (EXERCISES 41–46)

Work Exercises 41–46 in order. Refer to the figure as necessary.

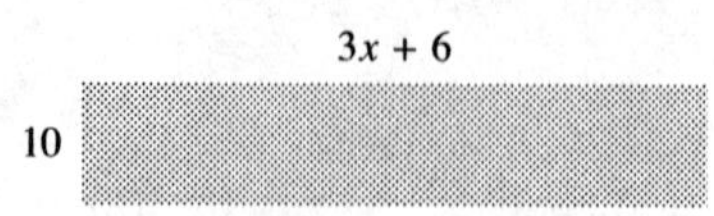

41. Find a polynomial that represents the area of the rectangle.

42. Suppose you know that the area of the rectangle is 600 square yards. Use this information and the polynomial from Exercise 41 to write an equation and solve it for x.

43. What are the dimensions of the rectangle (assume units are all in yards)?

44. Suppose the rectangle represents a lawn and its costs \$3.50 per square yard to lay sod on the lawn. How much will it cost to sod the entire lawn?

45. Use the result of Exercise 43 to find the perimeter of the lawn.

46. Again, suppose the rectangle represents a lawn and it costs \$9.00 per yard to fence the lawn. How much will it cost to fence the lawn?

6 *Products of Binomials*

OBJECTIVE 1 We can use the methods introduced in the last section to find the product of any two polynomials. They are the only practical methods for multiplying polynomials with three or more terms. However, many of the polynomials to be multiplied are binomials, with only two terms, so in this section we discuss a shortcut that eliminates the need to write all the steps. To develop this shortcut, let us first multiply $x + 3$ and $x + 5$ using the distributive property.

$$\begin{aligned}(x + 3)(x + 5) &= (x + 3)x + (x + 3)5 \\ &= (x)(x) + (3)(x) + (x)(5) + (3)(5) \\ &= x^2 + 3x + 5x + 15 \\ &= x^2 + 8x + 15\end{aligned}$$

The first term in the second line, $(x)(x)$, is the product of the first terms of the two binomials.

$(x + 3)(x + 5)$ Multiply the **first** terms: $(x)(x)$.

The term $(x)(5)$ is the product of the first term of the first binomial and the last term of the second binomial. This is the **outer product.**

$(x + 3)(x + 5)$ Multiply the **outer** terms: $(x)(5)$.

The term $(3)(x)$ is the product of the last term of the first binomial and the first term of the second binomial. The product of these middle terms is called the **inner product.**

$(x + 3)(x + 5)$ Multiply the **inner** terms: $(3)(x)$.

Finally, $(3)(5)$ is the product of the last terms of the two binomials.

$(x + 3)(x + 5)$ Multiply the **last** terms: $(3)(5)$.

In the third step of the multiplication above, the inner product and the outer product are added. This step should be performed mentally, so that the three terms of the answer can be written without extra steps as

$$(x + 3)(x + 5) = x^2 + 8x + 15.$$

WORK PROBLEM 1 AT THE SIDE.

A summary of these steps is given below. This procedure is sometimes called the **FOIL method,** which comes from the initial letters of the words *First, Outer, Inner,* and *Last.*

Multiplying Binomials by the FOIL Method

Step 1 Multiply the two **F**irst terms of the binomials to get the first term of the answer.

Step 2 Find the **O**uter product and the **I**nner product and add them (mentally, if possible) to get the middle term of the answer.

Step 3 Multiply the two **L**ast terms of the binomials to get the last term of the answer.

F $= x^2$ **L** $= 15$

$(x + 3)(x + 5)$

I $= 3x$

O $= 5x$

$8x$ Add.

OBJECTIVES

1. Multiply binomials by the FOIL method.
2. Square binomials.
3. Find the product of the sum and difference of two terms.

1. For the product $(2p - 5)(3p + 7)$, find the following.

(a) Product of first terms

(b) Outer product

(c) Inner product

(d) Product of last terms

(e) Complete product in simplified form

ANSWERS

1. (a) $2p(3p) = 6p^2$
(b) $2p(7) = 14p$
(c) $-5(3p) = -15p$
(d) $-5(7) = -35$
(e) $6p^2 - p - 35$

2. Use the FOIL method to find each product.

(a) $(m + 4)(m - 3)$

(b) $(y + 7)(y + 2)$

(c) $(r - 8)(r - 5)$

EXAMPLE 1 Using the FOIL Method

Use the FOIL method to find the product $(x + 8)(x - 6)$.

Step 1 **F** Multiply the **first** terms.

$$x(x) = x^2$$

Step 2 **O** Find the product of the **outer** terms.

$$x(-6) = -6x$$

I Find the product of the **inner** terms.

$$8(x) = 8x$$

Add the outer and inner products mentally.

$$-6x + 8x = 2x$$

Step 3 **L** Multiply the **last** terms.

$$8(-6) = -48$$

The product of $x + 8$ and $x - 6$ is found by adding the terms found in the three steps above, so

$$(x + 8)(x - 6) = x^2 + 2x - 48.$$

As a shortcut, this product can be found in the following manner.

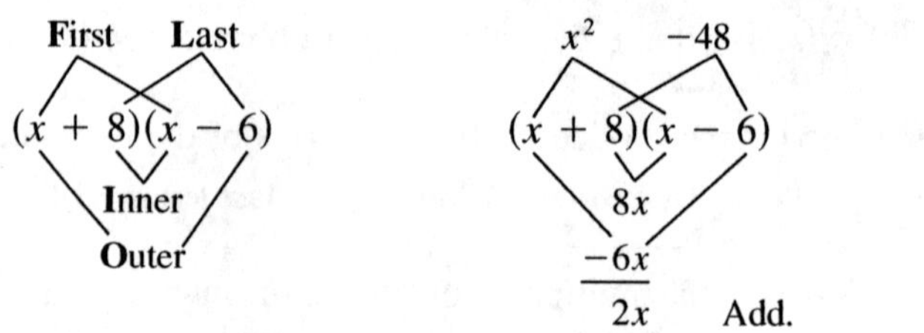

◀◀ WORK PROBLEM 2 AT THE SIDE.

EXAMPLE 2 Using the FOIL Method

Multiply $9x - 2$ and $3x + 1$.

First	$(\mathbf{9x} - 2)(\mathbf{3x} + 1)$	$27x^2$
Outer	$(\mathbf{9x} - 2)(3x + \mathbf{1})$	$9x$
Inner	$(9x \mathbf{- 2})(\mathbf{3x} + 1)$	$-6x$
Last	$(9x \mathbf{- 2})(3x \mathbf{+ 1})$	-2

$$\begin{aligned}(9x - 2)(3x + 1) &= \overset{\text{F}}{27x^2} + \overset{\text{O}}{9x} - \overset{\text{I}}{6x} - \overset{\text{L}}{2} \\ &= 27x^2 + 3x - 2\end{aligned}$$

ANSWERS

2. (a) $m^2 + m - 12$
(b) $y^2 + 9y + 14$
(c) $r^2 - 13r + 40$

EXAMPLE 3 Using the FOIL Method

Find the following products.

$$\begin{aligned}(2k + 5y)(k + 3y) &= \overset{\textbf{F}}{(2k)(k)} + \overset{\textbf{O}}{(2k)(3y)} + \overset{\textbf{I}}{(5y)(k)} + \overset{\textbf{L}}{(5y)(3y)} \\ &= 2k^2 + 6ky + 5ky + 15y^2 \\ &= 2k^2 + 11ky + 15y^2\end{aligned}$$

(a) (above)

(b) $(7p + 2q)(3p - q) = 21p^2 - pq - 2q^2$

WORK PROBLEM 3 AT THE SIDE.

OBJECTIVE 2 Certain special types of binomial products occur so often that the form of the answers should be memorized. For example, to find the square of a binomial quickly, use the method shown in Example 4.

EXAMPLE 4 Squaring a Binomial

Find $(2m + 3)^2$.

Squaring $2m + 3$ by the FOIL method gives

$$(2m + 3)(2m + 3) = 4m^2 + 6m + 6m + 9 = 4m^2 + 12m + 9.$$

The result has the square of both the first and the last terms of the binomial:

$$4m^2 = (\mathbf{2m})^2 \qquad \text{and} \qquad 9 = \mathbf{3}^2.$$

The middle term is twice the product of the two terms of the binomial, that is,

$$12m = 2(6m) = 2(\mathbf{2m})(\mathbf{3}).$$

This example suggests the following rule.

Square of a Binomial

The square of a binomial is a trinomial made up of the square of the first term, plus twice the product of the two terms, plus the square of the last term of the binomial. For a and b,

$$(a + b)^2 = a^2 + 2ab + b^2.$$

Also, $$(a - b)^2 = a^2 - 2ab + b^2.$$

EXAMPLE 5 Squaring Binomials

Use the formula to square each binomial.

$$\begin{array}{ccccccc}(a - b)^2 & = & a^2 & - & 2 \cdot a \cdot b & + & b^2 \\ \downarrow\ \downarrow & & \downarrow & & \downarrow\ \ \downarrow\ \ \downarrow & & \downarrow\end{array}$$

(a) $(\mathbf{5z} - \mathbf{1})^2 = (\mathbf{5z})^2 - 2(\mathbf{5z})(\mathbf{1}) + (\mathbf{1})^2$

$= 25z^2 - 10z + 1$

Recall that $(5z)^2 = 5^2z^2 = 25z^2$.

(b) $(\mathbf{3b} + \mathbf{5r})^2 = (\mathbf{3b})^2 + 2(\mathbf{3b})(\mathbf{5r}) + (\mathbf{5r})^2$

$= 9b^2 + 30br + 25r^2$

CONTINUED ON NEXT PAGE

3. Find each product.

(a) $(4k - 1)(2k + 3)$

(b) $(6m + 5)(m - 4)$

(c) $(8y + 3)(2y + 1)$

(d) $(3r + 2t)(3r + 4t)$

ANSWERS

3. (a) $8k^2 + 10k - 3$
(b) $6m^2 - 19m - 20$
(c) $16y^2 + 14y + 3$
(d) $9r^2 + 18rt + 8t^2$

4. Find each square by using the pattern for the square of a binomial.

(a) $(t + u)^2$

(b) $(2m - p)^2$

(c) $(4p + 3q)^2$

(d) $(5r - 6s)^2$

(e) $\left(3k - \frac{1}{2}\right)^2$

5. Find each product by using the pattern for the sum and difference of two terms.

(a) $(6a + 3)(6a - 3)$

(b) $(10m + 7)(10m - 7)$

(c) $(7p + 2q)(7p - 2q)$

(d) $\left(3r - \frac{1}{2}\right)\left(3r + \frac{1}{2}\right)$

Answers

4. (a) $t^2 + 2tu + u^2$
(b) $4m^2 - 4mp + p^2$
(c) $16p^2 + 24pq + 9q^2$
(d) $25r^2 - 60rs + 36s^2$
(e) $9k^2 - 3k + \frac{1}{4}$

5. (a) $36a^2 - 9$ (b) $100m^2 - 49$
(c) $49p^2 - 4q^2$ (d) $9r^2 - \frac{1}{4}$

(c) $(2a - 9x)^2 = 4a^2 - 36ax + 81x^2$

(d)
$$\left(4m + \frac{1}{2}\right)^2 = (4m)^2 + 2(4m)\left(\frac{1}{2}\right) + \left(\frac{1}{2}\right)^2$$
$$= 16m^2 + 4m + \frac{1}{4}$$

Caution

A common error in squaring a binomial is forgetting the middle term of the product. In general,

$$(a + b)^2 \neq a^2 + b^2.$$

WORK PROBLEM 4 AT THE SIDE.

Objective 3 Binomial products of the form $(a + b)(a - b)$ also occur frequently. In these products, one binomial is the sum of two terms, and the other is the difference of the same two terms. As an example, the product of $x + 2$ and $x - 2$ is

$$(x + 2)(x - 2) = x^2 - 2x + 2x - 4$$
$$= x^2 - 4.$$

As this example suggests, the product of $a + b$ and $a - b$ is the difference between two squares.

Product of the Sum and Difference of Two Terms

$$(a + b)(a - b) = a^2 - b^2$$

EXAMPLE 6 Finding the Product of the Sum and Difference of Two Terms

Find each product.

(a) $(\mathbf{5m} + \mathbf{3})(\mathbf{5m} - \mathbf{3})$ (with $(a + b)(a - b)$ written above, arrows pointing down to $5m$, 3, $5m$, 3)

Use the pattern for the sum and difference of two terms.

$$(5m + 3)(5m - 3) = (5m)^2 - 3^2$$
$$= 25m^2 - 9$$

(b) $(4x + y)(4x - y) = (4x)^2 - y^2$
$= 16x^2 - y^2$

(c) $\left(z - \frac{1}{4}\right)\left(z + \frac{1}{4}\right) = z^2 - \frac{1}{16}$

WORK PROBLEM 5 AT THE SIDE.

The product formulas of this section will be very useful in later work, particularly in the chapter on factoring. Therefore, it is important to memorize these formulas and practice using them.

6 Exercises

1. Consider the product $(2x + 3)(x - 5)$.
(a) What is the product of the first terms, $2x(x)$? ______
(b) What is the product of the outer terms, $2x(-5)$? ______
(c) What is the product of the inner terms, $3(x)$? ______
(d) What is the product of the last terms, $3(-5)$? ______
(e) What is the sum of the outer and inner products found in parts (b) and (c)? ______
(f) Write the complete product, which is a trinomial, using your results in parts (a), (e), and (d). ______

2. Repeat Exercise 1 for the product $(3y - 8)(2y + 5)$.

Find each product using the FOIL method. See Examples 1–3.

3. $(r + 1)(r + 3)$

4. $(p + 2)(p + 3)$

5. $(w - 4)(w - 6)$

6. $(y - 3)(y - 7)$

7. $(x - .3)(x + .8)$

8. $(y + .7)(y + .4)$

9. $\left(x - \frac{2}{3}\right)\left(x + \frac{1}{4}\right)$

10. $\left(z - \frac{5}{6}\right)\left(z + \frac{3}{4}\right)$

11. $(9x + 2)(3x + 7)$

12. $(6x + 7)(2x + 3)$

13. $(3m + 7)(3m + 5)$

14. $(7x - 1)(7x - 8)$

15. $(3 - 2x)(5 - 3x)$

16. $(4 - 3r)(5 - 2r)$

17. $(-5 + 6z)(3 - z)$

18. $(-4 + 5w)(5 - w)$

19. $\left(-\frac{8}{3} + 3k\right)\left(-\frac{2}{3} - k\right)$

20. $\left(-\frac{5}{4} + 2r\right)\left(-\frac{3}{4} - r\right)$

21. $(3w + 2z)(9w - z)$

22. $(2x - y)(5x + 3y)$

23. $(5y - 3z)(6y + 5z)$

24. $(4x + 5y)(7x - y)$

25. $(-.8p + .3s)(.2p + s)$

26. $(.7h + k)(-.4h + .3k)$

27. Consider the square $(2x + 3)^2$.
(a) What is the square of the first term, $(2x)^2$? ______
(b) What is twice the product of the two terms, $2(2x)(3)$? ______
(c) What is the square of the last term, 3^2? ______
(d) Write the final product, which is a trinomial, using your results in parts (a)–(c). ______

28. Repeat Exercise 27 for the square $(3x - 2)^2$.

Find each square. See Examples 4 and 5.

29. $(a - c)^2$

30. $(p - y)^2$

31. $(p + 2)^2$

32. $(r + 5)^2$

33. $(4x - 3)^2$

34. $(5y + 2)^2$

35. $(.8t + .7s)^2$

36. $(.7z - .3w)^2$

37. $\left(5x + \frac{2}{5}y\right)^2$

38. $\left(6m - \frac{4}{5}n\right)^2$

39. Consider the product $(7x + 3y)(7x - 3y)$.
   (a) What is the product of the first terms, $(7x)(7x)$? ______
   (b) Multiply the outer terms, $(7x)(-3y)$. Then multiply the inner terms, $(3y)(7x)$. Add the results. What is this sum? ______
   (c) What is the product of the last terms, $(3y)(-3y)$? ______
   (d) Write the complete product using your answers in parts (a) and (c). __________
   Why is the sum found in part (b) omitted here?

40. Repeat Exercise 39 for the product $(5x + 7y)(5x - 7y)$.

Find the following products. See Example 6.

41. $(q + 2)(q - 2)$

42. $(x + 8)(x - 8)$

43. $(2w + 5)(2w - 5)$

44. $(3z + 8)(3z - 8)$

45. $(10x + 3y)(10x - 3y)$

46. $(13r + 2z)(13r - 2z)$

47. $(2x^2 - 5)(2x^2 + 5)$

48. $(9y^2 - 2)(9y^2 + 2)$

49. $\left(7x + \frac{3}{7}\right)\left(7x - \frac{3}{7}\right)$

50. $\left(9y + \frac{2}{3}\right)\left(9y - \frac{2}{3}\right)$

MATHEMATICAL CONNECTIONS (EXERCISES 51–54)

To understand how the special product $(a + b)^2 = a^2 + 2ab + b^2$ can be applied to a purely numerical problem, work Exercises 51–54 in order.

51. Evaluate 35^2 using either traditional paper-and-pencil methods or a calculator.

52. The number 35 can be written as $30 + 5$. Therefore, $35^2 = (30 + 5)^2$. Use the special product for squaring a binomial with $a = 30$ and $b = 5$ to write an expression for $(30 + 5)^2$. Do not simplify at this time.

53. Use the rule for order of operations to simplify the expression you found in Exercise 52.

54. How do the answers in Exercises 51 and 53 compare?

7 Division of a Polynomial by a Monomial

OBJECTIVE 1 We add two fractions with a common denominator as follows.

$$\frac{a}{c} + \frac{b}{c} = \frac{a+b}{c}.$$

Looking at this statement in reverse gives us a rule for dividing a polynomial by a monomial.

Dividing a Polynomial by a Monomial

To divide a polynomial by a monomial, divide each term of the polynomial by the monomial:

$$\frac{a+b}{c} = \frac{a}{c} + \frac{b}{c} \quad (c \neq 0).$$

The quotient rule for exponents is used to reduce each fraction as shown in the following examples.

EXAMPLE 1 Dividing a Polynomial by a Monomial

Divide $5m^5 - 10m^3$ by $5m^2$.

Use the rule above, with + replaced by −.

$$\frac{5m^5 - 10m^3}{\mathbf{5m^2}} = \frac{5m^5}{\mathbf{5m^2}} - \frac{10m^3}{\mathbf{5m^2}} = m^3 - 2m$$

Check by multiplication.

$$5m^2(\mathbf{m^3 - 2m}) = 5m^5 - 10m^3$$

Because division by 0 is undefined, the quotient

$$\frac{5m^5 - 10m^3}{5m^2}$$

is undefined if $m = 0$. In the rest of the chapter, we assume that no denominators are 0.

WORK PROBLEM 1 AT THE SIDE.

EXAMPLE 2 Dividing a Polynomial by a Monomial

Divide: $\dfrac{16a^5 - 12a^4 + 8a^2}{4a^3}$.

Divide each term of $16a^5 - 12a^4 + 8a^2$ by $4a^3$.

$$\frac{16a^5 - 12a^4 + 8a^2}{4a^3} = \frac{16a^5}{\mathbf{4a^3}} - \frac{12a^4}{\mathbf{4a^3}} + \frac{8a^2}{\mathbf{4a^3}}$$

$$= 4a^2 - 3a + \frac{2}{a}$$

CONTINUED ON NEXT PAGE

OBJECTIVE

1 Divide a polynomial by a monomial.

Tutorial Tape 6 SSM, Sec. 3.7

1. Divide.

(a) $\dfrac{6p^4 + 18p^7}{3p^2}$

(b) $\dfrac{12m^6 + 18m^5 + 30m^4}{6m^2}$

(c) $(18r^7 - 9r^2) \div (3r)$

ANSWERS

1. (a) $2p^2 + 6p^5$
(b) $2m^4 + 3m^3 + 5m^2$
(c) $6r^6 - 3r$

The quotient is not a polynomial because of the expression $\frac{2}{a}$, which has a variable in the denominator. While the sum, difference, and product of two polynomials are always polynomials, the quotient of two polynomials may not be.

Again, check by multiplying.

$$4a^3\left(4a^2 - 3a + \frac{2}{a}\right) = 4a^3(4a^2) - 4a^3(3a) + 4a^3\left(\frac{2}{a}\right)$$
$$= 16a^5 - 12a^4 + 8a^2$$

◀◀ WORK PROBLEM 2 AT THE SIDE.

EXAMPLE 3 Dividing a Polynomial by a Monomial

Divide.

$$\frac{12x^4 - 7x^3 + 4x}{4x} = \frac{12x^4}{4x} - \frac{7x^3}{4x} + \frac{4x}{4x}$$
$$= 3x^3 - \frac{7x^2}{4} + 1$$

Check by multiplication.

Caution

In Example 3, notice that the quotient $\frac{4x}{4x} = 1$. It is a common error to leave the 1 out of the answer. Checking by multiplication will show that the answer $3x^3 - \frac{7}{4}x^2$ is not correct.

EXAMPLE 4 Dividing a Polynomial by a Monomial

Divide the polynomial

$$180y^{10} - 150y^8 + 120y^6 - 90y^4 + 100y$$

by the monomial $-30y^2$.

Using the methods of this section,

$$\frac{180y^{10} - 150y^8 + 120y^6 - 90y^4 + 100y}{-30y^2}$$
$$= \frac{180y^{10}}{-30y^2} - \frac{150y^8}{-30y^2} + \frac{120y^6}{-30y^2} - \frac{90y^4}{-30y^2} + \frac{100y}{-30y^2}$$
$$= -6y^8 + 5y^6 - 4y^4 + 3y^2 - \frac{10}{3y}.$$

To check, multiply this answer and $-30y^2$.

◀◀ WORK PROBLEM 3 AT THE SIDE.

2. Divide.

(a) $\dfrac{20x^4 - 25x^3 + 5x}{5x^2}$

(b) $\dfrac{50m^4 - 30m^3 + 20m}{10m^3}$

3. Divide.

(a) $\dfrac{8y^7 - 9y^6 - 11y - 4}{y^2}$

(b) $\dfrac{12p^5 + 8p^4 + 3p^3 - 5p^2}{3p^3}$

(c) $\dfrac{45x^4 + 30x^3 - 60x^2}{-15x^2}$

ANSWERS

2. (a) $4x^2 - 5x + \frac{1}{x}$ (b) $5m - 3 + \frac{2}{m^2}$

3. (a) $8y^5 - 9y^4 - \frac{11}{y} - \frac{4}{y^2}$

(b) $4p^2 + \frac{8p}{3} + 1 - \frac{5}{3p}$

(c) $-3x^2 - 2x + 4$

7 Exercises

1. Explain why the division problem $\frac{16m^3 - 12m^2}{4m}$ can be performed using the methods of this section, while the division problem $\frac{4m}{16m^3 - 12m^2}$ cannot.

2. If the area of a rectangle is $24x^6$ and the width is $4x^2$, what is the length?

3. Evaluate $\frac{5y + 6}{2}$ when $y = 2$. Evaluate $5y + 3$ when $y = 2$. Does $\frac{5y + 6}{2}$ equal $5y + 3$?

4. Evaluate $\frac{10r + 7}{5}$ when $r = 1$. Evaluate $2r + 7$ when $r = 1$. Does $\frac{10r + 7}{5}$ equal $2r + 7$?

Divide each polynomial by 2m. See Examples 1–4.

5. $10m^5 - 16m^4 + 8m^3$

6. $6m^5 - 4m^3 + 2m^2$

7. $8m^5 - 4m^3 + 4m^2$

8. $8m^4 - 4m^3 + 6m^2$

9. $m^5 - 4m^2 + 8$

10. $m^3 + m^2 + 6$

Divide each polynomial by $3x^2$. See Examples 1–4.

11. $12x^5 - 9x^4 + 6x^3$

12. $24x^6 - 12x^5 + 30x^4$

13. $3x^2 + 15x^3 - 27x^4$

14. $3x^2 - 18x^4 + 30x^5$

15. $36x + 24x^2 + 6x^3$

16. $9x - 12x^2 + 9x^3$

17. $4x^4 + 3x^3 + 2x$

18. $5x^4 - 6x^3 + 8x$

Perform each division. See Examples 1–4.

19. $\frac{27r^4 - 36r^3 - 6r^2 + 3r - 2}{3r}$

20. $\frac{8k^4 - 12k^3 - 2k^2 - 2k - 3}{2k}$

21. $\frac{2m^5 - 6m^4 + 8m^2}{-2m^3}$

22. $\frac{6r^5 - 8r^4 + 10r^2}{-2r^4}$

23. $(20a^4 - 15a^5 + 25a^3) \div (5a^4)$

24. $(16y^5 - 8y^2 + 12y) \div (4y^2)$

25. $(120x^{11} - 60x^{10} + 140x^9 - 100x^8) \div (10x^{12})$

26. $(120x^{12} - 84x^9 + 60x^8 - 36x^7) \div (12x^9)$

27. The quotient in Exercise 17 is $\frac{4x^2}{3} + x + \frac{2}{3x}$. Notice how the third term is written with x in the denominator. Would $\frac{2}{3}x$ be an acceptable form for this term? Explain why or why not. Is $\frac{4}{3}x^2$ an acceptable form for the first term? Why or why not?

28. If the area of a rectangle is represented by $12x^2 - 4x + 2$ and the width is $2x$, what expression represents the length?

29. What polynomial, when divided by $5x^3$, yields $3x^2 - 7x + 7$ as a quotient?

30. The quotient of a certain polynomial and $-12y^3$ is $6y^3 - 5y^2 + 2y - 3 + \frac{7}{y}$. Find the polynomial.

MATHEMATICAL CONNECTIONS (EXERCISES 31–34)

Our system of numeration is called a decimal system. It is based on powers of ten. In a whole number such as 2846, each digit is understood to represent the number of powers of ten for its place value. The 2 represents two thousands (2×10^3), the 8 represents eight hundreds (8×10^2), the 4 represents four tens (4×10^1), and the 6 represents six ones (or units) (6×10^0). In expanded form we write

$$2846 = (2 \times 10^3) + (8 \times 10^2) + (4 \times 10^1) + (6 \times 10^0).$$

Keeping this information in mind, work Exercises 31–34 in order.

31. Divide 2846 by 2, using paper-and-pencil methods: $2\overline{)2846}$.

32. Write your answer in Exercise 31 in expanded form.

33. Use the methods of this section to divide the polynomial $2x^3 + 8x^2 + 4x + 6$ by 2.

34. Compare your answers in Exercises 32 and 33. How are they similar? How are they different? For what value of x does the answer in Exercise 33 equal the answer in Exercise 32?

8 *The Quotient of Two Polynomials*

OBJECTIVE

1 Divide a polynomial by a polynomial.

OBJECTIVE 1 A method of "long division" is used to divide a polynomial by a polynomial (other than a monomial). This method is similar to the method of long division used for two whole numbers. For comparison, the division of whole numbers is shown alongside the division of polynomials. The polynomial must be in descending order.

Division of Whole Numbers	Division of Polynomials
Step 1	
Divide 27 into 6696.	Divide $2x + 3$ into $8x^3 - 4x^2 - 14x + 15$.
$27\overline{)6696}$	$2x + 3\overline{)8x^3 - 4x^2 - 14x + 15}$
Step 2	
27 divides into 66 **2** times; $2 \cdot 27 =$ **54**.	$2x$ divides into $8x^3$ $\mathbf{4x^2}$ times; $4x^2(2x + 3) = \mathbf{8x^3 + 12x^2}$.
$\begin{array}{r} \mathbf{2} \\ 27\overline{)6696} \\ \underline{\mathbf{54}} \end{array}$	$\begin{array}{r} \mathbf{4x^2} \\ 2x + 3\overline{)8x^3 - 4x^2 - 14x + 15} \\ \underline{\mathbf{8x^3 + 12x^2}} \end{array}$
Step 3	
Subtract; then bring down the next digit.	Subtract; then bring down the next term.
$\begin{array}{r} 2 \\ 27\overline{)6696} \\ \underline{54}\downarrow \\ 129 \end{array}$	$\begin{array}{r} 4x^2 \\ 2x + 3\overline{)8x^3 - 4x^2 - 14x + 15} \\ \underline{8x^3 + 12x^2} \quad \downarrow \\ -16x^2 - \mathbf{14x} \end{array}$
	(To subtract two polynomials, change the sign of the second and then add.)
Step 4	
27 divides into 129 **4** times; $4 \cdot 27 =$ **108**.	$2x$ divides into $-16x^2$ $\mathbf{-8x}$ times; $-8x(2x + 3) = \mathbf{-16x^2 - 24x}$.
$\begin{array}{r} 2\mathbf{4} \\ 27\overline{)6696} \\ \underline{54} \\ 129 \\ \underline{\mathbf{108}} \end{array}$	$\begin{array}{r} 4x^2 - \mathbf{8x} \\ 2x + 3\overline{)8x^3 - 4x^2 - 14x + 15} \\ \underline{8x^3 + 12x^2} \\ -16x^2 - 14x \\ \underline{\mathbf{-16x^2 - 24x}} \end{array}$
Step 5	
Subtract; then bring down the next digit.	Subtract; then bring down the next term.
$\begin{array}{r} 24 \\ 27\overline{)6696} \\ \underline{54} \\ 129 \\ \underline{108}\downarrow \\ 216 \end{array}$	$\begin{array}{r} 4x^2 - 8x \\ 2x + 3\overline{)8x^3 - 4x^2 - 14x + 15} \\ \underline{8x^3 + 12x^2} \\ -16x^2 - 14x \\ \underline{-16x^2 - 24x} \downarrow \\ 10x + \mathbf{15} \end{array}$

Step 6

27 divides into 216 **8** times; $8 \cdot 27 = \mathbf{216}$.

$$\begin{array}{r} \mathbf{248} \\ 27\overline{)6696} \\ \underline{54} \\ 129 \\ \underline{108} \\ 216 \\ \underline{\mathbf{216}} \end{array}$$

6696 divided by 27 is 248. There is no remainder.

$2x$ divides into $10x$ **5** times; $5(2x + 3) = \mathbf{10x + 15}$.

$$\begin{array}{r} 4x^2 - 8x + \mathbf{5} \\ 2x + 3\overline{)8x^3 - 4x^2 - 14x + 15} \\ \underline{8x^3 + 12x^2} \qquad\qquad \\ -16x^2 - 14x \qquad \\ \underline{-16x^2 - 24x} \qquad \\ 10x + 15 \\ \underline{\mathbf{10x + 15}} \end{array}$$

$8x^3 - 4x^2 - 14x + 15$ divided by $2x + 3$ is $4x^2 - 8x + 5$. There is no remainder.

Step 7

Check by multiplication.

$$27 \cdot 248 = 6696$$

Check by multiplication.

$$(2x + 3)(4x^2 - 8x + 5) = 8x^3 - 4x^2 - 14x + 15$$

E X A M P L E 1 Dividing a Polynomial by a Polynomial

Divide $4x^3 - 4x^2 + 5x - 8$ by $2x - 1$.

$$\begin{array}{r} \mathbf{2x^2 - x + 2} \\ 2x - 1\overline{)4x^3 - 4x^2 + 5x - 8} \\ \underline{4x^3 - 2x^2} \qquad\qquad \\ -2x^2 + 5x \qquad \\ \underline{-2x^2 + x} \qquad \\ 4x - 8 \\ \underline{4x - 2} \\ -6 \end{array} \quad \leftarrow \text{Remainder}$$

Step 1 $2x$ divides into $4x^3$ $\mathbf{2x^2}$ times; $2x^2(2x - 1) = 4x^3 - 2x^2$.

Step 2 Subtract; bring down the next term.

Step 3 $2x$ divides into $-2x^2$ $\mathbf{-x}$ times; $-x(2x - 1) = -2x^2 + x$.

Step 4 Subtract; bring down the next term.

Step 5 $2x$ divides into $4x$ **2** times; $2(2x - 1) = 4x - 2$.

Step 6 Subtract. The remainder is **−6**. Thus $2x - 1$ divides into $4x^3 - 4x^2 + 5x - 8$ with a quotient of $2x^2 - x + 2$ and a remainder of -6. Write the remainder as the numerator of a fraction that has $2x - 1$ as its denominator. The answer is not a polynomial because of the remainder.

$$\frac{4x^3 - 4x^2 + 5x - 8}{2x - 1} = 2x^2 - x + 2 + \frac{-6}{2x - 1}$$

CONTINUED ON NEXT PAGE

Step 7 Check by multiplication.

$$(\mathbf{2x-1})\left(2x^2 - x + 2 + \frac{-6}{2x-1}\right)$$
$$= (\mathbf{2x-1})(2x^2) + (\mathbf{2x-1})(-x) + (\mathbf{2x-1})(2) + (\mathbf{2x-1})\left(\frac{-6}{2x-1}\right)$$
$$= 4x^3 - 2x^2 - 2x^2 + x + 4x - 2 - 6$$
$$= 4x^3 - 4x^2 + 5x - 8$$

WORK PROBLEM 1 AT THE SIDE.

EXAMPLE 2 Dividing into a Polynomial with Missing Terms

Divide $x^3 - 1$ by $x - 1$.

Here the polynomial $x^3 - 1$ is missing the x^2 term and the x term. When terms are missing, use 0 as the coefficient for the missing terms. (Zero acts as a placeholder here, just as it does in our number system.)

$$x^3 - 1 = x^3 + \mathbf{0x^2} + \mathbf{0x} - 1$$

Now divide.

$$\begin{array}{r} x^2 + x + 1 \\ x - 1\overline{)x^3 + \mathbf{0x^2} + \mathbf{0x} - 1} \\ \underline{x^3 - x^2 } \\ x^2 + 0x \\ \underline{x^2 - x } \\ x - 1 \\ \underline{x - 1} \\ 0 \end{array}$$

The remainder is 0. The quotient is $x^2 + x + 1$. Check by multiplication.

$$(x^2 + x + 1)(x - 1) = x^3 - 1$$

WORK PROBLEM 2 AT THE SIDE.

EXAMPLE 3 Dividing by a Polynomial with Missing Terms

Divide $x^4 + 2x^3 + 2x^2 - x - 1$ by $x^2 + 1$.

Since $x^2 + 1$ has a missing x term, write it as $x^2 + 0x + 1$. Then go through the division process as follows.

$$\begin{array}{r} x^2 + 2x + 1 \\ x^2 + \mathbf{0x} + 1\overline{)x^4 + 2x^3 + 2x^2 - x - 1} \\ \underline{x^4 + 0x^3 + x^2 } \\ 2x^3 + x^2 - x \\ \underline{2x^3 + 0x^2 + 2x } \\ x^2 - 3x - 1 \\ \underline{x^2 + 0x + 1} \\ -3x - 2 \end{array}$$

CONTINUED ON NEXT PAGE

1. Divide.

(a) $(x^3 + x^2 + 4x - 6) \div (x - 1)$

(b) $\dfrac{p^3 - 2p^2 - 5p + 9}{p + 2}$

2. Divide.

(a) $\dfrac{r^2 - 5}{r + 4}$

(b) $(x^3 - 8) \div (x - 2)$

ANSWERS

1. (a) $x^2 + 2x + 6$

(b) $p^2 - 4p + 3 + \dfrac{3}{p + 2}$

2. (a) $r - 4 + \dfrac{11}{r + 4}$

(b) $x^2 + 2x + 4$

When the result of subtracting ($-3x - 2$, in this case) is a polynomial of smaller degree than the divisor ($x^2 + 0x + 1$), that polynomial is the remainder. Write the answer as

$$x^2 + 2x + 1 + \frac{-3x - 2}{x^2 + 1}.$$

Verify by multiplication that this is the correct quotient.

WORK PROBLEM 3 AT THE SIDE.

3. Divide.

(a)
$(2x^4 + 3x^3 - x^2 + 6x + 5) \div (x^2 - 1)$

(b)
$$\frac{2m^5 + m^4 + 6m^3 - 3m^2 - 18}{m^2 + 3}$$

ANSWERS

3. (a) $2x^2 + 3x + 1 + \frac{9x + 6}{x^2 - 1}$

(b) $2m^3 + m^2 - 6$

8 Exercises

1. In the division problem $(4x^4 + 2x^3 - 14x^2 + 19x + 10) \div (2x + 5) = 2x^3 - 4x^2 + 3x + 2$, which polynomial is the divisor? Which is the quotient?

2. When dividing one polynomial by another, how do you know when to stop dividing?

3. In dividing $12m^2 - 20m + 3$ by $2m - 3$, what is the first step?

4. In the division in Exercise 3, what is the second step?

Perform each division. See Example 1.

5. $\dfrac{x^2 - x - 6}{x - 3}$

6. $\dfrac{m^2 - 2m - 24}{m - 6}$

7. $\dfrac{2y^2 + 9y - 35}{y + 7}$

8. $\dfrac{2y^2 + 9y + 7}{y + 1}$

9. $\dfrac{p^2 + 2p + 20}{p + 6}$

10. $\dfrac{x^2 + 11x + 16}{x + 8}$

11. $(r^2 - 8r + 15) \div (r - 3)$

12. $(t^2 + 2t - 35) \div (t - 5)$

13. $\dfrac{4a^2 - 22a + 32}{2a + 3}$

14. $\dfrac{9w^2 + 6w + 10}{3w - 2}$

15. $\dfrac{8x^3 - 10x^2 - x + 3}{2x + 1}$

16. $\dfrac{12t^3 - 11t^2 + 9t + 18}{4t + 3}$

MATHEMATICAL CONNECTIONS (EXERCISES 17–20)

We can find the value of a polynomial in x for a given value of x by substituting that number for x. Surprisingly, we can accomplish the same thing by division. For example, to find the value of $2x^2 - 4x + 3$ *for* $x = -3$*, we would divide* $x - (-3)$ *into* $2x^2 - 4x + 3$*. The remainder will give the value of the polynomial for* $x = -3$*. Work Exercises 17–20 in order.*

17. Find the value of $2x^2 - 4x + 3$ for $x = -3$ by substitution.

18. Divide $2x^2 - 4x + 3$ by $x + 3$. Give the remainder.

19. Compare your answers to Exercises 17 and 18. What do you notice?

20. Choose another polynomial and evaluate it both ways at some value of the variable. Do the answers agree?

Perform each division. See Examples 2 and 3.

21. $\dfrac{3y^3 + y^2 + 2}{y + 1}$

22. $\dfrac{2r^3 - 6r - 36}{r - 3}$

23. $\dfrac{3k^3 - 4k^2 - 6k + 10}{k^2 - 2}$

24. $\dfrac{5z^3 - z^2 + 10z + 2}{z^2 + 2}$

25. $(x^4 - x^2 - 2) \div (x^2 - 2)$

26. $(r^4 + 2r^2 - 3) \div (r^2 - 1)$

27. $\dfrac{6p^4 - 15p^3 + 14p^2 - 5p + 10}{3p^2 + 1}$

28. $\dfrac{6r^4 - 10r^3 - r^2 + 15r - 8}{2r^2 - 3}$

29. $\dfrac{2x^5 + 9x^4 + 8x^3 + 10x^2 + 14x + 5}{2x^2 + 3x + 1}$

30. $\dfrac{4t^5 - 11t^4 - 6t^3 + 5t^2 - t + 3}{4t^2 + t - 3}$

31. $\dfrac{x^4 - 1}{x^2 - 1}$

32. $\dfrac{y^3 + 1}{y + 1}$

KEY TERMS

3	**scientific notation**	A number written as $a \times 10^n$, where $1 \leq \lvert a \rvert < 10$ and n is an integer, is in scientific notation.
4	**polynomial**	A polynomial is a term or the sum of a finite number of terms.
	descending powers	A polynomial is written in descending powers if the degrees of its terms are in decreasing order.
	degree of a term	The degree of a term with one variable is the exponent on the variable.
	degree of a polynomial	The degree of a polynomial in one variable is the highest exponent found in any term of the polynomial.
	trinomial	A trinomial is a polynomial with three terms.
	binomial	A binomial is a polynomial with two terms.
	monomial	A monomial is a polynomial with one term.

NEW SYMBOLS

x^{-n} x to the negative n

QUICK REVIEW

Concepts	*Examples*
1 The Product Rule and Power Rules for Exponents For any integers m and n: **Product rule** $a^m \cdot a^n = a^{m+n}$ **Power rules (a)** $(a^m)^n = a^{mn}$ **(b)** $(ab)^m = a^m b^m$ **(c)** $\left(\frac{a}{b}\right)^m = \frac{a^m}{b^m}$ $(b \neq 0)$	$2^4 \cdot 2^5 = 2^9$ $(3^4)^2 = 3^8$ $(6a)^5 = 6^5 a^5$ $\left(\frac{2}{3}\right)^4 = \frac{2^4}{3^4}$
2 Integer Exponents and the Quotient Rule If $a \neq 0$, for integers m and n: **Zero exponent** $a^0 = 1$ **Negative exponent** $a^{-n} = \frac{1}{a^n}$ **Quotient rule** $\frac{a^m}{a^n} = a^{m-n}$ $\frac{a^{-m}}{b^{-n}} = \frac{b^n}{a^m}$ $\left(\frac{a}{b}\right)^{-m} = \left(\frac{b}{a}\right)^m$	$15^0 = 1$ $5^{-2} = \frac{1}{5^2} = \frac{1}{25}$ $\frac{4^8}{4^3} = 4^5$ $\frac{6^{-2}}{7^{-3}} = \frac{7^3}{6^2}$ $\left(\frac{5}{3}\right)^{-4} = \left(\frac{3}{5}\right)^4$
3 An Application of Exponents: Scientific Notation To write a number in scientific notation (as $a \times 10^n$), move the decimal point to the right of the first nonzero digit. If the decimal point is moved n places, and this makes the number smaller, n is positive; otherwise, n is negative. If the decimal point is not moved, n is 0.	$247 = 2.47 \times 10^2$ $.0051 = 5.1 \times 10^{-3}$ $4.8 = 4.8 \times 10^0$

Concepts	*Examples*
4 Addition and Subtraction of Polynomials	
Addition: Add like terms.	Add: $\begin{array}{r} 2x^2 + 5x - 3 \\ 5x^2 - 2x + 7 \\ \hline 7x^2 + 3x + 4 \end{array}$
Subtraction: Change the signs of the terms in the second polynomial and add to the first polynomial.	$(2x^2 + 5x - 3) - (5x^2 - 2x + 7)$ $= (2x^2 + 5x - 3) + (-5x^2 + 2x - 7)$ $= -3x^2 + 7x - 10$
5 Multiplication of Polynomials	
Multiply each term of the first polynomial by each term of the second polynomial. Then add like terms.	Multiply: $\begin{array}{rrrrr} & 3x^3 - & 4x^2 + & 2x - & 7 \\ & & & 4x + & 3 \\ \hline & 9x^3 - & 12x^2 + & 6x - & 21 \\ 12x^4 - & 16x^3 + & 8x^2 - & 28x & \\ \hline 12x^4 - & 7x^3 - & 4x^2 - & 22x - & 21 \end{array}$
6 Products of Binomials	
FOIL Method	Find $(2x + 3)(5x - 4)$.
Step 1 Multiply the two first terms to get the first term of the answer.	$2x(5x) = \mathbf{10x^2}$
Step 2 Find the outer product and the inner product and mentally add them, when possible, to get the middle term of the answer.	$2x(-4) + 3(5x) = \mathbf{7x}$
Step 3 Multiply the two last terms to get the last term of the answer.	$3(-4) = \mathbf{-12}$ $(2x + 3)(5x - 4) = \mathbf{10x^2 + 7x - 12}$
Square of a Binomial $(a + b)^2 = a^2 + 2ab + b^2$ $(a - b)^2 = a^2 - 2ab + b^2$	$(3x + 1)^2 = 9x^2 + 6x + 1$ $(2m - 5n)^2 = 4m^2 - 20mn + 25n^2$
Product of the Sum and Difference of Two Terms $(a + b)(a - b) = a^2 - b^2$	$(4a + 3)(4a - 3) = 16a^2 - 9$
7 Division of a Polynomial by a Monomial	
Divide each term of the polynomial by the monomial: $\frac{a + b}{c} = \frac{a}{c} + \frac{b}{c}$	Divide: $\frac{4x^3 - 2x^2 + 6x - 8}{2x}$ $= 2x^2 - x + 3 - \frac{4}{x}$
8 The Quotient of Two Polynomials	
Use "long division."	Divide: $\begin{array}{r} 2x - 5 + \frac{-1}{3x + 4} \\ 3x + 4 \overline{)6x^2 - 7x - 21} \\ \underline{6x^2 + 8x} \quad\quad \\ -15x - 21 \\ \underline{-15x - 20} \\ -1 \end{array}$

[1] *Use the product rule to simplify each expression. Write each answer in exponential form.*

1. $4^3 \cdot 4^8$ **2.** $(-5)^6(-5)^5$ **3.** $(-8x^4)(9x^3)$ **4.** $(2x^2)(5x^3)(x^9)$

Use the power rules to simplify each expression. Write each answer in exponential form.

5. $(19x)^5$ **6.** $(-4y)^7$ **7.** $5(pt)^4$ **8.** $\left(\frac{7}{5}\right)^6$

Use a combination of rules to simplify each expression. Leave answers in exponential form.

9. $(3x^2y^3)^3$ **10.** $(t^4)^8(t^2)^5$ **11.** $(6x^2z^4)^2(x^3yz^2)^4$

12. Explain why the product rule for exponents does not apply to the expression $7^2 + 7^4$.

[2] *Evaluate each expression.*

13. $5^0 + 8^0$ **14.** 2^{-5} **15.** $\left(\frac{6}{5}\right)^{-2}$ **16.** $4^{-2} - 4^{-1}$

Simplify. Write each answer in exponential form, using only positive exponents. Assume all variables are nonzero.

17. $\frac{6^{-3}}{6^{-5}}$ **18.** $\frac{x^{-7}}{x^{-9}}$ **19.** $\frac{p^{-8}}{p^4}$ **20.** $\frac{r^{-2}}{r^{-6}}$

21. $(2^4)^2$ **22.** $(9^3)^{-2}$ **23.** $(5^{-2})^{-4}$

24. $(8^{-3})^4$ **25.** $\frac{(m^2)^3}{(m^4)^2}$ **26.** $\frac{y^4 \cdot y^{-2}}{y^{-5}}$

27. $\frac{r^9 \cdot r^{-5}}{r^{-2} \cdot r^{-7}}$ **28.** $(-5m^3)^2$ **29.** $(2y^{-4})^{-3}$

30. $\frac{ab^{-3}}{a^4b^2}$ **31.** $\frac{(6r^{-1})^2 \cdot (2r^{-4})}{r^{-5}(r^2)^{-3}}$ **32.** $\frac{(2m^{-5}n^2)^3(3m^2)^{-1}}{m^{-2}n^{-4}(m^{-1})^2}$

[3] *Write each number in scientific notation.*

33. 48,000,000

34. 28,988,000,000

35. .000065

36. .0000000824

Write each number without exponents.

37. 2.4×10^4

38. 7.83×10^7

39. 8.97×10^{-7}

40. 9.95×10^{-12}

Perform the indicated operation and write the answer without exponents.

41. $(2 \times 10^{-3}) \times (4 \times 10^5)$

42. $\dfrac{8 \times 10^4}{2 \times 10^{-2}}$

43. $\dfrac{12 \times 10^{-5} \times 5 \times 10^4}{4 \times 10^3 \times 6 \times 10^{-2}}$

44. $\dfrac{2.5 \times 10^5 \times 4.8 \times 10^{-4}}{7.5 \times 10^8 \times 1.6 \times 10^{-5}}$

[4] *For each polynomial, first simplify, if possible, and write it in descending powers of the variable. Then give the degree of the resulting polynomial, and tell whether it is a monomial, a binomial, a trinomial, or none of these.*

45. $9m^2 + 11m^2 + 2m^2$

46. $-4p + p^3 - p^2 + 8p + 2$

47. $2r^4 - r^3 + 8r^4 + r^3 - 6r^4 + 8r^5$

48. $12a^5 + 19a^4 + 8a^3 + 2a^3 - 9a^4 + 3a^5$

Add or subtract as indicated.

49. Add.
$$\begin{array}{r} -2a^3 + 5a^2 \\ \underline{-3a^3 - a^2} \end{array}$$

50. Add.
$$\begin{array}{r} 4r^3 - 8r^2 + 6r \\ \underline{-2r^3 + 5r^2 + 3r} \end{array}$$

51. Subtract.
$$\begin{array}{r} 6y^2 - 8y + 2 \\ \underline{-5y^2 + 2y - 7} \end{array}$$

52. Subtract.
$$\begin{array}{r} -12k^4 - 8k^2 + 7k - 5 \\ \underline{k^4 + 7k^2 + 11k + 1} \end{array}$$

53. $(2m^3 - 8m^2 + 4) + (8m^3 + 2m^2 - 7)$

54. $(-5y^2 + 3y + 11) + (4y^2 - 7y + 15)$

55. $(6p^2 - p - 8) - (-4p^2 + 2p + 3)$

56. $(12r^4 - 7r^3 + 2r^2) - (5r^4 - 3r^3 + 2r^2 + 1)$

Name ______________________

[5] *Find each product.*

57. $5x(2x + 14)$

58. $-3p^3(2p^2 - 5p)$

59. $(m - 9)(m + 2)$

60. $(3k - 6)(2k + 1)$

61. $(3r - 2)(2r^2 + 4r - 3)$

62. $(2y + 3)(4y^2 - 6y + 9)$

63. $(r + 2)^3$

64. Explain why $(a + b)^2$ is not equal to $a^2 + b^2$.

[6] *Find each product.*

65. $(3k + 1)(2k + 3)$

66. $(a + 3b)(2a - b)$

67. $(6k - 3q)(2k - 7q)$

68. $(a + 4)^2$

69. $(3p - 2)^2$

70. $(2r + 5s)^2$

71. $(6m - 5)(6m + 5)$

72. $(2z + 7)(2z - 7)$

73. $(5a + 6b)(5a - 6b)$

74. $(2x^2 + 5)(2x^2 - 5)$

[7] *Perform each division.*

75. $\dfrac{-15y^4}{-9y^2}$

76. $\dfrac{-12x^3y^2}{6xy}$

77. $\dfrac{6y^4 - 12y^2 + 18y}{-6y}$

78. $\dfrac{2p^3 - 6p^2 + 5p}{2p^2}$

79. $(5x^{13} - 10x^{12} + 20x^7 - 35x^5) \div (-5x^4)$

80. $(-10m^4n^2 + 5m^3n^3 + 6m^2n^4) \div (5m^2n)$

[8] *Perform each division.*

81. $(2r^2 + 3r - 14) \div (r - 2)$

82. $\dfrac{12m^2 - 11m - 10}{3m - 5}$

83. $\dfrac{10a^3 + 5a^2 - 14a + 9}{5a^2 - 3}$

84. $\dfrac{2k^4 + 4k^3 + 9k^2 - 8}{2k^2 + 1}$

Mixed Review Exercises

Simplify. Perform the indicated operations. Write with positive exponents. Assume that no denominators are equal to zero.

85. $19^0 - 3^0$

86. $(3p)^4(3p^{-7})$

87. 7^{-2}

88. $(-7 + 2k)^2$

89. $\dfrac{2y^3 + 17y^2 + 37y + 7}{2y + 7}$

90. $\left(\dfrac{6r^2s}{5}\right)^4$

91. $-m^5(8m^2 + 10m + 6)$

92. $\left(\dfrac{1}{2}\right)^{-5}$

93. $(25x^2y^3 - 8xy^2 + 15x^3y) \div (5x)$

94. $(6r^{-2})^{-1}$

95. $(2x + y)^3$

96. $2^{-1} + 4^{-1}$

97. $(a + 2)(a^2 - 4a + 1)$

98. $(5y^3 - 8y^2 + 7) - (-3y^3 + y^2 + 2)$

99. $(2r + 5)(5r - 2)$

100. $(12a + 1)(12a - 1)$

Name ____ Date ____ Class ____

Evaluate each expression.

1. 5^{-4}

2. $(-3)^0 + 4^0$

3. $4^{-1} + 3^{-1}$

Simplify, and write each answer using only positive exponents. Assume that variables represent nonzero numbers.

4. $\dfrac{8^{-1} \cdot 8^4}{8^{-2}}$

5. $\dfrac{(x^{-3})^{-2}(x^{-1}y)^2}{(xy^{-2})^2}$

Write each number in scientific notation.

6. (a) 344,000,000,000
   (b) .00000557

Write each number without exponents.

7. (a) 2.96×10^7
   (b) 6.07×10^{-8}

For each polynomial, combine like terms when possible, and write the polynomial in descending powers of the variable. Give the degree of the simplified polynomial. Decide whether the simplified polynomial is a monomial, binomial, trinomial, or none of these.

8. $5x^2 + 8x - 12x^2$

9. $13n^3 - n^2 + n^4 + 3n^4 - 9n^2$

Perform the indicated operations.

10. $(5t^4 - 3t^2 + 7t + 3) - (t^4 - t^3 + 3t^2 + 8t + 3)$

11. $(2y^2 - 8y + 8) + (-3y^2 + 2y + 3) - (y^2 + 3y - 6)$

12. Subtract.

$$\begin{array}{l} 9t^3 - 4t^2 + 2t + 2 \\ \underline{9t^3 + 8t^2 - 3t - 6} \end{array}$$

1. ____

2. ____

3. ____

4. ____

5. ____

6. (a) ____

(b) ____

7. (a) ____

(b) ____

8. ____

9. ____

10. ____

11. ____

12. ____

13. ____________________

13. $3x^2(-9x^3 + 6x^2 - 2x + 1)$

14. ____________________

14. $(t - 8)(t + 3)$

15. ____________________

15. $(4x + 3y)(2x - y)$

16. ____________________

16. $(5x - 2y)^2$

17. ____________________

17. $(10v + 3w)(10v - 3w)$

18. ____________________

18. $(2r - 3)(r^2 + 2r - 5)$

19. ____________________

19. $(x + 1)^3$

20. ____________________

20. $\dfrac{8y^3 - 6y^2 + 4y + 10}{2y}$

21. ____________________

21. $(-9x^2y^3 + 6x^4y^3 + 12xy^3) \div (3xy)$

22. ____________________

22. $\dfrac{2x^2 + x - 36}{x - 4}$

23. ____________________

23. $(3x^3 - x + 4) \div (x - 2)$

24. ____________________

24. What polynomial expression represents the area of this square?

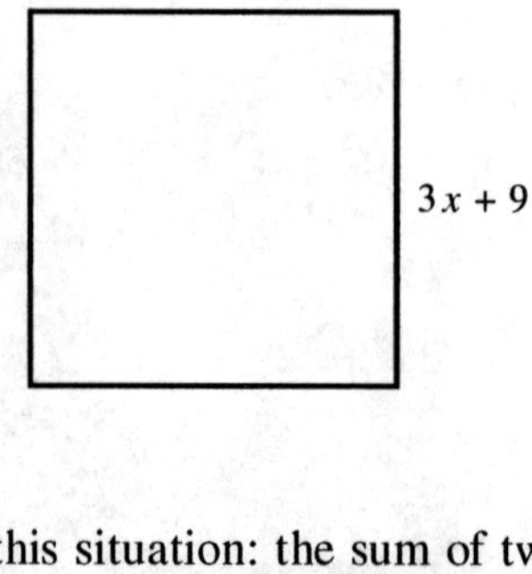

25. ____________________

25. Give an example of this situation: the sum of two fourth degree polynomials in x is a third degree polynomial in x.

Chapter R

R.1 Addition of Whole Numbers

There are 4 triangles at the left and 2 at the right. In all, there are 6 triangles.

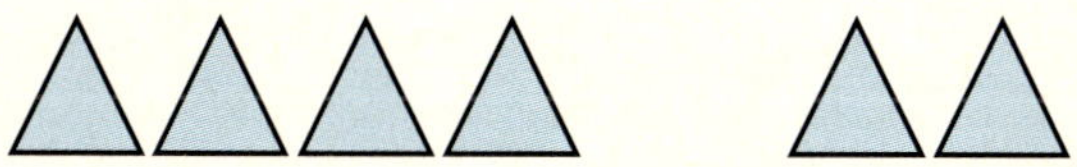

The process of finding the total is called *addition.* Here 4 and 2 were added to get 6. Addition is written with a + sign, so that

$$4 + 2 = 6.$$

OBJECTIVE 1 In addition, the numbers being added are called **addends,** (AD-ends), and the resulting answer is called the **sum** or **total.**

$$\begin{array}{r} 4 \leftarrow \text{Addend} \\ +\ 2 \leftarrow \text{Addend} \\ \hline 6 \leftarrow \text{Sum (answer)} \end{array}$$

Addition problems can also be written horizontally as follows.

$$\underset{\text{Addend}}{\underset{\uparrow}{4}} + \underset{\text{Addend}}{\underset{\uparrow}{2}} = \underset{\text{Sum}}{\underset{\uparrow}{6}}$$

Commutative Property of Addition

The **commutative** (cuh-MUE-tuh-tiv) **property of addition** states that changing the *order* of the addends in an addition problem does not change the sum.

For example, the sum of 4 + 2 is the same as the sum of 2 + 4. This allows the addition of the same numbers in a different order.

EXAMPLE 1 Adding Two Single-Digit Numbers

Add. Then use the commutative property to write another addition problem and find the sum.

(a) 6 + 2 = 8 and 2 + 6 = 8

(b) 5 + 9 = 14 and 9 + 5 = 14

(c) 8 + 3 = 11 and 3 + 8 = 11

(d) 8 + 8 = 16 (No change occurs when the commutative property is used.)

WORK PROBLEM 1 AT THE SIDE.

INTERNET
www.mathnotes.com

OBJECTIVES

1. Add two single-digit numbers.
2. Add more than two numbers.
3. Add when carrying is not required.
4. Add with carrying.
5. Solve application problems with carrying.
6. Check the answer in addition.

FOR EXTRA HELP

Tutorial | Tape 17 | SSM, Sec. R.1

1. Add. Then use the commutative property to write another addition problem and find the sum.

(a) 3 + 4

(b) 9 + 9

(c) 7 + 8

(d) 6 + 9

ANSWERS

1. (a) 7; 4 + 3 = 7 **(b)** 18; no change **(c)** 15; 8 + 7 = 15 **(d)** 15; 9 + 6 = 15

2. Add the following columns of numbers.

(a)

$$\begin{array}{r} 5 \\ 4 \\ 6 \\ 9 \\ +\ 2 \\ \hline \end{array}$$

(b)

$$\begin{array}{r} 7 \\ 5 \\ 1 \\ 2 \\ +\ 6 \\ \hline \end{array}$$

(c)

$$\begin{array}{r} 9 \\ 2 \\ 1 \\ 3 \\ +\ 4 \\ \hline \end{array}$$

(d)

$$\begin{array}{r} 3 \\ 8 \\ 6 \\ 4 \\ +\ 8 \\ \hline \end{array}$$

Associative Property of Addition

By the **associative** (uh-SOH-shuh-tiv) **property of addition,** changing the *grouping* of addends does not change the sum.

For example, the sum of $3 + 5 + 6$ may be found as follows.

$$(3 + 5) + 6 = 8 + 6 = 14$$ Parentheses tell what to do first.

Another way to add the same numbers is shown below.

$$3 + (5 + 6) = 3 + 11 = 14$$

Either method gives the answer 14.

OBJECTIVE 2 To add several numbers, first write them in a column. Add the first number to the second. Add this sum to the third digit; continue until all the digits are used.

EXAMPLE 2 Adding More Than Two Numbers

Add 2, 5, 6, 1, and 4.

$$\begin{array}{r} 2 \\ 5 \\ 6 \\ 1 \\ +\ 4 \\ \hline 18 \end{array}$$

$2 + 5 = 7$

$7 + 6 = 13$

$13 + 1 = 14$

$14 + 4 = 18$

Note

By the commutative and associative properties of addition, you may also add numbers by starting at the bottom of a column. Adding from the top or adding from the bottom will give the same answer.

◀◀ WORK PROBLEM 2 AT THE SIDE.

OBJECTIVE 3 If numbers have two or more digits, first you must arrange the numbers in columns so that the ones digits are in the same column, tens are in the same column, hundreds are in the same column, and so on. Next, you add column by column, starting at the right.

EXAMPLE 3 Adding Without Carrying

Add $511 + 23 + 154 + 10$.

First line up the numbers in columns, with the ones column at the right.

Hundreds in a column
Tens in a column
Ones in a column

$$\begin{array}{r} 511 \\ 23 \\ 154 \\ +\ 10 \\ \hline \end{array}$$

Ones digits at the right

CONTINUED ON NEXT PAGE

ANSWERS

2. (a) 26 (b) 21 (c) 19 (d) 29

Now start at the right and add the ones digits. Add the tens digits next, and finally, the hundreds digits.

```
    5 1 1
      2 3
    1 5 4
 +    1 0
 ---------
    6 9 8
    ↑ ↑ ↑
    │ │ └──── Sum of ones
    │ └────── Sum of tens
    └──────── Sum of hundreds
```

The sum of the four numbers is 698.

WORK PROBLEM 3 AT THE SIDE.

Objective 4 If the sum of the digits in a column is more than 9, use **carrying.**

EXAMPLE 4 Adding with Carrying

Add 47 and 29.

Add ones.

```
   47
 + 29
 ----
    ↑── Sum of ones is 16.
```

Because 16 is 1 ten plus 6 ones, place 6 in the ones column and carry 1 to the tens column.

```
   1 ←──────┐
   47       │  7 + 9 = 16
 + 29       │
 ----       │
    6 ←─────┘
```

Add the tens column.

```
   1
   47
 + 29
 ----
   76
   ↑── Sum of digits in tens column
```

WORK PROBLEM 4 AT THE SIDE.

EXAMPLE 5 Adding with Carrying

Add 324 + 7855 + 23 + 7 + 86.

Step 1 Add the digits in the ones column.

```
      2 ←──────────┐
    324            │ Carry 2 to the
   7855            │ tens column.
     23   Sum of the ones column is 25.
      7            │
 +   86            │ Write 5 in the
 ------            │ ones column.
      5 ←──────────┘
```

In 25, the 5 represents 5 ones and is written in the ones column, while 2 represents 2 tens and is carried to the tens column.

CONTINUED ON NEXT PAGE

3. Add.

(a)
```
   25
 + 73
```

(b)
```
   364
 + 532
```

(c)
```
   42,305
 + 11,563
```

4. Add by using carrying.

(a)
```
   69
 + 26
```

(b)
```
   76
 + 18
```

(c)
```
   56
 + 37
```

(d)
```
   34
 + 49
```

Answers

3. **(a)** 98 **(b)** 896 **(c)** 53,868

4. **(a)** 95 **(b)** 94 **(c)** 93 **(d)** 83

5. Add by carrying as necessary.

(a)
```
   481
    79
    38
+  395
```

(b)
```
  4271
   372
  8976
+  162
```

(c)
```
   57
    4
  392
  804
   51
+  27
```

(d)
```
  7821
   435
    72
   305
+ 1693
```

6. Add with mental carrying.

(a)
```
   278
   825
    14
     3
     7
+ 9275
```

(b)
```
  3305
   650
   708
    29
    40
     6
+    3
```

(c)
```
  15,829
     765
      78
      15
       9
       7
+ 13,179
```

Answers

5. **(a)** 993 **(b)** 13,781 **(c)** 1335 **(d)** 10,326

6. **(a)** 10,402 **(b)** 4741 **(c)** 29,882

Step 2 Now add the digits in the tens column, including the carried 2.

```
   12
   324
  7855
    23
     7
+   86
    95
```

Carry 1 to the hundreds column.

Sum of the tens column is 19.

Write 9 in the tens column.

Step 3 Add the hundreds column, including the carried 1.

```
  112
   324
  7855
    23
     7
+   86
   295
```

Carry 1 to the thousands column.

Sum of the hundreds column is 12.

Write 2 in the hundreds column.

Step 4 Add the thousands column, including the carried 1.

```
  112
   324
  7855
    23
     7
+   86
  8295
```

Sum of the thousands column is 8.

Finally, 324 + 7855 + 23 + 7 + 86 = 8295.

WORK PROBLEM 5 AT THE SIDE.

Note

For additional speed, try to carry mentally. Do not write the number carried, but just carry the number mentally to the top of the next column being added. Try this method. If it works for you, use it.

WORK PROBLEM 6 AT THE SIDE.

OBJECTIVE 5 The next two examples have application problems that require adding.

EXAMPLE 6 Applying Addition Skills

On this map, the distance in miles from one location to another is written alongside the road. Find the shortest distance from Altamonte Springs to Clear Lake.

CONTINUED ON NEXT PAGE

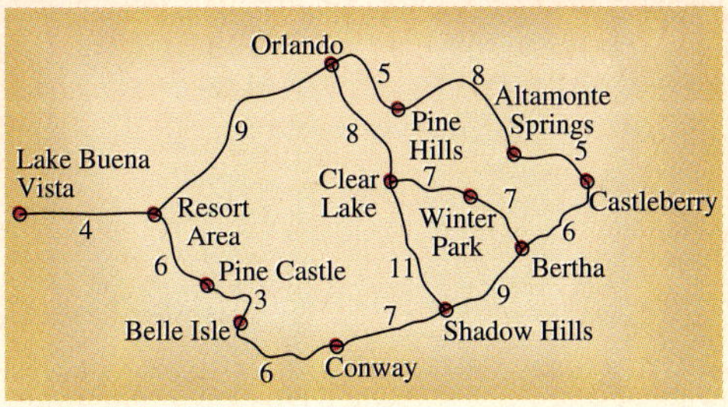

Approach Add the mileage along various routes to determine the distances from Altamonte Springs to Clear Lake. Then select the shortest route.

Solution One way from Altamonte Springs to Clear Lake is through Orlando. Add the mileage numbers along this route.

8	Altamonte Springs to Pine Hills
5	Pine Hills to Orlando
+ 8	Orlando to Clear Lake
21 →	miles from Altamonte Springs to Clear Lake, going through Orlando

Another way is through Bertha and Winter Park. Add the mileage numbers along this route.

5	Altamonte Springs to Castleberry
6	Castleberry to Bertha
7	Bertha to Winter Park
+ 7	Winter Park to Clear Lake
25 →	miles from Altamonte Springs to Clear Lake through Bertha and Winter Park

The shortest way from Altamonte Springs to Clear Lake is through Orlando.

WORK PROBLEM 7 AT THE SIDE.

EXAMPLE 7 Finding a Total

Using the map in Example 6, find the total distance from Shadow Hills to Castleberry to Orlando and back to Shadow Hills.

Approach Add the mileage from Shadow Hills to Castleberry to Orlando and back to Shadow Hills to find the total distance.

Solution Use the numbers from the map.

9	Shadow Hills to Bertha
6	Bertha to Castleberry
5	Castleberry to Altamonte Springs
8	Altamonte Springs to Pine Hills
5	Pine Hills to Orlando
8	Orlando to Clear Lake
+ 11	Clear Lake to Shadow Hills
52 →	miles from Shadow Hills to Castleberry to Orlando and back to Shadow Hills

WORK PROBLEM 8 AT THE SIDE.

7. Use the map to find the shortest distance from Lake Buena Vista to Conway.

8. The road is closed between Orlando and Clear Lake, so this route cannot be used. Use the map to find the next shortest distance from Orlando to Clear Lake.

ANSWERS

7. 19 miles

8.

Orlando to Pine Hills	5
Pine Hills to Altamonte Springs	8
Altamonte Springs to Castleberry	5
Castleberry to Bertha	6
Bertha to Winter Park	7
Winter Park to Clear Lake	+ 7
	38 miles

9. Check the following additions. If an answer is incorrect, give the correct answer.

(a)

32
8
5
+ 14
59

(b)

872
539
46
+ 152
1609

(c)

79
218
7
+ 639
953

(d)

21,892
11,746
+ 43,925
79,563

Objective 6 Checking the answer is an important part of problem solving. A common method for checking addition is to re-add from bottom to top. This is an application of the commutative and associative properties of addition.

EXAMPLE 8 Checking Addition

Check the following addition.

(Add down) ↓	**1428**	↑
	738	Adding down and adding up should give the same answer.
	63	
	125	
	17	(Add up)
	+ 485	Check
	1428	

Here the answers agree, so the sum is probably correct.

EXAMPLE 9 Checking Addition

Check the following additions. Are they correct?

		1033	Correct, because both answers are the same
(a)	785	785	
	63	63	(Add up)
	+ 185	+ 185	Check
	1033	**1033**	

		2454	Error, because answers are different
(b)	635	635	
	73	73	
	831	831	(Add up)
	+ 915	+ 915	Check
	2444	**2444**	

Re-add to find that the correct answer is 2454.

◀◀ WORK PROBLEM 9 AT THE SIDE.

Answers

9. (a) correct **(b)** correct
(c) incorrect, should be 943
(d) incorrect, should be 77,563

R.1 Exercises

Add. Then use the commutative property to write another addition problem and find the sum.

1. 3213 + 5715

2. 6344 + 1655

3. 38,204 + 21,020

4. 63,251 + 36,305

Add by carrying as necessary.

5. 67 + 83	**6.** 78 + 36	**7.** 746 + 905	**8.** 621 + 359	**9.** 798 + 206
10. 172 + 156	**11.** 7968 + 1285	**12.** 1768 + 8275	**13.** 7896 + 3728	**14.** 9382 + 7586
15. 3705 3916 + 9037	**16.** 6629 6076 + 8218	**17.** 32 + 4977	**18.** 402 + 9938	**19.** 3077 8 + 421
20. 56 7721 + 172	**21.** 9056 78 6089 + 731	**22.** 4022 709 8621 + 37	**23.** 18 708 9286 + 636	**24.** 1708 321 61 + 8926

Check each sum by adding from bottom to top. If an answer is incorrect, find the correct sum.

25. 179 214 + 376 759	**26.** 17 296 713 + 94 1220	**27.** 4713 28 615 + 64 5420	**28.** 6 215 744 36 + 4 284 11,279

Writing Calculator Small Group

Using the map below, find the shortest distance between the following cities.

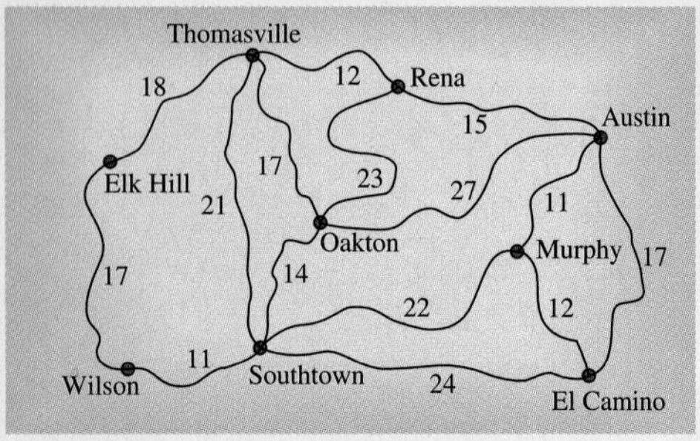

29. Southtown and Rena

30. Elk Hill and Oakton

31. Thomasville and Murphy

32. Murphy and Thomasville

Solve the following application problems.

33. A tune-up costs $65 and a tire rotation is $14. Find the total cost for both services.

34. Jane Lim ordered 68 large sprinkler valves and 47 small valves for her hardware store. How many valves did she order altogether?

35. There are 413 women and 286 men on the sales staff. How many people are on the sales staff?

36. One department in an office building has 283 employees while another department has 218 employees. How many employees are in the two departments?

37. At a charity bazaar, a library has a total of 9792 books for sale, while a book dealer has 3259 books for sale. How many books are for sale?

38. A plane is flying at an altitude of 5924 feet. It then increases its altitude by 7284 feet. Find its new altitude.

Find the perimeter of (or total distance around) each of the following figures.

39.

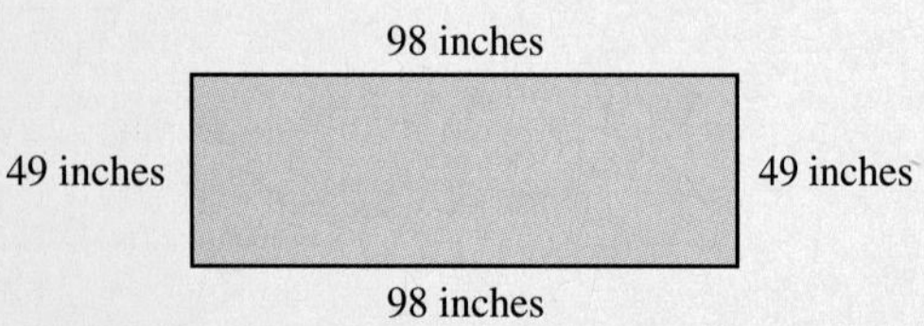

40.

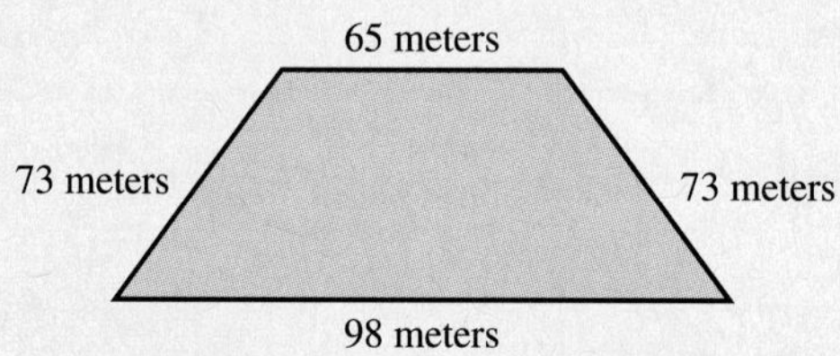

41.

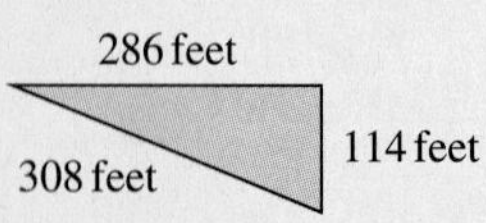

42.

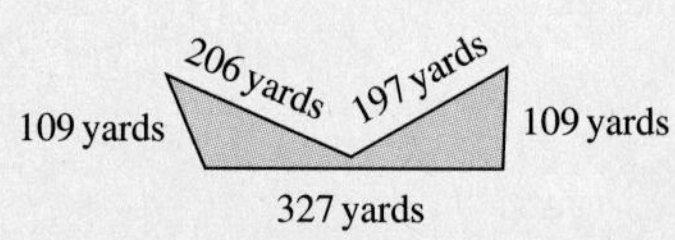

R.2 *Subtraction of Whole Numbers*

Suppose you have \$8, and you spend \$5 for gasoline. You then have \$3 left. There are two different ways of looking at these numbers.

As an addition problem:

\$5	+	\$3	=	\$8
↑		↑		↑
Amount spent		Amount left		Original amount

As a subtraction problem:

\$8	−	\$5	=	\$3
↑	↑	↑		↑
Original amount	Subtraction symbol	Amount spent		Amount left

OBJECTIVE 1 As this example shows, an addition problem can be changed to a subtraction problem and a subtraction problem can be changed to an addition problem.

EXAMPLE 1 Changing Addition Problems to Subtraction

Change each addition problem to a subtraction problem.

(a) $4 + 1 = 5$

Two subtraction problems are possible:

$$5 - 1 = 4 \quad \text{or} \quad 5 - 4 = 1$$

These figures show each subtraction problem.

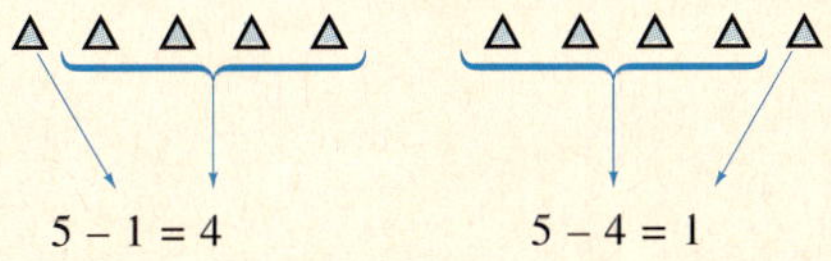

(b) $8 + 7 = 15$

$$15 - 7 = 8 \quad \text{or} \quad 15 - 8 = 7$$

WORK PROBLEM 1 AT THE SIDE.

EXAMPLE 2 Changing Subtraction Problems to Addition

Change each subtraction problem to an addition problem.

(a) $8 - 3 = 5$

↓ ↙ ↓ ↓

$8 = 3 + 5$

It is also correct to write $8 = 5 + 3$ (by the commutative property).

CONTINUED ON NEXT PAGE

OBJECTIVES

1. Change addition problems to subtraction and subtraction problems to addition.
2. Identify the minuend, subtrahend, and difference.
3. Subtract when no borrowing is needed.
4. Check answers.
5. Subtract by borrowing.
6. Solve application problems with subtraction.

FOR EXTRA HELP

Tutorial Tape 17 SSM, Sec. R.2

1. Write two subtraction problems for each addition problem.

(a) $4 + 3 = 7$

(b) $6 + 5 = 11$

(c) $15 + 22 = 37$

(d) $23 + 55 = 78$

ANSWERS

1. (a) $7 - 3 = 4$ or $7 - 4 = 3$
(b) $11 - 5 = 6$ or $11 - 6 = 5$
(c) $37 - 22 = 15$ or $37 - 15 = 22$
(d) $78 - 55 = 23$ or $78 - 23 = 55$

2. Write an addition problem for each subtraction problem.

(a) 5 − 3 = 2

(b) 8 − 3 = 5

(c) 21 − 15 = 6

(d) 58 − 42 = 16

3. Subtract.

(a) $\begin{array}{r} 56 \\ -\ 31 \\ \hline \end{array}$

(b) $\begin{array}{r} 38 \\ -\ 14 \\ \hline \end{array}$

(c) $\begin{array}{r} 378 \\ -\ 235 \\ \hline \end{array}$

(d) $\begin{array}{r} 3927 \\ -\ 2614 \\ \hline \end{array}$

(e) $\begin{array}{r} 5464 \\ -\ 324 \\ \hline \end{array}$

Answers

2. (a) 5 = 3 + 2 (b) 8 = 3 + 5 (c) 21 = 15 + 6 (d) 58 = 42 + 16

3. (a) 25 (b) 24 (c) 143 (d) 1313 (e) 5140

(b) 19 − 14 = 5
19 = 14 + 5

(c) 29 − 13 = 16
29 = 13 + 16

WORK PROBLEM 2 AT THE SIDE.

Objective 2 In subtraction, as in addition, the numbers in a problem have names. For example, in the problem, 8 − 5 = 3, the number 8 is the **minuend** (MIN-yoo-end), 5 is the **subtrahend** (SUB-truh-hend), and 3 is the **difference** or answer.

8 − 5 = 3 ← Difference (answer)
8 ↑ Minuend; 5 ↑ Subtrahend

8 ← Minuend
− 5 ← Subtrahend
3 ← Difference

Objective 3 Subtract two numbers by lining up the numbers in columns so that the digits in the ones place are in the same column. Next subtract by columns, starting at the right with the ones column.

EXAMPLE 3 Subtracting Two Numbers Without Borrowing

Subtract.

(a) Ones digits are lined up in the same column.

$\begin{array}{r} 53 \\ -\ 21 \\ \hline 32 \end{array}$

3 − 1 = 2
5 − 2 = 3

(b) Ones digits are lined up.

$\begin{array}{r} 385 \\ -\ 161 \\ \hline 224 \end{array}$

5 − 1 = 4
8 − 6 = 2
3 − 1 = 2

(c)

$\begin{array}{r} 9431 \\ -\ 210 \\ \hline 9221 \end{array}$

1 − 0 = 1
3 − 1 = 2
4 − 2 = 2
9 − 0 = 9

WORK PROBLEM 3 AT THE SIDE.

Objective 4 Use addition to check your answer to a subtraction problem. For example, check 8 − 3 = 5 by *adding* 3 and 5:

3 + 5 = 8, so 8 − 3 = 5 is correct.

EXAMPLE 4 Checking Subtraction

Check each answer.

(a)
$$\begin{array}{r} 89 \\ -\ 47 \\ \hline 42 \end{array}$$

Rewrite as an addition problem, as shown in Example 2.

Subtraction problem
$$\begin{array}{r} 89 \\ -\ 47 \\ \hline 42 \\ 89 \end{array}$$
Addition problem
$$\begin{array}{r} 47 \\ +\ 42 \\ \hline 89 \end{array}$$

Because 47 + 42 = 89, the subtraction was done correctly.

(b) 72 − 41 = 21

Rewrite as an addition problem.

$$72 = 41 + 21$$

But, 41 + 21 = 62, not 72, so the subtraction was done incorrectly. We rework the original subtraction to get the correct answer, 31.

(c)
$$\begin{array}{r} 374 \\ -\ 141 \\ \hline 233 \end{array}$$
374 ← Match → 141 + 233 = 374

The answer checks.

WORK PROBLEM 4 AT THE SIDE.

Objective 5 If a digit in the minuend is less than the one directly below it we cannot subtract, so **borrowing** will be necessary.

EXAMPLE 5 Subtracting with Borrowing

Subtract 19 from 57.

Write the problem.

$$\begin{array}{r} 57 \\ -\ 19 \\ \hline \end{array}$$

In the ones column, 7 is less than 9, so, in order to subtract, we must borrow 1 ten from the 5 tens.

5 tens − 1 ten = 4 tens → 4 17 ← 1 ten = 10 ones and 10 + 7 = 17

$$\begin{array}{r} \overset{4}{\cancel{5}}\ \overset{17}{\cancel{7}} \\ -\ 1\ 9 \\ \hline \end{array}$$

Now we can subtract 17 − 9 in the ones column and then 4 − 1 in the tens column,

$$\begin{array}{r} \overset{4}{\cancel{5}}\ \overset{17}{\cancel{7}} \\ -\ 1\ 9 \\ \hline 3\ 8 \end{array}$$ Difference

Finally, 57 − 19 = 38. Check by adding 19 and 38. You should get 57.

WORK PROBLEM 5 AT THE SIDE.

4. Decide whether these answers are correct. If incorrect, what should they be?

(a)
$$\begin{array}{r} 65 \\ -\ 23 \\ \hline 42 \end{array}$$

(b)
$$\begin{array}{r} 46 \\ -\ 32 \\ \hline 24 \end{array}$$

(c)
$$\begin{array}{r} 374 \\ -\ 251 \\ \hline 113 \end{array}$$

(d)
$$\begin{array}{r} 7531 \\ -\ 4301 \\ \hline 3230 \end{array}$$

5. Subtract.

(a)
$$\begin{array}{r} 67 \\ -\ 38 \\ \hline \end{array}$$

(b)
$$\begin{array}{r} 97 \\ -\ 29 \\ \hline \end{array}$$

(c)
$$\begin{array}{r} 31 \\ -\ 17 \\ \hline \end{array}$$

(d)
$$\begin{array}{r} 863 \\ -\ 47 \\ \hline \end{array}$$

(e)
$$\begin{array}{r} 762 \\ -\ 157 \\ \hline \end{array}$$

Answers

4. (a) correct (b) incorrect, should be 14 (c) incorrect, should be 123 (d) correct

5. (a) 29 (b) 68 (c) 14 (d) 816 (e) 605

6. Subtract.

(a) $\begin{array}{r} 354 \\ -\ \ 82 \\ \hline \end{array}$

(b) $\begin{array}{r} 457 \\ -\ \ 68 \\ \hline \end{array}$

(c) $\begin{array}{r} 874 \\ -\ 486 \\ \hline \end{array}$

(d) $\begin{array}{r} 1437 \\ -\ \ 988 \\ \hline \end{array}$

(e) $\begin{array}{r} 8739 \\ -\ 3892 \\ \hline \end{array}$

EXAMPLE 6 Subtracting with Borrowing

Subtract by borrowing as necessary.

(a) $\begin{array}{r} 7856 \\ -\ \ 137 \\ \hline \end{array}$

Borrow 1 ten. → 1 ten = 10 ones and 10 + 6 = 16.

$$\begin{array}{r} 4\ 16 \\ 7\ 8\ \not{5}\ \not{6} \\ -\ \ \ 1\ 3\ 7 \\ \hline 7\ 7\ 1\ 9 \end{array} \leftarrow \text{Difference}$$

(b) $\begin{array}{r} 635 \\ -\ 546 \\ \hline \end{array}$

Borrow 1 ten. → 1 ten = 10 ones and 10 + 5 = 15.

$$\begin{array}{r} 2\ 15 \\ 6\ \not{3}\ \not{5} \\ -\ 5\ 4\ 6 \\ \hline 9 \end{array}$$

Need to borrow farther because 2 is less than 4.

Borrow 1 hundred. → 1 hundred = 10 tens and 10 tens + 2 tens = 12 tens.

$$\begin{array}{r} 5\ 12\ 15 \\ \not{6}\ \not{3}\ \not{5} \\ -\ 5\ 4\ 6 \\ \hline 8\ 9 \end{array} \leftarrow \text{Difference}$$

(c) $\begin{array}{r} 412 \\ -\ 225 \\ \hline \end{array}$

$$\begin{array}{r} 0\ 12 \\ 4\ \not{1}\ \not{2} \\ -\ 2\ 2\ 5 \\ \hline 7 \end{array}$$

Need to borrow farther because 0 is less than 2.

$$\begin{array}{r} 3\ 10\ 12 \\ \not{4}\ \not{1}\ \not{2} \\ -\ 2\ 2\ 5 \\ \hline 1\ 8\ 7 \end{array} \leftarrow \text{Difference}$$

◀◀ WORK PROBLEM 6 AT THE SIDE.

Sometimes a minuend has zeros in some of the positions. In such cases, borrowing may be a little more complicated than what we have shown so far.

EXAMPLE 7 Borrowing with Zeros

Subtract.

$$\begin{array}{r} 4607 \\ -\ 3168 \\ \hline \end{array}$$

It is not possible to borrow from the tens position. Instead we must first borrow from the hundreds position.

Borrow 1 hundred → 1 hundred is 10 tens.

$$\begin{array}{r} 5\ 10\ \ \\ 4\ \not{6}\ \not{0}\ 7 \\ -\ 3\ 1\ 6\ 8 \\ \hline \end{array}$$

Now we can borrow from the tens position.

9 ← 10 tens − 1 ten = 9 tens.

17 ← 1 ten = 10 ones and 10 + 7 = 17.

$$\begin{array}{r} 9\ \ \ \ \\ 5\ \not{10}\ 17 \\ 4\ \not{6}\ \not{0}\ \not{7} \\ -\ 3\ 1\ 6\ 8 \\ \hline 9 \end{array}$$

CONTINUED ON NEXT PAGE

ANSWERS

6. (a) 272 **(b)** 389 **(c)** 388 **(d)** 449 **(e)** 4847

Complete the problem.

$$\begin{array}{r} \;\;9\;\; \\ 5\;\not{10}\;17 \\ 4\;\not{6}\;\not{0}\;\not{7} \\ -\;3\;1\;6\;8 \\ \hline 1\;4\;3\;9 \end{array} \quad \text{Difference}$$

As above, check by adding 1439 and 3168; you should get 4607.

WORK PROBLEM 7 AT THE SIDE. ▶▶

EXAMPLE 8 Borrowing with Zeros

Subtract.

(a) 708
− 149

Borrow 1 ten.
1 hundred is 10 tens.
1 ten is 10 ones and 10 + 8 = 18.
Borrow 1 hundred.

$$\begin{array}{r} 9\; \\ 6\;\not{10}\;18 \\ \not{7}\;\not{0}\;\not{8} \\ -\;1\;4\;9 \\ \hline 5\;5\;9 \end{array}$$

(b) 380
− 276

Borrow 1 ten. 1 ten is 10 ones.

$$\begin{array}{r} 7\;10 \\ 3\;\not{8}\;\not{0} \\ -\;2\;7\;6 \\ \hline 1\;0\;4 \end{array}$$

(c) 9000
− 6999

$$\begin{array}{r} 9\;\;9\;\; \\ 8\;\not{10}\;\not{10}\;10 \\ \not{9}\;\not{0}\;\not{0}\;\not{0} \\ -\;6\;9\;9\;9 \\ \hline 2\;0\;0\;1 \end{array}$$

WORK PROBLEM 8 AT THE SIDE. ▶▶

As explained above, an answer to a subtraction problem can be checked by adding.

EXAMPLE 9 Checking Subtraction

Check the following answers.

(a) **613**
− 275
338

Check
275
+ 338
613 Correct

Match

CONTINUED ON NEXT PAGE

7. Subtract.

(a) 308
− 285

(b) 206
− 148

(c) 5073
− 1632

8. Subtract.

(a) 405
− 267

(b) 370
− 163

(c) 1570
− 983

(d) 7001
− 5193

(e) 4000
− 1782

ANSWERS

7. **(a)** 23 **(b)** 58 **(c)** 3441

8. **(a)** 138 **(b)** 207 **(c)** 587 **(d)** 1808 **(e)** 2218

(b) 1915 − 1635 = 280

Check: 1635 + 280 = 1915 (Match) Correct

(c) 15,803 − 7 325 = 8 578

Check: 7 325 + 8 578 = 15,903 (No match) Error

Rework the original problem to get the correct answer, 8478.

◀◀ WORK PROBLEM 9 AT THE SIDE.

Objective 6 As shown in the next example, subtraction can be used to solve an application problem.

EXAMPLE 10 Applying Subtraction Skills

Diana Lopez drives a United Parcel Service delivery truck. Using the table below, decide how many more deliveries were made by Lopez on Monday than on Thursday.

PACKAGE DELIVERY (Lopez)

Day	Number of Deliveries
Sunday	0
Monday	137
Tuesday	126
Wednesday	119
Thursday	89
Friday	147
Saturday	0

Lopez made 137 deliveries on Monday, but had only 89 deliveries on Thursday. Find how many more deliveries were made on Monday than on Thursday by subtracting 89 from 137.

	137	Deliveries on Monday
−	89	Deliveries on Thursday
	48	More deliveries on Monday

Lopez made 48 more deliveries on Monday than she made on Thursday.

◀◀ WORK PROBLEM 10 AT THE SIDE.

9. Check the answers in the following problems. If the answer is incorrect, give the correct answer.

(a) 425 − 368 = 57

(b) 670 − 439 = 241

(c) 14,726 − 8 839 = 5 887

10. Using the table from Example 10, how many more deliveries did Lopez make on

(a) Friday than on Tuesday?

(b) Tuesday than on Wednesday?

Answers

9. **(a)** correct
(b) incorrect, should be 231
(c) correct

10. **(a)** 21 **(b)** 7

R.2 Exercises

Use addition to check each subtraction. If an answer is incorrect, find the correct answer.

1. $\begin{array}{r} 89 \\ -\ 27 \\ \hline 63 \end{array}$ **2.** $\begin{array}{r} 47 \\ -\ 35 \\ \hline 13 \end{array}$ **3.** $\begin{array}{r} 382 \\ -\ 261 \\ \hline 131 \end{array}$ **4.** $\begin{array}{r} 838 \\ -\ 516 \\ \hline 322 \end{array}$

Subtract by borrowing as necessary.

5. $\begin{array}{r} 36 \\ -\ 28 \\ \hline \end{array}$ **6.** $\begin{array}{r} 97 \\ -\ 39 \\ \hline \end{array}$ **7.** $\begin{array}{r} 83 \\ -\ 58 \\ \hline \end{array}$ **8.** $\begin{array}{r} 65 \\ -\ 28 \\ \hline \end{array}$ **9.** $\begin{array}{r} 45 \\ -\ 29 \\ \hline \end{array}$

10. $\begin{array}{r} 93 \\ -\ 37 \\ \hline \end{array}$ **11.** $\begin{array}{r} 719 \\ -\ 658 \\ \hline \end{array}$ **12.** $\begin{array}{r} 916 \\ -\ 618 \\ \hline \end{array}$ **13.** $\begin{array}{r} 771 \\ -\ 252 \\ \hline \end{array}$ **14.** $\begin{array}{r} 973 \\ -\ 788 \\ \hline \end{array}$

15. $\begin{array}{r} 9861 \\ -\ 684 \\ \hline \end{array}$ **16.** $\begin{array}{r} 6171 \\ -\ 1182 \\ \hline \end{array}$ **17.** $\begin{array}{r} 9988 \\ -\ 2399 \\ \hline \end{array}$ **18.** $\begin{array}{r} 3576 \\ -\ 1658 \\ \hline \end{array}$ **19.** $\begin{array}{r} 38,335 \\ -\ 29,476 \\ \hline \end{array}$

20. $\begin{array}{r} 61,278 \\ -\ 3,559 \\ \hline \end{array}$ **21.** $\begin{array}{r} 40 \\ -\ 37 \\ \hline \end{array}$ **22.** $\begin{array}{r} 80 \\ -\ 73 \\ \hline \end{array}$ **23.** $\begin{array}{r} 60 \\ -\ 37 \\ \hline \end{array}$ **24.** $\begin{array}{r} 70 \\ -\ 27 \\ \hline \end{array}$

25. $\begin{array}{r} 6020 \\ -\ 4078 \\ \hline \end{array}$ **26.** $\begin{array}{r} 7050 \\ -\ 6045 \\ \hline \end{array}$ **27.** $\begin{array}{r} 8503 \\ -\ 2816 \\ \hline \end{array}$ **28.** $\begin{array}{r} 16,004 \\ -\ 5\ 087 \\ \hline \end{array}$ **29.** $\begin{array}{r} 80,705 \\ -\ 61,667 \\ \hline \end{array}$

30. $\begin{array}{r} 72,000 \\ -\ 44,234 \\ \hline \end{array}$ **31.** $\begin{array}{r} 66,000 \\ -\ 444 \\ \hline \end{array}$ **32.** $\begin{array}{r} 77,000 \\ -\ 308 \\ \hline \end{array}$ **33.** $\begin{array}{r} 20,080 \\ -\ 96 \\ \hline \end{array}$ **34.** $\begin{array}{r} 80,056 \\ -\ 69 \\ \hline \end{array}$

Use addition to check each subtraction. If an answer is incorrect, find the correct answer.

35. $\begin{array}{r} 3070 \\ -\ 576 \\ \hline 2596 \end{array}$ **36.** $\begin{array}{r} 1439 \\ -\ 1169 \\ \hline 270 \end{array}$ **37.** $\begin{array}{r} 27,600 \\ -\ 807 \\ \hline 26,793 \end{array}$ **38.** $\begin{array}{r} 34,021 \\ -\ 33,708 \\ \hline 727 \end{array}$

Solve the following application problems.

39. A man burns 103 calories during 30 minutes of bowling while a woman burns 88 calories. How many fewer calories does a woman burn than a man?

40. Lynn Couch had $553 in her checking account. She wrote a check for $308 for school fees. How much is left in her account?

41. An airplane is carrying 254 passengers. When it lands in Atlanta 133 passengers get off the plane. How many passengers are left on the plane?

42. On Tuesday, 5822 people went to a soccer game, and on Friday, 7994 people went to a soccer game. How many more people went to the game on Friday?

The table below shows the average annual earnings of various types of medical doctors. Refer to the table to answer Exercises 43 and 44.

Type	Yearly Income
Anesthesiologist	$228,500
Radiologist	$253,300
Surgeon	$244,600
Family Doctor	$111,800
Pediatrician	$121,700

43. How much more does a radiologist earn than a surgeon?

44. How much more does an anesthesiologist earn than a pediatrician?

45. Downtown Toronto's skyline is dominated by the CN Tower which rises 1821 feet. The Sears Tower in Chicago is 1454 feet high. Find the difference in height between the two structures.

46. The fastest animal in the world, the peregrine falcon, dives at 217 miles per hour while a Boeing 747 cruises at 580 miles per hour. How much faster does the plane fly than the falcon?

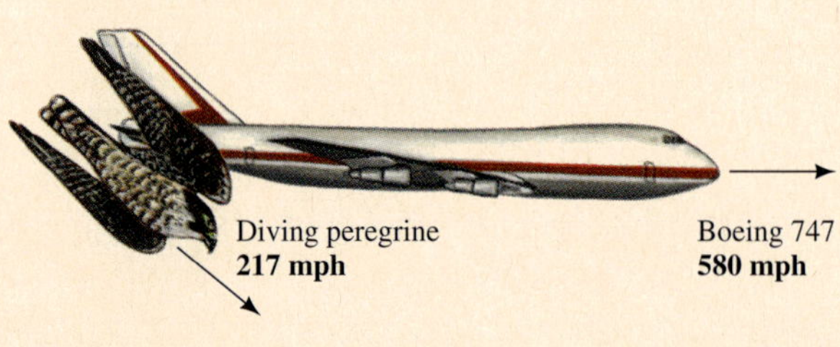

R.3 Multiplication of Whole Numbers

Adding the number 3 a total of 4 times gives 12.

$$3 + 3 + 3 + 3 = 12$$

This result can also be shown with a figure.

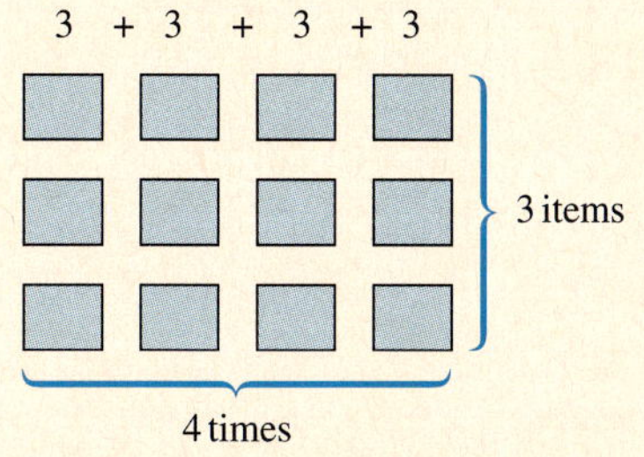

OBJECTIVES

1. Identify the parts of a multiplication problem.
2. Do chain multiplications.
3. Multiply by single-digit numbers.
4. Multiply quickly by numbers ending in zeros.
5. Multiply by numbers having more than one digit.
6. Solve application problems with multiplication.

FOR EXTRA HELP

Tutorial | Tape 17 | SSM, Sec. R.3

OBJECTIVE 1 Multiplication is a shortcut for repeated addition. The numbers being multiplied are called **factors.** The answer is called the **product.** For example, the product of 3 and 4 can be written with the symbol ×, a raised dot, parentheses, or, in computer work, an asterisk.

$$\begin{array}{rl} 3 & \leftarrow \text{Factor (also called } \textit{multiplicand}\text{)} \\ \times\ 4 & \leftarrow \text{Factor (also called } \textit{multiplier}\text{)} \\ \hline 12 & \leftarrow \text{Product (answer)} \end{array}$$

$3 \times 4 = 12$ $\quad 3 \cdot 4 = 12$ $\quad (3)(4) = 12$ $\quad 3 * 4 = 12$ (In computer work)

WORK PROBLEM 1 AT THE SIDE.

Commutative Property of Multiplication

By the **commutative property of multiplication,** changing the *order* of two factors does not change the product.

For example:

$$3 \times 5 = 15 \quad \text{and} \quad 5 \times 3 = 15$$

Both products are 15.

Note

Recall that addition also has a commutative property. Remember that $4 + 2$ is the same as $2 + 4$. Subtraction, however, is *not* commutative.

EXAMPLE 1 Multiplying Two Numbers

Multiply. (Remember that a raised dot means to multiply.) Do the work mentally.

(a) $3 \times 4 = 12$ By the commutative property, $4 \times 3 = 12$ also.

(b) $6 \cdot 0 = 0$ The product of any number and 0 is 0; if you give no money to each of 6 relatives, you give no money.

(c) $(4)(8) = 32$ By the commutative property, $(8)(4) = 32$ also.

WORK PROBLEM 2 AT THE SIDE.

1. Identify the factors and the product in each multiplication problem.

(a) $3 \times 6 = 18$

(b) $8 \times 4 = 32$

(c) $5 \cdot 7 = 35$

(d) $(3)(9) = 27$

2. Multiply. Do the work mentally. Then use the commutative property to write another multiplication problem and find the product.

(a) 4×7

(b) 0×9

(c) $8 \cdot 6$

(d) $5 \cdot 5$

(e) $(3)(8)$

ANSWERS

1. **(a)** factors: 3, 6; product: 18
(b) factors: 8, 4; product: 32
(c) factors: 5, 7; product: 35
(d) factors: 3, 9; product: 27

2. **(a)** 28; $7 \times 4 = 28$ **(b)** 0; $9 \times 0 = 0$
(c) 48; $6 \cdot 8 = 48$ **(d)** 25; no change
(e) 24; $(8)(3) = 24$

3. Multiply.

(a) $2 \times 3 \times 4$

(b) $6 \cdot 1 \cdot 5$

(c) $(8)(3)(0)$

(d) $3 \times 3 \times 7$

(e) $4 \cdot 2 \cdot 8$

(f) $(2)(2)(9)$

Answers

3. **(a)** 24 **(b)** 30 **(c)** 0 **(d)** 63 **(e)** 64 **(f)** 36

Objective 2 Some multiplications contain more than two factors.

Associative Property of Multiplication

By the **associative property of multiplication,** changing the *grouping* of factors does not change the product.

EXAMPLE 2 Multiplying Three Numbers

Multiply: $2 \times 3 \times 5$.

$$(2 \times 3) \times 5 \quad \text{Parentheses tell what to do first.}$$
$$6 \times 5 = 30$$

Also,

$$2 \times (3 \times 5)$$
$$2 \times 15 = 30$$

Either grouping results in the same answer.

Calculator Tip: The calculator approach to Example 2 uses chain calculations.

2 [×] 3 [×] 5 [=] 30

A problem with more than two factors, such as the one in Example 2, is called a *chain multiplication.*

WORK PROBLEM 3 AT THE SIDE.

Objective 3 Carrying may be needed in multiplication problems with larger factors.

EXAMPLE 3 Carrying with Multiplication

Multiply.

(a)
$$\begin{array}{r} 53 \\ \times\ 4 \\ \hline \end{array}$$

Start by multiplying in the ones column.

$$\begin{array}{r} 1 \\ 53 \\ \times\ 4 \\ \hline 2 \end{array} \quad 4 \times 3 = 12$$

Carry the 1 to the tens column.
Write 2 in the ones column.

Next, multiply 4 ones and 5 tens.

$$\begin{array}{r} 1 \\ 53 \\ \times\ 4 \\ \hline 2 \end{array} \quad 4 \times 5 \text{ tens} = 20 \text{ tens}$$

Add the 1 that was carried to the tens column.

$$\begin{array}{r} 1 \\ 53 \\ \times\ 4 \\ \hline 212 \end{array} \quad 20 \text{ tens} + 1 \text{ ten} = 21 \text{ tens}$$

CONTINUED ON NEXT PAGE

(b) $\begin{array}{r} 724 \\ \times \quad 5 \\ \hline \end{array}$

Work as shown.

$$\begin{array}{r} {}^{1\,2} \\ 724 \\ \times \quad 5 \\ \hline 3620 \end{array}$$

$\leftarrow$ 5 × 4 = **20** ones; write 0 ones and carry 2 tens.

5 × 2 tens = **10** tens; add the 2 tens to get 12 tens; write 2 tens and carry 1 hundred.

5 × 7 hundreds = **35** hundreds; add the 1 hundred to get 36 hundreds.

WORK PROBLEM 4 AT THE SIDE.

OBJECTIVE 4 The product of two whole-number factors is also called a **multiple** of either factor. For example, since 4 • 2 = 8, the whole number 8 is a multiple of both 4 and 2. Multiples of 10 are very useful when multiplying. A *multiple of 10* is a whole number that ends in zero, such as 10, 20, or 30; 100, 200, or 300; 1000, 2000, or 3000. There is a short way to multiply by these multiples of 10. Look at the following examples.

$$26 \times 1 = 26$$
$$26 \times 10 = 260$$
$$26 \times 100 = 2600$$
$$26 \times 1000 = 26{,}000$$

Do you see a pattern in the multiplications using multiples of 10? These examples suggest the following rule.

Multiplying by Multiples of 10

Multiply a whole number by 10, 100, or 1000, by attaching one, two, or three zeros to the right of the whole number.

EXAMPLE 4 Using Multiples of 10 to Multiply

Multiply.

(a) 59 × 10 = 590
Attach 0.

(b) 74 × 100 = 7400
Attach 00.

(c) 803 × 1000 = 803,000 ← Attach 000.

WORK PROBLEM 5 AT THE SIDE.

You can also find the product of other multiples of ten by attaching zeros.

4. Multiply.

(a) $\begin{array}{r} 52 \\ \times \quad 5 \\ \hline \end{array}$

(b) $\begin{array}{r} 79 \\ \times \quad 0 \\ \hline \end{array}$

(c) $\begin{array}{r} 862 \\ \times \quad 9 \\ \hline \end{array}$

(d) $\begin{array}{r} 2831 \\ \times \quad 7 \\ \hline \end{array}$

(e) $\begin{array}{r} 4714 \\ \times \quad 8 \\ \hline \end{array}$

5. Multiply.

(a) 45 × 10

(b) 102 × 100

(c) 571 × 1000

ANSWERS

4. **(a)** 260 **(b)** 0 **(c)** 7758 **(d)** 19,817 **(e)** 37,712

5. **(a)** 450 **(b)** 10,200 **(c)** 571,000

6. Multiply.

(a) 14×50

(b) 68×400

(c)
```
  180
×  30
```

(d)
```
  6100
×   90
```

(e)
```
  800
× 200
```

7. Complete each multiplication.

(a)
```
   35
×  54
  140
 175
```

(b)
```
   76
×  49
  684
 304
```

ANSWERS

6. **(a)** 700 **(b)** 27,200 **(c)** 5400 **(d)** 549,000 **(e)** 160,000

7. **(a)** 1890 **(b)** 3724

EXAMPLE 5 Multiplying by Using Other Multiples of Ten

Multiply.

(a) 75×3000

Multiply 75 by 3 and attach 3 zeros.

```
  75
×  3
 225
```

$75 \times 3000 = 225{,}000$ ← Attach 000.

(b) 150×70

Multiply 15 by 7, and then attach 2 zeros.

```
  15
×  7
 105
```

$150 \times 70 = 10{,}500$ ← Attach 00.

◀◀ WORK PROBLEM 6 AT THE SIDE.

OBJECTIVE 5 The next example shows multiplication when both factors have more than one digit.

EXAMPLE 6 Multiplying with More Than One Digit

Multiply 46 and 23.

First multiply 46 by 3.

```
  1
  46
×  3
 138  ← 46 × 3 = 138
```

Now multiply 46 by 20.

```
  1
  46
× 20
 920  ← 46 × 20 = 920
```

Add the results.

```
   46
×  23
  138  ← 46 × 3
+ 920  ← 46 × 20
 1058
  ↑ Add.
```

Both 138 and 920 are called *partial products.* To save time, the zero in 920 is usually not written.

```
   46
×  23
  138
  92   ← 0 not written. Be very careful to
 1058     place the 2 in the tens column.
```

◀◀ WORK PROBLEM 7 AT THE SIDE.

EXAMPLE 7 Using Partial Products

Multiply.

(a)

```
      2 3 3
    × 1 3 2
    -------
      4 6 6
    6 9 9       (Tens lined up)
  2 3 3         (Hundreds lined up)
  ---------
  3 0,7 5 6  ←  Product
```

(b)

```
   538
×   46
------
```

First multiply by 6.

```
    24
   538
×   46
------
  3228     Carrying is needed here.
```

Now multiply by 4, being careful to line up the tens.

```
    1 3
    2 4
    5 3 8
  ×   4 6
  -------
  3 2 2 8 ┐
2 1 5 2   ┘ Finally, add the results.
---------
2 4,7 4 8
```

WORK PROBLEM 8 AT THE SIDE.

When zero appears in the multiplier, be sure to move the partial products to the left to account for the position held by the zero.

EXAMPLE 8 Multiplication with Zeros

Multiply.

(a)

```
      1 3 7
    × 3 0 6
    -------
      8 2 2
    0 0 0       (Tens lined up)
  4 1 1         (Hundreds lined up)
  ---------
  4 1,9 2 2
```

(b)

```
          1 4 0 6
        × 2 0 0 1
        ---------
          1 4 0 6
        0 0 0 0      ← (0 to line up tens)
      0 0 0 0        ← (0 to line up hundreds)
  2 8 1 2
  -------------
  2,8 1 3,4 0 6
```

```
        1 4 0 6
      × 2 0 0 1
      ---------
        1 4 0 6
  2 8 1 2 0 0    ← Zeros are written so this partial product starts in the thousands column.
  -------------
  2,8 1 3,4 0 6
```

Note

In Example 8(b) in the solution on the right, zeros were inserted so that thousands were placed in the thousands column.

8. Multiply.

(a) 38 × 15

(b) 31 × 43

(c) 67 × 59

(d) 234 × 73

(e) 835 × 189

ANSWERS

8. **(a)** 570 **(b)** 1333 **(c)** 3953 **(d)** 17,082 **(e)** 157,815

9. Multiply.

(a) $\begin{array}{r} 28 \\ \times\ 60 \\ \hline \end{array}$

(b) $\begin{array}{r} 817 \\ \times\ 30 \\ \hline \end{array}$

(c) $\begin{array}{r} 481 \\ \times\ 206 \\ \hline \end{array}$

(d) $\begin{array}{r} 3526 \\ \times\ 6002 \\ \hline \end{array}$

10. Find the total cost of the following items.

(a) 289 redwood planters at $12 per planter

(b) 180 cordless drills at $42 per drill

(c) 15 forklifts at $8218 per forklift

Answers

9. **(a)** 1680 **(b)** 24,510 **(c)** 99,086 **(d)** 21,163,052

10. **(a)** $3468 **(b)** $7560 **(c)** $123,270

WORK PROBLEM 9 AT THE SIDE.

OBJECTIVE 6 The next example shows how multiplication can be used to solve an application problem.

EXAMPLE 9 Applying Multiplication Skills

Find the total cost of 24 cordless telephones that cost $54 each.

Approach To find the cost of all the telephones, multiply the number of telephones (24) by the cost of one telephone ($54).

Solution Multiply 24 by 54.

$$\begin{array}{r} 24 \\ \times\ 54 \\ \hline 96 \\ 120 \\ \hline 1296 \end{array}$$

The total cost of the cordless telephones is $1296.

Calculator Tip: If you are using a calculator for Example 9 you will do this calculation

$$24 \boxed{\times} 54 \boxed{=} 1296.$$

WORK PROBLEM 10 AT THE SIDE.

Name Date Class

R.3 Exercises

Work each of the following chain multiplications. Try to do the work mentally.

1. $3 \times 1 \times 3$ **2.** $2 \times 8 \times 2$ **3.** $9 \times 1 \times 7$ **4.** $2 \times 4 \times 5$ **5.** $9 \cdot 5 \cdot 0$

6. $6 \cdot 0 \cdot 8$ **7.** $4 \cdot 1 \cdot 6$ **8.** $1 \cdot 5 \cdot 7$ **9.** (2)(3)(6) **10.** (4)(1)(9)

Multiply.

11. 35×7 **12.** 76×9 **13.** 28×6 **14.** 83×5 **15.** 3182×6

16. 7326×5 **17.** $36{,}921 \times 7$ **18.** $28{,}116 \times 4$ **19.** 125×30 **20.** 246×50

21. 1485×30 **22.** 8522×50 **23.** 900×300 **24.** 400×700 **25.** $43{,}000 \times 2\,000$

26. $11{,}000 \times 9\,000$ **27.** 68×22 **28.** 82×32 **29.** 83×45 **30.** (43)(27)

31. (32)(475) **32.** (67)(218) **33.** (729)(45) **34.** (681)(47) **35.** 538×342

36. 3228×751 **37.** 8162×198 **38.** 528×106 **39.** 6310×3078 **40.** 3533×5001

Writing Calculator Small Group

Solve the following application problems.

41. An encyclopedia has 30 volumes. Each volume has 800 pages. What is the total number of pages in the encyclopedia?

42. A hospital has 20 bottles of thyroid medication, with each bottle containing 2500 tablets. How many of these tablets does the hospital have in all?

43. There are 12 tomato plants to a flat. If there are 18 flats, find the total number of tomato plants.

44. A hummingbird's wings beat about 65 times per second. How many times do the hummingbird's wings beat in 30 seconds?

45. A new Saturn automobile gets 38 miles per gallon on the highway. How many miles can it go on 11 gallons of gas?

46. Find the total cost of 16 gallons of paint at $18 per gallon.

Use addition, subtraction, or multiplication, as needed, to solve each of the following.

47. Alison Leow counted 53 joggers one day and 122 joggers the next day. How many joggers did she count altogether during the two days?

48. The largest living land mammal is the African elephant, and the largest mammal of all time is the blue whale. If an African elephant weighs 15,225 pounds and a blue whale weighs 28 times that amount, find the weight of the blue whale.

49. A large meal contains 1406 calories, while a small meal contains 348 calories. How many more calories are in the large meal than the small one?

50. The distance from Reno, Nevada, to the Atlantic Ocean is 2695 miles, while the distance from Reno to the Pacific Ocean is 255 miles. How much farther is it to the Atlantic Ocean than it is to the Pacific Ocean?

51. Dannie Sanchez bought 4 tires at $110 each, 2 seat covers at $49 each, and 6 socket wrenches at $3 each. Find the total amount that he spent.

R.4 *Division of Whole Numbers*

Suppose \$12 is to be divided into 3 equal parts. Each part would be \$4, as shown here.

\$12 total

\$4	\$4	\$4

3 equal parts

OBJECTIVE 1 Just as $3 \cdot 4$, 3×4, and $(3)(4)$ are different ways of indicating the multiplication of 3 and 4, there are several ways to write 12 divided by 3.

Being divided ↓	Divided by ↓	Being divided ↓	Being divided ↓
$12 \div 3 = 4$	$\begin{array}{r}4\\3\overline{)12}\end{array}$	$\dfrac{12}{3} = 4$	$12/3 = 4$
↑ Divided by	↑ Being divided	↑ Divided by	↑ Divided by

We will use all three division symbols, $\div$, $\overline{)\ }$, and —. In algebra the bar, —, is frequently used. In computer science the slash, /, is used.

EXAMPLE 1 Using Division Symbols

Write each division by using two other symbols.

(a) $12 \div 4 = 3$

This division can also be written as

$$\begin{array}{r}3\\4\overline{)12}\end{array} \quad \text{or} \quad \frac{12}{4} = 3.$$

(b) $\dfrac{15}{5} = 3$

$$15 \div 5 = 3 \quad \text{or} \quad \begin{array}{r}3\\5\overline{)15}\end{array}$$

(c) $\begin{array}{r}4\\5\overline{)20}\end{array}$

$$20 \div 5 = 4 \quad \text{or} \quad \frac{20}{5} = 4$$

WORK PROBLEM 1 AT THE SIDE.

OBJECTIVE 2 In division, the number being divided is the **dividend** (DIV-uh-dend), the number divided by is the **divisor** (div-EYE-zer), and the answer is the **quotient** (KWOH-shunt).

$$\text{dividend} \div \text{divisor} = \text{quotient}$$

$$\begin{array}{r}\text{quotient}\\\text{divisor}\overline{)\text{dividend}}\end{array} \qquad \frac{\text{dividend}}{\text{divisor}} = \text{quotient}$$

EXAMPLE 2 Identifying the Parts in a Division Problem

Identify the dividend, divisor, and quotient.

(a) $35 \div 7 = 5$

$35 \div 7 = 5$ ← Quotient

Dividend ↗ (35) Divisor ↖ (7)

CONTINUED ON NEXT PAGE

OBJECTIVES

1. Write division problems in three ways.
2. Identify the parts of a division problem.
3. Divide zero by a number.
4. Recognize that a number cannot be divided by zero.
5. Divide a number by itself.
6. Use short division.
7. Check the answer to a division problem.
8. Use tests for divisibility.

FOR EXTRA HELP

Tutorial | Tape 17 | SSM, Sec. R.4

1. Write each division problem using two other symbols.

(a) $48 \div 6 = 8$

(b) $24 \div 6 = 4$

(c) $\begin{array}{r}4\\9\overline{)36}\end{array}$

(d) $\dfrac{42}{6} = 7$

ANSWERS

1. (a) $\begin{array}{r}8\\6\overline{)48}\end{array}$ and $\dfrac{48}{6} = 8$

(b) $\begin{array}{r}4\\6\overline{)24}\end{array}$ and $\dfrac{24}{6} = 4$

(c) $36 \div 9 = 4$ and $\dfrac{36}{9} = 4$

(d) $\begin{array}{r}7\\6\overline{)42}\end{array}$ and $42 \div 6 = 7$

2. Identify the dividend, divisor, and quotient.

(a) $10 \div 2 = 5$

(b) $30 \div 5 = 6$

(c) $\frac{28}{7} = 4$

(d) $2\overline{)36}$ with quotient 18

3. Divide.

(a) $0 \div 9$

(b) $\frac{0}{8}$

(c) $\frac{0}{36}$

(d) $57\overline{)0}$

ANSWERS

2. **(a)** dividend: 10; divisor: 2; quotient: 5
(b) dividend: 30; divisor: 5; quotient: 6
(c) dividend: 28; divisor: 7; quotient: 4
(d) dividend: 36; divisor: 2; quotient: 18

3. all 0

(b) $\frac{100}{20} = 5$

$$\frac{100}{20} = 5$$

Dividend: 100 (↓); Divisor: 20 (↑); ← Quotient: 5

(c) $12\overline{)72}$ with quotient 6

$$12\overline{)72} \text{ with quotient } 6$$

6 ← Quotient; 72 ← Dividend; Divisor: 12 (↑)

WORK PROBLEM 2 AT THE SIDE.

OBJECTIVE 3 If no money, or \$0, is divided equally among five people, each person gets \$0. There is a general rule for dividing zero.

Dividing Zero

Zero divided by any nonzero number is *zero.*

EXAMPLE 3 Dividing Zero by a Number

Divide.

(a) $0 \div 12 = 0$

(b) $0 \div 1728 = 0$

(c) $\frac{0}{375} = 0$

(d) $129\overline{)0}$ with quotient 0

WORK PROBLEM 3 AT THE SIDE.

Just as a subtraction such as $8 - 3 = 5$ can be written as the addition $8 = 3 + 5$, any division can be written as a multiplication. For example, $12 \div 3 = 4$ can be written as

$$3 \times 4 = 12 \quad \text{or, by the commutative property,} \quad 4 \times 3 = 12$$

EXAMPLE 4 Converting Division to Multiplication

Convert each division to a multiplication.

(a) $\frac{20}{4} = 5$ becomes $4 \cdot 5 = 20$

(b) $8\overline{)48}$ with quotient 6 becomes $8 \cdot 6 = 48$

(c) $72 \div 9 = 8$ becomes $9 \cdot 8 = 72$

WORK PROBLEM 4 AT THE SIDE.

OBJECTIVE 4 Division by zero cannot be done. To see why, try to find

$$9 \div 0.$$

As we have just seen, all division problems can be converted to a multiplication problem so that

$$\text{divisor} \cdot \text{quotient} = \text{dividend}.$$

If you convert the problem $9 \div 0 = ?$ to its multiplication counterpart, it reads

$$0 \cdot ? = 9.$$

You already know that zero times any number must always equal zero. Try any number you like to replace the "?" and you'll always get 0 instead of 9. Therefore, the division problem $9 \div 0$ cannot be done. Mathematicians say it is *undefined* and have agreed never to divide by zero. However, $0 \div 9$ *can* be done. Check by rewriting it as a multiplication problem.

$$0 \div 9 = 0 \quad \text{because} \quad 0 \cdot 9 = 0 \text{ is true.}$$

Dividing by Zero

Since dividing by zero cannot be done, we say that division by *zero* is undefined. It is impossible to compute an answer.

EXAMPLE 5 Dividing by Zero Is Undefined

All the following are undefined.

(a) $\frac{6}{0}$ is undefined

(b) $0\overline{)8}$ is undefined

(c) $18 \div 0$ is undefined

(d) $\frac{0}{0}$ is undefined

Division involving 0 is summarized below.

$$\frac{0}{\text{nonzero number}} = 0 \qquad \frac{\text{number}}{0} \text{ is undefined}$$

Note

When "0" is the divisor in a problem you write undefined. You can never divide by zero.

WORK PROBLEM 5 AT THE SIDE.

Calculator Tip: Try these two problems on your calculator. Jot down your answers.

9 [÷] 0 [=] ______ 0 [÷] 9 = ______

When you try to divide by zero, the calculator cannot do it, so it shows the word "Error" in the display, or the letter "E" (for "error").

4. Write each division problem as a multiplication problem.

(a) $6\overline{)18}$ with quotient 3

(b) $\frac{28}{4} = 7$

(c) $48 \div 8 = 6$

5. Work the following problems whenever possible.

(a) $\frac{8}{0}$

(b) $\frac{0}{8}$

(c) $0\overline{)32}$

(d) $32\overline{)0}$

(e) $100 \div 0$

(f) $0 \div 100$

ANSWERS

4. **(a)** $6 \cdot 3 = 18$ **(b)** $4 \cdot 7 = 28$ **(c)** $8 \cdot 6 = 48$

5. **(a)** undefined **(b)** 0 **(c)** undefined **(d)** 0 **(e)** undefined **(f)** 0

OBJECTIVE 5 What happens when a number is divided by itself? For example, $4 \div 4$ or $97 \div 97$?

Dividing a Number by Itself

Any nonzero number divided by itself is *one.*

EXAMPLE 6 Dividing a Nonzero Number by Itself

Divide.

(a) $16 \div 16 = 1$

(b) $32\overline{)32}$ = 1

(c) $\dfrac{57}{57} = 1$

WORK PROBLEM 6 AT THE SIDE.

OBJECTIVE 6 **Short division** is a method of dividing a number by a one-digit divisor.

EXAMPLE 7 Using Short Division

Divide: $3\overline{)96}$.

First, divide 9 by 3.

$$3\overline{)96} \to 3 \quad \leftarrow \frac{9}{3} = 3$$

Next, divide 6 by 3.

$$3\overline{)96} \to 32 \quad \leftarrow \frac{6}{3} = 2$$

WORK PROBLEM 7 AT THE SIDE.

When two numbers do not divide exactly, the leftover portion is called the **remainder.**

EXAMPLE 8 Using Short Division with a Remainder

Divide 147 by 4.

Write the problem.

$$4\overline{)147}$$

Because 1 cannot be divided by 4, divide 14 by 4.

$$4\overline{)14^{2}7} \to 3 \qquad \frac{14}{4} = 3 \text{ with 2 left over}$$

CONTINUED ON NEXT PAGE

6. Divide.

(a) $5 \div 5$

(b) $14\overline{)14}$

(c) $\dfrac{37}{37}$

7. Divide.

(a) $2\overline{)18}$

(b) $3\overline{)39}$

(c) $4\overline{)88}$

(d) $2\overline{)462}$

ANSWERS

6. all 1

7. **(a)** 9 **(b)** 13 **(c)** 22 **(d)** 231

Next, divide 27 by 4. The final number left over is the remainder. Write the remainder to the side. "R" stands for remainder.

$$\begin{array}{r} 3\ 6\ \mathbf{R}3 \\ 4\overline{)14^{2}7} \end{array} \qquad \frac{27}{4} = 6 \text{ with 3 left over}$$

WORK PROBLEM 8 AT THE SIDE.

EXAMPLE 9 Dividing with a Remainder

Divide 1809 by 7.

Divide 7 into 18.

$$\begin{array}{r} 2 \\ 7\overline{)18^{4}09} \end{array} \qquad \frac{18}{7} = 2 \text{ with 4 left over}$$

Divide 7 into 40.

$$\begin{array}{r} 2\ 5 \\ 7\overline{)18^{4}0^{5}9} \end{array} \qquad \frac{40}{7} = 5 \text{ with 5 left over}$$

Divide 7 into 59.

$$\begin{array}{r} 2\ 5\ 8\ \mathbf{R}3 \\ 7\overline{)18^{4}0^{5}9} \end{array} \qquad \frac{59}{7} = 8 \text{ with 3 left over}$$

WORK PROBLEM 9 AT THE SIDE.

OBJECTIVE 7 **Check** the answer to a division problem as follows.

Checking Division

(divisor × quotient) + remainder = dividend

Parentheses tell you what to do first. In this case, multiply the divisor by the quotient first and then add the remainder.

EXAMPLE 10 Checking Division by Using Multiplication

Check each answer.

(a) $\begin{array}{r} 91\ \mathbf{R}3 \\ 5\overline{)458} \end{array}$

(divisor × quotient) + remainder = dividend

$$(5 \times 91) + 3$$

$$455 + 3 = 458$$

Matches original dividend so the division was done correctly.

CONTINUED ON NEXT PAGE

8. Divide.

(a) $2\overline{)225}$

(b) $3\overline{)275}$

(c) $4\overline{)538}$

(d) $\dfrac{819}{5}$

9. Divide.

(a) $5\overline{)937}$

(b) $\dfrac{675}{7}$

(c) $3\overline{)1885}$

(d) $8\overline{)1135}$

ANSWERS

8. **(a)** 112 R1 **(b)** 91 R2 **(c)** 134 R2 **(d)** 163 R4

9. **(a)** 187 R2 **(b)** 96 R3 **(c)** 628 R1 **(d)** 141 R7

10. Check each division. If an answer is incorrect, give the correct answer.

(a) $3\overline{)115}$ with quotient 38 R1

(b) $8\overline{)739}$ with quotient 92 R2

(c) $4\overline{)1312}$ with quotient 328

(d) $5\overline{)2033}$ with quotient 46 R3

Answers

10. **(a)** correct
(b) incorrect; should be 92 R3
(c) correct
(d) incorrect; should be 406 R3.

(b) $6\overline{)1437}$ with quotient 239 **R**4

(divisor × quotient) + remainder = dividend

$$(6 \times 239) + 4$$

$$1434 + 4 = 1438$$

Does not match original dividend

The answer does not check. Rework the original problem to get the correct answer, 239 **R**3.

Note
A common error when checking division is forgetting to add the remainder. Be sure to add any remainder when checking a division problem.

◀◀ WORK PROBLEM 10 AT THE SIDE.

Objective 8 It is often important to know whether a number is divisible by another number. You will find this useful in Chapter 4 when writing fractions in lowest terms.

Divisibility

One whole number is *divisible* by another if the remainder is zero.

Decide whether one number is exactly divisible by another by using the following tests for divisibility.

Tests for Divisibility

A number is divisible by

2 if it ends in 0, 2, 4, 6, or 8.
3 if the sum of its digits is divisible by 3.
4 if the last two digits make a number that is divisible by 4.
5 if it ends in 0 or 5.
6 if it is divisible by both 2 and 3.
8 if the last three digits make a number that is divisible by 8.
9 if the sum of its digits is divisible by 9.
10 if it ends in 0.

The most commonly used tests are those for 2, 3, 5, and 10.

Divisibility by 2

A number is divisible by **2** if the number ends in 0, 2, 4, 6, or 8.

EXAMPLE 11 Testing for Divisibility by 2

Are the following numbers divisible by 2?

(a) 986
↑ Ends in 6

Because the number ends in 6, which is in the list on the previous page, the number 986 is divisible by 2.

(b) 3255 is not divisible by 2.
↑ Ends in 5, and not in 0, 2, 4, 6, or 8

WORK PROBLEM 11 AT THE SIDE.

Divisibility by 3

A number is divisible by **3** if the sum of its digits is divisible by **3**.

EXAMPLE 12 Testing for Divisibility by 3

Are the following numbers divisible by 3?

(a) 4251

Add the digits.

$$\mathbf{4} + \mathbf{2} + \mathbf{5} + \mathbf{1} = 12$$

Because 12 is divisible by 3, the number 4251 is divisible by 3.

(b) 29,806

Add the digits.

$$\mathbf{2} + \mathbf{9} + \mathbf{8} + \mathbf{0} + \mathbf{6} = 25$$

Because 25 is not divisible by 3, the number 29,806 is not divisible by 3.

Note

Be careful when testing for divisibility by adding the digits. This method works only for the numbers 3 and 9.

WORK PROBLEM 12 AT THE SIDE.

11. Which are divisible by 2?

(a) 612

(b) 315

(c) 2714

(d) 36,000

12. Which are divisible by 3?

(a) 836

(b) 7545

(c) 242,913

(d) 102,484

ANSWERS

11. all but (b)

12. (b) and (c)

Divisibility by 5 and by 10

A number is divisible by **5** if it ends in 0 or 5.
A number is divisible by **10** if it ends in 0.

EXAMPLE 13 Determining Divisibility by 5

Are the following numbers divisible by 5?

(a) 12,900 ends in 0 and is divisible by 5.

(b) 4325 ends in 5 and is divisible by 5.

(c) 392 ends in 2 and is not divisible by 5.

WORK PROBLEM 13 AT THE SIDE.

EXAMPLE 14 Determining Divisibility by 10

Are the following numbers divisible by 10?

(a) 700 and 9140 end in 0 and are divisible by 10.

(b) 355 and 18,743 do not end in 0 and are not divisible by 10.

WORK PROBLEM 14 AT THE SIDE.

13. Which are divisible by 5?

(a) 160

(b) 635

(c) 3381

(d) 108,605

14. Which are divisible by 10?

(a) 290

(b) 218

(c) 2020

(d) 11,670

Answers

13. all but (c)

14. all but (b)

R.4 Exercises

Divide.

1. $\frac{12}{12}$ **2.** $\frac{9}{0}$ **3.** $24 \div 0$ **4.** $4 \div 4$ **5.** $\frac{0}{4}$

6. $0 \div 8$ **7.** $0 \div 12$ **8.** $\frac{0}{7}$ **9.** $0\overline{)21}$ **10.** $2 \div 0$

Divide by using short division.

11. $4\overline{)108}$ **12.** $5\overline{)135}$ **13.** $9\overline{)324}$ **14.** $8\overline{)176}$

15. $6\overline{)9137}$ **16.** $9\overline{)8371}$ **17.** $6\overline{)1854}$ **18.** $8\overline{)856}$

19. $4024 \div 4$ **20.** $16{,}024 \div 8$ **21.** $15{,}018 \div 3$ **22.** $32{,}008 \div 8$

23. $\frac{26{,}684}{4}$ **24.** $\frac{16{,}398}{9}$ **25.** $\frac{74{,}751}{6}$ **26.** $\frac{72{,}543}{5}$

27. $\frac{71{,}776}{7}$ **28.** $\frac{77{,}621}{3}$ **29.** $\frac{128{,}645}{7}$ **30.** $\frac{172{,}255}{4}$

Check each answer. If an answer is incorrect, find the correct answer.

31. $7\overline{)4692}$ = 67 **R**2 **32.** $9\overline{)5974}$ = 663 **R**5 **33.** $6\overline{)21{,}409}$ = 3,568 **R**2 **34.** $4\overline{)103{,}516}$ = 25,879

35. $6\overline{)18{,}023}$ = 3,003 **R**5 **36.** $8\overline{)33{,}664}$ = 4,208 **37.** $6\overline{)69{,}140}$ = 11,523 **R**2 **38.** $3\overline{)82{,}598}$ = 27,532 **R**1

Solve each application problem.

39. A carton of antifreeze holds 4 one-gallon jugs. Find the number of cartons needed to package 624 one-gallon jugs.

40. Eight people invested a total of $244,224 to buy a condominium. Each person invested the same amount of money. How much did each person invest?

41. If 6 identical service vans cost a total of $99,600, find the cost of each service van.

42. One gallon of beverage will serve 9 people. How many gallons are needed for 3483 people?

43. An estate of $127,400 is divided equally among 7 family members. Find the amount received by each family member.

44. How many 5-pound bags of rice can be filled from 8750 pounds of rice?

45. If 36 gallons of fertilizer are needed for each acre of land, find the number of acres that can be fertilized with 7380 gallons of fertilizer.

46. A roofing contractor has purchased 2268 squares (10 feet by 10 feet) of roofing material. If each home needs 21 squares of material, find the number of homes that can be roofed.

47. The Super Lotto payout of $8,100,000 will be divided equally by 36 people who purchased the winning ticket. Find the amount received by each person.

48. Oprah Winfrey reportedly earned $171,000,000 over the past two years. Find her monthly income for that time period.

49. Kaci Salmon, a supervisor at Albany Electric, earns $36,540 per year. Find the amount of her earnings in a three-month period.

50. A worker assembles 168 light diffusers in an 8-hour shift. Find the number assembled in 3 hours.

Put a ✓ mark in the blank if the number at the left is divisible by the number at the top of each column. Put an X in the blank if the number is not divisible by the number at the top.

	2	3	5	10		2	3	5	10
51. 30	____	____	____	____	**52.** 25	____	____	____	____
53. 184	____	____	____	____	**54.** 192	____	____	____	____
55. 445	____	____	____	____	**56.** 897	____	____	____	____
57. 903	____	____	____	____	**58.** 500	____	____	____	____
59. 5166	____	____	____	____	**60.** 8302	____	____	____	____
61. 21,763	____	____	____	____	**62.** 32,472	____	____	____	____

R.5 Long Division

Long division is used to divide by a number with more than one digit.

OBJECTIVE 1 In long division, estimate the various numbers by using a *trial divisor,* which is used to get a *trial quotient.*

EXAMPLE 1 Using a Trial Divisor and a Trial Quotient

Divide: $42\overline{)3066}$.

Because 42 is closer to 40 than to 50, use the first digit of the divisor as a trial divisor.

42
↑ Trial divisor

Try to divide the first digit of the dividend by 4. Because 3 cannot be divided by 4, use the first *two* digits, 30.

$$\frac{30}{4} = 7 \text{ with remainder } 2$$

$$\begin{array}{r} 7 \\ 42\overline{)3066} \end{array} \leftarrow \text{Trial quotient}$$

7 goes over the 6, because $\frac{306}{42}$ is about 7.

Multiply 7 and 42 to get 294; next, subtract 294 from 306.

$$\begin{array}{r} 7 \\ 42\overline{)3066} \\ \underline{294} \\ 12 \end{array} \quad \begin{array}{l} \\ \\ \leftarrow 7 \times 42 \\ \leftarrow 306 - 294 \end{array}$$

Bring down the 6 at the right.

$$\begin{array}{r} 7 \\ 42\overline{)3066} \\ \underline{294}\downarrow \\ 126 \end{array} \quad \begin{array}{l} \\ \\ \\ \leftarrow 6 \text{ brought down} \end{array}$$

Use the trial divisor, 4.

First two digits of 126 → $\frac{12}{4} = 3$

$$\begin{array}{r} 73 \\ 42\overline{)3066} \\ \underline{294} \\ 126 \\ \underline{126} \\ 0 \end{array} \quad \begin{array}{l} \\ \\ \\ \\ \leftarrow 3 \times 42 = 126 \\ \\ \end{array}$$

Check the answer by multiplying 42 and 73. The product should be 3066.

Note
The first digit on the left of the answer in long division must be placed in the proper position over the dividend.

WORK PROBLEM 1 AT THE SIDE. ▶▶

OBJECTIVES

1. Do long division.
2. Divide numbers ending in zero by numbers ending in zero.
3. Check answers.

FOR EXTRA HELP

Tutorial | Tape 17 | SSM, Sec. R.5

1. Divide.

(a) $25\overline{)1775}$

(b) $26\overline{)2132}$

(c) $51\overline{)2295}$

(d) $\frac{6552}{84}$

ANSWERS

1. **(a)** 71 **(b)** 82 **(c)** 45 **(d)** 78

2. Divide.

(a) $56\overline{)2352}$

(b) $38\overline{)1599}$

(c) $65\overline{)5416}$

(d) $89\overline{)6649}$

Answers

2. **(a)** 42 **(b)** 42 R3 **(c)** 83 R21 **(d)** 74 R63

EXAMPLE 2 Dividing to Find a Trial Quotient

Divide: $58\overline{)2730}$.

Use 6 as a trial divisor, since 58 is closer to 60 than to 50.

$$\frac{27}{6} = 4 \text{ with 3 left over}$$

(27: First two digits of dividend)

```
     4      ← Trial quotient
58)2730
   232      ← 4 × 58 = 232
    41      ← 273 − 232 = 41 (smaller than 58,
              the divisor)
```

Bring down the 0.

```
     4
58)2730
   232↓
    410     ← 0 brought down
```

$$\frac{41}{6} = 6 \text{ with 5 left over}$$

(41: First two digits of 410)

```
    46      ← Trial quotient
58)2730
   232
    410
    348     ← 6 × 58 = 348
     62     ← Greater than 58
```

The remainder, 62, is greater than the divisor, 58, so 7 should be used instead of 6.

```
    47 R4
58)2730
   232
    410
    406     ← 7 × 58 = 406
      4     ← 410 − 406
```

◀◀ **WORK PROBLEM 2 AT THE SIDE.**

Sometimes it is necessary to insert a zero in the quotient.

EXAMPLE 3 Inserting Zeros in the Quotient

Divide: $42\overline{)8734}$.

Start as above.

```
     2
42)8734
   84       ← 2 × 42 = 84
    3       ← 87 − 84 = 3
```

Bring down the 3.

```
     2
42)8734
   84↓
    33      ← 3 brought down
```

CONTINUED ON NEXT PAGE

Since 33 cannot be divided by 42, place a 0 in the quotient as a placeholder.

```
      20    ← 0 in quotient
42)8734
   84
    33
```

Bring down the final digit, the 4.

```
      20
42)8734
   84 ↓
    334   ← 4 brought down
```

Complete the problem.

```
     207 R40
42)8734
   84
    334
    294
     40
```

The answer is 207 **R**40.

Note

There must be a digit in the quotient (answer) above every digit in the dividend once the answer has begun. Notice that in Example 3 a zero was used to assure an answer digit above every digit in the dividend.

WORK PROBLEM 3 AT THE SIDE.

OBJECTIVE 2 When the divisor and dividend both contain zeros at the far right, recall that these numbers are multiples of 10. There is a short way to divide these multiples of 10. Look at the following examples.

$$26{,}000 \div 1 = 26{,}000$$

$$26{,}000 \div 10 = 2600$$

$$26{,}000 \div 100 = 260$$

$$26{,}000 \div 1000 = 26$$

Do you see a pattern in these divisions using multiples of 10? These examples suggest the following rule.

Dividing a Whole Number by 10, 100, or 1000

Divide a whole number by 10, 100, or 1000 by dropping the appropriate number of zeros from the whole number.

EXAMPLE 4 Dividing by Multiples of 10

Divide.

(a) 60 ÷ 1**0** = 6 (1 zero in divisor; 0 dropped)

(b) 3500 ÷ 1**00** = 35 (2 zeros in divisor; 00 dropped)

CONTINUED ON NEXT PAGE

3. Divide.

(a) 24)3127

(b) 52)10,660

(c) 39)15,933

(d) 78)23,462

ANSWERS

3. (a) 130 R7 **(b)** 205 **(c)** 408 R21 **(d)** 300 R62

(c) 915,000 ÷ 1000 = 915

3 zeros in divisor

000 dropped

◀◀ WORK PROBLEM 4 AT THE SIDE.

Find the quotient for other multiples of 10 by dropping zeros.

EXAMPLE 5 Dividing by Multiples of 10

Divide.

(a) $40\overline{)11{,}000}$ Drop 1 zero from the divisor and the dividend.

```
     275
  4)1100
    8
    30
    28
     20
     20
      0
```

(b) $3500\overline{)31{,}500}$ 2 zeros dropped from the divisor and the dividend

```
      9
  35)315
     315
       0
```

Note

Dropping zeros when dividing by multiples of 10 does not change the answer (quotient).

◀◀ WORK PROBLEM 5 AT THE SIDE.

OBJECTIVE 3 Answers in long division can be checked just as answers in short division were checked.

EXAMPLE 6 Checking Division

Check the answer.

```
      114 R43
  48)5324

        114
     ×   48   ← Multiply the quotient and the divisor.
        912
       456
       5472
     +   43   ← Add the remainder.
       5515   ← Result does not match dividend.
```

The answer does *not* check. Rework the original problem to get 110 **R**44.

◀◀ WORK PROBLEM 6 AT THE SIDE.

4. Divide.

(a) 50 ÷ 10

(b) 1800 ÷ 100

(c) 305,000 ÷ 1000

5. Divide.

(a) $60\overline{)7200}$

(b) $130\overline{)131{,}040}$

(c) $2600\overline{)195{,}000}$

6. Decide whether the following divisions are correct. If the answer is incorrect, find the correct answer.

(a)
```
      43
  18)774
     72
      54
      54
       0
```

(b)
```
          42 R178
  426)19,170
      17 04
       1 130
         952
         178
```

ANSWERS

4. (a) 5 (b) 18 (c) 305

5. (a) 120 (b) 1008 (c) 75

6. (a) correct (b) incorrect; should be 45

Name _______________ Date _______ Class _______

R.5 Exercises

Without doing the actual division, circle the correct answer from the three choices given.

Example: $42\overline{)7560}$ 18 180 1800 **Solution:** $42\overline{)7560}$ (1 over the 5) 18 (180) 1800

1 goes over the 5, because $\frac{75}{42}$ is about 1, so the answer must be a three-digit number.

1. $24\overline{)768}$

3 32 320

2. $35\overline{)805}$

2 23 230

3. $18\overline{)4500}$

2 25 250

4. $28\overline{)3500}$

12 125 1250

5. $86\overline{)10{,}327}$

12 120 R7 1200

6. $46\overline{)24{,}026}$

5 52 522 R14

7. $52\overline{)68{,}025}$

13 130 R1 1308 R9

8. $12\overline{)116{,}953}$

974 R2 9746 R1 97,460

9. $21\overline{)149{,}826}$

71 713 7134 R12

10. $32\overline{)247{,}892}$

77 R1 7746 R20 77,460

11. $523\overline{)470{,}800}$

9 R100 90 R100 900 R100

12. $230\overline{)253{,}230}$

11 110 1101

Divide by using long division.

13. $42\overline{)8699}$

14. $58\overline{)2204}$

15. $47\overline{)11{,}121}$

16. $83\overline{)39{,}692}$

17. $26\overline{)62{,}583}$

18. $28\overline{)84{,}249}$

19. $63\overline{)78{,}072}$

20. $238\overline{)186{,}948}$

21. $153\overline{)509{,}725}$

22. $402\overline{)29{,}346}$

23. $420\overline{)357{,}000}$

24. $900\overline{)153{,}000}$

Check each answer. If an answer is incorrect, give the correct answer.

25. $56\overline{)5943}$ answer: 106 R17

26. $87\overline{)3254}$ answer: 37 R37

27. $28\overline{)18{,}424}$ answer: 658 R9

28. $191\overline{)88{,}604}$ answer: 463 R171

29. $614\overline{)38{,}068}$ answer: 62 R3

30. $557\overline{)97{,}286}$ answer: 174 R368

31. $72\overline{)32{,}465}$ answer: 450 R65

32. $47\overline{)9570}$ answer: 23 R29

Solve each application problem by using addition, subtraction, multiplication, or division.

33. A car travels 1350 miles at 54 miles per hour. How many hours did it travel?

34. The U.S. Government Printing Office uses 255,000 pounds of ink each year. If it does an equal amount of printing on each of 200 work days in a year, find the weight of the ink used each day.

35. There were 1838 medals made for the 1996 Olympic Games. The ancient Olympic stadium is shown on one side, the pictogram of the sport on the other side. Of the medals made, 604 were gold, 604 were silver and the remainder were bronze. Find the number of bronze medals.

36. Two divorced parents each share some of the education costs of their child which amount to $3718. If one parent pays $1880, find the amount paid by the other parent.

37. Judy Martinez owes $3888 on a loan. Find her monthly payment if the loan is to be paid off in 36 months.

38. A consultant charged $13,050 for studying a school's compliance with the Americans with Disabilities Act. If the consultant worked 225 hours, find the rate charged per hour.

39. Clarence Hanks can assemble 42 circuits in 1 hour. How many circuits can he assemble in a 5-day workweek of 8 hours per day?

40. There are two conveyer lines in a factory each of which packages 240 sacks of salt per hour. If the lines operate for 8 hours, find the total number of sacks of salt packaged by the two lines.

41. A youth soccer association raised $7588 in fund-raising projects. There were expenses of $838 that had to be paid first, with the balance of the money divided evenly among the 18 teams. How much did each team receive?

42. Feather Farms Egg Ranch collects 3545 eggs in the morning and 2575 eggs in the afternoon. If the eggs are packed in flats containing 30 eggs each, find the number of flats needed for packing.

KEY TERMS

R.1	**addends**	Addends are the numbers being added in an addition problem.
	sum (total)	The answer in an addition problem is the sum (total).
	commutative property of addition	The commutative property of addition states that changing the *order* of two addends in an addition problem does not change the sum.
	associative property of addition	The associative property of addition states that changing the *grouping* of addends does not change the sum.
	carrying	The process of carrying is used in an addition problem when the sum of the digits in a column is greater than 9.
R.2	**minuend**	In the subtraction problem $8 - 5 = 3$, the 8 is the minuend. It is the number from which another number is subtracted.
	subtrahend	In the subtraction problem $8 - 5 = 3$, the 5 is the subtrahend. It is the number being subtracted.
	difference	The answer in a subtraction problem is the difference.
	borrowing	The method of borrowing is used in subtraction if a digit is less than the one directly below it.
R.3	**factors**	The numbers being multiplied are factors. For example, in $3 \times 4 = 12$, both 3 and 4 are factors.
	product	The answer in a multiplication problem is the product.
	commutative property of multiplication	The commutative property of multiplication states that changing the *order* of two factors in a multiplication problem does not change the product.
	associative property of multiplication	The associative property of multiplication states that changing the *grouping* of factors does not change the product.
	multiple	The product of two whole-number factors is a multiple of those numbers.
R.4	**dividend**	In division, the number being divided is the dividend.
	divisor	In division, the number being used to divide another number is the divisor.
	quotient	The answer in a division problem is the quotient.
	short division	A method of dividing a number by a one-digit divisor.
	remainder	The remainder is the number left over when two numbers do not divide evenly.
R.5	**long division**	The process of long division is used to divide by a number with more than one digit.

QUICK REVIEW

Concepts	*Examples*
R.1 Addition of Whole Numbers Add from top to bottom, starting with the ones column and working left. To check, add from bottom to top.	$\begin{array}{r} \underline{1}\,\underline{1}\,\underline{4}\,\underline{0} \\ 687 \\ 26 \\ 9 \\ +\ 418 \\ \hline 1140 \end{array}$ 687, 26, 9, 418: Addends; 1140: Sum (Add up to check)

Concepts	*Examples*
R.1 Commutative Property of Addition Changing the *order* of two addends in an addition problem does not change the sum.	$2 + 4 = 6$ $4 + 2 = 6$ By the commutative property, the sum is the same.
R.1 Associative Property of Addition Changing the *grouping* of addends does not change the sum.	$(2 + 3) + 4 = 9$ $2 + (3 + 4) = 9$ By the associative property, the sum is the same.
R.2 Subtraction of Whole Numbers Subtract the subtrahend from minuend to get the difference by borrowing when necessary. To check, add the difference to the subtrahend to get the minuend.	Problem 6 12 18 4 7 3 8 ← Minuend − 6 4 9 Subtrahend 4 0 8 9 Difference Check 4 0 8 9 + 6 4 9 4 7 3 8
R.3 Multiplication of Whole Numbers The numbers being multiplied are called *factors.* The multiplicand is being multiplied by the multiplier, giving the product. When the multiplier has more than one digit, partial products must be used and added to find the product.	78 Multiplicand } Factors × 24 Multiplier } Factors 312 Partial product 156 Partial product (one position left) 1872 Product
R.3 Commutative Property of Multiplication Changing the *order* of two factors in a multiplication problem does not change the product.	$3 \times 4 = 12$ $4 \times 3 = 12$ By the commutative property, the product is the same.
R.3 Associative Property of Multiplication Changing the *grouping* of factors does not change the product.	$(2 \times 3) \times 4 = 24$ $2 \times (3 \times 4) = 24$ By the associative property, the product is the same.
R.4 Division of Whole Numbers $\div$ and $\overline{)}$ mean divide. Also a —, as in $\frac{25}{5}$, means to divide the top number (dividend) by the bottom number (divisor).	22 ← Quotient Divisor → $4\overline{)88}$ ← Dividend 88 0 $88 \div 4 = 22$ ↑ Dividend ↑ Divisor ↑ Quotient Dividend → 88, Divisor → 4: $\frac{88}{4} = 22$ ← Quotient

If you need help with any of these review exercises, look in the section indicated in brackets.

[R.1] *Add.*

1. $74 + 18$

2. $35 + 78$

3. $807 + 4606 + 51$

4. $8215 + 9 + 7433$

[R.2] *Subtract.*

5. $238 - 199$

6. $573 - 389$

7. $2210 - 1986$

8. $99{,}704 - 73{,}838$

[R.3] *Multiply. Do the work mentally.*

9. $2 \times 4 \times 6$

10. $9 \times 1 \times 5$

11. $6 \cdot 1 \cdot 8$

12. $7 \cdot 7 \cdot 0$

Multiply.

13. 43×4

14. 781×7

15. 5440×6

16. $93{,}105 \times 5$

Multiply by using multiples of ten.

17. 320×60

18. 280×90

19. 517×400

20. $16{,}000 \times 8{,}000$

Multiply.

21. 34×18

22. 52×36

23. 655×21

24. 392×77

[R.4] *Divide.*

25. $42 \div 7$

26. $18 \div 18$

27. $\frac{125}{0}$

28. $\frac{0}{35}$

[R.4–R.5] *Divide.*

29. $4\overline{)432}$

30. $9\overline{)216}$

31. $76\overline{)26{,}752}$

32. $2704 \div 18$

Mixed Review Exercises

Solve each application problem.

33. There are 52 cards in a deck. How many cards are there in 9 decks?

34. Your college bookstore receives textbooks packed 12 books per carton. How many textbooks are received in a delivery of 238 cartons?

35. "Push type" gasoline-powered lawn mowers cost $100 less than self-propelled mowers that you walk behind. If a self-propelled mower costs $380, find the cost of a "push-type" mower.

36. The Village School wants to raise $115,280 for a new library. If $87,340 has already been raised, how much more must be raised to reach the goal?

37. It takes 2000 hours of work to build 1 home. How many hours of work are needed to build 12 homes?

38. A Japanese bullet train travels 80 miles in 1 hour. Find the number of miles traveled in 5 hours.

39. If an acre needs 250 pounds of fertilizer, how many acres can be fertilized with 5750 pounds of fertilizer?

40. Each home in a subdivision requires 180 feet of fencing. Find the number of homes that can be fenced with 5760 feet of fencing material.

41. Susan Hessney had $382 in her checking account. She wrote a check for $135. How much does she have left in her account?

42. Find the total cost of 4 T-shirts at $14 each and 3 sweatshirts at $29 each.

43. The Houston Space Center charges $15 for each adult admission and $12 for each child. Find the total cost to admit a group of 18 adults and 26 children.

44. A newspaper carrier has 56 customers who take the paper daily and 23 customers who take the paper on weekends only. A daily customer pays $18 per month and a weekend-only customer pays $11 per month. Find the total monthly collections.

Name ______ Date ______ Class ______

CHAPTER R TEST

Fill in the blanks to complete each sentence.

1. In an addition problem, the numbers being added are called ______ and the answer is called the ______.

2. In a multiplication problem, the numbers being multiplied are called ______ and the answer is called the ______.

3. In a subtraction problem the answer is called the ______ and in a division problem the answer is called the ______.

Add, subtract, multiply, or divide, as indicated.

4. $984 + 65 + 7561$

5.
$$\begin{array}{r} 17{,}063 \\ 7 \\ 12 \\ 1\,505 \\ 93{,}710 \\ +\quad 333 \\ \hline \end{array}$$

6. $17{,}002 - 54$

7.
$$\begin{array}{r} 5062 \\ -\ 1978 \\ \hline \end{array}$$

8. $5 \times 7 \times 4$

9. $57 \cdot 3000$

10. $(85)(21)$

11.
$$\begin{array}{r} 7381 \\ \times\ \ 603 \\ \hline \end{array}$$

12. $6\overline{)1236}$

13. $\dfrac{791}{0}$

14. $38{,}472 \div 84$

15. $280\overline{)44{,}800}$

1. ______
2. ______
3. ______
4. ______
5. ______
6. ______
7. ______
8. ______
9. ______
10. ______
11. ______
12. ______
13. ______
14. ______
15. ______

Solve each application problem.

16. ______________

16. Find the cost of 48 shovels at $11 per shovel.

17. ______________

17. In a recent test of side-bagging lawn mowers, the most expensive model sold for $350 and the least expensive model sold for $179. Find the difference in price between the most expensive and the least expensive mowers tested.

18. ______________

18. A stamping machine produces 936 license plates each hour. How long will it take to produce 30,888 license plates?

19. ______________

19. Kenée Shadbourne paid $690 for tuition, $185 for books, and $68 for supplies. If this money was withdrawn from her checking account, which had a balance of $1108, find her new balance.

20. ______________

20. An appliance manufacturer assembles 118 self-cleaning ovens each hour for 4 hours and 139 standard ovens each hour for the next 4 hours. Find the total number of ovens assembled in the 8-hour period.

21. ______________

21. The monthly rents collected from the four units in an apartment building are $485, $500, $515, and $425. After expenses of $785 are paid, find the amount that remains.

22. ______________

22. Describe the divisibility tests for 2, 5, and 10. Also give two examples for each test: one number that is divisible and one number that is not divisible.

Appendix A

Scientific Calculators

Calculators are among the more popular inventions of the last three decades. There are many types, from the inexpensive basic calculator to the more complex *financial* and *graphing* calculators. The discussion here is confined to the common scientific calculator with keys for percent, exponents, square root, memory, fractions, and parentheses.

Note
For an explanation of specific calculator models or special function keys, refer to the booklet supplied with your calculator.

OBJECTIVES

1. Learn the basic calculator keys.
2. Understand the [C], [CE], and [ON/C] or [ON/AC] keys.
3. Understand the floating decimal point.
4. Use the [%] key.
5. Use the [x^2] and the [x^3] keys.
6. Use the [y^x] and [$\sqrt{x}$] keys.
7. Use the [$a^b/_c$] key.
8. Solve problems with negative numbers.
9. Use the calculator memory function.
10. Solve chain calculations using the order of operations.
11. Use the parentheses keys.

OBJECTIVE 1 Most calculators use *algebraic logic.* For example, enter 14 + 28 in the order it is written.

14 [+] 28 [=] Answer is 42.

Enter 387 − 62 as follows.

387 [−] 62 [=] Answer is 325.

If your calculator does not work problems in this way, check its instruction book to see how to proceed.

OBJECTIVE 2 All calculators have a [C], [ON/C], or [ON/AC] key. Pressing this key erases everything in the calculator and prepares the calculator to begin a new problem. Some calculators also have a [CE] key. Pressing this key erases *only* the number displayed and allows you to correct a mistake without having to start the problem over.

Many calculators combine the [C] key and the [CE] key and use an [ON/C] key. This key turns the calculator on and is also used to erase the calculator display. If the [ON/C] key is pressed after the [=] or one of the operation keys ([+], [−], [×], [÷]), everything in the calculator is erased. If you happen to press the wrong operation key, immediately press the correct key to cancel the error. For example, 7 [+] [−] 3 [=] 4. Pressing the [−] key cancels the previous [+] key entry.

OBJECTIVE 3 Most calculators have a *floating decimal* that locates the decimal point in the final result. For example, to buy 55.75 square yards of carpet at $18.99 per square yard, proceed as follows.

55.75 [×] 18.99 [=] Answer is 1058.6925.

The decimal point is automatically placed in the answer. You should *round* money answers to the nearest cent. Draw a cutoff line after the hundredths place.

Look only at the first digit being cut off.

1058.69 | 25

Cent position (hundredths)

Because the first digit being cut off is less than 5, the part you are keeping remains the same. The answer is rounded to \$1058.69. If the first digit being cut off had been 5 or greater, you would have rounded up by adding 1 to the cent position (see **Section 5.2**).

When using a calculator with a floating decimal, enter the decimal point as needed. For example, enter \$47 by pressing 47 with no decimal point, but enter 95¢ as [•] 95 with a decimal point.

When you add \$21.38 and \$1.22, the answer is \$22.60, but the calculator does *not* show the final 0.

21.38 [+] 1.22 [=]   Answer is 22.6.

You must remember that the problem dealt with money and write the final 0, making the answer \$22.60.

OBJECTIVE 4 The [%] key moves the decimal point two places to the left when pressed following multiplication or division. Calculate 8% of \$4205 as follows.

4205 [×] 8 [%] [=]   Answer is 336.4.

Because the problem involved money, write the answer as \$336.40.

OBJECTIVE 5 The squaring key, [x^2], squares the number in the display (multiplies the number by itself). For example, find 7^2 (which means 7×7) as follows.

7 [x^2]   Answer is 49.

The cubing key, [x^3], finds the cube of a number (the number is multiplied by itself three times). To find the cube of 6 (that is, $6 \times 6 \times 6$), follow these keystrokes.

6 [x^3]   Answer is 216.

OBJECTIVE 6 The [y^x] key raises a base to any desired *power*. In 3^5, 3 is the base and the exponent, 5, tells how many times the base is multiplied by itself ($3 \times 3 \times 3 \times 3 \times 3$).

3 [y^x] 5   Answer is 243.

Because $3^2 = 9$, the number 3 is called the *square root* of 9. Square roots are written with the symbol $\sqrt{}$. Use the [$\sqrt{x}$] key to find $\sqrt{9}$ and $\sqrt{20}$ as follows.

9 [$\sqrt{x}$]   Answer is 3.

20 [$\sqrt{x}$]   Answer is 4.472135955.
Round to desired position.

OBJECTIVE 7 The [a b/c] key is used when solving problems containing fractions and mixed numbers.

Enter $\frac{3}{4} + \frac{6}{11}$ as follows.

3 [a b/c] 4 [+] 6 [a b/c] 11 [=] 1⌐13⌐44 Answer is $1\frac{13}{44}$.

Enter the mixed number problem $4\frac{7}{8} \div 3\frac{4}{7}$ as

4 [a b/c] 7 [a b/c] 8 [÷] 3 [a b/c] 4 [a b/c] 7 [=] 1⌐73⌐200. Answer is $1\frac{73}{200}$.

Note

The calculator automatically shows fractions in lowest terms and as mixed numbers when possible.

OBJECTIVE 8 To enter a negative number, first enter the number and then press the [+/−] key. This changes the number to a negative number. For example, enter $-10 + 6 - 8$ as follows.

10 [+/−] [+] 6 [−] 8 [=] Answer is −12.

(10 [+/−] gives **−10**; [−] is Subtract)

OBJECTIVE 9 Many calculators have memory keys, which are a sort of electronic scratch paper. The memory keys store intermediate steps in a calculation. On some basic calculators, the [M] key is used to store numbers in the display, with the [MR] key used to recall the numbers from memory.

Other basic calculators have [M+] and [M−] keys. The [M+] key adds the number in the calculator display to the number already in memory. At the beginning of a problem, the memory contains the number 0. If the calculator display contains the number 29.4, for example, pressing [M+] will cause 29.4 to be stored in the memory (the result of adding 0 + 29.4). If 57.8 is then entered into the display, pressing [M+] will cause 87.2 to be stored (the result of adding 29.4 + 57.8). If 11.9 is then entered into the display, and [M−] is pressed, the memory will contain 75.3 (the result of subtracting 87.2 − 11.9). The [MR] key is used to recall the number in memory, and [MC] is used to clear the memory.

Scientific calculators typically have one or more *registers* in which to store numbers. The memory keys are usually labeled [STO] for store and [RCL] for recall. For example, you can store 25.6 in register 1 by pressing 25.6 [STO] 1, or you can store it in register 2 by pressing 25.6 [STO] 2, and so on for other registers. To recall numbers from a particular memory register, use the [RCL] key followed by the number of the register. For example, pressing [RCL] 2 recalls the number from register 2.

With a scientific calculator, a number stays in memory until it is replaced by another number or until the memory is cleared. With some calculators, the contents of the memory are saved even when the calculator is turned off.

Here is an example of a problem that uses the memory keys. An elevator technician wants to find the average weight of a person using an elevator. She counts the number of people entering an elevator and also measures the weight of each group of people.

Number of People	Weight
6	839 pounds
8	1184 pounds
4	640 pounds

First, find the total weight of all three groups and store the result in memory register 1.

839 [+] 1184 [+] 640 [=] [STO] 1 — Stores 2663 in register 1.

Then, find the total number of people and store the result in memory register 2.

6 [+] 8 [+] 4 [=] [STO] 2 — Stores 18 in register 2.

Finally, divide the contents of memory register 1 (total weight) by the contents of memory register 2 (18 people).

[RCL] 1 [÷] [RCL] 2 [=] 147.9444444 pounds — Round as needed.

OBJECTIVE 10 Chain calculations, which involve several different operations, must be done in a specific sequence called the *order of operations* (see **Section 1.8**). The logic of the order of operations is built into most scientific calculators. To check your calculator, try entering $3 + 5 \times 2$ in the order it is written.

3 [+] 5 [×] 2 [=] — Answer should be 13.

If your calculator uses the order of operations, it will automatically multiply 5×2 *before* adding 3. If your calculator gives the *incorrect* answer of 16, it does *not* follow the order of operations.

OBJECTIVE 11 The parentheses keys allow you to group numbers in a chain calculation. For example, $\frac{24}{5 + 7}$ can be written as $\frac{24}{(5 + 7)}$ and entered as follows.

24 [÷] [(] 5 [+] 7 [)] [=] — Answer is 2.

(Left parentheses key: [(]; Right parentheses key: [)])

Without the parentheses the calculator would have automatically divided 24 by 5 *before* adding 7, giving an *incorrect* answer of 11.8.

Here is a more complicated example.

$$\text{To solve } \frac{16 - 2.5}{55 - 29.4 \div 0.6} \text{ write it as } \frac{(16 - 2.5)}{(55 - 29.4 \div 0.6)}$$

Use parentheses to set off the numerator and the denominator.

[(] 16 [−] 2.5 [)] [÷] [(] 55 [−] 29.4 [÷] 0.6 [)] [=] — Answer is 2.25.

Appendix B

Inductive and Deductive Reasoning

OBJECTIVE 1 In many scientific experiments, conclusions are drawn from specific outcomes. After many repetitions and similar outcomes, the findings are generalized into statements that appear to be true. When general conclusions are drawn from specific observations, we are using a type of reasoning called **inductive reasoning.** In the next several examples, this type of reasoning will be illustrated.

OBJECTIVES

1. Use inductive reasoning to analyze patterns.
2. Use deductive reasoning to analyze arguments.
3. Use deductive reasoning to solve problems.

EXAMPLE 1 Using Inductive Reasoning

Find the next number in the sequence 3, 7, 11, 15, ...

To discover a pattern, calculate the difference between each pair of successive numbers.

$$7 - 3 = 4$$
$$11 - 7 = 4$$
$$15 - 11 = 4$$

As shown, the difference is 4. Each number is 4 greater than the previous one. Thus, the next number in the pattern is $15 + 4$, or 19.

WORK PROBLEM 1 AT THE SIDE.

1. Find the next number in the sequence 2, 8, 14, 20, ...

EXAMPLE 2 Using Inductive Reasoning

Find the number that comes next in the sequence.

7, 11, 8, 12, 9, 13, ...

The pattern in this example can be determined as follows.

$$7 + 4 = 11$$
$$11 - 3 = 8$$
$$8 + 4 = 12$$
$$12 - 3 = 9$$
$$9 + 4 = 13$$

To get the second number, we add 4 to the first number. To get the third number, we subtract 3 from the second number. To obtain subsequent numbers, this pattern is continued. The next number is $13 - 3$, or 10.

WORK PROBLEM 2 AT THE SIDE.

2. Find the next number in the sequence 6, 11, 7, 12, 8, 13, ...

ANSWERS

1. 26
2. 9

3. Find the next number in the sequence 2, 6, 18, 54, ...

EXAMPLE 3 Using Inductive Reasoning

Find the next number in the sequence 1, 2, 4, 8, 16, ...

Each number after the first is obtained by multiplying the previous number by 2. So the next number would be $16 \cdot 2 = 32$.

WORK PROBLEM 3 AT THE SIDE.

EXAMPLE 4 Using Inductive Reasoning

Find the next geometric shape in the following sequence.

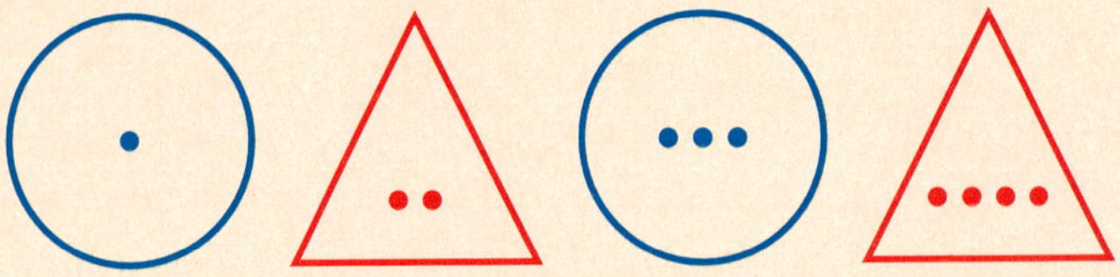

In this sequence, the figures alternate between a circle and a triangle. In addition, the number of dots increases by 1 in each subsequent figure. Thus, the next figure should be a circle with five dots contained in it, or

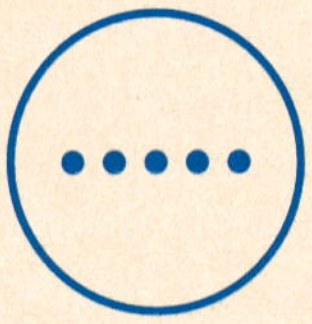

4. Find the next shape in the following sequence.

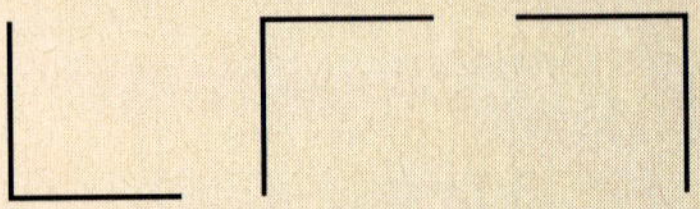

EXAMPLE 5 Using Inductive Reasoning

Find the next geometric shape in the following sequence.

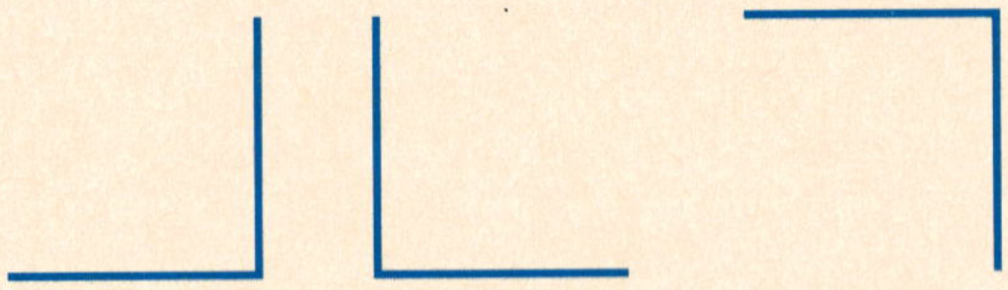

The first two shapes consist of vertical lines with horizontal lines at the bottom facing left and right. The third shape is a vertical line with a horizontal line at the top facing to the left. The fourth shape should be a vertical line with a horizontal line at the top facing to the right, or

WORK PROBLEM 4 AT THE SIDE.

ANSWERS

3. 162

4.

OBJECTIVE 2 In the previous discussion, specific cases were used to find patterns and predict the next event. There is another type of reasoning called **deductive reasoning,** which moves from general cases to specific conclusions.

EXAMPLE 6 Using Deductive Reasoning

Does the conclusion follow from the premises in this argument?

All Buicks are automobiles.
All automobiles have horns.
$\therefore$ All Buicks have horns.

In this example, the first two statements are called *premises* and the third statement (below the line) is called a conclusion. The symbol $\therefore$ is a mathematical symbol meaning "therefore." The entire set of statements is called an *argument.* The focus of deductive reasoning is to determine if the conclusion follows (is valid) from the premises. A series of circles called **Euler** (OI-ler) **circles** is used to analyze the argument. In Example 6, the statement "All Buicks are automobiles" can be represented by two circles, one for Buicks and one for automobiles.

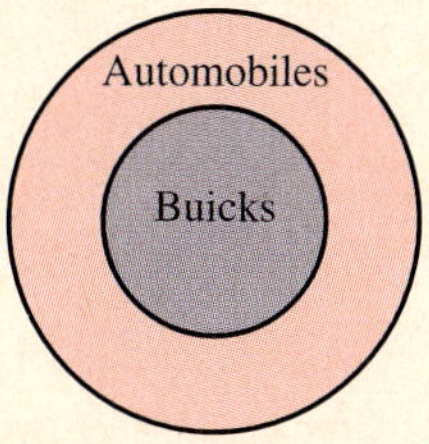

Note that the circle representing Buicks is totally inside the circle representing automobiles.

If a circle representing the second statement is added, a circle representing vehicles with horns must surround the circle representing automobiles.

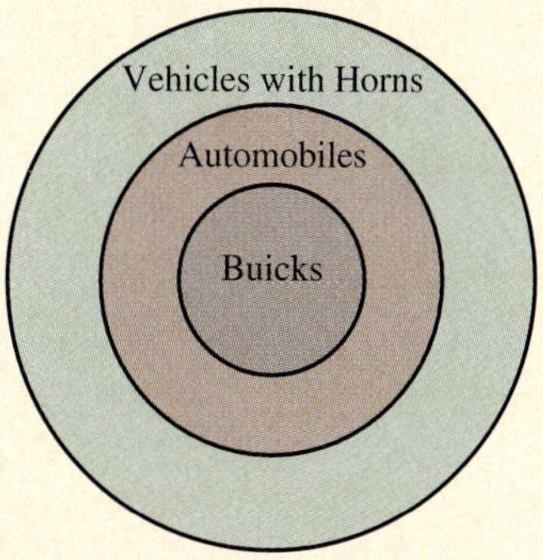

Notice that the circle representing Buicks is completely inside the circle representing vehicles with horns. It must follow that

all Buicks have horns.

WORK PROBLEM 5 AT THE SIDE.

5. Does the conclusion follow from the premises in the following argument?

All cars have four wheels.
All Fords are cars.
$\therefore$ All Fords have four wheels.

ANSWERS

5. The conclusion follows from the premises.

6. Does each conclusion follow from the premises?

(a) All animals are wild.
All cats are animals.
∴ All cats are wild.

(b) All students use math.
All adults use math.
∴ All adults are students.

Answers

6. (a) The conclusion follows from the premises.
(b) The conclusion does not follow from the premises.

EXAMPLE 7 Using Deductive Reasoning

Does the conclusion follow from the premises in this argument?

All tables are round.
All glasses are round.
∴ All glasses are tables.

Using Euler circles, a circle representing tables is drawn inside a circle representing round objects.

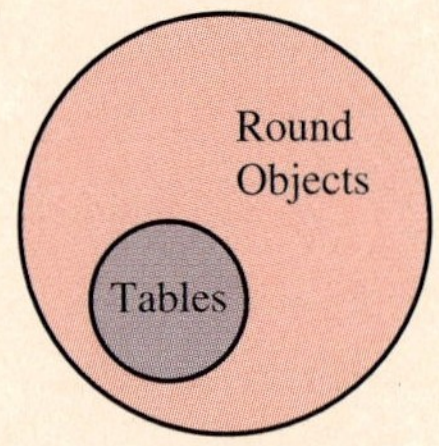

The second statement requires that a circle representing glasses must now be drawn inside the circle representing round objects but not necessarily inside the circle representing tables.

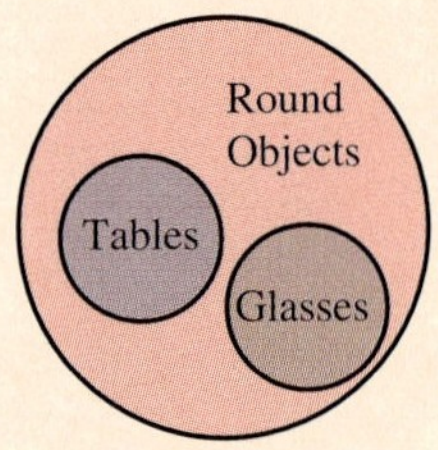

The conclusion does not follow from the premises. This means that the conclusion is invalid or untrue.

WORK PROBLEM 6 AT THE SIDE.

Objective 3 Another type of deductive reasoning problem occurs when a set of facts is given in a problem and a conclusion must be drawn by using these facts.

EXAMPLE 8 Using Deductive Reasoning

There were 25 students enrolled in a ceramics class. During the class, 10 of the students made a bowl and 8 students made a birdbath. Three students made both a bowl and a birdbath. How many students did not make either a bowl or a birdbath?

This type of problem is best solved by organizing the data using a device called a *Venn diagram.* Two overlapping circles are drawn, with each circle representing one item made by students.

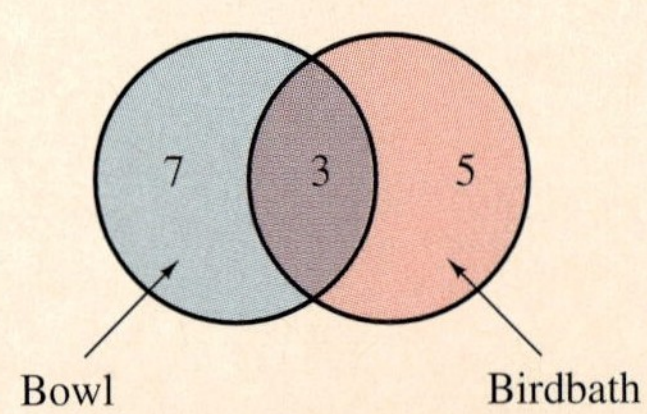

CONTINUED ON NEXT PAGE

In the region where the circles overlap, place the number that represents the number of students who made both items, namely 3. In the remaining portion of the birdbath circle, write the number 5, which when added to 3 will give the total number of students who made a birdbath, namely 8. In a similar manner, write 7 in the remaining portion of the bowl circle, since $7 + 3 = 10$, the total number of students who made a bowl. The three numbers that have been written in the regions total 15. Since there are 25 students in the class, this means $25 - 15$ or 10 students did not make either a birdbath or a bowl.

WORK PROBLEM 7 AT THE SIDE.

EXAMPLE 9 Using Deductive Reasoning

Four cars in a race finish first, second, third, and fourth. The following facts are known.

(a) Car A beat Car C.

(b) Car D finished between Cars C and B.

(c) Car C beat Car B.

In which order did the cars finish?

To solve this type of problem, it is helpful to use a line diagram.

1. *Write A before C,* since Car A beat Car C (fact a).

A C

2. *Write B after C,* since Car C beat Car B (fact c).

A C B

3. *Write D between C and B,* since Car D finished between Cars C and B (fact b).

So

A C D B

is the correct order of finish.

WORK PROBLEM 8 AT THE SIDE.

7. In a college class of 100 students, 35 take both math and history, 50 take history, and 40 take math. How many take neither math nor history?

8. A Chevy, BMW, Cadillac, and Oldsmobile are parked side by side.

(a) The Oldsmobile is on the right end.

(b) The BMW is next to the Cadillac.

(c) The Chevy is between the Oldsmobile and the Cadillac.

Which car is parked on the left end?

ANSWERS

7. 45

8. BMW

NUMBERS IN THE *Real World* *collaborative investigations*

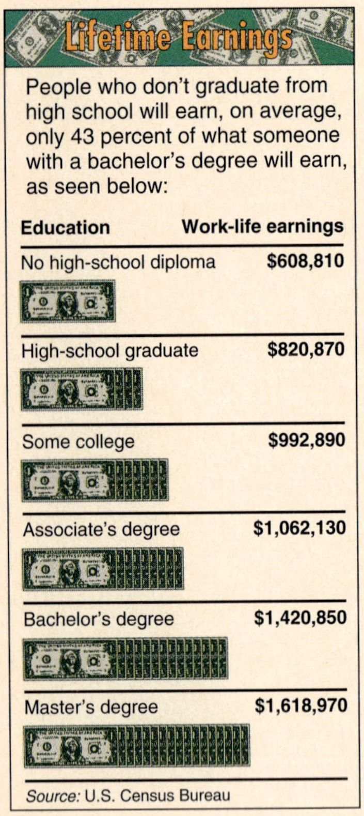

1. Read the information at the top of the graph. What does the phrase "on average" mean? How is an average calculated?

2. Is the 43 percent figure correct? Show how you would check it.

3. Find at least two other calculations you could do to compare "no high-school diploma" earnings with "bachelor's degree" earnings.

4. All of the earning amounts in the graph end in zero. This suggests that the amounts may have been rounded. If so, to what place were they rounded?

5. If you wanted to round the amounts even more, would it make sense to round them all to the nearest million? Explain why or why not.

6. To what place might you choose to round the earnings? Explain your choice.

7. Round each amount to the place you chose. Now look at the sequence of numbers from "no high-school diploma" to "master's degree." Suppose you wanted to add the next level to the graph, lifetime earnings of people with doctorate degrees. Can you use inductive reasoning to find the next number in the sequence of lifetime earnings? Explain why or why not.

8. Here is the deductive reasoning used by one student. Do you agree with the reasoning? If not, what would you change?

> A bachelor's degree guarantees lifetime earnings of exactly $1,420,850.
> I have a bachelor's degree.
> ---
> ∴ I will have lifetime earnings of $1,420,850.

9. If you wanted to verify that the earnings shown in the graph are correct for your city or state, how might you do that? Discuss some possibilities with several classmates, then share your ideas with the whole class.

Appendix B Exercises

Find the next number in each of the following sequences.

1. 2, 9, 16, 23, 30, ...

2. 5, 8, 11, 14, 17, ...

3. 1, 6, 11, 16, 21, ...

4. 3, 5, 7, 9, 11, ...

5. 1, 2, 4, 8, ...

6. 1, 8, 27, 64, ...

7. 1, 3, 9, 27, 81, ...

8. 3, 6, 12, 24, 48, ...

9. 1, 4, 9, 16, 25, ...

10. 6, 7, 9, 12, 16, ...

Find the next shape in each of the following sequences.

11. 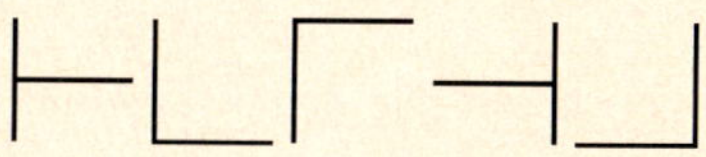

12. 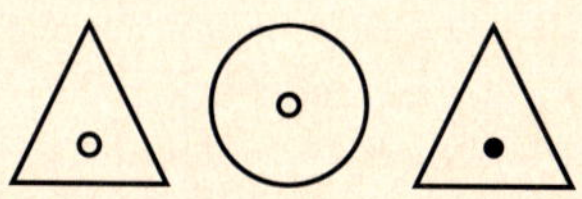

13. 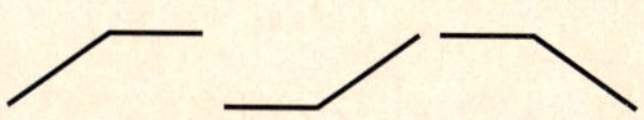

14. 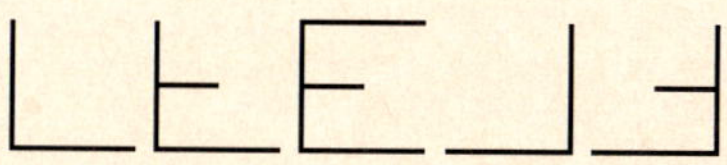

In each of the following, state whether or not the conclusion follows from the premises.

15. All animals are wild.
All lions are animals.
∴ All lions are wild.

16. All students are hard workers.
All business majors are students.
∴ All business majors are hard workers.

17. All teachers are serious.
All mathematicians are serious.
∴ All mathematicians are teachers.

18. All boys ride bikes.
All Americans ride bikes.
∴ All Americans are boys.

Solve the following application problems.

19. In a given 30-day period, a man watched television 20 days and his wife watched television 25 days. If they watched television together 18 days, how many days did neither watch television?

20. In a class of 40 students, 21 students take both calculus and physics. If 30 students take calculus and 25 students take physics, how many do not take either calculus or physics?

21. Tom, Dick, Mary, and Joan all work for the same company. One is a secretary, one is a computer operator, one is a receptionist, and one is a mail clerk.
(a) Tom and Joan eat dinner with the computer operator.
(b) Dick and Mary carpool with the secretary.
(c) Mary works on the same floor as the computer operator and the mail clerk.
Who is the computer operator?

22. Four cars—a Ford, a Buick, a Mercedes, and an Audi—are parked in a garage in four spaces.
(a) The Ford is in the last space.
(b) The Buick and Mercedes are next to each other.
(c) The Audi is next to the Ford but not next to the Buick.
Which car is in the first space?

Appendix C

Geometry: Lines and Angles

Geometry starts with the idea of a point. A **point** is a location in space. It has no length or width. A point is represented by a dot and is named by writing a capital letter next to the dot.

Point P

OBJECTIVE 1 A **line** is a straight row of points that goes on forever in both directions. A line is drawn by using arrowheads to show that it never ends. The line is named by using the letters of any two points on the line.

Line AB, written $\overleftrightarrow{AB}$

A piece of a line that has two endpoints is called a **line segment.** A line segment is named for its endpoints. The segment with endpoints P and Q is shown below. It can be named $\overline{PQ}$ or $\overline{QP}$.

Line segment PQ, written $\overline{PQ}$

A **ray** is a part of a line that has only one endpoint and goes on forever in one direction. A ray is named by using the endpoint and some other point on the ray. The endpoint is always mentioned first.

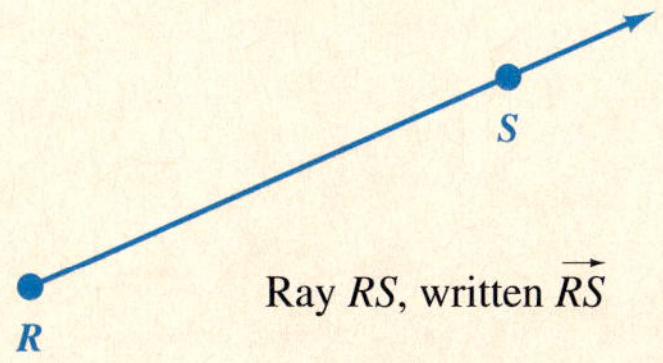

Ray RS, written $\overrightarrow{RS}$

OBJECTIVES

1. Identify lines, line segments, and rays.
2. Identify parallel and intersecting lines.
3. Identify and name an angle.
4. Classify an angle as right, acute, straight, or obtuse.
5. Identify perpendicular lines.
6. Identify complementary angles and supplementary angles.
7. Identify congruent angles and vertical angles.

1. Identify each of the following as a line, line segment, or ray.

(a)

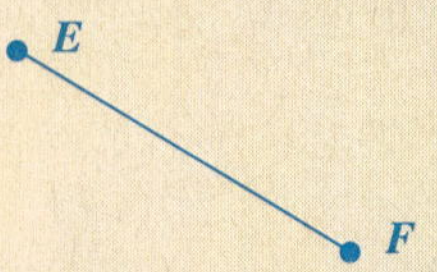

(b)

(c)

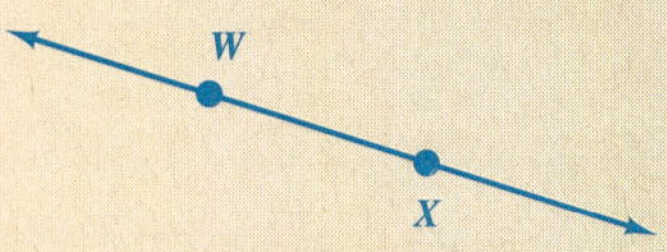

(d)

2. Label each pair of lines as parallel or intersecting.

(a)

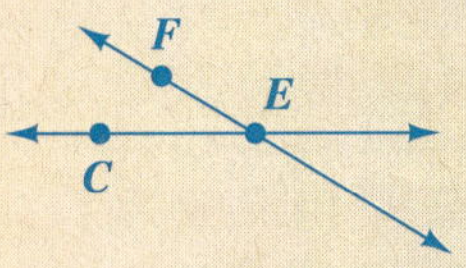

(b)

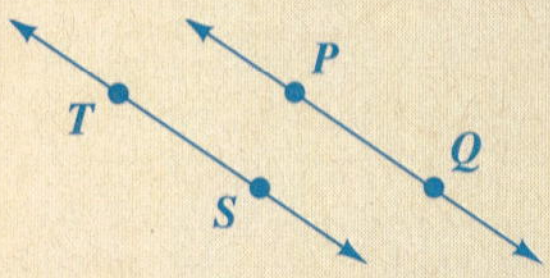

(c)

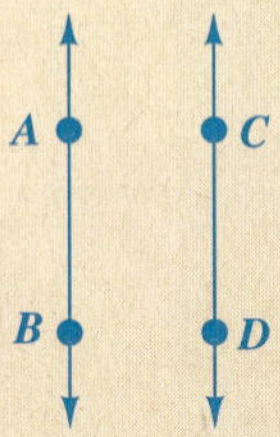

ANSWERS

1. **(a)** line segment **(b)** ray **(c)** line **(d)** line segment
2. **(a)** intersecting **(b)** parallel **(c)** parallel

EXAMPLE 1 Identifying Lines, Rays, and Line Segments

Identify each of the following as a line, line segment, or ray.

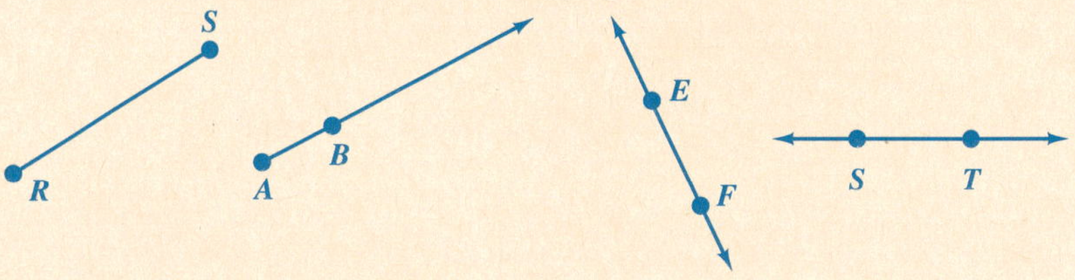

(a) **(b)** **(c)** **(d)**

Figure **(a)** has two endpoints, so it is a line segment.

Figure **(b)** starts at point *A* and goes on forever in one direction, so it is a ray.

Figures **(c)** and **(d)** go on forever in both directions, so they are lines.

WORK PROBLEM 1 AT THE SIDE.

OBJECTIVE 2 A *plane* is a flat surface, like a floor or a wall. Lines that are in the same plane but that never intersect (never cross) are called **parallel** (PAIR-uh-lell) **lines,** while lines that cross or merge are called **intersecting** (in-tur-SEKT-ing) **lines.** (Think of an intersection, where two streets cross each other.)

EXAMPLE 2 Identifying Parallel and Intersecting Lines

Label each pair of lines as parallel or intersecting.

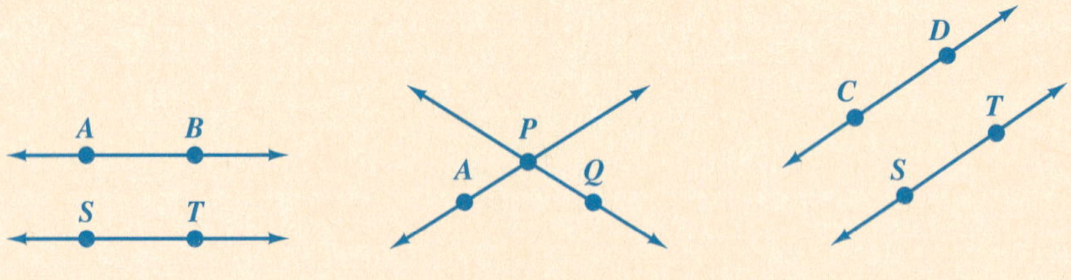

(a) **(b)** **(c)**

The lines in Figures **(a)** and **(c)** never intersect. They are parallel lines. The lines in Figure **(b)** cross at *P*, so they are intersecting lines.

WORK PROBLEM 2 AT THE SIDE.

OBJECTIVE 3 An **angle** (ANG-gul) is made up of two rays that start at a common endpoint. This common endpoint is called the *vertex.*

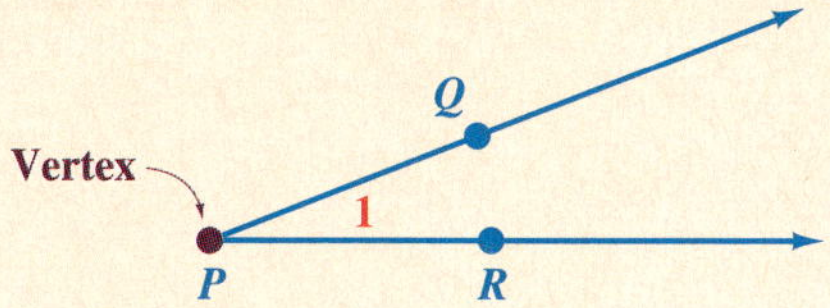

The rays *PQ* and *PR* are called *sides.* The angle can be named four ways:

$\angle 1$ $\angle P$ $\angle QPR$ $\angle RPQ$

Vertex alone — Vertex in the middle

Naming an Angle

To name an angle, write the vertex alone or write it in the middle of two other points. If two or more angles have the same vertex, as in Example 3, do not use the vertex alone to name an angle.

EXAMPLE 3 Identifying and Naming an Angle

Name the highlighted angle.

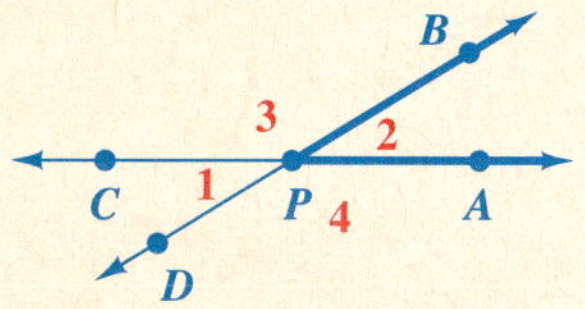

The angle can be named $\angle BPA$, $\angle APB$, or $\angle 2$. It cannot be named $\angle P$, using the vertex alone, because four different angles have *P* as their vertex.

WORK PROBLEM 3 AT THE SIDE.

OBJECTIVE 4 Angles can be measured in **degrees** (deh-GREEZ). The symbol for degrees is a small, raised circle °. Think of the minute hand on a clock as a ray of an angle. Suppose it is at 12:00. During one hour of time, the minute hand moves around in a complete circle. It moves 360 ***degrees***, or 360°. In half an hour, at 12:30, the minute hand has moved half way around the circle, or 180°. An angle of 180° is called a **straight angle.** (Notice that the two rays in a straight angle form a straight line.)

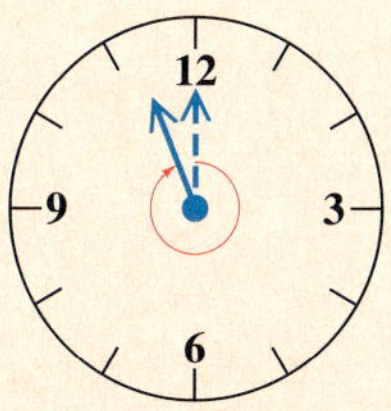

Complete circle
360°

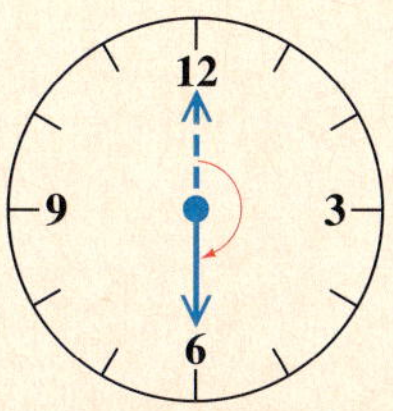

Straight angle
(half a circle)
180°

3. (a) Name the highlighted angle in three different ways.

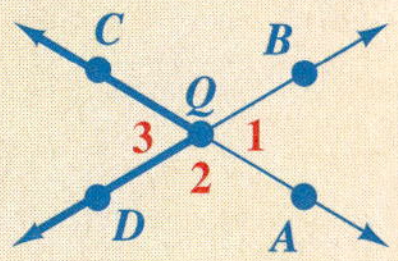

(b) Darken the lines that make up $\angle ZTW$.

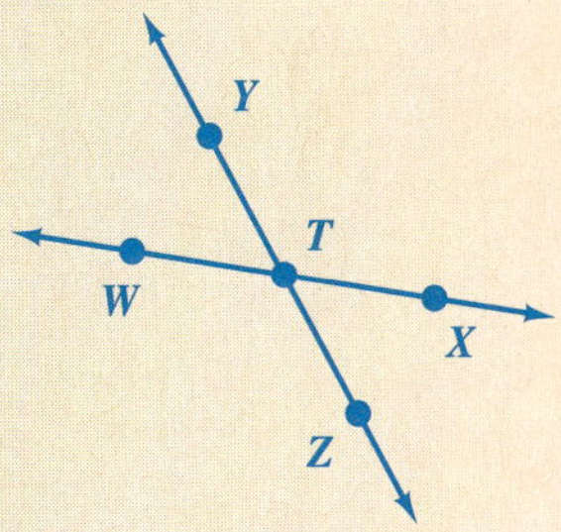

(c) Name this angle in four different ways.

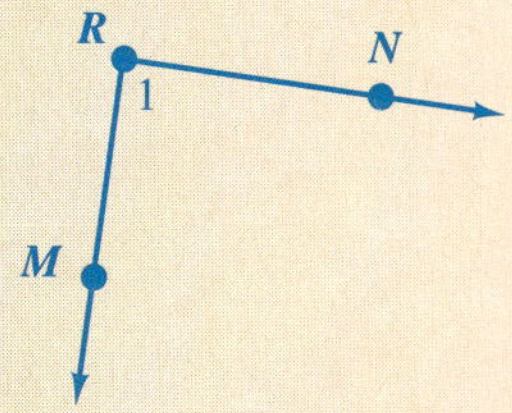

ANSWERS

3. (a) $\angle 3$, $\angle CQD$, $\angle DQC$

(b)

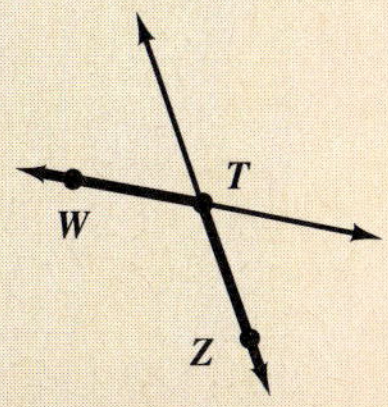

(c) $\angle 1$, $\angle R$, $\angle MRN$, $\angle NRM$

In a quarter of an hour, at 12:15, the minute hand has moved $\frac{1}{4}$ of the way around the circle, or 90°. An angle of 90° is called a **right angle.** Sometimes you hear it called a *square angle.* The minute hands at 12:00 and 12:15 form one corner of a square. So, to show that an angle is a **right angle**, we draw a **small square** at the vertex.

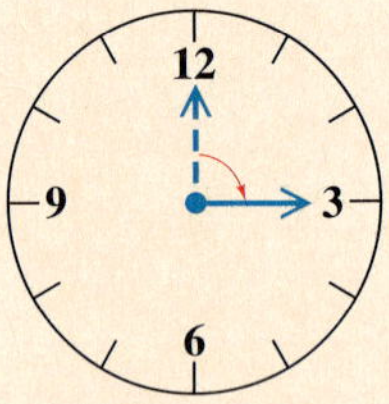

Right angle
($\frac{1}{4}$ of a circle)
90°

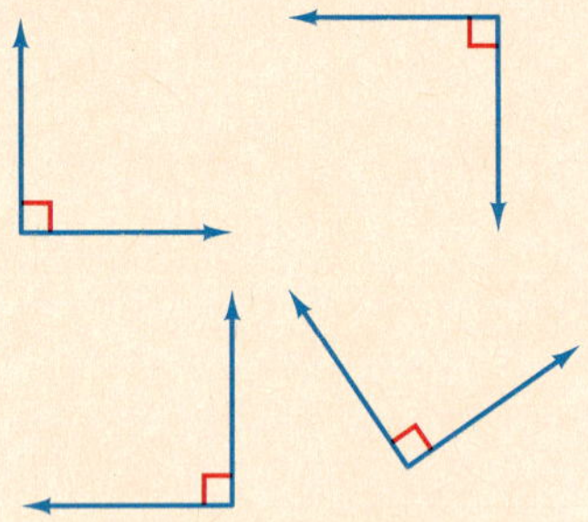

A small square at vertex identifies right angles.

You can see that an angle of 1° is very small. To be precise, it is only the distance that the *minute hand* moves in ten *seconds.*

Some other terms used to describe angles are shown below.

Acute (uh-CUTE) **angles** measure between 0° and 90°.

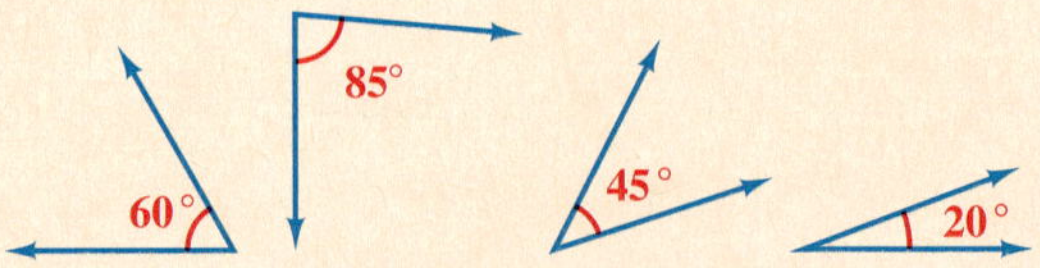

Examples of acute angles

Obtuse (ob-TOOS) **angles** measure between 90° and 180°.

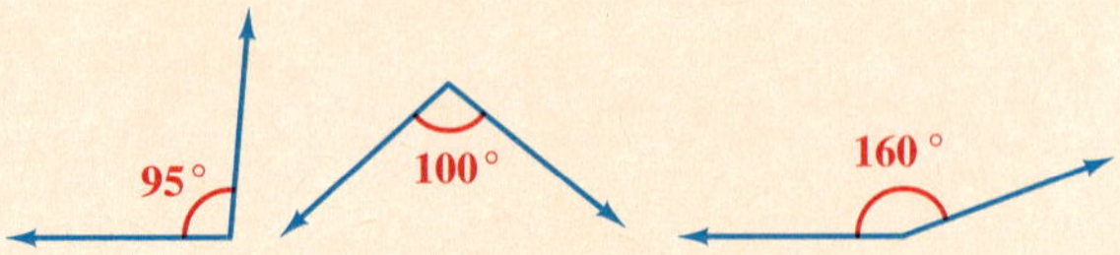

Examples of obtuse angles

Section 9.1 shows you how to use a tool called a *protractor* to measure the number of degrees in an angle.

Note

Angles can also be measured in radians, which you will learn about in a later math course.

EXAMPLE 4 Classifying an Angle

Label each of the following angles as acute, right, obtuse, or straight.

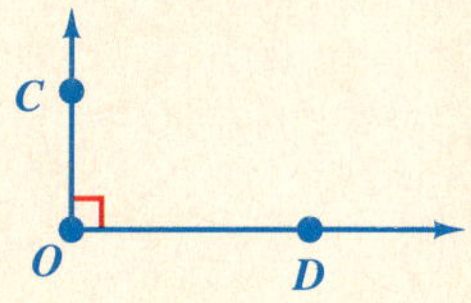

(a)

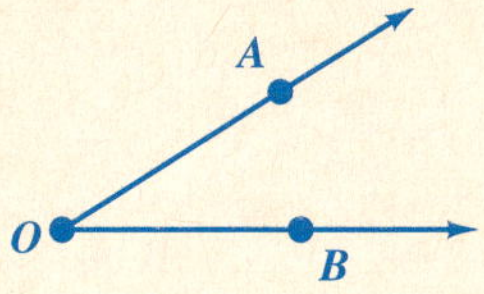

(b)

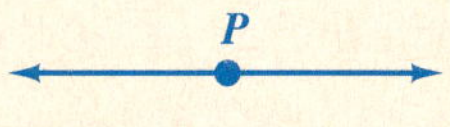

(c)

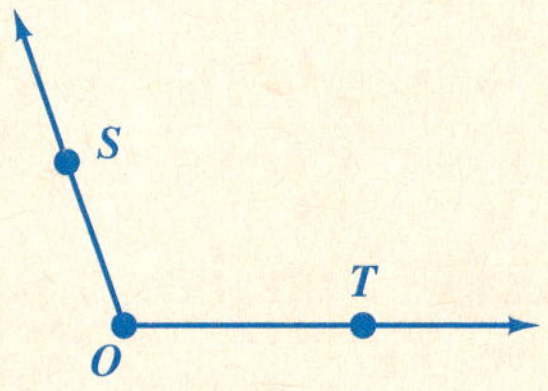

(d)

Figure **(a)** shows a right angle (exactly 90° and identified by a small square at the vertex).

Figure **(b)** shows an acute angle (between 0° and 90°).

Figure **(c)** shows a straight angle (exactly 180°).

Figure **(d)** shows an obtuse angle (between 90° and 180°).

WORK PROBLEM 4 AT THE SIDE.

Objective 5 Two lines are called **perpendicular** (per-pen-DIK-yoo-ler) **lines** if they intersect to form a right angle.

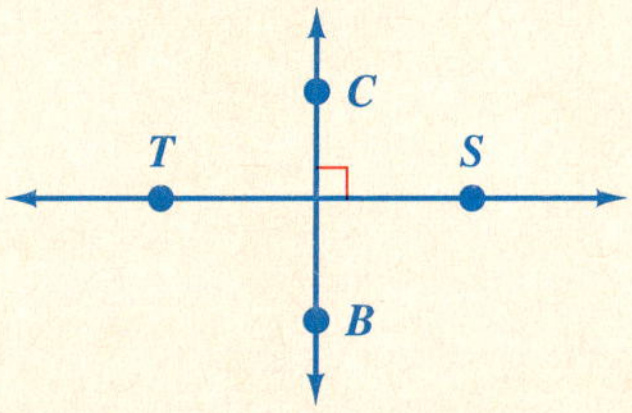

Lines *CB* and *ST* are **perpendicular**, because they intersect at right angles. This can be written in the following way: $\overleftrightarrow{CB} \perp \overleftrightarrow{ST}$.

EXAMPLE 5 Identifying Perpendicular Lines

Which of the following pairs of lines are perpendicular?

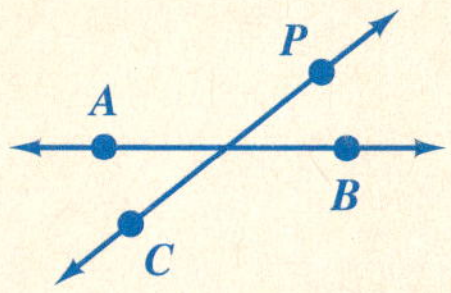

(a)

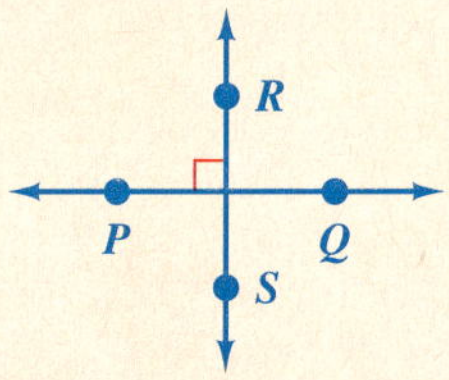

(b)

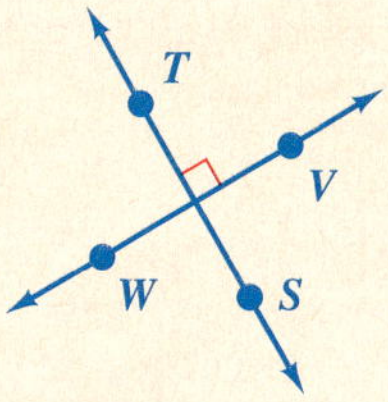

(c)

The lines in Figures **(b)** and **(c)** are perpendicular to each other, because they intersect at right angles. The lines in Figure **(a)** are intersecting lines, but are not perpendicular, because they do not form a right angle.

WORK PROBLEM 5 AT THE SIDE.

4. Label each of the following as an acute, right, obtuse, or straight angle.

(a)

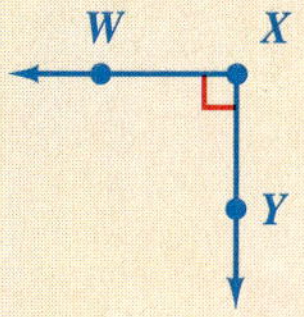

(b)

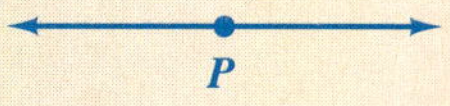

(c)

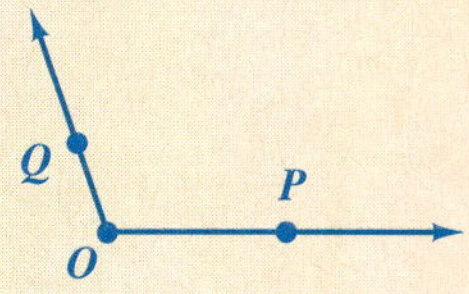

(d)

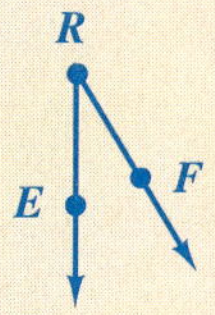

5. Which pair of lines is perpendicular? How can you describe the other pair of lines?

(a)

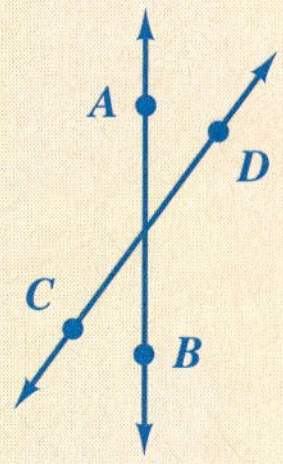

(b)

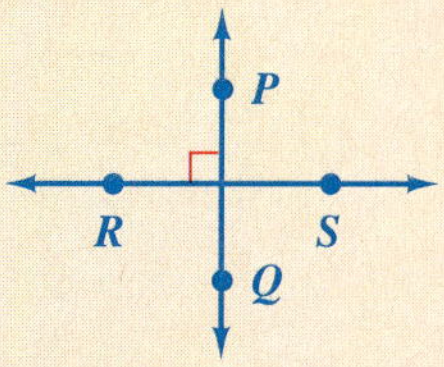

Answers

4. **(a)** right **(b)** straight **(c)** obtuse **(d)** acute

5. **(b)**; Figure **(a)** shows intersecting lines

6. Identify each pair of complementary angles.

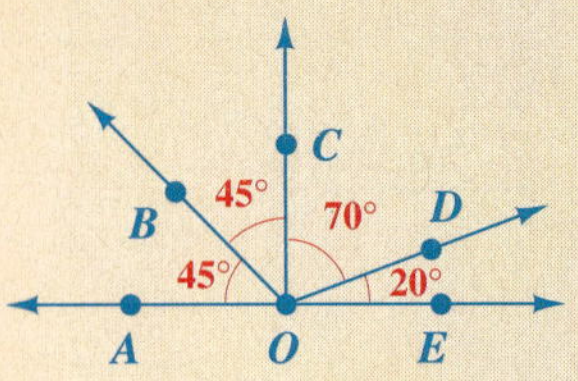

7. Find the complement of the following angles.

(a) 35°

(b) 80°

8. Identify each pair of supplementary angles.

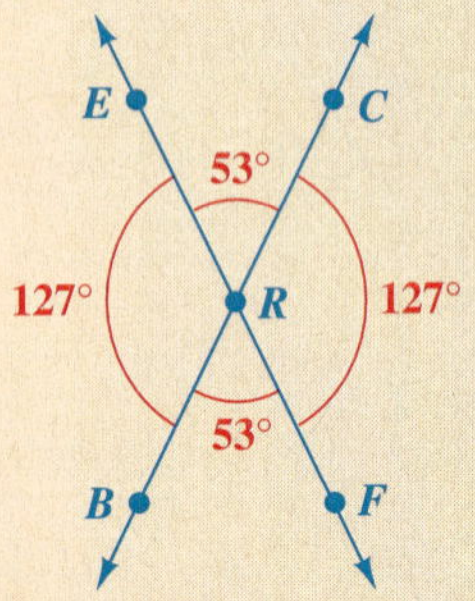

Answers

6. ∠*AOB* and ∠*BOC*; ∠*COD* and ∠*DOE*

7. (a) 55° **(b)** 10°

8. ∠*CRF* and ∠*BRF*
∠*CRE* and ∠*ERB*
∠*BRF* and ∠*BRE*
∠*CRE* and ∠*CRF*

Objective 6 Two angles are called **complementary** (kahm-pleh-MEN-tary) **angles** if their sum is 90°. If two angles are complementary, each angle is the *complement* of the other.

EXAMPLE 6 Identifying Complementary Angles

Identify each pair of complementary angles.

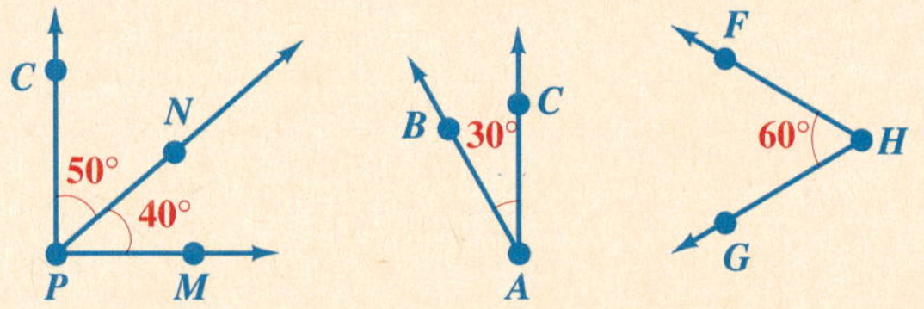

∠*MPN*(40°) and ∠*NPC*(50°) are complementary angles because

$$40° + 50° = \mathbf{90°}.$$

∠*CAB*(30°) and ∠*FHG*(60°) are complementary angles because

$$30° + 60° = \mathbf{90°}.$$

WORK PROBLEM 6 AT THE SIDE.

EXAMPLE 7 Finding the Complement of an Angle

Find the complement of each angle.

(a) 30°
The complement of 30° is 60°, because **90°** − 30° = 60°.

(b) 40°
The complement of 40° is 50°, because **90°** − 40° = 50°.

WORK PROBLEM 7 AT THE SIDE.

Two angles are called **supplementary** (sup-luh-MEN-tary) **angles** if their sum is 180°. If two angles are supplementary, each angle is the *supplement* of the other.

EXAMPLE 8 Identifying Supplementary Angles

Identify each pair of supplementary angles.

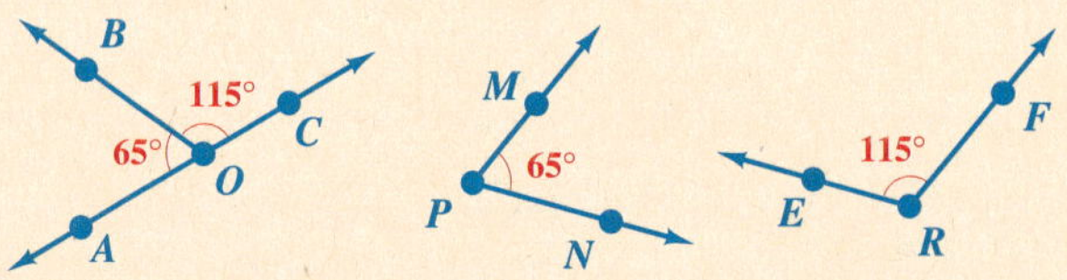

∠*BOA* and ∠*BOC*, because 65° + 115° = **180°**.
∠*BOA* and ∠*ERF*, because 65° + 115° = **180°**.
∠*BOC* and ∠*MPN*, because 115° + 65° = **180°**.
∠*MPN* and ∠*ERF*, because 65° + 115° = **180°**.

WORK PROBLEM 8 AT THE SIDE.

EXAMPLE 9 Finding the Supplement of an Angle

Find the supplement of the following angles.

(a) 70°

The supplement of 70° is 110°, because **180°** − 70° = 110°.

(b) 140°

The supplement of 140° is 40°, because **180°** − 140° = 40°.

WORK PROBLEM 9 AT THE SIDE.

OBJECTIVE 7 Two angles are called **congruent** (kuhn-GROO-ent) **angles** if they measure the same number of degrees. If two angles are congruent, this is written as

$$\angle A \cong \angle B$$

and read as, "angle *A* **is congruent to** angle *B.*"

EXAMPLE 10 Identifying Congruent Angles

Identify the angles that are congruent.

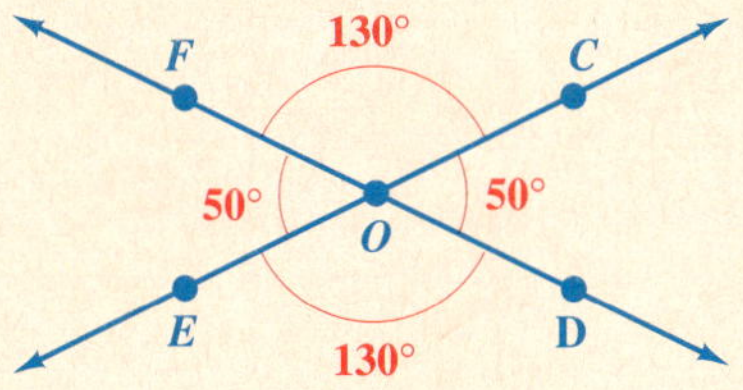

$\angle FOC \cong \angle EOD$ and $\angle COD \cong \angle EOF$

WORK PROBLEM 10 AT THE SIDE.

Angles that do not share a common side are called *nonadjacent* angles. Two nonadjacent angles formed by intersecting lines are called **vertical** (VUR-ti-kul) **angles.**

EXAMPLE 11 Identifying Vertical Angles

Identify the vertical angles in this figure.

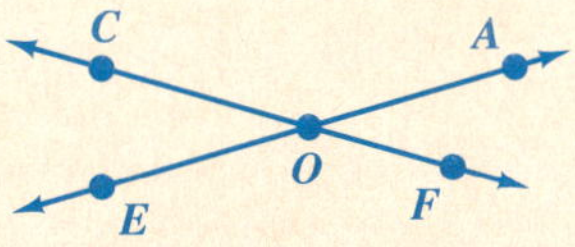

$\angle AOF$ and $\angle COE$ are vertical angles because they do not share a common side and they are formed by two intersecting lines ($\overleftrightarrow{CF}$ and $\overleftrightarrow{EA}$).

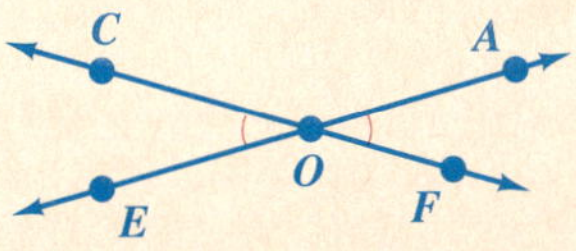

$\angle COA$ and $\angle EOF$ are also vertical angles.

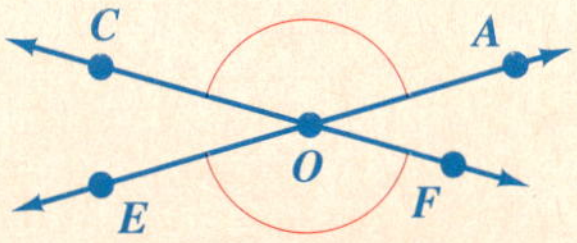

WORK PROBLEM 11 AT THE SIDE.

9. Find the supplement of each angle.

(a) 175°

(b) 30°

10. Identify the angles that are congruent.

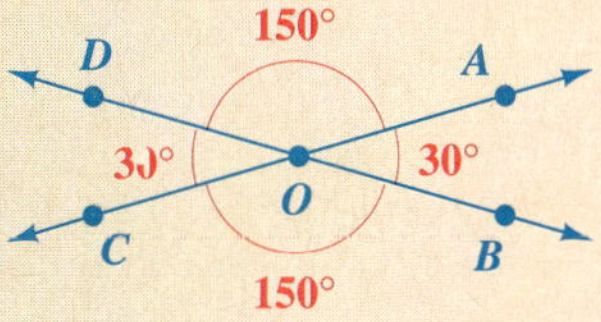

11. Identify the vertical angles.

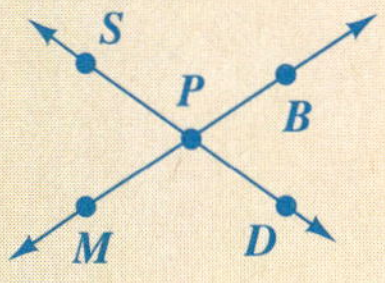

ANSWERS

9. **(a)** 5° **(b)** 150°

10. $\angle BOC \cong \angle AOD$, $\angle AOB \cong \angle DOC$

11. $\angle SPB$ and $\angle MPD$, $\angle BPD$ and $\angle SPM$

12. In the figure below, find the number of degrees in each of the following angles.

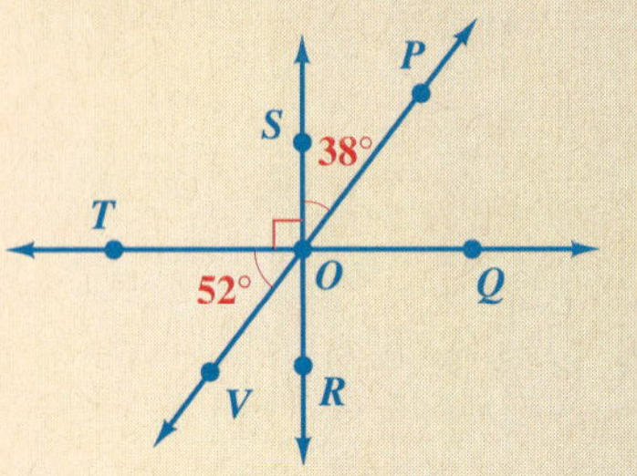

(a) $\angle VOR$

(b) $\angle POQ$

(c) $\angle QOR$

Look back at Example 10 on the previous page. Notice that the two congruent angles that measure 130° are also vertical angles. Also, the two congruent angles that measure 50° are vertical angles. This illustrates the following property.

Congruent Angles

If two angles are vertical angles, they are congruent; that is, they measure the same number of degrees.

EXAMPLE 12 Finding the Measures of Vertical Angles

In the figure below, find the measure of the following angles.

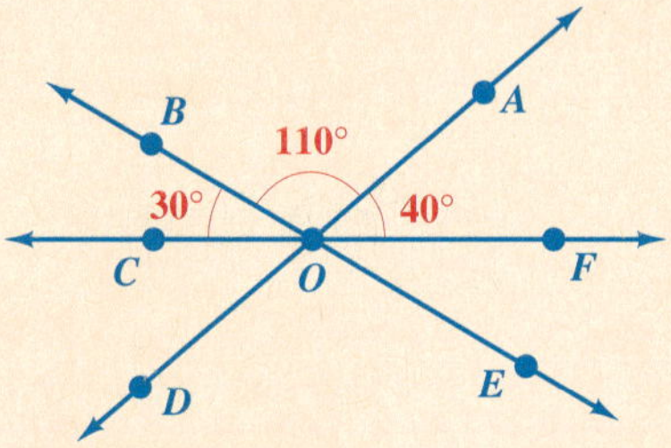

(a) $\angle COD$

$\angle COD$ and $\angle AOF$ are vertical angles so they are congruent. This means they measure the same number of degrees.

The measure of $\angle AOF$ is 40° so the measure of $\angle COD$ is 40° also.

(b) $\angle DOE$

$\angle DOE$ and $\angle BOA$ are vertical angles so they are congruent.

The measure of $\angle BOA$ is 110° so the measure of $\angle DOE$ is 110° also.

(c) $\angle EOF$

$\angle EOF$ and $\angle COB$ are vertical angles so they are congruent.

The measure of $\angle COB$ is 30° so the measure of $\angle EOF$ is 30° also.

WORK PROBLEM 12 AT THE SIDE.

ANSWERS

12. (a) 38° **(b)** 52° **(c)** 90°

Answers to Selected Exercises

In this section we provide the answers that we think most students will obtain when they work the exercises using the methods explained in the text. If your answer does not look exactly like the one given here, it is not necessarily wrong. In many cases, there are equivalent forms of the answer that are correct. For example, if the answer section shows $\frac{3}{4}$ and your answer is 0.75, you have obtained the right answer but written it in a different (yet equivalent) form. Unless the directions specify otherwise, 0.75 is just as valid an answer as $\frac{3}{4}$.

In general, if your answer does not agree with the one given in the text, see whether it can be transformed into the other form. If it can, then it is the correct answer. If you still have doubts, talk with your instructor.

Chapter 1

Section 1.1 (page 5)

1. 15; 0; 83,001 **3.** 7; 362,049 **5.** hundreds **7.** hundred thousands **9.** ten millions **11.** hundred billions **13.** ten trillions, hundred billions, millions, hundred thousands, ones **15.** eight thousand, four hundred twenty-one **17.** forty-six thousand, two hundred five **19.** three million, sixty-four thousand, eight hundred one **21.** eight hundred forty million, one hundred eleven thousand, three **23.** fifty-one billion, six million, eight hundred eighty-eight thousand, three hundred twenty-one **25.** three trillion, seven hundred twelve million **27.** 46,805 **29.** 5,600,082 **31.** 271,900,000 **33.** 12,417,625,310 **35.** 600,000,071,000,400 **37.** largest: 97,651,100 ninety-seven million, six hundred fifty-one thousand, one hundred; smallest: 10,015,679 ten million, fifteen thousand, six hundred seventy-nine **39.** six thousand, five hundred sixty-seven **41.** 101,280,000 **43.** Two million, twenty-one thousand, eighteen dollars **45.** 24,000,500 **47.** Answers will vary.

Section 1.2 (page 13)

1. $^{+}29{,}028$ feet or 29,028 feet **3.** $^{-}128.6$ degrees **5.** $^{-}18$ yards **7.** $^{+}\$100$ or \$100 **9.** $^{-}6\frac{1}{2}$ pounds

11. 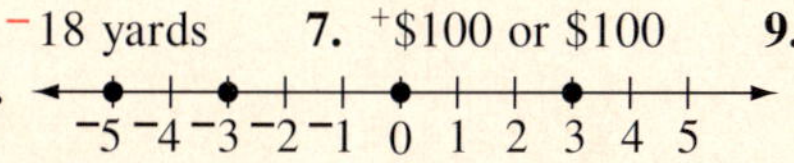

13. 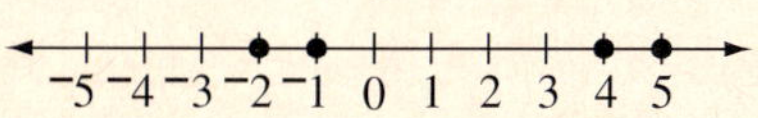

15. 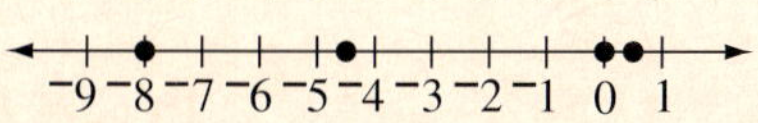

17. > **19.** < **21.** < **23.** > **25.** < **27.** > **29.** > **31.** < **33.** 15 **35.** 3 **37.** 0 **39.** 200 **41.** 75 **43.** 8042 **45.** Answers will vary. Some possibilities are overdrawn checkbook, winter temperatures, football, card games where you can end up with a negative score. **47.** The absolute value of a negative number is positive, for example, $|^{-}6|$ is 6. The absolute value of a positive number is the same number, for example, $|5|$ is 5. **49.** $^{+}3$ and $^{-}7$

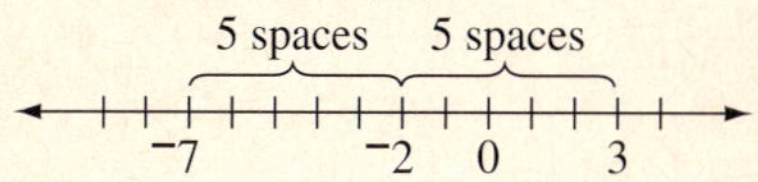

Section 1.3 (page 21)

1. 3

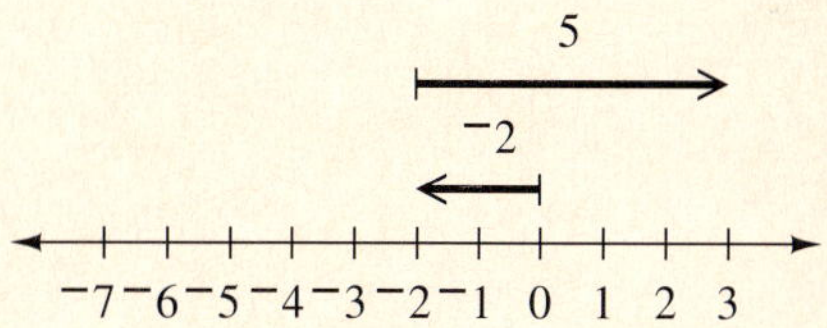

3. $^{-}7$

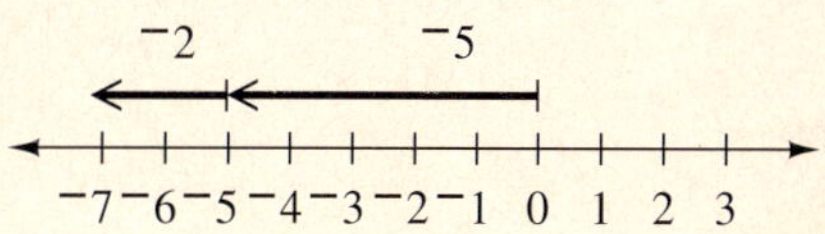

5. $^{-}1$

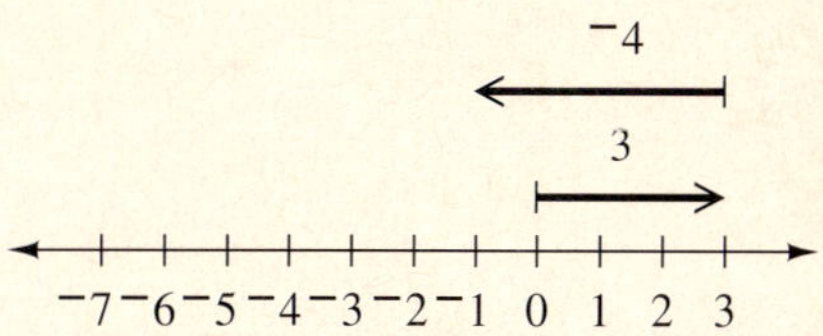

7. **(a)** $^{-}10$ **(b)** 10 **9.** **(a)** 12 **(b)** $^{-}12$ **11.** **(a)** $^{-}50$ **(b)** 50 **13.** **(a)** 158 **(b)** $^{-}158$ **15.** Each pair of answers matches except for the sign. This occurs because the absolute values are the same, so the only difference in the sums is the common sign. **17.** **(a)** 2 **(b)** $^{-}2$ **19.** **(a)** $^{-}7$ **(b)** 7 **21.** **(a)** $^{-}5$ **(b)** 5 **23.** **(a)** 150 **(b)** $^{-}150$ **25.** Each pair of answers differs only in the sign of the answer. This occurs because the signs of the addends are reversed. **27.** $^{-}3$ **29.** 7 **31.** $^{-}7$ **33.** 1 **35.** $^{-}8$ **37.** $^{-}20$ **39.** $^{-}17$ **41.** $^{-}22$ **43.** $^{-}19$ **45.** 6 **47.** $^{-}5$ **49.** 0 **51.** 5 **53.** $^{-}32$ **55.** $13 + {^{-}17} = {^{-}4}$ yards **57.** ${^{-}\$62} + \$50 = {^{-}\$12}$ **59.** $8 + {^{-}3} + {^{-}1} = 4$ feet **61.** ${^{-}20} + 75 + {^{-}55} = 0$ points **63.** $^{-}5$; $^{-}18$; $^{-}23$ **65.** 15; $^{-}4$; 11 **67.** $6 + ({^{-}14} + 14)$; $6 + 0 = 6$ **69.** $({^{-}14} + {^{-}6}) + {^{-}7}$; ${^{-}20} + {^{-}7} = {^{-}27}$ **71.** Some possibilities are: ${^{-}6} + 0 = {^{-}6}$; $10 + 0 = 10$; $0 + 3 = 3$ **73.** $^{-}4116$ **75.** 8686 **77.** $^{-}96{,}077$

Section 1.4 (page 27)

1. $^{-}6$; $6 + {^{-}6} = 0$ **3.** 13; ${^{-}13} + 13 = 0$ **5.** 0; $0 + 0 = 0$ **7.** 14 **9.** $^{-}2$ **11.** $^{-}12$ **13.** $^{-}25$ **15.** $^{-}23$ **17.** 5 **19.** 20 **21.** 11 **23.** $^{-}60$ **25.** 0 **27.** 0 **29.** $^{-}6$ **31.** **(a)** 8 **(b)** $^{-}2$ **(c)** 2 **(d)** $^{-}8$

33. **(a)** $^-3$ **(b)** 11 **(c)** $^-11$ **(d)** 3 **35.** Answers on left: $^-8$; 8. On right: $^-1$; 1. Subtraction is *not* commutative. The absolute value of the answer is the same, but the sign changes. **37.** $^-6$ **39.** $^-5$ **41.** 3 **43.** $^-10$ **45.** 12 **47.** $^-5$ **49.** The student forgot to change 6 to its opposite, $^-6$. It should be $^-6 + {}^-6 = {}^-12$ **51.** $^-11$ **53.** $^-5$ **55.** $^-10$

Section 1.5 (page 35)

1. ≈620 **3.** ≈$^-1090$ **5.** ≈7900 **7.** ≈$^-86{,}800$ **9.** ≈42,500 **11.** ≈$^-6000$ **13.** ≈15,800 **15.** ≈$^-78{,}000$ **17.** ≈6000 **19.** ≈53,000 **21.** ≈600,000 **23.** ≈$^-9{,}000{,}000$ **25.** ≈140,000,000 **27.** 30,000 miles **29.** $^-60$ degrees **31.** $10,000 **33.** 60,000,000 Americans **35.** $^-300$ feet **37.** 600,000 people **39.** Answers will vary but should mention looking only at the second digit, rounding first digit up when second digit is 5 or more, leaving first digit unchanged when second digit is 4 or less. Examples will vary. **41.** $^-40 + 90 = 50$; 47 **43.** $20 + {}^-100 = {}^-80$; $^-81$ **45.** $^-300 + {}^-400 = {}^-700$; $^-672$ **47.** 3000 + 7000 = 10,000; 9907 **49.** $20 + {}^-80 = {}^-60$; $^-58$ **51.** $^-40 + {}^-40 = {}^-80$; $^-78$ **53.** $^-100 + 30 + {}^+70 = 0$; 0 **55.** \$80,000 − \$50,000 = \$30,000; \$25,768 **57.** \$2000 − \$500 − \$300 − \$300 − \$200 − \$200 = \$500; \$458 **59.** $^-100 + 40 + 50 = {}^-10$ degrees; $^-13$ degrees **61.** \$400 + \$500 = \$900; \$905

Section 1.6 (page 45)

1. **(a)** 63 **(b)** 63 **(c)** $^-63$ **(d)** $^-63$ **3.** **(a)** $^-56$ **(b)** $^-56$ **(c)** 56 **(d)** 56 **5.** $^-35$ **7.** $^-45$ **9.** $^-18$ **11.** $^-50$ **13.** $^-40$ **15.** $^-56$ **17.** 32 **19.** 77 **21.** 0 **23.** 133 **25.** 13 **27.** 0 **29.** 48 **31.** $^-56$ **33.** $^-160$ **35.** Commutative property: changing the *order* of the factors does not change the product. Associative property: changing the *grouping* of the factors does not change the product. Examples will vary. **37.** Examples will vary. Some possibilities are: **(a)** $6 \cdot ({}^-1) = {}^-6$; $2 \cdot ({}^-1) = {}^-2$; $15 \cdot ({}^-1) = {}^-15$ **(b)** $^-6 \cdot ({}^-1) = 6$; $^-2 \cdot ({}^-1) = 2$; $^-15 \cdot ({}^-1) = 15$ The result of multiplying any nonzero number times $^-1$ is the number with the opposite sign. **39.** $9 \cdot {}^-3 + 9 \cdot 5$; Both products are 18. **41.** $8 \cdot 25$; Both products are 200. **43.** $({}^-3 \cdot {}^-2) \cdot {}^-5$; Both products are $^-30$. **45.** \$300 • 50 = \$15,000; \$324 • 52 = \$16,848 **47.** $^-\$10{,}000 \cdot 10 = {}^-\$100{,}000$; $^-\$9950 \cdot 12 = {}^-\$119{,}400$ **49.** \$200 • 10 = \$2000; \$182 • 13 = \$2366 **51.** 20 • 400 = 8000 hours; 24 • 365 = 8760 hours **53.** $^-512$ **55.** 0 **57.** $^-355{,}299$ **59.** \$247 **61.** $^-22$ degrees

Section 1.7 (page 55)

1. **(a)** 7 **(b)** 7 **(c)** $^-7$ **(d)** $^-7$ **3.** **(a)** $^-7$ **(b)** 7 **(c)** $^-7$ **(d)** 7 **5.** **(a)** 1 **(b)** 35 **(c)** $^-13$ **(d)** 1 **7.** **(a)** 0 **(b)** undefined **(c)** undefined **(d)** 0 **9.** $^-4$ **11.** $^-3$ **13.** 6 **15.** $^-11$ **17.** undefined **19.** $^-14$ **21.** 10 **23.** 4 **25.** $^-1$ **27.** 0 **29.** 191 **31.** $^-499$ **33.** 2 **35.** $^-4$ **37.** 40 **39.** $^-48$ **41.** 5 **43.** 0 **45.** 2 ÷ 1 = 2, but 1 ÷ 2 = 0.5 so division is not commutative. **47.** Similar: If the signs match, the result is positive. If the signs are different, the result is negative. Different: Multiplication is commutative, division is not. You can multiply by zero, but dividing by zero is not allowed. **49.** Some possibilities are: **(a)** $\frac{^-6}{^-1} = 6$; $\frac{^-2}{^-1} = 2$; $\frac{^-15}{^-1} = 15$ **(b)** $\frac{6}{^-1} = {}^-6$; $\frac{2}{^-1} = {}^-2$; $\frac{15}{^-1} = {}^-15$. When dividing by $^-1$, the sign of the number changes to its opposite. **51.** $^-40{,}000 \div 20 = {}^-2000$ feet; $^-36{,}198 \div 18 = {}^-2011$ feet **53.** $^-\$200 + \$500 = \$300$; $^-\$238 + \$450 = \$212$ **55.** 400 − 100 = 300 days; 365 − 106 = 259 days **57.** $^-700 \cdot 40 = {}^-28{,}000$ feet; $^-730 \cdot 37 = {}^-27{,}010$ feet **59.** 300 ÷ 5 = 60 miles; 315 ÷ 5 = 63 miles **61.** 168 average score **63.** The back shows 520 grams, which is 10 grams more than the front. **65.** $^-\$15$ **67.** 16 hours, with 40 minutes left over **69.** 13 pies, with 4 pieces of pie left over **71.** 48 computers, with \$1176 left over **73.** $^-10$ **75.** undefined

Section 1.8 (page 67)

1. 4 • 4 • 4; 4 cubed or 4 to the third power **3.** 2^7; 128; 2 to the seventh power **5.** 5^4; 625; 5 to the fourth power **7.** 7^2; 7 • 7; 49 **9.** 10^1; 10; 10 **11.** **(a)** 10 **(b)** 100 **(c)** 1000 **(d)** 10,000 **13.** **(a)** 4 **(b)** 16 **(c)** 64 **(d)** 256 **15.** 9,765,625 **17.** 4096 **19.** 4 **21.** 25 **23.** $^-64$ **25.** 81 **27.** $^-1000$ **29.** 1 **31.** 108 **33.** 225 **35.** $^-750$ **37.** $^-32$ **39.** The answers are 4, $^-8$, 16, $^-32$, 64, $^-128$, 256, $^-512$. When a negative number is raised to an even power, the answer is positive; when raised to an odd power, the answer is negative. **41.** $^-6$ **43.** 0 **45.** $^-39$ **47.** 16 **49.** 23 **51.** $^-43$ **53.** 7 **55.** $^-3$ **57.** 0 **59.** $^-38$ **61.** 41 **63.** $^-2$ **65.** 13 **67.** 126 **69.** 8 **71.** $\frac{27}{^-3} = {}^-9$ **73.** $\frac{^-48}{^-4} = 12$ **75.** $\frac{^-60}{^-1} = 60$ **77.** $^-4050$ **79.** 7 **81.** $\frac{27}{0}$ is undefined

Chapter 1 Review Exercises (page 77)

1. 86; 0; 35,600 **2.** eight hundred six **3.** three hundred nineteen thousand, twelve **4.** sixty million, three thousand, two hundred **5.** fifteen trillion, seven hundred forty-nine billion, six **6.** 504,100 **7.** 620,080,000 **8.** 99,007,000,356

9. 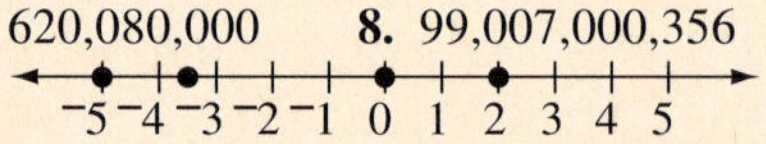

10. > **11.** < **12.** > **13.** < **14.** 5 **15.** 9 **16.** 0 **17.** 125 **18.** $^-1$ **19.** $^-13$ **20.** $^-3$ **21.** 0 **22.** 1 **23.** $^-24$ **24.** $^-7$ **25.** 3 **26.** 0 **27.** $^-17$ **28.** 5; $^-5 + 5 = 0$ **29.** $^-18$; $18 + {}^-18 = 0$ **30.** $^-7$ **31.** 17 **32.** $^-16$ **33.** 13 **34.** 18 **35.** $^-22$ **36.** 0 **37.** $^-20$ **38.** $^-1$ **39.** $^-3$ **40.** 14 **41.** 15 **42.** $^-16$ **43.** 3 **44.** $^-8$ **45.** 0 **46.** ≈$^-210$ **47.** ≈59,000 **48.** ≈85,000,000 **49.** ≈$^-3000$ **50.** ≈$^-7{,}060{,}000$ **51.** ≈400,000 **52.** $^-200$ pounds **53.** $^-1000$ feet **54.** 400,000,000 directories

55. 9,000,000,000 purchases **56.** -54 **57.** 56 **58.** -100 **59.** 0 **60.** 24 **61.** 17 **62.** -48 **63.** 125 **64.** -36 **65.** 50 **66.** -72 **67.** 9 **68.** -7 **69.** undefined **70.** 5 **71.** -18 **72.** 0 **73.** 15 **74.** -1 **75.** -5 **76.** 18 **77.** 0 **78.** 156 days and 2 extra hours **79.** 10,000 **80.** 32 **81.** 27 **82.** 16 **83.** -125 **84.** 8 **85.** 324 **86.** -200 **87.** -25 **88.** -2 **89.** 10 **90.** -28 **91.** $\frac{8}{-8} = -1$ **92.** $\frac{11}{0}$ is undefined **93.** Associative property of addition **94.** Commutative property of multiplication **95.** Addition property of zero **96.** Multiplication property of zero **97.** Distributive property **98.** Associative property of multiplication **99.** $10,000 • 200 = $2,000,000; $11,900 • 192 = $2,284,800 **100.** $200 + $400 − $700 = -$100; $185 + $428 − $706 = -$93 **101.** 800 ÷ 20 = 40 miles; 840 ÷ 24 = 35 miles **102.** ($40 • 20) + ($90 • 10) = $1700; ($39 • 19) + ($85 • 12) = $1761 **103.** -$700, $-100, $700, $900, $0, $700 **104.** Jan, April **105.** $2100 **106.** $1850

Chapter 1 Test (page 81)

1. twenty million, eight thousand, three hundred seven **2.** 30,000,700,005 **3.**

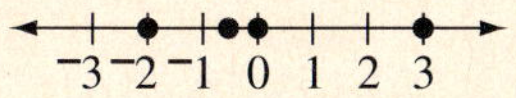

4. >; < **5.** 10; 14 **6.** -6 **7.** -5 **8.** 7 **9.** -40 **10.** 10 **11.** 64 **12.** -50 **13.** undefined **14.** -60 **15.** -5 **16.** -45 **17.** 6 **18.** 25 **19.** 0 **20.** 9 **21.** 128 **22.** -2 **23.** 8 **24.** -16 **25.** An exponent shows how many times to use a factor in repeated multiplication. Examples will vary. Some possibilities are: $(2)^4 = 2 \cdot 2 \cdot 2 \cdot 2 = 16$ and $(^-3)^2 = {^-3} \cdot {^-3} = 9$. **26.** Examples will vary. Some possibilities are: Commutative: 2 + 5 = 5 + 2; Associative: (-1 + 4) + 2 = -1 + (4 + 2) **27.** 900 **28.** 36,420,000,000 **29.** 350,000 **30.** *estimate:* 800,000 − 700,000 = 100,000 cars; *exact:* 80,991 cars **31.** *estimate:* $200 + $300 + -$500 = $0; *exact:* -$29 **32.** *estimate:* -1000 ÷ 10 = -100 yards; *exact:* -95 yards **33.** *estimate:* 30 • 100 = 3000 calories; *exact:* 3410 calories **34.** 27 cartons because 26 cartons would leave 28 pounds unpacked

Chapter 2

Section 2.1 (page 91)

1. *c* is variable; 4 is constant **3.** *h* is variable; 5 is coefficient **5.** *m* is variable; -3 is constant **7.** *c* is variable; 2 is coefficient; 10 is constant **9.** *x* and *y* are variables **11.** *g* is variable; -6 is coefficient **13. (a)** 654 + 10 is 664 robes **(b)** 208 + 10 is 218 robes **15. (a)** 3 • 11 is 33 inches **(b)** 3 • 3 is 9 feet **17. (a)** 3 • 12 − 5 is 31 brushes **(b)** 3 • 16 − 5 is 43 brushes **19. (a)** $\frac{332}{4}$ is 83 **(b)** $\frac{673}{7}$ is 91 **21.** 12 + 12 + 12 + 12 is 48; 4 • 12 is 48; 0 + 0 + 0 + 0 is 0; 4 • 0 is 0; -5 + -5 + -5 + -5 is -20; 4 • -5 is -20 **23.** -2 • -4 + 5 is 8 + 5 is 13; -2 • -6 + -2 is 12 + -2 is 10; -2 • 0 + -8 is 0 + -8 is -8 **25.** An expression expresses, or tells, the rule for doing something. A variable is a letter that represents the part of a rule that varies or changes depending upon the situation. For example, $c + 5$ is an expression, and c is the variable. **27.** $b \cdot 1 = b$ or $1 \cdot b = b$ **29.** $\frac{b}{0}$ is undefined or $b \div 0$ is undefined. **31.** $c \cdot c \cdot c \cdot c \cdot c \cdot c$ **33.** $x \cdot x \cdot x \cdot x \cdot y \cdot y \cdot y$ **35.** $^-3 \cdot a \cdot a \cdot a \cdot b$ **37.** $9 \cdot x \cdot y \cdot y$ **39.** $^-2 \cdot c \cdot c \cdot c \cdot c \cdot c \cdot d$ **41.** $a \cdot a \cdot a \cdot b \cdot c \cdot c$ **43.** 16 **45.** -24 **47.** -18 **49.** -128 **51.** -18,432 **53.** 311,040 **55.** 56 **57.** $\frac{36}{0}$ is undefined **59.** -1 **61.** -5 **63.** 0 **65.** 4

Section 2.2 (page 103)

1. $2b^2$; b^2; The coefficients are 2 and 1. **3.** ^-xy; $2xy$; The coefficients are -1 and 2. **5.** 7; 3; -4; The like terms are constants. **7.** $12r$ **9.** $6x^2$ **11.** ^-4p **13.** $^-3a^3$ **15.** 0 **17.** xy **19.** $6t^4$ **21.** $4y^2$ **23.** ^-8x **25.** $12a + 4b$ **27.** $7rs + 14$ **29.** $a + 2ab^2$ **31.** $^-2x + 2y$ **33.** $7b^2$ **35.** can't be simplified **37.** $^-15r + 5s + t$ **39.** $30a$ **41.** $^-8x^2$ **43.** $^-20y^3$ **45.** $18cd$ **47.** $21a^2bc$ **49.** $12w$ **51.** $6b + 36$ **53.** $7x - 7$ **55.** $21t + 3$ **57.** $^-10r - 6$ **59.** $^-9k - 36$ **61.** $50m - 300$ **63.** $8y + 16$ **65.** $6a^2 + 3$ **67.** $9m - 34$ **69.** -25 **71.** $24x$ **73.** $5n + 13$ **75.** $11p - 1$ **77.** A simplified expression still has variables, but is written in a simpler way. When evaluating an expression, the variables are all replaced by specific numbers. **79.** Like terms have matching variable parts, that is, matching letters and exponents. The coefficients do not have to match. Examples will vary. **81.** Keep the variable part unchanged when combining like terms. The correct answer is $5x + 8$. **83.** $^-2y + 9$ **85.** 0 **87.** ^-9x **89.** -7; 7 + -7 = 0 **91.** 12; -12 + 12 = 0

Section 2.3 (page 113)

1. 58 is the solution **3.** -16 is the solution **5.** -12 is the solution See Student's Solutions Manual for a sample of a check of answers to 7–39. **7.** $p = 4$ **9.** $r = 10$ **11.** $n = {^-8}$ **13.** $k = 18$ **15.** $y = 6$ **17.** $r = {^-6}$ **19.** $x = 11$ **21.** $t = {^-3}$ **23.** Does not balance. Correct solution is 8. **25.** Balances. -18 is correct solution. **27.** Does not balance. Correct solution is 0. **29.** $c = 6$ **31.** $y = 5$ **33.** $b = {^-30}$ **35.** $t = 0$ **37.** $z = {^-7}$ **39.** $w = 3$ **41.** $x = {^-10}$ **43.** $a = 0$ **45.** $y = {^-25}$ **47.** $x = 15$ **49.** $k = 113$ **51.** $b = 18$ **53.** $r = {^-5}$ **55.** $n = {^-105}$ **57.** $h = {^-5}$ **59.** An equation has an equal sign which shows that the two sides balance. An expression does not have an equal sign. Examples will vary. **61.** No, the solution is -14, the number you replace x with in the original equation. **63.** Equations will vary. Some possibilities are: $n - 1 = {}^-4 - 7$ and $3 - 8 = x - 3$ **65.** $m = {}^-19$ **67.** $x = 2$ **69.** **(a)** $^-42$ **(b)** 42 **(c)** $^-42$ **71.** **(a)** $^-3$ **(b)** 3 **(c)** 3 **73.** **(a)** undefined **(b)** 0 **75.** **(a)** $^-15$ **(b)** 1

Section 2.4 (page 123)

See Student's Solutions Manual for a sample of a check of answers to 1–16.
1. $z = 2$ **3.** $r = 4$ **5.** $y = 0$ **7.** $k = {}^-10$ **9.** $r = 6$ **11.** $b = {}^-5$ **13.** $r = 3$ **15.** $p = {}^-3$ **17.** $a = {}^-5$ **19.** $x = {}^-10$ **21.** $w = 0$ **23.** $t = 3$ **25.** $t = 0$ **27.** $m = 9$ **29.** $y = {}^-1$ **31.** $z = {}^-5$ **33.** $p = {}^-2$ **35.** $k = 7$ **37.** $b = {}^-3$ **39.** $x = {}^-32$ **41.** $w = 2$ **43.** $n = 50$ **45.** $p = {}^-10$ **47.** Each solution is the opposite of the number in the equation. So the rule is: change the number in the equation to its opposite. In $^-x = 5$, the opposite of 5 is $^-5$, so $x = {}^-5$.
49. Divide by the coefficient of x, which is 3, *not* by the opposite of 3. The correct solution is 5. **51.** $y = 27$ **53.** $x = 1$ **55.** 17 **57.** $^-18$ **59.** 24

Section 2.5 (page 131)

See Student's Solutions Manual for a sample of a check of answers to 1–16.
1. $p = 1$ **3.** $y = 1$ **5.** $m = 0$ **7.** $a = {}^-2$ **9.** $x = {}^-4$ **11.** $p = 4$ **13.** $k = {}^-2$ **15.** $a = 5$ **17.** $w = 6$ **19.** $y = {}^-9$ **21.** $t = {}^-5$ **23.** $x = 0$ **25.** $h = 1$ **27.** $y = {}^-2$ **29.** $m = {}^-3$ **31.** $w = 2$ **33.** $x = 5$ **35.** $a = 3$ **37.** $b = {}^-3$ **39.** $k = 4$ **41.** $c = 0$ **43.** $y = {}^-5$ **45.** $n = 21$ **47.** $c = 30$ **49.** $p = {}^-2$ **51.** $b = {}^-2$
53. The series of steps may vary. One possibility is:

$^-2t - 10 = 3t + 5$	Change subtraction to adding the opposite.
$^-2t + {}^-10 = 3t + 5$ $\underline{2t \qquad\quad 2t}$	Add $2t$ to both sides (addition property).
$0 + {}^-10 = 5t + 5$ $\underline{\quad {}^-5 \qquad {}^-5}$	Add $^-5$ to both sides (addition property).
$\dfrac{^-15}{5} = \dfrac{5t}{5}$	Divide both sides by 5 (division property).
$^-3 = t$	

55. **Check** $^-8 + \underbrace{4(3)} = \underbrace{2(3)} + 2$

$$\underbrace{^-8 + 12}_{4} = \underbrace{6 + 2}_{8} \qquad 4 \neq 8$$

Does not balance, so 3 is not the correct solution. The student added ^-2a to $^-8$ on the left side, instead of adding ^-2a to $4a$. The correct solution is 5.
57. $^-16$ **59.** 1 **61.** $\dfrac{-3}{0}$ is undefined

Chapter 2 Review Exercises (page 141)

1. Variable is k, constant is 3.
2. **(a)** 70 test tubes **(b)** 106 test tubes
3. **(a)** $x \bullet x \bullet y \bullet y \bullet y \bullet y$ **(b)** $5 \bullet a \bullet b \bullet b \bullet b$
4. **(a)** 9 **(b)** $^-27$ **(c)** $^-128$ **(d)** 720 **5.** $ab^2 + 3ab$
6. $^-4x + 2y - 7$ **7.** $16g^3$ **8.** $12r^2t$
9. $5k + 10$ **10.** $^-6b - 8$ **11.** $6y$ **12.** $20x + 2$
13. Expressions will vary. One possibility is $6a^3 + a^2 + 3a - 6$ **14.** $n = {}^-11$ **15.** $a = 4$ **16.** $m = {}^-8$ **17.** $k = 10$ **18.** $t = 0$ **19.** $p = {}^-6$ **20.** $r = 2$ **21.** $h = {}^-12$ **22.** $w = 4$ **23.** $c = {}^-2$ **24.** $n = {}^-3$ **25.** $a = {}^-5$ **26.** $p = 10$ **27.** $y = 5$ **28.** $m = 3$ **29.** $x = 9$ **30.** $b = 7$ **31.** $z = {}^-3$ **32.** $n = 4$ **33.** $t = 0$ **34.** $d = 5$ **35.** $b = {}^-2$

Chapter 2 Test (page 143)

1. $^-7$ is coefficient; w is variable; 6 is constant
2. Buy 177 hot dogs. **3.** $x \bullet x \bullet x \bullet x \bullet y \bullet y \bullet y$
4. $4 \bullet a \bullet b \bullet b \bullet b \bullet b$ **5.** $^-200$ **6.** $^-4w^3$ **7.** 0
8. c **9.** cannot be simplified **10.** $^-40b^2$ **11.** $15k$
12. $21t + 28$ **13.** $^-4a - 24$ **14.** $6x - 15$
15. $^-9b + c + 6$ **16.** $x = 5$ **17.** $w = {}^-11$
18. $p = {}^-14$ **19.** $a = 3$ **20.** $n = {}^-8$ **21.** $m = 15$
22. $x = {}^-1$ **23.** $m = 2$ **24.** $b = 54$ **25.** $c = 0$
26. Equations will vary. Two possibilities are: $x - 5 = {}^-9$ and $^-24 = 6y$

Cumulative Review 1–2 (page 145)

1. three hundred six billion, four thousand, two hundred ten **2.** 800,066,000 **3.** $>$; $<$
4. **(a)** Commutative property of addition
(b) Multiplication property of zero **(c)** Distributive property. **5.** **(a)** 9000 **(b)** 290,000 **6.** $^-8$
7. 10 **8.** 30 **9.** 25 **10.** 7 **11.** 0
12. $^-64$ **13.** undefined **14.** $^-60$ **15.** $^-40$
16. $^-9$ **17.** $^-25$ **18.** $^-28$ **19.** $\dfrac{11}{^-11} = {}^-1$
20. $600 \div 20 = 30$ miles; $616 \div 22 = 28$ miles
21. $^-50 + 20 = {}^-30$ degrees; $^-48 + 23 = {}^-25$ degrees
22. $\$2000 - (50 \bullet 10) = \$1500$; $\$2132 - (52 \bullet 8) = \$1716$ **23.** $10(600 + 40) = \$6400$; $12(552 + 35) = \$7044$ **24.** $^-4 \bullet a \bullet b \bullet b \bullet b \bullet c \bullet c$ **25.** 120
26. h **27.** 0 **28.** $5n^2 - 4n - 2$ **29.** $^-30b^2$
30. $28p - 28$ **31.** $^-9w^2 - 12$ **32.** $x = {}^-4$
33. $y = {}^-6$ **34.** $k = 7$ **35.** $m = 4$ **36.** $x = {}^-1$
37. $r = {}^-18$ **38.** $b = {}^-5$ **39.** $t = 1$ **40.** $y = {}^-12$

Chapter 3

Section 3.1 (page 153)

1. $P = 36$ cm **3.** $P = 100$ in.
5. $P = 4$ miles **7.** $P = 88$ mm

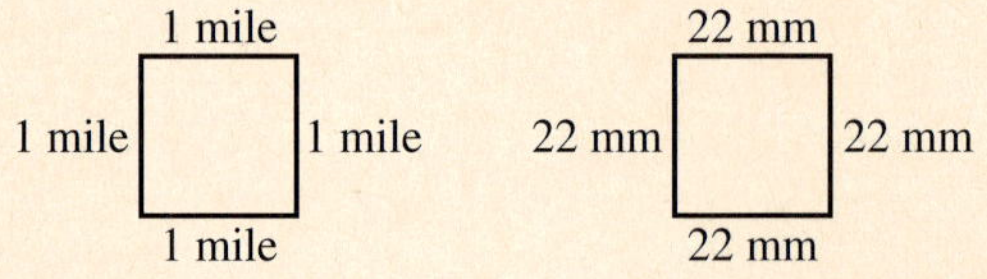

9. $s = 30$ ft **11.** $s = 1$ mm **13.** $s = 23$ yards
15. $s = 2$ ft
17. $P = 28$ yd
Check 8 yd + 8 yd + 6 yd + 6 yd = 28 yd
19. $P = 70$ cm
Check 25 cm + 25 cm + 10 cm + 10 cm = 70 cm

21. $P = 72$ ft

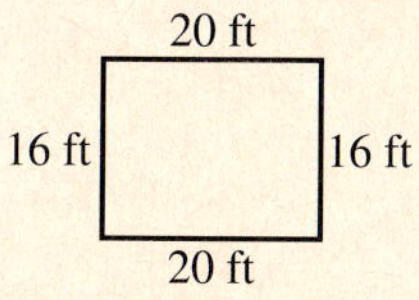

23. $P = 26$ in.

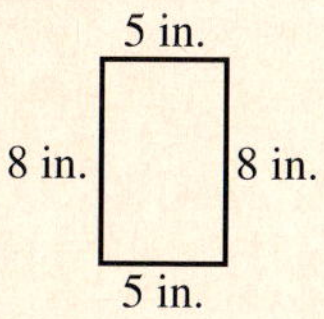

25. $l = 9$ cm

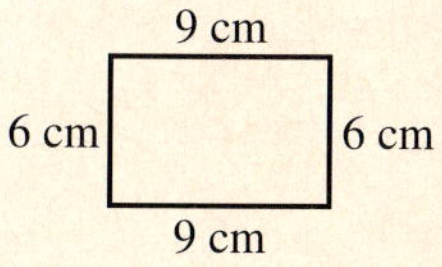

27. $w = 1$ mile

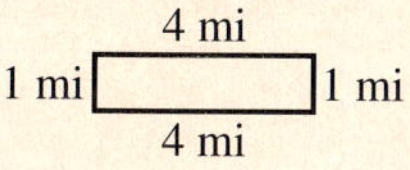

29. $w = 2$ ft

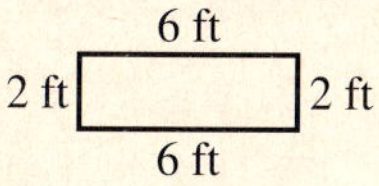

31. $l = 2$ m

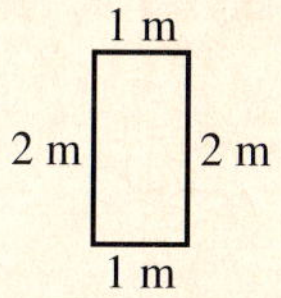

33. $P = 208$ m **35.** $P = 320$ ft **37.** $P = 54$ mm **39.** $P = 48$ ft **41.** $P = 78$ in. **43.** $P = 125$ m **45.** ? = 40 cm **47.** ? = 12 in.; $P = 78$ in. **49.** **(a)** Sketches will vary. **(b)** Use the formula $P = 2l + 2w$ **51.** **(a)** Sketches will vary. **(b)** Formula for perimeter of an equilateral triangle is $P = 3s$ where s is the length of one side. **(c)** The formula will *not* work for other kinds of triangles because the sides will have different lengths. **53.** 36 **55.** $^{-}8$ **57.** $x \cdot x$ **59.** $4^2 = 16$

SECTION 3.2 (page 163)

1. $A = 77\text{ ft}^2$ **3.** $A = 100\text{ m}^2$ **5.** $A = 775\text{ mm}^2$ **7.** $A = 36\text{ in}^2$ Sketches for Exercises 9–22 and 27–32 may vary; show them to your instructor. **9.** $A = 105\text{ cm}^2$ **11.** $A = 72\text{ ft}^2$ **13.** $A = 625\text{ mi}^2$ **15.** $A = 1\text{ m}^2$ **17.** $l = 6$ ft **19.** $w = 53$ yd **21.** $l = 14$ in. **23.** $s = 6$ m **25.** $s = 2$ ft **27.** $h = 20$ cm **29.** $b = 17$ in. **31.** $h = 1$ m **33.** Perimeter is the distance around the outside edges of a shape; area is the surface inside the shape measured in square units. Drawings will vary. **35.** $P = 32$ m; $A = 39\text{ m}^2$ **37.** $P = 34$ yd; $A = 56\text{ yd}^2$ **39.** $P = 36$ cm; $A = 80\text{ cm}^2$ **41.** Height is not part of perimeter; square units are used for area, not perimeter. $P = 25\text{ cm} + 25\text{ cm} + 25\text{ cm} + 25\text{ cm}$; $P = 100$ cm Sketches for Exercises 43–46 will vary. **43.** \$108 **45.** \$725 **47.** 42 m^2 **49.** $A = 6528\text{ ft}^2$ **51.** $y = 2$ **53.** $n = {}^{-}3$ **55.** $n = 11$

SECTION 3.3 (page 173)

1. $14 + x$ or $x + 14$ **3.** $^{-}5 + x$ or $x + {}^{-}5$ **5.** $20 - x$ **7.** $x - 9$ **9.** $x - 4$ **11.** $^{-}6x$ **13.** $2x$ **15.** $\frac{x}{2}$ **17.** $2x + 8$ or $8 + 2x$ **19.** $7x - 10$ **21.** $2x + x$ or $x + 2x$ **23.** $4n - 2 = 26$; $n = 7$ **25.** $2n + n = {}^{-}15$ or $n + 2n = {}^{-}15$; $n = {}^{-}5$ **27.** $5n + 12 = 7n$; $n = 6$ **29.** $30 - 3n = 2 + n$; $n = 7$ **31.** Let w be Ricardo's weight. $w + 15 - 28 + 5 = 177$. He weighed 185 pounds originally. **33.** Let c be the number of cookies the children ate. $18 - c + 36 = 49$. Her children ate 5 cookies. **35.** Let p be the number of pens in each box. $6p - 32 - 35 = 5$. There were 12 pens in each box. **37.** Let d be each member's dues. $14d + 340 - 575 = {}^{-}25$. Each member paid \$15. **39.** Let a be Tamu's age. $4a - 75 = a$. Tamu is 25 years old. **41.** Let m be the amount Brenda spent. $2m - 3 = 81$. Brenda spent \$42. **43.** Let b be the number of pieces in each bag. $5b - 3 \cdot 48 = b$. There were 36 pieces of candy in each bag. **45.** $k = 350$ **47.** $n = 19$

SECTION 3.4 (page 181)

1. a is my age; $a + 9$ is sister's age; $a + 9 + a = 51$; I am 21; my sister is 30. **3.** m is husband's earnings; $m + 1500$ is Lien's; $m + 1500 + m = 37{,}500$; Husband earned \$18,000; Lien earned \$19,500. **5.** m is printer's cost; $5m$ is computer's cost; $5m + m = \$1320$; Computer cost \$1100; printer cost \$220. **7.** shorter piece is x; longer piece is $x + 10$; $x + x + 10 = 78$; shorter piece is 34 cm; longer piece is 44 cm. **9.** longer piece is x; shorter piece is $x - 7$; $x + x - 7 = 31$; longer piece is 19 ft; shorter piece is 12 ft. **11.** first part is x; second part is x; third part is $x + 25$; $x + x + x + 25 = 706$; first part is 227 m; second part is 227 m; third part is 252 m.

13. length is 19 m

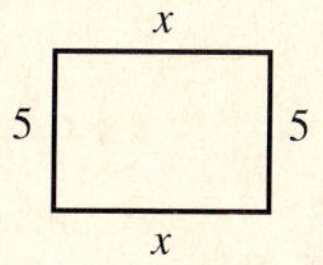

15. length is 12 ft width is 6 ft

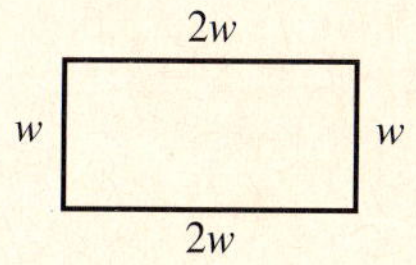

17. length is 13 in. width is 5 in.

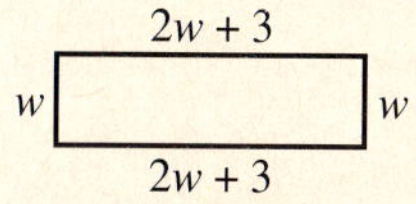

19. $A = 88\text{ in.}^2$ $P = 52$ inches

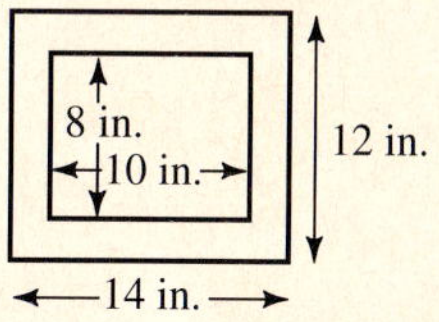

21.

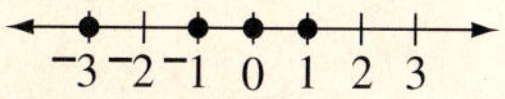

CHAPTER 3 REVIEW EXERCISES (page 187)

1. $P = 112$ cm **2.** $P = 22$ mi **3.** $P = 42$ yd **4.** $P = 141$ m **5.** 12 ft = $4s$; $s = 3$ ft **6.** 128 meters = $2l + 2$ (31 meters); $l = 33$ meters **7.** 72 in. = 2(21 in.) + $2w$; $w = 15$ in.

8. $A = 40\text{ ft}^2$

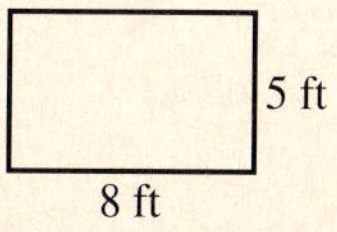

9. $A = 625\text{ m}^2$

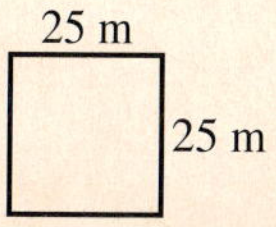

ANSWERS

10. $A = 208\ \text{yd}^2$

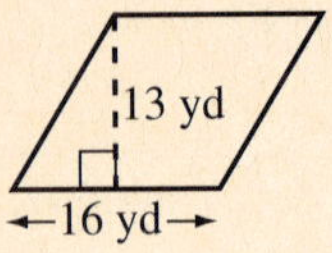

11. $126\ \text{ft}^2 = 14\ \text{ft} \bullet w;\ w = 9\ \text{ft}$ **12.** $88\ \text{cm}^2 = 11\ \text{cm} \bullet h;\ h = 8\ \text{cm}$ **13.** $100\ \text{mi}^2 = s \bullet s;\ s = 10\ \text{mi}$ **14.** $57 - x$ **15.** $15 + 2x$ or $2x + 15$ **16.** $^{-}9x$ **17.** $4n + 6 = {}^{-}30;\ n = {}^{-}9$ **18.** $10 - 2n = 4 + n;\ n = 2$ **19.** m is money originally in account; $m - 600 + 750 + 75 = 309$; \$84 was originally in Grace's account. **Check:** $84 - 600 + 750 + 75$ does equal 309. **20.** c is number of candles in each box; $4c - 25 = 23$; there were 12 candles in each box. **Check:** $4(12) - 25$ does equal 23. **21.** p is Reggie's prize money; $p + 300$ is Donald's prize money; $p + p + 300 = 1000$. Reggie gets \$350; Donald gets \$650. **Check:** \$650 is \$300 more than \$350; $\$350 + \$650 = \$1000$

22. x is the width;
$2x$ is the length;
$84 = 2(2x) + 2(x)$

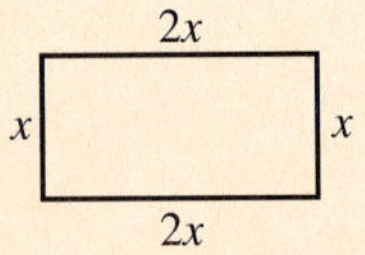

The width is 14 cm; the length is 28 cm.
Check: 28 cm is twice 14 cm;
$2(28\ \text{cm}) + 2(14\ \text{cm}) = 84\ \text{cm}$

23. **(a)** $36\ \text{ft} = 4s;\ s = 9\ \text{ft}$
(b) $A = 9\ \text{ft} \bullet 9\ \text{ft};\ A = 81\ \text{ft}^2$

24. Two possibilities are

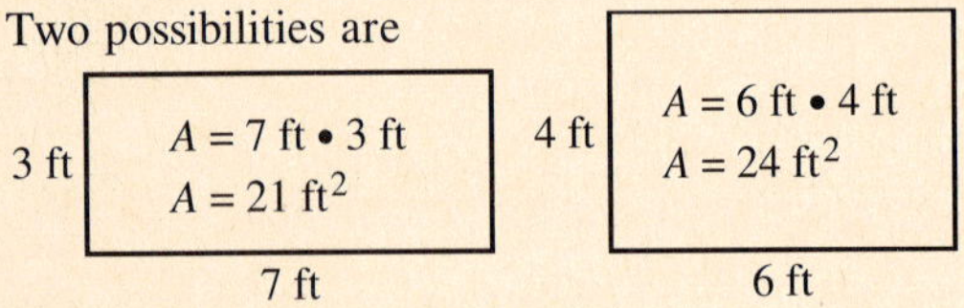

25. Let f be the fencing for the garden.
$36 - f + 20 = 41$; 15 ft of fencing was used on the garden.
Check: $36 - 15 + 20$ does equal 41

26. w is the width;
$w + 2$ is the length.

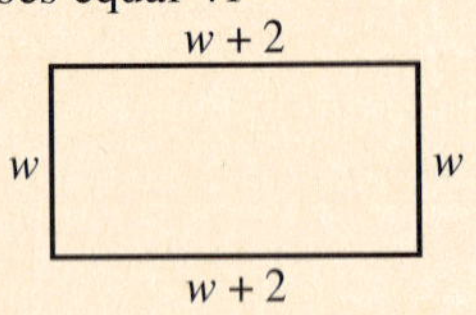

$36 = 2(w + 2) + 2 \bullet w$; width is 8 ft; length is 10 ft.
Check: 10 ft is 2 ft more than 8 ft;
$2(10\ \text{ft}) + 2(8\ \text{ft}) = 36\ \text{ft}$.

Chapter 3 Test (page 189)

1. $P = 262\ \text{m}$ **2.** $P = 40\ \text{in.}$ **3.** $P = 12\ \text{mi}$ **4.** $P = 12\ \text{ft}$ **5.** $P = 110\ \text{cm}$ **6.** $A = 486\ \text{mm}^2$ **7.** $A = 140\ \text{cm}^2$ **8.** $A = 3740\ \text{mi}^2$ **9.** $A = 36\ \text{m}^2$ **10.** $12\ \text{ft} = 4s;\ s = 3\ \text{ft}$ **11.** $34\ \text{ft} = 2l + 2(6\ \text{ft});\ l = 11\ \text{ft}$ **12.** $65\ \text{in}^2 = 13\ \text{in} \bullet h;\ h = 5\ \text{in.}$ **13.** $12\ \text{cm}^2 = 4\ \text{cm} \bullet w;\ w = 3\ \text{cm}$ **14.** $16\ \text{ft}^2 = s^2;\ s = 4\ \text{ft}$ **15.** Linear units like ft are used to measure length, width, height, perimeter, etc. Area is measured in square units like ft² (squares that measure 1 ft on each side). **16.** $4n + 40 = 0;\ n = {}^{-}10$ **17.** $7n - 23 = n + 7;\ n = 5$ **18.** son took \$15 **19.** daughter is 7 years old **20.** one piece is 57 cm; one piece is 61 cm **21.** length is 168 ft; width is 42 ft **22.** Marcella worked 11 hours; Tim worked 8 hours

Cumulative Review 1–3 (page 191)

1. four billion, two hundred six thousand, three hundred **2.** 70,005,489 **3.** $<;\ >$ **4.** **(a)** Multiplication property of 1 **(b)** Addition property of zero. **(c)** Associative property of multiplication **5.** **(a)** 3800 **(b)** 490,000 **6.** $^{-}24$ **7.** 27 **8.** $^{-}3$ **9.** $^{-}100$ **10.** 5 **11.** 0 **12.** $^{-}13$ **13.** 25 **14.** $^{-}32$ **15.** $\frac{27}{0}$ is undefined **16.** $10 \bullet w \bullet w \bullet x \bullet y \bullet y \bullet y \bullet y$ **17.** 240 **18.** $2k$ **19.** $3m^2 + 2m$ **20.** 0 **21.** $^{-}20a$ **22.** $^{-}2x^2 - 3$ **23.** $^{-}12n + 1$ **24.** $x = 2$ **25.** $y = {}^{-}1$ **26.** $b = 4$ **27.** $h = {}^{-}2$ **28.** $x = 0$ **29.** $a = {}^{-}3$ **30.** $P = 60\ \text{in.};\ A = 198\ \text{in}^2$ **31.** $P = 60\ \text{m};\ A = 225\ \text{m}^2$ **32.** $P = 24\ \text{ft};\ A = 32\ \text{ft}^2$ **33.** $5n + {}^{-}50 = 0;\ n = 10$ **34.** $10 - 3n = 2n;\ n = 2$ **35.** Let p be number of people originally. $p - 3 + 6 - 2 = 5$. There were 4 people in line. **36.** Let m be amount paid by each player. $12m - 2200 = {}^{-}40$. Each player paid \$180. **37.** Let g be one group; $3g$ the other; $g + 3g = 192$; 48 in one group; 144 in other group. **38.** w is the width; $w + 14$ is the length; $2(w + 14) + 2(w) = 92$; width is 16 ft; length is 30 ft

Chapter 4

Section 4.1 (page 201)

1. $\frac{5}{8};\ \frac{3}{8}$ **3.** $\frac{2}{3};\ \frac{1}{3}$ **5.** $\frac{7}{5};\ \frac{3}{5}$ **7.** $\frac{2}{11};\ \frac{3}{11};\ \frac{4}{11}$ **9.** $\frac{8}{25}$ **11.** $\frac{13}{71}$ **13.** 3; 4 **15.** 12; 7

17. proper: $\frac{1}{3}, \frac{5}{8}, \frac{7}{16}$ improper: $\frac{8}{5}, \frac{6}{6}, \frac{12}{2}$

19. proper: $\frac{3}{4}, \frac{9}{11}, \frac{7}{15}$ improper: $\frac{3}{2}, \frac{5}{5}, \frac{19}{18}$

21. One possibility is
3 ← Numerator
4 ← Denominator

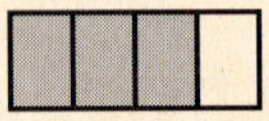

The denominator shows the number of equal parts in the whole and the numerator shows how many of the parts are being considered.

23.

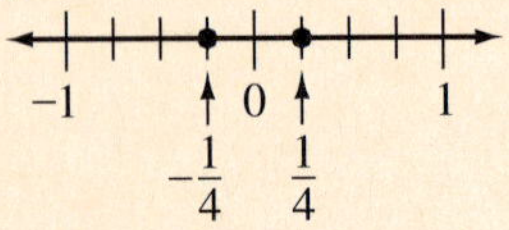

25.

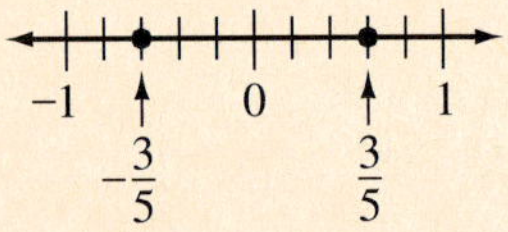

27.

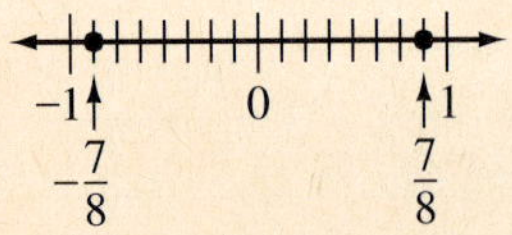

29. $-\frac{3}{4}$ **31.** $\frac{1}{2}$ **33.** $-\frac{5}{8}$ **35.** $\frac{2}{5}$ **37.** $\frac{9}{10}$

39. **(a)** $\frac{12}{24}$ **(b)** $\frac{8}{24}$ **(c)** $\frac{16}{24}$ **(d)** $\frac{6}{24}$ **(e)** $\frac{18}{24}$ **(f)** $\frac{4}{24}$ **(g)** $\frac{20}{24}$ **(h)** $\frac{3}{24}$ **(i)** $\frac{9}{24}$ **(j)** $\frac{15}{24}$ **41.** **(a)** $-\frac{1}{3}$ **(b)** $-\frac{2}{3}$ **(c)** $-\frac{2}{3}$ **(d)** $-\frac{1}{3}$ **(e)** $-\frac{2}{3}$ **(f)** Some possibilities are: $-\frac{4}{12} = -\frac{1}{3}$; $-\frac{8}{24} = -\frac{1}{3}$; $-\frac{20}{30} = -\frac{2}{3}$; $-\frac{24}{36} = -\frac{2}{3}$
43. $\frac{1467}{3912}$ **45.** $-\frac{1}{5}$ **47.** 7 **49.** 15
51. Multiply or divide the numerator and denominator by the same nonzero number. Some possibilities: $\frac{2}{3} = \frac{2 \cdot 4}{3 \cdot 4} = \frac{8}{12}$ and $\frac{10}{16} = \frac{10 \div 2}{16 \div 2} = \frac{5}{8}$
53. You cannot do it if you want the numerator to be a whole number, because 5 does not divide into 18 evenly. You could use multiples of 5 as the denominator, such as 10, 15, 20, etc. **55.** 10 **57.** −1 **59.** −6 **61.** 3 **63.** 1 **65.** 2 There are many correct ways to draw the answers for Exercises 67–82, so ask your instructor to check your work. **67.** $\frac{2}{5}$ is unshaded **69.** $\frac{5}{8}$ is unshaded **71.** $\frac{1}{4}$ is unshaded **73.** $\frac{2}{3}$ is unshaded **75.** $\frac{0}{6}$ is unshaded **83.** $x \cdot x \cdot x \cdot x$ **85.** $a \cdot a \cdot a \cdot a \cdot b \cdot b$ **87.** $b \cdot b \cdot b \cdot b$ **89.** $12 \cdot x \cdot y \cdot y$

SECTION 4.2 (page 215)

1. comp.; prime; comp.; prime; prime; comp.; comp.
3. $2 \cdot 3$ **5.** $2 \cdot 2 \cdot 5$ **7.** $5 \cdot 5$ **9.** $2 \cdot 2 \cdot 3 \cdot 3$ **11.** $2 \cdot 2 \cdot 11$ **13.** $2 \cdot 2 \cdot 2 \cdot 11$ **15.** $3 \cdot 5 \cdot 5$
17. A composite number has a factor(s) other than itself or 1. Examples include 4, 6, 8, 9, 10. A prime number is a whole number that has exactly two *different* factors, itself and 1. Examples include 2, 3, 5, 7, 11. The whole numbers 0 and 1 are neither prime nor composite.

19. $\frac{\overset{1}{\cancel{2}} \cdot \overset{1}{\cancel{2}} \cdot \overset{1}{\cancel{2}}}{\underset{1}{\cancel{2}} \cdot \underset{1}{\cancel{2}} \cdot \underset{1}{\cancel{2}} \cdot 2} = \frac{1}{2}$ **21.** $\frac{\overset{1}{\cancel{2}} \cdot \overset{1}{\cancel{2}} \cdot \overset{1}{\cancel{2}} \cdot \overset{1}{\cancel{2}} \cdot 2}{\underset{1}{\cancel{2}} \cdot \underset{1}{\cancel{2}} \cdot \underset{1}{\cancel{2}} \cdot \underset{1}{\cancel{2}} \cdot 3} = \frac{2}{3}$

23. $\frac{2 \cdot \overset{1}{\cancel{7}}}{3 \cdot \underset{1}{\cancel{7}}} = \frac{2}{3}$ **25.** $\frac{\overset{1}{\cancel{2}} \cdot 2 \cdot \overset{1}{\cancel{3}} \cdot 3}{\underset{1}{\cancel{2}} \cdot \underset{1}{\cancel{3}} \cdot 7} = \frac{6}{7}$ **27.** $\frac{3 \cdot 3 \cdot \overset{1}{\cancel{7}}}{2 \cdot 5 \cdot \underset{1}{\cancel{7}}} = \frac{9}{10}$

29. $\frac{\overset{1}{\cancel{3}} \cdot \overset{1}{\cancel{3}} \cdot 3}{\underset{1}{\cancel{3}} \cdot \underset{1}{\cancel{3}} \cdot 5} = \frac{3}{5}$ **31.** $\frac{\overset{1}{\cancel{2}} \cdot 2 \cdot \overset{1}{\cancel{3}}}{\underset{1}{\cancel{2}} \cdot \underset{1}{\cancel{3}} \cdot 3} = \frac{2}{3}$ **33.** $\frac{\overset{1}{\cancel{5}} \cdot 7}{2 \cdot 2 \cdot 2 \cdot \underset{1}{\cancel{5}}} = \frac{7}{8}$

35. $\frac{\overset{1}{\cancel{2}} \cdot \overset{1}{\cancel{3}} \cdot \overset{1}{\cancel{3}} \cdot \overset{1}{\cancel{5}}}{\underset{1}{\cancel{2}} \cdot 2 \cdot \underset{1}{\cancel{3}} \cdot \underset{1}{\cancel{3}} \cdot \underset{1}{\cancel{5}}} = \frac{1}{2}$ **37.** $\frac{2 \cdot \overset{1}{\cancel{3}} \cdot \overset{1}{\cancel{5}} \cdot \overset{1}{\cancel{7}}}{3 \cdot \underset{1}{\cancel{3}} \cdot \underset{1}{\cancel{5}} \cdot \underset{1}{\cancel{7}}} = \frac{2}{3}$

39. $\frac{\overset{1}{\cancel{3}} \cdot \overset{1}{\cancel{11}} \cdot 13}{\underset{1}{\cancel{3}} \cdot 3 \cdot 5 \cdot \underset{1}{\cancel{11}}} = \frac{13}{15}$ **41.** **(a)** $\frac{1}{4}$ **(b)** $\frac{1}{2}$ **(c)** $\frac{1}{10}$ **(d)** $\frac{60}{60} = 1$ **43.** **(a)** $\frac{1}{3}$ **(b)** $\frac{1}{5}$ **(c)** $\frac{7}{15}$
45. The result of dividing 3 by 3 is 1, so 1s should be written above and below the slashes. The numerator is $1 \cdot 1$, so the correct answer is $\frac{1}{4}$. **47.** Divide 437 by each of the prime numbers (2, 3, 5, 7, etc.) until you find one that divides evenly (no digits to the right of the decimal point). $437 = 19 \cdot 23$ **49.** $\frac{2c}{5}$ **51.** $\frac{4}{7}$ **53.** $\frac{6r}{5s}$ **55.** $\frac{1}{7n^2}$ **57.** already in lowest terms **59.** $\frac{7}{9y}$ **61.** $\frac{7k}{2}$ **63.** $\frac{1}{3}$ **65.** $\frac{c}{d}$ **67.** $\frac{6abd}{ce^2}$
69. **(a)** −28 **(b)** 27 **(c)** −80 **71.** **(a)** undefined **(b)** 0 **(c)** 9

SECTION 4.3 (page 227)

1. $-\frac{3}{16}$ **3.** $\frac{9}{10}$ **5.** $\frac{1}{2}$ **7.** −6 **9.** 36 **11.** $\frac{15}{4y}$ **13.** $\frac{1}{2}$ **15.** $\frac{6}{5}$ **17.** −9 **19.** $-\frac{1}{6}$ **21.** $\frac{11}{15d}$ **23.** b **25.** **(a)** Both products are $\frac{3}{20}$ because multiplication is commutative **(b)** The quotients are 12 and $\frac{1}{12}$. They are different because division is *not* commutative. **27.** **(a)** Answer is not in lowest terms; should be $\frac{\overset{1}{\cancel{3}} \cdot 2 \cdot \overset{1}{\cancel{4}}}{\underset{1}{\cancel{4}} \cdot \underset{1}{\cancel{3}} \cdot 3} = \frac{2}{3}$ **(b)** Divisor (4) must be changed to its reciprocal $\left(\frac{1}{4}\right)$. Correct answer is $\frac{1}{6}$ **(c)** Cannot divide out common factors until dividend is changed to its reciprocal. Should be $\frac{3}{10} \cdot \frac{3}{1} = \frac{9}{10}$ **(d)** Dividing by 0 is undefined. **29.** Rewrite division as multiplication. Leave the first number (dividend) the same. Change the second number to its reciprocal by "flipping" it. Then multiply. Divide out any common factors so the result will be in lowest terms. **31.** $\frac{4}{15}$ **33.** $-\frac{9}{32}$ **35.** 21 **37.** 15 **39.** undefined **41.** $-\frac{55}{12}$ **43.** $8b$ **45.** $\frac{3}{5d}$ **47.** $\frac{2x^2}{w}$ **49.** 48 vests **51.** Both lots have the same area, $\frac{3}{64}$ square mile **53.** 80 dispensers
55. 8 square miles **57.** 325 women; 455 men
59. **(a)** \$38,000 **(b)** \$7600 **61.** \$16,625
63. 9 trips **65.** −1 **67.** −6 **69.** −5 **71.** 4

SECTION 4.4 (page 239)

1. $\frac{7}{8}$ **3.** $-\frac{1}{2}$ **5.** $\frac{1}{2}$ **7.** $-\frac{9}{40}$ **9.** $-\frac{13}{24}$ **11.** $-\frac{3}{5}$ **13.** $-\frac{7}{18}$ **15.** $\frac{8}{7}$ **17.** $-\frac{3}{8}$ **19.** $\frac{3 + 5c}{15}$ **21.** $\frac{10 - m}{2m}$ **23.** $\frac{8}{b^2}$ **25.** $\frac{21 + bc}{7b}$ **27.** $\frac{-4 - cd}{c^2}$ **29.** $-\frac{44}{105}$ **31.** You cannot add or subtract until all the fractional pieces are the same size. For example, halves are larger than fourths, so you cannot add

$\frac{1}{2} + \frac{1}{4}$ until you rewrite $\frac{1}{2}$ as $\frac{2}{4}$. **33.** **(a)** Both sums are $\frac{1}{12}$ because addition is commutative. **(b)** The answers are $\frac{1}{3}$ and $-\frac{1}{3}$. They are different because subtraction is *not* commutative. **35.** **(a)** Cannot add fractions with unlike denominators. The LCD is 20 so should have $\frac{15}{20} + \frac{8}{20} = \frac{15+8}{20} = \frac{23}{20}$ **(b)** When rewriting fractions with 18 as denominator, you must multiply denominator *and numerator* by the same number. Should be $\frac{15}{18} - \frac{8}{18} = \frac{15-8}{18} = \frac{7}{18}$ **37.** $\frac{3}{4}$ acre **39.** $\frac{23}{24}$ cubic yard **41.** $\frac{41}{48}$ mile **43.** $\frac{3}{8}$ gallon **45.** $\frac{1}{4}$ of the day **47.** Work and travel 8 hours **49.** $\frac{1}{12}$ mile **51.** \$500 − 50 − 20 − 100= \$330; \$518 − 48 − 23 − 96 = \$351 **53.** 10 • 50 = 500 inches; 12 • 47 = 564 inches

SECTION 4.5 (page 253)

1. $7\frac{7}{8}$; 2 • 4 = 8 **3.** $1\frac{5}{21}$; 3 ÷ 3 = 1 **5.** $5\frac{1}{2}$; 4 + 2 = 6 **7.** $3\frac{2}{3}$; 4 − 1 = 3 **9.** $\frac{17}{18}$; 6 ÷ 6 = 1 **11.** $6\frac{1}{5}$; 8 − 2 = 6 **13.** P = 7 inches; $A = 3\frac{1}{16}$ square inches **15.** $P = 19\frac{1}{2}$ yards; $A = 21\frac{1}{8}$ square yards **17.** 13 + 9 = 22 feet; $21\frac{1}{6}$ feet **19.** 2 • 13 = 26 ounces; $21\frac{7}{8}$ ounces **21.** 4 − 2 = 2 miles; $2\frac{3}{10}$ miles **23.** 3 • 7 = 21 yards; $19\frac{1}{4}$ yards **25.** 10 − 2 − 3 = 5 cubic yards; $5\frac{1}{8}$ cubic yards **27.** 19 ÷ 5 is about 4 hours; $3\frac{3}{4}$ hours **29.** 3 + 8 + 2 = 13 tons; $12\frac{7}{8}$ tons **31.** 24 + 35 + 24 + 35 = 118 inches; $116\frac{1}{2}$ inches **33.** 25,730 ÷ 10 = 2573 anchors; 2480 anchors **35.**

12 • 19 = 228
24 • 36 = 864
16 • 74 = 1184
+ 12
\$2288; \$2296

37. −8 **39.** 144 **41.** −25 **43.** 3

SECTION 4.6 (page 261)

1. $\frac{9}{16}$ **3.** $\frac{8}{125}$ **5.** $-\frac{1}{27}$ **7.** $\frac{1}{32}$ **9.** $\frac{49}{100}$ **11.** $\frac{36}{25}$ or $1\frac{11}{25}$ **13.** $\frac{12}{25}$ **15.** $\frac{1}{100}$ **17.** $-\frac{3}{2}$ or $-1\frac{1}{2}$ **19.** The answers are $\frac{1}{4}, -\frac{1}{8}, \frac{1}{16}, -\frac{1}{32}, \frac{1}{64}, -\frac{1}{128}, \frac{1}{256}, -\frac{1}{512}$. When a negative number is raised to an even power, the answer is positive. When a negative number is raised to an odd power, the answer is negative. **21.** −4 **23.** $\frac{5}{16}$ **25.** $\frac{1}{3}$ **27.** $\frac{1}{6}$ **29.** $-\frac{17}{24}$ **31.** $-\frac{4}{27}$ **33.** $\frac{1}{36}$ **35.** $\frac{9}{64}$ square inch **37.** $\frac{19}{10}$ or $1\frac{9}{10}$ miles **39.** 4 **41.** $-\frac{25}{2}$ or $-12\frac{1}{2}$ **43.** $\frac{1}{14}$ **45.** $\frac{9}{100}$ **47.** $y = -8$ **49.** $n = 50$

SECTION 4.7 (page 269)

See Student's Solutions Manual for a sample check for Exercises 1–16.

1. $a = 30$ **3.** $b = -24$ **5.** $c = 6$ **7.** $m = \frac{3}{4}$ **9.** $d = -\frac{6}{5}$ **11.** $n = 12$ **13.** $r = -9$ **15.** $x = 24$ **17.** $y = 45$ **19.** $n = -18$ **21.** $x = \frac{1}{12}$ **23.** $b = -\frac{1}{8}$ **25.** **(a)** $\frac{1}{6}(18) + 1 = -2$; $3 + 1 = -2$; $4 \neq -2$. Does *not* balance. Correct solution is $x = -18$. **(b)** $-\frac{3}{2} = \frac{9}{4}\left(-\frac{2}{3}\right)$; $-\frac{3}{2} = -\frac{3}{2}$. Balances, so $-\frac{2}{3}$ is correct solution. **27.** There are many possibilities. Some examples are: $\frac{1}{2}x = 4$; $-\frac{1}{4}a = -2$; $\frac{3}{4}b = 6$. **29.** $109 = 100 + \frac{a}{2}$; 18 years old **31.** $122 = 100 + \frac{a}{2}$; 44 years old **33.** $\frac{11}{2}h - 220 = 209$; 78 inches **35.** $\frac{11}{2}h - 220 = 132$; 64 inches **37.** P = 58 in.; A = 154 in.^{2} **39.** P = 86 m; A = 297 m^2

SECTION 4.8 (page 279)

1. A = 1980 m^2 **3.** $A = \frac{115}{3}$ yd^2 or $38\frac{1}{3}$ yd^2 **5.** A = 198 m^2 **7.** A = 1196 m^2 **9.** Perimeter is the distance around the outside edges of a shape and is measured in linear units. Area is the space inside a flat shape and is measured in square units. Volume is the space inside a solid shape and is measured in cubic units. **11.** $A = \frac{63}{8}$ ft^2 or $7\frac{7}{8}$ ft^2 **13.** 127 m of curbing; 672 m^2

of sod **15.** $V = 528 \text{ cm}^3$ **17.** $V = 15\frac{5}{8} \text{ in.}^3$ **19.** $V = 800 \text{ cm}^3$ **21.** $V = 18 \text{ in.}^3$ **23.** $V = 651{,}775 \text{ m}^3$ **25.** $V = 513 \text{ cm}^3$ **27.** The correct answers are $A = 135 \text{ ft}^2$ (square feet, not squaring 135) and $5\frac{1}{2}$ in. (perimeter is in linear units not square units). **29.** hundreds **31.** thousands **33.** hundred thousands **35.** sixty million, three thousand, five hundred eight

Chapter 4 Review Exercises (page 289)

1. $\frac{2}{5}; \frac{2}{5}$ **2.** $\frac{3}{10}; \frac{7}{10}$
3.

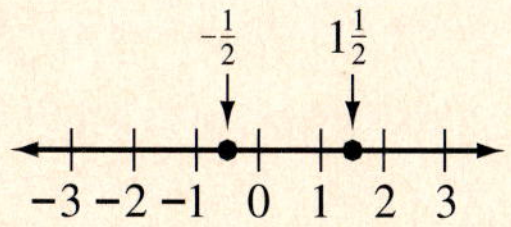

4. −4; 8; −1 **5.** $\frac{7}{8}$ **6.** $\frac{3}{5}$ **7.** already in lowest terms **8.** $\frac{3x}{8}$ **9.** $\frac{1}{5b}$ **10.** $\frac{4n}{7m^2}$ **11.** $\frac{1}{16}$ **12.** −12 **13.** $\frac{8}{27}$ **14.** $\frac{1}{6x}$ **15.** $2a^2$ **16.** $\frac{6}{7k}$ **17.** $\frac{5}{24}$ **18.** $-\frac{2}{15}$ **19.** $\frac{19}{6}$ **20.** $\frac{3}{2}$ **21.** $\frac{4n + 15}{20}$ **22.** $\frac{3y - 70}{10y}$ **23.** $2 \div 2 = 1; 1\frac{5}{13}$ **24.** $7 - 5 = 2; 2\frac{1}{2}$ **25.** $2 + 2 = 4; 4\frac{1}{20}$ **26.** $-\frac{27}{64}$ **27.** $\frac{1}{36}$ **28.** $-\frac{4}{5}$ **29.** $\frac{7}{6}$ or $1\frac{1}{6}$ **30.** 10 **31.** $-\frac{4}{27}$ **32.** $w = 20$ **33.** $r = -15$ **34.** $x = \frac{1}{2}$ **35.** $A = 14 \text{ ft}^2$ **36.** $V = 32\frac{1}{2} \text{ in.}^3$ **37.** $V = 93\frac{1}{3} \text{ m}^3$ **38.** $\frac{1}{4}$ pound; $7\frac{1}{2}$ pounds **39.** $\frac{5}{6}$ hour; $10\frac{11}{12}$ hours **40.** 12 preschoolers, 40 toddlers, 8 infants **41.** $P = 2\frac{1}{10}$ mi; $A = \frac{9}{40} \text{ mi}^2$

Chapter 4 Test (page 291)

1. $\frac{5}{6}; \frac{1}{6}$ **2.**

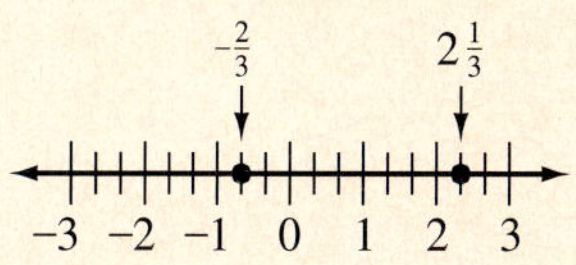

3. $\frac{1}{4}$ **4.** already in lowest terms **5.** $\frac{2a^2}{3b}$ **6.** $\frac{13}{15}$ **7.** −2 **8.** $-\frac{7}{40}$ **9.** 14 **10.** $-\frac{2}{27}$ **11.** $\frac{25}{8}$ or $3\frac{1}{8}$ **12.** $\frac{4}{9}$ **13.** $\frac{9}{16}$ **14.** $\frac{4}{7y}$ **15.** $\frac{24 - n}{4n}$ **16.** $\frac{10 + 3a}{15}$ **17.** $\frac{1}{18b}$ **18.** $-\frac{1}{18}$ **19.** $-\frac{31}{30}$ or $-1\frac{1}{30}$ **20.** $5 \div 1 = 5; \frac{64}{15}$ or $4\frac{4}{15}$ **21.** $3 - 2 = 1; 1\frac{1}{2}$ **22.** $d = 35$ **23.** $t = -\frac{15}{7}$ or $-2\frac{1}{7}$ **24.** $b = 8$ **25.** $x = -15$ **26.** $A = 48 \text{ m}^2$ **27.** $A = \frac{117}{2}$ or $58\frac{1}{2} \text{ yd}^2$ **28.** $V = 6480 \text{ m}^3$ **29.** $V = 16 \text{ yd}^3$ **30.** $14\frac{3}{4}$ hours; $1\frac{5}{6}$ hours **31.** $\frac{7}{2}$ or $3\frac{1}{2}$ days **32.** 7392 students work

Cumulative Review Exercises Chapters 1–4 (page 293)

1. five hundred five million, eight thousand, two hundred thirty-eight **2.** 35,600,000,916 **3. (a)** 60,700 **(b)** 100,000 **(c)** 3210 **4. (a)** Commutative property of multiplication **(b)** Associative property of addition **(c)** Distributive property **5.** −54 **6.** −20 **7.** undefined **8.** 81 **9.** 10 **10.** 4 **11.** −16 **12.** $\frac{27}{27} = 1$ **13.** −8 **14.** 35 **15.** $\frac{15a}{2}$ **16.** $\frac{3}{xy^2}$ **17.** $-\frac{8}{15}$ **18.** $\frac{5}{16}$ **19.** $\frac{14 - 3b}{21}$ **20.** $\frac{8n + 15}{5n}$ **21.** $\frac{13}{9}$ or $1\frac{4}{9}$ **22.** $\frac{13}{20}$ **23.** −1 **24.** $-\frac{5}{8}$ **25.** $a = 1$ **26.** $y = -5$ **27.** $k = -36$ **28.** $x = -4$ **29.** $m = 2$ **30.** $P = 18$ in.; $A = 20\frac{1}{4} \text{ in.}^2$ **31.** $P = 18\frac{1}{2}$ yd; $A = 15 \text{ yd}^2$ **32.** $P = 78$ mm; $A = 288 \text{ mm}^2$ **33.** Let b be the original weight of each bag; $3b - 16 - 25 = 79$; there were 40 pounds in each bag. **34.** Let w be the width; $2w$ is the length; $2(2w) + 2(w) = 102$; width is 17 yd; length is 34 yd

Chapter 5

Section 5.1 (page 301)

1. 7; 6; 3 **3.** 5; 1; 8 **5.** 4; 7; 0 **7.** 1; 6; 3 **9.** 1; 8; 9 **11.** 6; 2; 1 **13.** 410.25 **15.** 6.5432 **17.** 5406.045 **19.** $\frac{7}{10}$ **21.** $13\frac{2}{5}$ **23.** $\frac{7}{20}$ **25.** $\frac{33}{50}$ **27.** $10\frac{17}{100}$ **29.** $\frac{3}{50}$ **31.** $\frac{41}{200}$ **33.** $5\frac{1}{500}$ **35.** $\frac{343}{500}$ **37.** five tenths **39.** seventy-eight hundredths **41.** one hundred five thousandths **43.** twelve and four hundredths **45.** one and seventy-five thousandths **47.** 6.7 **49.** 0.32 **51.** 420.008 **53.** 0.0703 **55.** 75.030 **57.** Anne should not say "and" because that denotes a decimal point. **59.** 3-C **61.** 4-A **63.** one and six hundred two thousandths centimeters **65.** millionths, ten-millionths, hundred-millionths, billionths; these match the words on the left side of the chart with "ths" added. **67.** seventy-two million four hundred thirty-six thousand nine hundred fifty-five hundred-millionths **69.** eight thousand six and five hundred thousand one millionths **71.** 8240; 8200; 8000 **73.** 19,710; 19,700; 20,000

Section 5.2 (page 309)

1. ≈16.9 **3.** ≈0.956 **5.** ≈0.80 **7.** ≈3.661 **9.** ≈794.0 **11.** ≈0.0980 **13.** ≈49 **15.** ≈9.09 **17.** ≈82.0002 **19.** $0.82 **21.** $1.22 **23.** $0.50 **25.** ≈$17,250 **27.** ≈$310 **29.** ≈$379 **31.** Rounds to $0 (zero dollars) because $0.499 is closer to $0 than to $1. **33.** Round amounts less than $1.00 to nearest cent instead of nearest dollar. Round $0.4999 to $0.50. **35.** ≈$500 **37.** ≈$1.00 **39.** ≈$1000 **41.** 8000 + 6000 + 8000 = 22,000; 22,223 **43.** 80,000 + 100 + 800 = 80,900; 82,859

Section 5.3 (page 317)

1. 17.72 **3.** 11.98 **5.** 115.861 **7.** 59.323 **9.** 6 should be written 6.00; sum is 46.22. **11.** $0.3000 = \frac{3000 \div 1000}{10{,}000 \div 1000} = \frac{3}{10} = 0.3$ **13.** 89.7 **15.** 0.109 **17.** 0.91 **19.** 6.661 **21.** 15.32 should be on top; correct answer is 7.87 **23.** 23.013 **25.** −45.75 **27.** −6.69 **29.** −6.99 **31.** −4.279 **33.** −0.0035 **35.** 5.37 **37.** 0.275 **39.** 6.507 **41.** 1.81 **43.** 6056.7202 **45.** 40 − 20 = 20 hours; 26.15 hours **47.** $300 + $1 = $301; $311.09 **49.** 2 + 2 + 2 = 6 m; 6.2 m which is less than 6.4 m **51.** $20 − $9 = $11; $10.88 **53.** 8000 + 200 + 200 = 8400 miles; 8251.7 miles **55.** 10 + 5 = 15 hours; 14.3 hours **57.** 5 + 6 + 5 + 10 + 5 = 31 hours; 30 hours **59.** 20 + 6 + 20 + 6 = 52 inches; 52.1 inches **61.** $1105.75 **63.** $94.97 **65.** $219.22 **67.** $b = 1.39$ cm **69.** $k = 2.812$ inches **71.** 90.3 miles **73.** 9.372 gallons **75.** 19.486 **77.** 80 × 30 = 2400; 2324 **79.** 4000 × 200 = 800,000; 776,745

Section 5.4 (page 325)

1. 0.1344 **3.** −159.10 **5.** 15.5844 **7.** $34,500.20 **9.** −43.2 **11.** 0.432 **13.** 0.0432 **15.** 0.00432 **17.** 0.0000312 **19.** 0.000009 **21.** Multiplying by 10, decimal point moves one place to the right; by 100, two places to the right; by 1000, three places to the right. **23.** 40 × 5 = 200; 190.08 **25.** 40 × 40 = 1600; 1558.2 **27.** 7 × 5 = 35; 30.038 **29.** 3 × 7 = 21; 19.24165 **31.** unreasonable; $189.00 **33.** reasonable **35.** unreasonable; $3.19 **37.** unreasonable; 9.5 pounds **39.** $592.37 (rounded) **41.** $2.45 (rounded) **43.** $27.04 (rounded) **45.** $8771.00 **47.** $347.52 **49.** $76.50 **51.** $121.30 **53.** 46.3 gallons **55.** $388.34 **57.** $4.09 (rounded) **59.** 1000 ÷ 5 = 200; 190 R4 **61.** 20,000 ÷ 20 = 1000; 905 R15

Section 5.5 (page 335)

1. −3.9 **3.** 0.47 **5.** 400.2 **7.** 36 **9.** 0.06 **11.** 6000 **13.** 60 **15.** 0.0006 **17.** 25.3 **19.** ≈516.67 **21.** ≈−24.291 **23.** ≈10,082.647 **25.** Dividing by 10, decimal point moves one place to the left; by 100, two places to the left; by 1000, three places to the left. **27.** unreasonable; 40 ÷ 8 = 5; $8\overline{)37.8}$ = 4.725 **29.** reasonable; 50 ÷ 50 = 1 **31.** unreasonable; 300 ÷ 5 = 60; $5.1\overline{)307.02}$ = 60.2 **33.** unreasonable; 9 ÷ 1 = 9; $1.25\overline{)9.3}$ = 7.44 **35.** $4.00 (rounded) **37.** $19.46 **39.** $0.30 **41.** $5.89 per hour **43.** ≈21.2 miles per gallon **45.** ≈8.30 meters **47.** 0.05 meter **49.** 25.03 meters **51.** ≈$0.03 **53.** $237.25 **55.** 14.25 **57.** 3.8 **59.** −16.155 **61.** 3.714 **63.** $\frac{1}{2}$ **65.** $\frac{3}{5}$ **67.** > **69.** <

Section 5.6 (page 343)

1. 0.5 **3.** 0.75 **5.** 0.3 **7.** 0.9 **9.** 0.6 **11.** 0.875 **13.** 2.25 **15.** 14.7 **17.** 3.625 **19.** ≈0.333 **21.** ≈0.833 **23.** ≈1.889 **25.** $\frac{5}{9}$ means 5 ÷ 9 or $9\overline{)5}$ so correct answer is ≈0.556. **27.** Just add the whole number part to 0.375. So $1\frac{3}{8} = 1.375$; $3\frac{3}{8} = 3.375$; $295\frac{3}{8} = 295.375$. **29.** $\frac{2}{5}$ **31.** $\frac{5}{8}$ **33.** $\frac{7}{20}$ **35.** 0.35 **37.** $\frac{1}{25}$ **39.** $\frac{3}{20}$ **41.** 0.2 **43.** $\frac{9}{100}$ **45.** Shorter; 0.72 inch **47.** Too much; 0.005 gram **49.** 0.9991 cm, 1.0007 cm **51.** More; 0.05 inch **53.** 0.5399, 0.54, 0.5455 **55.** 5.0079, 5.79, 5.8, 5.804 **57.** 0.6009, 0.609, 0.628, 0.62812 **59.** 2.8902, 3.88, 4.876, 5.8751 **61.** 0.006, 0.043, $\frac{1}{20}$, 0.051 **63.** 0.37, $\frac{3}{8}$, $\frac{2}{5}$, 0.4001 **65.** ≈1.4 in. **67.** ≈0.3 in. **69.** ≈0.4 in. **71.** 0 **73.** 24 **75.** −19

Section 5.7 (page 351)

1. 7 months **3.** ≈69.8 **5.** $27,955 **7.** $58.24 **9.** $35,500 **11.** 6.7 **13.** ≈17.2 hours **15.** 15 messages **17.** 516 customers **19.** 48 calls **21.** 8 samples **23.** 68 and 74 years (bimodal) **25.** no mode **27.** mean = $223,500; median = $94,000. The mean is affected by the one very high value. The median gives a more representative picture of the home values because most of them are under $100,000. **29.** 2.60 **31.** **(a)** 2.80 **(b)** ≈2.93 **(c)** ≈3.13 **33.** 64 **35.** 225 **37.** 90 **39.** 23

SECTION 5.8 (page 359)

1. 4 **3.** 8 **5.** ≈ 3.317 **7.** ≈ 2.236 **9.** ≈ 8.544 **11.** ≈ 10.050 **13.** ≈ 13.784 **15.** ≈ 31.623 **17.** 30 is about halfway between 25 and 36 so $\sqrt{30}$ should be about halfway between 5 and 6, or ≈ 5.5. Using a calculator, $\sqrt{30} \approx 5.477$. Similarly $\sqrt{26}$ should be a little more than $\sqrt{25}$; by calculator it is ≈ 5.099. And $\sqrt{35}$ should be a little less than $\sqrt{36}$; by calculator it is ≈ 5.916. **19.** $\sqrt{1521} = 39$ ft **21.** $\sqrt{289} = 17$ in. **23.** $\sqrt{144} = 12$ mm **25.** $\sqrt{73} \approx 8.544$ in. **27.** $\sqrt{65} \approx 8.062$ yd **29.** $\sqrt{195} \approx 13.964$ cm **31.** $\sqrt{7.94} \approx 2.818$ m **33.** $\sqrt{65.01} \approx 8.063$ cm **35.** $\sqrt{292.32} \approx 17.097$ km **37.** $\sqrt{65} \approx 8.1$ ft **39.** $\sqrt{360{,}000} = 600$ m **41.** $\sqrt{135} \approx 11.6$ ft

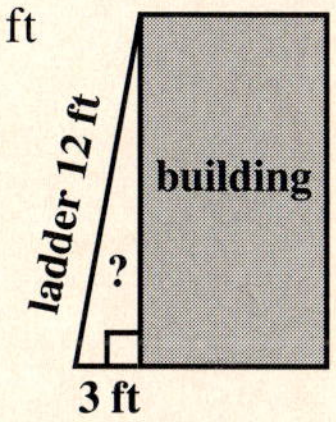

43. The student used the formula for finding the hypotenuse but the unknown side is a leg, so $? = \sqrt{(20)^2 - (13)^2}$. Also, the final answer should be m, not m^2. Correct answer is $\sqrt{231} \approx 15.2$ m. **45.** $\overline{BC} = 6.25$ ft; $\overline{BD} = 3.75$ ft **47.** $x = -12$ **49.** $b = -4$ **51.** $y = 4$

SECTION 5.9 (page 367)

See Student's Solutions Manual for a sample check for Exercises 1–6.
1. $h = 4.47$ **3.** $n = -1.4$ **5.** $b = 0.008$ **7.** $a = 0.29$ **9.** $p = -120$ **11.** $t = 0.7$ **13.** $x = -0.82$ **15.** $z = 0$ **17.** $k = -0.6$ **19.** $c = 0.45$ **21.** $w = 0$ **23.** $p = -40.5$ **25.** $y = -7$ **27.** Student should divide both sides by -4. Correct solution is -3.375. **29.** There are many possible equations. Two examples are $x + 2.5 = 2$ and $a - 3.1 = -3.6$.

SECTION 5.10 (page 377)

1. 18 mm **3.** 0.35 km **5.** $C \approx 69.1$ ft, $A \approx 379.9\ \text{ft}^2$ **7.** $C \approx 8.2$ m, $A \approx 5.3\ \text{m}^2$ **9.** $C \approx 47.1$ cm, $A \approx 176.6\ \text{cm}^2$ **11.** $C \approx 23.6$ ft, $A \approx 44.2\ \text{ft}^2$ **13.** $C \approx 27.2$ km, $A \approx 58.7\ \text{km}^2$ **15.** ≈ 219.8 cm **17.** ≈ 785.0 ft **19.** $\approx 70{,}650\ \text{mi}^2$ **21.** ≈ 3.5 m **23.** $\approx 44.2\ \text{in.}^2$ **25.** $\approx 201.0\ \text{in.}^2$ **27.** small $\approx \$0.084$; medium $\approx \$0.067$ Best Buy; large $\approx \$0.071$ **29.** π is the ratio of circumference of a circle to its diameter. If you divide the circumference of any circle by its diameter, the answer is always a little more than 3. The approximate value is 3.14 which we call π (pi). Your test question could involve finding the circumference or the area of a circle. **31.** $\approx 57\ \text{cm}^2$ **33.** $\approx 197.8\ \text{cm}^2$ **35.** $\approx 471\ \text{ft}^3$ **37.** $\approx 418.7\ \text{m}^3$ **39.** $\approx 3925\ \text{ft}^3$ **41.** $\approx 10.6\ \text{in.}^3$ **43.** Student used diameter of 7 cm; should use radius of 3.5 cm in formula. Units for volume are cm^3 not cm^2. Correct answer is $\approx 192.3\ \text{cm}^3$.

45. $5\frac{5}{6}$ **47.** -13 **49.** $2\frac{5}{11}$ **51.** $\frac{17}{50}$

CHAPTER 5 REVIEW EXERCISES (page 387)

1. 0; 5 **2.** 0; 6 **3.** 8; 9 **4.** 5; 9 **5.** 7; 6 **6.** $\frac{1}{2}$ **7.** $\frac{3}{4}$ **8.** $4\frac{1}{20}$ **9.** $\frac{7}{8}$ **10.** $\frac{27}{1000}$ **11.** $27\frac{4}{5}$ **12.** eight tenths **13.** four hundred and twenty-nine hundredths **14.** twelve and seven thousandths **15.** three hundred six ten-thousandths **16.** 8.3 **17.** 0.205 **18.** 70.0066 **19.** 0.30 **20.** ≈ 275.6 **21.** ≈ 72.79 **22.** ≈ 0.160 **23.** ≈ 0.091 **24.** ≈ 1.0 **25.** $\approx \$15.83$ **26.** $\approx \$0.70$ **27.** $\approx \$17{,}625.79$ **28.** $\approx \$350$ **29.** $\approx \$130$ **30.** $\approx \$100$ **31.** $\approx \$29$ **32.** -5.67 **33.** -0.03 **34.** 5.879 **35.** -6.435 **36.** $13 - 10 = 3$ hours; 2.75 hours **37.** $\$200 + \$40 = \$240$; \$260.00 **38.** $\$2 + \$5 + \$20 = \$27$; $\$30 - \$27 = \$3$; \$4.14 **39.** $2 + 4 + 5 = 11$ kilometers; 11.55 kilometers **40.** $6 \times 4 = 24$; 22.7106 **41.** $40 \times 3 = 120$; 141.57 **42.** 0.0112 **43.** -0.000355 **44.** reasonable; $700 \div 10 = 70$ **45.** unreasonable; $30 \div 3 = 10$; $\begin{array}{r}9.5\\2.8\overline{)26.6}\end{array}$ **46.** ≈ 14.467 **47.** 1200 **48.** -0.4 **49.** $\approx \$350$ **50.** $\approx \$1.99$ **51.** ≈ 133 shares **52.** $\approx \$3.12$ **53.** -4.715 **54.** 10.15 **55.** 3.8 **56.** 0.64 **57.** 1.875 **58.** ≈ 0.111 **59.** 3.6008, 3.68, 3.806 **60.** 0.209, 0.2102, 0.215, 0.22 **61.** $\frac{1}{8}, \frac{3}{20}$, 0.159, 0.17 **62.** 28.5 digital cameras (mean); 19.5 digital cameras (median) **63.** 44 claims (mean); 39 claims (median) **64.** \$51.05 **65.** **(a)** \$64 **(b)** 18 and 32 launchings (bimodal) **66.** 17 in. **67.** 7 cm **68.** ≈ 10.198 cm **69.** ≈ 7.211 in. **70.** ≈ 2.555 m **71.** ≈ 8.471 km **72.** $b = 0.25$ **73.** $x = 0$ **74.** $n = -13$ **75.** $a = -1.5$ **76.** $y = -16.2$ **77.** 137.8 m **78.** $1\frac{1}{2}$ in. or 1.5 in. **79.** $C \approx 6.3$ cm; $A \approx 3.1\ \text{cm}^2$ **80.** $C \approx 109.3$ m; $A \approx 950.7\ \text{m}^2$ **81.** $C \approx 37.7$ in.; $A \approx 113.0\ \text{in.}^2$ **82.** $\approx 549.5\ \text{cm}^3$ **83.** $\approx 1808.6\ \text{m}^3$ **84.** $\approx 512.9\ \text{m}^3$ **85.** 404.865 **86.** -254.8 **87.** ≈ 3583.261 **88.** 29.0898 **89.** 0.03066 **90.** 9.4 **91.** -15.065 **92.** 9.04 **93.** -15.74 **94.** 8.19 **95.** 0.928 **96.** 35 **97.** -41.859 **98.** 0.3 **99.** $\approx \$3.00$ **100.** $\approx \$2.17$ **101.** \$35.96 **102.** \$199.71 **103.** \$78.50 **104.** $y = -0.7$ **105.** $x = 30$ **106.** 15.7 ft of rubber striping; area $\approx 19.6\ \text{ft}^2$ **107.** 67.6 **108.** 20 miles **109.** $V \approx 87.9\ \text{in.}^3$

CHAPTER 5 TEST (page 393)

1. $18\frac{2}{5}$ **2.** $\frac{3}{40}$ **3.** sixty and seven thousandths **4.** two hundred eight ten-thousandths **5.** $8 + 80 + 40 = 128$; 129.2028 **6.** $-6(1) = -6$; -6.948

7. $-80 - 4 = -84$; -82.702 **8.** $-20 \div (-5) = 4$; 4.175 **9.** 669.004 **10.** 480 **11.** 0.000042 **12.** \$4.55 per meter **13.** Davida, by 0.441 minute **14.** \$5.35 (rounded) **15.** $y = -6.15$ **16.** $x = -0.35$ **17.** $a = 10.9$ **18.** $n = 2.85$ **19.** 0.44, $\frac{9}{20}$, 0.4506, 0.451 **20.** 32.09 **21.** 75 books **22.** 103° and 104° (bimodal) **23.** \$11.25 **24.** \$50.55 **25.** $\sqrt{85} \approx 9.220$ cm **26.** $\sqrt{279} \approx 16.703$ ft **27.** 12.5 in. **28.** ≈5.7 km **29.** ≈206.0 cm^2 **30.** ≈5086.8 ft^3 **31.** ≈169.6 $in.^3$

Cumulative Review Chapters 1–5 (page 395)

1. (a) forty-five and two hundred three ten-thousandths **(b)** thirty billion, six hundred fifty thousand, eight **2. (a)** 160,000,500 **(b)** 0.075 **3.** 46,900 **4.** 6.20 **5.** 0.661 **6.** 10,000 **7.** −13 **8.** −0.00006 **9.** undefined **10.** 44 **11.** 18 **12.** −6 **13.** 56.621 **14.** $\frac{3}{2}$ or $1\frac{1}{2}$ **15.** $\frac{3}{8b}$ **16.** 11.147 **17.** $-\frac{9}{20}$ **18.** $-\frac{1}{32}$ **19.** ≈1.3 **20.** $\frac{5x + 6}{10}$ **21.** −8 **22.** −32 **23.** $\frac{18}{18} = 1$ **24.** −5.76 **25.** $\frac{55}{24}$ or $2\frac{7}{24}$ **26.** ≈24.2 years **27.** 23.5 years **28.** 20 and 26 years (bimodal) **29.** $h = -4$ **30.** $x = 5$ **31.** $r = 10.9$ **32.** $y = -2$ **33.** $n = 4$ **34.** $P = 54$ ft; $A = 140$ ft^2 **35.** $C \approx 40.8$ m; $A \approx 132.7$ m^2 **36.** $x \approx 12.6$ km; $A \approx 37.8$ km^2 **37.** ≈3.4 $in.^3$ or $3\frac{3}{8}$ $in.^3$ **38.** \$20 − \$8 − \$1 = \$11; \$11.17 **39.** 2 + 4 = 6 yards; $6\frac{5}{24}$ yards **40.** \$3 ÷ 3 = \$1 per pound; ≈\$0.95 per pound **41.** \$80,000 ÷ 100 = \$800; ≈\$729

Chapter 6

Section 6.1 (page 403)

1. $\frac{8}{9}$ **3.** $\frac{2}{1}$ **5.** $\frac{1}{3}$ **7.** $\frac{8}{5}$ **9.** $\frac{3}{8}$ **11.** $\frac{9}{7}$ **13.** $\frac{6}{1}$ **15.** $\frac{5}{6}$ **17.** $\frac{8}{5}$ **19.** $\frac{1}{12}$ **21.** $\frac{5}{16}$ **23.** $\frac{4}{1}$ **25.** $\frac{22}{3}$ **27.** $\frac{5}{4}$ **29.** $\frac{1}{100}$ **31.** $\frac{11}{1}$ **33.** $\frac{3}{10}$ **35.** A ratio of 3 to 1 means your income is 3 times your friend's income. **37.** $\frac{2}{1}$ **39.** $\frac{3}{8}$ **41.** $\frac{7}{5}$ **43.** $\frac{6}{1}$ **45.** $\frac{38}{17}$ **47.** $\frac{1}{2}$ **49.** $\frac{34}{35}$ **51.** Answer varies. Some possibilities are:

$$\frac{4}{5} = \frac{8}{10} = \frac{12}{15} = \frac{16}{20} = \frac{20}{25} = \frac{24}{30} = \frac{28}{35}$$

53. ≈0.093 **55.** 1.025 **57.** ≈9.465

Section 6.2 (page 411)

1. $\frac{5 \text{ cups}}{3 \text{ people}}$ **3.** $\frac{3 \text{ feet}}{7 \text{ seconds}}$ **5.** $\frac{1 \text{ person}}{2 \text{ dresses}}$ **7.** $\frac{5 \text{ letters}}{1 \text{ minute}}$ **9.** $\frac{\$21}{2 \text{ visits}}$ **11.** $\frac{18 \text{ miles}}{1 \text{ gallon}}$ **13.** \$12 per hour or \$12/hour **15.** 5 eggs per chicken or 5 eggs/chicken **17.** 1.25 pounds/person **19.** \$103.30/day **21.** 325.9; ≈21.0 **23.** 338.6; ≈20.9 **25.** 4 oz for \$0.89 **27.** 17 ounces for \$2.89 **29.** 18 ounces for \$1.41 **31.** You might choose Brand B because you like more chicken, so the cost per chicken chunk may actually be the same or less than Brand A. **33.** 1.75 pounds/week **35.** \$12.26/hour **37.** \$11.50/share **39.** 0.1 second/meter or $\frac{1}{10}$ second/meter; 10 meters/second **41.** \$11.25/yard **43.** One battery for \$1.79; like getting 3 batteries, so \$1.79 ÷ 3 ≈ \$0.597 per battery **45.** Brand P with the 50¢ coupon is the best buy. \$3.39 − \$0.50 = \$2.89 ÷ 16.5 ounces ≈ \$0.175/ounce **47.** $6\frac{1}{5}$ **49.** $\frac{1}{4}$ **51.** 6

Section 6.3 (page 421)

1. $\frac{\$9}{12 \text{ cans}} = \frac{\$18}{24 \text{ cans}}$ **3.** $\frac{200 \text{ adults}}{450 \text{ children}} = \frac{4 \text{ adults}}{9 \text{ children}}$ **5.** $\frac{120}{150} = \frac{8}{10}$ **7.** $\frac{2.2}{3.3} = \frac{3.2}{4.8}$ **9.** True **11.** True **13.** False **15.** True **17.** False **19.** False **21.** True **23.** False **25.** True **27.** $\frac{16 \text{ hits}}{50 \text{ at bats}} = \frac{128 \text{ hits}}{400 \text{ at bats}}$ $50 \cdot 128 = 6400$, $16 \cdot 400 = 6400$ Cross products are equal so the proportion is true; they hit equally well. **29.** 4 **31.** 2 **33.** 88 **35.** 91 **37.** 5 **39.** 10 **41.** ≈24.44 **43.** 50.4 **45.** ≈17.64 **47.** $\frac{6.67}{4} \approx \frac{5}{3}$ or $\frac{10}{6} = \frac{5}{3}$ or $\frac{10}{4} = \frac{7.5}{3}$ or $\frac{10}{4} = \frac{5}{2}$ **49.** 1 **51.** $3\frac{1}{2}$ **53.** $\frac{1}{5}$ **55.** 0.005 or $\frac{1}{200}$ **57.** $\frac{25 \text{ feet}}{18 \text{ sec}} = \frac{15 \text{ feet}}{10 \text{ sec}}$; False **59.** $\frac{\$14.75}{2 \text{ hours}} = \frac{\$33.25}{4.5 \text{ hours}}$; False

Section 6.4 (page 427)

1. 22.5 hours **3.** \$7.20 **5.** 7 pounds **7.** \$153.45 **9.** 14 feet, 10 feet **11.** 14 feet, 8 feet **13.** 3 runs **15.** 665 students (reasonable); ≈1357 students with incorrect setup (only 950 students in the group). **17.** ≈190 people (reasonable); ≈298 people with incorrect setup (only 238 people attended). **19.** \$5.50 **21.** 625 stocks **23.** ≈4.6 hours **25.** ≈4.06 meters **27.** ≈10.53 meters **29.** You cannot solve this problem using a proportion because the ratio of age to weight is not constant. As Jim's age increases, his weight may decrease, stay the same, or increase. **31.** $4\frac{3}{8}$ cups water; $7\frac{1}{2}$ tbsp margarine; $1\frac{7}{8}$ cups milk; 5 cups potato flakes **33.** $2\frac{1}{3}$ cups water; 4 tbsp margarine; 1 cup milk; $2\frac{2}{3}$ cups potato flakes **35.** 3800 students **37.** $P = 32$ yd; $A = 60\text{ yd}^2$ **39.** $P = 36$ ft; $A = 54\text{ ft}^2$

Section 6.5 (page 435)

1. similar **3.** not similar **5.** similar **7.** $\angle B$ and $\angle Q$, $\angle C$ and $\angle R$, $\angle A$ and $\angle P$, $\overline{AB}$ and $\overline{PQ}$, $\overline{BC}$ and $\overline{QR}$, $\overline{AC}$ and $\overline{PR}$ **9.** $\angle P$ and $\angle S$, $\angle N$ and $\angle R$, $\angle M$ and $\angle Q$, $\overline{MP}$ and $\overline{QS}$, $\overline{MN}$ and $\overline{QR}$, $\overline{NP}$ and $\overline{RS}$ **11.** $\frac{3}{2}; \frac{3}{2}; \frac{3}{2}$ **13.** $a = 5$ mm; $b = 3$ mm **15.** $a = 6$ cm; $b = 15$ cm **17.** $x = 24.8$ m, perimeter = 72.8 m; $y = 15$ m, perimeter = 54.6 m **19.** Perimeter = 8 cm + 8 cm + 8 cm = 24 cm; Area = 0.5 • 8 cm • 5.6 cm = 22.4 cm^2 **21.** $h = 24$ ft **23.** One dictionary definition is, "Resembling, but not identical." Examples of similar objects are sets of different size pots or measuring cups; small and large size cans of beans; child's tennis shoe and adult tennis shoe. **25.** $x = 50$ m **27.** $n = 110$ m **29.** 6 **31.** 2870 **33.** 0.0193

Chapter 6 Review Exercises (page 443)

1. $\frac{3}{11}$ **2.** $\frac{19}{7}$ **3.** $\frac{3}{2}$ **4.** $\frac{9}{5}$ **5.** $\frac{2}{1}$ **6.** $\frac{2}{3}$ **7.** $\frac{5}{2}$ **8.** $\frac{1}{6}$ **9.** $\frac{3}{1}$ **10.** $\frac{3}{8}$ **11.** $\frac{4}{3}$ **12.** $\frac{1}{9}$ **13.** $\frac{10}{7}$ **14.** $\frac{7}{5}$ **15.** $\frac{5}{6}$ **16.** $\frac{5}{9}$ **17.** $\frac{\$11}{1\text{ dozen}}$ **18.** $\frac{12\text{ children}}{5\text{ families}}$ **19.** Both compare two things. In a ratio the common units cancel, but in a rate the units are different and must be written. Examples are:

$$(\text{ratio})\ \frac{5\ \not{\text{feet}}}{10\ \not{\text{feet}}} = \frac{1}{2} \qquad \frac{55\text{ miles}}{1\text{ hour}}\ (\text{rate})$$

20. A unit rate has 1 in the denominator. Examples are 55 miles in 1 hour, \$440 in 1 week, or 30 miles on 1 gallon of gas. We usually write them using "per" or a slash mark: 55 miles per hour, etc. **21.** 0.2 page/minute or $\frac{1}{5}$ page/minute; 5 minutes/page **22.** \$8/hour; 0.125 hour/dollar or $\frac{1}{8}$ hour/dollar **23.** 13 ounces for \$2.29 **24.** 25 pounds for \$10.40 − \$1 coupon **25.** $\frac{5}{10} = \frac{20}{40}$ **26.** $\frac{7}{2} = \frac{35}{10}$ **27.** $\frac{1\frac{1}{2}}{6} = \frac{2\frac{1}{4}}{9}$ **28.** True **29.** False **30.** False **31.** True **32.** True **33.** True **34.** 1575 **35.** 20 **36.** 400 **37.** 12.5 **38.** ≈14.67 **39.** ≈8.17 **40.** 50.4 **41.** ≈0.57 **42.** ≈2.47 **43.** 27 cats **44.** 46 hits **45.** ≈\$22.06 **46.** ≈3299 students **47.** 68 feet **48.** 14.7 milligrams **49.** 80 hours **50.** $27\frac{1}{2}$ hours or 27.5 hours **51.** $y = 30$ ft; $x = 34$ ft **52.** $y = 7.5$ m; $x = 9$ m **53.** $x = 12$ mm; $y = 7.5$ mm **54.** 105 **55.** 0 **56.** 128 **57.** ≈23.08 **58.** 6.5 **59.** ≈117.36 **60.** False **61.** False **62.** True **63.** $\frac{8}{5}$ **64.** $\frac{33}{80}$ **65.** $\frac{15}{4}$ **66.** $\frac{4}{1}$ **67.** $\frac{4}{5}$ **68.** $\frac{37}{7}$ **69.** $\frac{3}{8}$ **70.** $\frac{1}{12}$ **71.** $\frac{45}{13}$ **72.** ≈24,900 fans **73.** $\frac{8}{3}$ **74.** 75 feet for \$1.99 − \$0.50 coupon **75.** 16.5 feet **76.** $\frac{1}{2}$ teaspoon or 0.5 teaspoon **77.** ≈21 points **78.** 7.5 hours or $7\frac{1}{2}$ hours

Chapter 6 Test (page 447)

1. $\frac{4}{5}$ **2.** $\frac{20\text{ miles}}{1\text{ gallon}}$ **3.** $\frac{\$1}{5\text{ minutes}}$ **4.** $\frac{15}{4}$ **5.** $\frac{1}{80}$ **6.** $\frac{9}{2}$ **7.** 18 ounces for \$1.89 − \$0.25 coupon **8.** You earned less this year. An example is:

$$\begin{array}{l}\text{Last year} \rightarrow \\ \text{This year} \rightarrow\end{array} \frac{\$15{,}000}{\$10{,}000} = \frac{3}{2}$$

9. False **10.** True **11.** 25 **12.** ≈2.67 **13.** 325 **14.** $10\frac{1}{2}$ **15.** 576 words **16.** 6.4 hours **17.** ≈87 students **18.** No, 4875 cannot be correct because there are only 650 students in the whole school. **19.** ≈23.8 grams **20.** 60 feet **21.** $y = 12$ cm; $z = 6$ cm **22.** $x = 12$ mm; $P = 46.8$ mm; $y = 14$ mm; $P = 39$ mm

Cumulative Review Chapters 1–6 (page 449)

1. (a) seventy-seven billion, one million, eight hundred five **(b)** two hundredths **2. (a)** 3.040 **(b)** 500,037,000 **3.** 0 **4.** 14 **5.** $2\frac{11}{12}$ **6.** $\frac{2h-3}{10}$ **7.** 99.9905 **8.** $\frac{3}{0}$ is undefined **9.** −18 **10.** −14 **11.** $\frac{n}{4m}$ **12.** −0.00042 **13.** $\frac{1}{2y^2}$ **14.** $\frac{27+2n}{3n}$ **15.** −64 **16.** 5.4 **17.** $-\frac{1}{18}$ **18.** $y = -1$ **19.** $x = 15$ **20.** $x = 10$ **21.** Let t be the starting temperature;

ANSWERS

$t + 15 - 23 + 5 = 71$; the starting temperature was 74 degrees.
22. Let l be the length; $l - 14$ is the width;
$2(l) + 2(l - 14) = 100$
Length is 32 ft; width is 18 ft.
23. $P = 7$ km; $A \approx 2.0$ km^2 **24.** $d = 17$ m; $C \approx 53.4$ m; $A \approx 226.9$ m^2 **25.** $V \approx 339.1$ ft^3
26. 2000 + 2000 + 2000 + 2000 = 8000 students; 8400 students **27.** \$200,000 ÷ 2000 = \$100; ≈\$78
28. $\frac{2}{7}$; $\frac{13}{11}$ **29.** \$4 • 8000 = \$32,000; \$31,500
30. 4 + 3 = 7; 7 • 1 = 7 miles; $7\frac{3}{20}$ miles
31. 900 miles ÷ 50 gallons = 18 miles per gallon; ≈18.0 miles per gallon **32.** 250 pounds
33. $1\frac{1}{4}$ teaspoons

Chapter 7

Section 7.1 (page 459)

1. 0.25 **3.** 0.30 or 0.3 **5.** 0.06 **7.** 1.40 or 1.4 **9.** 0.078 **11.** 1.00 or 1 **13.** 0.005 **15.** 0.0035 **17.** 50% **19.** 62% **21.** 3% **23.** 12.5% **25.** 62.9% **27.** 200% **29.** 260% **31.** 3.12% **33.** $\frac{1}{5}$ **35.** $\frac{1}{2}$ **37.** $\frac{11}{20}$ **39.** $\frac{3}{8}$ **41.** $\frac{1}{16}$ **43.** $\frac{1}{6}$ **45.** $1\frac{3}{10}$ **47.** $2\frac{1}{2}$ **49.** 25% **51.** 30% **53.** 60% **55.** 37% **57.** $37\frac{1}{2}$% or 37.5% **59.** 5% **61.** $55\frac{5}{9}$% or ≈55.6% **63.** $14\frac{2}{7}$% or ≈14.3% **65.** 0.08 **67.** 0.65 **69.** 3.5% **71.** 200% **73.** $\frac{95}{100}$ or 95%; $\frac{5}{100}$ or 5% **75.** $\frac{3}{10}$ or 30%; $\frac{7}{10}$ or 70% **77.** $\frac{3}{4}$ or 75%; $\frac{1}{4}$ or 25% **79.** 0.01; 1% **81.** $\frac{1}{5}$; 20% **83.** $\frac{3}{10}$; 0.3 **85.** 0.5; 50% **87.** $\frac{9}{10}$; 0.9 **89.** $1\frac{1}{2}$; 150% **91.** $\frac{19}{25}$; 0.76; 76% **93.** $\frac{1}{5}$; 0.2; 20% **95.** $\frac{3}{5}$; 0.6; 60% **97.** $\frac{1}{5}$; 0.20; 20% **99.** $\frac{1}{10}$; 0.1;10% **101.** $\frac{1}{4}$; 0.25; 25% **103.** $\frac{2}{5}$; 0.4; 40%
105. (a) The student forgot to move the decimal point in 0.35 two places to the right. So $\frac{7}{20} = 35\%$
(b) The student did the division in the wrong order. Enter 16 ÷ 25 to get 0.64. So $\frac{16}{25} = 64\%$
107. \$78 **109.** \$39 **111.** 20 children **113.** 90 miles **115.** \$142.50 **117.** 4100 students **119.** Since 100% means 100 parts out of 100 parts, 100% is all of the number.
121. 30 **123.** 5 **125.** 10

Section 7.2 (page 471)

1. $\frac{10}{100} = \frac{n}{3000}$; 300 runners **3.** $\frac{4}{100} = \frac{n}{120}$; 4.8 feet **5.** $\frac{p}{100} = \frac{16}{32}$; 50% **7.** $\frac{p}{100} = \frac{16}{200}$; 8% **9.** $\frac{90}{100} = \frac{495}{n}$; 550 students **11.** $\frac{12.5}{100} = \frac{3.50}{n}$; \$28 **13.** $\frac{250}{100} = \frac{n}{7}$; 17.5 hours **15.** $\frac{p}{100} = \frac{32}{172}$; ≈18.6% **17.** $\frac{110}{100} = \frac{748}{n}$; 680 books **19.** $\frac{14.7}{100} = \frac{n}{274}$; ≈\$40.28 **21.** $\frac{p}{100} = \frac{105}{54}$; ≈194.4% **23.** $\frac{4}{100} = \frac{0.33}{n}$; \$8.25
25. 150% of \$30 cannot be less than \$30 because 150% is greater than 100%. The answer must be greater than \$30. 25% of \$16 cannot be greater than \$16 because 25% is less than 100%. The answer must be less than \$16.
27. The correct proportion is $\frac{p}{100} = \frac{14}{8}$. The answer should be labeled with the % symbol. Correct answer is 175%. **29.** 15 miles **31.** $7\frac{1}{2}$ miles **33.** \$310

Section 7.3 (page 479)

1. 1500 patients **3.** \$15 **5.** 4.5 pounds **7.** \$7.00 **9.** 52 students **11.** 10% means $\frac{10}{100}$ or $\frac{1}{10}$. The denominator tells you to divide the whole by 10. The shortcut for dividing by 10 is to move the decimal point one place to the left. Once you find 10% of a number, multiply the result by 2 for 20% and by 3 for 30%. **13.** 231 programs **15.** 50% **17.** 680 circuits **19.** 1080 people **21.** 845 species **23.** \$20.80 **25.** 76% **27.** 125% **29.** 4700 employees **31.** 83.2 quarts **33.** 2% **35.** 700 tablets **37.** 325 salads **39.** 1.5% **41.** 5 gallons **43.** 1029.2 meters **45. (a)** Multiply the solution by 100 to change it from a decimal to a percent. So, 0.20 = 20% **(b)** The correct equation is $50 = p \cdot 20$ so the solution is 250%.
47. (a) $\frac{1}{3} \cdot 162 = n$; solution is \$54.
(b) $0.333333333 \cdot 162 = n$; depending upon how your calculator rounds numbers, the solution is either \$54 or \$53.99999995. **(c)** Depending on your calculator, there is no difference, or the difference is insignificant.
49. 0.03 **51.** 0.125 **53.** 70% **55.** 180%

Section 7.4 (page 489)

1. \$19.80 **3.** 62.5% or $62\frac{1}{2}$% **5.** 836 drivers **7.** ≈15.5% **9.** 248.0 million **11.** 138% **13.** 40 problems **15. (a)** \$145 **(b)** \$1740 **17.** ≈29.4 miles per gallon **19.** March; 24.5 million cans, or 24,500,000 **21.** 52.5 million cans (January); 38.5 million cans (February) **23.** ≈5.6%

25. ≈18.0% **27.** 40% **29.** George ate more than the total recommended amount, so the percent must be >100%. Correct equation: $p \cdot 65 = 78$; Correct solution is 120% **31.** The brain could not weigh 375 pounds because that is more than the entire person weighs. The error is in changing $2\frac{1}{2}\%$ to a decimal: $2\frac{1}{2}\% = 2.5\% = 0.025$. Correct equation: $0.025 \cdot 150 = n$. Correct solution is 3.75 pounds **33.** No. 100% is the entire price, so a decrease of 100% would take the price down to zero. So 100% is the maximum possible decrease in the price of something. **35.** $43.20 **37.** 0.8 mile **39.** $4.32 **41.** $0.08

Section 7.5 (page 497)

1. $6; $106 **3.** 3%; $70.04 **5.** $21.96 (rounded); $387.94 **7.** $0.12 (rounded); $2.22 **9.** $4\frac{1}{2}\%$; $13,167 **11.** $3 + $1.50 = $4.50; $4.83 (rounded); 2 • $3 = $6; $6.43 (rounded) **13.** $8 + $4 = $12; $11.75 (rounded); 2 • $8 = $16; $15.67 (rounded) **15.** $1 + $0.50 = $1.50; $1.43 (rounded); 2 • $1 = $2; $1.91 **17.** $15; $85 **19.** 30%; $126 **21.** $4.38 (rounded); $13.12 **23.** $3.79 (rounded); $34.09 **25.** $2310 **27.** $52.50 **29.** 5% **31.** $74.25 **33.** $25.13 **35.** $106.20; $483.80 **37.** $21 **39.** $92.38 **41.** $230.12 **43.** **(a)** $18.43 (rounded to nearest cent) **(b)** When calculating the discount, the *whole* is $18.50. But when calculating the sales tax, the *whole* is only $17.39 (the discounted price). **45.** 0.625 **47.** 5.25 **49.** ≈0.42

Section 7.6 (page 507)

1. $8 **3.** $144 **5.** $488.75 **7.** $763.75 **9.** $6 **11.** ≈$634.38 **13.** $342 **15.** $773.30 **17.** $1725 **19.** $18,801.25 **21.** The answer should include: Amount of principal—This is the amount of money borrowed or loaned. Interest rate—This is the percent used to calculate the interest. Time of loan—The length of time that money is loaned or borrowed. **23.** $291 **25.** $6750 **27.** $1025 **29.** $323.40; $8163.40 **31.** $159.50 **33.** ≈$12,254.69 **35.** $1102.50; $2.50 **37.** $4587.79; $107.79 **39.** ≈$899.89 **41.** ≈$2027.46 **43.** ≈$7758.05 **45.** **(a)** ≈$11,901.56 **(b)** $4401.56 **47.** **(a)** ≈$37,202.58 **(b)** $7202.58

Chapter 7 Review Exercises (page 515)

1. 0.25 **2.** 1.8 **3.** 0.125 **4.** 0.00085 **5.** 265% **6.** 2% **7.** 87.5% **8.** 0.2% **9.** $\frac{3}{25}$ **10.** $\frac{3}{8}$ **11.** $2\frac{1}{2}$ **12.** $\frac{1}{20}$ **13.** 75% **14.** 62.5% or $62\frac{1}{2}\%$ **15.** 325% **16.** 6% **17.** 0.125 **18.** 12.5% **19.** $\frac{3}{20}$ **20.** 15% **21.** $1\frac{4}{5}$ **22.** 1.8 **23.** $46 **24.** $23 **25.** 9 hours **26.** $4\frac{1}{2}$ hours **27.** 242 meters **28.** 17,000 cases **29.** 27 telephones **30.** 870 reference books **31.** ≈9.5% **32.** 225% **33.** ≈$2.60 **34.** 80 days **35.** 4% **36.** $575 **37.** 20 people **38.** 175% **39.** 3000 patients **40.** $364 **41.** 80% **42.** ≈10.4% **43.** $8.40; $218.40 **44.** $7\frac{1}{2}\%$; $838.50 **45.** $4 + $2 = $6; $6.41 (rounded); 2 • $4 = $8; $8.55 (rounded) **46.** $0.80 + $0.40 = $1.20; $1.21 (rounded); 2 • $0.80 = $1.60; $1.61 **47.** $3.75; $33.75 **48.** 25%; $189 **49.** $47.25 **50.** $183.60 **51.** ≈$4534.96 **52.** ≈$14,943.07 **53.** **(a)** ≈48.4% **(b)** 7161 dogs **54.** **(a)** ≈36.4% **(b)** ≈180.3% **55.** **(a)** ≈38.1% **(b)** 265 animals **56.** $351.45 **57.** $1960.20 **58.** ≈13.1% **59.** 145%

Section 7 Test (page 519)

1. 0.75 **2.** 60% **3.** 180% **4.** 7.5% or $7\frac{1}{2}\%$ **5.** 3.00 or 3 **6.** 0.02 **7.** $\frac{5}{8}$ **8.** $2\frac{2}{5}$ **9.** 5% **10.** 87.5% or $87\frac{1}{2}\%$ **11.** 175% **12.** 320 files **13.** 400% **14.** $19,500 **15.** $6049.20 **16.** ≈34% **17.** To find 50% of a number, divide the number by 2. To find 25% of a number, divide the number by 4. Examples will vary. **18.** Round $31.94 to $30. 10% of $30. is $3 and 5% of $30 would be half of $3 or $1.50. So a 15% tip would be $3 + $1.50 = $4.50. A 20% tip would be 2 times $3 = $6. **19.** Exact tip is ≈$4.79. Each person pays ≈$12.24 **20.** $3.84; $44.16 **21.** ≈$80.98; $98.97 **22.** $815.66 ÷ 6 ≈ $135.94 **23.** $262.50 **24.** $52 **25.** $4876

Cumulative Review 1–7 (page 521)

1. **(a)** ninety and one hundred five thousandths **(b)** one hundred twenty-five million, six hundred seventy **2.** **(a)** 30,005,000,000 **(b)** 0.0078 **3.** **(a)** 50,000 **(b)** 0.70 **(c)** 8900 **4.** **(a)** Commutative property of multiplication **(b)** Associative property of addition **5.** $<$; $>$ **6.** 0.7005; 0.705; $\frac{3}{4}$; 0.755 **7.** mean is $771.25; median is $725 **8.** 17 ounces for $2.89 **9.** 48.901 **10.** 17 **11.** $\frac{9b^2}{8}$ **12.** $\frac{1}{2}$ **13.** −10 **14.** $\frac{16 + 5x}{20}$ **15.** −40 **16.** −0.001 **17.** $\frac{12}{11}$ or $1\frac{1}{11}$ **18.** 123 **19.** 0 **20.** $\frac{7m - 16}{8m}$ **21.** −30 **22.** $\frac{4}{9n}$ **23.** $3\frac{1}{2}$ **24.** $3\frac{2}{9}$ **25.** −44 **26.** −0.02

ANSWERS

27. $\frac{3}{10}$ **28.** 19 **29.** −1 **30.** −24 **31.** 160
32. $-9x^2 + 5x$ **33.** 0 **34.** $-40w^3$ **35.** $3h - 10$
36. $n = 5$ **37.** $h = -6$ **38.** $a = 2.2$
39. $y = -\frac{14}{3}$ **40.** $b = 20$ **41.** $x = 162.5$
42. $h = 3$ **43.** $x = -1$ **44.** $4n - 5 = -17$; $n = -3$ **45.** $n + 31 = 3n + 1$; $n = 15$ **46.** Let d be the diapers in each package; $3d - 17 - 19 = 12$; there were 16 diapers in each package. **47.** Let p be Susanna's pay; let $2p$ be Neoka's pay; $p + 2p = 1620$; Susanna receives \$540; Neoka receives \$1080.
48. $P = 50$ yd; $A = 142$ yd^2 **49.** $C \approx 31.4$ ft; $A \approx 78.5$ ft^2 **50.** $x \approx 8.1$ ft **51.** ≈7.4 m^3
52. ≈1271.7 cm^3 **53.** $y \approx 24.2$ mm **54.** 6 buses
55. $P = 64$ ft; $A = 252$ ft^2 **56.** 2100 students
57. 1 pound under the limit **58.** $\frac{1}{6}$ cup more than the amount needed **59.** \$56.70; \$132.30 **60.** 93.6 km
61. \$2756.25; \$11,506.25 **62.** ≈88.6%
63. ≈\$4.87; \$69.82 **64.** ≈\$0.43

Chapter 8

Section 8.1 (page 531)

1. 3 **3.** 8 **5.** 5280 **7.** 2000 **9.** 60
11. 2 **13.** 2 **15.** 76,000 to 80,000 pounds
17. 4 **19.** 112 **21.** 10 **23.** $1\frac{1}{2}$ or 1.5
25. $\frac{1}{4}$ or 0.25 **27.** $1\frac{1}{2}$ or 1.5 **29.** $2\frac{1}{2}$ or 2.5
31. $\frac{1}{2}$ day or 0.5 day **33.** 5000 **35.** 17
37. 44 ounces **39.** 216 **41.** 28 **43.** 518,400
45. 48,000 **47.** **(a)** pound/ounces **(b)** quarts/pints or pints/cups **(c)** min/h or s/min **(d)** feet/inches **(e)** pounds/tons **(f)** days/weeks **49.** 174,240
51. 800 **53.** 0.75 or $\frac{3}{4}$ **55.** 1.5 or $1\frac{1}{2}$ **57.** <
59. > **61.** >

Section 8.2 (page 541)

1. 1000; 1000 **3.** $\frac{1}{1000}$ or 0.001; $\frac{1}{1000}$ or 0.001
5. $\frac{1}{100}$ or 0.01; $\frac{1}{100}$ or 0.01 **7.** answer varies—about 8 cm **9.** answer varies—about 20 mm
11. cm **13.** m **15.** km **17.** mm **19.** cm
21. m **23.** Examples include 35 mm film for cameras, track and field events, metric auto parts, and lead refills for mechanical pencils. **25.** 700 cm **27.** 0.040 m or 0.04 m **29.** 9400 m **31.** 5.09 m **33.** 40 cm
35. 910 mm **37.** less; 18 cm or 0.18 m
39. ________________ 35 mm = 3.5 cm

________________________________ 70 mm = 7 cm

41. 0.0000056 km **43.** $\frac{7}{8}$ **45.** $\frac{2}{25}$

Section 8.3 (page 549)

1. mL **3.** L **5.** kg **7.** g **9.** mL
11. mg **13.** L **15.** kg **17.** unreasonable
19. unreasonable **21.** reasonable **23.** reasonable
25. Some examples are 2-liter bottles of soda, shampoo bottles marked in mL, grams of fat listed on cereal boxes, vitamin doses in mg. **27.** Unit for your answer (g) is in numerator; unit being changed (kg) is in denominator so it will divide out. The unit fraction is $\frac{1000 \text{ g}}{1 \text{ kg}}$.
29. 15,000 mL **31.** 3 L **33.** 0.925 L
35. 0.008 L **37.** 4150 mL **39.** 8 kg
41. 5200 g **43.** 850 mg **45.** 30 g
47. 0.598 g **49.** 0.06 L **51.** 0.003 kg
53. 990 mL **55.** mm **57.** mL **59.** cm
61. mg **63.** 2000 mL **65.** 0.95 kg
67. 0.07 L **69.** 3000 g
71. greater; 5 mg or 0.005 g **73.** 200 nickels
75. 0.2 gram **77.** 7, 1, 8, 2

Section 8.4 (page 555)

1. \$1.33 rounded to the nearest cent **3.** 9.75 kg
5. ≈71 beats **7.** 5.03 m **9.** 22.5 L
11. 215 g; 4.3 g; 4300 mg **13.** 1.55; 1500 g; 3 g
15. \$358.50 **17.** \$18 case **19.** 0.63 **21.** 5.733

Section 8.5 (page 561)

1. ≈21.8 yards **3.** ≈262.4 feet **5.** ≈4.8 m
7. ≈5.3 ounces **9.** ≈111.6 kg **11.** ≈30.3 quarts
13. ≈60.5 liters **15.** 28 °F **17.** 40 °C **19.** 150 °C
21. More. There are 180 degrees between freezing and boiling on the Fahrenheit scale, but only 100 degrees on the Celsius scale, so each Celsius degree is a greater change in temperature. **23.** ≈16 °C **25.** −20 °C
27. ≈46 °F **29.** 23 °F **31.** ≈58 °C; ≈−89 °C
33. Cold weather, at and below freezing. ≈39 °F to ≈5 °F Varies; in Minnesota it's 0 °C to −40 °C; in California, 24 °C to 0 °C. **35.** ≈\$9.57 (liters converted to gallons)
37. −32 **39.** 81 **41.** −39

Chapter 8 Review Exercises (page 567)

1. 16 **2.** 3 **3.** 2000 **4.** 24 **5.** 60 **6.** 8
7. 4 **8.** 5280 **9.** 12 **10.** 7 **11.** 60
12. 2 **13.** 48 **14.** 45 **15.** 4 **16.** 3
17. $2\frac{1}{2}$ or 2.5 **18.** $5\frac{1}{2}$ or 5.5 **19.** $\frac{3}{4}$ or 0.75
20. $2\frac{1}{4}$ or 2.25 **21.** 78 **22.** 28 **23.** 112
24. 345,600 **25.** mm **26.** cm **27.** km
28. m **29.** cm **30.** mm **31.** 500 cm
32. 8500 m **33.** 8.5 cm **34.** 3.7 m
35. 0.07 km **36.** 930 mm **37.** mL
38. L **39.** g **40.** kg **41.** L
42. mL **43.** mg **44.** g **45.** 5 L
46. 8000 mL **47.** 4580 mg **48.** 700 g
49. 0.006 g **50.** 0.035 L **51.** 31.5 L

52. ≈357 g **53.** 87.25 kg **54.** ≈\$3.01
55. ≈6.5 yards **56.** ≈11.7 inches
57. ≈67.0 miles **58.** ≈1288 km **59.** ≈21.9 L
60. ≈44.0 quarts **61.** 0 °C **62.** 100 °C
63. 37 °C **64.** 20 °C **65.** 25 °C **66.** −15 °C
67. ≈28 °F **68.** 104 °F **69.** L **70.** kg
71. cm **72.** mL **73.** mm **74.** km **75.** g
76. m **77.** mL **78.** g **79.** mg **80.** L
81. 105 mm **82.** $\frac{3}{4}$ hour or 0.75 hour
83. $7\frac{1}{2}$ feet or 7.5 feet **84.** 130 cm **85.** 77 °F
86. 14 quarts **87.** 0.7 g **88.** 810 mL
89. 80 ounces **90.** 60,000 g **91.** 1800 mL
92. ≈−1 °C **93.** 36 cm **94.** 0.055 L
95. 1.26 m **96.** ≈\$48.06
97. ≈113 g; ≈177 °C
98. ≈178.0 pounds; ≈6.0 feet

Chapter 8 Test (page 571)

1. 36 quarts **2.** 15 yards **3.** 2.25 or $2\frac{1}{4}$ hours
4. 0.75 or $\frac{3}{4}$ foot **5.** 56 ounces **6.** 7200 minutes
7. kg **8.** km **9.** mL **10.** g **11.** cm
12. mm **13.** L **14.** cm **15.** 2.5 m
16. 4600 m **17.** 0.5 cm **18.** 0.325 g
19. 16,000 mL **20.** 400 g **21.** 1055 cm
22. 0.095 L **23.** 6.32 kg **24.** 6.75 m
25. 95 °C **26.** 0 °C **27.** ≈1.8 m
28. ≈56.3 kg **29.** ≈13 gallons
30. ≈5.0 miles **31.** ≈23 °C **32.** ≈10 °F
33. Possible answers: Use same system as rest of the world; easier system for children to learn.
34. Possible answers: People like to use the system they're familiar with; cost of new signs, scales, etc.

Cumulative Review 1–8 (page 573)

1. (a) six hundred three billion, five million, forty thousand **(b)** nine and forty thousandths **2. (a)** 80.08 **(b)** 200,065,004 **3. (a)** 1.0 **(b)** 500 **(c)** 306,000,000
4. mean is ≈26.1 hours; median is 21.5 hours **5.** −17
6. $1\frac{13}{20}$ **7.** −25 **8.** $\frac{3}{4xy}$ **9.** 0.00015
10. −41.917 **11.** $\frac{18-5c}{6c}$ **12.** −30 **13.** 13
14. undefined **15.** −240 **16.** $-\frac{5}{12}$
17. $\frac{27xy}{4}$ **18.** $\frac{2m+3n}{3m}$ **19.** $2\frac{5}{6}$ **20.** $-\frac{7}{45}$
21. 2 **22.** −3.2 **23.** 0 **24.** 22 **25.** −34
26. −6 **27.** $-3p^2 - p$ **28.** $-5x - 14$
29. $-14r^3$ **30.** $b = -1$ **31.** $a = 8.05$
32. $t = 0$ **33.** $n = 2$ **34.** $w = 20$
35. $x = 0.4$ **36.** $y = -5$ **37.** $k = 4$
38. $11n - 8n = -9$; $n = -3$
39. $2n - 8 = n + 7$; $n = 15$ **40.** Let l be the length. $2l + 2(25) = 124$; the length is 37 cm.
41. Let p be the longer piece; $p - 6$ is the shorter piece. $p + p - 6 = 90$; The pieces are 48 ft and 42 ft long.
42. $P = 9$ ft; $A = 5\frac{1}{16}$ or ≈5.1 ft^2
43. $C \approx 28.3$ mm; $A \approx 63.6$ mm^2
44. $P \approx 6.5$ cm; $A \approx 2.2$ cm^2 **45.** $y \approx 13.2$ yd
46. 96 cm^3 **47.** $x = 14$ in. **48.** 54 in.
49. 3 days **50.** 3700 g **51.** 0.6 m
52. 0.007 L **53.** −4 °F **54.** ≈59%
55. Brand T at 15.5 ounces for \$2.99 − \$0.30 coupon
56. ≈2255.3 cm^3 **57.** \$15.40 **58.** $1\frac{1}{12}$ yard
59. ≈\$9.74 **60.** 5.97 kg **61.** 120 rows
62. \$3631.25 **63.** 23 blankets; \$9 left over

Chapter 9

Section 9.1 (page 581)

1. (a) \$32,000 **(b)** carpentry, \$12,100
3. (a) $\frac{\$9800}{\$32{,}000} = \frac{49}{160}$ **(b)** $\frac{\$3000}{\$2000} = \frac{3}{2}$
5. $\frac{\$16{,}000}{\$32{,}000} = \frac{1}{2}$ **7. (a)** history, 700 **(b)** science, 1000
9. $\frac{4000}{12{,}000} = \frac{1}{3}$ **11.** $\frac{1800}{2000} = \frac{9}{10}$ **13.** $\frac{4000}{1000} = \frac{4}{1}$
15. \$522,000 **17.** \$174,000 **19.** \$261,000
21. ≈1923 people **23.** ≈175 people
25. ≈262 people **27.** First find the percent of the total that is to be represented by each item. Next, multiply the percent by 360° to find the size of each sector. Finally, use a protractor to draw each sector. **29.** 90° **31.** 10%; 36% **33.** 15% **35.** \$630; 15%
37. (a) \$200,000 **(b)** 22.5°; 72°; 108°; 90°; 67.5°
(c)

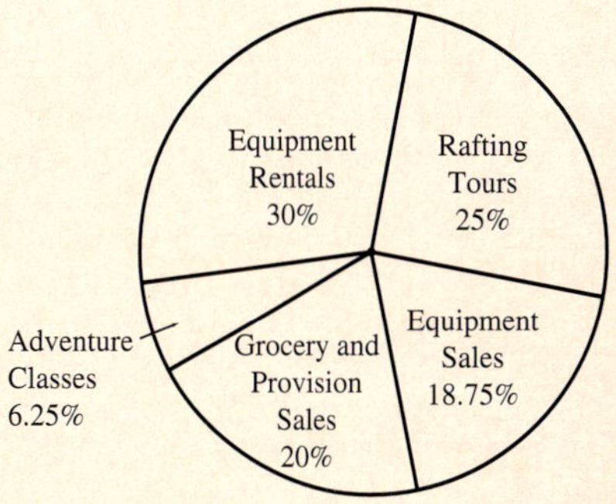

39. 30%, 108°; 20%, 72°; 15%, 54°; 10%, 36°; 5%, 18°; 5%, 18°; 15%, 54°; Total, \$32,000

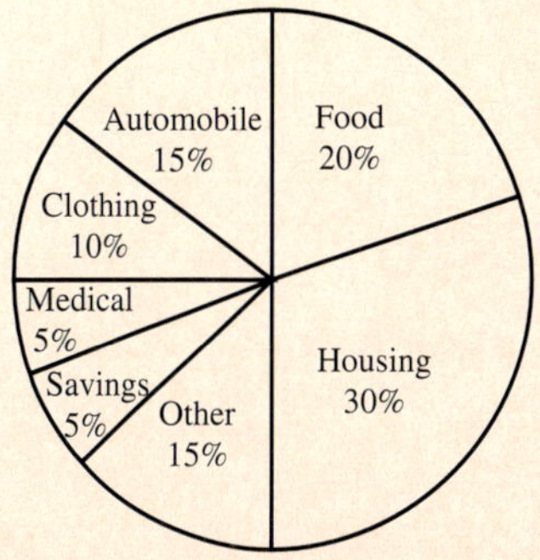

41. < **43.** > **45.** < **47.** >

SECTION 9.2 (page 591)

1. 8000 motorcycles **3.** 1996; 10,500 motorcycles **5.** 2500 motorcycles **7.** May; 10,000 unemployed **9.** 1500 workers **11.** 2500 workers **13.** 150,000 gallons **15.** 1994; 250,000 gallons **17.** 550,000 gallons **19.** April; 600 burglaries **21.** 200 burglaries **23.** Possibilities include extremely cold or snowy weather; greater police activity. **25.** 3,000,000 CDs **27.** 1,500,000 CDs **29.** 3,500,000 CDs **31.** Probably Store B with greater sales. Predicted sales might be 4,500,000 CDs to 5,000,000 CDs in 1999. **33.** A single bar or a single line must be used for each set of data. To show multiple sets of data, multiple sets of bars or lines must be used. **35.** \$40,000 **37.** \$25,000 **39.** \$5000 **41.** The decrease in sales may have resulted from poor service or greater competition. The increase in sales may have been a result of more advertising or better service. **43.** ≈\$597,000 **45.** ≈\$42,390,000 **47.** ≈\$11,941,000

SECTION 9.3 (page 601)

1.

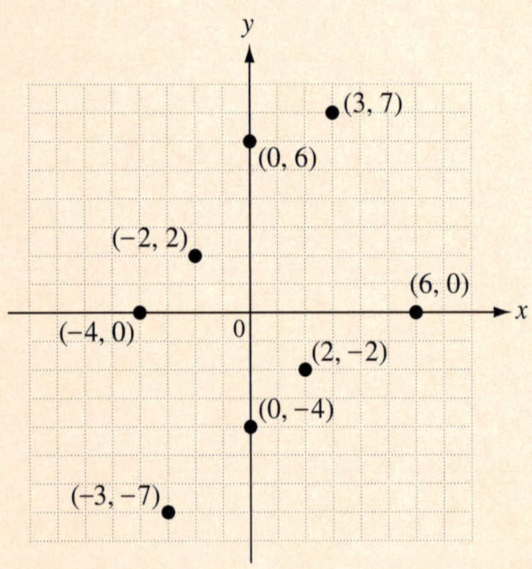

3.

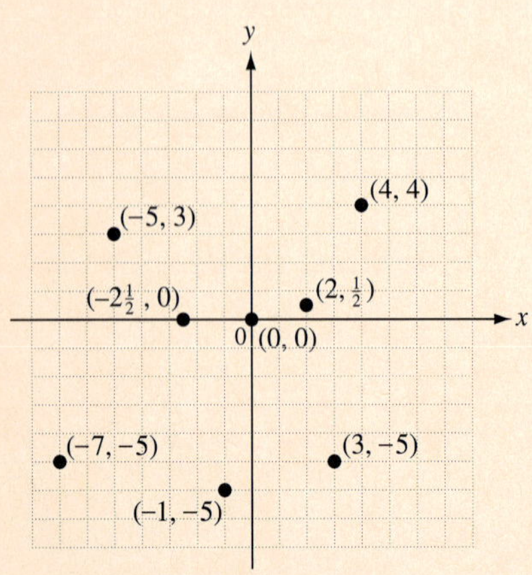

5. III, none, IV, II
7. A is (3, 4); B is (5, −5); C is (−4, −2); D is approximately $(4, \frac{1}{2})$; E is (0, −7); F is (−5, 5); G is (−2, 0); H is (0, 0).

9. Move left or right along the x-axis to the number a; then move up if b is positive or move down if b is negative.
11. **(a)** See graph. **(b)** Both (−2, −1) and (0, 0).

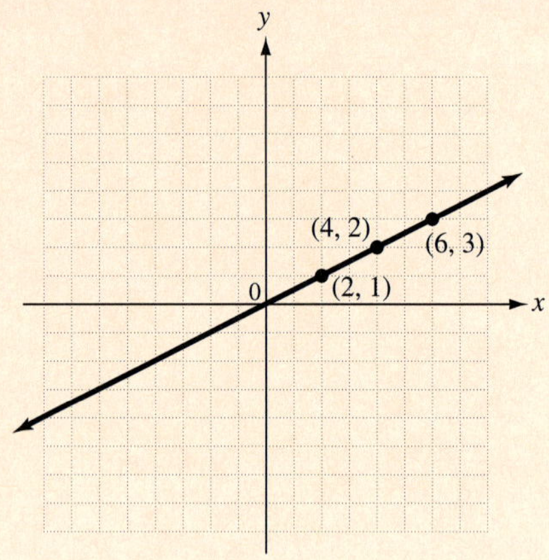

SECTION 9.4 (page 611)

1. 4; (0, 4); 3; (1, 3); 2; (2, 2)

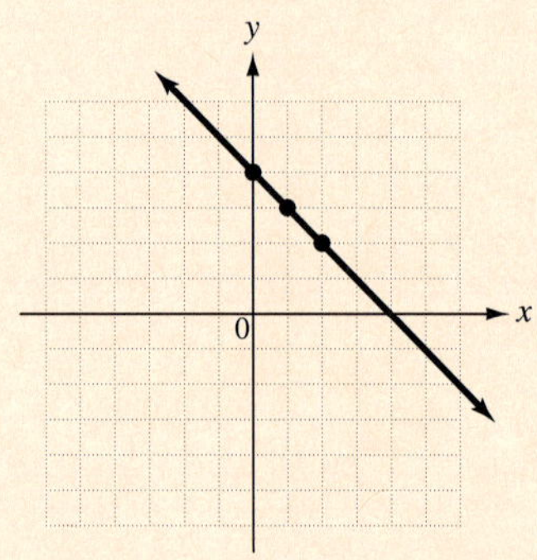

All points on the line are solutions. Some examples are (3, 1), (4, 0), and (−1, 5).
3. −1; (0, −1); −2; (1, −2); −3; (2, −3)

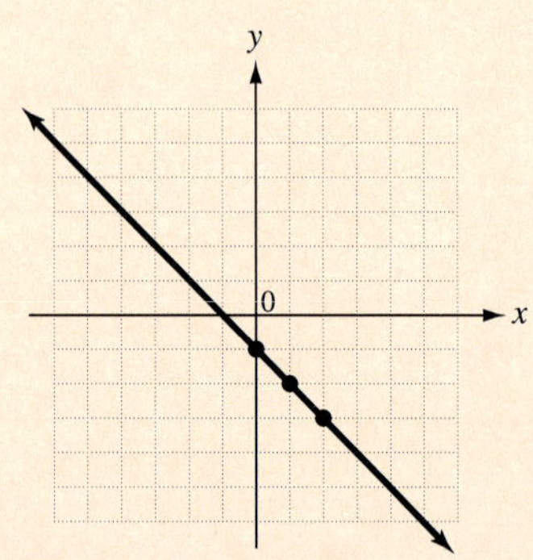

All points on the line are solutions. Some examples are (3, −4), (−1, 0), and (−2, 1).
5. 4; −1; −6; 99

7.

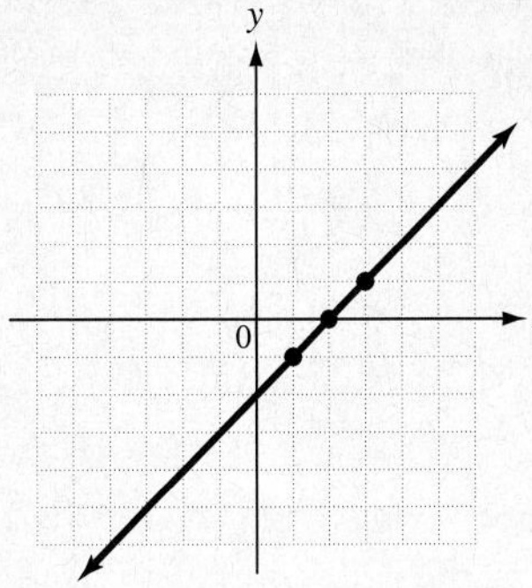

9.

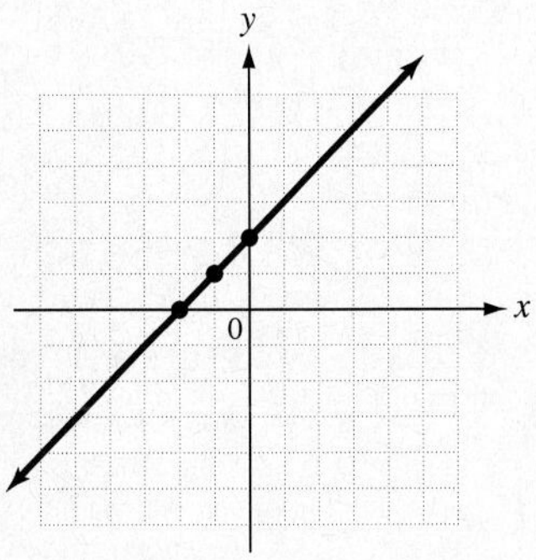

11.

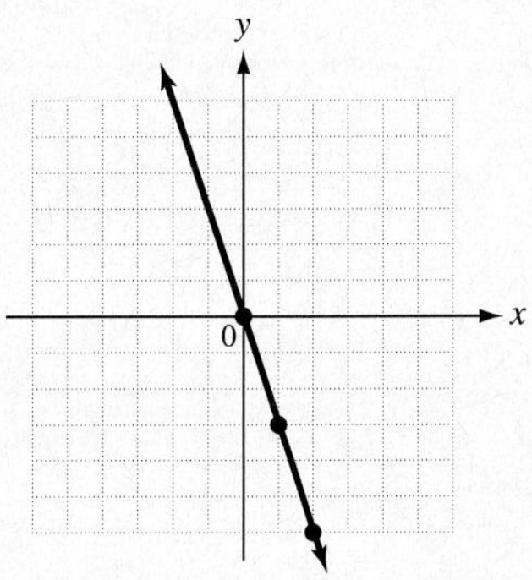

13. Exercises 7 and 9 show positive slope. Exercises 1, 3, and 11 show negative slope.

15.

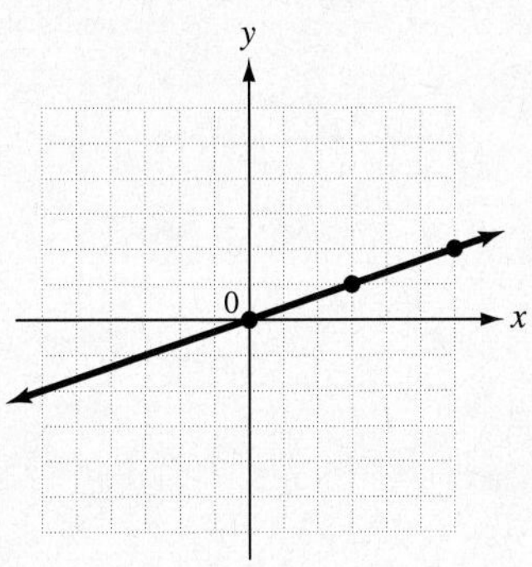

17.

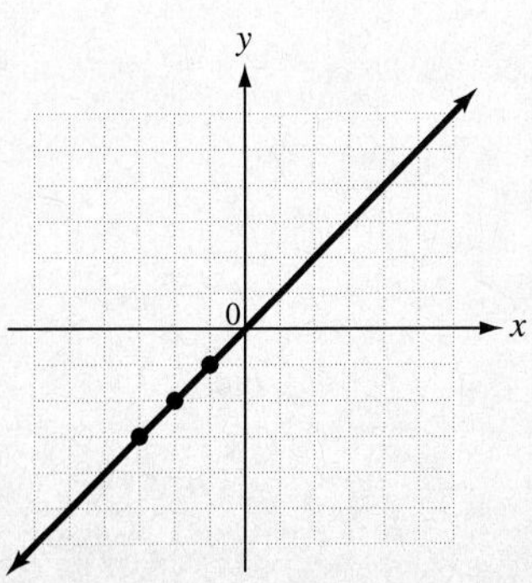

19.

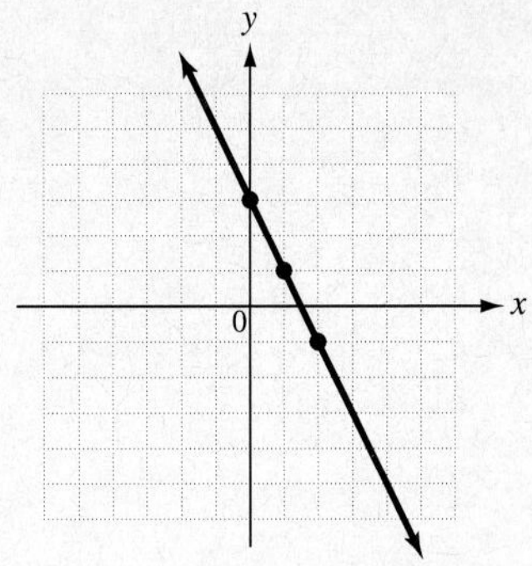

21.

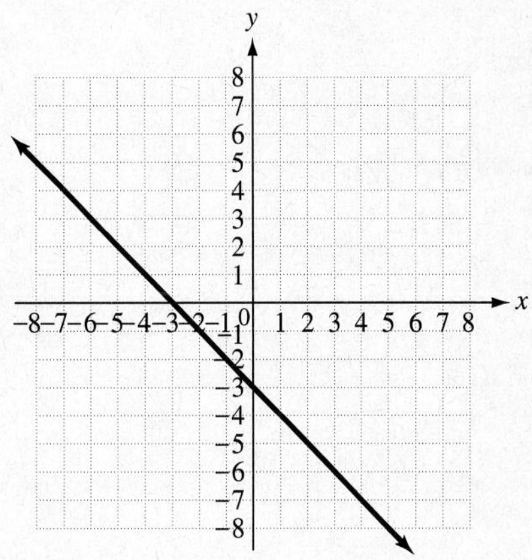

negative slope

23.

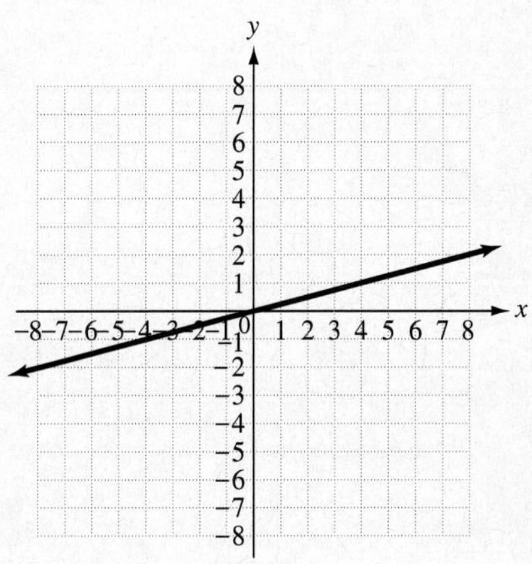

positive slope

25.

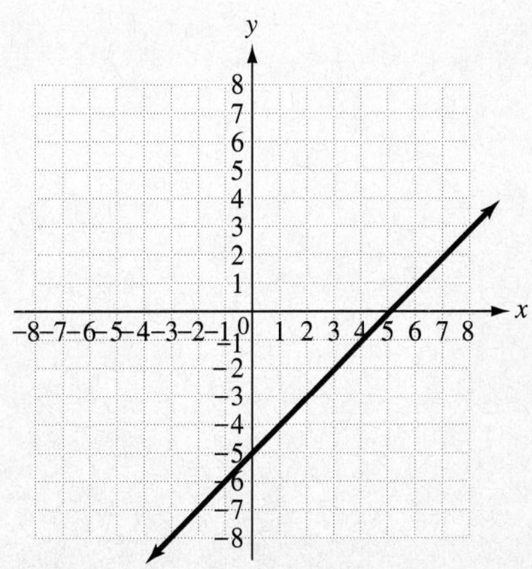

positive slope

ANSWERS

27.

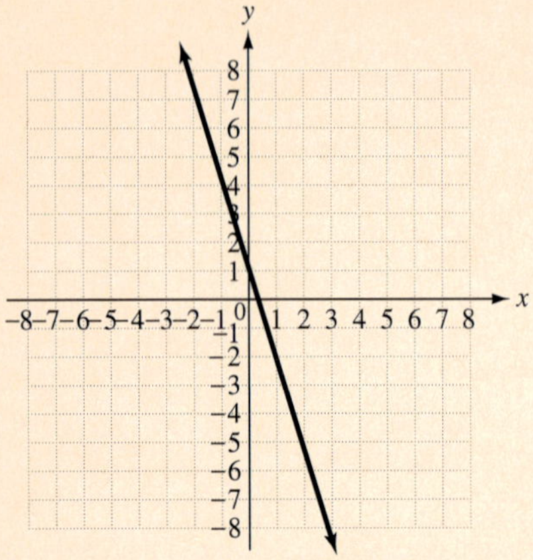

negative slope

28.

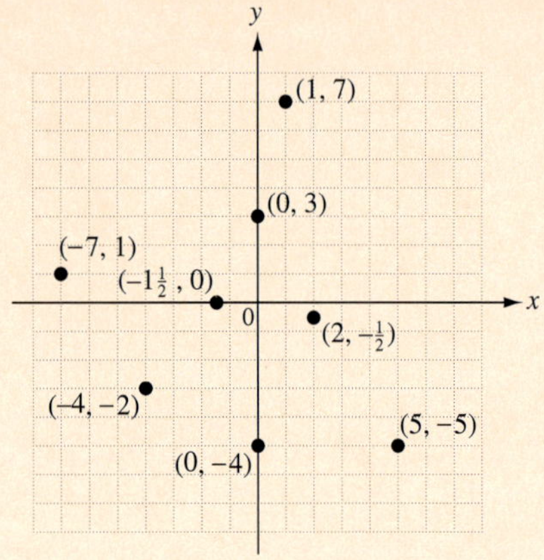

29. A is (0, 6); B is approximately $\left(-2, 2\frac{1}{2}\right)$; C is (0, 0); D is (−6, −6); E is (4, 3); F is approximately $\left(3\frac{1}{2}, 0\right)$; G is (2, −4)

30.

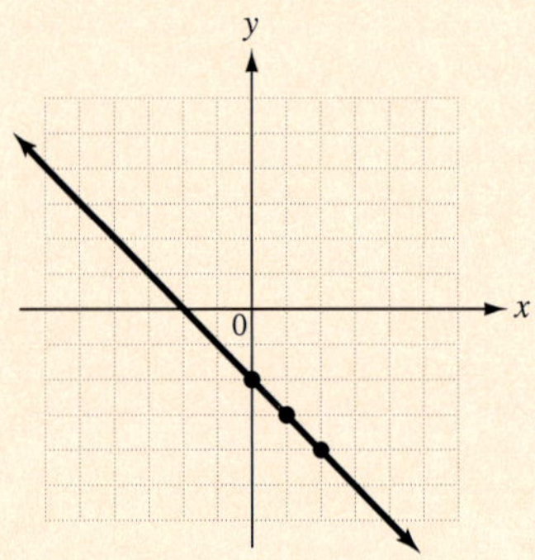

negative slope

31.

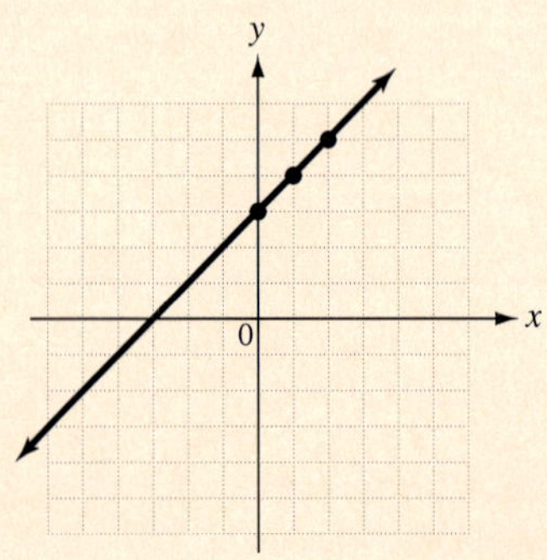

positive slope

Chapter 9 Review Exercises (page 621)

1. **(a)** Lodging; $560 **(b)** Food; $400 **2.** $\frac{4}{17}$ **3.** $\frac{3}{17}$ **4.** $\frac{14}{85}$ **5.** $\frac{15}{8}$ **6.** 3,300,000 phones **7.** 2,805,000 phones **8.** 1,650,000 phones **9.** 1,980,000 phones
10. Ages 12–17 and 35–44 buy more cellular phones. Explanations will vary. Perhaps they use a phone more often and feel the need to be reached by phone at all times. Or, it may be a status symbol among these age groups.
11. Ages 55–64 and 65 and over buy the fewest phones. Explanations will vary. Perhaps they use the phone less often and feel that they do not need to be reached by phone at all times and in different places. Or, those over 65 may not have enough income to afford cellular phones. **12.** **(a)** March; 8 million acre-feet **(b)** June; 2 million acre-feet
13. **(a)** 5 million acre-feet; **(b)** 4 million acre-feet
14. 5 million acre-feet **15.** 2 million acre-feet
16. $50,000 **17.** $20,000 **18.** $20,000
19. $40,000 **20.** The grocery purchases decreased for two years and then moved up slightly. Less children are attending the center or fewer children are eating at the childcare center. **21.** The grocery purchases are increasing. A greater number of children are attending the center or a greater number are eating at the childcare center.
22. 36° **23.** 35% **24.** 20% **25.** 25% **26.** 36°
27.

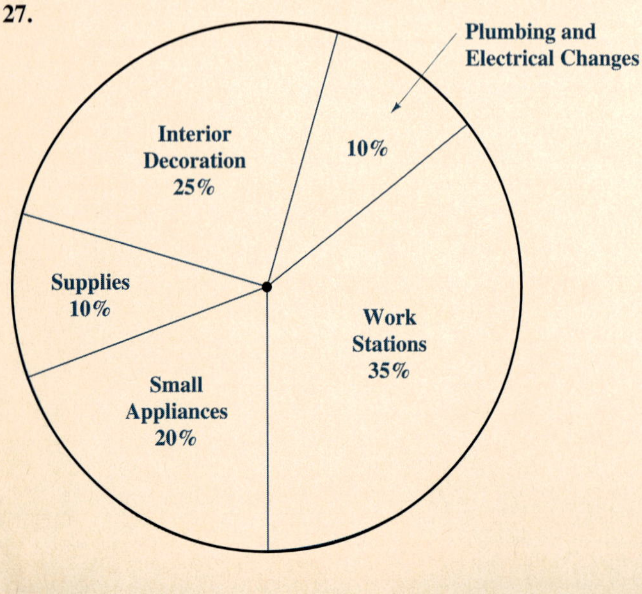

CHAPTER 9 TEST (page 625)

1. \$1,232,000 **2.** \$1,008,000 **3.** streets and gutters; \$1,680,000 **4.** miscellaneous; \$112,000 **5.** 1997; \$4000 **6.** \$9000 **7.** 1997; explanations will vary. Some possibilities are: laid off from work, changed jobs, was ill, cut down on hours worked. **8.** 5500 students; 3000 students **9.** College B; 1000 students **10.** Explanations will vary. For example, College B may have added new courses or lowered tuition or added child care. **11.** 108° **12.** 36° **13.** 72° **14.** 108° **15.** 10%

16.

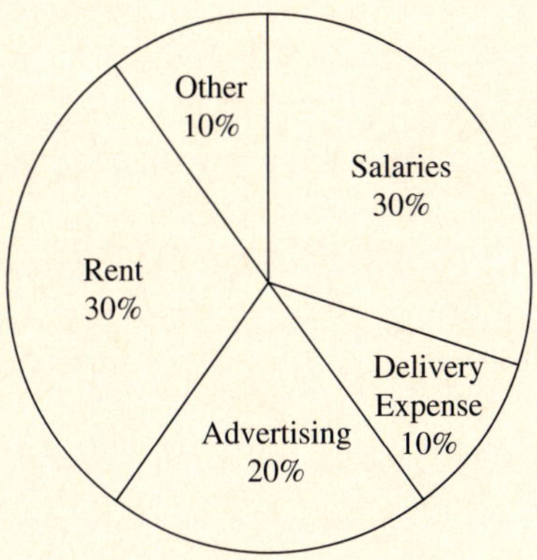

17–20.

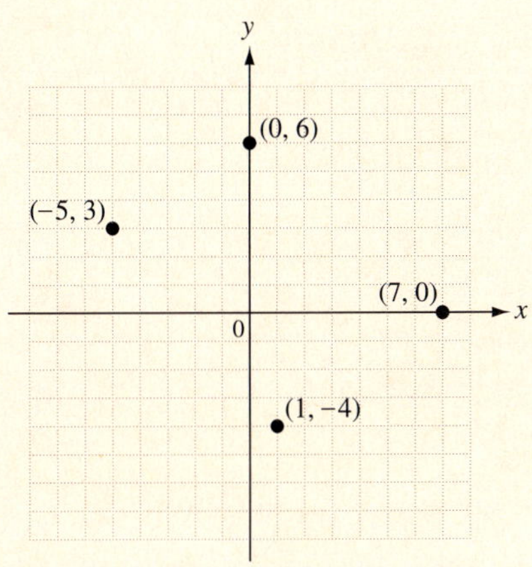

21. (0, 0) **22.** (−6, −5) **23.** (3, 3) **24.** (−2, 4) **25.** none **26.** III **27.** I **28.** II **29.** The order of the coordinates makes a difference. The first number tells where to move on the x-axis. (−4, 3) is in quadrant II and (3, −4) is in quadrant IV.

30.

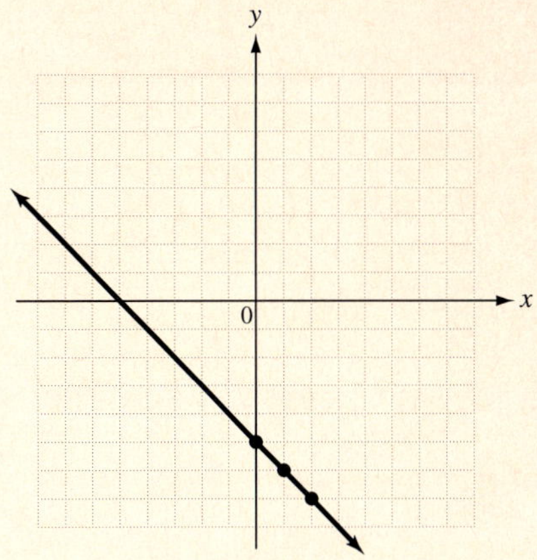

31.

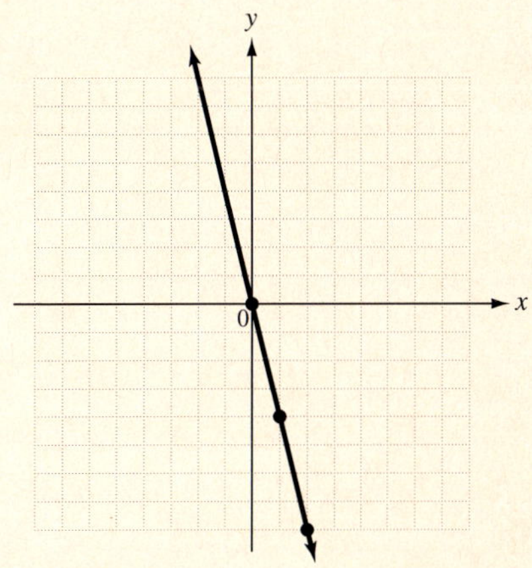

CUMULATIVE REVIEW 1–9 (page 629)

1. (a) six hundred two ten-thousandths **(b)** three hundred million, five hundred sixty **2. (a)** 70,005,000,043 **(b)** 18.09 **3. (a)** 3.0 **(b)** 0.80 **(c)** 69,000,000 **4.** mean is 29 years; median is 25 years; mode is 23 years **5.** $3\frac{2}{15}$ **6.** −0.25 **7.** −39 **8.** −0.00010 or −0.0001 **9.** $\frac{4b}{3a^2}$ **10.** $-2\frac{7}{12}$ **11.** −10.007 **12.** $\frac{2x - 56}{7x}$ **13.** −36 **14.** $\frac{25}{3mn}$ **15.** $3\frac{1}{3}$ **16.** undefined **17.** $7\frac{3}{8}$ **18.** $\frac{18 + wx}{6w}$ **19.** −4 **20.** 840 **21.** $\frac{0}{-8}$ is 0

22. −2.06 **23.** $-\frac{7}{20}$ **24.** 0 **25.** 41
26. −100 **27.** $x^2 - 5x$ **28.** $5y - 12$
29. $-40h^2$ **30.** $x = -7$ **31.** $y = -6$
32. $x = 2$ **33.** $m = -15$ **34.** $n = 16.2$
35. $w = 20$ **36.** $x = 7$ **37.** $h = -9$ **38.** $n = 2$
39. $n = -10$ **40.** Let m be the smaller amount; $m + 500$ the larger amount; $m + m + 500 = 1800$; the two amounts are \$650 and \$1150. **41.** Let x be the width; $3x$ the length; $2(3x) + 2x = 280$; the width is 35 ft; length is 105 ft. **42.** $P = 36$ ft; $A = 70$ ft^2
43. $P = 188$ m; $A = 1636$ m^2
44. $x = 25$ mi; $P = 56$ mi; $A = 84$ mi^2
45. $C \approx 18.8$ ft; $A \approx 28.3$ ft^2 **46.** $V \approx 3768$ m^3
47. $V = 30$ yd^3 **48.** 135 minutes **49.** 4.5 ft or $4\frac{1}{2}$ ft
50. 1850 mL **51.** 3.5 cm **52.** 0.01 kg
53. ≈−4 °C **54.** ≈556 nonsmoking employees
55. ≈\$116.78 **56.** ≈92.8% **57.** **(a)** 31 computers; \$333 left over **(b)** 17 calculators; \$10 left over
58. $1\frac{5}{12}$ pounds
59. $-2, 3\frac{1}{2}$ is in second quadrant; none in third

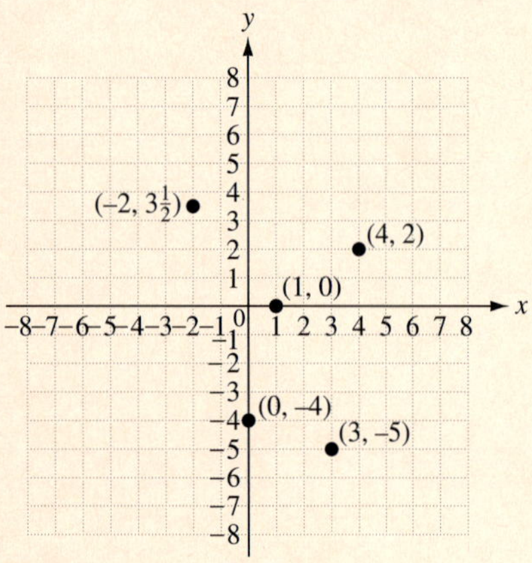

60.

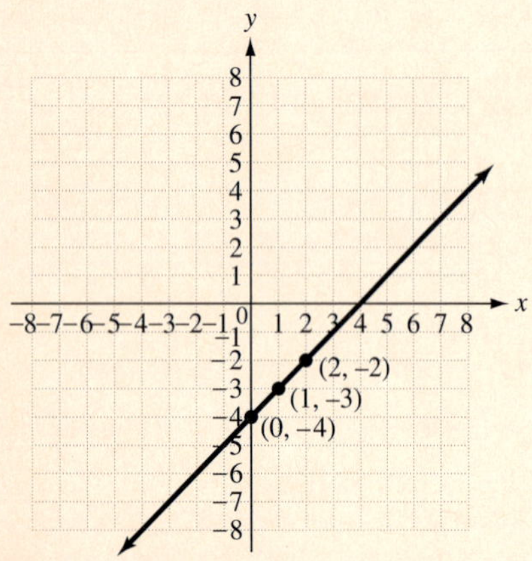

Chapter R

Section R.1 (page R-7)

1. 8928; 5715 + 3213 = 8928
3. 59,224; 21,020 + 38,204 = 59,224
5. 150 **7.** 1651 **9.** 1004
11. 9253 **13.** 11,624 **15.** 16,658
17. 5009 **19.** 3506 **21.** 15,954
23. 10,648 **25.** incorrect; should be 769
27. correct **29.** 33 miles
31. 38 miles **33.** \$79
35. 699 people **37.** 13,051 books
39. 294 inches **41.** 708 feet

Section R.2 (page R-15)

1. incorrect; should be 62 **3.** incorrect; should be 121
5. 8 **7.** 25 **9.** 16 **11.** 61 **13.** 519 **15.** 9177
17. 7589 **19.** 8859 **21.** 3 **23.** 23 **25.** 1942
27. 5687 **29.** 19,038 **31.** 65,556 **33.** 19,984
35. incorrect; should be 2494 **37.** correct
39. 15 calories **41.** 121 passengers
43. \$8700 **45.** 367 feet

Section R.3 (page R-23)

1. 9 **3.** 63 **5.** 0 **7.** 24 **9.** 36
11. 245 **13.** 168 **15.** 19,092 **17.** 258,447
19. 3750 **21.** 44,550 **23.** 270,000
25. 86,000,000 **27.** 1496 **29.** 3735
31. 15,200 **33.** 32,805 **35.** 183,996
37. 1,616,076 **39.** 19,422,180
41. 24,000 pages **43.** 216 plants **45.** 418 miles
47. 175 joggers **49.** 1058 calories **51.** \$556

Section R.4 (page R-33)

1. 1 **3.** undefined **5.** 0 **7.** 0 **9.** undefined
11. 27 **13.** 36 **15.** 1522 R5 **17.** 309
19. 1006 **21.** 5006 **23.** 6671 **25.** 12,458 R3
27. 10,253 R5 **29.** 18,377 R6 **31.** incorrect; should be 670 R2 **33.** incorrect; should be 3568 R1
35. correct **37.** correct **39.** 156 cartons
41. \$16,600 **43.** \$18,200 **45.** 205 acres
47. \$225,000 **49.** \$9135

	2	**3**	**5**	**10**
51.	✓	✓	✓	✓
53.	✓	×	×	×
55.	×	×	✓	×
57.	×	✓	×	×
59.	✓	✓	×	×
61.	×	×	×	×

Section R.5 (page R-39)

1. 32 **3.** 250 **5.** 120 R7 **7.** 1308 R9
9. 7134 R12 **11.** 900 R100 **13.** 207 R5
15. 236 R29 **17.** 2407 R1 **19.** 1239 R15
21. 3331 R82 **23.** 850 **25.** incorrect; should be 106 R7 **27.** incorrect; should be 658
29. incorrect; should be 62 **31.** correct
33. 25 hours **35.** 630 bronze medals
37. \$108 **39.** 1680 circuits **41.** \$375

Chapter R Review Exercises (page R-43)

1. 92 **2.** 113 **3.** 5464 **4.** 15,657 **5.** 39
6. 184 **7.** 224 **8.** 25,866 **9.** 48
10. 45 **11.** 48 **12.** 0 **13.** 172
14. 5467 **15.** 32,640 **16.** 465,525
17. 19,200 **18.** 25,200 **19.** 206,800
20. 128,000,000 **21.** 612 **22.** 1872
23. 13,755 **24.** 30,184 **25.** 6 **26.** 1
27. undefined **28.** 0 **29.** 108 **30.** 24
31. 352 **32.** 150 R4 **33.** 468 cards
34. 2856 textbooks **35.** $280 **36.** $27,940
37. 24,000 hours **38.** 400 miles
39. 23 acres **40.** 32 homes **41.** $247
42. $143 **43.** $582 **44.** $1261

Chapter R Test (page R-45)

1. addends; sum or total **2.** factors; product
3. difference; quotient **4.** 8610 **5.** 112,630
6. 16,948 **7.** 3084 **8.** 140 **9.** 171,000
10. 1785 **11.** 4,450,743 **12.** 206 **13.** undefined
14. 458 **15.** 160 **16.** $528 **17.** $171
18. 33 hours **19.** $165 **20.** 1028 ovens
21. $1140 **22.** A number is divisible by 2 if it ends in 0, 2, 4, 6, or 8. Examples: 312 is divisible by 2; 415 is not.
A number is divisible by 5 if it ends in 0 or 5. Examples: 75 is divisible by 5; 93 is not. A number is divisible by 10 if it ends in 0. Examples: 170 is divisible; 305 is not.

Whole Numbers Computation: Diagnostic Test

1. 390 **2.** 13,166 **3.** 3053 **4.** 117,510
5. 688,226

1. 350 **2.** 629 **3.** 24,894
4. 3909 **5.** 591,387

1. 0 **2.** 26,887
3. 1,560,000 **4.** 1846 **5.** 17,232 **6.** 519,477

1. 23 **2.** undefined **3.** 6259 **4.** 807 R6
5. 34 **6.** 50 **7.** 60 R20 **8.** 539 R62

CHAPTER 10

SECTION 1 (page 639)

1. 3^7 **3.** $(-2)^5$ **5.** w^6 **7.** $\left(-\frac{1}{4}\right)^5$ **9.** $(-7x)^4$
11. The expression $(5x)^3 = (5x)(5x)(5x) = 125x^3$, while $5x^3 = 5x \cdot x \cdot x \neq 125x^3$. **13.** base: 2; exponent: 7; 128 **15.** base: -2; exponent: 7; -128 **17.** base: $-8x$; exponent: 4 **19.** base: x; exponent: 4 **21.** The notation $a \mathbin{\char`\^} b$ means a^b. **23.** 1 **25.** 5^8 **27.** 4^{12} **29.** $(-7)^9$ **31.** t^{24} **33.** $-56r^7$ **35.** $42p^{10}$ **37.** The product rule applies only to exponentials with the same base.

$$3^2 \cdot 4^3 = 9 \cdot 64 = 576$$

39. $14x^4$; $45x^8$ **41.** $5a^2$; $-140a^6$ **43.** 4^6 **45.** t^{20} **47.** 7^3r^3 **49.** $5^5x^5y^5$ **51.** 5^{12} **53.** -8^{15} **55.** $8q^3r^3$ **57.** $\frac{1}{2^3}$ **59.** $\frac{a^3}{b^3}$ **61.** $\frac{9^8}{5^8}$ **63.** $\frac{5^5}{2^5}$ **65.** $\frac{9^5}{8^3}$ **67.** $2^{12}x^{12}$ **69.** $(-6)^5p^5$ **71.** $6^5x^{10}y^{15}$ **73.** x^{21} **75.** $2^2w^4x^{26}y^7$ **77.** $-r^{18}s^{17}$ **79.** $\frac{5^3a^6b^{15}}{c^{18}}$
81. No, $(10^2)^3 = 10^{(2)(3)} = 10^6 = 1{,}000{,}000$.

SECTION 2 (page 647)

1. 1 **3.** -1 **5.** 0 **7.** 1 **9.** 2 **11.** $\frac{1}{64}$ **13.** 16 **15.** $\frac{49}{36}$ **17.** $\frac{1}{81}$ **19.** $\frac{8}{15}$ **21.** negative **23.** negative **25.** positive **27.** zero **29.** 1 **30.** $\frac{5^2}{5^2}$ **31.** 5^0 **32.** $5^0 = 1$; this supports the definition of a zero exponent. **33.** $\frac{1}{9}$ **35.** $\frac{1}{6^5}$ **37.** 6^3 **39.** $2r^4$ **41.** $\frac{5^2}{4^3}$ **43.** $\frac{p^5}{q^8}$ **45.** r^9 **47.** $\frac{x^5}{6}$ **49.** 7^3 **51.** $\frac{1}{x^2}$ **53.** $\frac{4^3x}{3^2}$ **55.** $\frac{x^2z^4}{y^2}$ **57.** $6x$ **59.** $\frac{1}{m^{10}n^5}$ **61.** $\frac{5}{16x^5}$ **63.** $\frac{36q^2}{m^4p^2}$

SECTION 3 (page 651)

1. $\$3 \times 10^5$; $\$2.8 \times 10^6$ **3.** 7.52×10^8; 3.94×10^8 **5.** in scientific notation **7.** not in scientific notation; 5.6×10^6 **9.** not in scientific notation; 8×10^1 **11.** not in scientific notation; 4×10^{-3} **13.** A number is written in scientific notation if it is the product of a number whose absolute value is between 1 and 10 (inclusive of 1) and a power of 10. **15.** 5.876×10^9 **17.** 8.235×10^4 **19.** 7×10^{-6} **21.** -2.03×10^{-3} **23.** 750,000 **25.** 5,677,000,000,000 **27.** -6.21 **29.** .00078 **31.** .000000005134 **33.** 6×10^{11}; 600,000,000,000 **35.** 1.5×10^7; 15,000,000 **37.** 6.426×10^4; 64,260 **39.** 3×10^{-4} **41.** 4×10^1 **43.** 2.6×10^{-3} **45.** 635,000,000,000 **47.** $\$4.55 \times 10^7$ **49.** .00023; .006

SECTION 4 (page 659)

1. sometimes **3.** always **5.** never **7.** always **9.** 1; 6 **11.** 1; 1 **13.** 2; -19, -1 **15.** 2; 1, 8 **17.** $2m^5$ **19.** $-r^5$ **21.** cannot be simplified **23.** $-5x^5$ **25.** $5p^9 + 4p^7$ **27.** $-2y^2$ **29.** already simplified; 4; binomial **31.** already simplified; $6m^5 + 5m^4 - 7m^3 - 3m^2$; 5; none of these **33.** $x^4 + \frac{1}{3}x^2 - 4$; 4; trinomial **35.** 7; 0; monomial **37.** 4; \$5.00 **38.** 6; \$27 **39.** 2.5; 130 **40.** 3; 104 **41.** $5m^2 + 3m$ **43.** $4x^4 - 4x^2$ **45.** $\frac{7}{6}x^2 - \frac{2}{15}x + \frac{5}{6}$ **47.** $15m^3 - 13m^2 + 8m + 11$ **49.** $8r^2 + 5r - 12$ **51.** $5m^2 - 14m + 6$ **53.** $4x^3 + 2x^2 + 5x$ **55.** $-18y^5 + 7y^4 + 5y^3 + 3y^2 + y$ **57.** $-2m^3 + 7m^2 + 8m - 10$ **59.** $8x^2 + 8x + 6$ **61.** $8t^2 + 8t + 13$ **63.** $-11x^2 - 3x - 3$ **65.** The degree of a term is determined by the exponent on the *variable*, but 3^4 is not a variable. The degree of $3^4 = 3^4x^0$ is 0. **67.** $13a^2b - 7a^2 - b$; degree 3 **69.** $c^4d - 5c^2d^2 + d^2$; degree 5 **71.** $12m^3n - 11m^2n^2 - 4mn^2$; degree 4

SECTION 5 (page 667)

1. distributive **3.** $-6m^2 - 4m$ **5.** $6p - \frac{9}{2}p^2 + 9p^4$ **7.** $6y^5 + 4y^6 + 10y^9$ **9.** $n^2 + n - 6$ **11.** $8r^2 - 10r - 3$ **13.** $9x^2 - 4$ **15.** $9q^2 + 6q + 1$ **17.** $6t^2 + 23st + 20s^2$ **19.** $-.3t^2 + .22t + .24$ **21.** The answers are $x^2 - 16$, $y^2 - 4$, and $r^2 - 49$. All three products are the difference of the square of the first term and the square of the last term of the binomials. **23.** $12x^3 + 26x^2 + 10x + 1$ **25.** $20m^4 - m^3 - 8m^2 - 17m - 15$ **27.** $5x^4 - 13x^3 + 20x^2 + 7x + 5$ **29.** $x^2 + 24x + 144$ **31.** $9t^2 + 6t + 1$ **33.** $25x^2 - 20xy + 4y^2$ **35.** $h^3 - 15h^2 + 75h - 125$ **37.** $r^4 + 4r^3 + 6r^2 + 4r + 1$ **39.** $9x^4 + 6x^3 - 23x^2 - 8x + 16$ **41.** $30x + 60$ **42.** $30x + 60 = 600$; 18 **43.** 10 yards by 60 yards **44.** \$2100 **45.** 140 yards **46.** \$1260

SECTION 6 (page 673)

1. **(a)** $2x^2$ **(b)** $-10x$ **(c)** $3x$ **(d)** -15 **(e)** $-7x$ **(f)** $2x^2 - 7x - 15$ **3.** $r^2 + 4r + 3$ **5.** $w^2 - 10w + 24$ **7.** $x^2 + .5x - .24$ **9.** $x^2 - \frac{5}{12}x - \frac{1}{6}$ **11.** $27x^2 + 69x + 14$ **13.** $9m^2 + 36m + 35$ **15.** $15 - 19x + 6x^2$ **17.** $-15 + 23z - 6z^2$ **19.** $\frac{16}{9} + \frac{2}{3}k - 3k^2$ **21.** $27w^2 + 15wz - 2z^2$ **23.** $30y^2 + 7yz - 15z^2$ **25.** $-.16p^2 - .74ps + .3s^2$ **27.** **(a)** $4x^2$ **(b)** $12x$ **(c)** 9 **(d)** $4x^2 + 12x + 9$ **29.** $a^2 - 2ac + c^2$ **31.** $p^2 + 4p + 4$

33. $16x^2 - 24x + 9$ 35. $.64t^2 + 1.12ts + .49s^2$
37. $25x^2 + 4xy + \frac{4}{25}y^2$
39. (a) $49x^2$ (b) 0 (c) $-9y^2$ (d) $49x^2 - 9y^2$; Because 0 is the identity element for addition, it is not necessary to write "+0." 41. $q^2 - 4$
43. $4w^2 - 25$ 45. $100x^2 - 9y^2$
47. $4x^4 - 25$ 49. $49x^2 - \frac{9}{49}$
51. 1225 52. $30^2 + 2(30)(5) + 5^2$ 53. 1225
54. They are equal.

SECTION 7 (page 677)

1. To use the method of this section, the denominator must be just one term. This is true of the first problem, but not the second. 3. 8; 13; no 5. $5m^4 - 8m^3 + 4m^2$
7. $4m^4 - 2m^2 + 2m$ 9. $\frac{m^4}{2} - 2m + \frac{4}{m}$
11. $4x^3 - 3x^2 + 2x$ 13. $1 + 5x - 9x^2$
15. $\frac{12}{x} + 8 + 2x$ 17. $\frac{4x^2}{3} + x + \frac{2}{3x}$
19. $9r^3 - 12r^2 - 2r + 1 - \frac{2}{3r}$ 21. $-m^2 + 3m - \frac{4}{m}$
23. $4 - 3a + \frac{5}{a}$ 25. $\frac{12}{x} - \frac{6}{x^2} + \frac{14}{x^3} - \frac{10}{x^4}$
27. No, $\frac{2}{3}x$ means $\frac{2x}{3}$ which is not the same as $\frac{2}{3x}$. In the first case we multiply by x; in the second case we divide by x. Yes, $\frac{4}{3}x^2 = \frac{4x^2}{3}$. In both cases we are multiplying by x^2.
29. $15x^5 - 35x^4 + 35x^3$ 31. 1423
32. $(1 \times 10^3) + (4 \times 10^2) + (2 \times 10^1) + (3 \times 10^0)$
33. $x^3 + 4x^2 + 2x + 3$
34. They are similar in that the coefficients of the powers of ten are equal to the coefficients of the powers of x. They are different in that one is a number while the other is a polynomial. They are equal if $x = 10$.

SECTION 8 (page 683)

1. The divisor is $2x + 5$; the quotient is $2x^3 - 4x^2 + 3x + 2$.
3. Divide $12m^2$ by $2m$ to get $6m$. 5. $x + 2$
7. $2y - 5$ 9. $p - 4 + \frac{44}{p + 6}$ 11. $r - 5$
13. $2a - 14 + \frac{74}{2a + 3}$ 15. $4x^2 - 7x + 3$
17. 33 18. 33 19. They are the same.
20. The answers should agree. 21. $3y^2 - 2y + 2$
23. $3k - 4 + \frac{2}{k^2 - 2}$ 25. $x^2 + 1$
27. $2p^2 - 5p + 4 + \frac{6}{3p^2 + 1}$ 29. $x^3 + 3x^2 - x + 5$
31. $x^2 + 1$

CHAPTER 10 REVIEW EXERCISES (page 687)

1. 4^{11} 2. $(-5)^{11}$ 3. $-72x^7$ 4. $10x^{14}$
5. 19^5x^5 6. $(-4)^7y^7$ 7. $5p^4t^4$ 8. $\frac{7^6}{5^6}$
9. $3^3x^6y^9$ 10. t^{42} 11. $6^2x^{16}y^4z^{16}$
12. The product rule for exponents does not apply here because we want the sum of 7^2 and 7^4, not their product.
13. 2 14. $\frac{1}{32}$ 15. $\frac{5^2}{6^2}$ or $\frac{25}{36}$ 16. $-\frac{3}{16}$ 17. 6^2
18. x^2 19. $\frac{1}{p^{12}}$ 20. r^4 21. 2^8 22. $\frac{1}{9^6}$
23. 5^8 24. $\frac{1}{8^{12}}$ 25. $\frac{1}{m^2}$ 26. y^7 27. r^{13}
28. $(-5)^2m^6$ 29. $\frac{y^{12}}{2^3}$ 30. $\frac{1}{a^3b^5}$ 31. $2 \cdot 6^2 \cdot r^5$
32. $\frac{2^3n^{10}}{3m^{13}}$ 33. 4.8×10^7 34. 2.8988×10^{10}
35. 6.5×10^{-5} 36. 8.24×10^{-8} 37. 24,000
38. 78,300,000 39. .000000897
40. .00000000000995 41. 800 42. 4,000,000
43. .025 44. .01 45. $22m^2$; 2; monomial
46. $p^3 - p^2 + 4p + 2$; 3; none of these
47. $8r^5 + 4r^4$; 5; binomial
48. $15a^5 + 10a^4 + 10a^3$; 5; trinomial
49. $-5a^3 + 4a^2$ 50. $2r^3 - 3r^2 + 9r$
51. $11y^2 - 10y + 9$ 52. $-13k^4 - 15k^2 - 4k - 6$
53. $10m^3 - 6m^2 - 3$ 54. $-y^2 - 4y + 26$
55. $10p^2 - 3p - 11$ 56. $7r^4 - 4r^3 - 1$
57. $10x^2 + 70x$ 58. $-6p^5 + 15p^4$
59. $m^2 - 7m - 18$ 60. $6k^2 - 9k - 6$
61. $6r^3 + 8r^2 - 17r + 6$ 62. $8y^3 + 27$
63. $r^3 + 6r^2 + 12r + 8$
64. $(a + b)^2 = a^2 + 2ab + b^2$. The term $2ab$ is not in $a^2 + b^2$. 65. $6k^2 + 11k + 3$
66. $2a^2 + 5ab - 3b^2$ 67. $12k^2 - 48kq + 21q^2$
68. $a^2 + 8a + 16$ 69. $9p^2 - 12p + 4$
70. $4r^2 + 20rs + 25s^2$ 71. $36m^2 - 25$
72. $4z^2 - 49$ 73. $25a^2 - 36b^2$ 74. $4x^4 - 25$
75. $\frac{5y^2}{3}$ 76. $-2x^2y$ 77. $-y^3 + 2y - 3$
78. $p - 3 + \frac{5}{2p}$ 79. $-x^9 + 2x^8 - 4x^3 + 7x$
80. $-2m^2n + mn^2 + \frac{6n^3}{5}$ 81. $2r + 7$
82. $4m + 3 + \frac{5}{3m - 5}$ 83. $2a + 1 + \frac{-8a + 12}{5a^2 - 3}$
84. $k^2 + 2k + 4 + \frac{-2k - 12}{2k^2 + 1}$ 85. 0
86. $\frac{3^5}{p^3}$ 87. $\frac{1}{7^2}$ 88. $49 - 28k + 4k^2$
89. $y^2 + 5y + 1$ 90. $\frac{6^4r^8s^4}{5^4}$
91. $-8m^7 - 10m^6 - 6m^5$ 92. 2^5
93. $5xy^3 - \frac{8y^2}{5} + 3x^2y$ 94. $\frac{r^2}{6}$
95. $8x^3 + 12x^2y + 6xy^2 + y^3$ 96. $\frac{3}{4}$
97. $a^3 - 2a^2 - 7a + 2$ 98. $8y^3 - 9y^2 + 5$
99. $10r^2 + 21r - 10$ 100. $144a^2 - 1$

Chapter 10 Test (page 691)

1. $\frac{1}{625}$ **2.** 2 **3.** $\frac{7}{12}$ **4.** 8^5 **5.** x^2y^6
6. (a) 3.44×10^{11} **(b)** 5.57×10^{-6}
7. (a) 29,600,000 **(b)** .0000000607
8. $-7x^2 + 8x$; 2; binomial
9. $4n^4 + 13n^3 - 10n^2$; 4; trinomial
10. $4t^4 + t^3 - 6t^2 - t$ **11.** $-2y^2 - 9y + 17$
12. $-12t^2 + 5t + 8$
13. $-27x^5 + 18x^4 - 6x^3 + 3x^2$
14. $t^2 - 5t - 24$ **15.** $8x^2 + 2xy - 3y^2$
16. $25x^2 - 20xy + 4y^2$ **17.** $100v^2 - 9w^2$
18. $2r^3 + r^2 - 16r + 15$ **19.** $x^3 + 3x^2 + 3x + 1$
20. $4y^2 - 3y + 2 + \frac{5}{y}$ **21.** $-3xy^2 + 2x^3y^2 + 4y^2$
22. $2x + 9$ **23.** $3x^2 + 6x + 11 + \frac{26}{x - 2}$
24. $9x^2 + 54x + 81$ **25.** Answers will vary. One example is $(-4x^4 + 3x^3 + 2x + 1) + (4x^4 - 8x^3 + 2x + 7) = -5x^3 + 4x + 8$.

Index

D

E

F

G

N

O

P

Q

R

S

T

U

V

W

Z

Applications Index

Health/Life Sciences

Miscellaneous

Social Sciences

Sports/Hobbies/Entertainment

Statistics/Demographics

Transportation

* Denotes real data